MAIZE BREEDING AND GENETICS

MAIZE BREEDING AND GENETICS

Edited by
DAVID B. WALDEN
University of Western Ontario

A WILEY-INTERSCIENCE PUBLICATION

JOHN WILEY & SONS, New York • Chichester • Brisbane • Toronto

Library of Congress Cataloging in Publication Data:

Main entry under title:
Maize breeding and genetics.

  "A Wiley-Interscience publication."
  Includes index.
  1. Corn.   2. Corn—Breeding.   3. Corn—Genetics.
I. Walden, David B. 1932–
SB191.M2M325   633'.15'3   78-6779
ISBN 0-471-91805-9

Printed in the United States of America

10 9 8 7 6 5 4 3 2 1

INTERNATIONAL MAIZE SYMPOSIUM
GENETICS AND BREEDING
University of Illinois at Urbana—Champaign, September 1977

Advisory Committee

Dr. G. F. Sprague, United States, Chairman
Dr. D. E. Alexander, United States, Executive Secretary Department of
    Agronomy, Turner Hall, University of Illinois, Urbana, Illinois
Dr. H. H. Angeles A., Mexico
Dr. G. W. Beadle, United States
Dr. A. Bianchi, Italy
Dr. A. Brandolini, Italy
Dr. R. A. Brink, United States
Dr. W. L. Brown, United States
Dr. A. Cauderon, France
Dr. E. H. Coe, Jr., United States
Dr. R. Comstock, United States
Dr. G. S. Galeev, USSR
Dr. C. O. Gardner, United States
Dr. M. N. Harrison, Nigeria
Dr. P. H. Harvey, United States
Mr. R. Holland, United States
Mr. F. Ingersoll, United States
Dr. M. T. Jenkins, United States
Dr. R. W. Jugenheimer, United States
Dr. J. R. Laughnan, United States
Mr. R. C. Liebenow, United States
Dr. H. D. Loden, United States
Dr. J. H. Lonnquist, United States
Dr. B. McClintock, United States
Dr. O. E. Nelson, Jr., United States
Dr. F. Ogada, Kenya
Dr. E. Paterniani, Brazil
Dr. S. Rajki, Hungary
Dr. M. M. Rhoades, United States
Dr. H. F. Robinson, United States
Dr. W. A. Russell, United States
Dr. E. Sanchez-Monge, Spain
Ing. F. Scheuch, Peru
Dr. J. Singh, India

Dr. E. W. Sprague, Mexico
Dr. L. Steele, United States
Professor Dr. A. Tavcar, Yugoslavia
Dr. M. Torrigroza C., Columbia
Dr. V. Trifunovic, Yugoslavia
Dr. D. B. Walden, Canada

Program Committee

D. B. Walden, Chairman
E. H. Coe, Jr.
J. R. Harlan
A. L. Hooker
J. R. Laughnan
O. E. Nelson, Jr.
G. F. Sprague

Local Arrangements Committee

J. W. Dudley, Chairman
D. E. Alexander
J. M. J. de Wet
J. R. Harlan
A. L. Hooker
R. J. Jugenheimer
R. J. Lambert
J. R. Laughnan
E. B. Patterson
G. F. Sprague

Field Exhibits Committee

R. J. Lambert, Chairman
E. B. Patterson

INTRODUCTION

1975 marked the 75th anniversary of the rediscovery of Mendelism. Corn was involved in the rediscovery and continues to play an important role in advances in our knowledge of both genetics and cytogenetics. Furthermore, corn was the first agricultural species in which hybrid vigor was of commercial significance, a development which has had worldwide acceptance.

Scientific meetings are normally discipline oriented. The program developed for this symposium is different in that it is crop oriented, although the theme, of course, involves genetics. However, the object was to bring together the classical and quantitative geneticists, the cytogeneticists, the physiologists, the evolutionists, and the breeders who use or have used maize as their research organism. We believed it would be worthwhile to bring them together, with their diversity of interests and skills, to review that which is past and to develop projections for the future. The appropriate place for such a gathering was in the heart of the Corn Belt.

The desirability of such a meeting had been under sporadic discussion for several years. The 75th anniversary of the rediscovery of Mendelism provided a suitable opportunity, and a nuclear committee composed of University of Illinois faculty and Professor D. B. Walden of the University of Western Ontario performed the original feasibility study. An organizing committee of eminent corn researchers representing 14 countries were invited to assist in establishing general guidelines for the symposium. A smaller committee assumed responsibility for soliciting speakers and arranged the program.

The success of the Symposium depended on many people. We are particularly appreciative of the financial support we received. Without it, the Symposium as it was finally presented would have not been possible.

In addition, we must thank the Department of Agronomy and the Continuing Education Center of the University of Illinois for their support and efforts throughout the planning stages and during the execution of the Symposium itself.

G. F. SPRAGUE
Chairman, Advisory Committee

D. E. ALEXANDER
Executive Secretary

Financial support for the Symposium from the following sources is gratefully acknowledged.

Anderson Foundation
Continental Grain Company
Corn Refiner's Association, Inc.
DeKalb Ag Research, Inc.
Funk Seeds International, Inc.
Illinois Foundation Seeds, Inc.
Illinois Maize Genetics Laboratory
Moews Seed Company
Northrup, King and Company
Pioneer Hi-Bred International, Inc.
The University of Illinois
The University of Western Ontario

PREFACE

This volume is presented as a record of the Symposium. Each author was not constrained initially by style, length, or the usual format guidelines associated with a more formal review or monograph. All manuscripts were read by two reviewers, some by three. I am grateful to all the sessional chairpeople for their assistance in this task. Some of the alterations required reduction of the text and deletion of illustrative material. I attempted to retain the individual character and flavor of the contributions by leaving most of the decisions to the author. Finally, authors were given the opportunity to include in their contributions literature available up to late 1977.

After reading the manuscripts, I felt that the reordering and resequencing I have employed would be useful to the reader. In addition, several papers that were not read at the Symposium were invited as manuscripts to appear in this book. I am grateful to the chairpeople who prepared a script of remarks and views developed during their session.

There are some additional items, not to be found on the following pages, that deserve attention. My thoughts have often wandered back to that glorious Tuesday afternoon when so many Symposium participants visited the field exhibits and enjoyed the barbecue that Floyd Ingersoll arranged for us. It is difficult to imagine how Bob Lambert and his colleagues could have set up a better afternoon for us.

In keeping with the theme of mixing good times, good people, and good science, one additional event should be recorded. John Laughnan, on behalf of all those assembled, presented gifts to Marcus Rhoades and Ellen Dempsey in recognition of their valued editorial service and production of the Maize

Genetics Cooperation Newsletter and to Earl Patterson for his extensive efforts with the Maize Coop. It is entirely appropriate that the record shows the esteem and gratitude in which the three are held by all of us who have benefited from their labors.

I should like to extend my thanks to those who worked to make the Symposium a success. In addition to those who presented manuscripts and demonstration papers, the memberships of the working committees are listed elsewhere in this book. My special thanks go to George Sprague, Gene Alexander, Earl Patterson, John Dudley, Bob Lambert, Susan Gabay and John Laughnan—each of whom made my task considerably easier.

The Symposium and the attendant activities could not have occurred without the extensive financial support received from several sources. I draw your attention to the list of contributors on page viii. We are very grateful for this assistance.

D. B. WALDEN

London, Ontario
April 1978

CONTENTS

MAIZE BREEDING AND GENETICS

MEETING THE WORLD'S FOOD PROBLEMS

H. F. Robinson

Chancellor, Western Carolina University

Presented Thursday Evening, September 11, 1975

I want to express my appreciation to Dr. Sprague and other members of your organizing committee for allowing me to come and join with my friends—the workers in corn breeding and genetics at this, the 75th anniversary of the rediscovery of the laws of inheritance, with the great contributions by Gregor Mendel. No professional group is closer to me and to my thoughts than those who work in corn breeding and genetics programs in this country and abroad. I have kept reasonably informed about you and your work through your publications in the scientific journals. However, it is not easy for one who is devoting his full time to university administration to keep abreast of the rapid developments, even with one plant such as maize. Although my present assignment does not allow active participation in maize research, I have been and intend to continue to be involved in special assignments and committees concerned with the subjects important to genetics, plant breeding, and agriculture.

THE IMPORTANCE OF CORN GENETICS RESEARCH

Let me take a few minutes before we get into the main topic to recognize the efforts that all of you are making toward the improvement of this wonderful

1

plant with which we have labored and to which we owe so much. While working in the area of quantitative genetics I often failed to fully appreciate and recognize the significance of the contributions of those who worked on the inheritance of a specific gene, whether qualitative or quantitative in effect. We all know that it takes the full and combined efforts of all of the scientists working on the vast array of subjects to develop the complete story that is so very important as we endeavor to understand this complex organism. All of you, whether you are engaged in biochemical, quantitative, population, physiological, or other areas of genetics, are making significant contributions, each important in its own right, as you work continuously toward compiling the pieces of information required to achieve a total understanding of this, our most important seven billion dollar crop plant.

Speaking on the importance of corn, Allan Lenn has written in a recent issue of the *Smithsonian Magazine* that this is the new world's secret weapon. He begins his article with a statement to the effect that on Corn Hill in Truro, Massachusetts a statue stands with a bronze plaque dedicated to this important plant. On the plaque is the inscription in the words of William Bradford, Governor of the Plymouth colony, written in 1621. These words are, "and sure it was God's providence that we found this corn, for else we know not how we should have done."

No, neither do we, the more than 220 million people of this country and the four billion in the world tonight, know quite how we would have done had it not been for this miraculous plant. Neither do we know nor would we like to have learned what we would have done without the efforts of all of you and such great scientists as Emerson, East, Shull, Jones, Kiesselbach, Stadler, and many, many others no longer with us who have made outstanding contributions in the improvement of this organism.

THE CONTINUING WORLDWIDE NEED FOR FOOD

This brings me to the main topic about which Dr. Sprague suggested I speak to you tonight—namely, the continuing need for improving the food available for a hungry world and how this problem becomes increasingly serious with the worldwide population increase.

Some of you were involved with me in the 1966–1968 President's Science Advisory Committee (PSAC) study that we completed over a two-year span and that stands as one of the most comprehensive and authoritative studies ever undertaken on the world food problem. I have just recently served on another panel jointly sponsored by the National Academy of Sciences of the The National Science Foundation concerned with updating the PSAC study on the world food problem and determining the exact status of the problem as it exists today. The conclusions of this recent report, which have now gone to press but have not been released, contain the following statement, "The findings of the PSAC Study of the World Food Supply are as pertinent today as they were at the time they were written." Thus I cannot bring to you any *glowing* reports of major accomplishments in meeting our world food needs; neither can I be too optimistic with regard to the possibility of satisfying the needs of the people, particularly those in the underdeveloped countries as they face hunger, malnutrition, and starvation in the immediate years ahead.

I probably need not emphasize the seriousness of this situation as many of you either now work in those countries experiencing problems from lack of food and rapid population growth, or you may have visited some of these countries recently. Those of us from the United States have certainly read the increasingly frequent reports of the seriousness of the situation. In country after country such as Haiti, the Dominican Republic, the Honduras, most of the Central American area, much of South America, Kenya, Nigeria, Mali, all of middle Africa, the Far East, India, Bangladesh, Vietnam, and so many other countries that you would tire from the list, could be mentioned tonight as having very, very serious problems with hunger, malnutrition, and starvation. Some of these nations border on the brink of disaster. The question in the minds of many authorities involved in evaluating these problems is whether it may not already be too late for them to ever meet the food needs of their people and prevent starvation. We continue to have the reports of those who claim that a rather large number of the developing nations have crossed the brink of any hope of solution to their food problems, supporting the thesis that the crisis period of 1982–1985 predicted in the PSAC study appears ever so likely to occur. It is extremely regrettable that our pessimistic conclusions and predictions have arrived all too soon.

Let us examine for a moment the population growth with which we are certain to be faced. At the international level we know we have approximately four billion people in the world today and in just 25 years from now, or at the turn of the century, we will have somewhere between seven and eight billion, or almost double the number now inhabiting the world. This projected population growth presents a fantastic problem of even continuing the current average level of the diet, which in many areas is woefully inadequate.

Make no mistake about it; the population levels predicted will be reached. This will result because of the growth rates we are now experiencing and the likelihood that there will not be sufficient curbing of population growth to prevent this predicted number of people by the end of the century. In our nation, where we have almost a zero population growth in our birth rate, we will add approximately 50 million people during the next 25 crucial years. These will result, to a great degree, from extending the length of life at all levels, but particularly for older people. However, a report just published in

Science by two California demographers, June Sklar, of the University of California at Berkley and Beth Berkow, of the California State Department of Health, now predict that we will see a sudden change in the birth rate in the United States and a "baby boom" will occur. Women who have delayed having children now appear to wish to raise a family, a reversal of the "childless family" attitude of the late 1960s and early 1970s. This 20–25% growth in the population destined to occur in this country in such a short period of time is difficult to comprehend and promises to put almost unbearable pressure on our already strained economic and social systems.

As we face this population increase in the United States and a much greater growth at the worldwide level we find this country limited as to what we can and will do from the humanitarian standpoint to alleviate the hunger in foreign nations. This was brought to our attention in an emphatic manner with recently stated policies by our top officials from the Departments of both State and Agriculture at the International Food Conference in Rome.

THE U.S. POSITION AND THE DILEMMA WE FACE

It is interesting to examine the position of the United States with regard to our providing food to the hungry nations and to realize how dependent we are on the agricultural sector of this nation for the condition of our economy. Secretary of Agriculture Butz did have his very good reasons for insisting that the United States must sell its food grains rather than expand its foreign aid and food "give away program." On the other hand, we know that we cannot, as a nation, stand by and see others die from malnutrition and starvation. We have serious problems facing us in this country as we try to meet the expanding population and food requirements of this nation and at the same time continue with the outward shipment of food grains at the present and expected expanding levels for the future. We need the export trade to maintain our standard of living and to minimize our foreign trade deficit.

This seemingly selfish attitude by U. S. leaders is somewhat understandable when we realize that the grain from North America accounts for well over 90% of that grain that enters into world trade, and the United States accounts for more than 75% of this continent's grain exports. We in the United States, with some help from Canada, exported nearly 90 million metric tons of grain into the world trade market last year. Agriculture is the mainstay of our world trade, and it is largely on this tremendous grain export that the vital part of our economy rests. The increasing importance of the food grains to our total export trade and the indications that this will be increasingly important in the future provide some explanation for the attitude of our political leaders.

Today we cannot meet the demands of the foreign countries for feed grains. We have recently read where Russia needs 40 million metric tons of grain, and we had determined that we were prepared to ship to them about 25% of this requirement. We called a temporary halt to shipping grain to determine whether we could satisfy our own needs with crops yet to be harvested. We await the outcome of the present corn crop before making the final decision as to how far we can go toward meeting the needs of one of the major nations involved in buying the grain we produce.

We are rapidly approaching that situation that the PSAC Study indicated we might face during 1982–1985. We predicted that the United States would be faced with a decision as to whether to keep the food at home to feed our own people or send it abroad to assist in providing for the food needs of the hungry people throughout the world.

The alternatives with which we will soon be faced are whether we will: (a) be able to continue as the principal provider of the critical food supplies desired and needed by the rest of the world or (b) find other means whereby this country will meet its foreign trade deficit when we are no longer able to participate in the world grain trade at the level that we have reached and have come to depend on during the last few years.

In bringing these issues to you I do so with the realization that there is a danger of possible criticism of my suggestion that this country bear the major responsibility for alleviating the hunger and malnutrition in other parts of the world and for meeting the world food problem. I have not intended to imply that we should feed the world. We have a fantastic record with regard to the foreign aid that has been provided since 1949 and the development of the Truman Doctrine. However, for a variety of reasons—economic, biological, and agricultural—we cannot continue at the level of aid required to meet the increasing needs.

MEETING THE WORLD'S FOOD NEEDS

Realizing the fantastic problem to be faced, we are still hopeful that the critical food needs of most of the world population can and will be met. *How will it be done?*

Not through a miracle food—nor some ingenious development of a food substitute. Improvement in quality of the foods such as high-protein maize certainly could be a great benefit for some. However, the critical decade or two ahead will require that conventional food, principally food grains, must be provided from the land and will be produced by conventional agricultural methods.

Now we must determine how far we can expect to advance in increasing the productivity of the agricultural sector of our own nation, which will be extremely important in meeting the world's food needs. Various authorities see different outcomes with regard to the possibility of our continuing to increase productivity to meet these needs. For example, the Committee on Agricultural

Production Efficiency of the Board of Agriculture and Renewable Resources formed by the National Research Council of the National Academy of Sciences concluded recently:

> We end this report by indicating the clouds on the horizon do indeed cast doubts about our national ability to produce all the food we demand at reasonable prices. The clouds are yet small; we cannot know for sure how significant they may turn out to be. Agriculture output may not rise as fast as the demands of people and their pets, and those of foreign countries. The public must realize that new conditions such as supply and cost of energy and the rate of production, knowledge and technology are challenging our access to the plentiful supply of low-cost food that has dominated our attitude toward agriculture over the past quarter of a century. . . . For its own preservation, society must continue to seek understanding of the very complex agricultural system that produces its food. Support of research and extension and maintenance of a favorable balance between societal supports and constraints affecting agriculture and the environment are necessary concerns for all citizens now and for the future.

We will see some additional cultivated acres placed into production as a result of the draining of swamps, clearing of forests, and irrigating of dry lands. However, for the most part, the land now in production will provide the basis for meeting the nation's food needs and maintaining our export level.

What can we look for in terms of the productivity of the major cereal crops? Although I cannot speak as an authority in this area, it is my prediction that we will probably not find any major breakthroughs in our cereal crops such as the one that we experienced with the development of hybrid corn. My expectation is that within the next critical 25 years some 10–20% increase in production of wheat above the present national average of 32–33 bushels may be a rather realistic appraisal of the situation. As far as potential is concerned, this is a far distance from the reported top yield of 135 bushels and a record yield of 216 bushels. About the same may be forthcoming in soybeans, that is, 10–20% increase in yield over the next 15–25 years.

I would look to you for a prediction of what we can expect for the future from corn. Dr. W. A. Russell, of Iowa State University, in his paper on the comparative performance for maize hybrids representing different eras of maize breeding, made an experimental evaluation of the progress made in yield from corn breeding over a 40-year period (1930–1970). Russell indicated some 50–60% increase in yield in our most modern hybrids above the unimproved open-pollinated varieties. He does not think that we are yet on a "yield plateau."

In answer to the question as to whether maize breeders will continue to increase the genetic potential for grain yield he replied, "Yes, provided we do not experience some unforseen circumstances that will interrupt progress." I certainly would not argue with Dr. Russell in the claim that we will continue our progress, but I do not see a continuation of the progress that we have experienced in the past or any "major breakthroughs" in the foreseeable future in corn improvement. In my opinion, the major problem we are experiencing now in corn research is the lack of sufficient basic, fundamental information that should have come from increased research efforts, which have not been possible during these last 10–15 years. The financial resources have not been available for support of the needed research. With the increasing inflation and cost per unit of effort it can be shown at both the state and federal levels that the effective research support is less today than it was in years past. We are to a large measure drawing on the volume of research information created in the past, and we are not replenishing the information as rapidly as is necessary if continued progress is to result from our breeding and genetic efforts.

In my opinion, if we are to realize continued increases in yield of corn of the magnitude we have experienced in the past, we will require one or more of the following developments:

1. An entirely different kind of plant, possibly one that would have multiple stalks or grain borne on top of the plant, much as wheat, and most certainly a plant with much greater photosynthetic capacity than the present maize.

2. A required level of fertility, which is becoming increasingly difficult to maintain due to the lack of commercial fertilizer at a cost economically feasible to apply to corn (incidentally, the report of the discovery of "Nitrogen fixation in maize," *Science,* August 1, 1975 could be of great economic value).

3. A total agronomic program geared to maximum production with optimum utilization of all required inputs.

4. A much greater use of exotic germplasm that has been properly introduced into interbreeding germplasm pools and adequately studied to determine its usefulness and the appropriate breeding methods for realizing the full potential of the genetically diverse material.

CONCLUDING REMARKS

There are various ways that we must find to meet the vast food problems that the world is facing today. A most significant contribution has been the development of research information from the international centers that are dedicated to the improvement of our cereal crops. These centers have been jointly financed, built, and operated by the Rockefeller and Ford Foundations and participated in by scientists representing many nations.

We have been able to provide technology when called on to do so and, when resources are available, to finance the efforts of those of you who are willing and able to assist in person in these international developments. However, we are, as the PSAC study on the world food supply predicted, requiring an increasing commitment from both our scientists and our agricultural production personnel to meet the domestic needs as well as the requirements of the foreign trade of U. S. agriculture. To a great extent the productivity of your efforts will depend on resources available for an expanded research effort by you and your colleagues in the immediate years ahead. I hope that these resources will be forthcoming, but I have seen no indication that we will see any major change in the attitudes of the leaders at the state and national level with regard to providing for these needs.

This pessimism comes from experiences such as that of attempting to establish a National Plant Genetics Resources Commission and the development of a national plan for preservation and use of plant genetic resources. Those of us who have been involved have been discouraged at the tremendous amount of time required to initiate a plan that is so obviously needed to maintain present production levels and is required for increasing production of our principal cereal crops. Be that as it may, we cannot be discouraged even though we may be somewhat pessimistic.

We honor a great man by honoring Gregor Mendel and his tremendous contributions. We do so realizing that he worked for many years with an idea and with very limited resources and certainly without any appreciable encouragement from many of his colleagues both within and without the monastery where he resided.

The situation is vastly different today, but there is a similarity. The world depended on Gregor Mendel and other geneticists who followed him, and without these contributions the tremendous advances that we have seen in corn and other crops would have been impossible. The extent to which we in the United States will meet our commitments abroad and continue with our present level of foreign trade, and the extent to which vast numbers of people in many countries will have a sustaining diet in the years ahead, will depend to a great degree on your continuing efforts toward deriving the basic information and applying it to the improvement of the yield of corn. I hope that Dr. Russell is right in stating that we have not reached a yield plateau level of corn. If we are to meet the needs that I foresee, we will have to accomplish a steeper climb in the yield curve for corn in the future than we have had in the recent past. We and the rest of the world look to and depend on you for this most important accomplishment.

Section One

HISTORY

Chapter 1

INTRODUCTORY REMARKS TO THE SESSION ON THE HISTORY OF HYBRID CORN

G. F. Sprague
University of Illinois,
Urbana

Corn was one of the species used in the trio of papers confirming the earlier findings of Gregor Mendel. The species has played a continuing role in both basic and applied studies. The correlation between cytologic and genetic crossing over was first established in corn, as was the mechanics for the utilization of heterosis on a commercial scale, one of the major contributions to increased agricultural productivity in this century.

One of the important factors leading to continued genetic and plant-breeding progress within this species has been the prevailing spirit of cooperation. Beginning some time before 1920 most of the corn geneticists and breeders in the United States either did their Ph.D. research under Professor R. A. Emerson of Cornell University or were exposed to his influence during their postdoctoral training. Under his influence there developed a spirit of willingness to exchange both genetic stocks and information. The Maize Genetics Cooperation, which Professor Emerson founded and supervised during his lifetime, was a tangible expression of this cooperative spirit. This spirit within

11

the corn fraternity has become international in scope as evidenced by the membership of the advisory committee of this symposium and the contributions they have made to the development of the program of the next few days.

The interests of scientists using corn as a test organism are so diverse as to cover many different disciplines. In consequence their paths cross only infrequently at various scientific meetings. One major objective of this symposium is to provide an opportunity for those diverse groups to meet personally, to review the progress achieved in the separate areas in the hope that even more productive cooperation may be established.

Our first session covers some of the early developments in corn breeding. Many of the early corn breeders received training under Professor Emerson. It is not surprising, therefore, that the same spirit of cooperation exemplified in genetics carried over to applied breeding. Support under the Purnell Act and the establishment of the Corn Conference under the guidance of Dr. F. D. Richey provided the necessary vehicle for exchange of both materials and information during these early years.

In 1933 approximately 0.4% of our corn acreage was planted to hybrids. Adoption was rapid, and by 1945 approximately 90% of the acreage was hybrid. Hybrid corn technology has now spread throughout most of the important corn-growing areas of the world.

The speakers selected for the first session review some of these important developments.

Chapter 2

MAIZE BREEDING DURING THE DEVELOPMENT AND EARLY YEARS OF HYBRID MAIZE

Merle T. Jenkins

Formerly Principal Agronomist in Charge of Corn Investigations
USDA Agricultural Research Service

Hybrid maize is a product of the 20 century. The idea was outlined by G. H. Shull in 1909. It was implemented by suitable breeding programs beginning about 1916 and expanded during the 1920s. Commercial use began in the early 1930s and by 1950 most of the maize-growing area of the United States was being planted with hybrid maize. The detailed history of hybrid maize in the United States has been reviewed by several writers (Hayes, 1963; Jenkins, 1936; Richey, 1950; Sprague, 1959), and I do not attempt to cover it in detail here. Instead I have chosen to review and emphasize some of the contributions that, in my opinion, facilitated progress in the development and improvement of hybrid maize during its early years.

    G. H. Shull deserves full credit for suggesting the pure-line method of maize breeding. As a result of maize-inheritance studies begun in 1905 at the Carnegie Institute of Washington, Cold Spring Harbor, New York, he published "The Composition of a Field of Maize" in 1908 (Shull, 1908) and

13

"A Pure-line Method of Corn Breeding" in 1909 (Shull, 1909). Shull's 1908 paper was the first suggestion that the deterioration that results from self-fertilization was due to the isolation of homozygous biotypes and that the individual plants in a field of maize are complex hybrids among such biotypes. In his 1909 paper he outlined the three essential steps in his pure-line method of maize breeding. These included; (a) large-scale inbreeding to obtain many homozygous or near-homozygous lines, (b) testing the selected pure lines in all possible crosses, and (c) utilizing the best crosses for practical corn production. In later publications, Shull (1910, 1911) urged the various agricultural experiment stations to try his pure-line method of maize breeding.

In spite of Shull's urging, maize breeders were slow in adopting his pure-line method of breeding. Shull envisioned using single crosses involving only two inbred lines. The reduced vigor of the lines with the resulting low yields of small kernels introduced difficulties that made the method seem less practical to breeders than the mass selection and ear-to-row selection procedures then in common use. To circumvent these difficulties, Jones (1918) suggested the use of the double cross, which is the product obtained by crossing two single crosses. Interest in the pure-line method of breeding increased with additional research on quantitative inheritance and on the genetic explanation of hybrid vigor.

At the time Shull outlined his pure-line method of maize breeding, he believed that hybrid vigor was due to the physiological stimulation of heterozygosis (Hayes, 1963, p. 15). In 1910, Bruce suggested on purely mathematical considerations that hybrid vigor might be explained on the basis of the complementary action of favorable dominant growth factors. Shortly thereafter Keeble and Pellow (1910) reported data on a pea hybrid illustrating the action of two genes in increasing the plant height of the hybrid.

Two main objections were raised to the complementary action of favorable dominants as the explanation of hybrid vigor. The first related to the assumption that it should be possible in later generations to collect all of the dominants in one line that would equal the F_1 hybrid in vigor. The second objection was based on the assumption that the F_2 distributions should be skewed. Jones (1917) suggested that linkage between favorable dominant growth factors would account for failure to meet these two theoretical requirements. Collins (1921) demonstrated that for as many as 20 pairs of genes, neither of these assumptions was valid.

In 1913 Emerson and East reviewed previous work on the inheritance of quantitative characters and reported data on their extensive experiments with maize. They concluded that the "multiple-factor hypothesis," that is, the assumption of numerous independently mendelizing genes, "furnished a satisfactory and simple interpretation not only of all of the results secured in these experiments but also of the experiments previously reported for other plants and animals." This explanation of the inheritance of quantitative characters undoubtedly aided in the acceptance of Bruce's (1910) suggestion of the complementary action of favorable dominant growth factors as the explanation of hybrid vigor. In fact, East (1936) states, "The key to heterosis is in the inheritance of quantitative characters." Of course, size factors are not necessarily dominants, but the mechanism of numerous genes mendelizing independently is similar for both the explanation of hybrid vigor and of quantitative inheritance.

SOURCES OF INBRED LINES

When inbreeding projects first were initiated, the adapted local varieties served as source material for line development. In the central Corn Belt, many of the local dent varieties such as Reid Yellow Dent and similar strains had been developed from crosses between early northern flints and late southern dents such as the Gourdseed variety, followed by some 50 years of intercrossing and selection. In Minnesota and Wisconsin, north of the central Corn Belt, the local varieties represented a much wider range of genetic types than in the central Corn Belt. In the areas south and west of the central Corn Belt, especially where white varieties were popular, fewer varietal types were included in the early inbreeding experiments.

Early crossing experiments indicated the importance of genetic diversity between the parents of productive hybrids. Crosses between lines from different varieties were likely to be more productive than crosses between lines from the same variety. The early cooperation and free exchange of breeding material among state and federal maize breeders provided all breeders with a genetic diversity of material that aided the overall program. This cooperation was promoted first informally by F. D. Richey of the U. S. Department of Agriculture (USDA) and later by the Purnell and Regional Corn Improvement Conferences.

One important contribution to the early success of hybrid maize in the Corn Belt was the fortunate introduction of the Pennsylvania variety Lancaster Surecrop into the Corn Belt breeding programs through the efforts of F. D. Richey. As Richey was much impressed with the performance of Lancaster Surecrop on his father's farm in LaSalle County, Illinois, he began inbreeding it himself and supplied seed for the Iowa and Ohio breeding projects. The Lancaster variety was developed by Isaac Hershey in Pennsylvania (Anderson, 1944) from the cross of an early flint with a large, late, rather rough local variety resembling Golden Queen. The variety is very subject to root and stalk lodging, and the lines developed from it tend to carry these weaknesses. The outstanding performance over the past 40 years of numerous hybrids involving lines from Lancaster in combination with lines from the Corn Belt dents would indicate a substantial genetic difference between the two strains. The Lancaster lines have been unusually prepotent in their reaction with Corn Belt dent lines, one Lancaster line in a double cross being sufficient to exert a marked influence on the performance of the hybrid.

Some of the more important Lancaster lines are CI.4-8 developed by Richey, L289 and L317 developed in Iowa, Oh40B developed in Ohio, and C103 developed in Connecticut. The Iowa line L289 was one parent of Iowa 939, the most widely grown hybrid in its day, and the first of the widely grown hybrids. The L317 line was one parent of U. S. 13, which succeeded Iowa 939 as the most widely grown hybrid of its time, and is possibly the most widely grown hybrid we have had. The Oh40B and C103 lines appeared on the scene a little later than the Iowa lines. The Oh40B line has not been used as widely as its second-cycle recoveries Oh43 and Oh45. The Oh43 and C103 lines have been used as parents of many outstanding hybrids, and both they and their related recoveries still are being used extensively.

The first hybrids widely used in Kentucky and Tennessee are interesting exceptions to the intervarietal hybrids mentioned above. In Kentucky, the Boone County White variety was popular, and the first lines released were all from this variety. The first hybrids released in Kentucky all involved lines from Boone County White and by the mid 1940s, Ky72B, the most popular of these early releases, was being grown on close to 200,000 ha. A very similar situation existed in Tennessee with respect to the Neal Paymaster variety. Neal Paymaster was a popular variety in Tennessee, and inbreeding in the beginning years was concentrated on this variety. Several hybrids involving only Neal Paymaster lines were released in the early 1940s, of which Tenn. 10 became the most popular. Nearly 3.7 million kilos of seed of Tenn. 10 was produced in 1949, enough to plant more than 230,000 ha in 1950.

In the late 1920s and early 1930s several synthetics were initiated, utilizing the better inbred lines available at that time. The most successful of these was the Stiff Stalked Synthetic, so named because the parent lines were especially selected for resistance to root and stalk lodging. This synthetic has yielded several outstanding inbred lines, B37 being the most important. It was 25 years from the time the synthetic was started before lines from it became available for general use.

CORRELATION STUDIES AND SELECTION CRITERIA

With the expansion of inbreeding programs in the 1920s increased interest developed in identifying characters of the inbred lines associated with their productivity and the productivity and other desirable attributes of the hybrids involving them. Nilsson-Leissner (1927) studied four groups of lines, two dent and two flint, and computed coefficients of correlation between the mean values of each of six characters in the two parental lines and the values of the same character in the F_1 cross between them. All correlations were positive. The combined multiple correlation of characters in the parents with yield of the hybrids for the two groups of dent lines was 0.6687 and for the two groups of flint lines, 0.8240. Jorgenson and Brewbaker (1927) computed similar parent–hybrid correlations for 10 lines from the Silver King variety. Their correlations also were all positive and the multiple correlation with yield was 0.6074.

Jenkins (1929) studied five groups of lines and the F_1 crosses among the lines within each group. He reported positive correlations between 19 characters in the inbred parent lines and the means of the same characters in all of the F_1 crosses of the individual lines. Yield of the F_1 crosses was positively and significantly correlated with date of tasseling, date of silking, plant height, number of nodes per plant, number of nodes below the ear, number of ears per plant, ear length, and ear diameter of the parental inbred lines. Correlations between the yield of the parent lines and the mean yields of their F_1 crosses for the five groups of lines studied were 0.67, 0.64, 0.45, 0.41, and 0.25.

Hayes and Johnson (1939) reported positive and significant correlations between five characters of 110 inbred lines and the same characters in their inbred-variety crosses. They also reported positive and significant correlations between twelve characters of the inbred parents and the yield of the inbred-variety crosses. The correlation between the yield of the parent lines and yield of their crosses was 0.2470 and the multiple correlation between twelve characters of the inbred parents and yield of the crosses was 0.6660.

In general, the correlation studies all seem to indicate that effective selection may be made in the inbred lines for characters desired in the hybrids and that the characters indicating vigor in the parental lines are associated with increased productivity in the hybrids.

EVALUATING COMBINING ABILITY

In the early years when relatively few lines were available, it was common to evaluate the combining ability of lines in single crosses. Lines frequently were tested in groups in diallel crosses among members of the group. This procedure involved much labor and expense and a simpler procedure was required. Davis (1929) suggested the use of inbred-variety crosses to evaluate S_2 lines, but the suggestion was overlooked. Jenkins and Brunson (1932) reported coefficients of correlation between the rankings of inbred lines by their average performance in a number of single crosses and their performance in inbred-variety crosses. The experiments involved eight groups of lines and different testing locations. The correlations between the average single-cross yield and the inbred-variety cross yield ranged from 0.53 to 0.90, the larger correlations being obtained for lines selfed for six generations. For comparison, the correlations between the average yields of three groups of lines tested in two series of single crosses were 0.65, 0.69, and 0.82. It was concluded that inbred-variety crosses may be used efficiently in the preliminary testing of new lines to eliminate perhaps 50% of the material without serious danger of losing superior material.

Johnson and Hayes (1936) compared inbred-variety crosses or, more briefly, topcrosses, to evaluate the combining ability of 11 inbred lines of Golden Bantam sweet corn. The correlation between the topcross yields of the 11 lines and the average yield of their single crosses with the other 10 lines in the group was 0.7835. Cowan (1943) obtained highly significant correlations of 0.4872 and 0.3278 between topcross yields of unrelated inbred lines and their yields in single crosses and predicted double crosses, respectively.

Topcrosses have been used fairly widely for the preliminary evaluation of new inbred lines. They give information on general combining ability only and need to be followed by tests in single crosses for information on nicking with individual lines. Crossed seed may be produced by using the variety as male parent and detasseling the inbred lines, or it may be made by hand pollination, using pollen from the inbred lines on silks of the variety. Sprague (1939) has shown that 10 plants are needed to adequately represent the variety.

Any technique for the evaluation of combining ability requires the choice of an appropriate tester. Federer and Sprague (1947) found that the line X tester interaction was nearly 40% as great as the line effect when both effects were measured by means of the variance component. For maximum efficiency the tester should be tailored for the specific information desired. They also concluded that increasing the number of testers will improve the estimate of combining value more than increasing the number of replications. Sprague (1946) reported effective selection for resistance to stalk breaking using a susceptible double-cross tester. The writer has obtained similar results (unpublished) using a susceptible single cross as tester. Susceptible testers for other specific characters should promote critical testing for these characters.

Green (1948a,b) compared the high-yielding double cross U. S. 35 and the open-pollinated variety Black Yellow Dent as testers for evaluating 83 S_2 plants from each of three single crosses. The single crosses were between lines designated as high × high, high × low, and low × low in combining ability. Using U. S. 35 as tester, no difference could be established in the segregation for combining ability among plants from the high × high versus the high × low series. Using the variety as the tester, the difference between the two series was highly significant. Both U. S. 35 and Black Yellow Dent were able to establish significant segregations for root lodging and stalk breaking in 1945, a year of severe damage. In 1946, a year of little damage, Black Yellow Dent, the more susceptible tester, gave the best classification. Thus the detected segregation for combining ability was a function of the genotype of the lines and the genetic characteristics of the tester parent.

Sprague and Tatum (1942) partitioned combining ability, as expressed in single crosses into two portions, general and specific, on the basis of the type of gene action involved. They interpreted the variance of general combining ability as a measure of additive genetic variance and that of specific combining ability as a measure of the nonadditive genetic variance.

EARLY TESTING

Common practice in the earlier years of inbred line development was to self-pollinate for three to five generations before evaluating the lines for combining ability. As breeding programs expanded and more lines were developed, it was advantageous to discard unpromising lines as early as possible. Jenkins (1935) reported data on topcrosses of a number of inbred lines from the Lancaster and Iodent varieties made after S_1 through S_8, except for S_7. The variance due to differences between lines was significantly greater than that due to the interaction between lines and generations. It was concluded that the lines exhibited their individuality early in the inbreeding period, justifying early evaluation for combining ability and early discarding of unpromising lines. Chance sampling in following generations should result in choosing individuals from the modal classes and maintain the general combining ability of the lines. Beard (1940) compared single crosses and an open-pollinated variety for the preliminary testing of new inbred lines and found them comparable in accuracy. A high-yielding, heterozygous single cross seemed to give the most accurate evaluation. Jenkins (1940) in a second experiment tested the topcrosses of 16 plants within seven different S_1 lines of the Krug variety. The average standard deviation among the topcrosses of individual plants within the seven S_1 populations was 2.7 bushels per acre, indicating only one chance in 10 of obtaining a plant within this generation whose topcross would yield 5.6 bushels or 8.9% above the mean for the line. Sprague (1946) tested 167 S_0 plants from a stiff-stalked synthetic in topcrosses on the double cross, Iowa Hybrid 13. Highly significant differences were obtained among the yields of these topcrosses. Selfed progenies from the parents of the upper 10% of the topcrosses were grown, the S_1 plants were selfed and crossed on Iowa Hybrid 13, and these topcrosses were compared for yield. Their mean yield equaled the mean yield of the topcrosses of their selected parents. A second sample of selected parental lines represented a seriated sampling of the entire frequency distribution. The selfed progenies from these lines were grown and the S_1 plants also were selfed and crossed on Iowa Hybrid 13 and the topcrosses tested. The correlation between the topcross yields of the S_0 and S_1 plants was 0.85.

LINE IMPROVEMENT

In the beginning years all inbred lines were isolated from varieties. As good lines became increasingly available the chance of isolating a new line superior to those already available was greatly reduced. One of the first methods tried to obtain superior lines was to self in good single crosses and double crosses. Usually the single crosses were between lines that complemented the weaknesses they each carried. The most extensive data on the extraction of new lines from single crosses, a procedure they designated as "pedigree selection," came from Hayes and Johnson (1939), Wu (1939), and Johnson and Hayes (1940). Lines were obtained which were superior to their parents in combining ability, lodging resistance, and smut resistance. The largest proportion of the high-combining lines was obtained from single crosses with at least one parent that had high combining ability.

Hayes, Rinke, et al. (1946a) have reported extensive data on line improvement for yield, moisture content, and combining ability following two and three generations of backcrossing. Backcrossing also has been used extensively to transfer individual genes such as wx, $o2$ and the Ht gene for resistance to the leaf blight caused by $Helminthosporium$ $turcicum$.

Convergent improvement, suggested by Richey (1927), is a system of double backcrossing. It was suggested originally as a method of improving inbred lines and of evaluating the complementary dominant theory of hybrid vigor suggested by Bruce (1910). Richey and Sprague (1931) reported six comparisons of yields of progeny lines and crosses after one to six successive generations of crossing to the same recurrent parent. Yields of the progeny lines were somewhat above expectation on the basis of theory, indicating some contributions from the nonrecurrent parent had been retained by selection. The yields of the F_1 crosses between the selected backcrossed lines and the nonrecurrent parents also were in excess of the theoretical values for similar crosses with unselected backcrossed lines. As the F_1 crosses involving the backcrossed lines had fewer genes heterozygous than the original crosses, the results were interpreted in favor of the complementary dominant theory of hybrid vigor. Murphy (1942) reported data on the convergent improvement of the four inbred parents of two single crosses. The recovered lines showed marked improvement in vigor, in resistance to smut and lodging, and in soundness of ear in those instances where the original lines were lacking in these characters. For each of the two crosses it was possible to obtain F_1 crosses between recovered parental lines that yielded significantly higher than the original F cross. These data further support the complementary dominant theory of hybrid vigor. Hayes, Rinke, et al. (1946b) reported the results of efforts to improve Minhybrid 403 by convergent improvement. Two recovered versions of the hybrid were produced and the better one was slightly earlier than Minhybrid 403 and was significantly higher yielding.

Stadler (1944) suggested gamete selection as a more efficient method of sampling open-pollinated varieties or other heterozygous populations than selecting plants. The best 1% among gametes represent a level expected to occur among zygotes at the rate of only 1 in 10,000. Pinnell, Rinke, et al. (1952) tested gamete selection for improving the combining ability of three inbred lines, the lowest combining parents of each of three established double crosses. Each line was crossed by gametes from a different variety. Almost 50% of the gametes showed evidence of having combining ability in excess of the sampler line. Lonnquist and McGill (1954) sampled gametes from South American popcorn and the dent variety Hays Golden. Gametes were obtained from Hays Golden, which exceeded in combining ability the gametes from 16 elite lines.

Line improvement resulting from the intercrossing of superior lines and reselection gradually limits the genetic base of the breeding material. Jenkins (1936) listed some 350 lines developed by State Agricultural Experiment Stations and the USDA. Only eight of the lines listed were second-cycle lines. In 1948 Clarion Henderson of Illinois Foundation Seeds assembled a list of "Inbred lines released to private growers from state and federal agencies," which he has updated every four years. In 1948, 20% of the lines released by agencies in the Corn Belt were second-cycle lines. In 1952 the proportion of second-cycle lines had increased to 26%, in 1956 to 40%, and in 1960 to 50%. Since 1960 most of the new inbred lines released in the Corn Belt have been second-cycle, recovered, or backcrossed lines. In view of this trend we might be wise to devote more attention to broadening the genetic base of new breeding material.

PREDICTING HYBRID PERFORMANCE

With the general use of double crosses for commercial maize production and the rapid increase in the number of new lines being developed, the testing of the many possible double-cross combinations among them soon become burdensome. The number of new lines increases arithmetically, whereas the number of possible double-cross combinations among them increases exponentially. Excluding reciprocals, there are 14,535 double-cross combinations among 20 inbred lines and 11,765,675 among 100 lines! Obviously, some procedure for estimating the most promising combination was highly desirable.

Jenkins (1934) reported on four methods of estimating double-cross performance from single-cross and topcross data. The four prediction procedures compared were the average performance of the: (a) six possible single crosses among the four parental lines, (b) four nonparental single crosses, (c) four parental lines over all of their single-cross combinations, and (d) topcrosses of the four parental lines. Data were available on 53 of the 55 possible single crosses among 11 parental lines, the topcrosses of the 11 lines, and 42 double crosses involving the lines. Coefficients of correlation were computed between predicted and actual performance of the double crosses.

Methods a, b, and c gave rather similar estimates and correlations with actual performance. Method d gave the poorest estimate. From the genetic standpoint, Method b estimates should be the most reliable since as Jenkins states, "In any double cross, the genes of each of the four parental lines are united only with allelomorphs of the two lines whch enter the double cross from the opposite parent." Doxtator and Johnson (1936), Anderson (1938), Hayes, Murphy, et al. (1943), and Hayes, Rinke, et al. (1946) all have reported data indicating good agreement between method b predictions and actual double-cross performance. Peterson, Eldridge, et al. (1949) reported data indicating that the performance of popcorn three-way crosses may be predicted satisfactorily from single-cross data. Millang and Sprague (1940) and Combs and Zuber (1949) have outlined procedures for the use of punched-card equipment to facilitate routine prediction computations.

Method b predictions imply that there may be significant differences in performance among the three double crosses that may be made among four parental inbred lines. This is fully demonstrated in the data reported by the Minnesota workers mentioned above. Eckhardt and Bryan (1940a, b) have presented data on the order of pairing the four parents of a double cross. In comparing double crosses among inbreds from two varietal sources, the highest yielding combinations were obtained when the two lines from one variety came into the cross from one side and the two lines from the other variety came in from the other side. Crosses of this kind also were less variable for plant and ear height, ear length, and ear weight. In similar crosses involving early (E) and late (L) lines the variance for silking data, ear height, ear weight, ear diameter, and ear length were significantly less for the (E × E) (L × L) combinations than for the (E × L) (E × L) combinations. Pinnell (1943) studied the variability of eight plant characters of four unrelated inbred lines, their six single crosses, and three double crosses. He was not able to predict the relative variability of the double crosses from the character means of either the inbreds or the single crosses.

RECURRENT SELECTION

The general procedure or recurrent selection was suggested by Jenkins (1940) as a method of developing and improving synthetic varieties. Hull (1945) suggested a modified version adapted to selection for specific combining ability and named the procedure "recurrent selection." The general procedure is suitable for selection of quantitative characters of all types. It consists of selfing once or twice, evaluating the selfed lines for the character under considera-tion, intercrossing the superior lines, and repeating the process using the reconstituted population as the source of the new lines.

Collins (1921) showed that with as many as 20 factor pairs affecting a character, extremely large F_2 populations must be grown to provide an even chance of recovering all of the dominant alleles in one F_2 plant. If such a plant should occur, it would be extremely difficult or impossible to recognize. The alternative is to grow smaller F_2 populations, select several F_2 plants with as many favorable dominant alleles as possible, intercross the selected plants to increase the frequency of the selected genes within a smaller population, and repeat the process until an acceptable frequency of selected genes has been attained.

Sprague and Brimhall (1950) and Sprague, Miller, et al. (1952) compared the effectiveness of inbreeding with selection versus recurrent selection for increasing the oil content of the maize kernel. In both experiments the recur-rent selection procedure was the more efficient by a factor of 2–5 and still retained enough variability to indicate that maximum progress had not been achieved.

Jenkins, Robert, et al. (1954) reported results of recurrent selection for resistance to *Helminthosporium turcicum* leaf blight. The plants were inocu-lated and resistant plants could be identified before pollination. Pollen was collected from resistant plants, mixed, and applied to the silks of the same plants, thus permitting a cycle of selection in each generation. More progress was made in the first and second cycles of selection than in the third. The method was effective in isolating populations with acceptable resistance in each of the nine breeding lines included in the experiment.

Recurrent selection for general combining ability implies the production of a synthetic or population developed from intercrosses among the selected lines. The tester should have a rather broad genetic base such as a variety, selected population, double cross, or several single crosses. If the selected plants are selfed and crossed on the tester by hand, care must be exercised to ensure that the tester is adequately sampled. Sprague and Brimhall (1950) reported that one cycle of selection for general combining ability increased the mean of the selected group about 7 bushels per acre.

Lonnquist (1951) reported the results of one generation of selection for combining ability in the Krug variety and McGill and Lonnquist (1955) sum-marized the results following the second cycle of selection. Three synthetic varieties were developed from Krug, two were selected for high yield, and one for low yield, using WF9 × M14 as the tester. Following the second cycle, the three synthetics and Krug were evaluated by topcrosses of S_0 plants from each population made on the tester WF9 × M14. The mean yields for the topcrosses of the two high yield synthetics were 5.1 and 5.5 bushels per acre above that of the topcrosses for the Krug variety, and the mean yield of the topcrosses for the low yield synthetic was 2.3 bushels below that for Krug. It was evident that selection had been effective in modifying the combining ability of the Krug plants with WF9 × M14. Variability tests indicated that the variability of the synthetics was below that of the Krug variety and that it had been reduced more than expected.

Lonnquist and McGill (1956) compared the yields of four synthetics after one and two cycles of selection with the yield of U. S. Hybrid 13. Their average yield increased 14% of the yield of U. S. Hybrid 13 from the first to the second cycle of selection.

Hull (1945) suggested a process of selecting for specific combining ability. The plan involved choosing an outstanding inbred line that would be used as the tester and a suitable heterozygous population as the source of the S_1 lines to be crossed with the tester. As yet the method has not been tried to any extent. Its most serious disadvantage is that with the continuous shift and replacement of lines, few have remained in use long enough to warrant such a breeding program dependent on the continued use of a single line.

Comstock, Robinson, et al. (1949) suggested another modification of recurrent selection that they designated as *reciprocal recurrent selection.* Two heterozygous populations are used and each population serves as the tester for the other. On the basis of theory the authors state that reciprocal recurrent selection should give results superior to recurrent selection for general combining ability and for all loci exhibiting overdominance. It should be superior to recurrent selection for specific combining ability for all loci exhibiting partial dominance and should be no more than slightly inferior to the better of the other procedures. Much experimental evidence will be required to fully evaluate the relative efficiency of the three recurrent selection procedures.

Recurrent selection is without question a very useful procedure for increasing the frequency of desirable genes for any quantitative character or combination of quantitative characters in a selected population. It would seem to be most useful in the development of populations that would later be sampled by selfing, gamete selection, or other methods.

SUMMARY

By 1950 hybrid maize was well established in the United States and was beginning to attract attention in other areas. Almost 90% of the maize growing area of the Corn Belt was being planted with hybrid seed, and hybrids had been developed that were superior to local varieties in most of the other maize-growing areas of the country. The substitution of hybrids for adapted varieties was resulting in increased yields of at least 25%.

It had been well established that a diversity of germplasm was desirable for the production of good hybrids. Some inbred lines are outstanding in their combining ability, and others are mediocre. Methods had been developed and types of testers suggested to efficiently evaluate the combining ability of inbred lines for yield, lodging resistance, and other characteristics.

Double crosses were well established as the most efficient type of cross for general commercial production. A method of estimating the expected performance of new double-cross combinations had been developed and was being used to increase the frequency of superior combinations among the new experimental hybrids produced for testing.

From the standpoint of maize breeding it had been reasonably well demonstrated that hybrid vigor was due to the complementary action of favorable dominant or semidominant growth factors.

Procedures had been suggested and evaluated for improving existing inbred lines and developing superior new lines through the use of pedigree selection, backcrossing, gamete selection, convergent improvement, recurrent selection, and reciprocal recurrent selection. Several new synthetic populations were in the process of development for future sampling.

All of these important developments along with others not specifically reviewed here formed a substantial foundation for the additional progress that has been made in maize improvement in more recent years.

REFERENCES

Anderson, D. C. 1938. The relation between single and double cross yields in corn. *J. Am. Soc. Agron.* **30:** 209–211.

Anderson E. 1944. The sources of effective germ-plasm in hybrid maize. *Ann. Mo. Bot. Garden* **31:** 355–361.

Beard, D. F. 1940. Relative values of unrelated single crosses and an open pollinated variety as testers of inbred lines of corn. *Diss. Abstr.* **33:** 19–18.

Bruce, A. B. 1910. The mendelian theory of heredity and the augmentation of vigor. *Science, N.S.* **32:** 627–628.

Collins, G. N. 1921. Dominance and the vigor of first generation hybrids. *Am. Natur.* **55:** 116–133.

Combs, J. B. and M. S. Zuber. 1949. Further use of punched card equipment in predicting the performance of double-crossed corn hybrids. *Agron. J.* **41:** 485–486.

Comstock, R. E., H. F. Robinson, and P. H. Harvey. 1949. A breeding procedure designed to make maximum use of both general and specific combining ability. *Agron J.* **41:** 360–367.

Cowan, J. Richie. 1943. The value of double cross hybrids involving inbreds of similar and diverse genetic origin. *Sci. Agr.* **23:** 287–296.

Davis, R. L. 1929. Report of the plant breeder. *Puerto Rico Agr. Exp. Sta. Annu. Rep.* **1927:** 14–15.

Doxtator, C. W. and I. J. Johnson. 1936. Prediction of double cross yields in corn. *J. Am. Soc. Agron.* **28:** 460–462.

East, E. M. 1936. Heterosis. *Genetics* **21:** 375–397.

Eckhardt, R. C. and A. A. Bryan. 1940a. Effect of the method of combining the four inbreds of a double cross of maize upon the yield and variability of the resulting hybrid. *J. Am. Soc.Agron.* **32:** 347–353.

Eckhardt, R. C. and A. A. Bryan. 1940b. Effect of the method of combining two early and two late lines of corn upon the yield and variability of the resulting double crosses. *J. Am. Soc. Agron.* **32:** 645–656.

Emerson, R. A. and E. M. East. 1913. The inheritance of quantitative characters in maize. *Nebr. Agr. Exp. Sta. Res. Bul.* 2.

Federer, W. T. and G. F. Sprague. 1947. A comparison of variance components in corn yeild trials. I, Error, tester x line and line components in top-cross experiments. *J. Am. Soc. Agron.* **39:** 453–463.

Green, J. M. 1948a. Relative value of two testers for estimating top cross performance in segregating maize progenies. *J. Am. Soc. Agron.* **40:** 45–57.

Green, J. M. 1948b. Inheritance of combining ability in maize hybrids. *J. Am. Soc. Agron.* **40:** 58–63.

Hayes, H. K. 1963. A professor's story of hybrid corn. Minneapolis: Burgess.

Hayes, H. K. and I. J. Johnson. 1939. The breeding of improved selfed lines of corn. *J. Am. Soc. Agron.* **31:** 710–724.

Hayes, H. K., R. P. Murphy, and E. H. Rinke. 1943. A comparison of the actual yield of double crosses of maize with their predicted yield from single crosses. *J. Am. Soc. Agron.* **35:** 60–65.

Hayes, H. K., E. H. Rinke, and Y. S. Tsiang. 1946a. The relationship between predicted performance of double crosses of corn in one year with predicted and actual performance of double crosses in later years. *J. Am. Soc. Agron.* **38:** 60–67.

Hayes, H. K., E. H. Rinke, and Y. S. Tsaing. 1946b. Experimental study of covergent improvement and backcrossing in corn. *Minn. Agr. Exp. Sta. Tech. Bull.* No. 172.

Hull, F. H. 1945. Recurrent selection for specific combining ability in corn. *J. Am. Soc. Agron.* **37:** 134–145.

Jenkins, M. T. 1929. Correlation studies with inbred and crossbred strains of maize. *J. Agr. Res.* **39:**677–721.

Jenkins, M. T. 1934. Methods of estimating the performance of double crosses in corn. *J. Am. Soc. Agron.* **26:** 199–204.

Jenkins, M. T. 1935. The effect of inbreeding and of selection within inbred lines of maize upon the hybrids made after successive generations of selfing. *Iowa State Coll. J. Sci.* **9:** 429–450.

Jenkins, M. T. 1936. Corn improvement. In *U. S. Department of Agriculture Yearbook of Agriculture 1936,* pp. 455–522.

Jenkins, M. T. 1940. The segregation of genes affecting yield of grain in maize. *J. Am. Soc. Agron.* **32:** 55–63.

Jenkins, M. T. and A. M. Brunson. 1932. Methods of testing inbred lines of maize in crossbred combinations. *J. Am. Soc. Agron.* **24:** 523–530.

Jenkins, M. T., A. L. Robert, and W. R. Findley, Jr. 1954. Recurrent selection as a method of concentrating genes for resistance to *Helminthosporium turcicum* leaf blight in corn. *Agron. J.* **46:** 89–94.

Johnson, I. J. and H. K. Hayes. 1936. The combining ability of inbred lines of Golden Bantam sweet corn. *J. Am. Soc. Agron.* **28:** 246–252.

Johnson, I. J. and H. K. Hayes. 1940. The value in hybrid combination of inbred lines of corn selected from single crosses by the pedigree method of breeding. *J. Am. Soc. Agron.* **32:** 479–485.

Jones, D. F. 1917. Dominance of linked factors as a means of accounting for heterosis. *Genetics* **2:** 466–479.

Jones, D. F. 1918. The effects of inbreeding and cross-breeding upon development. *Conn. Agr. Expt. Sta. Bul.* 207.

Jorgenson, L. and H. E. Brewbaker. 1927. A comparison of selfed lines of corn and first generation crosses between them. *J. Am. Soc. Agron.* **19:** 819–830.

Keeble, F. and C. Pellew. 1910. The mode of inheritance of stature and time of flowering in peas (*Pisum sativum*). *J. Genetics* **1:** 47–56.

Lonnquist, J. H. 1951. Recurrent selection as a means of modifying combining ability in corn. *Agron. J.* **43:** 311–315.

Lonnquist, J. H. and D. P. McGill. 1954. Gametic sampling from selected zygotes in corn breeding. *Agron. J.* **46:** 147–150.

Lonnquist, J. H. and D. P. McGill. 1956. Performance of corn synthetics in advanced generations of synthesis and after two cycles of recurrent selection. *Agron. J.* **48:** 249–253.

McGill, D. P. and J. H. Lonnquist. 1955. Effects of two cycles of recurrent selection for combining ability in an open-pollinated variety of corn. *Agron. J.* **47:** 319–323.

Millang, A. and G. F. Sprague. 1940. The use of punched card equipment in predicting the performance of corn double crosses. *J. Am. Soc. Agron.* **32:** 815–816.

Murphy, R. P. 1942. Convergent improvement with four inbred lines of corn. *J. Am. Soc. Agron.* **34:** 138–150.

Nilsson-Leissner, G. 1927. Relation of selfed strains of corn to F_1 crosses between them. *J. Am. Soc. Agron.* **19:** 440–454.

Peterson, E. L., J. C. Eldridge, and I. J. Johnson. 1949. Predicting the performance of popcorn hybrids from single cross data. *Agron J.* **41:** 104–106.

Pinnell, E. L. 1943. The variability of certain quantitative characters of a double cross hybrid in corn as related to the method of combining the four inbreds. *J. Am. Soc. Agron.* **35:** 508–514.

Pinnell, E. L., E. H. Rinke, and H. K. Hayes. 1952. Gamete selection for specific combining ability. In *Heterosis*. Ames: Iowa State College Press, pp. 378–388.

Richey, F. D. 1927. The convergent improvement of selfed lines of corn. *Am. Nat.* **61:** 430–449.

Richey, F. D. 1950. Corn Breeding. *Adv. Genet.* **3:** 159–192.

Richey, F. D. and G. F. Sprague. 1931. Experiments on hybrid vigor and convergent improvement in corn. *USDA Tech. Bull.* **267:** 1–22.

Shull, G. H. 1908. The composition of a field of maize. *Am. Breeders Assoc. Rep.* **4:** 296–301.

Shull, G. H. 1909. A pure line method of corn breeding. *Am. Breeders Assoc. Rep.* **5:** 51–59.

Shull, G. H. 1910. Hybridization methods of corn breeding. *Am. Breeders Mag.* **1:** 98–107.

Shull, G. H. 1911. The genotypes of maize. *Am. Nat.* **45:** 234–252.

Sprague, G. F. 1939. An estimation of the number of top-crossed plants required for adequate representation of a corn variety. *J. Am. Soc. Agron.* **31:** 11–16.

Sprague, G. F. 1946. Early testing of inbred lines of corn. *J. Am. Soc. Agron.* **38:** 108–117.

Sprague, G. F. 1959. Mais (*Zea mays*). I. General considerations and American breeding work. *Handbuck der Pflanzuchtung 2 Auflage.* **2:** 103–143.

Sprague, G. F. and B. Brimhall. 1950. Relative effectiveness of two systems of selection for oil content of the corn kernel. *Agron. J.* **42:** 83–88.

Sprague, G. F., P. A. Miller, and B. Brimhall. 1952. Additional studies on the relative effectiveness of two systems of selection for oil content of the corn kernel. *Agron. J.* **44:** 329–331.

Sprague, G. F. and L. A. Tatum. 1942. General vs. specific combining ability in single crosses of corn. *J. Am. Soc. Agron.* **34:** 923–932.

Stadler, L. J. 1944. Gamete selection in corn breeding. *J. Am. Soc. Agron.* **36:** 988–989.

Wu, Shao-Kwei. 1939. The relationship between the origin of selfed lines of corn and their value in hybrid combination. *J. Am. Soc. Agron.* **31:** 131–140.

Chapter 3

THE HYBRID CORN INDUSTRY IN THE UNITED STATES

Leon Steele

Vice President and Research Director,
Funk Seeds International,
Bloomington, Illinois

> Since the early days of the Twentieth Century, hybrid corn has gone through the pilot plant stage. It has evolved into an industry which has affected agriculture in America more than any single thing since the steel plow came to make possible the farming of the boundless prairies. Like many other great contributions to our way of living, hybrid corn has crept quietly into our economy with only a few realizing what has happened.

These words are copied directly from an article (Holbert, 1944) that had a substantial impact on the business community, for most earlier articles about hybrid corn had appeared in farm magazines. Here was a success story that bankers and businessmen could appreciate—an industry that had grown from practically nothing in 1934 to a 60–70 million-dollar business in 1944.

What were those early days of the hybrid corn industry like? My personal experience began in 1925, when for three weeks I worked as a water boy on the USDA corn-breeding station on Funk Farms. At that time hybrid corn was

29

an idea and a few people had faith that the idea could grow into something useful. Henry Wallace was one such man.

Mr. Wallace promoted the idea of hybrid corn in public addresses (Wallace, 1955) and through his family's publication "Wallaces Farmer." But his vision went further than just writing and speaking, and in 1926 he founded the first company organized for the sole purpose of producing hybrid seed corn. This company was named the "Hi-Bred Corn Company," and a few years later the name "Pioneer" was added.

Another man with the vision to foresee that hybrid seed corn would be a viable business was E. D. Funk, Sr. The Funk family had been farmers, cattle raisers, and suppliers of seed to their neighbors for over a hundred years, when Mr. Funk looked at the prospects, decided the time was right, and organized a hybrid corn division of Funk Brothers Seed Company in 1927.

Tom Roberts and Charles Gunn began their first attempts at commercial hybrid corn breeding in 1925 as a part of the corn improvement work Gunn had been doing for many years for the DeKalb Agricultural Association. Roberts had enough faith and vision to commit resources to the new kind of corn and produced 75 acres (30 ha) of double-cross hybrid seed corn in 1934. Despite the meager seed crop of 1934 caused by drought and heat, the DeKalb organization decided to go all out in 1935 and produced 15,000 bushels (378,000 kg) of hybrid seed.

In 1925 Lester Pfister began inbreeding corn starting with his own version of Krug open pollinate. Pfister had no organization to back him up, but he stuck to his original dream and by 1933 had put together a hybrid from his own breeding and produced and sold 225 bushels to his neighbors. Getting advice from Holbert, Pfister combined his inbreds and some from the USDA and began producing and selling Pfister 360 and 366, and by 1937 had 37,000 bushels (932,400 kg) to sell.

Universities, experiment stations, and the USDA promoted the concept of hybrid corn to farmers through meetings, publications, and demonstrations. Dozens of companies and farmers were attracted to the idea of hybrid corn and with help from their local university or experiment station became hybrid seed corn producers. Sometimes a system was developed where a university might supply a hybrid seed producer with information, pedigrees, and inbred lines and direct him to a source of foundation seed. Certification was available through crop-improvement associations. Some pedigrees such as U. S. Hybrid 13 developed by the USDA were very widely used in the Corn Belt and were produced by many seed firms in various versions and marketed under a variety of names and numbers.

In 1934 only 0.4% of the corn acreage in the United States was planted with hybrid seed; by 1944 this percentage had grown to 59%, and in the Corn Belt 90% of the corn grown was hybrid. What had happened in this decade? How could farmers be persuaded to give up their cherished seed corn that had been so painstakingly selected over so many years?

Farmers could see the differences and of all of the advantages of hybrid corn, one was so easy to spot that everyone noticed, namely, that hybrids *stood up*! Any farmer still planting open pollinated seed corn who had a neighbor with hybrids became a believer in one season as he watched his neighbor able to pick his corn with a corn picker, while he struggled to pick 80 bushels (2016 kg) a day by hand.

Farmers knew a bargain when they saw one—in seed costs. The USDA credited 600 million bushels (15 million metric tons) of the 1943 crop to increased yields caused by hybrids. And this increase was from an expenditure of 60 million dollars for seed. A 10-dollar return for each dollar spent.

But yield and standability were only two features of these new seeds. Another great advantage evident in this decade was that hybrid corn had more stability than open-pollinates. Drought, insects, diseases all had an effect on hybrids but these stress conditions caused less damage on hybrids than on the open pollinates. With more stable yields, land values were enhanced, credit was easier, rotations and soil management practices could be improved.

Wider use of labor saving machinery was made possible by hybrid corn, particularly the corn picker. Corn that stood with good-quality grain began to change storage methods, grain-handling systems, and farm-labor requirements and contributed to an increasing usage of tractors and the disappearance of horses.

In 1935, with 1.1% of U. S. corn acreage planted to hybrids, 100 million acres (40 million ha) of corn were planted and 2 billion 400 million bushels (60 million metric tons) harvested, a yield of 24 bushels per acre (1494 kg/ha). By 1944 acreage had dropped to 95 million acres (38 million ha), 59% was hybrid and production was 3 billion 130 million bushels (79 million metric tons), an average of 32.8 bushels per acre (2042 kg/ha). Hybrid corn made a significant contribution to the feeding of the United States and her allies during World War II!

No one can accurately assess the contributions of inbreds and pedigrees from the USDA and various other institutions during this early period. It is fair to say that most of the inbreds used in this first decade did come from public sources, but that by 1944 significant numbers of useful inbreds and pedigrees were coming from research subsidized by private funds.

By 1944 there were hundreds of enterprises of various kinds producing and marketing hybrid seed corn. Many were family concerns producing a few hundred to a few thousand bags and selling mostly to their neighbors. Other firms, like Pioneer, DeKalb, Funk and Associates, Pfister and Associates, Michael-Leonard, and Northrup King were moderate-sized businesses, many selling other seeds in addition to hybrid corn. Investments for these businesses were not excessive; many small firms used crib dryers, had a building for processing, sizing, and storage, and sold most of what they produced. Practically all of the hybrid seed was double cross. Seed production was dependable and labor for detasseling was plentiful and cheap. Seed prices in 1944 were $9.00 for a bushel of flat grades of seed ($.36 per kg) and corn was selling for $1.15 per bushel ($.046/kg).

The year 1944 was something of a turning point in the history of hybrid corn in the United States. With most of the Corn Belt planted to hybrid seed, expansion of the hybrid seed corn business had to take place in the southern, western, and northwestern areas where open pollinates were still competitive with the hybrids then available. Both private and public research attacked the problem of developing hybrids suited for these fringe areas, and by 1956 the USDA stated that over 90% of all corn in the United States was hybrid.

Shortly after 1944 other events were taking place that would alter the patterns of corn farming and reshape the aims and methods of the hybrid-seed corn industry. Following World War II cheap nitrogen fertilizer became available from excess manufacturing plants that had been making explosives during the war. Farmers began to use more fertilizer, particularly nitrogen, and yields began to increase moving from 32.8 bu/acre (2042 kg/ha) in 1944 to 38.1 (2371 kg/ha) in 1954. Many hybrids in use in 1944 were not suited for the extra fertility and responded by developing weak stalks, susceptibility to various diseases, particularly stalk rots, and were in general unsatisfactory. The hybrid corn industry met the demands of farmers for something better with new hybrids bred to better utilize higher rates of fertilizer.

Great advances in research were made during this second decade (1945–1954), and both public and private sources were prolific in new inbreds and hybrids.

Advertising by the larger seed companies became an increasingly effective tool for informing farmers about new and better hybrids. Competition became more intense and many companies turned to state yield tests as devices for proving their hybrids were good ones. Most seed marketed was still double cross, but an occasional single cross was sold by a few companies.

Yield levels on a national basis had tended to level off at 37 (2300 kg/ha) to 40 bu/acre (2500 kg/ha) during 1948–1955. Some people wondered whether the ceiling had been reached and yields would move up only fractionally in years to come. It was obvious to a few people that a new approach had to be made to start yields on an upward trend. This new approach was furnished by part of the hybrid-seed industry when both inbreeding and testing procedures for new inbreds and hybrids were done on highly fertilized fields and under high populations. These studies began in 1948, and by 1955 a few hybrids were being marketed as being especially bred for and adapted to higher plant populations and higher fertility levels.

More hybrids suited for even higher populations and higher levels of fertilization appeared in the late 1950s and early 1960s and as farmers realized that yields of 120–160 bu/acre (7500–10,000 kg/ha) were possible on many soils and locations, pressure on the hybrid-seed industry to develop hybrids capable of producing these yields was intense. Not every company nor indeed every experiment station and university endorsed the "high profit trio idea"— use special hybrids, plant them thicker, and fertilize heavier. It sounded too simple to work, but it did work if the right hybrids were used. As the idea spread and farmers demonstrated in their own fields that the concepts were workable, most of the new hybrids introduced from about 1965 on were suitable for use in the high profit trio.

Ideas were changing rather quickly in these exciting years of the late 1950s. Corn had always been considered a soil-depleting crop. Now the hybrid seed corn industry began to promote corn as a nondepleting crop if it were planted thickly, and adequately fertilized for grain yields and if adequate fertilizer, especially nitrogen, were applied to the crop residues as they were being plowed under. Farmers found that on many soils, continuous corn at high yield levels was possible and was more profitable than the older systems of rotation. But there were problems of insect, disease, and weed buildups, and some new inputs began to have a wide impact.

Crop insecticides were not new, but with the introduction of DDT and other synthetic insecticides farmers found they could manage continuous corn, reduce losses from such insects as corn rootworm and European corn borer, and reduce harvest losses because their corn stood better and dropped fewer ears. Added to the benefits of useful insecticides was a growing family of herbicides, with 2-4-D being the first widely used broadleaf weed killer.

All of these imputs, better hybrids, increased populations, heavier fertilization, and the increasing use of insecticides and herbicides, coupled with better farming methods, began to push yields up in an almost uninterrupted climb. Yields moved from the 1955 level of 41 bu/acre (2553 kg/ha) to 53 bu/acre (3300 kg/ha) in 1960, 73 bu/acre (4545 kg/ha in 1965 and to 84 bu/acre (5231 kg/ha) in 1969. Then came the "year of the blight" (1970) (a full discussion of industry role in the blight appears later in this chapter), and yields plummeted to 71.5 bu/acre (4452 kg/ha). However, in 1971 yields were back to 86.5. bu/acre (5386 kg/ha) and in 1972 reached an all time high of 96.5 bu/acre (6009 kg/ha), dropped some in 1973 to 91.5 bu/acre (5698 kg/ha), and in the very poor growing year of 1974 dropped to 72.1 bu/acre (4490 kg/ha).

Improved hybrids from industry research and public research contributed to an indeterminable but substantial part of the gains in productivity. Not only improved and better inbred lines were produced but the very type of pedigree underwent a radical change in the four decades from 1934 through 1975. Most of the early hybrids that were marketed were double crosses, but surprisingly a number of hybrids in the late 1930s and early 1940s were three-way crosses. It was difficult to find four inbreds that would combine together to make a good double cross.

The bulk of the hybrids produced until 1960 were double crosses, including a few hybrids that were made with two related inbreds as a pollinator. Several companies tried to market single crosses before 1960, but acceptance was limited. However, when DeKalb introduced XL45 about 1963, the whole picture began to change. Farmers liked the uniformity, beauty, and performance of XL45, and other companies were forced to add single crosses to their lineup of hybrids. Today, my estimate is that 60% of the hybrids grown in the United States are single crosses or modified single crosses (produced with related inbred parents), about 20% are three-way crosses or something like a three-way, and less than 20% are double crosses.

Production of single crosses and various modified single crosses confronts the hybrid-seed corn industry with substantially more risk than encountered with double crosses. Obviously, with both costs and risks much higher for single crosses, seed costs to the farmer are much higher. As the hybrid corn industry has grown and matured, the taking of sound risks and the development of better methods of producing and handling high priced single cross seed have become a way of life for this highly sophisticated industry. In the United States the current annual dollar volume of hybrid-seed corn sales is estimated at $450–$500 million dollars.

A modern hybrid-seed corn company depends on research—without research, no company can remain competitive very long. New hybrids and concepts are an absolute must in this highly competitive business. Research inputs can come from several sources. A few public institutions continue to develop good new inbred lines, and as these are released, all companies, large or small, have access to them. Companies with their own research departments will usually test out all new inbred releases from public institutions, many times using their own proprietary inbreds as testers.

Most large companies have their own research staff and depend on them to create new proprietary inbred lines and hybrid pedigrees. Research and development of inbreds and pedigrees by firms that are in the business of selling foundation seed is becoming more important and many seed companies depend to varying degrees on these research outputs.

Research expenditures for the total hybrid seed industry in the United States can be only estimated since not many companies publish these figures. Based on figures made public, my estimate would be that 15–20 million dollars annually are spent by the industry. The USDA and other public institutions spend substantial sums, much of which goes into research that does not result in the creation of new inbreds but is extremely valuable to the industry. The industry generally has supported the use of public funds for continuing research by various institutions. Most publicly funded research is directed toward basic investigations and less toward applied research.

After research has created inbreds and pedigrees, the next step in the hybrid seed corn industry is the increase of inbreds for use as parents in commercial seed fields and/or the crossing of inbred lines for the same use. This step is foundation seed. Most larger companies produce the major part of their foundation seed, some companies purchase part of their needs, and some purchase all of their needs. There are a number of companies whose principal business is the growing of foundation seed for sale to companies who produce commercial seed for farming use. These foundation seed companies provide an invaluable aid to the total industry and probably produce 25–30% of the total foundation seed used.

Great care must be used in the production of foundation seed. Most companies forecast their needs for foundation seed 2 or 3 years in advance and make every effort to have foundation seed supplies on hand well in excess of current needs. A safe supply of foundation seed would suffice to carry a company through one disastrous seed-production year and one poor year. Precautions are taken to maintain genetic purity of inbreds and single crosses. Many companies "grow out" a sample of each foundation seed lot by seed size before using the seed.

Some companies minimize genetic drift of inbreds by using a base population that is increased once, kept in cold storage, and drawn on for subsequent increases. Numbers of generations from the base population are held to only one, allowing better preservation of the original genetic makeup of each inbred.

Most companies that raise foundation seed usually contract with farmers for land for increase and crossing fields, which may range from 1 to 100 acres in size (0.4–40 ha). These isolations must be well isolated from other corn— usually a minimum of 640 ft (200 m). Land is preferred that has been in crops other than corn for the previous year so that problems of contamination from volunteer plants can be avoided.

Trained technicians plant, inspect, rogue, detassel, and care for these fields throughout the growing season, and seed ears are carefully sorted and dried at harvest. Foundation seed is usually stored in climate controlled buildings since it is not uncommon to continue to use the same seed lot for a period of 1–5 years.

Most of the estimated 400,000 acres (162,000 ha) of commercial seed production of hybrid corn grown in the United States is grown on land owned by farmers and under contract with seed companies. Usually a contract calls for payment by the company of a fee usually based on some multiple of the Chicago futures market for grain corn. Single cross and other risky productions usually carry a minimum guarantee payment by the company. Often a yield incentive is built into the contract to encourage the farmer to produce best possible seed yield.

Usually ground preparation, fertilizer, insecticide, and herbicide are the responsibility of the farmer with the company specifying what should be used. Planting is done by the farmer under supervision by trained company personnel.

Roguing and detasseling are the company's responsibility, although many farmers contract with the company to recruit labor and to detassel on a part or all of their seed acreage. With the current shortage of labor and high minimum wage, a substantial part of the seed acreage is being detasseled with a mechanical cutter or puller at least once and then "cleaned up" by hand detasseling crews. In 1975 only a small percentage of the seed acres in the United States were planted using a cytoplasmic male-sterile seed parent such as *cms*C.

Harvest of the seed crop is usually the farmer's job and seed on the ear goes through an inspection and sorting before it goes into the dryer. After drying the seed is shelled, cleaned, and stored until sized. In some areas and with some female parents, seed can be shelled directly in the field and dried as shelled corn. Sizing separates seed into various shapes and dimensions, enabling farmers to plant accurately. However, the advent and widening use of plateless planters will reduce the necessity for extremely accurate sizing. After sizing, seed is treated with a combination fungicide and insecticide and is bagged, tagged, and stored for shipment.

Before shipment there are numerous quality-control checks. As seed is sized, each size is carefully sampled and a representative sample is evaluated in a warm germination test where it must meet certain standards of percentage of germinating seeds. Most sizes are also evaluated under cold test germination conditions. Cold tests measure general vigor and adequacy of seed treatment. Before sale, some companies require an actual grow-out of each sample, usually in Florida, Hawaii, or Mexico during the fall or winter seasons. In most years, germinations on warm tests will be over 90% with most lots exceeding 95%. The 1974 seed was an exception with some hybrid-seed corn being sold at 85% germination with a little seed germinating only 80%.

In many areas hybrid-seed corn sales begin campaigns in August and continue through planting season. The bulk of seed corn used in the Corn Belt is sold by farmer–dealers who are actively engaged in farming and can demonstrate their wares on their own farm or who are retired or part-time farmers. Seed stores and elevators are also outlets for hybrid-seed corn in many areas, particularly where corn is not the dominant crop. In the midsouthern and southern part of the United States much seed corn moves from the seed company to jobbers or wholesalers who will have salesmen on the road selling to stores or directly to customers.

Most seed corn is sold with a return privilege; if a customer has more seed than he needs, he can return the unbroken packages to his dealer or store and get credit. The dealer or store can return such seed to the company. Returned seed is carefully checked for germination, age, vigor, and general quality. If returned seed meets quality standards, it may be repackaged, placed in good storage, and offered for sale a second time. Usually seed is not kept beyond this time.

Since return seed is always a factor in determining the size of inventory a company must have to service its customers, most seed companies plan on having an inventory of 25–40% over their anticipated sales. This inventory reserve provides some flexibility in planning seed production and some insurance of supply in case a bad seed production year (e.g., 1974) comes along.

The response of the U. S. hybrid corn industry to the poor year of 1974 was a good illustration of a mature and dynamic industry at work. In the Corn Belt a poor spring followed by hot dry weather in the summer, with a series of disastrous freezes, reduced the seed crop to 60–75% of projected yield. In addition, significant quantities of the seed crop harvested were not high quality in germination and seedling vigor. Seed companies moved quickly to remedy their shortages. Thousands of acres of commercial seed production and foundation seed isolations were planted in the late summer and fall of 1974 in Florida, Hawaii, Mexico, and Central and South America. Some seed was purchased in Europe and from Canada. The USDA figures show that over 15,000 metric tons of hybrid-seed corn were imported into the United States during this past season. Sufficient good-quality seed was made available to American farmers to plant intended acreage in 1975.

Perhaps even more dramatic was the response of the hybrid seed corn industry to the blight year of 1970.

Corn scientists had discovered and reported cytoplasmic male sterility (CMS) in corn as early as 1933, but it was in the mid 1940s before seed companies began to convert their seed parent inbreds to CMS. Two sources of CMS were available, cmsS and cmsT. The cmsS was the first to be used on a commercial scale in 1951. Pollen supplies for farmers fields were assured by blending a percentage (25–50%) of seed produced on normal cytoplasm with seed produced on male sterile cytoplasm.

As scientists and seed producers gained experience, the T source of cytoplasmic male sterility was recognized as being superior in several respects. Male sterility caused by cmsT was much less affected by environment variation and was more definite in its response to restoring genes.

Genes for restoration of pollen shedding to cytoplasmic male-sterile plants were discovered and incorporated into seed-production pollen parents. With dependable male sterile seed parents and good restoring pollen parents, the hybrid seed corn producer could eliminate the tedious and expensive operation of detasseling. In 1970 80–90% of the hybrid-seed corn used in the United States was made on females carrying the Texas type of male-sterile cytoplasm (cmsT). As the growing season of 1970 developed, it became evident, starting in Florida and moving northward, that plants with cmsT were highly susceptible to a devastating leaf blight that was quickly attributed to a subrace of *Helminthosporium maydis* and was christened *Race T*.

Race T not only devastated the grain fields of corn but seed fields as well, including foundation seed production fields. By fall 1970 seed companies found themselves short of seed to sell, short of foundation seed, and in many cases, unable to use reserve supplies of foundation seed because it was in T cytoplasm. Some foundation seed in "normal" (N) cytoplasm was available, and seed companies quickly moved to get more.

Beginning in August 1970, foundation-seed-production fields were planted in Florida, Hawaii, Mexico, and Central and South America. Commercial seed production fields were planted in the same areas. Hybrid seed was purchased and imported from Europe and South America. Farmers had seed to plant in the spring of 1971, even though much of it was either 100% T cytoplasm or various blends of T and N cytoplasm.

Seed companies planted their 1971 seed-production fields using all of the N foundation seed available and by the spring of 1972 almost all seed was in normal cytoplasm. This shift in only one year from 80–90% use of T cytoplasm to practically 0% use again illustrates the responsiveness of the industry and its willingness to make the sacrifices necessary to serve their customers.

Corn farmers in the United States are accustomed to having new hybrids continually being offered to them and many farmers will look for and plant the best hybrids from several different companies. Since competition among companies is severe, farmers are assured of having good products and good service. Service included not only providing the customer with seed but giving him the latest information on corn growing through publications, meetings, and services of agronomists and other experts.

The industry has sponsored and supported many research projects at various institutions and through the American Seed Trade Association has organized and conducted an annual corn research conference. This conference, first held in 1947, brought together scientists from universities, agricultural experiment stations, the USDA, and industry. Speakers and programs presented have contributed substantial information to the industry on subjects such as genetics, breeding, physiology, insects, diseases, fertilizer practices, and many others. Proceedings are published and the total constitutes a valuable library of information.

In the four decades of the hybrid corn industry's life span, many changes have taken place. Growth in dollar volume is quite evident—from nothing to 450 million dollars a year in sales. Packaging was originally in 56-lb bushels;

today some companies market hybrid-seed corn in 50-lb bags, some in bushels, and a growing number in packages containing 80,000 seeds. Plant populations in corn fields have increased from 12,000 to 24,000 per acre and more during these 40 years. In the 1930s, 1940s, and 1950s most hybrids were double crosses with seed sizes like medium flat, regular flat, large flat, and three round sizes being marketed and farmers generally were rather reluctant to use rounds. Today, particularly in single-cross hybrids, all sorts of sizes are marketable with 15–20 sizes in many hybrids.

Almost all of the first companies raising and selling hybrid seed corn were family owned. Today, several companies are big business, with Pioneer, DeKalb, and Northrup-King listed on the stock exchange, and with other companies changing ownership. Funk was originally a family-owned company, and then owned by CPC International, it became a public company and was purchased by CIBA-GEIGY. Trojan was acquired by Fuqua Industries and then by Pfizer, Asgrow was acquired by Upjohn, Farmers Hybrid Co. by Monsanto, and Tewles by Kent Feeds, and Cargill has its own hybrids and has purchased PAG. Many other shifts in ownership have taken place and are still going on. The hybrid-seed corn industry has had good profit margins throughout its four decades of existence.

The industry is continually upgrading itself not only in research but in physical facilities as well. An estimated 20–30 million dollars will be spent in 1975 on improved and enlarged facilities. These newer facilities represent improvements in drying, sizing, and storage methods.

There are an estimated 300 companies producing and marketing the 18–20 million bags (25 kg) of hybrid-seed corn the farmers of United States need annually. These companies purchase machinery and supplies, buy 25 million bags and tags each year, spend 2–3 million dollars annually for advertising, employ some 6000 people permanently with thousands (mostly teenagers) employed on a part-time basis during detasseling time.

The hybrid-seed corn industry in the United States has often been pointed out as one of the best examples of scientific research being put to practical use. The successful hybridization of corn has stimulated research in other crops and we now have hybrid sorghum and some hybrids in wheat, barley, sunflower, and in many vegetables and flowers.

The success of this industry has stimulated activity in other countries. Hybrid-seed industries now exist in Canada, Mexico, South America, Europe, Africa, Asia, and Australia.

With exciting new research going on within the industry and in institutions around the world, methodology of breeding, insecticides, herbicides, fertilizers, farming concepts will all be changed and improved. The hybrid corn industry in the United States has an exciting and dramatic history—it has a future full of promise.

ADDITIONAL READING

Airy, J. M. 1955. Production of hybrid seed corn. In G. F. Sprague (ed.), *Corn and corn improvement.* New York: Academic Press.

Airy, J. M., L. A. Tatum, and J. W. Soreson, Jr. 1961. Producing seed of hybrid corn and sorghum. In *Seeds. Year book of agriculture.* Washington, D. C.: Government Printing Office.

Aldrich, S. R. and E. R. Leng. 1965. *Modern corn production.* Urbana, Illinois: F & W.

Anonymous. 1973. Hybrid grains and the green revolution. In *Interactions of Science and technology in the innovative process: some case studies.* Columbus, Ohio: Battelle Columbus Laboratories.

Crabb, A. R. 1948. *The hybrid corn makers.* New Brunswick, New Jersey: Rutgers University Press.

Duvick, D. M. 1959. The use of cytoplasmic male sterility in hybrid seed production. *Econ. Bot.* **13:** 167–195.

Fowler, W. 1967. Cultural practices for today's seed fields. In *Proc. 22nd Annu. Hybrid Corn Res. Conf.* Washington, D. C.: ASTA.

Holbert, J. R. 1944. A great industry of one decade. In *Finance:* September 25.

Huey, J. R. 1971. Experiences and results of mechanical topping versus hand detasseling in 1971. In *Proc. 26th Annu. Corn and Sorghum Res. Conf.* Washington, D. C.: ASTA.

Jones, D. F. and P. C. Manglesdorf. 1951. The production of hybrid corn seed without detasseling. *Conn. Agr. Exp. Sta. Bul.* 550.

Jones, D. F., H. T. Stinson, and A. Khoo. 1957. Pollen Restoring Genes. *Conn. Agr. Exp. Sta. Bul.* 610.

Jugenheimer, R. W. 1958. *Hybrid maize breeding and seed production.* FAO Agr. Dev. Paper 62.

Tatum, L. A. 1972. The southern corn leaf blight epidemic. *Science* **171:** 1113–1116.

Wallace, H. A. 1955. Public and private contributions to hybrid corn—past and future. *Proc. 10th Annu. Hybrid Corn Industry Res. Conf.* Washington, D. C.: ASTA.

Welch, F. J. 1955. Hybrid corn—a symbol of American agriculture. *Proc. 16th Annu. Hybrid Corn Industry Res. Conf.* Washington, D. C.: ASTA.

Chapter 4

MAIZE PRODUCTION AND MAIZE BREEDING IN EUROPE

Vladimir Trifunovic
Department of Maize Breeding,
Maize Research Institute,
Belgrade, Yugoslavia

For Europeans, as for the entire Old World, maize was unknown until the discovery of America. Historians are not unanimous as to the year in which maize was brought from America to Europe. The first European to mention maize was Columbus, who wrote in his report in 1492 after discovering America, "I have seen a grain known as Mahyz." This is why some Europeans have referred to American corn as *maize*.

Whether Columbus and his sailors brought maize to Europe on their return or after their second voyage is not so important—the fact is that with the discovery of America, maize had begun its voyage toward the Old World, and this crop has since been expanding in Europe with variable intensity on an ever-increasing area and at an ever-increasing production rate.

If the historians are not unanimous about the year maize was introduced into Europe, they are unanimous in stating that maize growing began in the countries of southern Europe—Spain, Portugal, and Italy—and from there expanded to the north into southern France and into the southeastern part of

41

the Mediterranean, namely, Turkey. The Turks propagated maize further to the north into the Balkans and the Danubian Basin, and from there it was imported by countries of central and north Europe. This is why in many European countries corn or maize is referred to by terms such as *grano turko*, *türkischer Weizen*, *tzarevica*, and *turkinja*, *kukuruz*.

We present more detailed information about the origin and expansion of maize in Europe in the section on maize breeding. However, let it be mentioned here that maize is today known in all European countries. In the majority of those it represents a significant, and in some, the most significant, crop. The agriculture of France, Rumania, Yugoslavia, Italy, and Hungary could not be considered today without maize. Even today the trend in maize expansion in European countries still continues. There is a further tendency to expand to the north into Belgium, Holland, and Denmark. Even Sweden and Finland have in recent years been investigating possibilities of growing this crop for silage.

AREA, PRODUCTION, AND AGROECOLOGICAL CONDITIONS

To illustrate the area under maize and its production in Europe for the past two to three decades, we have used FAO statistical data for the years 1969–1973.

Area

The total area under maize in Europe in comparison with the other continents is shown in Table 1. According to these data, approximately 12% of the total world area under maize cultivation is in Europe. It can also be seen that for the 5-year periods, the variations in area under maize in Europe range between 11,700,000 ha in 1934–1938 and 11,856,000 ha in 1969–1973. Although there is an increase in area, it is slight compared to the area increase in Asia, Africa, and South America.

From the data in Table 2 a few general conclusions can be drawn regarding the characteristics of areas under maize in Europe. The largest portion of the area under maize is found in countries of the Danubian Basin—Rumania, Yugoslavia, Hungary, and Bulgaria. These four countries occupy approximately 7.5 million ha or 65% of total area under maize in Europe. If we add to this the area of the Danube region in Austria and Czechoslovakia, the Danubian Plain with valleys of rivers flowing into it can rightfully be called the *European Corn Belt*.

In addition to these countries, considerable areas under maize cultivation are found in France, averaging 1,629,000 ha for the 5-year period 1969–1973 and

TABLE 1.   Area under maize for the 5-year periods 1934–1938, 1959–1963, 1964–1968, and 1969–1973 for Europe and the other continents and percent of total world area[a]

Continents		Acreage in 1000 ha and percent of total world area under maize cultivation			
		1934–1948	1959–1963	1964–1968	1969–1973
Europe	Acreage	11,700	11,488	11,098	11,856
	(%)	(13.8)	(13.7)	(12.3)	(12.2)
Asia	Acreage	13,500	12,244	14,681	19,518
	(%)	(15.9)	(14.6)	(16.3)	(20.0)
Africa	Acreage	7,700	12,606	15,847	16,613
	(%)	(9.1)	(15.1)	(15.6)	(17.1)
North and	Acreage	41,900	35,226	33,464	32,744
Central	(%)	(49.4)	(42.1)	(37.1)	(33.6)
America					
South	Acreage	10,000	12,138	15,031	16,553
America	(%)	(11.8)	(14.5)	(16.7)	(17.0)
Oceania	Acreage	0	0	83,800	90,200
	(%)	(0)	(0)	(0.1)	(0.1)
Total		84,800	83,702	90,204	97,374

[a] Grain production only.

in Italy nearly 1,000,000 ha, 13.8% and 8.7% of European corn hectarage, respectively.

The largest increase in area under maize cultivation of all European countries in the last 10 years is to be found in France, followed by Austria and Czechoslovakia. In France, areas under maize cultivation increased in the five-year period 1969–1973, approximately 5 times that of 1934–1938.

Taking Europe as a whole, the greatest concentration in area under maize cultivation is in the plains and river valleys situated between 43° and 48° north latitude. Located at this latitude, are the Danubian Basin, the Po River Valley, and the most important areas under maize in France. Considering the altitude, maize is primarily grown in Europe on areas between sea level and approximately 1200 m above sea level, although areas beyond 600–700 m above sea level are of little significance.

Production

Data on maize production are given in Tables 3 and 4. In Table 3 data are presented on total production, yield per ha, and percent of total production. In Table 4 yield increase is given in percent for every analyzed 5-year period,

TABLE 2. Area under maize cultivation in various European countries in 1000 ha and percent of total area in Europe

Countries	Acreage in 1000 ha				% of Total area in Europe			
	1934–1938	1959–1963	1964–1968	1969–1973	1934–1938	1959–1963	1964–1968	1969–1973
Austria	66	59	58	130	0.6	0.5	0.5	1.1
Albania	—	152	156	143	—	1.3	1.4	1.2
Bulgaria	811	663	582	651	7.6	5.8	5.3	5.5
Belgium	—	—	1	3	—	—	—	—
East Germany	—	1	1	6	—	—	—	0.1
West Germany	—	9	35	104	—	0.1	0.3	0.9
Czechoslovakia	115	191	159	138	1.0	1.7	1.5	1.2
France	342	866	952	1,629	3.2	7.5	8.7	13.8
Greece	256	197	146	163	2.4	1.7	0.7	1.4
Netherlands	2	—	—	2	—	—	—	—
Italy	1,458	1,164	1,014	966	14.1	10.1	9.3	8.2
Hungary	1,156	1,335	1,242	1,322	11.1	12.1	11.4	11.3
Poland	—	9	6	5	—	0.1	0.1	—
Portugal	—	494	465	396	—	4.3	4.3	3.4
Romania	3,879	3,408	3,096	3,190	36.1	30.0	28.4	27.1
Spain	—	455	495	526	—	4.0	4.5	4.5
Switzerland	—	3	4	16	—	—	—	0.1
Yugoslavia	2,600	2,506	2,490	2,386	24.1	21.8	22.8	20.2
Total	10,685	11,512	10,903	11,787	—	—	—	—

TABLE 3. Total average production for the 5-year periods 1934–1938, 1959–1963, 1964–1968, and 1969–1973 for Europe and the other continents in 1000 tons and percent of world production

Continents		Production in 1000 tons			
		1934–1938	1959–1963	1964–1968	1969–1973
Europe	Total	17,400	25,126	30,272	41,631
	(%)	(15.8)	(14.4)	(14.7)	(15.9)
Asia	Total	15,300	12,386	16,306	30,087
	(%)	(13.9)	(7.1)	(7.9)	(11.5)
Africa	Total	6,200	13,310	17,937	20,313
	(%)	(5.6)	(7.6)	(8.7)	(7.8)
North and	Total	55,900	107,814	118,944	143,104
Central	(%)	(50.8)	(61.6)	(57.8)	(54.8)
America					
South	Total	15,300	18,464	22,024	25,931
America	(%)	(13.9)	(9.4)	(10.7)	(9.9)
Oceania	Total	0	0	195	277
	(%)	(0)	(0)	(0.1)	(0.1)
Total		110,100	175,100	205,678	261,342

TABLE 4. Maize yield increase in Europe and the other continents for the four 5-year periods as percent of the 1934–1938 average

Continents		Years			
		1934–1938	1959–1963	1964–1968	1969–1973
Europe	Production	100	144	174	239
	mc/ha	100	138	184	243
Asia	Production	100	81	106	197
	mc/ha	100	89	100	132
Africa	Production	100	215	289	328
	mc/ha	100	130	139	151
North and	Production	100	193	213	257
Central	mc/ha	100	230	251	320
America	Production				
South	mc/ha	100	108	144	170
America		100	89	96	102

TABLE 5. Total maize production in Europe

Country	Production in 1000 metric tons				Percentage (approx.)			
	1934–1938	1959–1963	1964–1968	1969–1973	1934–1938	1959–1963	1964–1968	1969–1973
Austria	170	189	278	721	1	1	1	2
Albania	—	164	213	254	—	1	1	1
Bulgaria	913	1,545	1,848	2,626	6	6	6	6
Belgium	—	2	3	16	—	—	—	—
East Germany	—	3	2	18	—	—	—	—
West Germany	—	38	154	528	—	—	—	1
Czechoslovakia	225	517	443	573	1	2	1	1
France	541	2,571	3,884	8,198	3	10	13	20
Greece	246	284	308	535	2	1	1	1
Netherlands	3	—	—	7	—	—	—	—
Italy	2,960	3,720	3,727	4,776	19	15	12	11
Hungary	2,306	3,314	5,694	5,036	15	13	12	12
Poland	—	22	15	12	—	—	—	—
Portugal	—	582	549	542	—	2	2	1
Rumania	4,032	5,581	6,911	8,148	26	22	23	20
Spain	—	1,052	1,233	1,873	—	4	4	5
Switzerland	—	11	20	83	—	—	—	—
Yugoslavia	4,300	5,606	6,974	7,676	27	22	23	18
Total	15,696	25,201	30,256	41,620	—	—	—	—

taking the average 5-year yield in 1934–1938 as the base. According to these data, approximately 16% of the total world maize is produced in Europe.

Data on maize production in various European countries are given in Tables 5 and 6. The largest maize producer was, until recently, Rumania. It also had the largest area under this crop. This country, with 27% of the European area under maize cultivation, has produced approximately 20% of the total European production during the period 1969–1973. In second place is Yugoslavia, with 18%. Rumania, Yugoslavia, Hungary, Bulgaria, and part of south Czechoslovakia and southeast Austria produce 60% of Europe's maize. Considerable maize production is to be found in France and Italy as well.

Although total production, as well as yield per hectare, has considerably increased in all European countries, France has made by far the greatest progress. This country has, as we have already mentioned, increased its maize-cultivation area approximately 5 times and at the same time achieved a 3.2-fold increase in yield per hectare. In 1973 France became the greatest maize producer in Europe with a total production of 10,675,000 tons (FAO Produc-

TABLE 6. Average maize yield in specified European countries and increase in percent related to 1934–1938

Country	Average yield in mc/ha				Percent increase (approx.)		
	1934–1938	1959–1963	1964–1968	1969–1973	1959–1963	1964–1968	1969–1973
Austria	26	36	47	56	142	185	218
Albania	—	11	14	18	—	—	—
Bulgaria	11	23	32	40	206	280	356
Belgium	—	47	46	52	—	—	—
East Germany	—	19	23	27	—	—	—
West Germany	27	32	42	51	118	158	187
Czechoslovakia	20	27	29	41	139	146	211
France	16	30	40	50	187	256	318
Greece	10	17	21	33	180	221	342
Netherlands	15	—	39	53	—	262	171
Italy	20	32	37	—	157	181	244
Hungary	20	25	30	38	125	149	189
Poland	—	24	24	27	—	—	—
Portugal	—	12	12	14	—	—	—
Rumania	10	16	21	26	158	202	246
Spain	—	23	25	36	—	—	—
Switzerland	—	47	50	55	—	—	—
Yugoslavia	16	22	28	32	136	171	196

tion Yearbook, 1968–1973). In this way it changed relatively quickly from a large-scale importer to a maize exporter.

It is also interesting to mention that the highest average yield in the north European maize production region—in Austria, Belgium, West Germany, Switzerland, and Holland—is nearly equal to yields in the United States and considerably higher than those in Rumania, Yugoslavia, Hungary, and Bulgaria.

Agroecological Production Conditions

For a better explanation and understanding of the state and trends in European production, it is necessary to become familiar with some of the basic agroecological conditions for growing maize. Although the soil is a significant factor in production, we do not dwell on it here. Temperature and precipitation during the growing season are more significant factors, so we are concerned mainly with them.

The Southern, Coastal Mediterranean Zone. This zone consists of southern Italy, southern Spain, the southeastern part of the Mediterranean coast, and almost all of Greece. The basic characteristics of this region are: (a) relatively poor soil, (b) spring temperatures higher than those in Iowa (12°C), (c) very hot summers with an average temperature above 21°C in June and July, and (d) total precipitation for the 5-month vegetative period (April–August) less than 250 mm. This zone is the least productive one, both in area under maize and in average yield. Very early to early varieties are grown and their critical vegetative season takes place before high summer temperatures set in. In this zone a profitable and stable high maize production can be attained only under irrigation. Maize can also be grown as a second crop following small grains.

The Main or Central Zone. The central European maize growing region is at 45° north latitude. The 20°C June isotherm is approximately the same. The following regions fall into this zone; the Danube basin with adjoining valleys and hills up to 300 m above sea level, the Po valley in Italy and southern France. This zone covers 70% of the area under maize cultivation in Europe. The soil is quite good and naturally fertile. Springs are somewhat warmer than those in the U. S. Corn Belt making earlier planting possible. June and July temperatures are 20–23°C, slightly lower than those in Iowa. Total precipitation for the five-month vegetative period (April–August) ranges within 200–300 mm, approximately 150 mm less than in the American Corn Belt. In addition to this, the distribution is unfavorable so that drought occurs quite frequently during the critical vegetative period.

These conditions influenced the creation of varieties relatively resistant to drought. They also had an effect on the adoption of cultural practices such as conserving moisture, early planting, and low rate of planting (25,000–40,000 plants/ha). For stable high yields, most of this region should have 30–100 mm of irrigation during the growing season. The basic crop rotation in this region is maize and wheat. The best suited hybrids are those of FAO maturity groups 5, 6, and 7.

Northern Production Zone. The remaining areas under maize cultivation fall into this zone; central and northern France, most of Austria, central and northern Czechoslovakia, West and East Germany, Poland, Belgium, and other northern European countries. In this production region the degree days are less than 2500, the springs are colder than in the American Corn Belt, and the average monthly temperature in April is below 9°C. The average monthly temperature in June and July is 18°C or lower, with profuse precipitation in the vegetative season, particularly during the critical maize vegetative period. It is a relatively new, intensive maize-producing region with such basic requirements as very early hybrids of the FAO 300 or earlier group, high mineral fertilizer requirements, and a very high rate of planting (above 50,000–80,000 plants/ha). As already mentioned, the greatest progress in maize production in the past 15 years has been made in this region. Let us once again mention that the division into three climatically different zones is very general and that there are numerous smaller specific microzones within which are regions considered as intermediate or narrow specific and very different from the principal production zones.

MAIZE BREEDING

For a better understanding of the program and problems in maize breeding in Europe besides the knowledge on climate and soil, it is necessary to be acquainted with the origin, introduction, evolution, and major characteristics of European maize varieties.

Maize Introduction into Europe

Although historians are not unanimous, most of them agree that the first field of maize in Europe was planted in Spain in the vicinity of Seville in 1494 (Brandolini, 1968). It is almost certain that it was planted to one of the West Indian flints (Hatheway, 1957). Due to poor adaptation, these initially imported forms could not appreciably be expanded to the agriculture of south

Europe. It was not until later that varieties and populations from Andean slopes and valleys and the Mexican Highlands were brought to Europe and germplasms of different origin were crossed on a large scale under the new climatic and soil conditions. These new introductions increased variability and created improved adaptation capabilities. These somewhat better adapted populations were grown in Portugal, Spain, and Italy. From there, Spanish merchants carried them to Dalmatia and the Ionian Islands in 1572 (Djordjević, 1956), from whence they were spread to the Balkan Penninsula by the Greeks where, according to Piper (1965), maize was planted in the spring of 1576. By 1590 some very early types from the Italian province of Venetto were carried into Austria (Brandolini, 1968).

In the 17th century maize was already grown on a large scale in almost all countries of southern Europe, the Balkans, the Danube Basin, and southern France. During that period maize became the main food crop of peasants and became a trade item between Turkey, Venice, Spain, and other countries. It is believed that during that time the Turks had been intensively introducing maize into the agriculture of countries under their domination, namely, Bulgaria, Rumania, Serbia, and parts of present Hungary (personal communication). This period in the history of maize expansion in Europe was unquestionably a period of intensive crossing and natural hybridization and the beginning of the differentiation of ecotypes adapted to diverse European growing conditions.

During the 18th century new germplasm was introduced. French and English settlers brought a number of very early forms from Canada and New England to northern France, England, Holland, and southern Germany. Some time later, eight-rowed varieties were introduced into Austria and Czechoslovakia and from there spread into the border regions of Poland, Hungary, Rumania, Slovenia, and Croatia (Brandolini, 1968). This new introduction of early germplasm and its consequent crossing with earlier grown forms contributed to the creation of early ecotypes adapted to growing conditions in colder regions.

Until the end of the 19th century only flints were grown in Europe. In the second half of the century the well-known dent varieties were created in the American midwest. The most progressive farmers of the Danube Basin countries, particularly Hungary, imported a number of these varieties between 1880 and 1930. The names of these varieties can even today be heard among older farmers of Hungary, Yugoslavia, and Rumania namely, the Silvermine, Goldmine, Queen of Prairie, Lester Pfister, and others. Some of them are simply known as "American," and for some the origin and time of import is known. For example, the well-known Hungarian variety *Szegedi sarga lofogu* (Szeged Yellow dent) originated from the imported "Funk's Yellow Dent" (Edwards and Leng, 1965).

The introduction of the Corn Belt dents into the Danubian Basin and the Balkan regions and their natural crossing and hybridization with existing flint populations was of exceptional significance for the establishment of European maize dent types similar to the American Corn Belt dents. This was the last major natural crossing and hybridization of old already formed European ecotypes with the new germplasm imported from the North American continent.

This information, of course, comprises only a part of the history of the origin and evolution of European maize varieties. However, the following conclusion can be drawn. European varieties originated from diverse germplasms brought into Europe from totally different maize-growing centers of North, Central, and South America. Maize traveled from the new continent to the Old World for more than four centuries, that is, from the discovery of America until present times. The growing of this diverse material under different conditions in Europe, accompanied by multiple crossing and selection, resulted in the creation of special varieties adapted to many climatic regions. Photoperiodism, thermal, and moisture conditions were basic factors in the differentiation and creation of European varieties. However, the new types retained some of the principal characteristics of the original types, but they also acquired something new and different.

In the past two decades, many papers on the classification and evolution of European maize varieties have been published. Not all of the authors are unanimous regarding both the origin and evolution of European varieties. In view of this, it seems obvious that further research might be worthwhile. I take the liberty to propose a general classification of European varieties and other maize forms on the basis of what has already been the subject of this paper so far.

Corn Belt Dent-like Forms.   This group of dents is similar to those grown in the Corn Belt of the United States. They were created by crossing Corn Belt dents and old flint varieties between 1800 and 1900. They range from genuine Corn Belt dents to dents with more pronounced flint features. The majority of these varieties have a somewhat shorter vegetative period than typical Corn Belt dents, greater cold tolerance, and more seedling vigor and are slightly better adapted to drought. Most of these varieties are adapted to the Danubian Plain. Varieties of this group are considered to have highest yield capacity of all European maize varieties. In the establishment of some of them, well-known European breeders and scientific institutions of Hungary, Rumania, Yugoslavia, and Bulgaria have taken part. Apart from being mainly diffused in the Danube Plain, representatives of this group are also found in other parts of the European corn belt as in the Po River Valley in Italy and regions of southern France.

Southeastern European Flints.   This group of varieties draws its origin mainly from Northeastern flints imported into Europe in the 19th and beginning of the 20th century from northeastern America. This group of 8–12 row flints retained most of the original characteristics. They are grown in Balkan hills, Carpathian Mountains and Highlands, and partially in Czechoslovakia on the margins of the Danubian Plain. They are also found in northern Italy, in some maize-growing regions of northern Spain, and sporadically in southern France. In short, they are grown in countries where soil and climatic conditions are unfavorable for Corn Belt dent varieties.

Northern European and Alpine Flint Types.   This group of varieties originated from crosses of Canadian and New England flints with earlier imported West Indian, Peruvian highland, and Mexican highland flints. These derived varieties have to the greatest extent retained basic features of very early flints of the Canadian and New England regions. They are the earliest maturing maize grown in Europe and are adapted to cold springs, short cool summers, and heavy summer precipitation. These varieties are diffused in parts of Austria, northern Italy, northwestern Yugoslavia, Switzerland, parts of central and northern France, Germany, northern Czechoslovakia, and southern Poland, that is, mainly in regions of predominantly Alpine, Atlantic, and similar climates of the so-called Northern Cold Maize Growing Zone.

Southern European or Mediterranean Types.   Maize varieties adapted to arid spring and summer conditions fall into this group. They are found in the very narrow southern Mediterranean maize-growing region. They are mostly flints, but a few early and medium maturity dents originating mainly from Central and South America also belong (Edwards and Leng, 1965; Sanchez-Monge y Parellada, 1962). The maize varieties of this group have not been introgressed by North American varieties, that is, northeastern flints and southern dents. They have largely retained features of the original varieties and differ greatly from all other European maize forms.

It is understandable that among these distinct principle groups numerous intermediate forms exist and that unique varieties and populations exist that are adapted to very limited specific conditions of small regions.

Breeding of Open-pollinated Varieties

At the end of the last century and in the beginning of this one, particularly between the two World Wars, work was carried out in the majority of European countries to create open-pollinated varieties that would be better adapted and of greater yield capacity than the previous ones established by natural crossing and subsequent selection. This phase occurred somewhat later than the same phase of breeding in the United States. The principal method used in creating these varieties was mass selection.

The objectives in creating these new varieties differed depending on regional conditions where the selection was carried out, but they were basically the same as today, that is, to increase yield and to improve quality and earliness. As a result, a great number of varieties were created in Europe during that time. Most of them have been used as basic material in the development of hybrids.

How successful these programs were in improving yield is still an open question. However, it can be stated with great certainty that the substitution of dents for flints has proven to be most useful. The breeding of early maturity varieties was of significant success, particularly in those regions where the growing season was short and where it was necessary to harvest maize early in order to prepare the seedbed for wheat planting (September–October). Breeding for earliness was also desirable in view of autumn rainy periods and early frosts. The creation of maize varieties adapted to cool climates and short growing seasons was of exceptional significance for the later period of breeding inbred lines and the establishing of high-yielding maturing hybrids.

Varietal Hybridization

Maize varietal hybridization was already well known by European breeders and biologists at the time of Darwin. More progressive farmers started "mixing" two or more varieties when they noticed that a certain "degeneration" of some of their varieties took place. However, especially after World War I and during 1946–1950 attempts were made in several European countries to quickly and inexpensively improve yield by varietal hybridization. Some of the best-known European geneticists and breeders were engaged in this work. A number of varietal hybrids were registered as approved hybrids and were used commercially. We do not go into a detailed estimate of their value, but rather mention why they were not produced over a longer period of time:

1. It was difficult to maintain parental varieties unchanged.
2. The majority of these hybrids did not give remarkably better yields than the parental varieties.
3. Relatively soon after their introduction, American hybrids were imported and the first European–American hybrids were created. Both had greater yield capacity and better agronomic characteristics than varietal hybrids.

Let it be mentioned here that some varietal hybrids were successful and contributed to yield increases in regions of intensive maize production. In the majority of these cases the best varietal hybrids were crosses between late maturity dents and early and medium maturity flints (personal communication; Savulescu, 1957; Tavčar, 1956).

Inbreeding and Hybrids

Although the creation of new varieties led to some improvement in European maize production, the rate of increase was slight so that, for example, the average maize yield in the first three decades of the 20th century in Italy was 14.5 mc/ha, 15.4 mc/ha, and 16.9 mc/ha, respectively (Bunting, 1968). However, the increase in yield in other parts of the world at the same time was also insignificant. However, experiments with inbreeding and hybridization in the United States had already gone beyond the experimental stage and, in the early 1940s, new hybrids resulted in a rapid increase in yield. In Europe, too, work in inbreeding was initiated immediately after World War I but was discontinued during World War II, so that results were not applied until after.

Initial experiments with hybrids of American origin were sponsored by UNRRA in 1947. An international conference on maize breeding was sponsored by the FAO in the summer of 1947 (Trifunović, personal communication) at Bergamo, Italy and attended by specialists from 11 countries. As a result of this conference FAO supplied seed of 30 open-pedigree hybrids to be tested in various European countries in 1948. A second conference was held in Rome in 1949, and results of the tests were discussed. An "Association for Maize Improvement in Europe and the Middle East" was founded during this conference. Experiments with American hybrids were continued in the following years. At the same time, several American hybrids were introduced into different European maize-improvement institutions to be used in testing and seed-improvement programs. Annual meetings to discuss progress and problems in maize improvement and to coordinate cooperative research were held under FAO auspices until 1958. Subsequent meetings were organized by the European Association for Research on Plant Breeding (EUCARPIA) every second year.

From the first meeting held in Bergamo in 1947 to the 1975 meeting in Paris, maize production in Europe changed drastically. Jointly with the introduction of hybrids, improved cultural practices were applied, resulting in a large increase in yield. The increase was brought about after 20 years delay, but thanks to the experience and assistance of American breeders, the process of introduction and application of modern hybrid maize varieties in European agriculture was significantly accelerated. Let us add here that the introduction of American hybrids and lines into European maize-growing regions has been a most significant stage in the four centuries of voyages of American germplasm to Europe. Further introgression of American germplasm into European local types created improved forms adapted to European maize-growing conditions.

The most significant stages during the postwar period probably are: (a) testing of American hybrids and their substitution for local varieties, (b) development of inbreds from local open-pollinated varieties, (c) creation of hybrids involving American and European inbreds and the substitution of American hybrids with the newly developed European hybrids, (d) education of European breeders, and (e) exchange of material and information through EUCARPIA.

Jugenheimer and Silow (1952) summarized 141 experiments carried out in 13 European countries after World War II and found that the best hybrids produced on the average 60% more grain than the best local open-pollinated variety. The majority of European breeders agree that the best hybrids gave on the average 20–30% better yield compared to previously grown local open-pollinated varieties. As a consequence, American hybrids were introduced into western European, Mediterranean, and somewhat later, southeast European agriculture. The most widely used medium maturity hybrids were Iowa 4417, Wisconsin 641AA, Nebraska 301, U. S. 13, Ohio C92, and Kansas 1859. Late maturing hybrids were also widely adopted at this time.

The performance of American hybrids encouraged European breeders to initiate their own breeding programs. These programs were initiated in the period 1946–1957 by state institutions and private enterprises. The first European registered hybrids appeared between 1953 and 1955, and hundreds of hybrids have since been released and registered.

During the same period many European maize breeders spent some time in the United States specializing in genetics, methods of breeding, and production of hybrid maize seed. For this reason, methods applied in obtaining inbred lines, testing, evaluation of hybrids, techniques, and so on were acquired from American breeders.

The first series of uniquely European hybrids was created by crossing American inbreds, developed from Corn Belt dent varieties, and inbreds originated from European local varieties. The American lines were in most cases carriers of high-yield capacity and good standability, whereas the inbred lines developed from local European varieties contributed to a better adaptation to diverse European maize-growing conditions. Hence it is understandable that the new hybrids were usually superior to introduced hybrids of American origin.

Among American lines introduced into Europe's prime production areas, for example, the following were used in hybrid combinations with European inbreds: WF9, 38-11, C103, Hy, Oh7, Oh51, W22, K148, K150, N6, and M14. Early maturity lines, mainly from Wisconsin, Iowa, and Minnesota, W153, W37A, W19A, W41A, W59E, W9, WD, A374, A375, M13, and I205, were used in the northern production zone.

In northern maize-growing regions, excellent early maturing hybrids were created by crossing American dent inbreds with flint inbreds developed from varieties adapted to the cooler climate of northern Europe. Thus the basic characteristics of these hybrids were cold tolerance and adaptation to cool growing conditions.

In the main maize-production zone new European hybrids are predominantly of the dent type. The principal characteristics of the first European hybrids of this production zone were their better adaptation to drought, especially of those bred in the eastern region of Yugoslavia, Hungary, Romania, and Bulgaria. These countries belong to the semiarid part of the main European maize production zone. Apart from the adaptation to drought, hybrids of this zone have good early vigor as well. Hybrids of the Mediterranean coastal zone are characterized by earliness and adaptation to drought.

The principal breeding method applied by the majority of European breeders in the development of the first cycle of inbred lines was the ear-to-row method. The combining ability of the new European inbreds was determined by crossing with the best adapted parental single crosses of imported American hybrids already in production. For evaluation of general combining ability, some of the better local open-pollinated varieties were used as testers.

The first stage in breeding European hybrids was concluded in most European countries in the period 1955-1965. Up to that time many young European agronomists and biologists attended training and education courses to study maize genetics, contemporary breeding methods, characteristics of hybrids and gene pools, and other topics. The next stage in breeding, therefore, is likely to be more diverse in methodology. Contemporary European maize-breeding programs are quite typical of the majority of programs of the main U. S. breeding centers. Perhaps European programs are still deficient in theoretical research, particularly in quantitative genetics and population genetics.

In general, problems now being worked on in most European breeding centers, aside from yield, include:

1. Breeding for improvement in quality, primarily improvement of quantity and quality of protein. The *opaque*-2 (*o*2) and other mutants are widely used. Furthermore, breeding for increased oil content and quality used to be and still is part of many European breeding programs. Breeding of waxy and low-lignin hybrids is in its initial stages. Breeding is continuing on flint types and on semidents with an increased content of carotenoids.

2. Plant geometry, specifically upright leaves, and other novelties also are of interest at some stations.

3. Classification and preservation of local European maize germplasms is still of great interest. A lot has been done about it, but there are many problems yet to be solved.

4. Corn borer tolerance, *Helminthosporium* sp. tolerance, and resistance to stalk and ear rot are important items in most European breeding programs.

Typical problems in each of the three principal maize production zones are described in the text that follows.

Northern European Maize-growing Zone. The most significant problem is still earliness and cold tolerance. Apart from this, maize-breeding centers of these regions devote much attention to research on high population fitness, standability, and response to heavy fertilization.

The European "Corn Belt." Because of the maize–wheat rotation, the development of hybrids with earlier and relatively high yield are of prime importance. The development of hybrids suitable for growing at a relatively low rate of population (under drought conditions) are essential for the eastern and southern part of this zone. Prolific hybrids are of special value for the same reason. Very early hybrids to be used as a second crop after wheat and barley, under irrigation for both grain and silage or dehydration, also receive attention.

Southern Europe. Adaptation to drought together with exceptional earliness is still one of the important goals of breeders of the southern European maize growing area. By introducing irrigation, full-season hybrids are becoming increasingly important, however. In particular, the development of early and medium early hybrids to be used as a second crop after small grains is also of great interest.

Let me conclude with the brief observation that European hybrid maize development related to the practical exploitation of the heterosis phenomenon was initiated two decades later than similar programs in the United States. In the last 20–25 years European breeders have attempted to make up for the late start. Thanks to the assistance and experience of American colleagues, their efforts soon proved to be successful. Their endeavors resulted in double and triple increases in yield in most European countries in the postwar period. In our opinion they have enriched, through their work, the world pool of inbreds. They have increased variation and thereby widened the possibility of an even greater utilization of the heterosis phenomenon, an important factor in the struggle to increase food production. I believe that most of them will be happy to see the results of their work used anywhere in the world. They would be exceptionally pleased if their achievements could be of use to American colleagues.

I think that the journey of maize from the New to the Old World is no longer a one-way process as it was until recent times. In the last few years an intensive exchange of breeding material has taken place so that maize presently journeys in both directions.

REFERENCES

Brandolini, G. A. 1968. European races of corn. *Proc. 24th Annu. Corn and Sorghum Res. Conf.* Washington, D. C.: ASTA, Publ. No. **24**: 36–48.

Broekhuizen, S. 1969. *Atlas of the cereal-growing areas in Europe,* Vol. 2. *Production Centre for Agricultural Publication and Documentation, Wageningen.* Amsterdam: Elsevier.

Bunting, I. S. 1968. Maize in Europe. *Field Crop Abstr.* University of Oxford. Vol. **21**: 1–9.

Djordjević, Dj. V. 1956. *Kukuruz.* Belgrade: Zadružna Knjiga.

Edwards, J. R., and E. R. Leng. 1965. Classification of some indigenous maize collections from southern and southeastern Europe. *Euphytica* **14**: 161–169.

European Association for Research on Plant Breeding, Eurcarpia. Freising-Weihenstephan, September 14–17, 1971.

FAO Production Yearbook, 1968–1973. Rome. 1969–1974.

Hatheway, H. W. 1957. Races of maize in Cuba. Washington, D. C.: National Academy of Sciences, National Research Council, Publ. 453.

Jugenheimer, R. W. and R. A. Silow. 1954. *Co-operative Hybrid Maize Tests in European and Mediterranean Countries,* 1952. Food and Agr. Org. of the U. N.

Personal communication with: W. Schuster, Institut für Pflanzenbau und Pflanzenzüchtung, Justus Liebig, Universität Giessen; E. K. Rohringer, Saatzucht und Versuchsanstalt Landeskammer für Land und Forstwirtschaft, Gleisdorf; A. Piršel, Vyskumneho Ustavu kukurice V Trnave; I. Kovacs, Department of Maize Breeding, Agricultural Research Institute, Hungarian Academy of Sciences, Martonvasar, Fejer; O. M. Cosmin, Institut de Recherches pour Cereales et Plantes Industrielles, Fundulea, Bucharest; N. Tomov and A. Sevov, Naučno-issledovatelskii institute kukuruzi, Kneja.

Piper, M. 1965. Značaj, poreklo i stanje proizvodnje kukuruza. In *Kukuruz.* Belgrade: Zadružna knjiga, pp. 7–40.

Sanchez-Monge y Parellada, E. 1962. Races de maiz en Espana. Madrid: Ministerio de Agricultura.

Savulescu, T. 1957. *Porumbul (studiu monografic).* Bucarest: Editura Academiei Republicii populare Romine.

Tavčar, A. 1956. *II. Züchtung für europäische Verhältnisse Handbuch der Pflanzenzüchtung,* 6 vols. 2. *Auflage* Berlin and Hamburg: Paul Parey, pp. 144–169.

Chapter 5

RECENT DEVELOPMENTS IN MAIZE BREEDING IN THE TROPICS

E. J. Wellhausen

Special Staff Member, Rockefeller Foundation

According to the FAO, the total worldwide planted maize area in 1973 was 111 million ha, with a total production of 312 million tons (average yield 2010 kg/ha). Over half of the total world area planted is in Latin America, Africa, and South and Southwest Asia—in the tropics and semitropics south of 30°N latitude—but this vast region produces only about one fourth of the total world tonnage. The distribution of this acreage and tonnage among the different maize-growing areas is shown in Table 1. About 50% of all the maize produced in the tropics and semitropics is produced in Latin America.

With the exception of Argentina, parts of Brazil, South Africa, and Thailand, most of the maize in the tropics and subtropics is produced by small farmers as a subsistence food crop. It is grown under a wide range of temperature, soil fertility, and rainfall conditions. Little chemical fertilizer is used. Total rainfall varies from area to area and is often poorly distributed. Average yields per unit area are low. Most of the indigenous varieties that evolved under these conditions over the last 5000 years are low yielding but tough. They are well adapted to a primitive, traditional type of agriculture, but not well suited for the more productive, sophisticated agriculture that is now being promoted throughout the tropics.

TABLE 1.   World maize production (1973)

Region	Area harvested (million ha)	Production (million metric tons)	Percentage of world total	
			Area	Pro-duction
United States	26.5	146.0	23.7	47.0
Europe–North Asia	25.6	85.8	22.8	27.3
Latin America	27.8	41.4	25.0	13.3
South and Southeast Asia	13.6	14.8	12.3	4.7
Africa minus South Africa	12.4	14.4	11.2	4.6
South Africa	5.6	9.6	5.0	3.1
Total	111.5	312.0		

As part of the overall effort in accelerating maize production in the tropics, the breeders are aiming at the development of higher yielding, widely adapted, fertilizer-responsive, biologically efficient varieties that are highly tolerant to the vagaries of weather, prevailing diseases, and insect pests and have more and better quality proteins—for a more modern scientific agriculture. One of the major problems is that most high-yielding varieties grow too tall under better fertility conditions and are susceptible to lodging. Reduction of plant height in most breeding programs has high priority.

In this chapter I try to briefly summarize the accomplishments and progress in maize improvement in the tropics and subtropics during the last 25 years. Although about 15% of the 80 million tons of maize produced in the tropics is produced in the cool highlands, I concentrate my remarks primarily on recent developments in the lowland tropics below 1000–1500 m elevation.

IDENTIFICATION OF SUPERIOR BREEDING MATERIALS

Maize improvement in the lowland tropics really began in the late 1940s and early 1950s with the systematic collection, classification and evaluation of indigenous varieties of maize in Latin America. There is a tremendous variation in varieties. Over 300 races have been described (Wellhausen, Roberts, et al. 1952; Brown, 1953; Wellhausen, Fuentes, et al. 1957; Roberts, Grant, et al. 1957; Brieger, Gurgel, et al. 1958; Brown, 1960; Ramirez, Timothy, et al. 1960; Grobman, Salhuana, et al. 1961; and Timothy, Pena, et al. 1961). These studies have provided an inventory of the kinds of germplasm available and in many cases have revealed the relationships and paths of its origin.

The first attempt at a systematic identification of elite racial complexes and heterosis patterns for the lowland tropics was made by Wellhausen in the early 1960s in collaboration with maize breeders in Mexico and Central and South America. In this study 10 dent or semident varieties were crossed with each of 10 flint or semiflint varieties, and the resulting crosses were tested in six locations in Latin America, along with local hybrids as checks. The results are summarized in Table 2.

These results, together with those obtained from subsequent studies, (Wellhausen, 1965), identified four outstanding basic racial complexes for the immediate improvement of maize in the tropics; namely, Tuxpeño and its related Caribbean and U. S. A. dents, Cuban Flint (CF), Coastal Tropical Flint (CTF), and ETO. These four complexes are highly resistant to the common maize diseases and possess excellent yield potential. They have become extremely attractive to breeders and provide the basic resources for the further improvement of maize throughout the tropics. Their origin and relationships have been adequately described in the series of publications on Races of Maize in Latin America. Origin of each is distinctly different and crosses among them exhibit considerable hybrid vigor, especially combinations of Tuxpeño with each of the others.

Tuxpeño, a pure dent type, is particularly outstanding. It originated on the gulf coast of Mexico, with a complex pedigree very distinct from the other three races. It possesses exceptional vigor and yield capacity and is one of the most productive of all maize races. It has influenced the development of the Cuban and West Indian dent races and is one of the putative parents of the U. S. A. Corn Belt dents. Crosses of Tuxpeño with ETO, CF, and CTF are strikingly vigorous, high-yielding, and exhibit an exceptionally strong heterosis pattern.

The three flint complexes, Cuban Flint, Coastal Tropical Flint, and ETO, although of diverse origin, are more closely related to each other than with Tuxpeño, yet crosses among them also exhibit considerable hybrid vigor.

According to Brown (1953, 1960) Cuban Flint is the only true flint found in the West Indies. Its distribution until recently was limited, primarily to the Oriente Province of Cuba, where it is commonly known as *Maiz Argentino.* Hatheway (1957) describes this race as Argentino and postulated that it originated from the accidental crossing of the Catetos of Argentina (imported during 1914–1930) with local *Criollo* Cuban varieties. Cuban Flint has contributed greatly, together with an introduced dent, to the development of the widely distributed *Maiz Criollo* (Cuban Yellow Flint) that Brown (1960) describes as one of the varieties of Coastal Tropical Flints. Hybrids and varieties developed by C. G. del Valle, one of the early corn breeders of Cuba, from the combinations of Cuban Flint and Cuban Dent, two distinct, unrelated races, were excellent yielders in Cuba and Central America. Del Valle's outstanding suc-

TABLE 2. Average yield (kg/ha) of each variety, its 10 crosses F_1, and of the F_1 in percentage of the average yield of the parents, the superior parent, the constant parent, and the checks (average of six locations)

| | Yield | | | | | | |
| | kg/ha | | F_1 in percent of | | | | |
	Parent	Average F_1	Average of parents	Superior parent	Constant parent	Check
Group A Dents						
Sin. Bl. (Tuxpeño)	5559	5810	116.3	104.5	104.5	103.8
Mix. 1 (Tuxpeño)	6032	5770	110.5	95.6	95.6	103.1
Capitaine (Tuxpeño)	5537	5646	114.1	102.0	102.0	101.0
D. V.-101 (Tuxpeño × Común)	5145	5354	112.0	104.0	104.0	95.6
Cuba 28 (dent)	4615	5297	117.2	115.0	115.0	94.6
Paulista Dent[a]	4374	5213	118.5	118.3	119.2	93.1
Azteca (Tuxpeño)	4758	5155	112.5	108.3	108.3	92.1
Venezuela 3 (CTF)	4953	5123	109.4	103.4	103.4	91.5
I-452 (Costa Rica)[b]	4130	4953	115.8	112.4	119.9	88.5

Carmen (Early Tuxpeño)	4068	4662	110.7	107.1	114.6	84.1
(Average)	(4920)	(5298)	(113.7)	(106.5)	(107.7)	(94.7)
Group B Flints						
N. 330 × P. 330 (CF)	4405	5787	123.9	117.6	131.4	103.4
ETO Blanco	5573	5773	110.0	103.6	103.6	103.1
ETO Am.	5392	5516	106.8	102.3	102.3	98.5
S. C. Florida[c]	4339	5390	116.7	110.1	124.2	96.3
N. 330 (CF)[d]	3303	5371	130.3	109.1	99.6	95.9
Ven. 1 CTF	4690	5255	109.3	106.8	112.0	93.9
Am. Sal. (CTF)	4670	5139	107.5	105.2	110.0	92.7
PD (MS) 6 (CTF)	5023	5155	104.2	102.6	102.6	92.8
Cateto	3845	5003	114.2	101.7	130.1	89.4
Cristal Bl.	2775	4598	119.5	93.4	165.7	82.1
(Average)	(4407)	(5299)	(113.7)	(105.2)	(120.2)	(94.7)

[a] Dent from State of Sao Paulo, Brazil—probably introduced from Southern U. S. A.

[b] Variety synthesized from Cuban materials by the Interamerican Institute of Agricultural Sciences at Turrialba, Costa Rica.

[c] Variety synthesized from combination of CF and CTF.

[d] Propagated from a single ear.

cess can be attributed in part to the hybrid vigor resulting from crossing these two types. Good sources of Cuban Flint in the CIMMYT maize bank are Cuba Group I, Cuba Group II, Cuba 11J, Nariño 330, and Nariño 330 × Peru 330. Of these the purest forms are Cuba Group I, Cuba 11J, and Nariño 330. Nariño 330 was propagated from a single ear found in a market in Colombia and has been the source of many good inbred lines in the Colombian hybrid program.

Coastal Tropical Flint as described by Hatheway (1957) and Brown (1960) is widely distributed in the West Indies and along the east coast of South America. Its center of origin appears to be Venezuela and in some way is associated with development of the early maize culture of the Arawak Indians in northern South America. Its precursors are unrelated, and distinct from those of Tuxpeño.

The more typical and less contaminated forms of CTF today are found in the small islands of Antigua, Dominica, Saint Vincent, Barbados, and Guadaloupe. These tend to be earlier in maturity, two-eared, and shorter in plant height. One of the typical varieties now widely used in tropical breeding programs is Antigua 2D and Antigua Group 2, the latter representing a composite of several collections. Crosses between Tuxpeño and Antigua 2D or Group 2 were consistently among the highest-yielding crosses tested in Latin America. The Antigua variety, however, has certain defects. In the lowland, high-rainfall areas of Veracruz, it is very susceptible to a number of leaf diseases. It is also susceptible to root lodging. Early forms of CTF were introduced into Europe during the time of Columbus and from there were distributed to South and Southeast Asia in ancient times.

Although distinctly different in their origin, the CF and CTF complexes have intermixed and have become commonly known throughout the tropics as the Caribbean Flints.

A variety synthetized in the 1940s by Eduardo Chavarriaga at the Estación Experimental Tulio Ospina at Medellín, Colombia, 1500 m above sea level was ETO, which was basically a mixture of Venezuela I (mixture of CF and CTF) and Blanco Común, the predominant race in Colombia at elevation 1000–2000 m, plus selected improved lines and varieties from Mexico, Puerto Rico, Venezuela, Brazil, Argentina, and the United States. Since many of the improved lines added to the mixture were derived from the Cuban and Coastal Tropical Flints and since Común itself was influenced by Costeño derived from the intercrossing of Coastal Tropical Flint and Tuxpeño, it appears that ETO was developed primarily from a complex of Caribbean germplasm. This variety has not only proven to be one of the highest-yielding ones in the temperate climates of Colombia, but also an excellent source of inbred lines for the development of hybrids. Seed of the original variety of ETO is available in the Colombian Seed Bank maintained by the Instituto Colombiano Agropecuario.

The four complexes discussed above have become extremely attractive to tropical maize breeders and have completely revolutionized the breeding and production of maize throughout the lowland tropics in the last 20 years. Some very outstanding open-pollinated varieties and hybrids have been produced and have become widely distributed thoughout the tropics through CIMMYT and the International Maize Program of the Rockefeller Foundation before CIMMYT.

DEVELOPMENT AND DISTRIBUTION OF IMPROVED OPEN-POLLINATED VARIETIES

In Mexico three outstanding open-pollinated populations of Tuxpeño have been developed in a cooperative program between CIMMYT and the National Institute of Agricultural Research (INIA): Crema I, Tuxpeño La Posta, and Tuxpeño 1 (Tuxpeñito).

Tuxpeño Crema I was derived from the intercrossing of eight collections selected as the best performers out of 200 in yield trials of individual collections and their top crosses on two testers, conducted at several locations in Mexico and Central America. The collections composited were Veracruz 48, 143, 174, V-520-C from the Gulf coast, and Michoacán 137, 166, and Colima Group 1 from the West Coast of Mexico. The eighth collection Mixeño 1 was collected in a lowland area of Guatemala.

Tuxpeño La Posta was derived from the intercrossing of 15 elite Tuxpeño inbred lines. Tuxpeño 1, commonly referred to as *Tuxpeñito,* is a variety with shorter plants developed from Tuxpeño Crema I through recurrent full sib selection.

During the last few years tropical maize breeders have given considerable emphasis to shortening the height of Tuxpeño and other tall high-yielding tropical varieties. Johnson (1974) points out that CIMMYT has put high a priority on the development of tropical breeding materials with reduced plant height, using three different approaches: (a) use of genetic dwarfs (*brachytic*-2), (b) selection in advanced generation progenies of crosses between tall and very early, low yielding, short varieties, and (c) recurrent selection within tall, elite, high-yielding tropical materials. The CIMMYT found that all three approaches provided shorter plants; however, recurrent selection within elite tall materials appeared to be the best. The genetic dwarf gene (*br2*) produced certain undesirable side effects, such as excessive leaf width, erratic height and uneven development.

Johnson (1974) greatly reduced the height of both Tuxpeño Crema I and ETO through recurrent selection. Yield in both varieties was maintained, if not improved, and lodging and number of days to flowering were slightly reduced.

These results have stimulated CIMMYT to use the same procedure in reducing the height of many other elite populations with which its maize breeders and collaborators are working.

In El Salvador, Ing. Roberto Vega Lara of the Ministry of Agriculture compared the yields of Tuxpeñito with the original Tuxpeño Crema I and the *br2* conversion at densities of 40,000, 65,000, 90,000, and 115,000 plants per hectare. The yield of Tuxpeñito was greatly superior to the *br2* cultivars at all densities and slightly superior to the original variety at all densities except at 40,000. Furthermore, Johnson (1974) reported that in farm trials conducted in the state of Veracruz, Mexico, in 1973, Tuxpeñito was the highest yielding of all entries, including the commercial hybrid H-507, at 75,000 plants/ha and still in second place at a density of 50,000.

Tuxpeñito is being widely distributed throughout the tropics. Seven thousand tons of seed were produced in Mexico in 1975, about one-third of the total seed produced. Some of this will probably be used in other areas of the tropics and will rapidly substitute for the taller elite varieties of Tuxpeño in breeding programs.

The Caribbean area is particularly rich in outstanding breeding materials for the lowland tropics. It is where the outstanding germplasm complexes—the Dents and Coastal Tropical Flints (and more recently the Cuban Flints, or their precursors)—were brought together. Some very elite varieties have come out of the interhybridization of these complexes. Cuprico, a synthetic formed by CIMMYT from the intermixture of Cuba Group I and Puerto Rican flint–dents, is one of the highest yielding open-pollinated Caribbean varieties. In Southeast Asia, where hard yellow flints are highly desired, an excellent high yielding population has been developed from a combination of Cuprico × Cuba Group I.

About 15 years ago the late I. E. Melhus (Professor Emeritus, Iowa State University) developed a variety that he named *Tiquisate Golden Yellow* from a composite of Caribbean collections. It has become very popular in Southeast Asia, under the name of *Guatemala* in Thailand and *Metro* in Indonesia and the Philippines.

These more productive Tuxpeño dents, along with certain open-pollinated varieties, developed directly from the flint complexes or from combinations of dent and flint by various tropical breeders, have made a tremendous impact throughout the lowland tropics. Although only relatively small amounts of seed have been produced and distributed commercially, these elite germplasm complexes have spread rapidly from farmer to farmer with little effort on the part of research and extension workers. Today, open-pollinated varieties common 20 years ago in the lowland areas have been replaced, either directly by improved varieties developed from the elite materials or indirectly by varieties derived by farmers from the random hybridization between introduced materials and the previously predominating types.

In general, the initial impact of the farmer-developed varieties has been much greater than the direct utilization of the improved materials themselves. This presents a striking example of what can be accomplished with little effort, merely through the exploitation of two main evolutionary forces: (a) introduction and migration of elite germplasm complexes and (b) random hybridization and selection by the farmers themselves.

In the lowlands of Central America, the formerly prevailing local indigenous varieties of the Salvadoreño race have been replaced by varieties derived from the introgression of Tuxpeño, ETO, and the Caribbean Flints. In this area, Tuxpeño, although a little late in maturity, is rapidly replacing all other open-pollinated varieties.

In the lowlands of Venezuela and Colombia, indigenous varieties are being replaced directly by Tuxpeño or varieties derived from the random interhybridization of the local, and introduced Tuxpeño, ETO, and Caribbean complexes.

Varieties developed from mixtures of Caribbean flint–dents and Tuxpeño now predominate in the lowlands of Ecuador and Bolivia.

In Brazil the Azteca, developed by Pinto Viegas (1957) from collections of Yellow Tuxpeño introduced from Mexico, and subsequently improved versions are rapidly replacing all indigenous open-pollinated varieties in the vast region from the state of Santa Catalina to the Amazon Basin.

In tropical humid West Africa, the more productive Tuxpeños and varieties derived from interracial crosses of West Indian dents and Caribbean flints are rapidly taking over. Tuxpeñito, the shorter-growing Tuxpeño developed by CIMMYT, is at the top in most yield trials.

In the East African highlands the introduction of Ecuador 573 from the Colombian Maize Bank, has made a striking impact. With this introduction, varietal yield potentials were raised by 40%. Tuxpeño is also becoming widely distributed in the lowland more humid area.

In India, varietal yield levels were raised 100–200% through a combination of southern United States dents with the Caribbean Flints, both recently introduced.

In Southeast Asia, local varieties are being replaced by varieties formed primarily with the Caribbean fling–dent complexes. The flint population developed from an intermixture of Cuprico and Cuba Group I has been especially outstanding. It has become widely distributed in South and Southeast Asia and is widely used in breeding programs in this area of the world. Thailand's 2.5-million-ton corn export business is based entirely on the open-pollinated variety Guatemala (Tiquisate Golden Yellow) and its derivatives.

The influx of the superior Coastal Tropical and Cuban Flints, and to some extent Tuxpeño and related Caribbean dents, has greatly improved the yield potentials of the formerly predominating early white and yellow flint varieties of the Philippines and Indonesia. Throughout Southeast Asia the currently available synthetics developed from combinations of Tuxpeño, CF, and CTF outyield the native varieties by 100% in the absence of downy mildew and under good fertility and moisture conditions. However, they are a bit late in maturity for most cropping systems; earlier maturing varieties are badly needed.

Downy Mildew (*Sclerospora philippinensis, S. sacchari,* and *S. maydis*) is a serious disease in the high rainfall areas throughout Southeast Asia. The main source of resistance is from the local native Philippine varieties (College White and Yellow Flints). A composite of College White Flint and Tuxpeño seems to be a good source of resistant lines. The CIMMYT is concentrating on the development of a downy mildew resistant Tuxpeño, in collaboration with the maize program at Los Baños, using College White Flint as a source of resistance. Varieties resistant to downy mildew in the Philippines are also generally resistant in other areas of the humid tropics. The reverse is not true. Philippine maize breeders have been fairly successful in incorporating a high degree of tolerance to downy mildew into some of the new high-yielding synthetics.

IMPORTANCE OF HYBRID MAIZE TECHNOLOGY

Simultaneously with the improvement and distribution of superior exotic, open-pollinated materials, there has been considerable emphasis on the use of hybrid maize technology in the further improvement of yield potentials, especially in Latin America, East Africa, and India. In the early 1950s the Mexican Ministry of Agriculture, in a cooperative program with the Rockefeller Foundation, developed some very outstanding hybrids, utilizing principally the races Tuxpeño, Celaya, Chalqueño, and Bolita as basic breeding materials. Due to political difficulties in seed production and distribution, hybrid seed never gained much momentum. In the early 1960s emphasis was shifted to the formation of superior open-pollinated composites and their further improvement through recurrent selection. In this effort some of the resulting synthetics soon overshadowed the hybrids produced earlier. Nevertheless, about 70% of the 23,000 tons of seed produced in 1975 by the Mexican National Seed Producing Agency, PRONASA, continues to be hybrid. The remaining third is mostly Tuxpeñito. Together, this is sufficient seed to plant about 18% of the total maize area. With favorable pricing policies and heavy emphasis on increasing use of improved seed, fertilizer, and improved agronomic practices, plus the relaxation of seed-production policies that discouraged private producers, it is likely that there will be a new surge in the production of hybrid maize in Mexico by private seed companies. Genetic materials and hybrid patterns are well identified for the production of hybrids yielding significantly above that of present synthetics.

After the first big impact with superior open-pollinated varieties, breeders in Central America are beginning to concentrate on hybrids. Guatemala is successfully producing and distributing the intervarietal cross Tuxpeño × ETO (using shorter varieties developed by the CIMMYT). According to data compiled by Villena (1975), the 10 highest-yielding varieties in the 1974 cooperative yield tests in Central America (Table 3) were hybrids yielding an average of 8–9% more than the best Tuxpeño synthetic. As evident in Table 3, hybrids made from a combination of Tuxpeño × ETO or West Indian dents × West Indian flints tend to be higher yielding than hybrids made solely with lines from Tuxpeño. In addition, some of the hybrids are earlier maturing and shorter in plant height. El Salvador has been particularly successful in the use of the hybrid-maize technology. Over 60% of the maize area in El Salvador (consisting mainly of small farmers) is planted to hybrids, and much of the seed is produced by private producers. In addition, El Salvador supplies considerable seed to other countries. Demand is greater than supply.

Venezuela is successfully exploiting the Tuxpeño × ETO hybrid-vigor pattern. In 1975, through a combination of Government and private enter-

TABLE 3.   Origin, yields, and agronomic characteristics of the 11 highest-yielding entries in the cooperative Central American yield trials (average of 12 locations) (Villena, 1975)

Entry	Yield tons/ha	Days to flower	Plant height (cm)	Ear height (cm)	Germplasm
Dekalb H-4[a]	5324	59	241	135	Tuxpeño × (ETO)[b]
Desarrural HB 105	4963	62	245	136	Tuxpeño × ETO
Pioneer 304A (×)[a]	4928	58	225	122	WI Dents × WI Flints[b]
Desarrural HB 104	4826	62	241	141	Tuxpeño × ETO
Poey T31	4819	62	241	142	Tuxpeño × Tuxpeño
E S-H B 1	4794	60	246	130	Tuxpeño × Tuxpeño
H 5[a]	4724	60	236	129	Tuxpeño × Tuxpeño
E S-H A 1	4633	59	234	134	Tuxpeño × Tuxpeño
Mexico H-507[a]	4584	64	253	148	Tuxpeño × Tuxpeño
Pioneer 304 B[a]	4583	58	215	114	WI Dents × WI Flints[b]
Centa 1 B (syn)[a]	4576	59	225	118	Tuxpeño Synthetic

[a] Seed is being produced.

[b] The source of germplasm for these hybrids is a guess based on their appearance and performance and may not be exact.

prise, Venezuela produced about 6200 tons of hybrid seed, enough to plant over 50% of total maize area. In addition, about 350 tons of certified seed of improved open-pollinated varieties were produced. Only about 5.3% of total seed production is open pollinated.

Colombia is principally exploiting the hybrid-vigor pattern among Tuxpeño, the Caribbean flints, and ETO, in areas below 1800 m elevation. Although the hybrids are excellent, the demand for seed in the past has not been great, due to the absence of effective programs for the acceleration of maize production throughout the country. In 1973, through the combined efforts of government and private enterprise, Colombia produced approximately 2885 tons of hybrid seed, enough to plant 165,000 ha in the tropical lowlands (ca. 20% of total maize area). This was more than produced in any previous year. Seed production in 1974 diminished about 10%.

In Peru, 90% of the maize in the lowland coastal area under irrigation is planted with hybrid seed (2200 tons) made primarily from a combination of lines from Perla (the formerly predominating local flint variety) and the exotic Caribbean flint–dent complex. The remaining 10% is planted mostly with advanced generation hybrid seed.

In Argentina approximately 80% of the 4–6 million hectares planted to maize is planted with hybrid seed. Since the dark orange Cateto flint type of grain continues to be the preferred type for export by the Argentinian Grain Board, most of the hybrids have been formed from the local Cateto flints. Hybrids made from a combination of U. S. dent lines with lines from Cateto, CF, and CTF, have proven to be greatly superior in yield, but grain types of these do not meet present export requirements.

Brazil is number one in maize production in Latin America, with about 12 million hectares planted annually and a total production around 16 million tons. About 40% of the total maize area in Brazil is planted to hybrid seed. However, in the central maize belt 50–80% of the maize planted is now hybrid. Seed is produced by both private companies and associated seed growers. Sementes Agroceres, S. A., one of the largest seed producers, produced about 25,000 tons in 1975 (personal communication from Gladstone Drummond, Director of Research).

The most striking example of the effect of exotic germplasm in increasing potential yields of hybrids is in Brazil. Dr. Ernesto Paterniani, Director of the Institute of Genetics in Piracicaba, Brazil, has kindly provided me with a summary of the progress made in the dynamic maize improvement program in Brazil since its beginning.

Thirty years ago the two most common maize races in Brazil were Cateto and Paulista Dents. Both of these races were described by Brieger, Gurgel, et al. (1958). Paulista Dent was derived from the interhybridization of Cateto and dents introduced from U. S. A. These varieties were inbred, beginning in 1932, and resulting lines were evaluated. In 1946 the first double-cross H-3531, developed by Carlos Krug and collaborators, with four Cateto inbreds (C1 × C2) × (C3 × C4) was released for commercial production by the Institute of Agronomy at Campinas (IAC). It yielded about 22% more than Cateto and somewhat less in comparison to Paulista Dent. The relative yield of this hybrid and of hybrids produced subsequently and of two open-pollinated varieties representative of Cateto and Paulista Dent, are illustrated in Figure 1. Although the graph only indicates comparative yield potentials, the hybrids are also superior to the open-pollinated varieties in agronomic characters and disease resistance. The curves on the far right in Figure 1 show the progress made in yield potential through half sib selection in four synthetic populations. These are discussed later.

In 1953 the semident hybrid H-4624 developed by IAC was released for commercial production to replace H-3531. In this double cross two Cateto lines C1 × C2 were combined with two dent lines (T1 × PD1); T1 is a line derived from Tuxpeño germplasm by the Texas Agricultural Experiment Station, and PD1 was derived by IAC from Paulista Dent. This flint–dent hybrid was strikingly superior to the pure Cateto hybrid H3531 and yielded about 43% more than the better varieties of Paulista Dent.

Also in 1953 the IAC received seed of 21 yellow Tuxpeño samples collected in lowlands of the state of San Luis Potosi, Mexico, by the Cooperative Mexican–Rockefeller Foundation Agricultural Program. A synthetic made by Pinto Viegas at IAC (1957) from these introductions was named *Azteca*. In yield trials Azteca was found to yield the same as H-4624. As mentioned earlier, Azteca has become widely distributed throughout Brazil and has provided a source of Tuxpeño germplasm for a new series of inbred lines and the development of still higher-yielding synthetics.

In 1956 the IAC released its third hybrid H-6999A. The only difference between the pedigree of this hybrid and the previous one is that the Paulista Dent line PD1 was substituted for a line derived from Azteca. Its yield was only slightly better.

In 1958 the IAC released its fourth hybrid H-6999B, a double cross between two Cateto lines and two Azteca lines (C1 × C2) (T1 × T2). In performance trials it yields about 9% better than H-6999A and 50% better than Paulista Dent. Since its appearance, this hybrid has become the standard to which all other flint–dent hybrids developed by private, federal, and state breeders are compared. Yields of most hybrids produced commercially in Brazil today are similar to H-6999B.

C. G. del Valle was among the first to develop commercial hybrids and improved open-pollinated varieties from the flint–dent heterosis patterns in the

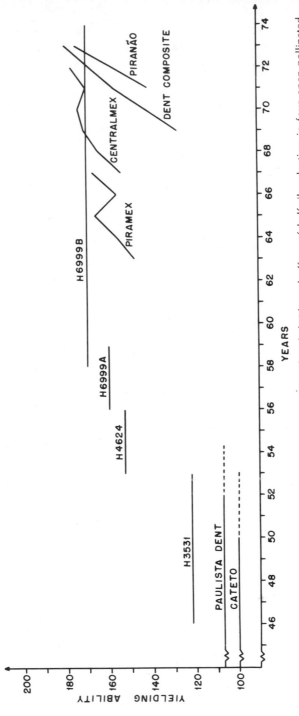

Figure 1. Progressive improvement of yield potential in Brazilian hybrids and effect of half-sib selection in four open-pollinated composites.

Caribbean. The Cornelli flint–dent hybrids developed by del Valle for Cornelli Seed Company were the first hybrids commercially available in the area. They were also used in Central America. Similar hybrids were later successfully produced in Cuba by Poey Seed Company. The open-pollinated synthetic produced by del Valle [PD(MS6)] is still used in breeding programs.

More recently, Pioneer International Hi-bred Seed Company has begun to exploit the flint–dent Caribbean heterosis pattern in the formation of hybrids for the Caribbean area and lowland tropics of Central America. In Central America, as indicated early in this chapter, some of the Pioneer hybrids are similar in yield to the best Tuxpeño or Tuxpeño × ETO hybrids but are shorter in plant-height growth and are earlier maturing.

Poey Seed Company has moved its base to Mexico, where it has become an affiliate of Northrup and King and is continuing to successfully exploit the flint–dent hybrid-vigor pattern in the production of hybrid seed for the lowland tropics.

Prior to 1969, 85–90% of the 432,000 ha of maize planted annually in Egypt was planted to two varieties: (a) American Early (a derivative of Boone County White U. S. A.) and Nabal Gamal (isolated from Hickory King U. S. A.). An earlier attempt to substitute these with double-cross hybrids made from U. S. A. inbred lines failed, primarily because the resulting hybrids yielded very little more than the prevailing open-pollinated varieties.

In 1969, at the beginning of the CIMMYT/Ford Foundation/Egyptian cooperative project, varieties of Coastal Tropical Flints, Tuxpeño and mixtures of Tuxpeño–U. S. A. dents developed by CIMMYT, were introduced into Egypt, and the program was reoriented toward the development of new elite heterozyous maize populations through interracial crosses. This approach has resulted in the formation of a new series of synthetics and intervarietal hybrids with a much higher yield potential.

Two interracial crosses—American Early × a Composite of Tuxpeño–U. S. A. Dents and American Early × Tuxpeño La Posta—and one interracial composite have been released for commercial production. The two interracial crosses in F_1 yield at the level of 12–14 tons/ha, 40–50% more than the better open-pollinated varieties and double crosses made with U. S. A. lines. Sufficient seed of this new material was produced in 1973 to plant about 7,000 ha in 1974.

In Kenya the major part of the hybrid seed industry is based on interracial hybrids made from improved populations of KII and EC. 573. According to the 1974 EAAFRO Maize Report, about 375,000 ha are planted to hybrids in the highlands of Kenya. This is about 30% of the total maize planted. Of the maize planted, about 89% is planted on small farms.

Breeders in India have concentrated on the formation and distribution of hybrids since the beginning of the maize-breeding program in 1954. A number

TABLE 4.   Commercial seed production in Latin America, 1974

	Production (tons)			
Country	Hybrid	Open-pollinates	Percentage[a]	Kind of Enterprise
Mexico	16,000	7,000	18	Mostly government
Central America	4,500	1,000	25[b]	Government and private
Venezuela	6,200	400	60	Government and private
Colombia	2,900	?	20	Government and private
Peru	2,200	200	90[b]	Agr. univ. and private
Argentina	80,000	0	80	Private
Brazil	90,000	1,000	40[c]	Mostly private
Total	201,800	9,600		

[a] Percentage of total maize area planted with improved seed.

[b] Percentage of area in which improved varieties are adpated. In El Salvador 60% of total maize area is planted with hybrid seed.

[c] Eighty percent in State of São Paulo.

of outstanding hybrids has been made, with a yield potential of two to three times greater than the local varieties prevailing in the 1950s. These hybrids were formed in a short period of five years, from the combination of inbred lines from U. S. A. with lines developed from the Caribbean flint–dent complexes. However, due to a myriad of problems, mainly a lack of farmer incentives to produce more maize, these hybrids never made significant impact. Nevertheless, the hybrids served to introduce new germplasm into the hands of subsistence farmers from which they developed a new set of more productive varieties.

In spite of the many difficulties with the hybrid-maize technology in the tropics, much more hybrid seed is commercially produced than seed of open-pollinated varieties. In Latin America (Table 4), it is estimated that 202,000 tons of hybrid seed were produced in 1974–1975, sufficient to plant about 10 million ha, or about 37% of the total land area planted annually to maize. Only about 10,000 tons of open-pollinated seed were produced commercially (not including seed that might be sold by individual farmers). Countries that have encouraged private enterprise in hybrid-seed production have progressed more rapidly in increasing seed supplies. Private seed producers have not only helped to increase the volume of seed, but have been a big factor in its distribution and use with improved agronomic practices on all kinds of farms, large and small. As maize production becomes more sophisticated in the tropics, the trend definitely will be toward the formation and greater use of hybrids.

BREEDING METHODOLOGY FOR FURTHER IMPROVEMENT OF YIELD POTENTIAL OF HYBRIDS

In the attempt to raise yields of hybrid varieties still further, the Brazilian maize team is devoting a major effort to further improvement of a number of open-pollinated flint and dent composites (see Table 5).

All of these composites, with the exception of America Central, are being improved mainly through recurrent selection among and within half-sib families (modified ear-to-row selection). Very substantial increases in yield capacity are being obtained and yields of certain populations, as shown in Figure 1, are now equal or superior to hybrid H-6999B. Seed of the related dent composites, Azteca, Maya, Piramex, Centralmex, and IAC1, has been widely distributed in the vast region from northern Rio Grande do Sul to the Amazon basin. These new higher yielding varieties are rapidly spreading from farmer to farmer and gradually replacing open-pollinated varieties formerly distributed in this area.

The broad-based dent and flint composites (numbers 7 and 8, Table 5) are extremely promising for the development of higher-yielding hybrids and have

TABLE 5. Principal composites being improved through recurrent selection (Brazil)

Name	Year formed	Institution	Germplasm
Azteca	1953	IAC[a]	Yellow Tuxpeño dent
America Central	1961	ESALQ[b]	Mixture of West Indies and Tuxpeño Dents
Piramex	1963	ESALQ	Tuxpeño (Azteca) Dent (combination of 20 S_1 lines)
Maya	1965	IAC	Mixture of lines and collections of Tuxpeño (Azteca)
IAC 1	1965	IAC	Tuxpeño × Caribbean Flint
Centralmex	1967	ESALQ	Piramex × America Central
Dent Composite	1969	ESALQ	Wide-base Tuxpeño dent
Flint Composite	1969	ESALQ	Wide-base flint mixture (Cateto, ETO, CF, and CTF)
Piranão	1971	ESALQ	Piramex × Tuxpeño brachytic
Cateto–Colombia			Wide-base flint mixture of inbreds from Cateto, Venezuela 1, Nariño 330, ETO, and Cuba 23

[a] Instituto Agronómico Campinas.

[b] Escola Superior de Agricultura Luis de Queiroz, Instituto Genética, Piracicaba.

been widely distributed among public and private breeders. Recurrent selection is being continued within each, and a program for the further improvement of yield of both composites and their specific combining ability is underway at IAC.

Composite Cateto–Colombia is another very promising flint type. It is being further improved at the National Maize Center, Sete Lagoas, M. G., and seed is being extensively distributed in northern Minas Gerais, where the farmers still prefer the Cateto-type flints.

Piranão is one of the most interesting populations of all. It is a short plant population derived by Paterniani (1974) through recurrent selection within the advance generation progeny of the cross between Piramex and Tuxpeño brachytic received from CIMMYT. Reduced height is primarily due to the shortening of the internodes below the ear. Above the ear, internode length is normal, avoiding a compaction of leaves. It is equal in yield capacity to the dent composite Centralmex, but its plants are one meter shorter. It is being enthusiastically received by farmers and seed producers. One seed producer had about 800 tons available for planting in 1975. This new short plant variety presents a breakthrough for Brazil and other tropical maize programs, along with the short plant Tuxpeñito developed in Mexico.

All of these new, higher-yielding materials, in addition to their direct use as open-pollinated varieties, will be extremely valuable to the hybrid maize breeders in Brazil. Until a new series of inbred lines can be developed from these new materials, hybrid seed producers in certain cases are resorting to the production of intervarietal crosses or the substitution of one of the high-yielding composites for one of the single crosses in a double-cross hybrid. Some of the new flint × dent intervarietal crosses yield up to 12% better than H-6999B. Thus the Brazilian breeders now have the basic ingredients for the formation of a new series of double-cross hybrids that can be expected to greatly surpass the present ones in yield capacity and plant type. Breeding activities leading to higher yielding, open-pollinated varieties and hybrids in the fuller exploitation of certain heterosis patterns are also underway in other areas of the tropics.

In Venezuela the Ministry of Agriculture breeders have launched a population improvement program within two populations and a reciprocal, recurrent selection program between them. One of the two populations is a flint, synthesized from a mixture of ETO, Nariño 330, and Venezuela 1, and the other is a dent synthesized from a mixture of three local varieties of Tuxpeño. These two populations should offer the commercial breeders a new source of inbred lines. It might be highly beneficial to add CIMMYT's short ETO to the flint population and Tuxpeñito to the dent. During the last 3 years the prevalence of stunt virus and downy mildew diseases has increased. Venezuelan breeders are very much aware of this and are beginning to emphasize selection for resistance to these two potentially destructive diseases in their breeding programs.

The flint–dent hybrid-vigor pattern is also being further improved in the lowlands of Colombia, Peru, and Central America through the formation and improvement of separate wide-based flint and dent composites.

In Colombia, Torregroza, head of the ICA maize program, has established certain interracial heterosis patterns for the Andean highlands. Yield capacities of the component populations of these patterns are being further improved through recurrent selection for prolificacy.

Breeders at the Argentine Agricultural Research Center at Pergamino, looking forward to a time when the semiflints may be acceptable for export, have formed two populations: (a) a composite of local and Caribbean flints and (b) a mixture of U. S. A., Tuxpeño, and West Indian dents. These populations have now been grown for a number of years with the intention that once they become fairly well adapted to the Argentine environment, a reciprocal recurrent selection scheme between them would be initiated.

In the highlands of Kenya, the Kitale II × Ecuador 573 heterosis pattern has yielded some excellent hybrids and open-pollinated synthetics. Attempts are being made to further improve the yield of the two components and their combining ability.

Harrison, of the International Institute of Tropical Agriculture at Ibadan, is working on the improvement of a hybrid vigor pattern for West Africa (Harrison, 1970). Egypt is distributing seed of interracial crosses of American Early (Boone County White) and Tuxpeño and is concerned with the further improvement of these combinations.

In Thailand the dark orange-yellow flint population derived from a mixture of Cuba Group 1 and Cuprico is being further improved through recurrent selection for direct use in commercial maize production. It is the highest-yielding population available in Southeast Asia today. Although it is not specifically being improved for use in a hybrid pattern with Tuxpeño or related dents, it might be an excellent source of flint inbred lines for breeders interested in the production of superior flint × dent hybrids.

For years, certain breeders have dreamed about the formation of a high-yielding dwarf corn that could be planted at high densities and machine harvested. Mario Castro Gil (1973), working in Mexico with initial encouragement and support from CIMMYT, produced a dwarf hybrid adapted to 1000–1800 m elevation from inbreds developed from an interracial mixture segregating for *brachytic*-2, derived from a composite of Puebla Group I, Tuxpeño (V-520-C), and a variety from Argentina, with a square stalk (*Tallo cuadrado*), two ears from the same node, and erect leaves. The hybrid is currently commercially produced as a three-way cross. It grows about 1.5 m tall.

Yields of 15–18 tons/ha have been obtained with a density of 130,000 plants/ha in experimental plots under high fertility irrigated conditions, at elevations of 1100–1800 m. This is about double the yield of current hybrids in this area of Mexico. Commercial seed production of the three-way cross, as presently constituted, is difficult, because the inbred used as a pollinator is too short for good pollination of the female single cross. About 40 tons of seed were available for planting in 1976.

DEVELOPMENT OF MORE NUTRITIVE VARIETIES

Thus far very little impact has been made with high-quality-protein varieties in Latin America, but interest in the development of such varieties continues to be high. Work underway, accomplishments and prospects for the future, were recently reviewed in the worldwide CIMMYT–Purdue Symposium on protein quality in Maize (1975). This year Guatemala is starting a program in collaboration with CIMMYT and USAID, based on the modified endosperm, high lysine–trytophan varieties developed by the CIMMYT–UNDP project. These varieties are also being tested in other countries of Central and South America.

Colombia has had two high lysine hybrids (H-208 and H-205) in commercial production for a number of years; they yield about 10% less than their normal counterparts. Seed production of these hybrids has stabilized around 60–70 tons annually.

Peru has four hybrids converted to *opaque*-2 that are well adapted to the low and medium Sierra. However, like the Colombian hybrids, they yield 10–15% less than their normal counterparts but still considerably more than the common open-pollinated varieties. Very little seed is being produced.

Most of the high-lysine maize produced in Latin America is produced in Brazil. The best variety is Agroceres 504. About 800 tons of seed of this variety are produced and sold annually by Agroceres. This variety has also performed consistently well in CIMMYT's international yield trials. Another 800–1000 tons of seed are produced annually of the converted intervarietal cross Maya × IAC by various smaller producers. Practially all of the *opaque*-2 maize produced is fed directly to hogs by farmers not using commercially prepared, balanced rations. Some is being used in a nutrition project, in part sponsored by the USAID at Vicosa.

Seed producers everywhere, including Brazil, claim there is little demand for the better protein-quality maize in its present state of development. Breeders and extension workers talk about the need for special promotion campaigns. Even with these, it is doubtful that much progress can be made. I believe that introducing the more nutritive protein qualities in the form of special varieties is the hard way. In my opinion, the only way to attain the impact we all desire and know is feasible is to treat the high-quality protein character as one more attribute to add to all of the improved open-pollinated varieties and hybrids released for production. Wheat breeders have been operating this way for years, in the development of superior varieties, with the desired milling and baking qualities.

THE CIMMYT PROGRAM

This program is discussed by Johnson (1974) and in the CIMMYT 1973 annual report on maize improvement. As indicated in these descriptions, CIMMYT is devoting its major efforts to the formation, development, and initial improvement of broad-based gene pools, composited with emphasis on the combination of desirable attributes for different ecological situations, without regard to racial origins or heterosis patterns. Fourteen highland, 12 tropical and eight temperate pools have been formed. Breeding efforts are concentrated on the development of widely adapted, high-yielding, open-pollinated varieties, with desirable attributes as may be needed for different prevailing environmental situations, through recurrent, full-sib selection within a given pool. The full-sib progenies are widely tested within the environment for which a better variety is sought. The pool selected for improvement in a given area depends on the end product desired.

The CIMMYT procedure with 12 populations in the lowland tropics will certainly produce some outstanding varieties and probably some good hybrids; there is, however, another approach that I believe warrants special attention and would result in a fuller exploitation of the natural heterosis patterns existing between the flint and dent (and other) germplasm complexes. Instead of the formation of 12 populations, in which racial complexes and hybrid patterns are disregarded, it seems to me that tropical maize breeders might better focus their main cooperative efforts on the development of *two* broad-based, high-yielding, widely adapted, fertilizer-responsive, more nutritive, biologically efficient populations, as follows: (a) a dent composite, consisting of the combination of Tuxpeño and related dents, such as the Cuban, West Indian, and U. S. A. dents and their precursors and (b) a flint composite, consisting mainly of the Cuban, Coastal Tropical, and Cateto flints and their precursors. Each of these might be further divided into three subpopulations: (a) early, (b) medium, and (c) late maturity. For the short range, it might be better to start with more narrowly based populations constituted with only the best materials. For example, the late dent population might primarily be Tuxpeñito, and the medium and early maturing ones might be comprised of the earlier tropical dents. Each population would be constituted without regard to grain color. The color desired could readily be sorted out as needed, when varieties are formed for commercial production.

Desirable attributes, such as reduced plant height, high lysine, wide adaptation, elasticity and flexibility in performance under varying moisture conditions, and more comprehensive resistance to disease and insect pests, could be incorporated into each, along with a steady improvement of yield through indicated recurrent selection procedures. Once the desired performance level of each of the individual flint and dent populations has been attained, reciprocal recurrent selection procedures could be initiated to further enhance their combining ability.

Where open-pollinated synthetics are desired, the two populations could be combined and the resulting progeny further improved through recurrent selection. This, like the formation of hybrids, can best be done at the local level by local breeders. With our present knowledge of available materials, the combination of the flint and dent germplasm complexes into a single population should provide the best possible base for synthesizing open-pollinated varieties with a high yield capacity. However, where hybrids are feasible and practical and where maximum yield capacity is desirable, the best way to exploit the strong natural heterosis pattern existing between the flint and dent complexes is in the formation of hybrids. The yield capacity of the best hybrids, produced as modern technology provides, would be at least 15–30% higher than that of the best synthesized open-pollinated varieties in the same maturity class. Where wide adaptability is a desirable trait, one might resort to the use of multiline hybrids.

This kind of worldwide cooperative program would set the stage for the formation of both super open-pollinated varieties and hybrids at the national level. The Brazilian maize team is basically using this approach in the further improvement of the yield potential of hybrids and varieties in Brazil.

DISCUSSION AND CONCLUSIONS

Breeding and production of maize in the lowland tropics has been revolutionized during the last 20 years by the isolation, introduction, and widespread dissemination of four outstanding and distinct germplasm complexes: (a) Tuxpeño and related dents (Mexican, West Indian, Cuban, and Southern U. S. dents), (b) Cuban Flint (CF), (c) Coastal Tropical Flint (CTF), and (d) ETO Flint.

Today, open-pollinated varieties that prevailed in the lowland tropics 20 years ago have been replaced by: (a) improved open-pollinated varieties or composites, developed by breeders directly from the individual complexes or combinations of them, (b) intervarietal, interracial, or inbred-line hybrids (usually dent–flint crosses), or (c) varieties developed by the farmers themselves, either directly from the introduced materials or from the random hybridization of introduced materials with prevailing local, indigenous types.

The impact from the farmer-developed varieties has been much greater than from the direct utilization of the breeder-developed materials. In the early stages of most tropical maize improvement programs, the farmer-developed varieties often were very good, and in the absence of effective seed production and distribution systems, these varieties spread very rapidly from farmer to farmer, with little effort on the part of research and extension workers. This is a striking example of what can be accomplished with relatively little effort, merely through the exploitation of the two evolutionary factors of introduction and migration of elite germplasm complexes through random hybridization with prevailing less productive types.

Although a tremendous impact has been made with the development and distribution of open-pollinated varieties, as maize production becomes more sophisticated in the hands of both large and small farmers, and as reliable sources of good seed develop, the trend toward the formation and greater use of hybrids will definitely continue. Over 200 thousand tons of hybrid seed were commercially produced in 1974–1975 in Latin America alone, sufficient to plant about 10 million ha or about 37% of the area planted annually to maize. In the same year only about 10,000 tons of seed of improved open-pollinated varieties were produced. Use of hybrid seed is not limited to the large farmers. In Central America, where 80% or more of the maize is produced by small farmers with 1–5 ha, the demand for hybrid seed is growing rapidly. It has been demonstrated in this area that the small subsistence farmer, as his maize production moves from a low-input subsistence operation to a highly profitable operation with the use of fertilizer and improved agronomic practices, will abandon his former custom of saving his own seed and resort to the annual purchase of good seed with a higher yield potential for his conditions, provided he is assured of a constant and reliable supply. In most cases this will be hybrid, because the other is in limited supply commercially.

During my 32 years of promotion of maize production in the tropics, I have been able to interest neither the public nor the private sector in the production of large volumes of seed of open-pollinated varieties. Where it is produced, it is produced by individual farmers, or as a stopgap by commercial seed-producing agencies, until some kind of hybrid can be developed. In Central America, several good, improved open-pollinated varieties are available; yet the trend in both the public and private sectors is toward hybrid seed production, much of it initially in the form of intra- or interracial variety crosses. In Venezuela, a high-yielding open-pollinated variety derived from a combination of Tuxpeño by ETO is available; yet practically all of the seed commercially produced is hybrid, either as crosses between open-pollinated varieties of Tuxpeño and

ETO or as crosses between inbred lines developed from these races. Similarly in Brazil, in the maize area of Rio Grande do Sul, an open-pollinated variety is available that outyields the best of the hybrids, but no one seems to be interested in producing large volumes of commercial seed of this variety.

Hence it appears to me that the wise maize breeders in the lowland tropics are those who devote their major resources and efforts toward the further improvement of the components of the flint–dent heterosis pattern, and the combining ability between them, in the attempt to more fully exploit these outstanding germplasm complexes for the further improvement of hybrid yields. At any point along the line of progress, the two components may be combined to form an open-pollinated population from which outstanding open-pollinated varieties could be developed. Actual development of the open-pollinated varieties might best be left to the farmers themselves, especially those who save seed for their next planting. Experience has shown that some of these farmers will come up with open-pollinated varieties from this elite material, which for their purposes will be as good or better than any breeders may produce.

Many of the hybrid maize breeders in the tropics, especially in Latin America, are devoting major efforts to the next major increase in the yield potential of their hybrids by a more fuller exploitation of the flint–dent heterosis pattern. In Brazil, yield levels are being raised in both the recently formed wide-based flint and dent open-pollinated populations through recurrent half-sib selection. At the same time, reciprocal recurrent selection programs are underway to enhance the combining ability between them. Yield capacities of the dent composites are now higher than the best hybrids, and major progress is being made in increasing yield potential of the flint composites combining Cateto and Caribbean flints. In the early cycles of selection, yield increases of up to 8% per cycle have been obtained in the dent composites. Rate of increase in the flint composites has been equally striking. These new populations, in addition to being directly useful as open-pollinated varieties, have set the stage for the development of more productive hybrids.

The CIMMYT is devoting its major efforts to the formation and improvement of broad-based gene pools, consisting of a combination of different varieties with desirable attributes. However, varieties are pooled without regard to racial origin or heterosis patterns. Undoubtedly, some excellent open-pollinated varieties are being formed, but in view of the trends in commercial seed production, more attention to population improvement for the fuller exploitation of specific, natural heterosis patterns would be highly desirable. This new orientation would in no way hinder the development of immediately useful open-pollinated varieties, but it would be much more helpful to tropical maize breeders in general.

As a final comment, I would like to stress that the limiting factor in the acceleration of maize production in the lowland and highland tropics today is not varieties. Good varieties (open-pollinates or hybrids), highly stable in their performance and highly responsive to fertilizer and improved agronomic practices, are now available for almost all areas in which maize is grown. What is lacking are effective production campaigns in which good seed is combined with the use of fertilizer, improved agronomic practices, and input-product pricing policies that will allow the farmer to make a profit.

ACKNOWLEDGMENTS

The author is deeply indebted to Drs. Hermilo Angeles A. and Ramón Covarrubias of Mexico, Drs. Ernesto Paterniani, Glauco Pinto Viegas, William da Silva, Luis Torres de Miranda, and Gladstone Drummond of Brazil, Drs. Federico Scheuch and Alejandro Grobman of Peru, Drs. Manuel Torregroza and Charles Francis of Colombia, Drs. Pedro Obregón and Pedro Elias Marcano of Venezuela, Dr. Hugo Córdova of Central America, Drs. E. C. Johnson, Willy Villena, and N. L. Dahwan of CIMMYT, Drs. Sujin Jinahyon and Bobby Renfro of Thailand, and Dr. V. R. Caranjal of the Philippines, for information in the preparation of this chapter.

REFERENCES

Annonymous. 1974. *Ann. Rep. Maize Genetics Division.* Kitale, Kenya: EAAFRO.

Brieger, F. G., J. T. A. Gurgel, E. Paterniani, A. Blumenschein, and M. R. Alleoni. 1958. Races of maize in Brazil and other eastern South American countries. *Nat. Acad. Sci., Nat. Res. Coun.,* No. 593.

Brown, W. L. 1953. Maize of the West Indies. *Trop. Agr.* **30:** 141–170.

Brown, W. L. 1960. Races of maize in the West Indies. *Nat. Acad. Sci., Nat. Res. Coun.,* No. 792.

Castro Gil, Mario. 1973. Maíces "super enanos" para el Bajio. *Boletín Técnico,* Universidad de Coahuila, ESA, Mexico.

CIMMYT Report on Maize Improvement 1973. 1974. Mexico D. F., Mexico: CIMMYT.

Grobman, A., W. Salhuana, and R. Sevilla, in collaboration with P. C. Mangelsdorf, 1961. Races of maize in Peru. Their origins, evolution and classification. *Nat. Acad. Sci., Nat. Res. Coun.,* No. 915.

Harrison, M. N. 1970. Maize improvement in East Africa. In C. L. A. Leakey (ed.), Crop improvement in East Africa. Commonwealth Agr. Bur., Farnham Royal, Burks, U.K.

Hatheway, W. H. 1957. Races of maize in Cuba. *Nat. Acad. Sci., Nat. Res. Coun.,* No. 453.

Johnson, E. C. 1974. Maize improvement. *Proc. World-Wide Symp. Maize Improvement in the 70's,* Mexico, D. F.: CIMMYT.

Paterniani, E. 1974. O factor braquitico o melhoramento do milho. X Reuniao Brasileira do milho e sorgo. Piracicaba, Brasil: Inst. Genetica, E.S.A.L.Q.

Pinto Viegas, G. 1957. Realizacoes do Instituto Agronomico o Milho Azteca. Campinas, Brasil: Inst. Agronomico.

Proceedings of the CIMMYT-Purdue International Symposium on Protein Quality in Maize. 1975. Dowden, Hutchinson and Ross Inc., Stroudsburg, Pa.

Ramírez, E. R., D. H. Timothy, B. E. Díaz, and U. J. Grant, in collaboration with G. E. Nicholson, E. Anderson, and W. L. Brown. 1960. Races of maize in Bolivia. *Nat. Acad. Sci., Nat. Res. Coun.,* No. 747.

Roberts, L. M., U. J. Grant, E. R. Ramírez, W. H. Hatheway, D. L. Smith, in collaboration with P. C. Mangelsdorf. 1957. Races of maize in Colombia. *Nat. Acad. Sci., Nat. Res. Coun.,* No. 510.

Timothy, D. H., V. B. Pena, and E. R. Ramírez, in collaboration with W. L. Brown and E. Anderson. 1961. Races of maize in Chile. *Nat. Acad. Sci., Nat. Res. Coun.,* No. 847.

Villena, D. W. 1975. Summary of results 1974 PCCMCA cooperative yield trials (mimeo.). Mexico, D. F.: CIMMYT.

Wellhausen, E. J. 1965. Exotic germ plasm for improvement of Corn Belt maize. *Proc. 20th Annu. Hybrid Corn Ind., Res. Conf.,* pp. 31–45.

Wellhausen, E. J., O. Fuentes, and A. Hernández Corzo, in collaboration with P. C. Mangelsdorf. 1957. Races of maize in Central America. *Nat. Acad. Sci., Nat. Res. Coun.,* No. 511.

Wellhausen, E. J., L. M. Roberts, and X. E. Hernández, in collaboration with P. C. Mangelsdorf. 1952. Races of maize in Mexico. Cambridge, Mass.: The Bussey Institution, Harvard University, pp. 1–223.

EVOLUTION

Chapter 6

INTRODUCTORY REMARKS TO THE SESSION ON EVOLUTION

William L. Brown
Pioneer Hi-Bred International, Inc.,
Des Moines, Iowa

A number of theories and hypotheses relating to origin have been proposed over the years. As a brief background to the chapters that follow, it seems desirable to enumerate the more important of these. The first is the pod corn theory of Saint Hilaire offered in 1829. This hypothesis simply held that cultivated maize arose from pod corn that St. Hilaire presumably thought to represent wild maize.

The teosinte theory and its various modifications goes back to Ascherson who in the late 1800s suggested that maize arose either directly from teosinte or from the hybridization of teosinte and some unknown species of the *Andropogoneae.* More modern versions of the teosinte theory have been proposed by some of the authors contributing to this section, whose views are explained in the chapters that follow.

The theory of common ancestry, first proposed by Montgomery in 1906 and later expanded by Weatherwax, suggests that maize, teosinte, and *Tripsacum* arose from a common ancestor and evolved independently by divergent evolution.

Finally, the tripartite theory of Mangelsdorf and Reeves postulates that: (a) pod corn represents the ancestral form of cultivated maize, (b) teosinte is not the progenitor of maize but is instead the progeny of a hybrid of maize and *Tripsacum,* and (c) teosinte has played an important role in the evolution of modern maize through its introgression with maize. More recently Mangelsdorf has abandoned the hypothesis relating to the hybrid origin of teosinte and has suggested that it is probably a derivative of maize.

Chapters 7–9 and 11 deal specifically with questions of origin, and Chapter 10, treats the end products of evolution, that is, modern races and their relationships.

Each of the contributors to this section is a proponent of the teosinte hypothesis and each has presented evidence, both old and new, supporting that position. Perhaps the most convincing evidence in support of the teosinte hypothesis is the close genetic relationship between teosinte and maize. The two species hybridize readily, both experimentally and in nature. The hybrids exhibit normal chromosome pairing and essentially normal fertility. As Dr. Beadle points out, teosinte has survived as a wild plant for at least 7000 years, whereas maize as we know it requires the aid of man for its survival. For these and other reasons, he believes it reasonable to assume that teosinte was ancestral to maize and that the transformation from teosinte to maize occurred as a result of human selection. Beadle emphasizes the fact that despite extensive searches for wild maize, none has been found. Furthermore, he believes the earliest archaeological specimens from Tehuacán not to be "wild," since the maize represented by those remains has no means of seed dispersal.

Chapter 11, which deals with the chromosomal constitution in maize, is a brief summary of extensive data accumulated over a period of years by Dr. McClintock and her collaborators. These data provide the best information available on the geographic centers of important maize races, the routes of migration of races into new territories, and the evolution of new races resulting from introgression between maize originating in different centers.

Numerous clearly defined maize complexes have been identified based on chromosome constitution. These are associated with specific geographic areas that probably represent secondary centers of concentrated propagation. The migration paths of certain complexes are clear-cut, and the influence of others is noted at locations distantly removed from the complex center. This information should be of special interest not only to students of evolution but to the anthropologist as well.

In addition, the data presented by Dr. McClintock suggest that much of the maize of the Caribbean Islands reached there from Central America, that highland Guatemala was the source of the initial introductions of maize into the Andes, and that eastern and central Mexican maize in combination with that from the northern United States was an important source of southeastern U. S.

varieties. Chromosome constitutions also indicate a distinct relationship between the maize of highland Guatemala and that of the northern United States. This is particularly interesting in that it is directly opposed to the previously postulated role of the Mexican race, Harinosa de Ocho, in the evolution of the Northern Flint Corns.

Regarding the role of teosinte in the origin of maize, McClintock points out that Kato-Y's studies of the chromosomal components of teosinte "suggests that the Mexican teosintes are the source of all the basic germplasms of maize." It is further suggested that Guatemalan teosinte has, at best, played a minor role in the evolution of maize.

Galinat considers the hypothesis that ancestral populations of teosinte, which eventually evolved into maize, became established accidentally near camp sites from seed spilled or lost following gathering from wild populations. He suggests that the semiwild-type camp site populations diverged from the wild populations in the direction of increased condensation and spike clustering that led eventually to the highly condensed rachis and shank of maize. Galinat believes the natural tendency of preagricultural gatherers was to harvest seed from those plants exhibiting a high degree of spike clustering, thereby enhancing the efficiency of the gathering process. He further theorizes that early man recognized the advantage of those populations that were reproduced from his accidentally lost seed. Consequently, he deliberately repeated the process and eventually saved such seed for sowing. Although this may provide a reasonable explanation of the processes affecting the change from noncondensed to highly condensed inflorescences, it does not shed light on the nasty problem of differences in induration in teosinte and primitive maize. Without exception, the fruit case of teosinte is highly indurated (lignified). In contrast, the earliest archaeological maize from Tehuacán and Bat Cave exhibits little induration. Indeed the older cobs recovered from the rock shelters of Tehuacán are less indurated than those of more recent origin.

It seems reasonable to assume that if maize evolved from teosine in the manner described by Galinat, the female inflorescences of the earliest forms of such maize should logically be expected to be highly indurated. Galinat recognizes this problem and offers as an explanation an earlier suggestion by Beadle that a single mutation to a higher tunicate allele could block the process of induration and thereby produce a phenotype similar to that found in the oldest cobs from Tehuacán and Bat Cave. It is unlikely that this explanation will satisfy all students of maize.

Chapter 9 provides a much needed, revised taxonomic treatment of the genus *Tripsacum* L. The authors recognize eleven species, two of which, *T. andersonii* Gray and *T. bravum* Gray, are described for the first time. On the basis of chromosome behavior, it is suggested that *T. andersonii* may have arisen as a natural hybrid between maize and *Tripsacum*. It is further suggested that the putative hybrid genome probably comprises 54 *Tripsacum* and 10 *Zea* chromosomes.

Students of evolution will find particularly interesting the discussion on cytogenetic affinities between *Zea* and *Tripsacum* and the morphology of maize–*Tripsacum* introgression.

The authors emphasize the lack of close relationship between teosinte and *Tripsacum* by referring to the thousands of crosses that have been attempted between the two genera without the production of a single successful hybrid. In contrast, *Tripsacum* and cultivated maize have been crossed many times. Chromosomes of the hybrid *T. dactyloides* × maize remain as univalents during meiosis, except for the presence of two or three open bivalents, indicating a lack of close genomic relationship. The hybrids are male sterile but when backcrossed to *Zea mays*, produce some offspring. The backcross progeny have 18 *Tripsacum* + 20 *Zea* chromosomes. Although most of the *Zea* chromosomes pair as bivalents among themselves, the authors have observed as many as four *Tripsacum* chromosomes associated with *Zea* chromosomes to form bivalents or trivalents. Hybrids with 10 *Zea* and 36 *T. dactyloides* chromosomes when repeatedly backcrossed to *Zea* produce some plants with $2n = 20$ *Zea* chromosomes. Most plants so derived resemble pure maize in phenotype, yet a few plants exhibit a number of "Tripsacoid" characteristics and are similar in phenotype to maize–teosinte hybrids, suggesting the incorporation of some *Tripsacum* genetic material into the maize genome. The authors do not rule out the possibility of natural introgression of *Tripsacum* into *Zea* but consider more likely the introgression of maize into *Tripsacum,* an example of which may be represented by *T. andersonii* Gray.

Chapter 10 is a summary of more extensive data contained in several of Goodman's recent publications.

The more important racial complexes of economic importance are identified. These include the Mexican Dents, U. S. Corn Belt Dents, Caribbean Dents and Flints, the Cateto Flints of South America, and the Northern Flints and Flours. In addition, a number of other regionally important broad racial complexes are recognized from North, South, and Central America.

A section describing the current status of racial studies provides an account of work currently in progress (pp. 143–148). A final section describes in considerable detail some of the more recent methods being utilized in attempts to more clearly define racial relationships (pp. 148–156). These include chromosome knob-pattern data, numerial taxonomy, morphological studies of racial F_2 populations, and the use of isozyme frequencies.

As clearly indicated in these papers, a considerable body of cytological, genetic, and morphological evidence points to teosinte as a probable progenitor of maize. Yet if maize arose directly from teosinte as a result of selection by early man, a number of questions arise for which we still lack satisfactory answers:

1. If teosinte was ancestral to maize and if the transformation to primitive maize were a gradual process, why have intermediate forms not been recovered either from natural populations of teosinte or from prehistoric remains?

2. If teosinte were used as food by primitive man in the premaize era (as suggested by Galinat), would not one expect to find an abundance of teosinte remnants in the premaize strata of primitive man's shelters and dump heaps located in the areas of teosinte distribution?

3. If maize arose directly from teosinte through selection, would not the extensive induration of the teosinte fruit case be expected to occur in the cobs of early maize?

Chapter 7

THE INHERITANCE OF SOME TRAITS ESSENTIAL TO MAIZE AND TEOSINTE

Walton C. Galinat

Suburban Experiment Station, Waltham,
Massachusetts Agricultural Experiment Station,
College of Food and Natural Resources,
University of Massachusetts, Amherst

Maize (*Zea mays*) was the cereal that permitted the development of the major New World civilizations. The process of its domestication is still controversial. We have not agreed as to whether the wild ancestor of maize is extinct or if it is represented by its closest relative, teosinte (*Z. mexicana*). The position that teosinte or a teosinte-like plant is the ancestor of maize has been maintained by Beadle (1939, 1972, 1975) and others. The reverse, that maize is the ancestor of teosinte, is now supported by Mangelsdorf (1974).

A compromising opinion supporting a common ancestry for maize and teosinte is held by a few others (Randolph, 1975; Weatherwax, 1935). Implicit by a common ancestor is that it be somewhat intermediate between its derivatives. In the case of maize and teosinte, it is doubtful that a truly intermediate form would be adaptive for natural survival or that it would be consistent with the evolutionary trends in this part of the grass family. Rather, one might ask

93

whether such an ancestor was more teosinte-like than maize-like, or vice versa. Did man have a role in the divergence of teosinte and maize, or was it a result of a natural process of speciation?

The question of which came first, teosinte or maize, may be resolved in part by measuring the evolutionary distance between them. The distance would be much greater if teosinte diverged from a wild maize by natural evolution over a period of millions of years (Mangelsdorf, 1974), than if maize was domesticated from teosinte in less than 10,000 years (Galinat, 1971a).

The morphology of the fruit case in *Tripsacum,* the second closest relative of maize, may reflect the more primitive condition of the genus *Zea.* That is, *Tripsacum,* like teosinte, has single female spikelets enclosed in cupulate fruit cases arranged in two-ranked spikes that become indurated during maturation. Because of its apparent great age, *Tripsacum* has had time to undergo an extensive differentiation in the architecture and number of its chromosomes during the speciation process since its divergence from *Zea. Tripsacum* has diverged into at least five diploid ($2n = 36$) and four tetraploid ($2n = 72$) species, all of which are perennial, but of a diverse morphology and adapted to a wide range of habitats throughout much of the New World (Cutler and Anderson, 1941; Randolph, 1970; Tantravahi, 1968, 1971; Wilkes, 1972; Galinat, 1973b). The diploid species regularly produce 18 bivalents at meiosis, whereas the tetraploids always form at least some quadrivalents. Because the tetraploids may segregate for some of the morphological characteristics that distinguish their putative parents, they are regarded as segmental allotetraploids (Tantravahi, 1968, 1971).

The genus *Tripsacum* probably originated as an allopolyploid following doubling of the chromosomes from a wide-cross hybrid between two other genera, one of which is presumed to be *Manisuris* of the tribe Andropogoneae (Anderson, 1944; Stebbins, 1950). The other parent of the allopolyploid was probably of the genus *Zea* (of the tribe Maydeae), at first identified as maize (Galinat, Chaganti, et al, 1964) and later modified as being more similar to teosinte (Galinat, 1970). In this case the teosinte-like ancestor would be older than both maize and *Tripsacum.* Teosinte has not undergone the same degree of speciation as that found in *Tripsacum,* and this may be explained as a result of the broader base of variability in *Tripsacum* resulting from its allopolyploid origin. Teosinte is cytologically more variable than maize in having certain knob positions unknown in maize and in occurring in both diploid and tetraploid forms, the latter of which happens to be perennial. In having both annual and perennial plant habits, teosinte is not only more variable than maize but also more variable than *Tripsacum.* Maize has the least cytological variation of the American Maydeae, despite the fact that more than 200 races have been described, primarily on a basis of the ear. Mangelsdorf (1974) has suggested that a wild maize differentiated into at least six wild races that were spread to

North, Central, and South America, whereas teosinte has remained confined to Mexico, Guatemala, and Honduras.

In contrast to the relatively wide cytological differences between *Zea* and *Tripsacum,* the chromosomes of teosinte and maize are essentially identical. They have the same number of chromosomes ($n = 10$), their F_1 hybrids are usually fertile, and their chromosomes pair closely at pachytene (Beadle, 1932a,b). For the loci tested, maize–teosinte hybrids have about the same frequencies of crossing over as in maize (Emerson and Beadle, 1932), except in the presence of rare heterozygous inversions (Beadle, 1932b). Furthermore, the close similarity of maize and teosinte chromosomes, not only in their size, but also in their basic organization such as the chromomere pattern, indicates that these species have the same chromosome complements (Kato-Y, 1975; Longley, 1937, 1941). Both species are polymorphic for size and number of chromosome knobs. Of the 36 knob positions known in the genus *Zea*, only 22 have been found in maize, whereas all occur in a composite of teosinte (Kato-Y, 1975).

Before considering the inheritance of the traits essential to demark maize and teosinte, a discussion of the quantitative traits of condensation and lignification is necessary. Because these polygenic traits interact or are epistatic with the expression of the more simple traits, they complicate their inheritance in a background of modern maize. Therefore, an understanding of how these traits became adaptive will help to explain how they became polygenic and how they fit into evolutionary trends with known relatives.

CAMP-SITE COLONIES AS CRADLES LEADING TO ZEA DOMESTICATION

It is now generally recognized that man had been gathering and influencing food plants for at least 10,000 years in the New World (Bryan, 1973). Even in the premaize period of the middle of the El Riego phase in the Tehuacan Valley, there are indications that a cereal, *Setaria,* was being domesticated and, based on seed size, evidence of man's selection in certain tree crops (avocado, chupandilla, and cosahuico) has been pointed out by Smith (1967). Without animal species amenable to domestication in the New World comparable to the species of cattle, sheep, and goats available in the Old World, man may have been forced to turn to plant domestication in the New World at an earlier time than that for which he is usually given credit, on the basis of his lower level of cultural evolution by 1492 (lack of the wheel, arch, etc.).

For the sake of this discussion of the probable steps involved in the evolutionary emergence of maize, it is assumed that a teosinte-like plant was the ancestor of maize and that the influence of man transformed it into maize (Galinat, 1970, 1971a). In the absence of maize, the harvesting of teosinte is a productive source of nutritious food. On the basis of harvesting semiwild teosinte in Guerrero, Mexico, Beadle (Flannery, 1973) estimates that a man could harvest one liter of teosinte per hour (fruit cases threshed into blankets, skins). As with the first semidomesticated colonies of other cereals (Hawkes, 1970), the ancestral populations leading to maize became established by accident near the camp sites from teosinte seed that man has spilled or otherwise lost during transport home from wild populations. These camp-site populations diverged from the wild type because the greater part of the seed harvested was of the harvestable types.

In harvesting natural stands of teosinte in Guerrero, Mexico, my son and I noted that most of our harvested seed was automatically coming from those plants with some conspicuous degree of spike clustering. The tight compaction of many husk-enclosed spikes causes a retention of their fruit cases by interfering with the natural process of seed dispersal by the action of wind lashing on elongate branches.

Despite the tardy seed dispersal from spikes that are condensed into clusters, this trait is maintained in about 25% of the plants in natural stands, because it seems to have a counterbalancing adaptive advantage in the form of some additional protection against the feeding of birds. Thus variation for condensation probably occurred in ancient teosinte populations and made the origin of maize possible. As a result of the process of harvesting, populations of teosinte with a higher-than-average degree of condensation tended to become grouped about the camp sites.

In recapitulating the process of *Zea* domestication, seed spilled of the harvested teosinte easily attained a foothold in the disturbed and often moist soil at the threshing and grinding sites near man's habitations. Various genetic factors for increased levels of condensation accumulated and recombined in these evolutionary cradles, perhaps for thousands of years, until divergent types with superior harvestability emerged. With this passage of time, multigenic assemblages for increased condensation in the lateral branches and their female spikes became the primary factors separating the wild teosinte from the protodomestic types of teosinte that represented primordial maize. An understanding of how this condensation moved teosinte in the direction of maize is the key to visualizing how a subsequent program of selection by man completed the transformation of the spike of teosinte into the cob of maize.

Man recognized the advantage of these populations that grew from his accidentally lost seed, so he deliberately repeated the process by saving their seed for sowing. When man learned to select and sow the most useful types, a dramatic transformation in the female spike followed. As the condensed teosinte became dependent on man for seed dispersal, there was a corresponding loss in its natural means for disarticulation by abscission layers in the rachis.

The effects of the condensation factors that had first concentrated their expression in the shank were either switched or extended into the rachis. As a result, the cupulate interspace representing the barren side of the alternate internode became reduced to zero length, the internodes assumed alternate triangular shapes, and the nodes bearing solitary spikelets zigzagged back and forth across the rachis. Compression caused the apex of cupules to become fused to the glume cushion of the spikelet borne at the second node above and, thereby, the rachis became rigid. Although the abscission layers across the pith and through the rind may have continued to develop at first, as shown in some of the older Tehuacán cobs, they were nonfunctional in these fused cobs and were soon lost. Other specializations that further adapted the emerging ear to man's use were acquired through selection. The raw material for this selection came both from within the population's variability and by the introduction of new cultivars of maize derived through trade and human migration. These domestic traits included softer chaff, longer rachillas for easier shelling, pairing of female spikelets, increases in kernel size, and increases in the level of ranking in spikelets, all of which resulted in higher-grain yields per ear and made the ear more useful to man.

Man's selections, then as now, tended to give the uppermost spike apical dominance as it became the prime energy sink. The uppermost ear has become the largest if not the only one to develop. The uppermost tassel branch became so conspicuous that it has been given a special term, the *central spike*. The process has been compared by Iltis (1973) to the origin of the monocephalic head and stout single stem of the cultivated sunflower through a suppression of the many small lateral heads of the wild sunflower.

Concurrent with a concentration of nourishment into the uppermost female spike, a suppression of both the lower spikes and the internodes that separate them caused their subtending leaf sheaths to remain concentrated in a single multihusk system that enclosed and protected the terminal ear. The suppressor genes (sleeper genes) involved sometimes reveal their presence in modern maize as they awake under abnormal environmental conditions, leaving an unprotected ear or cluster of ears. The development of this multihusk enclosure of the ear cannalized both man and maize into a symbiotic partnership.

TWO-RANKED VERSUS MANY-RANKED

The many-ranked arrangement of spikelets on the ear and central spike in the tassel of maize, in contrast to the strictly two-ranked spikes of teosinte, logically follows the discussion on the adaptive values, under semidomestication, of high condensation in the ancestor of maize. When increased condensation became the primary step leading toward domestication, instead of a nonshattering rachis as in the other cereals, it opened the way for new patterns of domestic specializations, unknown in other grasses. As a result, we now have a many-ranked maize ear enclosed and protected by a system of multiple husks.

Within the ear, condensation is generally more reduced or relaxed toward the tip with a gradient of increasing levels toward the butt. When a threshold in compression created by the condensation is reached, relaxation and its associated space for the differentiation of spikelets is gained by pinching and twisting the expanding primordia off to one side. As a result of this slippage, higher orders of ranking become associated with increasing levels of condensation, a correlation studied in detail by Anderson and Brown (1948). During this process modifying genes accumulated that expanded the cob (rachis), thereby permitting condensation without fasciation (Anderson and Brown, 1948).

Earlier, Collins (1919) had speculated on a basis of the derivatives from maize–teosinte hybrids that condensation, twisting, and yoking of rachis segments could have resulted in an origin of the many-ranked condition of maize. Unfortunately, the arguments of Collins became confused when he claimed erroneously that drop rowing (abortion of two kernel rows) occurred simultaneously in two single rows borne on opposite sides of the ear, a mistake that disregards the paired nature of female spikelets in maize and, thereby, the fact that drop rowing must occur by members of a pair (Weatherwax, 1935; Mangelsdorf, 1945).

A significant number of the oldest Tehuacán cobs were two-ranked at the tip of the ear, and a few were two-ranked throughout the entire ear. A similar phenotypic result has been produced experimentally by selection for lower levels of condensation among the F_2 segregates of a cross between a string cob sweet corn inbred (MW401) that derived its slender cob from the race Confite Morocho with another similar inbred derived in part from Coroico "interlocked" maize. When these stocks were carried into the F_3 and F_4, it was found that the percent of two-ranking along the ear was both a heritable trait and related to the overall level of rachis condensation.

Because this elongate type of induction of two-ranking depends on a polygenic complex controlling the level of condensation even within a background of four-ranked (eight-rowed) maize, its inheritance is complex and it has not been pinpointed to known loci. Several loci are involved in higher orders of many-ranking (Emerson and Smith, 1950). Since condensation operates directly on ranking in the central spike of the tassel and ear, the control of two-ranks versus many-ranks in this case is *not* by means of suppression versus activation of the central spike. The lateral branches of the tassel and ear (ramosa type) are usually two-ranked in modern maize, whereas the central spike is many-ranked.

Probably the most significant type of control over development of a two-ranked spike is an incompletely dominant factor first discovered as one of the effects of a *Tripsacum* chromosome (Tr9) when present as an addition monosome on a background of eight-rowed (four ranks of paired spikelets) chromosome 2 (M2) tester maize. This *Tripsacum* chromosome is marked by a terminal knob, dominant *Lg*1 and *Gl*2, as well as a recessive *b*. In a substitution stock for an interchange chromosome (M2-Tr9) derived from two independent crossovers between M2 and Tr9, a derivative was obtained that helps to localize the position of the locus controlling ranking. In the first M2–Tr9 crossover the interchange acquired the long arm of maize chromosome 2. A subsequent derivative of this stock showed the effects of a second crossover between the *Lg*1 and *Gl*2 loci. In this stock the terminal knob and the dominant *Lg*1 allele from *Tripsacum* were replaced by their counterparts from maize. The stocks remained two-ranked throughout development of the stock carrying primary and secondary interchanges. Because this *Tripsacum* chromosome (Tr9) appears to carry the same loci as the short arm of maize chromosome 2 (known coverage of *ws*3, *lg*1, *gl*2, *b*, *sk*, *fl*1) and as these appear to be in essentially the same order, although closer together than in maize (Galinat, 1973b), it is concluded that the position of the locus controlling ranking is between *Lg*1 and the centromere. The factor from *Tripsacum* for two-ranks produces fertile compounds with a mutant gene (*pd*) for single female spikelets. The dominance of both the *Tripsacum* and *pd* genes is relative to the genetic and environmental background.

This semi-dominant factor from *Tripsacum* appears to be allelic to a gene for two-ranking from teosinte, as shown by derivatives from a hybrid between the two-ranked trait from teosinte and the Tr9–M2 interchange stock. This hybrid has failed to segregate a stable many-ranked condition through the F_3, thereby satisfying the allelism test for the teosinte and *Tripsacum* factors for ranking control.

Prior to hybridization with the teosinte derivative stock, the point of expression of the two-ranking factor from the Tr9–M2 interchange chromosome was often variable along the length of the ear. After crossing with the teosinte derivative, several Tr9–M2 selections were made in which the expression for two-ranking was synchronized with the onset of rachis formation. The nature and position of this regulating factor is as yet unknown.

It should be pointed out that eight-rowed races of maize in general, when grown under stress, frequently produce two-ranked ears. The tendency for this type of two-ranking is stronger in eight-rowed strains having an inherited reduction in the vascular system of the cob. When such a reduction in vascularization is combined with a mutant gene (*pd*) for single female spikelets, the styles may fail to elongate, producing a "silkless" type of expression. This defective and sometimes lethal system of producing a two-ranked spike appears to be evolutionarily unimportant in comparison to the other two systems, which are fully viable.

Recessive mutations to two-ranking have been reported, but it is not clear at this time whether all of them are of the "depauperate" type. The one reported by Daniel (1973) is of a defective type because when he combined this trait with the factor for single female spikelets, only silkless ears resulted. The two-ranked mutant discovered by Tavcar (1935) was controlled by a single recessive gene, whereas that isolated by Lindstrom and studied by Burdick (1951) required the interaction of two complementary genes.

That two genes may control the "switch" between the two-ranked condition of teosinte and the four-ranked state of the primitive maize is supported by some of the linkage data from maize–teosinte segregations. Rogers (1950) found evidence for genes controlling the two-ranked condition on chromosomes 3 and 6 of Nobogame teosinte. Both Rogers (1950) and Langham (1940) had also found a gene for single spikelets (*pd*) on chromosome 3. While Langham found the two genes, *tr* and *pd*, to be linked by about 20 crossover units, this was not confirmed by Rogers. With Durango teosinte, Rogers (1950) found that the two genes for two-ranking are on chromosomes 1 and 2. The chromosome 2 location agrees with our results with *Tripsacum* chromosome 9, which is homeologous to the short arm of maize chromosome 2. The chromosome 1 location for two-ranking agrees with the earlier results of Langham with the same teosinte. Mangelsdorf (1947) found two-ranking from teosinte to show linkage to marker genes on maize chromosomes 1, 2, 3, 6, 7, 8, and 9. These variable findings on linkage of two-ranking probably result from its segregation on a genetic background distinguishing various levels of condensation. That is the segregation of many genes for condensation would confound the linkage results with two-ranking.

INDURATION IN THE FEMALE SPIKE

A high degree of induration (largely lignification) of the cupulate fruit case occurs in all known races of teosinte. It appears to be a polygenic trait, because the experimental introduction of any of the various teosinte chromosomes into modern maize is known to increase the induration of the cob (Mangelsdorf, 1974). An induration of the teosinte fruit case is adaptive in protecting the grain. In contrast, all of the oldest archaeological maize cobs from Tehuácan, Mexico and from Bat Cave, New Mexico are soft.

The post domestication archeological maize cobs, like modern cobs, include some indurated specimens. In some cases the induration of such cobs may be ascribed to teosinte introgression, especially teosinte chromosome 4, as is indicated by associated teosinte traits such as short rachillas, upward inclination of the spikelets, and relaxed condensation. Other indurated cobs, such as in some Northern Flint specimens, could derive their hardness from the greater efficiency of the ear (kernels and cob) as an energy sink. Maize cobs may become indurated when the growing point of the ear is aborted (blasted) before the cob achieves its full potential length, causing the available energy to become concentrated into a smaller storage organ, which may in turn result in lignification.

If the original wild teosinte were of extreme induration, a single genetic change, such as to a higher tunicate allele, could block this deposition of lignin (and elongate the rachilla), producing a phenotype similar to that of the ancient soft cobs from Tehuácan, Mexico (Beadle, 1972). One difficulty in accepting Mangelsdorf's (1974) suggestion of an evolutionary change from a hypothetical wild maize to the phenotype of teosinte is the prerequisite for a multigenic assemblage channeling energy into the rachis. Therefore, in the case of induration, the direction of simplest change would favor teosinte as the older species.

THE CHROMOSOME 4 COMPLEX

Teosinte chromosome 4 carries a genetic complex that is primarily concerned with the development of its specialized fruit case. An induration of the glumes and rachis is one of the most obvious effects of this critical genetic assemblage. Other important traits of the fruit case affected are an inclination of the female spikelets toward their cupules, the development of abscission layers in the rind and across the pith, and, according to Mangelsdorf (1947), one of the two genes for single female spikelets [the other pd gene being on chromosome 8–or chromosome 3 according to Langham (1940), or chromosomes 3 and 6, according to Rogers (1950)].

The position of this fourth chromosome complex has been estimated to include the whole short arm, as marked by the Su locus (Mangelsdorf and Reeves, 1939; Mangelsdorf, 1947; Rogers, 1950), and it is incompletely dominant in expression (Mangelsdorf, 1952). Separate genes for pith abscission (Ph) and rind abscission (Ri) are about 27 crossover units apart, with the Ri gene near the gametophyte allele (Ga) on the short arm of chromosome 4 (Galinant, 1975). According to Mangelsdorf (1947), genes on at least three different chromosomes are involved in rigid versus shattering rachis, but apparently his data refer to cupule–glume cushion fusion from compaction as well as abscission-layer development.

Because Ga provides one-way cross sterility, it may help to protect the integrity of the chromosome 4 complex when coupled in linkage to it. The frequency of *Ga* in present-day teosinte is unknown. There is a small inversion in the long arm of chromosome 4 in certain races of teosinte (Nobogame), but it is not clear if it is functioning to protect the adaptive value of this chromosome.

In the process of intergenomic mapping between the chromosomes of maize and *Tripsacum* it was found that *Tripsacum* does not carry an assemblage of loci corresponding to maize chromosome 4 (Galinat, 1973b). This is considered as evidence that teosinte could not have acquired its essential fourth chromosome assemblage from *Tripsacum*, in contradiction to a basic assumption of the hypothesis that teosinte originated as a maize–*Tripsacum* hybrid.

This conclusion that teosinte is not a maize–*Tripsacum* hybrid has now been accepted in a reversal of position by Mangelsdorf (1974) on different grounds involving pollen morphology of the maize–*Tripsacum* derivatives developed by Galinat and by the identical fine structure of the pollen-grain ektexine of maize and teosinte, in contrast to the clearly different ektexine of *Tripsacum* (Banerjee and Barghoorn, 1972).

The cross-mapping studies revealed that more than two different *Tripsacum* chromosomes carry M4 loci; *Su1* is located on Tr7 and *Gl3* is located on Tr13. At least four other M4 loci (*La, Bm3, Ra3, J2*) not covered by either Tr7 or Tr13 are present in the *Tripsacum* complement but have not as yet been assigned to particular *Tripsacum* chromosomes.

It should also be pointed out that *Tripsacum* does not develop abscission layers across the pith in the manner of teosinte. Rather, during rind abscission and disarticulation, a loose mass of parenchyma cells from the pith breaks loose and clings to the underside of the fruit case forming a "ball-and-socket" type of separation. Thus *Tripsacum* carries only one of the two types of abscission factors found in teosinte. On this basis, if teosinte was a parent of *Tripsacum*, it was a less specialized form of teosinte than that occurring today. The same conclusion may be drawn from the fact that *Tripsacum* does not carry the chromosome 4 complex of modern teosinte.

SINGLE VERSUS PAIRED FEMALE SPIKELETS

In maize, paired spikelets occur normally on both tassels and ears. In the cob each pair is carried outside an associated cupule. In teosinte, only the male spikelets are paired. The female spikelets are borne singly and within the protective confines of their individual cupules. The cupule, together with the outer glume of the enclosed spikelet, forms the fruit case. The single condition is a result of reduction of the pedicellate spikelet of each pair, as shown by the occasional remains of the second spikelet. A similar condition occurs in *Tripsacum*, the second closest relative of maize and teosinte. Further back in *Manisuris* of the Andropogoneae, only the pedicel of the second spikelet remains. The pedicel is adaptive as a functional part of the fruit case by adhering to one wing of the rachis cavity, thereby contributing to the formation of a cupule-like structure.

The mutant gene, *pd,* for single female spikelets discovered by Hepperly (1949) may be recessive, incompletely dominant, or fully dominant depending on the maize background involved. In most maize stocks, expression of *pd* is not at first synchronized with the onset of development of the female spike, or its expression may extend into the male rachis. The point of *pd* expression along the ear and/or in the tassel may be altered by selection for regulatory genes until the stable system of teosinte is ultimately established. Langham (1940) reports that a mutation to *pd* that he discovered is allelic to one of the genes for single spikelet in teosinte. We have subsequently obtained similar results with the *pd* mutant gene discovered by Hepperly.

When the single spikelet gene is extracted directly out of teosinte, it may segregate as a single recessive in some families (Collins and Kempton, 1920), sometimes on chromosome 3 (Langham, 1940), and sometimes segregating from chromosome 7 with Nobogame teosinte (Rogers, 1950). In another segregation involving Durango teosinte, Rogers (1950) found linkage of *pd* on both chromosomes 3 and 7. These results do not agree with those of Mangelsdorf (1947), who reported genes for single spikelets on chromosomes 4 and 8 of these two teosintes. Later, however, Mangelsdorf (1952) did report a block of genes on chromosome 3 of Florida teosinte that affects the expression of single spikelets. The results with El Valle or Florida teosinte of Rogers (1950) are also at variance in which *pd* was found to be linked with *tr*, which in turn showed linkage with chromosomes 8 and 10.

The somewhat variable results in trying to pin down linkage relations of the *pd* gene or genes from teosinte in a segregation with marker genes on a background of modern maize is probably due to its incongruous nature, resulting in problems of variable expression similar to those experienced with the mutant gene for single spikelets.

When the *tr* and *pd* genes are transferred to a background of the string cob (*Sg*) trait in combination with the teosinte chromosome 4 complex, the rachis is vascularized more like that of teosinte, and these mutant genes acquire a stable phenotype that is fully fertile in the homozygous compound (*tr tr, pd pd*). The linkage association of single spikelets with teosinte chromosome 4 reported by Mangelsdorf (1947) may have as its basis an indirect interaction through its effect on increasing vascular development in the rind of the rachis (Galinat, 1974).

SESSILE VERSUS PEDICELLATE FEMALE SPIKELETS

Whereas all races of teosinte must have sessile female spikelets to contain them and their enclosed grain within the cavity of their associated cupulate fruit case, the various races of maize have various degrees of pedicel development. Some of the more primitive races of maize, such as Confite Morocho and its derivatives from South America, carry sessile spikelets controlled by the interaction of two complementary genes (Galinat, 1969). These same races of maize have reduced cupules or, in some selections, they may become totally lacking in cupule development. As a result, the kernels (and spikelets) are emergent and threshable on the surface of the rachis.

Although cupule reduction in South American maize may possibly reflect an independent pathway to domestication from a teosinte-like ancestor, the absence of any known teosinte in South America seems to rule this out. Rather, it may be that man transported a primitive cultivar of maize to South America, where it was geographically isolated from teosinte and further divergence from its fruit case by cupule reduction became possible. When reduced couples were combined with a slight relaxation in condensation it became possible for adjacent members from alternate whorls of spikelets to interlock and, thereby, to give structural stability to the Piricinco–Coroico complex of South American races.

In the maize of Central and North America the cupule was not reduced, but highly compressed, and pedicels of the spikelets were elongated sufficiently to carry the paired spikelets out of their associated cupule. Thus the adaptive function of the cupule became transferred from one of enclosing and protecting the grain to one of providing the mechanical strength required for the agronomic utility of a large ear. During this process the elongate pedicel (foot stalk) became adnate to the floor of a deep upward sloping cupule. When condensation collapsed the cupule its roof became fused to the floor and its apex to the glume cushion borne at the second node above. The whole lignified structure is so solid and strong that an industry of manufacturing smoking pipes has been built using the corn cob as the tobacco fire pit.

THE GENETIC DISTANCE AND SELECTIVE GENETIC ISOLATION OF MAIZE AND TEOSINTE

The essential genetic differences that demark the female spikes of teosinte and maize are of no greater magnitude than might be expected to separate the one form selected and nurtured by man from its counterpart that survives in nature. Disruptive selection between man and nature, acting in opposite directions over the millennia, would be expected to create such extreme differences between the domestic and wild forms of crops (Doggett, 1965). Because unconscious mass selection during a preagricultural period of gathering food by man has apparently accumulated high polygenic levels of condensation in the shank and rachis of maize, and, because high condensation interferes with the expression of other traits essential to teosinte, especially two-ranking and single female spikelets, the genetic distance between these wild and domestic forms of *Zea* may in some segregations be exaggerated. When present day segregations involve low condensation maize such as occurred in the oldest archaeological maize of Tehuacán, the two-ranking and single female spikelet traits are controlled by only one or two genes each.

The two-rank and single female spikelet traits not only have a relatively simple inheritance, but they also occur in the more distant relatives of maize, namely *Tripsacum* and *Manisuris*. Thus they could easily mutate to their counterpart traits in maize, and they would be expected in the ancestor of maize (Galinat, 1956). Teosinte fulfills these morphological qualifications as an ancestor of maize, and it is the closest relative of maize on cytological grounds. The complex inheritance of condensation and lignification does not rule out the possibility that maize originated from teosinte by a shift in allelic frequencies in response to the predomestication influence of man on the seeds gathered for food.

In contrast to the genetic differences between teosinte and maize, the structure of their chromosomes has remained relatively conservative. The few cytological differences occurring between the species appear to have acquired adaptive values during their coevolution, perhaps as follows. When man learned to cultivate his domestic teosinte or emerging maize in large stands, the original wild populations became threatened by genetic swamping. To protect and stabilize their system of disarticulation and natural seed dispersal the wild types had to coevolve systems of selective genetic isolation (Galinat, 1973a,b). As summarized by Galinat (1975), wild individuals carrying chromosome knobs, small inversions and gametophyte alleles that are located by chance in or near their essential combinations of genes, would be resistant to the swamping effects of introgression from the domestic types. With possibly 30–40 essential genes of teosinte now tied up in only four or five blocks (Collins and Kempton, 1920; Mangelsdorf and Reeves, 1939; Beadle, 1972) or supergenes, destabilizing introgression by their nonadaptive corresponding segments from maize is rapidly cast off. As a result, the integrity of the essential features of teosinte is protected from genetic dissection. Thus teosinte survives the continued genetic onslaught by maize despite the array of fertile F_1 hybrids in about 5–10% of the combined maize-teosinte populations in the Chalco area near Mexico City, as recorded by Wilkes (1972).

These crossover inhibitors and factors for preferential fertilization would tend to maintain the integrity of their associated blocks of genes with the result that the wild types of teosinte could more readily cope with and recover from outcrossing to the nonadaptive types of domestic teosinte representing the emerging maize. Opposite differences in the presence or absence of knobs as well as small inversions between sympatric partners of maize and teosinte as would be necessary to maintain adaptive blocks of genes (supergenes) have been observed in the extensive cytological studies of Kato-Y (1975), who also showed a negative correlation between the presence of knobs and the presence of inversions as if they were an alternate way to perform the same coadaptive function of segment protection.

Whereas short inversions may be relatively numerous and important to the coexistence of maize and teosinte, they are difficult to demonstrate cytologically or genetically. They are fully fertile and reduce crossing over through loose pairing. Such crossover inhibitors may be responsible for the rapid recovery of parental types from maize–teosinte hybrids that segregate in frequencies, indicating that only four to five inherited units distinguish the species. The mechanisms for genetic protection are more important to the ability of teosinte to cope with maize than vice versa, because the selective powers of man would tend to prevent deleterious introgression of teosinte into maize.

Close linkage alone without the benefit of knobs or small inversions may also be important in maintaining block inheritance of the genes that are essential to demark maize and teosinte. According to Beadle (personal communication), a process of intrachromosomal translocations within inversion heterozygotes could assemble these essential genes, which then would be maintained through close linkage.

Studies of the speciation process in nature (Carson, 1975) provide some insights toward understanding what probably occurred through disruptive selection between man and nature during the divergence of maize and teosinte. The initial linkage disequilibrium need not be unusually tight if epistasis (interaction in this case via condensation) between the genes involved is strong. The present-day crossover inhibitors that provide block inheritance would be selected for in the later stages. Once achieved, there would be an accelerated response to selection for the maize-ear phenotype because the coadaptations and supergenes would be less vulnerable to disintegration through crossing over. With man now clearly in charge of maintaining the phenotype of the maize ear, introgression from teosinte is more of an asset (especially as a source of heterosis) than a liability. Meanwhile, the coevolved allelic counterparts of these essential blocks in teosinte began to ensure its survival against genetic swamping from an artificially expanded maize population.

When primitive maize was carried by man into geographic isolation from teosinte, its previous systems of obligatory epistasis between blocks were broken down. Thus a greater amount of variability was released to recombination. This was followed by bizarre combinations of genes resulting in a flush of hundreds of diverse races at the Peruvian and Guatemalan–Mexican centers of variation. When these expanding populations of maize with their altered segments were moved by man or otherwise spread into a reencounter with teosinte, the latter was forced to readjust its epistatic supergenes in relation to that of its newly acquired maize partner.

The Guatemalan races of teosinte that have not had to cope with maize introgression in recent years (Galinat, 1973) may be more open to the free release of genetic variability to maize and may thereby be more amenable for maize improvement.

CONCLUSION

From the various clues provided by genetics, cytology, morphology, anthropology and archaeology, one may postulate the sequence of events during an origin of maize. A high degree of significance is attached to the cytological method of measuring the evolutionary distance between teosinte and maize, because this feature has been relatively free from direct human selection. On a basis of their close cytological homology and because there are apparently no compelling genetic reasons for considering maize and teosinte to result from natural evolutionary divergence, the one possible exception is examined in terms of how it might have originated through the influence of man.

The major polygenic trait separating the female spikes of teosinte and maize, which at first glance might appear to have taxonomic significance, is the high level of condensation in the rachis and shank of maize in comparison to the relaxed condition of teosinte. But the origin of this high condensation was probably an unintentional result of harvesting. The gathering of teosinte seed at random would automatically and unconsciously increase the degree of condensation in material carried to the camp sites for threshing and grinding. Here, in the disturbed and often moist soil, divergent colonies of more highly condensed teosinte would gain a foothold from spilled seed. Multigenic assemblages for increased levels of condensation would accumulate until protodomestic types of teosinte, requiring the intervention of man for their propagation, emerged. An understanding of how this condensation moved teosinte in the direction of maize during a preagricultural period is the key to visualizing how a later program of deliberate selection by man could be successful in transforming the spike of teosinte into the cob of maize.

The other polygenic trait, that of induration of the fruit case, poses a more serious problem if we attempt to derive teosinte from maize than vice versa. A single genetic change, such as to a high tunicate allele, can block this deposition of lignin and elongate the rachilla, producing a phenotype similar in these respects to the ancient soft cobs from Tehuacán, Mexico, and from Bat Cave, New Mexico. On a basis of simplest change, in indurated versus soft rachis, teosinte would be older.

The accumulation of factors for increased condensation and their association with reduced shattering during the preagricultural process of seed gathering, followed by disruptive selection between man and nature, were basic to the evolutionary emergence of maize. This also requires that present-day teosinte,

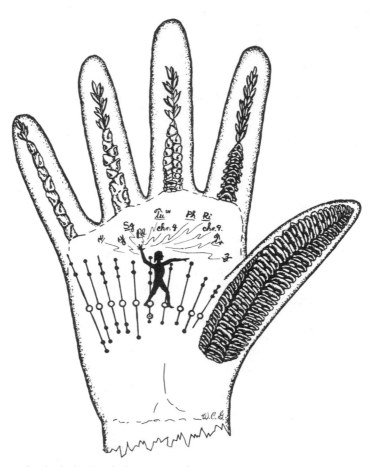

Plate 1. What hath the hand of man wrought?

This symbolic diagram illustrates how a teosinte-like spike (age 10,000 years) shown in the little-finger position may have been transformed by man into a Nal-Tel type maize cob (age 6000 years, paired spikelets removed) shown in the thumb position. About 1000 years are presumed to have passed between each finger. The ring and middle finger illustrate increasing degrees of condensation that were unconsciously accumulated by the mere act of harvesting. The index finger with the first maize cob (paired spikelets removed) has become four-ranked (eight rows of kernels). It is similar to some of the oldest archaeological cobs from Tehuacán, Mexico (age, 7000 years). The Indian illustrates his role in directing genetic changes leading to maize. He treads a single Zea complement of chromosomes. Of the 36 knob positions indicated, only 22 are known in maize. All the knobs occur in various frequencies in the different races of teosinte (Kato-Y, 1975). Solid circles are knobs and open circles are centromeres. The hand traced from that of the artist (W. C. G.) is relative to the size of the spikes.

108

which developed under a sympatric relationship with maize, had to undergo a process of coevolution to protect its system of natural seed dispersal. The ultimate archaeological proof of a transformation from teosinte to maize is still lacking, but the experimental evidence available supports it.

The possible process of the domestication of a maize cob from a teosinte-like spike is summarized pictorially in Plate 1.

ACKNOWLEDGMENT

Paper No. 2206, Massachusetts Agricultural Experiment Station, University of Massachusetts at Amherst. This research was supported in part from Experiment Station Hatch Projects Nos. 258R and 352-NE66, as well as a grant (GB-15767 #1) from the National Science Foundation and by the Bussey Institution of Harvard University.

REFERENCES

Anderson, E. 1944. Cytological observations on *Tripsacum dactyloides*. *Ann. Mo. Bot. Gard.* **31:** 317–323.

Anderson, E. and W. L. Brown. 1948. A morphological analysis of row number in maize. *Ann. Mo. Bot. Gard.* **35:** 323–336.

Banerjee, U. C. and E. S. Barghoorn. 1972. Fine structure of pollen grain ektexine of maize, teosinte and *Tripsacum*. *30th Annu. Proc. Electron Microscopy Soc. Am.,* pp. 226–227.

Beadle, G. W. 1932a. Studies of *Euchlaena* and its hybrids with *Zea*. 1. Chromosome behavior in *Euchlaena mexicana* and its hybrids with *Zea mays*. *Z. Abstam. Vererbungsl.* **62:** 291–304.

Beadle, G. W. 1932b. The relation of crossing-over to chromosome association in *Zea–Euchlaena* hybrids. *Genetics* **17:** 481–501.

Beadle, G. W. 1939. Teosinte and the origin of maize. *J. Heredity* **30:** 245–247.

Beadle, G. W. 1972. The mystery of maize. *Field Mus. Nat. Hist. Bull.* **43:** 1–11.

Beadle, G. W. 1975. Of maize and men. (Review of *Corn: its origin, evolution and improvement,* P. C. Mangelsdorf.) *Quart. Rev. Biol.* **50:** 67–69.

Bryan, A. L. 1973. Paleoenvironments and cultural diversity in late pleistocene South America. *Quatern. Res.* **3:** 237–256.

Burdick, A. B. 1951. Dominance as a function of within organism environment in kernel-row number in maize. *Genetics* **36:** 652–666.

Carson, H. L. 1975. The genetics of speciation at the diploid level. *Am. Nat.* **109:** 83–92.

Collins, G. N. 1919. Structure of the maize ear as indicated in *Zea–Euchlaena* hybrids. *J. Agr. Res.* **17:** 127–135.

Collins, G. N. and J. H. Kempton. 1920. A teosinte–maize hybrid. *J. Agr. Res.* **19:** 1–37.

Cutler, H. C. and E. Anderson. 1941. A preliminary survey of the genus *Tripsacum*. *Ann. Mo. Bot. Gard.* **28:** 249–269.

Daniel, L. 1973. The synthesis of two-rowed maize ears. *Acta Agron. Acad. Sci. Hungaricae, Tomus* **22:** 13–18.

Doggett, H. 1965. Disruptive selection in crop development. *Nature* **206:** 279–280.

Emerson, R. A. and G. W. Beadle. 1932. Studies of *Euchlaena* and its hybrids with *Zea*. II. Crossing-over between the chromosomes of *Euchlaena* and those of *Zea*. *Z. Abstam. Vererbungsl.* **62:** 305–315.

Emerson, R. A. and H. H. Smith. 1950. Inheritance of number of kernel rows in maize. *Cornell Univ. Agr. Exp. Sta. Mem.* No. 296, pp. 1–30.

Flannery, K. V. 1973. The origins of agriculture. *Annu. Rev. Anthro.* **2:** 271–310.

Galinat, W. C. 1956. Evolution leading to the formation of the cupulate fruit case in the American Maydeae. *Bot. Mus. Leafl., Harvard Univ.* **17:** 217–239.

Galinat, W. C. 1969. The evolution under domestication of the maize ear: string cob maize. *Mass. Agr. Exp. Sta. Bull.* No. 577, pp. 1–19.

Galinat, W. C. 1970. The cupule and its role in the origin and evolution of maize. *Mass. Agr. Exp. Sta. Bull.* No. 585, pp. 1–20.

Galinat, W. C. 1971a. The origin of maize. *Annu. Rev. Genet.* **5:** 447–478.

Galinat, W. C. 1971b. The morphological nature of "string-cob" corn. *Maize Genet. Coop. Newsl.* **45:** 95–96.

Galinat, W. C. 1973a. Preserve Guatemalan teosinte, a relict link in corn's evolution. *Science* **180:** 323.

Galinat, W. C. 1973b. Intergenomic mapping of maize, teosinte and *Tripsacum*. *Evolution* **27:** 644–655.

Galinat, W. C. 1974. A congruous background for the *tr* and *pd* genes. *Maize Genet. Coop. Newsl.* **48:** 93.

Galinat, W. C. 1975. The evolutionary emergence of maize. *Bull. Torrey Bot. Club.* **102:** 313–324.

Galinat, W. C., R. S. K. Chaganti, and F. D. Hager. 1964. *Tripsacum* as a possible amphidiploid of wild maize and *Manisuris*. *Bot. Mus. Leafl., Harvard Univ.* **20:** 289–316.

Hawkes, J. G. 1970. The origins of agriculture. *Econ. Bot.* **24:** 131–133.

Hepperly, I. W. 1949. A corn with odd-rowed ears. *J. Hered.* **40:** 62–64.

Iltis, H. H. 1973 (ms. distrib.). The maize mystique: A taxonomic-geographic interpretation of the origin of corn. *Proc. Symp. on Origin of Zea mays and its relatives*. Botanical Museum, Harvard Univ.

Kato-Y, T. A. 1975. Cytological studies of maize and teosinte in relation to their origin and evolution. Ph.D. thesis. Amherst: Univ. Massachusetts.

Langham, D. G. 1940. The inheritance of intergeneric differences in *Zea–Euchlaena* hybrids. *Genetrics* **25:** 88–107.

Longley, A. E. 1937. Morphological characters of teosinte chromosomes. *J. Agr. Res.* **54:** 835–862.

Longley, A. E. 1941. Chromosome morphology in maize and its relatives. *Bot. Rev.* **7:** 263–289.

Mangelsdorf, P. C. 1945. The origin and nature of the ear of maize. *Bot. Mus. Leafl., Harvard Univ.* **12:** 33–75.

Mangelsdorf, P. C. 1947. The origin and evolution of maize. *Adv. Genet.* **1:** 161–207.

Mangelsdorf, P. C. 1952. Hybridization in the evolution of maize. In *Heterosis*, J. W. Gowen (ed.). Ames: Iowa State College Press.

Mangelsdorf, P. C. 1974. Corn: Its origin, evolution, and improvement. Cambridge, Mass.: Harvard Univ. Press, 262 pp.

Mangelsdorf, P. C. and R. G. Reeves. 1939. The origin of Indian corn and its relatives. *Texas Agr. Exp. Sta. Bull.* No. 574, pp. 1–315.

Mangelsdorf, P. C., R. S. MacNeish, and W. C. Galinat. 1967. Prehistoric wild and cultivated maize. In *The prehistory of the Tehuacan Valley*, Vol. 1, D. S. Byers (ed.). Austin: Univ. Texas Press, pp. 178–200.

MacNeish, R. S. 1972. Summary of the cultural sequence and its implications in the Tehuacan Valley. In *The prehistory of the Tehuacan Valley*, Vol. 5, R. S. MacNeish (ed.). Austin: Univ. Texas Press, pp. 496–504.

Randolph, L. F. 1970. Variation among *Tripsacum* populations of Mexico and Guatemala. *Brittonia* **22:** 305–337.

Randolph, L. F. 1975. Contributions of wild relatives of maize to the evolutionary history of domesticated maize: A synthesis of divergent hypotheses. *Proc. Symp. on Origin of Zea mays and its relatives*. Botanical Museum, Harvard Univ.; *Econ. Bot.* **30:** 321–345.

Rogers, J. S. 1950. The inheritance of inflorescence characters in maize–teosinte hybrids. *Genetics* **35:** 541–558.

Smith, Jr., C. E. 1967. Plant remains. In *The prehistory of the Tehuacan Valley*, Vol. 1, D. S. Byers (ed.). Austin: Univ. Texas Press, pp. 220–255.

Stebbins, Jr., G. L. 1950. *Variation and evolution in plants*. New York: Columbia Univ. Press, 643 pp.

Tantravahi, R. V. 1968. Cytology and crossability relationships of *Tripsacum*. Cambridge, Mass.: Bussey Inst., Harvard Univ., 123 pp.

Tantravahi, R. V. 1971. Multiple character analysis and chromosome studies in the *Tripsacum lanceolatum* complex. *Evolution* **25:** 38–50.

Tavcar, A. 1935. Beitrag zur Vererbung der Körnereihenanzahl on Maiz-Kolben. *Zeitschr. f. Zuchtung, Reihe A. Pflanzenzuchtg.* **20:** 364–376.

Weatherwax, P. 1935. The phylogeny of *Zea mays*. *Am. Midl. Natur.* **16:** 1–71.

Wilkes, H. G. 1972. Maize and its wild relatives. *Science* **177:** 1071–1077.

Chapter 8

TEOSINTE AND THE ORIGIN OF MAIZE

George W. Beadle
Department of Biology,
The University of Chicago

Among the world's major cultivated plants used primarily as human food, maize (corn) is the only one with but a single existing wild relative that can reasonably be considered its immediate ancestor. Despite this, the origin of maize has in many ways remained the most confused and controversial of all. In searches for plausible ancestors of major human food crops, plant explorers and taxonomists found and described for each except maize half a dozen or more species congeneric with their cultivated counterparts. But for maize, in many ways the most important of all cultivated crop plants of the Western Hemisphere prior to Columbus, not a single wild relative was found that a majority of early plant taxonomists judged could reasonably be assigned to *Zea*, the genus Linnaeus established for maize.

Finally in 1790, almost three centuries after the discovery of maize by Europeans, and this on the island of Cuba by men of Columbus, a short account of an obviously close relative of maize was published by Hernández (Wilkes, 1967). This Mexican maize-like plant, now called *teosinte*, was by some considered a plausible ancestor of maize. Whether in his earlier more extensive writings, which were destroyed by fire in 1651, Hernández had given a more substantial description of teosinte than that "The plant looks like maize but its seed is triangular," we cannot know now. It was another four decades after the surviving observation of Hernández before a detailed taxonomic description of teosinte was given by Schrader (1833). He assigned it to the newly created genus *Euchlaena*, species *mexicana*. Knowing this as the obviously closest relative known of maize, emphasized especially by Ascherson, who in 1877 wrote "In effect *Euchlaena* is a *Zea* . . ." (Wilkes, 1967), one can only wonder why teosinte was not more widely regarded as the ancestor of maize. The writer believes the answer is clear, that plant taxonomists and others found it most difficult to conceive that a simple female spike of teosinte could have evolved into the relatively monstrous ear of maize, even under the powerful influence of human selection.

Kempton (1938) was well aware that hybrids of maize and teosinte were fully fertile, that in such hybrids meiosis is essentially normal (Beadle, 1932), and that genetic crossing over occurs with frequencies closely similar to those in maize controls in the nine chromosome pairs for which suitable genetic markers were available (Emerson and Beadle, 1932). Yet he stated the difficulties as he saw them in the following dramatic passage:

> Before man could initiate the selection which some authorities believe eventually led to maize [from teosinte] there would have to be some encouraging characteristic, some useful part, on the plant with which he started. No more useless grasses from the standpoint of human consumption could be devised than the American relatives of maize. The seeds of teosinte, the most maize-like relative, are no larger than a No. 4 shot, are firmly imbedded in a brittle articulated rachis and are further protected by a hard shell-like outer glume. They have the disadvantage of being distributed throughout the plant in spikes of only six or seven that ripen over an extended period. The labor of gathering and the diffficulty of separating enough seeds from that covering to furnish a day's food supply would be disheartening, especially with other more promising grasses from which to choose a cereal food.

Kempton further reveals his emotional bias by granting that "maize could arise from teosinte by selection just as a troupe of monkeys pounding at random on a typewriter would in time reproduce the works of Shakespeare."

More recently we find an equally unjustified position taken by Duncan (1975), who categorically asserts that "The origin of maize poses a mystery because no wild ancestor from which it could have originated has ever been found."

What are the differences between maize and teosinte that convinced so many taxonomists and others that maize could not have been directly derived from teosinte? As has often been pointed out, for example, by Weatherwax (1954), Galinat (1971) and Mangelsdorf (1974), those of greatest taxonomic significance have to do with the laterally borne pistillate inflorescences of the two.

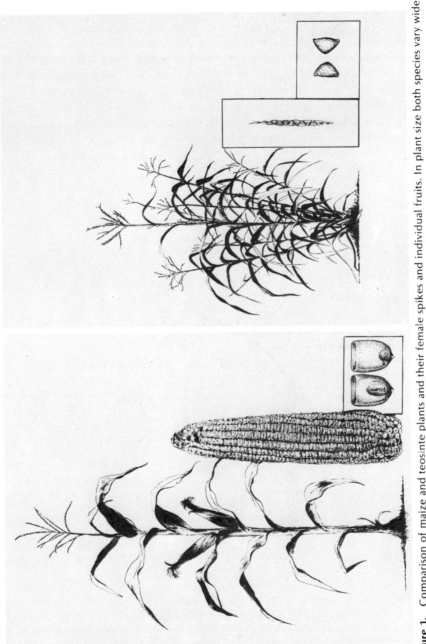

Figure 1. Comparison of maize and teosinte plants and their female spikes and individual fruits. In plant size both species vary widely depending on genetic and environmental factors. Kernel sizes likewise vary, especially in maize.

As Figure 1 indicates, maize plants typically bear one or a few four- or more-ranked pistillate spikes (ears) with two rows of pedicellate spikelets developing per rank. Thus at maturity each resulting ear consists of a massive complex rachis (cob), usually with an even number of rows of naked kernels arranged in two regular rows per rank and oriented at right angles to the axis of the cob. Each spikelet has the potential of producing two fertile florets, but in all but a few varieties one floret of each pair aborts. Individual ears are typically tightly enclosed in several leaf-sheath husks.

In contrast to maize, the two-ranked pistillate inflorescences of teosinte, also laterally borne, are many, often in scores, and at many nodes. One spikelet and

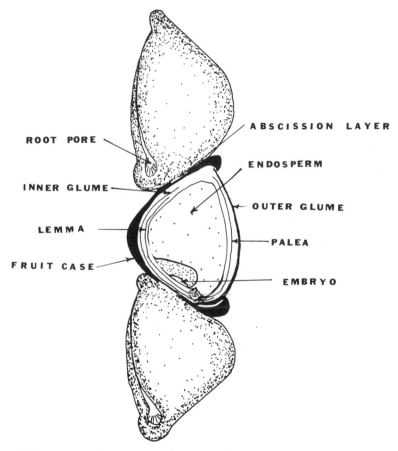

Figure 2. Diagrammatic segment of a normal teosinte spike including a section through a rachis segment showing the heavily lignified fruit case. The suppressed lower floret is not clearly indicated. A fruit case consists of a rachis segment plus the outer glume.

one of its two potential florets differentiate and mature at each rachis segment, resulting in a simple mature two-ranked pistillate spike in which alternately facing heavily lignified triangular rachis segments partially enclose single small kernels. The outer side of each rachis segment is sealed by a specialized corneous outer glume, with the result that each kernel is tightly enclosed in a hard nut-like "fruit case," usually black or gray in color (Figure 2). Pistillate spikes are individually enclosed in single leaf-sheath husks. At maturity the individual fruits, separated through the development of abscission layers, are readily disseminated by falling or being shaken out of their loose dry husks.

The terminal staminate inflorescences or tassels of maize and teosinte are basically similar with the exception that those of maize usually have polystichous central tassel spikes and distichous branches, whereas those of teosinte rarely show such differentiation, the central tassel spike being distichous, as are their pistillate spikes.

THE TEOSINTE HYPOTHESIS

Despite the substantial morphological and anatomical differences between teosinte and maize, the two hybridize freely, both experimentally, as well as in nature whenever they grow in close proximity and overlap in flowering times. Meiosis in the hybrids is essentially normal, with the frequency of genetic crossing over about the same as that in maize controls. Pollen abortion in such hybrids may be little or no greater than in pure maize, again indicating a very close genetic relationship of the parents, with one being ancestral to the other.

Since teosinte is a highly successful wild plant known to have existed in its present form some 7000 years ago (Lorenzo and Quintero, 1970), whereas modern maize is quite incapable of surviving without human intervention, the most reasonable assumption would seem to be that teosinte was ancestral to maize and the transformation was the result of human selection.

Kempton's contention, fully or partially shared by many others, that the morphological and anatomical differences between maize and teosinte are so great that the transition is essentially impossible requires serious consideration. The most persuasive evidence to the contrary consists in the essentially complete fertility of their hybrids, plus the fact that in their second and subsequent generations a detailed and complete array of viable and fertile offspring appears. It is most remarkable that Kempton seems not to have recognized the significance of this, for he was for many years an associate of G. N. Collins of the USDA, who as early as 1919 showed that parental and transitional types between teosinte and maize are recovered in modest sized second generation populations of their hybrids and that these are viable and fertile (Collins, 1919).

More extensive investigations of teosintes and their hybrids with maize were not made in Collins' time or for several years later, as most investigators who might have been interested worked in temperate Corn Belt areas where teosinte and many segregants of its hybrids with maize do not mature because of the then unrecognized short photoperiod requirement for initiation of flowering.

With the advice and cooperation of Edwin J. Wellhausen and Mario Gutiérrez of the Mexican-based International Maize and Wheat Improvement Center (CIMMYT) and Walton C. Galinat of the University of Massachusetts, the writer has for several years investigated larger second-generation and backcross populations of maize-teosinte hybrids grown at the CIMMYT El Batán station near Mexico City where summer photoperiods permit maturity of all genotypes.

The results are generally similar to those of Collins and others. Thus in a second generation population of 15,464 F_2 plants of the cross Chapalote maize by Chalco teosinte, a complete array of intermediates and parental types was noted. Good maize-like and teosinte-like plants were recovered, each with a frequency of approximately one in 500. Since we know few of the detailed linkage and dominance relations of the differentiating genes and have not determined the proportions of parental phenotypes that were true-breeding, we cannot estimate with any precision the number of genes that differentiate the two parental types. We do know, however, that two of the important differentiating traits, paired versus solitary fertile pistillate spikelets and four- or more-ranked pistillate spikes, can each be differentiated by what appear to be single discrete genetic units (Galinat, 1971).

In subsequent generations of this and other hybrid populations, many true-breeding maize and teosinte lines have been established, some of which are discussed in other contexts.

THE WILD MAIZE HYPOTHESIS

With the view of many early taxonomists and others that teosinte could not have been directly ancestral to maize, extensive searches were made for a more promising candidate. The result has been negative. No other living plant has ever been found that even begins to qualify.

However, in 1960 preserved and intact cobs of maize were found in a dry cave in the Tehuacán Valley of Mexico, the oldest of which were found by radiocarbon dating to have lived some 7000 years ago. These were classified and evaluated by Mangelsdorf, MacNeish, et al. (1964) and judged to have grown wild and been harvested for food by the occupants of the cave. Collections from this and other caves of the same valley and elsewhere include large numbers of specimens that provide detailed and invaluable records of the subsequent evolution of maize through human cultivation and selection (Mangelsdorf, MacNeish, et al. 1964).

Remarkable and revealing as these findings and evaluations are, the conclusion that the earliest archaeological specimens are the remains of wild maize cannot be accepted.

THE TRIPARTITE HYPOTHESIS

Largely since teosinte is in a number of taxonomic traits intermediate between maize and species of the genus *Tripsacum,* a third member of the American *Maydeae,* Mangelsdorf and Reeves (1939) long ago postulated that teosinte was not ancestral to maize but instead was a derivative of hybrids between their postulated wild maize and a species of *Tripsacum.* Despite seemingly persuasive arguments against this hypothesis (Beadle, 1939; Galinat, 1971; de Wet, Harlan, et al. 1972), it continued to be defended by its principal advocates for more than thirty years. Recently however, Mangelsdorf (1974) has stated that he no longer regards it as defensible.

REVIVAL OF THE WILD MAIZE HYPOTHESIS

With the demise of the tripartite hypothesis, Mangelsdorf (1974) now favors the view that a postulated wild maize of the 7000-year-old Tehuacán type antidated teosinte and was its direct ancestor. This would seem to be untenable for several reasons. Tehuacán maize has no means of dispersing its seeds. This is clear from the fact that the recovered intact cobs have survived harvesting, transportion to the cave, removal of kernels, being trampled on the cave floor prior to burial in debris, and the final process of excavation.

The kernels of this alleged wild corn are naked and thus highly vulnerable to destruction in the fields by birds and rodents, and in storage by various grain destroying insects. John Fox, a graduate student at The University of Chicago, has collaborated with the writer in the tests of the comparative vulnerability to predators of such maize kernels and teosinte fruits. Chicago tree squirrels, for example, will readily remove and eat the germ of a corn kernel but not that of teosinte, for its germ is well protected by the fruit case. Rats and mice make the same distinction.

As for insect damage, Warren (1954) has shown that in striking contrast to teosinte fruits, those of maize are highly susceptible to destruction by grain and flour beetles as well as by grain weevils, borers, and moths.

Examination of the second-generation population of maize–teosinte hybrids grown at the CIMMYT station adds substantial support to the earlier finding that the intermediate phenotypes vary in morphology, including many closely approaching the parental types and that substantially all are potentially fertile. It is equally clear that all phenotypes except those very close to the teosinte parent have no effective means of seed dispersal. One cannot easily overestimate the importance of this, for it indicates clearly that under the influence of natural selection maize could not have given rise to teosinte. Thus Mangelsdorf's current hypothesis that maize of the earliest Tehuacán type was ancestral to teosinte must be judged untenable. In contrast, the reverse transformation, teosinte to maize, is entirely plausible, for in this direction of change seed dispersal of all intermediates and the maize end product could have been facilitated by man.

IS TEHUACÁN MAIZE TRANSITIONAL?

As a result of the foregoing considerations, the earliest Tehuacán maize seems most reasonably interpreted as a fairly advanced stage in the transformation of teosinte to maize through human selection. If so, the process by which it was brought about must have begun long before 5000 B.C. probably several millennia before.

POLLEN EVIDENCE OF ANCIENT MAIZE

The contention that maize of the Tehuacán type could not have been a wild plant is inconsistent with pollen evidence purported to show that wild maize existed in the area of present Mexico City as long ago as 70,000–80,000 years (Barghoorn, Wolfe et al. 1954). This evidence consists in the discovery and characterization of fourteen analyzable pollen grains recovered in a construction drill core sample taken some 70 m below the site of modern Mexico City and estimated to have been deposited some 80,000 years ago and judged to be those of maize on the basis of size.

The writer and some others (Beadle, 1972; Galinat, 1971; Kurtz, Liverman, et al., 1960) regard this evidence as inconclusive on at least two counts. Engineers experienced in drill-core analysis point out that unless very special precautions are taken of a kind not usual in commercial drill coring, contamination of deeper samples by microscopic quantities of surface materials from higher levels cannot be entirely prevented or detected. It is known that at upper levels of the Belles Artes site under consideration modern maize pollen was present. Furthermore, it is known that maize pollen can be airborne for long distances, especially during dust devils and other updrafts. In addition, it is known that maize pollen varies considerably in size, that there is an overlap in the sizes of it and that of teosinte, that there is a positive correlation between ear size in maize and pollen size, and that modern maize pollen is larger than that of more primitive types such as Chapalote maize (Galinat, 1971). Thus we would expect the pollen of an 80,000-year-old maize to be smaller than that of any existing maize. Since the drill-core pollen in question is larger than that of some existing maize varieties, and that sophisticated electron microscope techniques cannot differentiate pollen of the maize and teosinte (Banerjee and Barghoorn, 1972; Grant, 1972), substantial skepticism is indicated.

PRODUCTIVITY AND USE OF TEOSINTE

It is unfortunate that Kempton was not more adventurous and experimentally oriented, for otherwise he might well have determined that teosinte is not inedible, that its yields in wild populations are not markedly less than those of some other progenitors of cultivated food plants, and that harvesting it not nearly as difficult and unrewarding as he believed.

In 1971, with the support of the National Science Foundation, a party of 16 persons including the writer collaborated for a 2-week period in investigating various aspects of wild teosinte in several separate areas of Mexico in which it grows in some abundance. Among the party were specialists in plant genetics, cytology, taxonomy, morphology, anatomy, agronomy, plant breeding, archaeology, anthropology, and ethnobotany.

Estimates of yields of wild populations did not differ much from those of other plants such as wild grain crops in the Near East (Flannery, 1973). It was also found that the efficiency of hand collection of seed from individual plants was greater for plants with clustered female spikes, than for those with single or few spikes per node, suggesting that this could have been a significant factor in the earliest transitional stages of the origin of maize from teosinte (Galinat and Galinat, 1972).

In 1972, and again with support of the National Science Foundation plus participation by Mario Gutiérrez of CIMMYT, H. Garrison Wilkes of the University of Massachusetts, Robert D. Drennan of the University of Michigan, Comisario Daniel Santos of the village of Mazatlán, Guerrero, Mexico, and Rafael Ortega of the Escuela Nacional De Agricultura, Chapingo, it was determined that ripe teosinte could be rather efficiently collected by bending groups of ripe wild plants over a blanket or plastic sheet and shaking and beating out the ripe seeds. Obviously an animal skin, such as was available in preagricultural times, could have served equally well. It was determined that in this way one person could collect about a kilo of seed per hour. Thus it is reasonable to estimate that a family of five could in a 3-week period collect as much as a metric ton of teosinte seeds, probably more at optimal harvest times and by better adapted and less naive collectors.

Unless teosinte were a tenable human food, there would of course have been little incentive to collect it in preagricultural times. In 1939 the writer reported that properly dried teosinte will pop in the same way as does modern popcorn, with the result that the edible part of the fruit is exploded free of the fruit case (Beadle, 1939). Thus in this way teosinte could have been a source of food in preceramic times, for it is well known that the popping of maize can be accomplished in or near hot coals or hot sand.

Teosinte fruits can be ground or broken with primitive grinding or pounding stones known to have been used during and prior to the first beginnings of agriculture (Beadle, 1972). Such ground teosinte can be eaten with no ill effects, as the writer has demonstrated by consuming 75 g of such meal for 2 successive days and 150 grams for each of 2 following days. In this connection it should be pointed out that many modern diets are deficient in celluloses, hemicelluloses, and lignins, with the result that colonic disorders such as diverticulitis and appendicitis may be favored rather than otherwise as an uninformed layman might well assume (Reilly and Kirsner, 1975; Robson, Ford, et al., 1976).

In the course of simple experiments in preparing teosinte for human consumption it was found that whole or broken kernels of stone-crushed or stone-ground teosinte meal can be more or less effectively separated from their cases by a primitive flotation process such as can readily be carried out with the aid of an animal skin.

It is also readily demonstrable that the immature fruits of teosinte, up to and including the "stiff dough" stage, can be eaten directly without special preparation or processing. Soaked mature fruits are likewise edible.

Until recently there has been no direct evidence that teosinte was ever extensively collected for food or other purposes. Now, however, the writer is kindly permitted to report that Richard I. Ford and Robert D. Drennan of the University of Michigan have recovered carbonized teosinte fruits from a refuse area of ancient living sites near Oaxaca, Mexico. These are estimated to have been originally collected some 3500 years ago. Were they used for food, possibly by popping? We cannot at present know, but it seems reasonable to assume that they were not harvested for any trivial reason.

CHROMOSOME MORPHOLOGY

Consistent with the fact that maize-teosinte hybrids and their offspring are highly fertile and show little pollen sterility, the chromosomes are much alike morphologically. Heterochromatic chromosome knobs are found in both (Longley and Kato-Y, 1965; Kato-Y, 1975). The Mexican teosinte chromosomes bear many knobs, terminal and interstitial, large and small, while the Guatemala teosinte knobs are terminal. Maize varieties in general have fewer knobs than do the Annual Mexican teosintes. Maize knobs appear to be samples of the larger diversity of Mexican teosinte knobs and knob positions and thus to favor the hypothesis that teosinte was ancestral to maize rather than the reverse and that maize was derived from Mexican teosintes, not those of Guatemala.

PROTEIN COMPARISONS

The proteins of teosinte and maize that have so far been investigated are much alike, whereas those of *Tripsacum* show substantial differences. This is true for storage proteins (Gray, Grant, et al., 1972; Waines, 1972) as well as those of enzymes. Senadhira (1946), a graduate student working under the direction of Robert A. Allard of the University of California at Davis, has investigated the electrophoretic characteristics of enzymes of teosinte, maize, and *Tripsacum* specified by genes at 19 different loci, these concerned with five classes of enzymes. The results confirm the similarity of Mexican teosinte and maize and the dissimilarity of these to the *Tripsacum* species.

LINGUISTICS

Webster's Third New International Dictionary defines the word *teosinte* as derived from Nahatl *teotl* (God) and *centli* (ear of maize). If correct, this would strongly suggest that early natives of that linguistic group believed maize had been derived from teosinte.

Throughout its 1700-mi range from lower Chihuahua south and east to Guatemala, teosinte's designation in Spanish is *madre de maiz* (mother of maize). If, as seems plausible, this had been derived separately from more than one of the preconquest native languages spanning that 1700-mi range, it would strengthen the evidence for a kind of cultural recollection that teosinte was derived from maize.

RECONSTRUCTION OF PRIMITIVE MAIZE

Whether or not teosinte was directly ancestral to maize, the earliest true maize was surely as primitive as the earliest Tehuacán specimens. In attempts by the writer to determine the genetic relation of primitive maize types to

Figure 3. Above, ear of U.S. Corn Belt dent for scale. Below, left to right: ear of primitive type maize, rachis of same, F_1 hybrid ear of primitive type maize with teosinte, rachis of same, recovered maize ear from an F_2 generation hybrid population, rachis of same, a teosinte spike. Coin diameter 16 mm.

teosinte, a number of such lines have been established through several generations of selection in maize–teosinte hybrid populations. In terms of anatomy, morphology, and size of their pistillate spikes, these equivalents of the earliest known archaeological maize are obviously very much closer to teosinte than are any of the living races of maize.

Over a period of several years Professor Walton C. Galinat and the writer have selected from these populations early maturing teosintes that are not dependent on short daily photoperiods and are otherwise adapted to Corn Belt latitudes. Second-generation hybrid populations of these and primitive maize lines are now being grown in the hope that the most significant genetic differences required to differentiate the two can be characterized (Figure 3).

A PLAUSIBLE FIRST STEP IN THE TEOSINTE–MAIZE TRANSITION

We can do little more than speculate as to the precise sequence of genetic changes by which teosinte could have given rise to maize through human selection. One seemingly plausible hypothesis assumes that a dominant tunicate mutation was a first step in the transition (Beadle, in press).

Professor Galinat has provided the writer with a number of teosinte lines to which such a dominant tunicate genetic determinant had been transferred by repeated backcrossing of subsequent tunicate types to normal teosintes, and the writer has continued to backcross these to other teosintes. Many such lines have markedly reduced rachis segments and a lesser tendency of their rachises to shatter at maturity (Figure 4). Naked kernels are readily threshed free in many of the lines derived in this way.

Had such a mutant type been found during the early stages of plant cultivation, it could well have been purposely perpetuated. As a dominant mutation, no less than half the progeny would have been of the new type, in contrast to the situation for a recessive mutant in a predominantly cross-pollinated species.

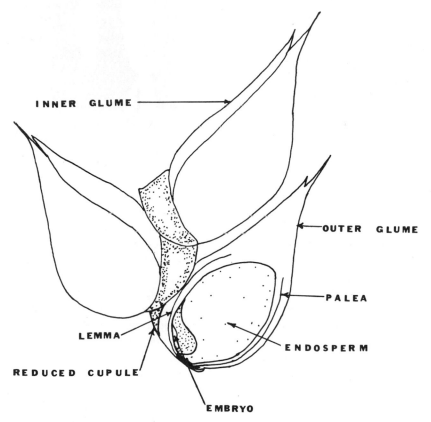

Figure 4. Segment of tunicate teosinte spike to indicate differences from normal teosinte of Figure 2. Note membranous outer glumes and reduction of much less lignified cupules.

Continuing the speculation, further mutations to paired spikelets per rachis segment, a four-ranked rachis and multiple husk protection would have gone a long way toward deriving a primitive maize type much like that of the earliest known Tehuacán specimens.

Interestingly, and perhaps highly significantly, Mangelsdorf, MacNeish, et al. (1967) have described a tunicate-like teosinte spike with a nonfragile rachis, this from Romero's cave in Tamaulipas, Mexico. Its age, however, is indicated to be thousands of years later than that of any reasonable estimate of the earliest postulated cultivation of teosinte and the beginnings of its transformation into maize. It is also important to point out that this tunicate-like teosinte find is far outside the range of existing teosinte populations and was thus presumably perpetuated by man, for all modern reconstructed tunicate teosintes have low survival potential in the wild due to their high susceptibility to avian, rodent, and insect predation. Thus it seems probable that this Tamaulipas tunicate teosinte and the normal teosintes found with it were deliberately planted and perpetuated by the inhabitants of the region more than 3000 years ago, possibly for ceremonial purposes, for by then they had maize as a far superior food plant.

ACKNOWLEDGMENTS

I gratefully acknowledge advice, encouragement, and other help generously given in connection with the studies of the teosinte–maize relation reported here. H. Garrison Wilkes, whose monograph (1967) inspired me to return to earlier studies of almost 40 years ago, has over a period of several years introduced me to the major surviving teosinte populations of Mexico and Guatemala. Walton C. Galinat, with whom I have also had many helpful discussions, has generously provided me Corn Belt-adapted teosintes, normal and tunicate and has collaborated in classifying hybrid populations grown in Mexico. The University of Chicago has provided extensive greenhouse and field facilities and financial support. The Mexican-based CIMMYT, especially Edwin J. Wellhausen, Mario G. Gutiérrez, and Haldore Hanson, have made available field facilities and other help in growing extensive experimental populations not adapted to United States Corn Belt latitudes. The University of Illinois at Urbana and the Chicago Horticultural Society Botanic Garden have also made available supplemental field facilities for special plantings. The National Science Foundation has provided grant support through the Chicago Horticultural Society for teosinte collecting in Mexico and Guatemala, as well as for searches for mutations in native teosinte populations. These grants have enabled the renewal and expansion of seed-bank stocks of teosintes at the United States Department of Agriculture's National Seed Storage Laboratory at Fort Collins, Colorado, its Regional Plant Introduction Station at Experiment, Georgia and the Genetics Cooperation Center's seed-bank at the University of Illinois, Urbana, as well as by direct dissemination of seed stocks to interested investigators.

Much of the substance of this account was given at the Connecticut Agricultural Experiment Station in the initial Donald Forsha Jones Memorial Lecture on September 14, 1972.

REFERENCES

Banerjee, U. C. and E. S. Barghoorn. 1972. Fine structure of pollen grain ektexine of maize, teosinte, and Tripsacum. *30th Annu. Proc. Electron Microscopy Soc. Am.*, pp. 226–227.

Barghoorn, E. S., M. K. Wolfe, and K. H. Clisby. 1954. Fossil maize from the Valley of Mexico. *Bot. Leafl., Harvard Univ.* **16**: 229–240.

Beadle, G. W. 1932. Studies of *Euchlaena* and its hybrids with *Zea*. I. Chromosome behavior in E. mexicana and its hybrids with *Zea mays. Z. Abstam. Vererbungsl.* **62**:291–304.

Beadle, G. W. 1939. Teosinte and the origin of maize. *J. Heredity* **30**: 245–247.

Beadle, G. W. 1972. The mystery of maize. *Field Mus. Nat. Hist. Bull.* **43**: 2–11.

Beadle, G. W. (in press). The Origin of *Zea mays*. In *The origins of agriculture*, Charles Reed (ed.). Moulton.

Collins, G. H. 1919. Structure of the maize ear as indicated in *Zea–Euchlaena* hybrids. *J. Agr. Res.* **17**: 127–135.

de Wet, J. M. J., J. R. Harlan, and C. A. Grant. 1972. Origin and evolution of teosinte [*Zea mexicana* (Shrad) Kuntz]. *Euphytica,* **20**: 255–265.

Duncan, W. G. 1975. In *Crop physiology,* L. T. Evans (ed). New York: Cambridge Univ. Press, p. 23.

Emerson, R. A. and G. W. Beadle. 1932. Studies of *Euchlaena* and its hybrids with *Zea*. II. Crossing over between the chromosomes of *Euchlaena* and those of *Zea. Z. Abstam. Vererbungsl.* **62**: 305–315.

Flannery, K. 1973. The origins of agriculture. *Annu. Rev. Anthropol.* **2**: 271–310.

Galinat, W. C. 1971. The origin of corn. *Ann. Rev. Genet.* **5**: 447–478.

Galinat, W. C. and D. W. Galinat. 1972. A possible role of condensation in a domestication of teosintes. *Maize Genet. Coop. Newsl.* **46**: 109–110.

Grant, C. A. 1972. A scanning electron microscopy survey of some Maydeae pollen. *Grana* **12**: 177–184.

Gray, J. R., C. A. Grant, and J. M. J. de Wet. 1972. Protein electrophoresis as an indicator of relationship among maize, teosinte and tripsacum. *Ill. State Acad. Sci. Newsl.* **4**: 5.

Kato-Y, T. A. 1975. Cytological studies of maize and teosinte in relation to their origin and evolution. Ph.D. thesis. Amherst: University of Massachusetts. *Univ. Mass. Agric. Exp. Sta. Bull.* No. 635, pp. 1–200.

Kempton, J. H. 1938. Maize—our heritage from the Indian. Washington, D. C.: *Smithsonian Report 1937:* 385–408.

Kurtz, E. B., J. R. Liverman, and H. Tucker. 1960. Some problems concerning fossil and modern corn pollen. *Torrey Bot. Club Bull.* **87:** 85–94.

Longley, A. E. and T. A. Kato-Y. 1965. Chromosome morphology of certain races of maize in Latin America. *Internat. Conf. for the Improvement of Maize and Wheat, Chapingo, State of Mexico, Mexico. Res. Bull.* No. 1, p. 112.

Lorenzo, J. L. and L. Gonzalez Quintero. 1970. El mas antiguo teosinte. *INAH Boletin* **42:** 41–43.

Mangelsdorf, P. C. 1974. In *Corn, its origin, evolution and improvement.* Cambridge, Mass.: Belknap Press, Harvard Univ., pp. 1–262.

Mangelsdorf, P. C., R. S. MacNeish, and W. C. Galinat. 1964. Domestication of corn. *Science* **143:** 538–545.

Mangelsdorf, P. C., R. S. MacNeish, and W. C. Galinat. 1967. Prehistoric maize, teosinte and *Tripsacum* from Tamaulipas, Mexico. *Bot. Mus. Leafl., Harvard Univ.* **22:** 33–62.

Mangelsdorf, P. C. and R. G.Reeves. 1939. The origin of Indian corn and its relatives. *Texas Agr. Exp. Sta. Bull.* **574:** 1–315.

Reilly, R. W. and J. B. Kirsner (eds.). 1975. *Fiber deficiency and colonic disorders.* New York: Plenum.

Robson, J. R. K., R. I. Ford, K. V. Flannery, and J. E. Konlande. 1976. The nutritional significance of maize and teosinte. *Ecol. Food Nutr.* **4:** 243–249.

Schrader, H. 1833. In *Litteratur Linnaea (Berlin) 1833,* pp. 25–26.

Senadhira, D. 1976. Genetic variation in corn and its relatives. Ph.D. thesis. Davis: University of California, 126 pp.

Waines, J. G. 1972. Protein electrophoretic patterns of maize, teosinte and *Tripsacum dactyloides. Maize Genet. Coop. Newsl.* **46:** 164–165.

Warren, L. O. 1954. Teosinte as a host of stored grain insects. *J. Econ. Entomol.* **47:** 630–632.

Weatherwax, P. 1954. In *Indian Corn in Old America.* New York: Macmillan, pp. 1–253.

Wilkes, H. G. 1967. *Teosinte: The closest relative of maize.* Cambridge, Mass.: Bussey Inst., Harvard Univ., pp. 1–159.

Chapter 9

TRIPSACUM AND THE ORIGIN OF MAIZE

J. M. J. de Wet and J. R. Harlan

Crop Evolution Laboratory,
Department of Agronomy,
University of Illinois,
Urbana

Tripsacum L. is the only genus with which domesticated maize (*Zea mays* L. ssp. *mays*) is known to cross to produce viable hybrids. The basic genomes of *Tripsacum* ($x = 18$) and *Zea* ($x = 10$) are distantly related (Rao and Galinat, 1974). Gene exchange is possible between them (Maguire, 1964; Reeves and Brockholt, 1964), and Mangelsdorf (1961, 1968) suggested that introgression from *Tripsacum* may have played a role in the evolution of domesticated maize. These studies involved hybrids between maize and diploid *T. dactyloides* (L.) L. ($2n = 36$) and *T. floridanum* Porter ex Vasey ($2n = 36$). Attempts by Randolph (1952) to cross maize with other species of *Tripsacum* failed, but de Wet and Harlan (1974) reported successful crosses with tetraploid *T. dactyloides*, tropical *T. lanceolatum* Rupr. ex Fourn., and *T. pilosum* Scribn. and Merrill. In this chapter we present a new classification for *Tripsacum*, discuss the

Research financed by the Illinois Agricultural Experiment Station and grant GB-40136-X from the National Science Foundation.

cytogenetic affinities between *Zea* and *Tripsacum,* and indicate the phenotypic effects of maize–*Tripsacum* introgression.

SYSTEMATICS OF TRIPSACUM

Tripsacum and *Zea* are monoecious New World grass genera. They are usually classified with the Old World monoecious genera *Coix* L., *Chionachne* R. Br., *Sclerachne* R.Br., *Polytoca* R.Br., and *Trilobachne* Henr. into the tribe Maydeae. Celarier (1957) demonstrated that these monoecious taxa belong more naturally in the Andropogoneae, and Clayton (1973) classified *Zea* and *Tripsacum* into the subtribe Tripsacinae and the Old World genera into the subtribe Coicinae. Phylogenetically the Coicinae are allied to the subtribe Rottboellinae rather than the Tripsacinae.

The genus *Zea* is characterized by male terminal inflorescences with paired staminate spikelets and lateral female inflorescences with single or paired pistillate spikelets. A section of paired staminate spikelets sometimes terminates the branches of female inflorescences. *Zea* includes two species, *Z. mays* and *Z. perennis* (Hitchc.) Reeves and Mangelsdorf. *Zea mays* is recognized to include domesticated maize (ssp. *mays*) and teosinte [ssp. *mexicana* (Schrad.) Iltis].

The genus *Tripsacum* is characterized by bisexual terminal and lateral inflorescences. The lower section of each inflorescence branch is female, and the upper part is male. Female spikelets are arranged in two alternate rows of cupules that each contains a single fertile spikelet. Male spikelets are paired and arranged on one side of the terminal portion of the rachis.

Tripsacum is a small but complex genus (Cutler and Anderson, 1941; Randolph, 1970). We recognize 11 species, two of which (*T. andersonii* Gray and *T. bravum* Gray) were recently described. The species recognized are distinguished from each other by the following key characters:

1. One spikelet of each staminate pair pedicelled, the other sessile; pedicel slender, 2–5 mm long; staminate section of racemes pendulous.

    A. Leaf blades at most 3.5 cm wide; sheaths hispid at least at the base of the plant; racemes of terminal inflorescence 1–9; fruitcases about as long as broad; $2n = 72$: *T. lanceolatum.*

    B. Leaf-blades up to 10 cm wide; sheaths variously pubescent; racemes of terminal inflorescence 2–50; fruit cases longer than broad.

    C. Sheaths coarsely pilose; leaf blades 3–10 cm wide.

        i. Leaf blades coriaceous, essentially glabrous, 4–10 cm wide; terminal inflorescence strongly branched with 12–50 racemes; $2n = 36$: *T. maizar.*

   ii.   Leaf blades not coriaceous, more or less pilose, 3–7 cm wide; terminal inflorescence with 2–15 racemes; $2n = 72$: *T. pilosum.*

  D. Sheaths glabrous; leaf blades 3–6 cm wide, glabrous; terminal inflorescence with 5–25 racemes; $2n = 36$: *T. laxum.*

2. Staminate spikelets sessile, or one of a pair with a short pedicel, pedicel usually almost as broad as long; racemes stiffly ascending.

  A. One staminate spikelet of each pair shortly pedicelled; fruit cases 4–6 mm in diameter; leaf blades 6–10 cm wide; $2n = 64$: *T. andersonii.*

  B. Staminate spikelets essentially sessile; fruit cases 2–5 mm in diameter; leaf blades 0.3–6 cm wide.

    i.   Sheaths, blades, and culm with a wooly tomentum.

       a. Staminate glumes coriaceous, 5–9 mm long; terminal inflorescence with 1–5 racemes; $2n = 36$: *T. australe.*

       b. Staminate glumes membranaceous, 3–5 mm long; racemes usually solitary; $2n = 36$: *T. bravum.*

    ii. Sheaths, blades, and culm never tomentose; sheaths glabrous to pilose.

   iii. Leaf blades 3 cm wide; terminal inflorescences regularly with more than one raceme.

       a. Staminate spikelets 2–5 mm long; $2n = 72$: *T. latifolium.*

       b. Staminate spikelets 4–12 mm long; $2n = 36, 54, 72$: *T. dactyloides.*

   iv. Leaf blades less than 3 cm wide; terminal inflorescences with 1–3, usually solitary, racemes at the end of stiffly erect penduncles.

       a. Plants rhizomatous; Florida and Caribbean Islands; $2n = 36$: *T. floridanum.*

       b. Plants caespitose; Zopilote Canyon; $2n = 36$: *T. zopilotense.*

Tripsacum andersonii Gray is a robust, rhizomatous perennial. It is further characterized by glabrous leaf sheaths, terminal inflorescences with 1–7 ascending racemes, one spikelet of each staminate pair having a short and stout pedicel, and by $2n = 64$ chromosomes. This species is widely grown in Mesoamerica and in South America as a fodder. It is known to occur naturally in the Cordillera de Merida of Venezuela, where it occupies mesic woodlands. It is frequently confused with *T. latifolium* Hitchc. or *T. laxum* Nash (*T. fasciculatum* Trin. ex Aschers.) (Cutler and Anderson, 1941). These three species resemble one another in being robust, with broad leaves and glabrous sheaths. However, the staminate spikelets are sessile in *T. latifolium,* and one of each pair has a long and slender pedicel in *T. laxum,* whereas in *T. andersonii* the pedicel is short and stout. Staminate spikelet morphology of *T. andersonii* is similar to that of maize–*Tripsacum* hybrids combining 10 *Zea* and 72 *Tripsacum* chromosomes studied by Newell and de Wet (1974a). Indeed, 10 of the $2n = 64$ chromosomes of *T. andersonii* usually remain as univalents during meiotic prophase, and the remaining 54 chromosomes behave as if they represent three homologous sets of 18 *Tripsacum* chromosomes. The origin of *T. andersonii* has as yet not been studied in biosystematic detail. However, its cytology and sexual sterility suggest that *T. andersonii* may represent a natural hybrid between maize and *Tripsacum.* If this hybrid originated in Mesoamerica, the species involved could have been *T. latifolium.*

Tripsacum australe Cutler and Anderson is a medium robust, rhizomatous perennial with tomentose leaf sheaths and usually also blades, with terminal inflorescences having 1–5 ascending racemes and essentially sessile staminate spikelets and with $2n = 36$ chromosomes. This species is widely distributed in South America from Panama to Central Brazil and Paraguay. It is almost as diverse morphologically as is *T. dactyloides.* Specimens from Matto Grosso in Brazil appear nearly vine-like in growth habit. The species grades into *T. dactyloides* in inflorescence morphology, but can always be recognized by the wooly tomentum on its leaf sheaths. It typically occupies the ecotone between savanna and woodlands.

Tripsacum bravum Gray is a medium robust, rhizomatous perennial with a sparse wooly tomentum on the leaf sheaths. Its terminal inflorescences usually have solitary racemes, with essentially sessile staminate spikelets, and the plants are diploid with $2n = 36$ chromosomes. This species is known only from Valle de Bravo in the state of Mexico, where it occupies wooded slopes. It differs conspicuously from *T. australe* in inflorescence morphology. The terminal inflorescence of *T. australe* is composed of 1–5 racemes that are subdigitately arranged on a short primary axis. Terminal inflorescences of *T. bravum* consist of single racemes that are each partially enclosed by a spathe.

Tripsacum dactyloides (L.) L. is a small-to-medium robust, rhizomatous perennial with glabrous to pilose leaf sheaths, terminal inflorescences with 1–12 ascending or rarely somewhat pendulous racemes, with essentially sessile staminate spikelets, and with $2n = 36$ 54 or 72 chromosomes. This species occurs across the range of the genus in North and South America. It is extremely variable morphologically. Newell and de Wet (1974b) demonstrated that the species occurs in the United States as morphologically distinct populations that are isolated from one another but have no ecological or geographical unity and do not deserve subspecific rank. In Mesoamerica and South America a robust race occurs with somewhat pendulous racemes. Around Taxco, Mexico a pilose race occurs that is often recognized as ssp. *hispidum* Hitchcock.

Tripsacum floridanum Porter ex Vasey is a small, rhizomatous perennial with glabrous leaf sheaths, terminal inflorescence with 1–3 ascending racemes, with essentially sessile staminate spikelets, and with $2n = 36$ chromosomes. This species occurs in southern Florida and in Cuba. It is further characterized by narrow leaf blades that become pendulous above the middle. Inflorescence morphology is similar to that of small *T. dactyloides,* and these two species cross readily. Hybrids are sterile, although chromosome pairing is almost complete between the basic genomes of *T. dactyloides* and *T. floridanum.*

Tripsacum lanceolatum Rupr. ex Fourn. is a medium robust, rhizomatous perennial with hispid leaf sheaths. The terminal inflorescences have 1–9 pendulous racemes, with one spikelet of each staminate pair having a long and slender pedicel. The species is tetraploid with $2n = 72$ chromosomes. It occurs from southern Arizona to central Mexico, and occupies streambanks or moist cliffs. It differs from *T. pilosum* primarily in having narrower leaves and smaller stature.

Tripsacum latifolium Hitchcock is a robust, rhizomatous perennial with glabrous to pubescent leaf sheaths. The terminal inflorescences have 1–3 ascending racemes, with essentially sessile staminate spikelets, and the plants are tetraploid with $2n = 72$ chromosomes. The species occupies open habitats in mesic woodlands from Guatemala to Costa Rica. It resembles a large *T. dactyloides* but typically has smaller staminate spikelets and broader leaf-blades.

Tripsacum laxum Nash is a robust, rhizomatous perennial with glabrous leaf sheaths, having terminal inflorescence with 5–20 pendulous racemes and one of each staminate spikelet pair with a long and slender pedicel. Plants are diploid with $2n = 36$ chromosomes. The species occurs in mesic woodlands of Central America. Cutler and Anderson (1974) recognized it to include the cultivated "Guatemala grass" (*T. andersonii*). A study of the type, however, revealed that *T. laxum* is characterized by long pendulous racemes and a distinct pedicel to one spikelet of each staminate pair. It resembles *T. andersonii* only in size and general vegetative appearance.

Tripsacum maizar Hernandez and Randolph is an extremely robust, rhizomatous perennial with coarsely hispid leaf sheath, terminal inflorescences having 12–50 pendulous racemes, and one spikelet of each staminate pair having a long and slender pedicel. This is a diploid species with $2n = 36$ chromosomes. It is widely distributed in western Central America from Nayarit in Mexico to Guatemala. It resembles *T. pilosum* and *T. laxum* in having numerous pendulous racemes, but can always be distinguished by its coriaceous leaf blades and coarsely pilose sheaths.

Tripsacum pilosum Scribn. and Merrill is a medium to robust rhizomatous perennial with hispid leaf sheaths. Terminal inflorescences have 2–13 pendulous racemes, with one spikelet of each pair having a long and stiff pedicel, and the plants are tetraploid with $2n = 72$ chromosomes. The species ranges from

Tamaulipas to Zacatecas in Mexico and south to the Honduras. It often resembles pilose *T. dactyloides* in vegetative characteristics, but can always be distinguished by the stiff pedicel of the staminate spikelets and pendulous racemes.

Tripsacum zopilotense Hernandez and Randolph is a small, nonrhizomatous perennial with glabrous leaf sheaths having terminal inflorescences with a solitary erect raceme and essentially sessile staminate spikelets. The species is diploid with $2n = 36$ chromosomes. It is confined to talus slopes in Zopilote Canyon of Guerrero in Mexico and is unique among *Tripsacum* species in lacking a rhizome.

CYTOGENETIC AFFINITIES BETWEEN *ZEA* AND *TRIPSACUM*

Teosinte (*Zea mays* ssp. *mexicana*) and *Z. perennis* are reproductively isolated from *Tripsacum*. Thousands of attempted crosses between different species of *Tripsacum* and the wild forms of *Zea* failed to produce a single successful hybrid. However, domesticated maize (*Zea mays* ssp. *mays*) crosses with different *Tripsacum* collections when they are not isolated from one another by gametophytic and other genetic barriers. Pollen of the one genus frequently fails to germinate on the stigma of the other genus, or hybrid embryos fail to germinate even when excised from the endosperm. With the proper techniques and matching gametophytic factors, however, annual as well as perennial teosinte will probably eventually be crossed with at least some *Tripsacum* species. Their basic genomes can be introduced into maize–*Tripsacum* hybrids by pollinating such hybrids with the wild taxa of *Zea*.

The basic genomes of *Tripsacum* and *Zea* are not closely allied (Anand, 1966; Chaganti, 1965). Except for two or three open bivalents that tend to remain paired through diakinesis, the chromosomes are present as univalents during meiotic prophase in hybrids that combine 18 Td (*Tripsacum*) and 10 Zm (*Zea*) chromosomes. These hybrids are male sterile but somewhat female fertile. When they are pollinated by maize, offspring that were sexually produced have 18 Td + 20 Zm chromosomes. In these hybrids the 20 Zm chromosomes frequently associate into bivalents, but as many as four Td chromosomes can associate with them to form bivalents or trivalents (Newell and de Wet, 1973). Hybrids with 36 Td + 10 Zm chromosomes were obtained when maize was pollinated with tetraploid *T. dactyloides* ($2n = 72$). These hybrids exhibit relatively frequent multivalent associations, although the 36 Td chromosomes often associate autosyndetically into bivalents. When $2n = 46$ chromosome hybrids are backcrossed with maize, some offspring with 36 Td + 20 Zm chromosomes are regularly produced. Autosyndetic Zm and Td bivalents are commonly, but not always, present in these hybrids. Again, as many as four Zm chromosomes tend to associate with Td chromosomes to form trivalent or bivalent configurations. Hybrids with 72 Td + 10 Zm chromosomes were obtained when unreduced gametes of tetraploid *T. dactyloides* were fertilized by normal male gametes of *Z. mays*. Eight to ten maize univalents were commonly observed at metaphase, but four to eight Zm chromosomes are sometimes associated with Td chromosomes. The maize chromosomes that tend to pair with Td chromosomes were not positively identified but appear to be chromosomes 2, 4, 7, and 9 (Engle, de Wet, *et al.* 1974).

The basic genomes of maize and annual teosinte are essentially homologous (Beadle, 1932, 1972; de Wet and Harlan, 1972; de Wet, Harlan, et al., 1971). Hybrids with 20 Zp (*Zea perennis*) and 10 Zm chromosomes show more trivalents and fewer bivalents than do hybrids combining 20 Zp and 10 Ze (teosinte) chromosomes. This suggests that the chromosomes of *Z. perennis* are more compatible with those of *Z. mays* ssp. *mays* than with those of ssp. *mexicana* (Longley, 1934). In tetraploid hybrids with 20 Zm + 20 Zp chromosomes, Collins and Longley (1935) demonstrated almost complete autosyndetic pairing within the Zm and Zp genomes. Comparisons of hybrids with 18 Td + 20 Zm and with 18 Td + 10 Zm + 10 Ze chromosome combinations

TABLE 1. Chromosome configurations in hybrids involving *Tripsacum dactyloides* (Td), *Z. mays* ssp. *mays* (Zm), *Z. mays* ssp. *mexicana* (Ze), and *Z. perennis* (Zp).

Hybrid	2n	Average and range of chromosome configurations			
		I	*II*	*III*	*IV*
10 Zm + 18 Td	28	21.04	2.94	0.28	0.06
		14–28	0–7	0–1	0–1
20 Zm + 18 Td	38	14.99	11.22	0.11	0.06
		10–18	10–14	0–1	0–1
10 Zm + 36 Td	46	10.40	17.33	0.14	0.13
		8–13	15–19	0–1	0–1
20 Zm + 36 Td	56	3.65	25.64	0.13	0.17
		0–8	24–28	0–1	0–1
10 Zm + 10 Ze + 18 Td	38	16.64	9.78	0.60	
		10–24	6–13	0–2	
10 Zm + 10 Ze + 36 Td	56	3.00	25.61	0.22	0.22
		0–8	21–28	0–1	0–1
20 Zm + 10 Ze + 36 Td	66	7.46	25.14	1.62	0.85
		5–10	19–29	0–4	0–2
10 Zm + 20 Zp + 18 Td	48	14.81	12.26	2.45	0.33
		10–22	7–17	0–4	0–2
20 Zm + 20 Zp + 36 Td	76	5.94	32.96	0.54	0.63
		1–10	27–36	0–2	0–3

reveal no major differences in meiotic chromosome behavior. This is also true when hybrids with 36 Td + 20 Zm and 36 Td + 10 Zm + 10 Ze chromosomes are compared. Similarly, hybrids with 10 Zm + 20 Zp + 18 Td or 20 Zm + 20 Zp + 36 Td chromosomes show little homology between the Td and Zp genomes (Newell and de Wet, 1973). Trivalent configurations involving Zm and Zp chromosomes are commonly produced in the $2n = 48$ chromosome hybrids, and autosyndetic pairing within the Zm, Zp, and Td genomes is often complete in $2n = 76$ chromosome hybrids. Data on chromosome pairing are summarized in Table 1.

MORPHOLOGY OF MAIZE-*TRIPSACUM* INTROGRESSION

Hybrids with 10 *Z. mays* and 36 *T. dactyloides* chromosomes, in maize cytoplasm, are usually facultative gametophytic apomicts. When they are pollinated with maize their offspring mostly have the same genome constitution as the hybrid parent. These offspring originate both apomictically and sexually (de Wet, Harlan, et al. 1971). Functional sexual female gametes with 36 Td chromosomes are sometimes produced, and the backcrosses substitute the 10 parental Zm chromosomes for the haploid genome of the new maize pollen parent (de Wet and Harlan, 1974). Such a substitution of the haploid maize genome has been achieved for as many as eight successive backcrosses. The overall phenotype of these backcross individuals usually does not differ significantly from that of the first generation hybrid parent. Some individuals, however, become successively more maize-like in female inflorescence morphology (Newell and de Wet, 1974a). This is due to accumulation of maize genetic material in the *Tripsacum* genome. Such backcross individuals, when pollinated with maize, sometimes produce offspring with 36 Td + 20 Zm chromosomes. Hybrid derivatives with 36 Td + 20 Zm chromosomes in turn produce offspring with 18 Td + 10 Zm, and 36 Td + 20, 30 or various aneuploid Zm chromosome complements (de Wet, Harlan, et al. 1973). Several further generations of backcrossing with maize eventually yield plants with $2n = 20$ maize chromosomes (de Wet, Engel, et al. 1972).

Plants with $2n = 20$ chromosomes recovered from these backcrosses are for the greater part phenotypically pure maize. A small percentage of individuals, however, differ from their maize parents in several morphological traits. They tiller excessively, generate new growth after the inflorescences mature, and produce lateral inflorescences at as many as seven different nodes of a single stalk. Female inflorescences are often borne on elongated lateral branches with as many as five fertile cobs on each branch; inflorescences are commonly bisexual, and glume as well as rachis tissues are often strongly indurated. In gross morphology several of these tripsacoid plants resemble derivatives of maize-teosinte hybrids [cf. Grant (1973), Galinat (1970, 1971)].

The more maize-like individuals are usually fertile, whereas the most tripsacoid ones are sterile. When fertile, tripsacoid traits are often rapidly eliminated by selfing or mating among siblings. Strong selection seems to operate against gametes carrying haploid genomes that were excessively contaminated with *Tripsacum* genetic material when in competition with gametes carrying an essentially pure maize haploid genome. The amount of *Tripsacum* genetic material that can permanently be incorporated into the maize genome, and the phenotypic effects of these genetic transfers have as yet not been studied in detail. We have, however, several hundred lines of recovered maize that are breeding true for one or more of the obvious tripsacoid traits enumerated above. When these are crossed or allowed to mate within lines, their offspring are often extremely variable. Several characteristics not present in either the maize or *Tripsacum* parents are uncovered. The more common among these are: (a) large glumes that cover the grain as is typical of half-tunecate described by Mangelsdorf and Galinat (1964), (b) the presence of a fertile male and female spikelet in each cupule of the cobsection of lateral inflorescences, (c) grains that fall readily from the indurated glumes, and (d) four- or eight-rowed cobs, with these cobs often breaking into sections of yoked or cross-yoked rachis segments carrying four or eight grains respectively as is typical of some maize–teosinte hybrid derivatives described by Collins (1919) and by Mangelsdorf and coworkers (Mangelsdorf 1974; Mangelsdorf MacNeish, et al. 1964; Reeves and Mangelsdorf, 1959).

Mangelsdorf (1958) suggested that the teosinte genome may have a mutagenic effect on the maize genome in hybrids between these two taxa. Presence of *Tripsacum* genetic material similarly releases genetic variability within the maize genome that seems to have lain dormant, probably since the time of domestication. Attempts to match some of these phenotypes with known maize and teosinte alleles are now underway.

DISCUSSION

There seems to be no possibility that *Tripsacum* could be a progenitor of maize. If the genus has played any role at all in maize evolution, it would have been in some sort of introgression. In our experience, introgression, when it occurs, is rather conspicuous in the field. We have been able to study introgression hybridization between spontaneous and cultivated races of barley, wheat, pearl millet, sorghum, Asian rice, African rice, maize, and among wild races of *Bothriochloa, Dichanthium, Cynodon, Andropogon, Bromus,* and other grasses. Introgression between maize and teosinte is conspicuous in

Mexico and Guatemala. Hybrids and backcrosses can be found in the field. Maizoid characters such as fertile paired female spikelets, four ranked female inflorescences and central tassel spike, are detectable in plants that otherwise closely resemble teosinte.

We have searched fairly extensively for evidence of maize–*Tripsacum* interactions in Mexico and Guatemala without finding a trace of any kind. The situation may well be different in South America. We have only started our field work there and have little experience with the materials. However, certain bits and pieces of evidence suggest to us the possibility that interactions of some kind may have taken place.

1. Cytological evidence implies that *T. andersonii* carries 10 maize chromosomes. If so, hybridization is possible. There is some variation among clones of Guatemala grass, suggesting that such hybrids have been produced more than once and from different parents.

2. The race pollo as described and figured by Roberts, Grant, et al. (1957) has a strongly tripsacoid appearance. A number of our experimentally produced hybrid derivatives resemble it very closely. According to their description the race segregates for extreme tripsacoidy and they suggested this might be due to a virus. In our materials the same effect is obtained when a line carries one pair of *Tripsacum* chromosomes in addition to the maize complement.

3. The race *Chococeño* crosses naturally with *Tripsacum* (Roberts, Grant, et al., 1957). This remains to be verified experimentally and studied in detail. The race does have plants with flat stems and rough leaf margins, tillers heavily and has an extraordinary adaptation for maize. Annual rainfall in Choco is 8–10 m. Chococeño maize is not planted; the seeds are broadcast on the ground, sometimes in standing vegetation, and fields may be subject to frequent flooding. High tillering is not characteristic of lowland tropical races. Chococeño is a most unusual race, indeed, and some of its characteristics could have been derived from *Tripsacum*.

4. We have extracted a number of tunicate and tunicate-like lines from our maize × *Tripsacum* hybrid derivatives. Could the complex Tu gene be an insertion of a fragment of a *Tripsacum* chromosome into chromosome four of maize?

5. Our strongly tripsacoid plants, especially those carrying one or two pairs of *Tripsacum* chromosomes have string cobs. Could string cob represent a transfer from *Tripsacum* to maize?

6. The interlocking character is based on a relatively long distance between cupules within a row of cupules. Wide spacing of cupules is rather characteristic of tripsacoid plants.

7. Some sweetcorns and popcorns tend toward tripsacoidy in plant type with several ears at a node, several nodes with ears, long shank, male tips to the ears, tassel seed, and the like.

All of these bits and fragments of evidence need to be examined. Some of the questions can be attacked experimentally, and it is important for an understanding of maize that such studies be pursued.

If crosses occur in nature they are most likely to happen with *Tripsacum* as female. It is not absolutely necessary to open the husk and shorten the silk as recommended by Mangelsdorf and Reeves (1939), but it is necessary to place the *Tripsacum* pollen near the base of the silk. This can happen in nature under rather special circumstances, but the seeds tend to be weak with low viability due, in part, to poorly developed endosperm. On the other hand, if the cross is made in the other direction, maize pollen has no difficulty growing the length of *Tripsacum* styles, and the endosperm is likely to be well developed and the seeds near normal in size. In our experience almost all of the hybrids obtained from such crosses are derived from unreduced gametes of the *Tripsacum* parent. We do find difficulty in getting the hybrid seeds to germinate, and this constitutes one of the major barriers that separates the genera. Some seeds will germinate, however, and hybrid seedlings are more vigorous than *Tripsacum* seedlings so that they should compete well with the wild grass parent.

Tripsacum inflorescences are strongly protogynous, and stigmas are often excerted when no pollen is available, and they dry up before pollination. This is an ideal situation for *Tripsacum* × maize hybridization to take place.

Our interpretation of the origin of clones of *T. andersonii*, would follow this scheme: (a) a diploid $2n = 36$ *Tripsacum* species is pollinated by maize, resulting in a $2n = 46$ F_1 via unreduced female gamete, and (b) the F_1 is male sterile but backcrosses to female parent, also via unreduced gametes, and produces $2n = 64$ plants with three genomes of *Tripsacum* and one of maize.

If $2n = 46$ F_1 plants are produced, they could also be backcrossed to maize. Unreduced female gametes would yield $2n = 56$ plants. In such plants every chromosome has a partner (36 Td + 20 Zm), and meiosis is essentially regular. The next backcross is, therefore, likely to be derived from reduced gametes yielding $2n = 38$ plants. These we know to yield $2n = 20$ maize, and $2n = 22$–24 tripsacoid plants on further backcrossing. Note that $2n = 46, 56, 38$ plants are perennial and vigorous, so that opportunities for backcrossing would extend over considerable time.

At present we would have to consider these schemes as possible, even plausible, but undemonstrated. If this sort of interchange does occur in South America, it should be possible to find intermediate generations in the field. Until the field work is done or these pathways are verified, we must consider that introgression of maize into *Tripsacum* is likely and that introgression of *Tripsacum* into maize is speculative.

REFERENCES

Anand, S. C. 1966. Meiotic behaviour in hybrids between maize and *Tripsacum*. *Caryologia* **19**: 227–230.

Beadle, G. W. 1932. The relation of crossing-over to chromosome association in *Zea–Euchlaena* hybrids. *Genetics* **17**: 481–501.

Beadle, G. W. 1972. The mystery of maize. *Field Mus. Nat. Hist. Bull.* **43**: 2–11.

Celarier, R. P. 1957. Cytotaxonomy of the Andropogoneae. II. Subtribes Ischaeminae, Rottboelliinae, and Maydeae. *Cytologia* **22**: 160–183.

Changanti, R. S. K. 1965. Cytogenetic studies of maize–*Tripsacum* hybrids and their derivatives. Ph.D. thesis. Cambridge, Mass.: Bussey Inst., Harvard Univ.

Clayton, W. D. 1973. The awnless species of Andropogoneae. *Kew Bull.* **28**: 49–58.

Collins, G. N. 1919. Structure of the maize ear as indicated in *Zea–Euchlaena* hybrids. *J. Agr. Res.* **17**: 127–136.

Collins, G. N. and A. E. Longley. 1935. A tetraploid hybrid of maize and perennial teosinte. *J. Agr. Res.* **50**: 123–133.

Cutler, M. C. and E. Anderson. 1941. A preliminary survey of the genus *Tripsacum*. *Ann. Mo. Bot. Gard.* **28**: 249–269.

de Wet, J. M. J., L. M. Engle, C. A. Grant, and S. T. Tanaka. 1972. Cytology of maize-*Tripsacum* introgression. *Am. J. Bot.* **59**: 1026–1029.

de Wet, J. M. J. and J. R. Harlan. 1972. Origin of maize: The tripartite hypothesis. *Euphytica* **21**: 271–279.

de Wet, J. M. J. and J. R. Harlan. 1974. *Tripsacum*–maize interaction: A novel cytogenetic system. *Genetics* **78**: 493–502.

de Wet, J. M. J., J. R. Harlan, L. M. Engle, and C. A. Grant. 1973. Breeding behaviour of maize–*Tripsacum* hybrids. *Crop Sci.* **13**: 254–256.

de Wet, J. M. J., J. R. Harlan, and C. A. Grant. 1971. Origin and evolution of teosinte [*Zea mexicana* (Schrad.) Kuntze]. *Euphytica* **20**: 255–265.

Engle, L. M., J. M. J. de Wet, and J. R. Harlan. 1974. Chromosome variation among offspring of hybrid derivatives with 20 *Zea* and 36 *Tripsacum* chromosomes. *Caryologia* **27**: 193–209.

Galinat, W. C. 1970. The cupule and its role in the origin and evolution of maize. *Univ. Mass. Agr. Exp. Sta. Bull.* No. 585.

Galinat, W. C. 1971. The origin of maize. *Ann. Rev. Genet.* **5**: 447–478.

Grant, C. A. 1973. A scanning electron microscopy survey of some Maydeae pollen. *Grana* **12**: 177–184.

Longley, A. E. 1934. Chromosomes in hybrids between *Euchlaena perennis* and *Zea mays. J. Agr. Res.* **48**: 789–806.

Maguire, M. P. 1964. Chromatid interchange in allodiploid maize–*Tripsacum* hybrids. *Can. J. Genet. Cytol.* **5:** 414–420.

Mangelsdorf, P. C. 1958. The mutagenic effect of hybridizing maize and teosinte. *Cold Spr. Harb. Symp. Quant. Biol.* **23:** 409–421.

Mangelsdorf, P. C. 1961. Introgression in maize. *Euphytica* **10:** 157–168.

Mangelsdorf, P. C. 1968. Cryptic genes for tripsacoid characteristics in Miaz Amargo of Argentina and other Latin American varieties. Bol. Soc. Argent. Bot. **12:** 180–187.

Mangelsdorf, P. C. 1974. *Corn: its origin, evolution and improvement.* Cambridge, Mass.: Belknap Press, Harvard Univ.

Mangelsdorf, P. C. and W. C. Galinat. 1964. The tunicate locus in maize dissected and reconstituted. *Proc. Nat. Acad. Sci.* **51:** 147–150.

Mangelsdorf, P. C., R. S. MacNeish, and W. C. Galinat. 1964. Domestication of corn. *Science* **143:** 538–545.

Mangelsdorf, P. C. and R. G. Reeves. 1939. The origin of Indian corn and its relatives. *Texas Agr. Exp. Sta. Bull.* No. 575.

Newell, C. A. and J. M. J. de Wet. 1973. A cytological survey of *Zea–Tripsacum* hybrids. *Can. J. Genet. Cytol.* **15:** 763–778.

Newell, C. A. and J. M. J. de Wet. 1974a. Morphology of some maize–*Tripsacum* hybrids. *Am. J. Bot.* **61:** 45–53.

Newell, C. A. and J. M. J. de Wet. 1974b. Morphological and cytological variability in *Tripsacum dactyloides* (Gramineae) *Am. J. Bot.* **61:** 652–664.

Randolph, L. F. 1952. New evidence on the origin of maize. *Am. Nat.* **86:** 193–202.

Randolph, L. F. 1970. Variation among *Tripsacum* populations of Mexico and Guatemala. *Brittonia* **22:** 305–337.

Rao, B. G. S. and W. C. Galinat. 1974. The evolution of the American Maydeae. 1. The characteristics of two *Tripsacum* chromosomes (Tr7 and Tr13) that are partial homeologs to maize chromosomes 4. *J. Hered.* **65:** 335–340.

Reeves, R. G. and A. J. Bockholt. 1964. Modification and improvement of a maize inbred by crossing it with *Tripsacum. Crop Sci.* **4:** 7–10.

Reeves, R. G. and P. C. Mangelsdorf. 1959. The origin of corn II. Teosinte, a hybrid of corn and *Tripsacum. Bot. Mus. Leafl. Harvard Univ.* **18:** 357–387.

Roberts, L. M., U. J. Grant, R. Ramirez E., W. H. Hatheway, and D. L. Smith, in collaboration with P. C. Mangelsdorf. 1957. Races of maize in Colombia. *Nat. Acad. Sci., Nat. Res. Coun. Publ.* No. 510.

Chapter 10

A BRIEF SURVEY OF THE RACES OF MAIZE AND CURRENT ATTEMPTS TO INFER RACIAL RELATIONSHIPS

Major M. Goodman

North Carolina State University,
Raleigh, North Carolina 27607

Before the early 1900s little was known about the diversity of maize varieties, even for those grown within the United States. Effective characterization of maize in Latin America did not begin until World War II. In essence, the Rockefeller Foundation, National Academy of Sciences collections, dating from about 1945 in Mexico to about 1960 in Chile, numbering nearly 12,000 collections, form the basis for our current knowledge of maize diversity. The remnants of those collections, as well as data obtained from the original studies of the collections, are still serving as a basis for current attempts to study the diversity within the species *Zea mays* L. Those collections, supplemented

Paper number 4752 of the Journal Series of the North Carolina Agricultural Experiment Station, Raleigh, North Carolina. This investigation was supported in part by NIH research grant number GM 11546 from the National Institute of General Medical Sciences.

143

somewhat by more recent ones of lesser scope, reasonably span the range of variability of maize in Latin America. From the approximately 12,000 collections, some 250 or so races were named in a series of monographs describing the maize of Latin America, its uses, and its known and/or inferred history (Brown and Goodman, 1977). Since then archaeological studies of maize remains [reviewed by Mangelsdorf (1974)] have supplemented studies of currently existing maize varieties and races.

Despite such studies, we know less than we should about the interrelationships among the races of maize. Only some of the more obvious cases (limited largely to races of current economic importance) are discussed here, and the discussion is limited to the maize of the New World. It should be kept in mind that our current groupings of the maize collections into races and the subsequent groupings of races into groups or complexes of races are little more than refinements of Sturtevant's (1899) system of grouping corn varieties into flints, dents, popcorns, sweetcorns, and so on. Whereas Sturtevant arbitrarily used rather simply inherited differences, more recent workers have tried to use multiple (often quantitatively inherited) characters as well as geographic distributions.

RACIAL COMPLEXES OF CURRENT WORLDWIDE ECONOMIC IMPORTANCE

The Mexican dents (e.g., Tuxpeño, Vandeño, Tepecintle, the Zapalotes, Celaya) and their derivatives form the economically most important of the several groups of maize races of current widespread use (see Figure 1). Some of these dents apparently spread to the southeastern part of what is now the United States at about the time of European settlement. In the United States, descendants of the Mexican dents crossed with the Northern Flints from the Northeast to produce the widely used Corn Belt dents (Anderson and Brown, 1952).

Since the earliest descriptions of corn from the Caribbean are those of floury maize (Sauer, 1960), the Mexican dents apparently were widely introduced there also by early European colonists. In the Caribbean and along the northern edge of South America, the Mexican dents apparently combined with flints from northern or eastern South America to form two racial complexes that are now widely distributed throughout the tropics and subtropics. The Tusóns are cylindrical semidents, much flintier than the Mexican dents but similar to them in ear size and shape. The Caribbean flints, which include the Coastal Tropical Flints, Comúns, Costeños, and the northernmost Catetos, are usually more conical in ear shape than the Tusóns and are still more flinty. Their kernels, however, do usually contain some soft starch. In more recent

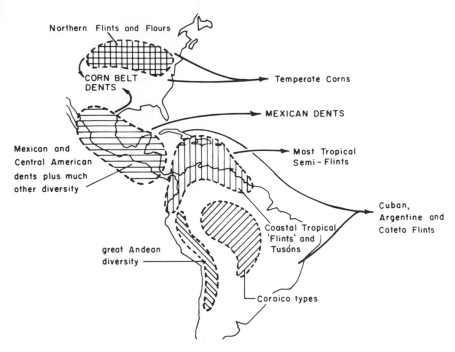

Figure 1. Groups of maize races of worldwide economic importance along with Andean and Coroico types from South America [adapted from Simmonds (1976)].

times, Mexican dents, as well as U. S. and Caribbean materials derived in large part from Mexican dent germ plasm, have been widely used in Brazil, Chile, and more recently, in Argentina.

The other races of widespread commercial importance are the orange–yellow Cateto flints of southern Brazil, Uruguay, and Argentina and the long, narrow-eared Northern Flints and Flours from the northern United States and southern Canada. These two races have spread throughout the temperate regions of the world, wherever flint types of corn have been preferred. They have been particularly important in Southern Europe and Argentina.

REGIONALLY IMPORTANT RACIAL COMPLEXES

Whereas only six racial complexes (the Mexican dents, the Corn Belt dents, the Tusóns, the Caribbean Flints, the Northern Flints and Flours, and the Catetos or Argentine Flints) have achieved worldwide importance, there are several racial groups that are regionally dominant, often almost to the exclusion of racial groups that have achieved much wider acceptance. The Indian corns of the U. S. Southwest and northwestern Mexico form one such complex. These races are typically floury with long, narrow ears that taper at both ends, but flint types are also common. Also included are the "popcorns," Reventador and Chapalote, from the same region. The influence of this racial group seems to extend as far south as the race Dzit-Bacal of Guatemala and as far north as Hickory King of the United States. In the central highlands of Mexico a group of conically eared corns with narrow and droopy-leaved plants, sparsely branched but highly condensed tassels, and very weak roots are predominant. Although the kernels are consistently pointed, they may be flinty (Palomero Toluqueño, from which "Japanese Hulless" popcorn was apparently derived), floury (Cacahuacintle), or slightly dented (Cónico, Cónico Norteño, Pepitilla).

The most distinctive corn in Guatemala is flinty and late maturing with large rounded kernels borne on long ears that are positioned high on the tall, thick stalks. This race, Olotón, is found at mid-to-high elevations in Guatemala, and it and related races are found in adjacent portions of Mexico. Morphologically similar races (Montaña, Amagaceño) are found at similar altitudes in Colombia and Ecuador.

Along the northern edge of South America and in the Caribbean, a narrow-eared type of maize with long, often flexible, ears and deep kernels contrasts with the more widely distributed Tusóns and Coastal Tropical Flints. It is known variously as Canilla (which is usually flinty) or Chandelle (more often dented). The elongate-eared Puyas from Venezuela and Colombia seem to be derived in part from Canilla or Chandelle types.

At mid-to-high elevations in the Andes there is a great diversity of maize. The variations in aleurone and pericarp colors and in kernel shapes and sizes are particularly striking as are differences in plant morphology. Nevertheless, much of the variation can be characterized by three racial complexes. The North Andean flints and flours are predominantly conically eared with large, rounded kernels and moderate, often irregular, row numbers. Perhaps most typical and widely distributed of these is the race Sabanero from northern Peru, Ecuador, Colombia, and Venezuela. Other closely similar races include Cacao and Cabuya from Colombia, Kcello from Ecuador, Pollo from Venezuela and Colombia, and Huevito from Venezuela.

In the Central Andes (southern Peru, western Bolivia) near the center of the influence of the Incas, the most distinctive group of races is usually floury, also, with conical ears, but with low row numbers and often flattened, almost coin-shaped kernels, which in extreme cases are about the size of nickels. The various Cuzco races are typical of this group.

The third common Andean complex consists of races with grenade-shaped ears having high row numbers of floury and often dramatically colored kernels. The size of the ears of these races, as well as the height of the plants, is inversely correlated with the altitude in which the plants are grown (such altitudes often exceed 10,000 ft). This group encompasses races such as the various Chulpi's, Paro, Huayleño, Capia, and Shajatu.

In the Amazon basin east of the Andes a single racial complex is dominant, although maize cultivation is sparse throughout the area. This maize, called Coroico, Piricinco, Pojoso, or Entrelaçado, is characterized by its long, narrow ears. It usually has a distinctive kernel arrangement that results in very low numbers of kernel rows, especially near the tip of the ear, as well as often having odd numbers of rows of kernels rather than the usual even numbers found in almost all other maize. Galinat (1970) has recently given a detailed analysis of the structure of the cob of this maize.

CURRENT STATUS OF RACIAL STUDIES WITH MAIZE

In addition to the racial groups mentioned here, there are many individual races and small groups of races that must be omitted from consideration in this chapter. Some of these are omitted for lack of time and space, others are omitted as a result of lack of known current importance, and still others are omitted due to my lack of knowledge about them. The popcorns, some of which may have played an important role in the early evolution of modern corn, have received little attention here, since they seem not to have played an important role in the recent (last 500 or so years) evolution of the more economically important maize races. They are treated in more detail by Brown and Goodman (1977).

A series of attempts to apply several techniques of numerical taxonomy to racial studies with maize have recently been completed and are in the process of publication (Bird, 1970; Goodman, 1972; Bird and Goodman, 1977; Goodman and Bird, 1977). Such techniques have been reasonably successful, but there have been suggestions of a few similarities between races that appear to contradict conclusions that would be reached using more traditional methods based largely on morphology and geographic distribution. Several methods for further study of such relationships are reviewed in a later section.

Mangelsdorf (1974) has described six lineages of races from which, he suggests, most other races have been derived. Three of his lineages (Kulli, Chulpi, and Confite Morocho) involve various Andean races that certainly do appear to have contributed to the three Andean complexes described here. A fourth lineage, Chapalote–Nal Tel, involves most lower altitude Mexican races. A fifth lineage (Palomero Toluqueño) involves the pointed popcorns discussed in a later section of this chapter. The sixth lineage (Pira Naranja) includes the Catetos and Argentine and Cuban Flints. Several of the lineages (Confite Morocho and Pira Naranja, especially) are quite speculative, and often the degree of speculation increases as the geographic distance of the lineage increases (notably for Chulpi and Confite Morocho, both lineages that extend from South to North America).

After the original series of almost 12,000 collections of Latin American maize was completed in about 1955, preparation of some of the races of maize reports continued into the early 1960s. Since 1955, apparently about 3000 additional collections have been made in Latin America. Some of these collections have been aimed at strengthening weak representation of certain races, others have been aimed at replacing collections that have been lost through attrition, and still others have filled gaps in the original geographic composition (which seems to have closely paralleled the existing Pan-American highway system in at least a few instances). In addition, there have been several attempts to specifically collect material of current interest to breeders. Examples of these include attempts to discover additional mutants rich in particular amino acids (e.g., as *opaque*-2 or *floury*-2) and the assemblage of over 700 collections of the Mexican race Tuxpeño.

One of the most important features of the races of maize reports was, with a few exceptions, their listings of specific collection numbers of collections considered typical of each race. This feature has enabled others to undertake further racial studies and has provided a rational basis for breeders and geneticists to sample the variability present in Latin American maize. In the past 15 years some of these typical collections have been lost. Others have been replaced or supplemented by more recent collections thought to be typical of certain of the races. In addition, a few new races have been named and, in most cases, collections typical of these races have been identified. Unfortunately, much of the work done within the last 15 years has not been published, except for the new races from northwestern Mexico described by Hernandez and Alanis (1970).

RECENT METHODS OF STUDYING INTERRELATIONSHIPS

Chromosome Knob Patterns

Whereas the races of maize reports were based largely on ear and plant morphology combined with the geographic distributions of the races, some consideration was given to the cytological characteristics of the collections, especially for the races of the Andean region. Dr. Barbara McClintock (1959, 1960) and her associates, especially Kato, have continued this work focusing especially on the geographic distributions of chromosome knob complexes (Longley and Kato, 1965; Kato and Blumenschein, 1967; Blumenschein, 1974). They appear to be able to trace the migrations of several knobs and knob complexes over quite large distances. Since historical and archaeological records are scarce throughout many areas of Latin America, especially the lowlands, their study, to be summarized at these meetings by Dr. McClintock, is especially significant and should be a classic in the area of crop-plant evolution. Their conclusions should encounter widespread interest in anthropology and archaeology as well.

Morphological Studies of Racial F_2 Populations

A complementary approach for the study of racial divergence is to ask how far a pair of races have diverged relative to the variation present in their common ancestral population (Goodman, 1969). In Figure 2 populations C and D have

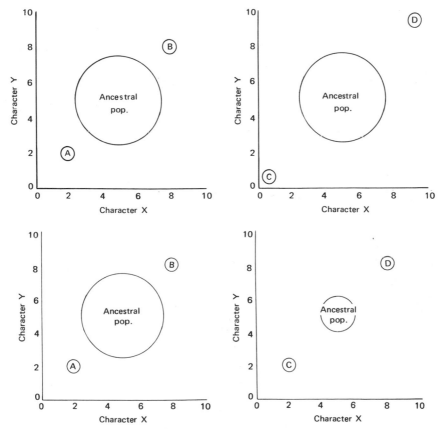

Figure 2. A comparison of divergence of pairs of populations from ancestral populations from which they arose [from Goodman (1969)].

clearly diverged more, relative to their ancestral population, than have populations A and B. For a specific pair of races the best estimate we can usually make of the variation in the ancestral population is the variation observable in the F_2 population derived from the racial cross. Thus if we examine the position of the parental populations relative to the F_2 variation, we can obtain an indication of how far they may have diverged relative to their hypothesized common ancestral population.

I should like to illustrate several such examples. The characters measured for each cross were ear length (cm), ear diameter (cm), row number, kernel length (mm), kernel thickness (mm), and the following ratios: (a) ear diameter/length, (b) kernel thickness/length, and (c) kernel length/ear length. These were chosen for their repeatability and ease of measurement (Goodman and Paterniani, 1969). For each specific pair of parental populations, the pair of characters having the highest ratios of the difference between the parental means to the estimate of the pooled standard deviation within parental populations (the estimate was actually obtained from observations on the F_1 population) was chosen with the following restrictions: (a) such pairs of characters were selected from those having correlations within the F_1 population of less than 0.35 and (b) characters having such a degree of transgressive segregation

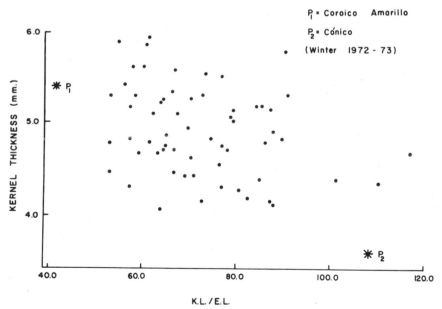

Figure 3. Comparison of F_2 variation of racial cross Cónico × Coroico Amarillo (grown in Florida, winter of 1972/1973) with its parental means. These races are quite clearly distantly related [K.L./E.L. = ratio of 100 (kernel length, mm)/ear length, cm].

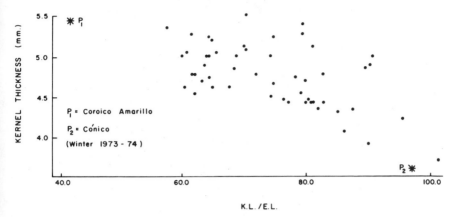

Figure 4. Comparison of F_2 variation of racial cross Cónico × Coroico Amarillo (grown in Florida, winter 1973/1974) with its parental means. Compare with Figure 3 [K.L./E.L. = ratio of 100 (kernel length, mm)/ear length, cm.].

or F_2 breakdown that the F_2 mean lay outside the range of the parental means were not used.

Figure 3 illustrates that Cónico from high altitude central Mexico and Coroico Amarillo from the low altitudes east of the central Andes in South America lie outside the range of variation of the F_2 population studied. Such a finding was not unexpected; this particular pair of races was deliberately chosen as an example of a pair of unquestionably distantly related races. Figure 4 shows similar results obtained when the experiment was grown in a different year.

It would be possible to use many characters simultaneously to obtain distances between the parents relative to the F_2. Perhaps the simplest appropriate multivariate procedure would be to measure the multivariate distance between the parents relative to one of several possible multivariate generalizations of Dempster's (1949) *parental-combination variance.* Here attention is restricted to the bivariate case.

Additional examples are illustrated in Figures 5 and 6, where it can be seen that Harinoso de Ocho Occidentales and Tabloncillo Perla from northwestern Mexico fall well within the range of F_2 variation, as do the two Catetos from Brazil. On this basis it would appear reasonable to conclude that the races within each of Figures 5 and 6 are closely related. Similar conclusions have been reached elsewhere on the basis of other evidence (Brown and Goodman, 1977).

In some cases reasonably critical evidence can be assembled in this manner. Goodman and Bird (1977) grouped the race Nal-Tel from the Yucatan in

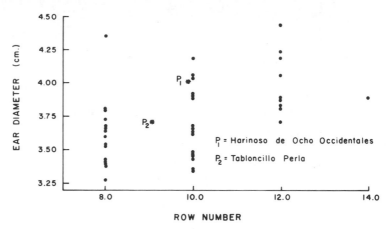

Figure 5. Comparison of F_2 variation of racial cross Harinoso de Ocho Occidentales × Tabloncillo Perla with the means of the parents.

Mexico with Tusilla, a long, narrow-eared popcorn or small flint from Ecuador. We regarded this grouping with doubt, since the result was neither consistent with the gross morphology of the ears nor the geographic distribution of the races and their possible intermediates. It was also inconsistent with some other types of analyses (Bird and Goodman, 1977). Figure 7 indicates that Nal-Tel and Tusilla do fall outside the range of distribution of the F_2 population studied and hence are probably not closely related. On the other

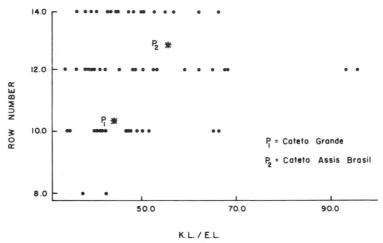

Figure 6. Comparison of F_2 variation of racial cross of two Brazilian Catetos with means of parental populations [K.L./E.L. = ratio of 100 (kernel length, mm)/ear length, cm].

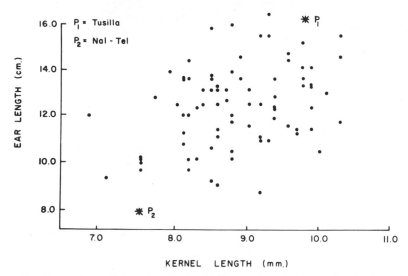

Figure 7. Comparison of F_2 variation of racial cross of Tusilla from Eduador by Nal-Tel from the Yucatan with parental means.

hand, the Colombian race Amagaceño and the Guatemalan and southern Mexican race Olotón have several times been considered to be closely related (Brown and Goodman, 1977), despite their geographic isolation. The results presented in Figure 8 would appear to support such a conclusion, since the racial means of the parents fall well within the range of the F_2 population.

Unfortunately, we do not always reach such clearcut results with our F_2 studies of racial crosses. Palomero Toluqueño, the pointed popcorn from near Mexico City, bears a striking resemblance to a number of pointed South American popcorns, one of which was apparently described as early as 1784 by Dobrizhoffer (1822). Mangelsdorf (1974) has perhaps best assembled the list of traits shared by these seemingly rare yet widely distributed popcorns. Despite an extensive list of shared characteristics, the reticulate nature of their distribution coupled with the great distances involved casts suspicion on the hypothesis that they are closely related. The results shown in Figure 9 suggest neither a very close nor a very distant relationship between Pisankalla, the pointed popcorn from Bolivia, and Palomero Toluqueño. The two parents fall just inside the ranges of the F_2 distribution observed.

Use of Isozyme Frequencies for Grouping Maize Races

Within the past 10 years the analyses of gene frequencies have greatly expanded as a result of the widespread use of electrophoretic techniques. These

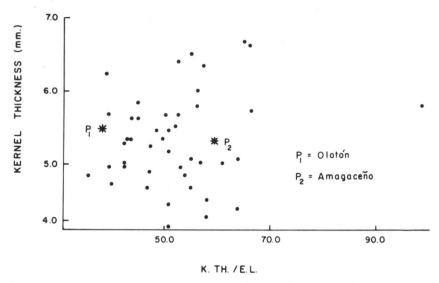

Figure 8. Comparison of F_2 variation of racial cross Olotón × Amagaceño with its parental means [K.TH./E.L. = ratio of 100 (kernel thickness, mm)/ear length, cm].

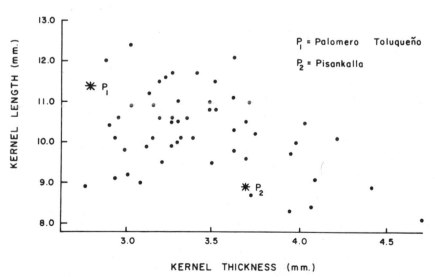

Figure 9. Comparison of F_2 variation of racial cross Palomero Toluqueno × Pisankalla with its parental means.

techniques are based on the fact that enzymes of different weights, conformations, and potentials for ionization migrate at different rates across an electrical field. Our work involves the migration of several enzymes across electrically charged horizontal starch gels followed by histochemical staining of the enzyme variants. The techniques we are using are slight modifications of those widely used elsewhere (Shaw, 1968; Brewbaker, Upadhya, et al, 1968; Selander, Smith, et al, 1971). The work with which we are currently involved was initiated by Dr. Frank Johnson, who has been active in similar work with *Drosophila* species since the mid-1960s, and by Dr. Charles Stuber, who has carried out much of this work since Dr. Johnson began working for the Research Triangle Institute. It is with Dr. Stuber's collaboration that I present a few very preliminary results here. Much of our data has just been acquired, and much remains to be done with both genetic and statistical analyses. Somewhat similar work is being carried out by Allard's group at the University of California at Davis, while others such as Scandalios (1974), Brewbaker, Upadhya, et al (1968), and Schwartz (Schwartz and Endo, 1966) have been more concerned with the developmental genetics and biochemistry of the enzyme systems. We are just now completing a study of the well-defined Mexican and Guatemalan races of maize. Our genetic analyses are far from complete, in part because of the discovery of new alleles and/or loci as a result of the survey. For present purposes it is simplest to divide the isozyme systems studied into two groups: (a) those for which we feel we can read genotypes (or groups of genotypes) from the banding patterns and (b) those for which we mainly read phenotypes. In the first set are included β-glucosidase (β-GLU) (Stuber, Goodman, and Johnson, 1977), acid phosphatase (ACPH), phosphohexose isomerase (PHI), alcohol dehydrogenase (ADH), glutamate-oxalacetate transaminase (GOT), an esterase (EST A), and malate dehydrogenase (MDH). In the second set are phosphoglucomutase (PGM) and 6-phosphogluconate dehydrogenase (6-PGD). Our data are currently being summarized, but it is apparent that at least for several of these systems, especially β-GLU, ACPH, and MDH, substantial intraspecific variation occurs. For each of these three systems the collections we have studied differ not only because of the presence of different alleles, but also because of different allelic frequencies. Although less variation among collections appears to be present for the other enzymes studied, it is apparent that there is appreciable intraspecific variation associated with differences among the collections we have studied. This conclusion may seem self-evident, but many intraspecific studies have encountered little such variation (Avise, 1974). Nevertheless, two widely studied enzyme systems, the esterases and ADH, show little variation among the Mexican collections.

We hope to be able to use the isozyme gene and band frequencies to group the collections and to compare such a grouping with that achieved using mor-

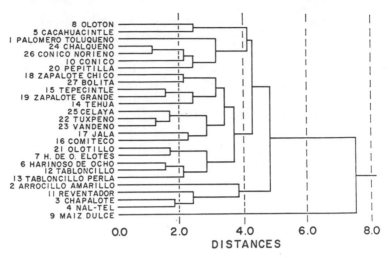

Figure 10. Dendrogram grouping most of the Mexican races of maize on the basis of distances calculated from standardized principal components derived from correlations among morphological and geographic data [from Goodman (1972)].

phological and geographic data (Figure 10). Since the electrophoretic data are in many respects similar to the cytological data of McClintock and her associates, we also hope to ultimately compare our racial groupings with their conclusions as well. They have covered a much broader geographic area than we have at present, however.

REFERENCES

Anderson, E. and W. L. Brown. 1952. Origin of Corn Belt maize and its genetic significance. In *Heterosis*, J. W. Gowen (ed.). Ames: Iowa State College Press, pp. 124–148.

Avise, J. C. 1974. Systematic value of electrophoretic data. *Syst. Zool.* **23:** 465–481.

Bird, R. McK. 1970. Maize and its cultural and natural environment in the sierra of Huánuco, Peru. Ph.D. thesis. Berkeley: University of California, 310 pp.

Bird, R. McK. and M. M. Goodman. 1977. The races of maize: V. Grouping maize races on the basis of ear morphology. *Econ. Bot., ***31:**471–481.

Blumenschein, A. 1974. Chromosome knob patterns in Latin American maize. In *Genes, enzymes, and populations*. A. M. Srb (ed.). New York: Plenum, pp. 271–277.

Brewbaker, J. L., M. D. Upadhya, Y. Mäkinen, and T. Macdonald. 1968. Isoenzyme polymorphism in flowering plants. III. Gel electrophoretic methods and applications. *Physiologia Plantarum* **21:** 930–940.

Brown, W. L. and M. M. Goodman. 1977. Races of corn. In: *Corn and corn improvement*, rev. ed., G. F. Sprague (ed.). Madison, Wisconsin: American Society of Agronomy, pp. 49–88.

Dempster, E. R. 1949. Effects of linkage on parental-combination and recombination frequencies in F_2. *Genetics* **34**: 272–284.

Dobrizhoffer, M. 1822. *An account of the Abipones, an equestrian people of Paraguay*, Vol. 2. London: John Murray, 435 pp. (Originally published in Latin in 1784.)

Galinat, W. C. 1970. The cupule and its role in the origin and evolution of maize. *Mass. Agr. Exp. Sta. Bull.* No. 585.

Goodman, M. M. 1969. Measuring evolutionary divergence. *Japanese J. Genet.* **44** (suppl. 1): 310–316.

Goodman, M. M. 1972. Distance analysis in biology. *Syst. Zool.* **21**: 174–186.

Goodman, M. M. and R. McK. Bird. 1977. The races of maize: IV. Tentative grouping of 219 Latin American races. *Econ. Bot.*, **31**: 204–221.

Goodman, M. M. and E. Paterniani. 1969. The races of maize: III. Choices of appropriate characters for racial classification. *Econ. Bot.* **23**: 265–273.

Hernandez X., E. and G. Alanis F. 1970. Estudio morfologico de cincos nuevas razas de maiz de la Sierra Madre Occidental de Mexico: Implicaciones filogeneticas y fitogeograficas. *Agrociencia* **5**: 3–30.

Kato Y., T. A. and A. Blumenschein. 1967. Complejos de nudos cromosómicos en los maíces de América. *Fitotecnia Latinoamericana* **4**: 13 24.

Longley, A. E. and T. A. Kato Y. 1965. Chromosome morphology of certain races of maize in Latin America. Chapingo, Mexico: *International Center for the Improvement of Maize and Wheat Res. Bull.* No. 1. 112 p.

Mangelsdorf, P. C. 1974. *Corn. its origin, evolution, and improvement*. Cambridge, Mass.: Harvard Univ. Press, 262 pp.

McClintock, B. 1959. Genetic and cytological studies of maize. *Carnegie Inst. of Washington Year Book* **58**: 452–456.

McClintock, B. 1960. Chromosome constitutions of Mexican and Guatemalan races of maize. *Carnegie Inst. of Washington Year Book* **59**: 461–472.

Sauer, C. O. 1960. Maize into Europe. Akten Internationalen Amerikanisten Kongresses (Vienna) **34**: 778–786. [Republished in 1969 in C. O. Sauer. *Agricultural origins and dispersals*, 2nd ed. Cambridge, Mass.: MIT Press, pp. 147–167.]

Scandalios, J. G. 1974. Isozymes in development and differentiation. *Ann. Rev. Plant Physiol.* **25**: 225–258.

Schwartz, D. and T. Endo. 1966. Alcohol dehydrogenase polymorphism in maize— simple and compound loci. *Genetics* **53**: 709–715.

Selander, R. K., M. H. Smith, S. Y. Yang, W. E. Johnson, and J. B. Gentry. 1971. IV. Biochemical polymorphism and systematics in the genus *Peromyscus*. I. Variation in the old field mouse. *Studies in Genetics* VI. *Univ. Texas Publ.* 7103, pp. 49–90.

Shaw, C. R. 1968. Starch gel electrophoresis of enzymes. In *Chromatographic and electrophoretic techniques*, Vol. II, ed. 2, New York: Interscience, pp. 325–364.

Simmonds, N. W. 1976. *Crop plant evolution*. London: Longman.

Stuber, C. W., M. M. Goodman, and F. M. Johnson. 1977. Genetic control and racial variation of β-glucosidase isozymes in maize (*Zea mays* L.). *Biochem. Genet.* **15:** 383–394.

Sturtevant, E. L. 1899. Varieties of corn. Washington, D. C.: *USDA Office of Exp. Sta. Bull. No.* 57.

Chapter 11

SIGNIFICANCE OF CHROMOSOME CONSTITUTIONS IN TRACING THE ORIGIN AND MIGRATION OF RACES OF MAIZE IN THE AMERICAS

Barbara McClintock

Carnegie Institution of Washington,
Cold Spring Harbor, New York.

Examination of chromosome constitution in plants of many races of maize has revealed extensive diversity in origin of their germplasms. Their chromosomes have distinctive structural components that derive from more than one initial source. The sources differed from each other in the presence or absence of a particular component and also in the morphological organization of the component, if present. The components of these initial sources are now distributed among the races, and in distinctive combinations. Recognition of them provides a means of determining degrees of relationship among the races, and estimates based on chromosome constitutions both confirm and supplement those based on morphological physiological, and genetic considerations.

The examinations have also revealed clearly defined geographic centers in which maize with a particular combination of initial germplasms was propa-

159

gated. This maize "migrated" into new territories, often along well-defined paths, or was introduced into a distant location where it could then initiate a new center of distribution. As a consequence of migrations and introductions, introgressions took place between maize originating in different centers, and these initiated new combinations of basic germplasms. Not only did new races arise through modification of genetic expression of maize with a particular germplasm combination, they also arose as a consequence of introgressions between maize strains having different combinations. In some territories an initial introduction was followed by later introductions, and in some instances it is possible to discern the sequence of arrivals. It is the purpose of this chapter to present briefly the nature of the evidence that is responsible for the above statements.

The studies reviewed here commenced with some initial observations of mine conducted with plants of races of maize derived from Ecuador, Bolivia, Chile, and Mexico, and with plants of a few additional races indigenous to the Central American countries, Venezuela, and the Caribbean Islands (McClintock, 1959, 1960). These studies revealed the highly conservative nature of the distinctive chromosomal components. This suggested that knowledge of the distribution of these components to the races of maize in the Americas could reveal much about the history of maize—the centers of origin of particular germplasm combinations, migration paths from these centers, introductions and their sources, and introgressions that have occurred. In these respects the fuller knowledge proved to be rewarding.

Extensive recordings of chromosomal constitutions of maize races began with the studies of Longley and Kato Y. (1965), who examined races collected in the Central American countries and from the Caribbean Islands. In addition, a few races derived from Mexico, Colombia, and Venezuela were examined. Subsequently, Blumenschein (1973) examined the chromosome constitutions of plants of various races collected in Venezuela and in central and eastern South America. His studies reveal much about the origin of races in these areas. Dr. Kato's extraordinary talents for unraveling complicated combinations of basic germplasms, as expressed in the pachytene chromosomes, has been of inestimable importance for this study. His examinations include many races of Mexico, some strains of Indian maize from the southwest and central United States, and maize of formerly important open-pollinated, commercial varieties from the southeastern parts of the United States. All of the maize from the United States was provided by Dr. William L. Brown, who has generously cooperated in these studies since their very beginning. Recently, Dr. Kato determined chromosomal similarities and differences among many teosintes collected in Mexico and Guatemala. The information he obtained from these teosintes makes it possible to appreciate just how close the relationship must be between germplasms of Mexican teosintes and those of maize. It suggests that

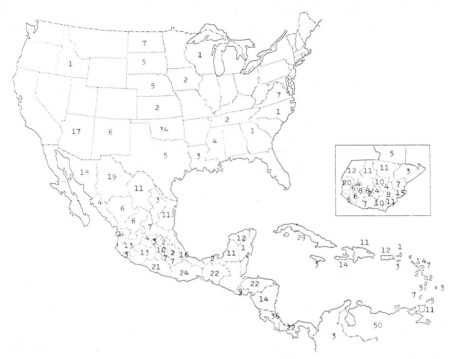

Figure 1. Number of collections, within each bounded area, of maize races of strains whose chromosome constitutions were determined. To indicate the number of examined collections from each Department of Guatemala, an enlarged outline of this country is inserted to the right. This same placement of Guatemala appears on each map of North America. It represents the format devised by Dr. Kato.

the Mexican teosintes are the sources of all the basic germplasms of maize. The Guatemalan teosintes appear to have contributed little or not at all initially.

The number of examined plants of a single collection from a given locality varied considerably. In the extended studies of Kato and Blumenschein, this averaged between four and six. The numbers were increased for collections of those races considered to be of particular importance in the history of maize, or among collections of different races coming from a restricted territorial region and considered to be of special significance in this regard. The number of examined plants from such collections ranged within 7–24. Examination of plants from different collections of the same race, either from the same general area or distantly located, was much increased for those races considered to have contributed to the origin of other races, or to have been extensively distributed. The total number of collections and the areas from which they were taken are shown in Figures 1 and 2.

Figure 2. Geographic locations in South America of those collections of maize races from which chromosome constitutions were determined. An enlarged outline of Ecuador is inserted just to the left of Chile to indicate clearly the locations of examined collections. This format was devised by Dr. Blumenschein and appears on all maps of South America.

This report is based largely on the observations of Drs. Longley, Kato and Blumenschein. Although we may share a common mode of interpreting them, I must take full responsibility for the conclusions drawn from them that are expressed here.

CHROMOSOMAL COMPONENTS CONTRIBUTING TO THE STUDY OF RACE RELATIONSHIPS

The basic chromosome constitution of maize consists of 10 chromosomes, numbered according to their relative lengths from the longest, chromosome 1, to the shortest, chromosome 10. Among the examined races, these length relationships were found to be constant, except that plants of some races carried an aberrant chromosome 10 (called *Abnormal-10*) having a heterochromatic extension of its long arm. Each chromosome is divided into two arms whose relative lengths are determined by the location of the centromere. These locations also appear to have remained relatively constant, as no strains were observed that had a readily detectable modification in the location of the centromere. Slight shifts, however, could have been overlooked. Each arm exhibits a specific pattern of chromatin compaction at the stage of the meiotic cycle, late pachytene, that best served the purpose of this study. The chromomere patterns so formed also are basically conservative; they are quite similar in most examined plants. There are, however, several locations in the chromosome complement that may have an enlarged chromomere. One or another of these enlarged chromomeres may be present in a strain and be either homozygous or heterozygous with a smaller homologous chromomere. Although such chromomere variations were noted, no effort was made in this study to systematically record them.

The nucleolus organizer is another chromosome component that exhibits variation among maize strains. Not only does the size vary, but the location within the organizer where the nucleolus is initiated at telophase also varies. Although notes were taken when exceptional modifications of the organizer were observed, no attempt was made to record the state of the organizer within each plant.

Another variable in chromosome organization relates to rearrangements of segments of chromosomes, such as inversions, duplications, and translocations. The examined races were free of gross rearrangements, with the exception of the inversion in the short arm of chromosome 8, whose presence in maize and teosinte has been known for many years (McClintock, 1933; Ting, 1958). Occasionally a modification was noted in an individual plant or in plants of a single collection, but none has spread throughout populations as has the inversion in chromosome 8.

The most detailed observations and recordings of variation of chromosome organization involve the knobs in terms of their presence or absence at any one knob-forming region and, if present, the size class to which a knob belongs. The locations of these knob-forming regions in each of the maize chromosomes are shown in Figure 3. At any one of these locations in any one plant of a race, a knob may be present or absent in one or both homologs, and this reveals an aspect of the ancestral history of the plant. If present, a knob might be small or large or some size in between, and in one or both homologs. The exceptions are the knobs designated 8L-2, 10L-1, and 10L-2 in Figure 3, which were always observed to be relatively small. Again, the size and shape as well as presence or absence of any one knob depends on the ancestral history of the plant. It is the distinctiveness of each of these many knobs that has provided the most detailed and useful data for judging degrees of relationship between races and for

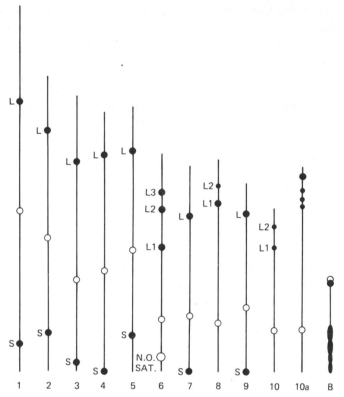

Figure 3. The distinctive components of the maize chromosome complement that were recorded in this study. From left to right the chromosomes are numbered according to relative size, 1 to 10: 10a refers to Abnormal-10; B refers to the B-type chromosome. The small open circles represent locations of the centromeres. The large open circle represents the nucleolus organizer, SAT refers to the satellite of this chromosome, knob locations are indicated by solid circles, S refers to the short arm of the chromosome, and L refers to the long arm. (Courtesy of Dr. T. A. Kato.)

reconstructing the probable past history of some of them. Examples illustrating this are given after two other sharply exhibited variables in chromosome constitution are discussed. These are the B-type chromosome and the Abnormal-10 chromosome.

DISTRIBUTION OF ABNORMAL-10 AND THE B-TYPE CHROMOSOME INITIAL CONSIDERATIONS

The heterochromatic component that lengthens the long arm of chromosome 10 allows ready detection of the presence of Abnormal-10 in a plant (see chromosome 10a, Figure 3). Although some variation in organization of the heterochromatic component was noted, only in one small area in the Argentine Andes was any conspicuous modification observed in the organization of the B-type chromosome (see chromosome B, Figure 3). It had an elongated euchromatic segment.

One or the other or both of these distinctive chromosomes appeared in plants derived from collections made in various parts of the Americas, as shown in Figures 4 and 5. Each circle refers to plants derived from a collection made at the marked geographic location. It is obvious at a glance that each of these chromosomes has a distinctive distribution pattern and that overlapping is confined to restricted areas. There can be no question that plants having a B-type chromosome share at least one ancestor in common, as do plants that have Abnormal-10. The question then turns to the degree of relatedness among plants having either one or the other of these chromosomes. This may be answered by comparing other components of their chromosome constitutions.

MIGRATION PATHS IN NORTH AMERICA OF COMPONENTS OF ORIGINAL GERMPLASMS

The distribution patterns of the B-type chromosome in Mexico and Guatemala (Figure 4a) illustrate both old and relatively recent migration paths of maize germplasms. These paths diverge from definable areas, and several paths may diverge from a single area. Some of the paths are sharply revealed by those knobs that have restricted distributions. The geographic distribution of several of these knobs will serve to illustrate one mode of detecting migration paths.

Paths to and along the Pacific Coast of Mexico

One of the older paths is along the Pacific coast of Mexico, from its border with Guatemala into Sonora. Figures 6 and 13a define this path. Another path starts in the area that includes northern–central Guerrero, which is adjacent to Morelos, and the state of Mexico and runs into southern Guanajuato and northern Michoacan, from which it enters Jalisco and Nayarit, where it may join the Pacific coastal path to the north. No one knob alone illustrates this path but its existence is recognizable on viewing a number of knob distribution maps. It can be discerned in Figures 4a, 7, 8, and 9.

The path from Central Mexico to the U. S. Border

A relatively recent path begins in the Bajío of Mexico, directly north of Mexico City, and extends into Chihuahua. The territory covered by this migration broadens toward the north, resulting in a cone-shaped distribution pattern with

166

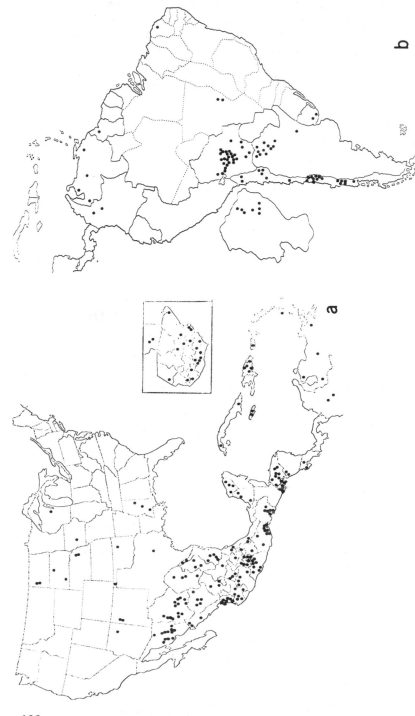

Figure 4. Locations of collections having plants with one or more B-type chromosomes: (a) North America and Caribbean area; (b) South America.

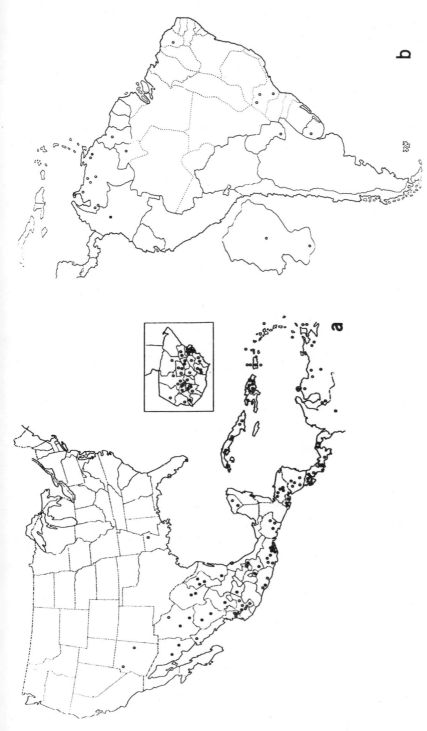

Figure 5. Locations of collections having plants that carry the Abnormal-10 chromosome (solid circles represent those collections in which the B-type chromosome was also present): (*a*) North America and Caribbean area; (*b*) South America. [In states located along the Pacific coast of Mexico the maize races are as follows: in Sonora: Onaveño-Cristalino and Cristalino de Chihuahua; in Sinaloa: Harinoso-de-Ocho; in Nayarit: Harinoso-de-Ocho, Jala, and one unclassified race; in Jalisco: Reventador, Celaya, Tabloncillo, Tuxpeño, and Pepitilla (two collections); in Guerrero: Nal-Tel (two collections) and Vandeño-Olotón; in Oaxaca: Bolita (two collections), Zapalote Grande (one collection), Zapalote Chico (four collections), and Nal-Tel (two collections); in Chiapas: Nal-Tel, Zapalote Grande, and Tepecintle.]

Figure 6. Locations of collections having plants with a large- and/or a medium-sized knob in the short arm of chromosome 4. Solid triangles indicate the large knob; open circles indicate the medium-sized knob. Underlined triangle with circle indicates that both types of knobs were distributed among plants of the same collection. [The races in Mexico are as follows. In Sonora: Chapalote with large knob, Dulcillo de Sonora, and Cristalino de Chihuahua; in Sinaloa: Chapalote and Harinoso-de-Ocho; in Nayarit: Harinoso-de-Ocho and Reventador; in Jalisco: Reventador and Pepitilla; in Oaxaca: Bolita with medium-sized knob (one collection), Zapalote Chico (all other collections); in Chiapas: Zapalote Chico (one collection with large knob), Zapalote Grande (three collections; one has both the large and the medium-sized knob), and Tuxpeño. The two collections from Guatemala are of Nal-Tel Ocho and Comiteco.] These knobs were not found elsewhere in the Americas.

its apex in central Mexico (see Figures 7–9 and their legends). The race contributing the major fraction of germplasm to the maize in this cone-shaped area is Cónico Norteño, itself a product of earlier racial crosses that combined and integrated distinctly different basic germplasms.

The origin of Cónico Norteño, as projected by Wellhausen, Roberts, et al. (1952), starts with a hybrid between two strikingly different races (Palomero Toluqueño and Cacahuacintle), both of which are grown within the state of

Mexico. Stabilization of this hybrid produced the race known as Cónico, also grown in central Mexico. Not only are the two parental races strikingly different morphologically; their chromosome constitutions also are strikingly different, emphatically demonstrating basic differences in their germplasms. The chromosome constitution of Cónico contributes direct support for Wellhausen's projection of its origin. Wellhausen, Roberts, et al. (1952) also projected that Cónico was taken north into the Bajío, where introgression occurred with a race from the west (Celaya). This series of mixing gave rise to the successful combination of germplasms that appears in the race designated Cónico Norteño. All of the projected racial combinations thought to be responsible for the origin of Cónico Norteño are fully supported by the chromosome constitutions of putative ancestral races and by Cónico Norteño itself. Clearly, it has germplasm components that are shared with each parental race but totally carried by no other single race.

Figure 7.   Distribution of the small knob at position 1 in the long arm of chromosome 6. This knob was found elsewhere only in plants of one collection from the Territory of Roraima, Brazil, adjacent to Venezuela.

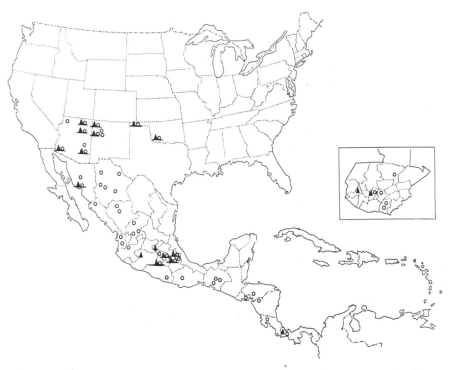

Figure 8. Locations in the Americas of collections having plants carrying the large- and/or the medium-sized knob in the long arm of chromosome 1. Solid triangles indicate the large knob; open circles indicate the medium-sized knob. Underlined triangle with circle indicates that both types of knobs were found among plants of the same collection. The two collections in Oklahoma are from the Kiowa and the Mescalero Indian tribes. Both of these tribes are related to the Apache tribe in Arizona. Chromosome constitutions of plants from collections of these two tribes indicate that their maize came mainly from the Apache tribe in the southwestern part of the United States, although the Kiowa tribe has some maize from the northcentral states. Note the relation of maize in the southwest United States to that in central Mexico. The two races in the highlands of Guatemala having the large knob are Nal-Tel Blanco and Imbricado.

A Path along the East Coast of Mexico

The distribution of the race Tuxpeño defines a relatively recent path along the eastern coast of Mexico. The path runs north, parallel to the coast, from the lower part of the state of Veracruz into the state of Tamaulipas. Toward the north the Tuxpeño germplasm has spread westward through the northern Mexican states. The spread was accomplished by introgressions with maize having other basic germplasm combinations, which had arrived in these states by other migration routes or by direct introductions. This path is illustrated by the distribution of the 9L large knob shown in Figure 10a. The triangles refer to collections of the race Tuxpeño proper. The distribution of this knob within the described area of Mexico should be compared with that of the race Tuxpeño and its introgressions shown on the map of Figure 90 in Wellhausen, Roberts, et al. (1952). They are the same except for the states of Campeche and Yucatan. Tuxpeño strains arrived in these two states by an entirely different route, and they contained a few distinctly different basic germplasm components.

Tuxpeño's initial germplasm undoubtedly was formulated from a succession of introgressions between maize strains having different combinations of basic germplasms. The successful combination that produced the race now known as Tuxpeño appears to have occurred in southwestern Guatemala or in the adjacent state of Chiapas, Mexico. From this general area of origin, its migration took several distinctly different paths. One of these carried it through

Figure 9. Distribution of the small knob at position 2 in the long arm of chromosome 10. The open triangles represent collections of the race Tuxpeño. The collection from Virginia (U. S. A.) is of the strain White Tuxpan. The three collections from the highlands of Guatemala are of the races Serrano, Nal-Tel, and Imbricado. This knob was not found elsewhere in the Americas.

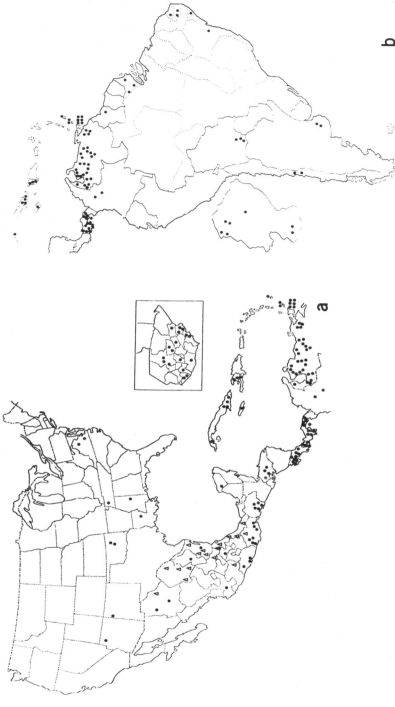

Figure 10. Distribution in the Americas of the large knob in the long arm of chromosome 9 (open triangles represent collections in Mexico of the race Tuxpeño): (a) North America and Caribbean area and (b) South America and Caribbean area.

Chiapas into and across Oaxaca, where it entered the state of Veracruz at Oaxaca's northeastern border. It then went north along the eastern part of Mexico, as just described and illustrated. In the course of its migration, it picked up bits of other basic germplasm components from maize growing along its route, without however suffering serious loss of character identity. Such an instance is illustrated in Figure 9.

Distribution Paths of B-Type Chromosome

The distribution of the B-type chromosome in the upper two-thirds of Mexico (Figure 4a) follows along the described western and central migration paths and to a lesser extent along the eastern path. This could be anticipated if the B-type chromosome originated in central Mexico and was integrated into the early maize races that were being propagated in this area. [The B-type chromosome is present in some races considered to be ancient (Palomero Toluqueño, Chapalote, Reventador, Pepitilla, Zapalote Chico, Nal-Tel). It is also present in teosintes located in western–central Mexico (Kato Y. 1975). Its origin in a teosinte is possible, if not probable].

From the southern border of the Mexican state of Nayarit on the western coast to the southern coast of Oaxaca, plants from only two collections along the coastal region had B-type chromosomes. One of the collections is of the race Nal-Tel, and the other is of the race Vandeño, and this is significant. In Oaxaca plants from nine collections had this chromosome. The collection from the far eastern part of this state is of the race Tuxpeño. Of the eight along the coast, seven are of the race Zapalote Chico, and the eighth is a hybrid of the related race, Zapalote Grande, with the race Bolita. In the adjacent state to the south, Chiapas, there are seven collections having plants with B-type chromosomes. Three are of Zapalote Chico, two are of Zapalote Grande, and there is one each of Vandeño and Comiteco. In these states the B-type chromosome is concentrated in the race Zapalote Chico and its close relative, Zapalote Grande. (In Oaxaca, the chromosome constitutions of seven races were examined: Bolita, Bolita-Zapalote Grande, Cónico, Nal-Tel, Tuxpeño, Zapalote Chico, and Zapalote Grande. In Chiapas, 13 were examined: Comiteco, Nal-Tel, Olotillo, Olotón, Tehua, Tepecintle, Tepecintle-Olotón, Tuxpeño, Tuxpeño-Tepecintle, Tuxpeño-Vandeño, Vandeño, Zapalote Chico, and Zapalote Grande.) In these studies Zapalote Chico proved to be an extraordinary race. It is a repository for components of germplasms of several races thought to be relics of early races (Nal-Tel, Chapalote, and Reventador). In addition, it has many of the components of a presumed early race that cannot yet be identified. It also may be viewed as a "point of departure" for the north coastal and southern distributions of some of the rare chromosomal components (e.g., see Figures 6, 11, and 13a, and their legends).

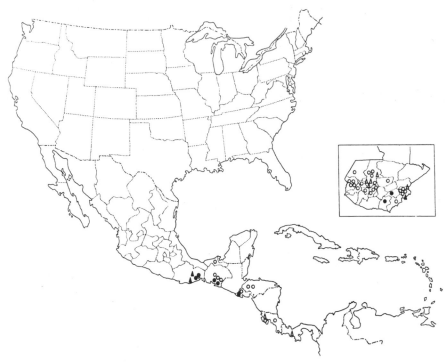

Figure 11. Distribution in the Americas of the small knob at position 1 in the long arm of chromosome 10. Solid circles represent collections having this knob and the B-type chromosome. Open triangles represent collections with this knob and also the Abnormal-10 chromosome. Solid triangles represent collections in which this knob, the B-type chromosome, and Abnormal-10 were distributed among the plants. Open circles represent collections in which this knob was found, but no plants had either the B-type or the Abnormal-10 chromosome.

For the anthropologist, the geographic path described by the B-type chromosome is of some significance. This chromosome, as well as other specific chromosomal components, depicts the likely mode of entrance of maize into the Mayan territories of Campeche and Yucatan. Its path in Guatemala is in the form of an arc, passing through the southern Departments. It then curves toward the northwest into Jalapa, El Progresso, and Baja Verapaz, proceeding north into El Peten. From El Peten it enters Campeche in Mexico and follows northeast into Yucatan (see Figures 12a and 14a).

From Guatemala south, the paths described by the B-type chromosome again are instructive. These paths are extensions from the southern Guatemalan arc. They enter El Salvador (B-type chromosome in six of nine collections) and Honduras (in eight of 24 collections). The path stops with the two

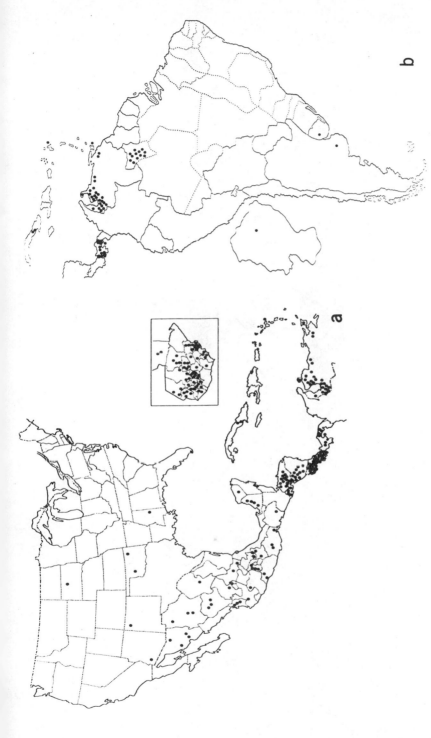

Figure 12. Distribution in the Americas of the small knob in the short arm of chromosome 2: (a) North America and Caribbean area; (b) South America and Caribbean area.

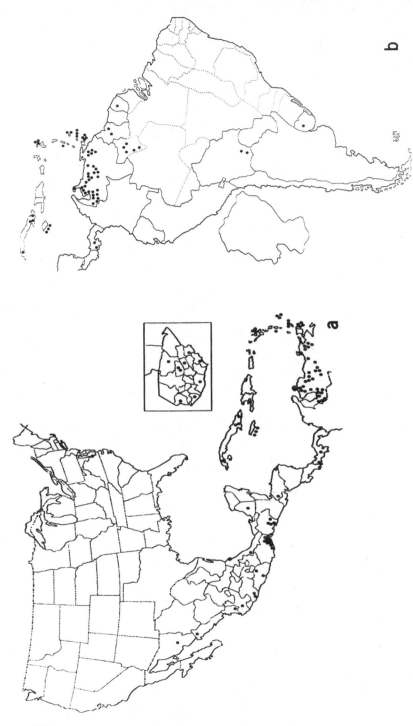

Figure 13. Distribution in the Americas of the large knob terminating the short arm of chromosome 7. [The races in states of Mexico are as follows. In Sonora: Cristalino de Chihuahua; in Sinaloa: Chapalote; in Nayarit: unclassified; in Jalisco: Celaya; in Michoacan: Reventador (both collections); in Guerrero: Nal-Tel and Pepitilla; in Oaxaca: all are from collections of Zapalote Chico; in Chiapas: Zapalote Chico (one collection), Zapalote Grande (two collections), Tuxpeño–Tepecintle (one collection); in Veracruz: Tuxpeño; in Campeche: Nal-Tel–Tuxpeño. The races in Guatemala are mostly Nal-Tel Blanco or hybrids with it.]

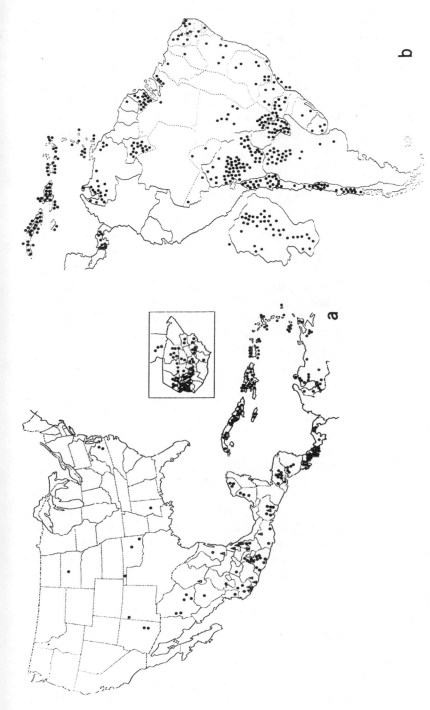

Figure 14. Distribution in the Americas of the small knob in the long arm of chromosome 7. The open triangles in Mexico represent collections of the race Tuxpeño: (a) North America and Caribbean area; (b) South America and Caribbean area.

collections located in the northwestern corner of Nicaragua, adjacent to Honduras. The appearance of the B-type chromosome south of Nicaragua reflects an introduction from the north of maize that had this chromosome in it. The significance of the scattered distributions of the B-type chromosome in the Caribbean area, the clustered distributions in the Andean regions of Ecuador, Bolivia, Argentina, and Chile, and its appearance in the race Cateto in Guyana, Uruguay, and eastern Argentina will become evident later during the discussion of other chromosomal components that suggest the geographic locations from which maize with the B-type chromosome could have been introduced into each of these areas.

Origin and Distribution of Abnormal-10

The distribution of Abnormal-10 (Figure 5) differs from that of the B-type chromosome. In states directly north and northeast of the central Mexican area the appearance of Abnormal-10 relates to the race Celaya, either directly in representatives of that race or through its contribution to the formation of the race Cónico Norteño, whose germplasm distributions were discussed. Celaya is grown in Guanajuato and Jalisco and has been introduced into northeastern Mexico. In fact, all three examined collections from Nuevo Leon are of this race, and all have Abnormal-10 (the legend to Figure 5a names the races on the western coast that have this chromosome).

I suspect that Abnormal-10 may have originated from some event occurring to an ancestor of the race Tepecintle, located at the time of the event in southwestern Mexico or adjacent Guatemala. (That Abnormal-10 may have entered the race via teosinte cannot be excluded, as it is present in some of the Mexican teosintes.) Its presence in Zapalote Chico and Zapalote Grande would not, then, be surprising as Tepecintle and Zapalote share many of the same basic germplasm components. Through subsequent and successive introgressions, Abnormal-10 was delivered to other races within the range of distribution of Tepecintle in Guatemala, and it now appears in some of them (Nal-Tel Ocho, Nal-Tel Blanco, Nal-Tel Amarillo, Serrano, Olotillo, and Comiteco). It also appears in some of the successful races or strains produced by hybridization with Tepecintle (Tepecintle-Salvadoreño, Tepecintle-Comiteco, and Nal-Tel–Tepecintle). These are the main races in Guatemala in which Abnormal-10 was found. The germplasms of these races are distributed along the previously described paths into El Salvador, Honduras, and Nicaragua, and into Yucatan by the described northern route.

Migration Path from the Guatemalan Highlands and Lowlands into Central America and Its Relation to Maize in the Caribbean Area

Another migration path, repeatedly recognized in these studies, takes the germplasm of maize of the highlands of Guatemala and also of two of the lowland

Departments (Jutiapa and Chiquimula) into the Central American countries. In Costa Rica and Panama introgressions then established new germplasm combinations. Some of these produced the races and strains that were taken to the Caribbean Islands. Also, some populated a large area in northwestern Venezuela, either directly as a racial type or indirectly after introgressions with maize previously introduced from the north (e.g., see Figure 12a). The general absence of the 2S small knob in the Caribbean Islands would postulate its general absence in eastern South America, because most of the maize in that part of South America relates to maize in the Islands. Its distribution in South America (Figure 12b) supports this postulate.

The map in Figure 12b conspicuously illustrates a correlation that was observed repeatedly. It shows a relationship between the maize of Venezuela and that of the Territory of Roraima, Brazil. The distribution of other knobs narrows this connection to maize located in northwestern Venezuela. These other knobs are mainly the small ones. Although they occupy the same locations in the chromosomes as the large knobs of the "Venezuelan complex" (see next section), the size differences are in marked contrast. The main center of these small knobs is in the highlands of Guatemala, and a group of them constitute the "small knob complex" (knobs in 2S, 2L, 3L, 4L, 5L, 7L, 8L-1, 9S, and 10L-1).

The distribution of the 2S small knob in Mexico (Figure 12a) also illustrates a pattern that is repeatedly expressed. This pattern relates to maize containing components of the small knob complex. Knobs of this complex appear in maize of the Central Mesa of Mexico. There can be no doubt that exchanges of maize occurred between inhabitants of the highlands of Guatemala and those of the Central Mesa of Mexico. Both chromosomal constitutions and racial characters give evidence of this. The distribution of these knobs from and about the Central Mesa follows the previously described paths diverging from this center, although the concentration of the knobs is much reduced within the migrant territories.

ORIGIN OF THE "VENEZUELAN COMPLEX" AND THE SIGNIFICANCE OF ITS DISTRIBUTION PATTERN

The knobs appearing in much of the maize of Venezuela suggest that there had been two early introductions preceding the one just described. One entered the western highlands (a spur of the Andes) and the other entered more easterly (Caribbean territory). The sources were related to each other although they differed in some of their basic germplasm components. Their commonly shared chromosomal components resemble some of those now commonly shared between the race Nal-Tel Blanco in Guatemala and Zapalote Chico in Oaxaca and Chiapas, Mexico. Both of these races have a wide assortment of large knobs, and it is these knobs that characterize the Venezuelan complex. They are the large knobs in 2L, 3L, 4L, 5L, 7S, 7L, 8L-1, 9S, and 9L.

Several of the enumerated large knobs have restricted distributions in the Americas, and this simplifies the process of detecting the probable source of the Venezuelan complex. One is the large knob in the long arm of chromosome 9. Its distribution in Mexico (Figure 10a) was discussed with reference to the race Tuxpeño. In South America (Figure 10b), its concentration within the boundaries of Venezuela is conspicuous. It is well represented in maize of the adjacent islands of Trinidad and Tobago, as is each member of this complex. It should be noted that it is present in Costa Rica and Panama, in the two examined races in Colombia (Pollo and Pira), and in several collections from Ecuador. It is also present in the race Perola of Bolivia, but only in this race. In the east it has been introduced into the race Cateto.

Another large knob, with a distribution in northern South America resembling that of 9L, terminates the short arm of chromosome 7 (Figure 13b). Clearly, it is concentrated in Venezuela and adjacent eastern islands. In eastern South America it appears only in four collections of the race Cateto, three from the Guianas, and one from Uruguay. Its near absence in South America below the Venezuela–Roraima territory probably relates to its near absence in Central America. Central America provided much of the maize of the Caribbean Islands, which, in turn, provided much of the maize of eastern Brazil. It also provided the examined maize that entered Ecuador along the Pacific coast. In western South America this 7S large knob is totally absent except in the two marked collections in Bolivia, both of which are of the race Perola. This race must have been introduced into Bolivia quite recently. It has the full complement of large knobs of the Venezuelan complex, and these knobs are not yet integrated into maize of other races grown in its vicinity. I suspect that this race may have been introduced into Bolivia from a location in the highlands of Colombia, possibly close to the border with Venezuela.

The two collections in Panama having plants with the 7S large knob are of related races: (a) Coastal Tropical Flint-Pira and (b) Coastal Tropical Flint–Tusón. The races in the Caribbean Islands that have this knob are related to the Panamanian races, sufficiently so to be given similar racial designations: Coastal Tropical Flint, Coastal Tropical Flint–Tusón, and Tusón. In Mexico it is concentrated in the race Zapalote Chico, but it also is present in the western Mexican Nal-Tel and in its close relatives. Thus, both the 9L and the 7S large knobs associate the germplasm components of the Venezuelan complex with those in these two races from the north.

It is evident that most of the knobs of this Venezuelan complex appear in maize that now is spread over much of Venezuela and the adjacent eastern islands of Trinidad and Tobago. In South America the distinctive germplasm of this complex is not distributed much beyond the territory of Venezuela proper.

This probably reflects the early introduction of maize into the Colombian and Venezuelan territories and the late arrival of maize in territories to the southeast. When the latter territories were populated with maize, it came mostly from the Caribbean Islands, and it entered along the eastern coast at least as far south as the examined collections extend. These introductions occurred at different times and places, bringing with them maize having various different combinations of the basic germplasms. Those having distinctive components of the Venezuelan complex are now distributed to the Catetos of the Guianas, Uruguay, and eastern Argentina, as the maps in Figures 10*b* and 13*b* show.

THE DISTRIBUTION AND INFLUENCE OF
THE "ANDEAN COMPLEX"

In the early study of chromosome constitutions of races of maize from Ecuador, Bolivia, Chile, and Venezuela a result was obtained that seemed extraordinary at the time. With few exceptions, all of the examined races from collections made at high elevations in Ecuador, Bolivia, and Chile, but *not* Venezuela, had the same or nearly the same chromosome constitution. It was a simple one; there was a small knob in the long arm of chromosome 7 that was homozygous, and a small knob at position 3 in the long arm of chromosome 6 that could be homozygous, heterozygous with a knobless chromosome 6, or totally absent. From north to south, the highland territory in which this "Andean complex" prevailed—either exclusively as in Chile, or with few exceptions, as in Ecuador and Bolivia—extended approximately 3000 km. The outstanding exception appeared in the race Perola of Bolivia. The germplasms of most of the other exceptional races were highly diluted by the Andean complex. One of these is the race Pisinkalla of Bolivia. Its plant characters suggest a connection with maize of central Mexico. Support for this relatedness is given not only by the types of its few non-Andean knobs, but also by the presence of the same inversion in the short arm of chromosome 8 that is known to be distributed among maize races of central and western Mexico. A second exception, the race Canguil of Ecuador, also must have a connection with maize from the north. It is one of the two races in Ecuador that were found to have the Abnormal-10 chromosome (Figure 5*b*). The other is Uchima, grown in southern Ecuador at lower elevations.

Other constitutions began to appear in the lower elevations in Ecuador and Bolivia. The races had more knobs distributed among their chromosomes, and these ranged in size from small to very large. In the lowest elevations along the Pacific coast of Ecuador, the races had many large knobs in their chromosomes. Some of them were very large indeed, resembling in this regard some of the knobs of the Venezuelan complex. In the lower elevations in Chile, chromosome constitutions other than those in Ecuador were noted, suggesting distinctly different sources for their germplasms. In each of these countries, some of the races collected from regions of intermediate elevation had chromosome constitutions that suggested origins initiated by introgressions between lowland and highland maize.

The very extended distribution of the Andean complex throughout the highlands of Ecuador, Bolivia, Chile, and Argentina points to an early introduction of one type of maize into the highlands. This maize was then distributed over a vast territory. Unlike the maize of Mexico, whose history is marked by many introgressions with teosintes having distinctly different knob constitutions, this maize suffered no extensive modification of its chromosome constitution until other maize was brought in very much later. Successive introgressions with the initially introduced maize diluted the constitutions of the secondarily introduced maize. Such dilutions are evident in the race Pisinkalla. Thus knowledge of chromosome constitutions of the Andean races makes it possible to project the probable sequence in time of arrivals of introduced maize—a very early introduction that gave rise to the Andean type maize, a later introduction eventuating in the race Pisinkalla, and a recent introduction, the race Perola, not yet modified by introgressions with maize carrying the Andean complex.

INFLUENCE OF THE ANDEAN COMPLEX ON MAIZE OF CENTRAL AND EASTERN SOUTH AMERICA

Andean maize has had an extraordinary influence on maize now growing in central and eastern South America. Its chromosome constitution appears unmodified, or nearly so, in some maize of Paraguay and of the Mato Grosso of Brazil. In eastern Bolivia, parts of Paraguay, and regions south and west of this central area, Andean maize has come in contact with maize migrating inland from the Atlantic coastal areas, or directly introduced into previously isolated areas populated only by maize with the Andean complex. It is evident that Andean maize also has influenced maize of northeastern Brazil and as far as the Atlantic coast, as if the Andean maize had been brought from the highlands to the lower lands along river routes. Introgressions that must have occurred of this Andean maize with maize from the east would be responsible for new character expressions, and thus for the appearance of incipient new races.

That much of the maize in Brazil, Paraguay, and Argentina has been influenced by Andean maize is revealed by the wide prevalence of the knobless state in all chromosomes except chromosomes 6 and 7, and by the exceptionally wide distribution of the Andean type small knob in 7L. The distribution of this knob is shown in Figure 14*b*.

Figure.14*a* reveals an uneven distribution of the 7L small knob in North America. It is concentrated in the highlands of Guatemala, in Costa Rica, in northwestern Venezuela, and in the Caribbean Islands. Since it is a component of the small knob complex, its distributions are typical of members of this complex. They include the introductions into central Mexico from the highlands of Guatemala, as previously described. On the map the triangles represent collections of the race Tuxpeño that exhibited this 7L small knob. It may be noted that triangles appear not only in the east, but also in west Mexico in the states of Michoacan and Colima. Many of the knob distribution maps suggest the source of these westerly located Tuxpeños. They relate to the area in northern Veracruz where the triangles are located (see also Figure 9.)

The Andean·complex is exceptional in having so few knobs—only two—each of which is small. Present information would suggest that the initial introduction into the Andes came from the highlands of Guatemala. Not only is this region the center of the small knob complex, as defined earlier, it also is the center of the knobless complex. Maize of the Guatemalan highlands strikingly exhibits both these complexes. Morphologically, Andean maize resembles some maize types peculiar to the Guatemalan highlands. Anthropological considerations, along with those of chromosome constitutions and morphological and genetic characters, could point directly to the source.

CONCLUDING STATEMENT

The purpose of this chapter is to indicate by specific examples the usefulness of knowledge of chromosome constitutions in helping to resolve some of the problems associated with the origin of maize and its present-day races. This has involved reconstructions of its biological history, which often must have been associated with man's migration patterns and his commerce. As the discussions indicate, anthropological projections are central to the reconstructions. This association applies equally well to the maize of the United States that has not been specifically mentioned here. Relationships are fascinatingly revealed by the chromosome constitutions—the southwest maize with races of central and western Mexico (see Figure 8 and its legend), the northern maize with that of the highlands of Guatemala, the southeastern maize that was constructed from combinations of northern United States maize with some of the developed races of central and eastern Mexico, and so on.

In this chapter emphasis is placed on the importance of Mexican teosintes in supplying the basic germplasm components that were followed in these studies.

It should be emphasized again that it is the Mexican teosintes, not the Guatemalan ones, that have played the major role in the history of maize, suggesting that maize started its expansive development in those areas of Mexico where teosintes prevailed, that is, central and southwestern Mexico.

ACKNOWLEDGMENTS

Those of us who participated in these studies are indebted to the Rockefeller Foundation and to the International Maize and Wheat Improvement Center (CIMMYT) for the support these organizations provided. We also are particularly indebted to Dr. E. J. Wellhausen, formerly Director of CIMMYT, for his continuing support and enthusiastic responses as the work progressed. Personally, I am indebted to Drs. David H. Timothy, Lewis M. Roberts, William H. Hatheway, and Ing. Ricardo Ramirez E. for their generous collaboration during the initial phase of the study.

REFERENCES

Blumenschein, A. 1973. Chromosome knob patterns in Latin American maize. In *Genes, enzymes and populations*, A. M. Srb (ed.). New York: Plenum.

Kato Y., T. A. 1975. Cytological studies of maize (*Zea mays* L.) and teosinte (*Zea mexicana* Schrader Kuntze) in relation to their origin and evolution. Amherst, Mass.: *Mass. Agric. Exp. Sta. Res. Bull.* No. 635.

Longley, A. E. and T. A. Kato Y. 1965. Chromosome morphology of certain races of maize in Latin America. Chapingo, Mexico: *Internat. Center for Improvement of Maize and Wheat, Res. Bull.* No. 1.

McClintock, B. 1933. The association of non-homologous parts of chromosomes in the mid-prophase of meiosis. *Zeitsch. f. Zellforsch. u. mikro. Anatomie* **19**: 191.

McClintock, B. 1959. Genetic and cytological studies of maize. *Carnegie Institution of Washington Year Book* No. 58: 452.

McClintock, B. 1960. Chromosome consitutions of Mexican and Guatemalan races of maize. *Carnegie Institution of Washington Year Book* No. 59: 461.

Ting, Y. C. 1958. Inversions and other characteristics of teosinte chromosomes. *Cytologia* **23**: 239.

Wellhausen, E. J., L. M. Roberts, E. Hernández X., in collaboration with P. C. Mangelsdorf. 1952. *Races of maize in Mexico*. Cambridge, Mass.: Bussey Inst., Harvard Univ., p. 223.

For extensive literature references to races of maize, see Brown, W. L. and M. M. Goodman. 1977. Races of Maize. In *Corn and corn improvement*, 2nd ed., G. F. Sprague (ed.). American Society of Agronomy.

BREEDING

Chapter 12

INTRODUCTORY REMARKS TO THE SESSION ON CORN BREEDING

J. H. Lonnquist

University of Wisconsin, Madison

The history of applied science is marked by successive steps forward as basic knowledge and the technology required for advance have been expanded and utilized. In retrospect, the forward steps often may appear to have been rather small in magnitude. Yet at the time of their initial exposure they were generally considered to have been highly significant. Advances in productivity potential in maize were almost always, I think, based upon increases in understanding of the underlying principles of the genetic and biological aspects of the crop as well as those of plot techniques and statistical refinements needed for the detection and evaluation of small differences among genotypes in field trials. There were times when important guidelines from the theoretical quantitative geneticists needed for advances in productivity potential were not available prior to realized advances. Successful development of widely used corn hybrids resulted more from the sheer weight of effort directed toward their development than from well-defined procedures based on application of theoretical developments from the quantitative geneticists.

Following the spectacular gains from the first cycle of hybrid development in the U. S. Corn Belt, progress slowed appreciably. Differences of opinion developed relative to the underlying reasons for the apparent impasse. A quantitative geneticist by the name of R. E. Comstock, heading a team of researchers at North Carolina State University (Raleigh) in the late 1940s, set out to characterize and quantify the genetic variances for various agronomic characters in maize. This group and their students subsequently provided the necessary theoretical development and guidelines for future trends in maize improvement as well as for many other crop species.

The trends in increased productivity potential of the maize crop have continued through further elucidation of issues that appeared to be important. At different research centers results of various selection procedures reflecting the utility of application of information gained from quantitative genetic studies have been of considerable interest. Progress has been made in the improvement of yield, resistance to major pests, and in modification of other agronomic traits of importance to corn producers. The end is still not in sight. Important research findings are being reported constantly. These developments generate continuing interest in breeders looking for materials and methods useful in the further improvement of the crop.

The use of cytoplasmically sterile stocks for production of hybrid seed was dealt a severe blow by the occurrence of southern corn leaf blight (*H. maydis*) in 1970. Uniformity in any major crop, whether genetic or cytoplasmic in nature, constitutes a serious risk. The use of cytoplasmically male-sterile cytoplasms, although extremely helpful to seed producers, was demonstrated to be a major risk to corn production generally, when limited to a single source. Uniformity in cytoplasms (cytosteriles) as well as uniformity in background genotypes leads to genetic vulnerability and can be a serious deterrent to continued maximum productivity. Methods of circumventing the potential dangers in the use of cytoplasmic male sterility are known, but needed materials have not been located or satisfactorily developed to provide immediate practical usage. Considerable effort is being directed toward the solution of problems in this area.

Nutritional and industrial quality in maize have been of interest over an extended period of time. The best-known example is the long-time selection for divergent quantities of oil and protein in the maize endosperm. The work begun by C. H. Hopkins at the University of Illinois toward the end of the 19th century continues today. It is the longest continuous selection program in history and is of considerable theoretical interest to quantitative geneticists.

The discovery of waxy endosperm in maize led to the development of hybrids for production of waxy maize to supply a specialized starch product for the food industry. More recently, interest has been shown in the possibilities for developing high amylose starch types for the production of edible films and fibers. With the increasing interest in protein malnutrition in the world, the discovery of mutants that altered the balance of essential amino acids in the maize endosperm created interest in genetic improvement for nutritional qualities in all major cereals. The search for new mutants of nutritional value in maize and the study of interactions of various mutants of long standing have received intensified emphasis in recent years. Much remains to be done in this area and intensified research programs are being directed toward solutions of the current problems. With continued advances in technology increased opportunities exist for the geneticist and breeder to study new types of variations in the endosperm of maize. Maize may soon be more than just an energy source in man's search for food.

Chapter 13

QUANTITATIVE GENETICS IN MAIZE BREEDING

Ralph E. Comstock

Department of Genetics and Cell Biology,
University of Minnesota,
St. Paul

The focus of this presentation is on quantitative genetics in relation to the decisions required in the design of a maize-breeding program. The purpose is to identify: (a) significant contributions made by quantitative genetics in the past and (b) questions important to maize breeding that are in the realm of quantitative genetics and remain to be solved.

The difference between quantitative and qualitative traits resides in the relative magnitudes of allele substitution effects. If the effect of substituting one allele for another is large relative to total phenotypic variation, the trait is qualitative. If such substitution effects are small, the trait is quantitative. If effects are small relative to total variation, it is because the trait is affected by numerous genes or because a portion of the phenotypic variation is environmental in origin. Thus the concepts, heritability and multiple factors, unique to quantitative genetics are inherent in the nature of quantitative characters.

The boundary between quantitative genetics and population genetics is obviously and properly indistinct. In this chapter (and in other published works)

191

I use the term *quantitative genetics* to include all aspects of population genetics required for understanding the variation of quantitative traits.

Historically, the first question of quantitative genetics was whether the inheritance of continuously distributed traits was Mendelian. Evidence obtained from maize by East and Hayes (1911), Emerson and East (1913), and others was an important portion of the total that led, in the second decade of this century, to the rejection of the "blending" inheritance hypothesis and to general agreement that Mendelian principles apply to quantitative as well as other traits. Then Jones (1917), a maize breeder, rationalized observations on heterosis and inbreeding depression in terms of a Mendelian model that involved genetic linkage. Thus maize geneticists made important contributions to the shaping of the general model that has been employed in quantitative genetics through the last six decades, the model that embraces the multiple-factor hypothesis, genes located in chromosomes and hence sometimes linked, and incomplete heritability.

To my knowledge the first work that related a theoretical concept from quantitative genetics to a maize-breeding problem was the estimation by Sprague and Tatum (1942) of the components of genetic variance, among crosses of inbred lines, due to general and specific combining ability. It was inevitable that more research involving ideas from quantitative genetics would follow, but it was Fred Hull (1945) who hurried the process by bringing a critical issue, namely, level of dominance, into sharp focus. He proposed, and argued persuasively for, overdominance at segregating loci pertinent to grain yield as the appropriate explanation for the total of things known and presumed known concerning grain yield genetics. In addition, he emphasized that if his proposal were correct, new approaches in maize breeding would be required, that at best the then current methods could not achieve the full genetic potential of maize and, at worst, they might achieve little beyond what had already been achieved. This concise delineation of the implications concerning effective breeding systems made testing of the overdominance hypothesis imperative and urgent. Quantitative genetic investigations came quickly and have since been strongly identified with maize breeding.

STATE OF KNOWLEDGE IN RELATION TO PROBLEMS OF THE BREEDER

Recurrent Selection as a Frame of Reference

The species, *Zea mays L.,* is a germplasm pool in which there may be various alleles of each of the thousands of genes in the genome. The number of possible genotypes is very, very large but, with respect to any specified measure of value, one will be the best, although quite a few others are likely to be almost as good. The synthesis, identification, and multiplication of such superior genotypes is the ultimate goal of maize breeding.

Perspective concerning this task is obtained by recalling that in general the probability of a genotype increases when the frequencies of its component alleles are increased but that because the number of genes with effects on value of genotype is large, the probability of a specific superior genotype will be infinitesimal unless the frequencies of almost all required alleles are close to one. (Consider the fact that 0.9^{200} is less than one in 1 billion.) Fortunately, allele frequencies can be modified by selection and, whereas expected allele frequency increments per generation are small when numerous segregating genes are involved, major changes can be affected by recurrent selection continued through a number of generations.

Because recurrent selection appears to be an essential element of any program for developing the best possible genotypes from the allele resource of a species, I have chosen to discuss quantitative genetics in terms of the issues having to do with the potential effectiveness of recurrent selection programs.

The more obvious things to be said about recurrent selection programs include the following.

1. The potential product is limited by the allele content of the foundation stock employed.

2. The selection criterion employed should be such that all useful alleles are favored by the selection practiced. This means that the selection criterion must be consistent with the goal, that is, with the specific definition of phenotypic value, for the program concerned.

3. Desired alleles present in the foundation stock can be lost by random drift, even though favored by selection, if effective population size is not adequate.

4. Given a substantial amount of overdominance, the recurrent selection program should involve selection for combining ability between two populations. Unless it were safe to assume no partial dominance, this should be reciprocal recurrent selection (Comstock, Robinson et al. 1949).

5. Simple epistasis has no special bearing on the effectiveness of recurrent selection. However, if the rank order values of alleles at any locus depend on genotypes present at one or more other loci (i.e., given multiple-peak epistasis), certain alleles required in the best possible genotype may never be favored by selection. The result would be that the ideal genotype would not be achieved even though population size were sufficient to minimize the consequences of genetic drift.

Major Specific Problems

The Goal.   The goal has two dimensions: (a) the definition of phenotypic value (net worth) and (b) the specification of the target population of environments, that is, the environments for which value of genotype will be sought.

The definition of phenotypic value reflects decisions concerning values to be attached to unit variations in yield, amounts of oil and amino acids in the grain, ear height, and other variables. These decisions flow from socio and economic, and not genetic, considerations.

The genetic issue involved in choice of the target population of environments is genotype–environmental interaction. The operational problem is the number of selection programs required for the development of genotypes well adapted to all environments that are significant for maize production. The target population of environments for each program should be homogeneous enough so that the best allele of each gene is the same for all environments of the population. The dilemma is to have enough programs (recurrent selection populations) to achieve that end and at the same time to have effective population size large enough in each program so that no favorable alleles are lost through random drift. The genetic basis for good decisions is information concerning: (a) adequate population size and (b) genotype–environment interactions. The first of these is discussed on pp. 198–201.

A great deal of information is available concerning genotype–environment interaction as a component of the deviation of phenotypic from genotypic values (Sprague and Federer, 1951; Rojas and Sprague, 1952; Comstock and Moll, 1963). However, very little is known about the kinds of environmental differences responsible for changes in the rank-order values of alleles of the same gene. It is apparent that the values of early and late maturity alleles are reversed by the differences between short- and long-season environments, but this may be the only clear case. Whether, for example, alleles responsible for disease or drought resistance become unfavorable when the pathogen is absent or moisture is plentiful is not nearly so certain.

Foundation Stocks.   Maize breeders understand this problem. The species is not a single panmictic population. It is composed instead of subpopulations that have evolved under diverse sorts of selection pressures and among which amounts of migration have been variable. From which of these many subpopulations should inputs to a foundation stock be made? Inbred lines with high general combining ability for the target environments and open-pollinated varieties well adapted to the target environments obviously carry useful alleles. But when these have been incorporated, will favorable alleles be lacking that are present elsewhere in the species? It is widely believed that any subpopulation may possess one or more favorable alleles that are absent in all or many other subpopulations. On the other hand, there is a genetic dogma that says all alleles arise repeatedly by mutation and hence will be present in frequencies that may be high or very low in every population for which effective size has been large through many generations.

A limited but growing body of direct information is available. I am thinking of such data as that on the frequencies of blood antigen alleles in diverse races of man and on various allele frequencies determined by electrophoresis in different populations of such cross-fertilizing species as *Drosophila* and mice. A comprehensive review of such data should obviously be interpreted with reference to the problem we have been discussing. Results from the electrophoretic exploration of allele frequencies in races and varieties of maize referred to by Goodman (1978) should be especially significant in this connection.

In practice, the conservative approach to ensuring the presence of most favorable alleles is to establish foundation stocks with inputs from numerous diverse sources. Unfavorable aspects of that approach are low performance at the outset, the correlated long interval before products having practical utility can be derived by recurrent selection, and the possible loss of favorable alleles gained from exotic sources before they can be disentangled from linkage association with unfavorable alleles at other loci. On the favorable side, theory and simulation information from quantitative genetics can now provide useful guides for the necessary decisions concerning population size and selection strategies.

The Selection System. The first and major decision that is required concerning selection systems is whether selection for combining ability with a specific population is to be employed. The critical genetic issue is the importance of overdominance. As first indicated by Hull (1945), such selection must be employed to achieve the full potential of a gene pool if overdominance is involved. The second and less critical decision concerns the specific selection system to be employed. For example, there are various ways in which either reciprocal recurrent selection or selection based on intrapopulation performance can be conducted.

Quantitative genetics has made two major contributions in relation to choice of selection system: (a) it established that overdominance is not a major aspect of the genetics of grain yield and (b) it has provided a theoretical basis for predicting rates of genetic improvement by almost any selection system.

In response to the challenge posed by Hull's overdominance hypothesis, Comstock and Robinson (1948, 1952) described experimental procedures for obtaining information on levels of dominance and set out the associated genetic logic. These procedures were employed in extensive research programs at the North Carolina and Nebraska agricultural experiment stations. Data published by Robinson, Comstock, et al. (1949, 1955), Gardner and Lonnquist (1959), Lindsey, Lonnquist, et al. (1962), Gardner (1963) and Moll, Lindsey, et al. (1964) were consistent and provided strong evidence that overdominance is not an overriding feature of grain yield genetics. Corollaries for the breeder were that:

1. If selection for specific combining ability is to be practiced, it should be reciprocal recurrent selection rather than selection for combining ability with an inbred line.

2. Selection based on intrapopulation performance should be effective rather than ineffective as many breeders then believed.

Substantial supporting evidence has since been obtained from recurrent selection programs (Gardner, 1961, 1968; Moll and Stuber, 1971).

Lush (1947) derived formulas for predicting response to individual selection, family selection, and combinations of the two. His derivation employed the idea of regression of "breeding value" on phenotype or family mean phenotype. Comstock, Robinson, et al. (1949) showed how approximate expressions for change in allele frequencies resulting from selection and change in genotypic mean per unit change in allele frequency can be obtained and employed together to obtain formulas for expected response to selection. Empig, Gardner, et al. (1972) applied this procedure to obtain prediction formulas for many selection systems that have been used or proposed for use in maize breeding. Comstock (1964) pointed out that many system comparisons can be made most easily using formulas for expected change in allele frequencies and Jones, Compton, et al. (1971) followed that procedure in their theoretical comparison of full-sib and half-sib reciprocal recurrent selection. We see later (pp. 199–200) that expressions for expected change in allele frequencies serve another significant purpose as well.

In addition to providing a basis for choice between selection systems, the available prediction formulas facilitate decisions concerning details of the chosen system. Specifically, they enable evaluation of: (a) variations having effects on phenotypic variance of the selection criterion and (b) inbreeding prior to progeny testing in systems that use progeny testing.

Empig, Gardner, et al. (1972) specified that their results assume random mating populations, no multiple alleles, no epistasis, and independent assortment of genes (linkage equilibrium would have been sufficient); much of the published prediction theory is subject to these same assumptions. It is thus reasonable to question whether this theory is useful to practice. Actually, more is known than has been published, although there is more theoretical work that is needed. On the basis of my own unpublished work and the indications from numerous selection experiments I consider that available theory has general applicability far beyond what is suggested by assumptions stated above.

It should be noted that theory discussed above enables a comparison of selection systems in terms of expected response per unit of time. However, in the context of recurrent selection programs we should probably be more interested in comparisons based on response limits (probabilities of eventual fixation of favorable alleles). Some such work has been done by Hill (1970) and by Vencovsky (1975), but more can and should be done in the context of maize breeding.

The Selection Criterion.   When net worth (phenotypic value) is a function of more than one trait, the rule in plant breeding, the relative weights to be placed on various traits in the selection process, pose a major problem. The importance of this problem is especially obvious when considered in the context of recurrent selection. More than optimum emphasis on one trait (or one set of traits) may cause unfortunate correlated responses in others. In the end, such correlated responses would doubtless lead to loss of alleles that are favorable with respect to net worth.

The idea and theory for an optimum selection criterion were set out by Smith (1936) and Hazel (1943), and an early example application for maize was offered by Robinson, Comstock, et al. (1951). Despite the fact that selection index theory has been available for a long time, it is my impression that it has not been applied very generally in maize breeding. The following are some probable reasons.

1. The total data required for construction of a Smith–Hazel index is costly to obtain.
2. The Smith-Hazel theory assumes a single stage selection procedure with all traits of the index measured for each selection unit whereas, in practice, there are cost reasons for using a two- or three-stage procedure in which selection units are screened first on the basis of inexpensive measures and last on traits most expensive to measure.
3. Uncertainty concerning the effects of sampling errors in genetic parameter estimates on the real utility of a selection index and the attendant uncertainty concerning amounts of data required as the base for a high-efficiency index.

Data collection costs can be reduced by use of long-cycle selection systems, although these may suffer when judged in terms of expected genetic progress per year. Comparisons in terms of progress per year per unit of cost seem desirable. Young (1964) considered multiple trait selection in two-stage systems, but I have not studied his manuscript and am not aware that practical implications have been interpreted to plant breeders. Harris (1964) studied the impact of sampling errors of parameter estimates in the case of two traits having equal heritabilities but stated that it was not clear to him on the basis of his results, that inferences to more complex cases would be justified.

It appears to me that selection criterion problems deserve more attention by quantitative genetics and that useful information can be produced. In particular, the long-term goal of recurrent selection suggests our need for a selection criterion that is optimal with respect to the probability of fixation of favorable alleles rather than genetic gain per unit of time.

Population Size.   The complete genetic potential of a foundation stock can only be realized when no favorable alleles are lost by random drift. We must ask, therefore, what population size is required to make the probability of ultimate fixation approach 1.0 for each favorable allele.

For the case of natural selection among individuals in a randomly cross-fertilizing population, Kimura (1957) obtained an expression that is general with respect to levels of dominance other than overdominance. In the special case where there is no intralocus interaction of alleles, that is, no dominance, his expression for the probability of fixation of an allele becomes

$$P = \frac{1 - e^{-2Nsq_0}}{1 - e^{-2Ns}} \tag{1}$$

where N = effective population size; s = the difference in relative fitness between homozygous genotypes at the single locus; q_0 = the initial frequency of the allele; and e = the base of the natural system of logarithms (exp).

In addition to no dominance, this expression assumes only two alleles per locus, that both N and s are constant from one generation to another and that the gene assorts independently from all other genes for which $s \neq 0$. It is worth adding that partial or complete dominance causes only quite small changes in P. Although developed for natural selection, the formula works as well for artificial selection when s-values are employed that are appropriate functions of the mode and intensity of selection to be practiced.

Whereas the assumptions to which the formula is subject may not be completely valid for any gene in a real live population, I will argue that Equation (1) is extremely useful as a guide to good decisions concerning population size. Consider a gene that meets all of the specifications (two alleles, no dominance, etc.) and for which the difference between homozygous genotypes in effect on net worth is 1% of the population mean. Now suppose that $N = 120$ is required to make $P = 0.95$ if $q_0 = 0.10$. My position is that this would be enough information to convince me, as a breeder, that population size in any recurrent selection program with a broad foundation stock should be at least 120.

The choice of reasonable values of s and q_0 deserves some attention. Values for s are considered first. Making the assumptions required for Equation (1), the approximate expected change in frequency of a favorable allele from one generation to the next is

$$E(\Delta q) = \frac{s}{2} q(1 - q) \qquad (2)$$

This familiar expression is found in many books. On the other hand, the procedure used by Comstock, Robinson, et al. (1949) can be applied to produce an approximation to $E(\Delta q)$ as a function of k, σ, u, and q, where k = the selection differential in standard units; σ^2 = the variance of the selection criterion; and u = one-half of the difference in genetic effect on the selection criterion of the two homozygous genotypes for the gene in question.

In some instances the inbreeding coefficient (F) may be a part of the expression. As an example, consider any of the systems in which individuals (represented by S_1 progenies) are selected on the basis of the performance of noninbred progenies. Then, making the same assumptions as for Equation (2), the following is obtained

$$E(\Delta q) \approx \frac{k}{2\sigma} q(1 - q)u(1 + F) \qquad (3)$$

where F = the inbreeding of the progeny-tested plants. Now if Equations (2) and (3) are equated and the resulting equation solved for s, we obtain

$$s \approx \frac{ku}{\sigma}(1 + F) \qquad (4)$$

The above procedure was used by Falconer (1960) to show that $s \approx (2ku/\sigma)$ in the case of mass selection (in both sexes, i.e., among both pollen and seed parents), but he did not emphasize its applicability to other sorts of selection. In order to obtain numerical values of s from such expressions ku/σ must be evaluated. Since the approximation of k as a function of selection intensity has become standard procedure, only u/σ requires special attention.

Let $\bar{x}$ represent the population mean of the selection criterion. Then $C = 100 \, \sigma/\bar{x}$ is the coefficient of variation of the selection criterion, and $E = 100(2u/\bar{x})$ is the effect as a percent of the mean of substituting the favorable for the unfavorable allele in homozygous genotypes. Dividing the latter by the former, we obtain

$$\frac{u}{\sigma} = \frac{E}{2C} \qquad (5)$$

Providing a value of C that is reasonable for the selection system being considered should never be a problem but realistic values for E are another matter. We know very little concerning the proportions of potential genetic gain that are inherent in genes for which E is in the range 0.5–1.0%, the range 1–1.5%, and so on.

However, there are some bits of rationale and evidence that can be brought to bear. First, the obvious physiological complexity of net worth suggests that it must be affected by a very large number of genes, and if a finite potential for genetic gain must be divided among many genes, the potential associated with single genes must be small. Second, the amounts of linkage indicated by yield data in the design III studies of Moll, Lindsey, et al. (1964) suggest that many loci were segregating in their populations originated from pure-line single crosses. Broader based populations should have larger numbers of segregating genes. In my view we would be distinctly imprudent if we did not assume that a substantial portion of the genetic gain potential resides in genes for which E is less than 1%.

In the case of q_0, our interest centers on the lowest values to be considered since the N required to achieve a specific probability of fixation, increases as q_0 decreases. The most obvious facts to be recognized are that foundation stocks are established by interbreeding materials from different sources, that the various sources are sampled because each may possess favorable alleles that the others lack, and hence that values of q_0 must be contemplated that are no higher than the lowest fraction of the foundation stock obtained from any single source. For example, in the case of a synthetic originated with equal contributions from n lines, q_0 values as low as $1/n$ should be considered.

To provide a little perspective concerning adequate effective population size, some values obtained using Equation (1) are shown in the table that follows. They apply for selection among noninbred plants on the basis of the performance of noninbred progenies in a system such that $C = 7.5$ and $k = 1.5$. Equation (4) provided the s values, shown in parentheses following E values.

		$E(s)$		
P	q_0	2(0.2)	1(0.1)	0.5(0.05)
0.9	0.1	58	115	230
	0.05	115	230	460
0.95	0.1	75	150	300
	0.05	150	300	600

It is worth noting that P is a function of Ns [see Equation (1)] and that s is increased by 50% when S_1 instead of S_0 plants are tested [see Equation (4)].

Therefore, all required $N's$ values would be $\frac{2}{3}$ of those listed if one generation of self-fertilization were introduced before making test crosses.

The foregoing is approximate and incomplete but suffices, I think, to demonstrate that quantitative genetics has supplied a very useful new way to examine the critical population-size problem. At the same time, important questions remain. Those that seem to me most important are touched on below.

Linkage is guaranteed when the number of genes is large. Robertson (1970) made a very extensive study of linkage effects given linkage equilibrium when selection is initiated. His evidence indicates that maximum reductions in fixation probabilities due to linkage are probably about 10%. On the other hand, simulation studies at Minnesota, Ho (1973) and Cardellino (1975), indicate, as many of you would have predicted, that the impact of linkage can be much greater when there is initial linkage disequilibrium. More information, together with interpretation in the context of population size requirements, is needed.

The logic and process for obtaining numerical values of N have been discussed at length (Crow and Kimura, 1970). However, appropriate working formulas for each of the selection systems potentially useful in maize breeding have not been made readily available. This can and no doubt will soon be accomplished. Hill (1970) pointed out that in reciprocal recurrent selection, s is not constant when there is any level of dominance. The impact of this relative to population size strategy seems (to me) worthy of further attention. Finally, more information concerning the distribution of u/σ in the context of net worth is sorely needed.

What to Do When Plateaus Have Been Reached.   Consider recurrent selection based on intrapopulation performance, assuming that the selection criterion used is optimum at all times. When there is no further response to selection (plateau reached) all genes should fit into one of four classes. These are: (a) nonoverdominance, the best allele fixed, (b) overdominance, segregation involving the alleles of the best heterozygous genotype, (c) nonoverdominance, a less favorable allele fixed, and (d) overdominance, one or both of the alleles of the best heterozygote absent from the population. Inbreeding can be used to test for class (b) genes; their presence will be indicated by inbreeding depression in net worth. With the exception of loci where one of the alleles of the superior heterozygote is lethal or very deleterious relative to seed production, class (b) genes can be dealt with by reciprocal recurrent selection. They are not considered further here.

If the genetic state of the plateaued population were such that all genes were in class (a) or (b), the breeder would have achieved his goal save for the RRS phase required to deal with class (b) genes. The tough problems posed by classes (c) and (d) are discussed in terms of class (c). How can the breeder judge whether a nontrivial number of useful alleles are missing from his plateaued population? Information is provided by phenotype; for example, an inadequate root system or susceptibility to significant pathogens or parasites. However, this must inevitably be incomplete information. In the past it has always been tacitly assumed, almost certainly correctly, that even the best improved populations were not very close to the full genetic potential of the species, that numerous useful alleles existing somewhere in the species were lacking in each of the improved populations. The two most obvious reasons why this would be true if an optimum selection criterion had been used are either that the alleles were not present in the foundation stock or were lost during the course of recurrent selection because of inadequate effective population size.

The addition of alleles not present in a plateaued population requires either outcrossing or mutation. It has been conventional to assume that the deficiencies of plateaued populations can be corrected by appropriate outcrosses, followed by renewed recurrent selection and that repetitions of this process will bring an ever closer approach to full genetic potential of the species.

The problem associated with the outcross-reselection process is that the better the plateaued population, the more difficult further improvement becomes. Let A be a high-performance plateaued population, B be a population or mixture of populations to which A will be outcrossed, and A′ be the product of the reselection program. Let n be the number of loci at which the best allele is homozygous in A but lacking or at low frequency in B, and let m be the number at which the best allele is lacking in A but present (although not necessarily homozygous) in B. Completely successful reselection would fix the favorable allele at all $n + m$ loci. No genetic gain in A′ (relative to A) would be achieved unless the number of losses at loci of the first set was less than the number of loci of the second set at which the best allele reached homozygosity in A′. If a net gain were made, the next time around the difference between n and m would be greater and the chance of success less. To add to the difficulty, as the difference in genetic value between the progressively improved population and those that are potential sources of additional useful alleles becomes greater, the problem posed by linkage disequilibrium becomes greater, and population size required for any degree of success becomes greater. It seems clear that a point will be reached where, although useful alleles exist in the species that are not present in the best population, the cost of incorporating them would be prohibitive. And it appears almost certain that the time required to achieve that limit would be much less if all germplasm sources considered worth including in a sequence of outcrosses were incorporated in the original foundation stock with population size made adequate to capitalize on the potential of such a broad foundation.

It goes without saying that the practical payoff from recurrent selection applied to so broad a base would be relatively slow in coming and that interim objectives would require parallel programs based on foundation stocks having less long range potential but higher average genotypic value at the outset.

It should be quite clear that there is much more that quantitative genetics can do to improve general understanding in this area.

Anticipation of Goal Changes.   The flaw in the frame of reference that I have employed is the probability of goal changes. These can occur for various reasons, but it will suffice to note that there are a number of possible reasons for change in the target population of environments, for example, new pathogens or new variants of old ones, or new optima for fertilization or plant density. If a goal change were responsible for changes in rank order values of alleles, then alleles fixed or brought to high frequency by recurrent selection in the past might not be "best" alleles relative to the new goal.

To my knowledge, quantitative genetics has produced no supplements to conventional wisdom concerning this issue. My own thoughts are as follows:

1. In total, the world needs for maize (different environments, variations in product quality requirements, etc.) are sufficiently diverse to require special goals in a substantial number of separate selection programs. Consequently, the genetic populations produced in these programs would be variable in alleles brought to fixation and taken together would constitute a considerable germplasm resource. In addition, if in each of the world regions in which different programs are conducted a sample of the foundation stock were maintained with large population size but no artificial selection, the total "store" of alternative alleles should be of very comfortable magnitude.

2. Spontaneous mutation in the total of populations envisaged in (1) cannot be discounted as a source of useful alleles. Beyond that, presently ongoing efforts to develop methods in molecular biology for increasing the potential of induced mutation seem likely to become fruitful relative to the production and selection of specific new special-value alleles.

SUMMARY

The most significant contributions of quantitative genetics relative to maize breeding have, in my view, been the following.

1. Development of a theoretical basis for comparing selection systems in terms of predicted response to selection and for obtaining the parameter estimates required by the prediction formulas.

2.   Demonstration that overdominance is not characteristic of a majority of the segregating genes that affect grain yield, and hence that selection for specific combining ability with a homozygous line cannot be expected to achieve the full genetic potential of maize. A related contribution was the description of a system, reciprocal recurrent selection, that should be effective relative to both overdominant and nonoverdominant genes.

3.   Development of approximate theory for the probability of fixation of favorable alleles that enables significant insights relative to optimum effective population size in recurrent selection programs.

4.   Theory relating to optimum weighting of different traits in a selection criterion. Despite problems encountered in the application of this theory, it provides important insights.

Areas in which further research is needed and significant new contributions appear most likely are first, improvement of the basis for decisions concerning population size. More information is required on the impacts of linkage and linkage disequilibrium in the sorts of situations significant to the maize breeder; some of this can be obtained by computer simulation. Better information concerning the number of genes affecting net worth and the magnitudes of associated allele effects is also needed. No close approach to complete information on these matters can be expected but more certain insights than we presently have should be possible. Second, some of the selection criterion problems deserve more attention, both because they are critical and because it appears that some useful things can be accomplished. Whether the criterion that is optimum in terms of genetic gain per cycle of selection is also optimum in terms of probabilities of fixation of useful alleles should be determined. Finally, the question of whether multiple peak epistasis is significant relative to potentials of the *Zea mays* germplasm pool deserves investigation by procedures effective for the purpose. If an important amount of such epistasis were discovered, procedures for coping with it would need to be developed. Such procedure has not yet been described.

Present insights and evidence from quantitative genetics suggest to me that the greatest present and future need in maize breeding is for sound recurrent selection programs designed to be effective in relation to the long-range goal of achieving very high frequencies of the most useful alleles. Effective population sizes substantially larger than those characteristic of most present programs and collection of more data to enable more nearly optimum selection criteria would both be required. In the interest of achieving such changes it appears desirable to reduce the total number of separate programs as much as justified by actual genotype–environment interaction and the truly appropriate variation in phenotypic goals. This would undoubtedly necessitate a higher level of cooperation and coordination of effort among maize breeders.

REFERENCES

Cardellino, Ricardo, 1975. Personal communication.

Comstock, R. E. and H. F. Robinson. 1948. The components of genetic variance in populations of biparental progenies and their use in estimating the average degree of dominance. *Biometrics* **4**: 254–266.

Comstock, R. E., H. F. Robinson, and P. H. Harvey. 1949. A breeding procedure designed to make maximum use of both general and specific combining ability. *Agron. J.* **41**: 360–367.

Comstock, R. E. and H. F. Robinson. 1952. Estimation of average dominance of genes. In *Heterosis*. Ames: Iowa State College Press, pp. 494–516.

Comstock, R. E. and R. H. Moll. 1963. Genotype–environment interaction. In *Statistical genetics and plant breeding*. National Research Council Publication No. 982, pp 164–194.

Comstock, R. E. 1964. Selection procedures in corn improvement. Washington, D. C.: *Proc. 19th Annu. Hybrid Corn Ind. Res. Conf.* pp. 1–8.

Crow, James F. and Motoo Kimura. 1970. *An introduction to population genetics theory*. New York, Evanston, and London: Harper and Row.

East, E. M. and H. K. Hayes. 1911. Inheritance in maize. *Conn. Agr. Exp. Sta. Bull.* No. 167.

Emerson, R. A. and E. M. East. 1913. The inheritance of quantitative characters in maize. *Nebr. Agr. Exp. Sta. Res. Bull.* No. 2.

Empig, L. T., C. O. Gardner, and W. A. Compton. 1972. Theoretical gains for different population improvement procedures. *Nebr. Agr. Exp. Sta. Bull.* No. M26 (revised).

Falconer, D. S. 1960. *Introduction to quantitative genetics*. New York: The Ronald Press.

Gardner, C. O. 1961. An evaluation of effects of mass selection and seed irradiation with thermal neutrons on yield of corn. *Crop Sci.* **1**: 241–245.

Gardner, C. O. 1963. Estimates of genetic parameters in cross-fertilizing plants and their implications in plant breeding. In *Statistical genetics and plant breeding*. National Research Council Publication No. 982, pp. 225–248.

Gardner, C. O. 1968. Mutation studies involving quantitative traits. *Gamma Field Symposium No. 7. The present state of mutation breeding*, pp. 57–77.

Gardner, C. O. and J. H. Lonnquist. 1959. Linkage and the degree of dominance of genes controlling quantitative characters in maize. *Agron. J.* **51**: 524–528.

Goodman, Major M. 1978. In *Maize breeding and genetics*. D. B. Walden (ed). New York: John Wiley & Sons. Chap. 10.

Harris, D. L. 1964. Expected and predicted progress from index selection involving estimates of population parameters. *Biometrics* **20**: 46–72.

Hazel, L. N. 1943. Genetic basis for selection indices. *Genetics* **28**: 476–490.

Hill, W. G. 1970. Theory of limits to selection with line crossing. In *Mathematical topics in population genetics*. New York: Springer-Verlag, pp. 210–245.

Ho, Y. T. 1973. Genetic improvement by combining superior alleles from different populations in a naturally cross-fertilizing species. Ph.D. thesis. St. Paul: University of Minnesota.

Hull, F. H. 1945. Recurrrent selection for specific combining ability in corn. *J. Am. Soc. Agron.* **37:** 134–145.

Jones, D. F. 1917. Dominance of linked factors as a means of accounting for heterosis. *Genetics* **2:** 466–479.

Jones, L. P., W. A. Compton, and C. O. Gardner. 1971. Comparison of full and half-sib reciprocal recurrent selection. *Theor. Appl. Genet.* **41:** 36–39.

Kimura, M. 1957. Some problems of stochastic processes in genetics. *Ann. Math. Stat.* **28:** 882–901.

Lindsey, M. F., J. H. Lonnquist, and C. O. Gardner. 1962. Estimates of genetic variance in open-pollinated varieties of cornbelt corn. *Crop Sci.* **2:** 105–108.

Lush, J. L. 1947. Family merit and individual merit as bases for selection. *Am. Natur.* **81:** 241–261.

Moll, R. H., M. F. Lindsey, and H. F. Robinson. 1964. Estimates of genetic variances and level of dominance in maize. *Genetics* **49:** 411–423.

Moll, R. H. and C. W. Stuber. 1971. Comparisons of response to alternative selection procedures initiated with two populations of maize (*Zea mays* L.). *Crop Sci.* **11:** 706–711.

Robertson, A. 1970. A theory of limits in artificial selection with many linked loci. In *Mathematical topics in population genetics.* New York: Springer-Verlag, pp. 246–288.

Robinson, H. F., R. E. Comstock, and P. H. Harvey. 1949. Estimates of heritability and the degree of dominance in corn. *Agron. J.* **41:** 353–359.

Robinson, H. F., R. E. Comstock, and P. H. Harvey. 1951. Genetic and phenotypic correlations in corn and their implications to selection. *Agron. J.* **43:** 282–287.

Robinson, H. F., R. E. Comstock, and P. H. Harvey. 1955. Genetic variances in open pollinated varieties of corn. *Genetics* **40:** 45–60.

Rojas, B. A. and G. F. Sprague. 1952. A comparison of variance components in corn yield trials. III. General and specific combining ability and their interaction with location and years. *Agron. J.* **44:** 462–466.

Smith, H. F. 1936. A discriminant function for plant selection. *Ann. Eugenics* **7:** 240–250.

Sprague, G. F. and W. T. Federer. 1951. A comparison of variance components in corn yield trials. II. Error, year × variety, location × variety and variety components. *Agron. J.* **43:** 535–541.

Sprague, G. F. and L. A. Tatum. 1942. General and specific combining ability in single crosses of corn. *J. Am. Soc. Agron.* **34:** 923–932.

Vencovsky, Roland. 1975. Personal communication.

Young, S. S. Y. 1964. Multi-stage selection for genetic gain. *Heredity* **19:** 131–145.

Chapter 14

POPULATION IMPROVEMENT IN MAIZE

C. O. Gardner

Professor of Agronomy,
University of Nebraska,
Lincoln

If population improvement through the use of well-designed cyclic selection and recombination procedures had been practiced in corn over the past half century along with inbred line and hybrid development, there is reason to believe that yields might be substantially higher than they are today. The concept of recurrent selection as a method for the improvement of yield of corn varieties was suggested by Hayes and Garber (1919) over 50 years ago. However, at that time, primary interest of breeders was being focused on inbreeding and hybridization. Selection methods then being used to improve populations were considered relatively ineffective and were generally abandoned.

Jenkins (1940) described a recurrent selection scheme in greater detail, which served as the basis for programs initiated in Nebraska by Lonnquist and in Iowa by Sprague in the early 1940s. The Nebraska program was initiated in

Published as Paper No. 5026, Journal Series, Nebraska Agricultural Experiment Station. Research reported was conducted under project 12-49.

1943, and empirical results published by Lonnquist (1949) clearly indicated the effectiveness of recurrent selection for yield improvement.

Work at North Carolina State University in the late 1940s and early 1950s showed that considerable additive genetic variance for grain yield existed in F_2 populations and in open-pollinated varieties of corn (Robinson, Comstock, et al., 1949, 1955; Robinson and Comstock, 1955). Progress from any of the intrapopulation selection schemes should have been possible if good techniques had been used.

By the mid-1950s several recurrent selection programs had been initiated that were designed to permit more effective evaluation of genotypes, and such programs are now an important component of almost every corn-breeding program.

Recently, the genetic vulnerability of commercial maize hybrids in the United States has been recognized, and the need for broadening the germplasm base has been emphasized. The populations already improved by recurrent selection procedures, although far from perfect, are promising sources of new germplasm. Recurrent selection systems provide an excellent way of utilizing exotic germplasm in maize-breeding programs to further broaden the genetic base.

A considerable body of evidence now available from quantitative genetic studies indicates that recurrent selection procedures are effective in improving both population performance and interpopulation hybrid vigor. Furthermore, inbred lines developed from improved populations are clearly superior in test-cross hybrids when compared to comparable hybrids of inbred lines developed from the original populations (Harris, Gardner, et al., 1972; Gardner, 1974).

The basic genetic principle involved in population improvement by recurrent selection procedures is simply the substitution of more favorable alleles for less favorable ones in the population. Sets of genes that interact favorably with one another will also be favored in such a program, even though some sets of genes are often broken in the recombination phase. Increasing the frequencies of more favorable alleles increases the population mean. If population improvement can be achieved without a significant reduction in genetic variability, our knowledge of quantitative genetic theory leads us to expect that the best double-cross and single-cross hybrids from a series of random lines derived from the improved population will exceed the population mean by at least 20–30% (Sprague and Eberhart, 1977).

Although it is too early to predict with certainty the effect of continued selection on genetic variability in grain yield, quantitative genetic theory indicates that increases in the frequencies of favorable alleles above 0.5 should cause a reduction in additive genetic variability.

At this point we might turn to the classical selection experiments, which were initiated by Hopkins (1899) at the University of Illinois prior to the turn of the century and before the rediscovery of Gregor Mendel's paper. These experiments were aimed at altering the chemical composition of the maize kernel in protein and oil content. Now after over 70 generations of selection, the population means of the high-protein, low-protein, high-oil, and low-oil strains are, respectively, 12, 8, 27, and 10 standard deviations away from the mean of the original population (Dudley, Lambert, et al., 1974). If 6000 ears of the original strain were examined, the most extreme ear expected would have been only 3.8 standard deviations from the population mean. Obviously, continued progress has been made in increasing protein and oil content without exhausting genetic variability. It is quite surprising that changes of the magnitude observed could result from selection in view of the original variability observed. Hopefully, we will achieve the same result with grain yield.

PROCEDURES FOR POPULATION IMPROVEMENT IN MAIZE

Several sophisticated procedures have been developed by different geneticists and breeders to more effectively identify the genetically superior individuals or families selected in the various recurrent selection schemes used for population improvement and to shorten the selection cycle. These include: (a) improved field-plot techniques, (b) improved field designs and use of irrigation to control environmental variation, (c) improved statistical procedures for better estimation, (d) testing in several environments to better evaluate genotypes, (e) use of winter nurseries for recombination phases, and (f) developing new families during the recombination phase to shorten the time required per cycle in family selection programs.

Population-improvement systems can be divided into two main categories: (a) intrapopulation selection systems and (b) interpopulation selection systems. The first is aimed at improving the population directly. The second is aimed at improving the interpopulation cross or at improving hybrids of lines derived from the two reciprocally selected populations. Some of the most widely used recurrent selection systems are listed in Table 1.

As Moll and Stuber (1971) point out, intrapopulation selection may also have an indirect effect of improving interpopulation crosses or of improving hybrids of lines derived from such populations. Interpopulation selection may have the indirect effect of improving the population *per se.*

PREDICTION EQUATIONS

Formulas for calculating response to be expected from many of the various recurrent selection systems used in population improvement in maize have been

TABLE 1.   Population-improvement systems

1. Intrapopulation selection systems.
    a. Mass selection as modified by Gardner (1961, 1968, 1969).
        i.   With pollen control.
        ii. Without pollen control.
    b. Full-sib family testing and selection.
    c. Half-sib family testing and selection.
        i.   Ear-to-row (no replication)
        ii. Replicated test.
    d. Modified ear-to-row.
        i.   Without pollen control as done by Lonnquist (1964).
        ii. With pollen control as proposed by Eberhart and now being done by Compton at Nebraska. Selection within as well as between families is involved.
    e. S_1 Family testing and selection.
    f. S_2 Family testing and selection.
    g. S_1 Family selection based on testcross family performance. Tester may be the population itself, a double cross, a single cross, or an inbred line.
2. Interpopulation selection systems.
    a. Reciprocal recurrent selection (Comstock, Robinson, et al., 1949).
    b. Full-sib reciprocal recurrent selection (Hallauer, 1967; Lonnquist and Williams, 1967).
    c. Full-sib reciprocal recurrent selection using an inbred line tester derived from the tester population (Russell and Eberhart, 1975).

developed. Comstock and Cockerham have made important contributions to plant breeders in developing such prediction equations, and several of them have been published by Sprague (1966) and by Empig, Gardner, et al. (1972). Sprague and Eberhart (1977) have also summarized several prediction equations.

Parameter estimates needed for calculating expected response from most intrapopulation selection systems are additive genetic variance in the population to be improved and the total phenotypic variance among individuals or among families being selected. Lush (1945) pointed out that expected gain is the product of the selection differential (superiority of selected group over the population mean) and heritability (proportion of total phenotypic variance that is additive genetic variance). In a normally distributed population, expected gain per year can be expressed as:

$$G_y = \frac{ck\ \sigma^2_{g*}}{y\ \sqrt{\sigma^2_g + \dfrac{\sigma^2_{ge}}{e} + \dfrac{\sigma^2}{re}}}$$

TABLE 2. Expected gain per cycle (G_c) from intrapopulation selection systems

System (Table 1)	Expected gain (G_c)[a]	Crop seasons per cycle
1. Mass selection		
a. Pollen control	$\dfrac{k\sigma_A^2}{\sqrt{\sigma_A^2 + \sigma_D^2 + \sigma_{AE}^2 + \sigma_{DE}^2 + \sigma_w^2}}$	1
b. No pollen control	$\dfrac{k(\frac{1}{2})\sigma_A^2}{\sqrt{\sigma_A^2 + \sigma_D^2 + \sigma_{AE}^2 + \sigma_{DE}^2 + \sigma_w^2}}$	
2. Full-sib family	$\dfrac{k(\frac{1}{2})\sigma_A^2}{\sqrt{(\frac{1}{2}\sigma_A^2 + \frac{1}{4}\sigma_D^2) + \dfrac{(\frac{1}{2}\sigma_{AE}^2 + \frac{1}{4}\sigma_{DE}^2)}{e} + \dfrac{\sigma^2}{re}}}$	3
3. Half-sib family		
a. Ear-to-row	$\dfrac{k(\frac{1}{4})\sigma_A^2}{\sqrt{\frac{1}{4}\sigma_A^2 + \frac{1}{4}\sigma_{AE}^2 + \sigma^2}}$	1
b. Replicated test	$\dfrac{k(\frac{1}{4})\sigma_A^2}{\sqrt{\frac{1}{4}\sigma_A^2 + \dfrac{\frac{1}{4}\sigma_{AE}^2}{e} + \dfrac{\sigma^2}{re}}}$	3
4. Modified ear-to-row		
a. No pollen control	$\dfrac{k_1(\frac{1}{8})\sigma_A^2}{\sqrt{\frac{1}{4}\sigma_A^2 + \frac{1}{4}\dfrac{\sigma_{AE}^2}{e} + \dfrac{\sigma^2}{re}}}$ $+ \dfrac{k_2(\frac{3}{8})\sigma_A^2}{\sqrt{\frac{3}{4}\sigma_A^2 + \sigma_D^2 + \frac{3}{4}\sigma_{AE}^2 + \sigma_{DE}^2 + \sigma_w^2}}$	
5. S_1 Family	$\dfrac{k\,\sigma_A^{2*b}}{\sqrt{\sigma_A^{2*} + \frac{1}{4}\sigma_D^2 + \dfrac{\sigma_{A\,E}^{2*} + \frac{1}{4}\sigma_{DE}^2}{e} + \dfrac{\sigma^2}{re}}}$	3
6. Testcross population as tester	$\dfrac{k(\frac{1}{2})\sigma_A^2}{\sqrt{\frac{1}{4}\sigma_A^2 + \dfrac{\frac{1}{4}\sigma_{AE}^2}{e} + \dfrac{\sigma^2}{re}}}$	3

[a] σ_A^2 is additive genetic variance, σ_D^2 is dominance variance, and σ_{AE}^2 and σ_{DE}^2 are their interactions with environments, respectively.

[b] σ_A^{2*} indicates that the definition of additive genetic variance changes slightly with inbreeding.

where G_y = gain per year; y = number of years per cycle; k = standardized selection differential; σ^2_{g*} = portion of the genetic variance among individuals or families that is additive; σ^2_g = genetic variance among individuals or families evaluated; σ^2_{ge} = genotype × environment interaction variance; σ^2 = experimental error; e = number of environments; and r = number of replications per environment.

$$\sigma^2 = \frac{\sigma^2_w + (\sigma^2_h - \sigma^2_g)}{n} + \sigma^2_e,$$

where σ^2_w = within-plot environment variance; σ^2_h = total heritable variance; n = number of plants per plot; and σ^2_e = plot-to-plot environmental variance within replications. Prediction equations are given in Tables 2 and 3 for some of the specific systems listed in Table 1.

YIELD IMPROVEMENT REALIZED FROM RECURRENT SELECTION SYSTEMS

In recent years results from several different recurrent selection programs have appeared in the literature. It is not possible to discuss these results in any comprehensive way, so only a few examples are cited for each system.

Mass Selection (M)

The most extensive mass-selection studies have been conducted at the University of Nebraska, where a program initiated in 1955 as part of a mutation breeding experiment is still in progress (Gardner, 1961, 1968, 1969, 1974). The complete program is illustrated in Figure 1. The control population has now undergone 19 generations of selection and a population that received thermal neutron seed irradiation twice in early generations is in its 20th cycle. Individual plant selection at harvest time has been for high grain yield in an isolated planting with approximately 19,000 plants/ha.

Since an increase in prolificacy was noted first in the irradiated population and later in the control, Lonnquist (1967) initiated the prolific mass-selected population, where selection is based on number of ears per plant.

At generation 14 in the irradiated selected population, a program was initiated to select for early flowering, lower ear and plant height, and improved stalk quality while maintaining or improving yield. Other programs indicated in Figure 1 are selection for grain yield at high plant populations (69,000 plants/ha) using generation 13 of the irradiated selected population as a starting base and recurrent X-irradiation and mass selection for grain yields using generation 9 of the control selected population as the base.

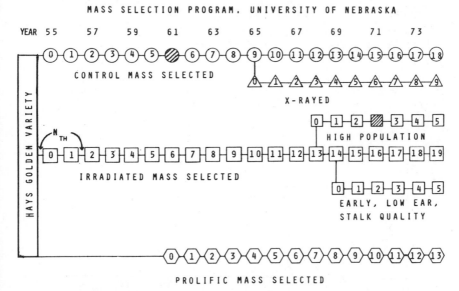

Figure 1. Diagram of the University of Nebraska mass selection program initiated in 1955 in the open-pollinated variety Hays Golden.

For the first 16 generations an excellent linear response of 3% per generation was realized from mass selection for high grain yield (Gardner, 1974), which is in excellent agreement with the 3.2% expected progress based on extensive genetic variance studies of the Hays Golden variety (Gardner, 1968) and under the assumption that genetic variability has not been decreased by selection. However, the addition of data collected on the three most recent generations tested in our 1973 and 1974 evaluation studies, suggest that a plateau may have been reached (Figure 2). Yields even appear to have decreased. This is believed to be the result of severe environmental stresses experienced at critical times that differentially affected the selected populations more severely than they did Hays Golden. Yields are all expressed in percent of Hays Golden. Also some shifts to less desirable and more variable isolated fields may have rendered selection less effective in recent years.

Mass selection for prolificacy has increased yield faster than selection for yield itself in early generations as reported by Lonnquist (1967). He indicated gains of 6.3% per generation for the first five generations. More recent data covering 13 generations indicate a gain of about 5% per year for seven or eight generations and then a leveling off (Figure 3). Perhaps there is some optimum number of ears, or perhaps the environmental stresses have caused the plateau effect noted.

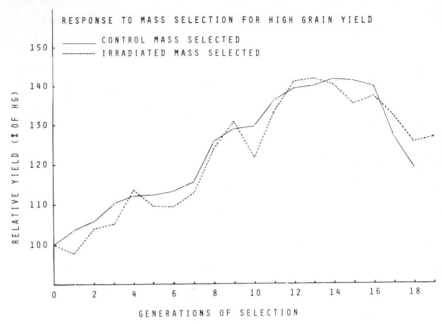

Figure 2. Response to mass selection for high-grain yield in a nonirradiated control population [NHG(M)C] and an irradiated one [NHG(M)I] derived from the open-pollinated variety Hays Golden.

Correlated responses to mass selection for yield have been a delay in maturity, more ears per plant, and higher ears and taller plants (Table 4). Selection for earlier maturity, lower ear height and improved stalk quality have been very effective in shifting maturity and height in a favorable direction.

Some other mass selection studies are summarized in Table 5. Hallauer and Sears (1969) had rather limited success with two populations, but Johnson (1963) made striking progress in tropical material in Mexico. In Nebraska, Mathema (1971) demonstrated that considerable success had been made using "adaptive mass selection" on two Corn Belt composites and eight composites involving exotics. The latter yield about as well as the two Corn Belt composites. In Colombia, Arboleda and Compton (1974) had excellent success from selection in the wet season, good success from selection in both seasons, and moderate success from selection in the dry season.

Full-sib Family Selection (FS)

The most extensive work involving full-sib family selection has been conducted at North Carolina State University and has been reported by Moll and

Robinson (1967) and Moll and Stuber (1971). Gains have been in the range of 3–4% per cycle (Table 5).

Modified Ear-to-row Selection (ER)

The most extensive study of this system has been the one involving Hays Golden at the University of Nebraska. It was initiated by Lonnquist in 1961 and has been continued by Compton. Webel and Lonnquist (1967) using a linear relationship reported gains of 9.4% per cycle for the first four cycles; however, even their early data indicated some decrease in gains with the advance in generations. Bahadur (1974), working under Compton, studied the first 10 selected generations simultaneously and reported a quadratic response. Other data·from the three most recent generations included in our mass selection evaluation tests each year are in agreement with that of Bahadur (Figure

TABLE 3.   Expected gain per cycle (G_c) for the population cross from interpopulation selection systems

System (Table 1)	Expected gain $(G_c)^a$	Crop seasons per cycle
1. Reciprocal recurrent selection	$\dfrac{k(\frac{1}{4})\,\sigma^2_{A(1)}}{\sqrt{\frac{1}{4}\,\sigma^2_{A(1)} + \frac{1}{4}\,\dfrac{\sigma^2_{AE(1)}{}^b}{e} + \dfrac{\sigma^2}{er}}}$ $+ \dfrac{k\frac{1}{4}\,\sigma^2_{A(2)}}{\sqrt{\frac{1}{4}\,\sigma^2_{A(2)} + \frac{1}{4}\,\dfrac{\sigma^2_{AE(2)}}{e} + \dfrac{\sigma^2}{er}}}$	3
2. Full-sib reciprocal recurrent selection	$\dfrac{k(\frac{1}{2})\,\sigma^2_{A'}{}^c}{\sqrt{(\frac{1}{2}\,\sigma^2_{A'} + \frac{1}{4}\,\sigma^2_{D'}) + \dfrac{(\frac{1}{2}\,\sigma^2_{A'E} + \frac{1}{4}\,\sigma^2_{D'E})}{e} + \dfrac{\sigma^2}{er}}}$	3

[a] $\sigma^2_{A(1)}$ and $\sigma^2_{A(2)}$ are the additive genetic variances among testcross families, that is, $4\sigma^2_{m(1)}$ and $4\sigma^2_{m(2)}$, respectively, where $\sigma^2_{m(1)}$ and $\sigma^2_{m(2)}$ are the testcross (half-sib) family components of variance for population 1 with population 2 as tester and for population 2 with population 1 as tester, respectively.

[b] $\sigma^2_{AE(1)}$ and $\sigma^2_{AE(2)}$ are the additive genetic × environment interaction variances and equal $4\sigma^2_{mE(1)}$ and $4\sigma^2_{mE(2)}$, respectively.

[c] $\sigma^2_{A'}$ and $\sigma^2_{D'}$ are additive genetic and dominance variances for the population cross. The variance among full-sib testcross families $= \frac{1}{2}\,\sigma^2_{A'} + \frac{1}{4}\,\sigma^2_{D'}$ and the variance among half-sib testcross families $= \frac{1}{4}\,\sigma^2_{A'} = (\sigma^2_{m(1)} + \sigma^2_{m(2)})/2$. $\sigma^2_{A'E}$ and $\sigma^2_{D'E}$ are the corresponding interactions with years.

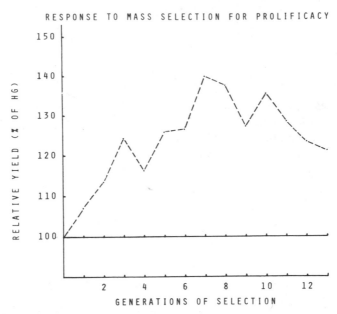

Figure 3. Response to mass selection for prolificacy (number of ears per plant) in a population [NHG(M)P] derived from the open-pollinated variety of Hays Golden.

TABLE 4. Correlated responses to mass selection for high grain yield or prolificacy and recovery observed from selection for early maturity, low ears, and stalk quality in the I-14 generation

Population	Changes relative to Hays Golden (average of three most recent generations tested in 1973 and 1974)				
	Days to flower	Percent moisture at harvest	Ears per plant	Ear height (cm)	Plant height (cm)[a]
Control selected (C)	+6.5	3.1	+.12	+43	+46
Irradiated selected (I)	+6.3	1.6	+.19	+47	+53
Prolific selected (P)	+4.4	0.6	+.42[b]	+31	+29
Early, low ear selection from I-14	+2.1	−0.1	+.13	+22	+21

[a] Only 1974 data.
[b] Direct response.

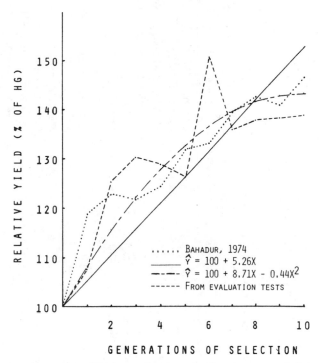

Figure 4. Response to modified ear-to-row selection in a population [NHG(ER)C] derived from the open-pollinated variety of Hays Golden. Linear and quadratic response curves are from Bahadur (1974).

4). Although rapid early gains were realized, the total gain over 10 cycles has been 40–45%, which is an average gain of 4.0–4.5% per generation.

Others have also had excellent success with modified ear-to-row selection (Table 5). Paterniani (1967) realized gains of 13.6% per cycle, and Dudley, Lambert, et al. (1974) realized gains of 2.4–4.8% per cycle in *opaque*-2 populations.

S_1 Family Selection (S_1)

Genter (1973) in Virginia has gotten an excellent yield response from S_1 family testing and selection in a Corn Belt southern composite. The improved population has proven to be 5.4% superior to the original one (Table 5) over a wide

TABLE 5. Average yield gains realized per cycle in some intrapopulation recurrent selection systems

Selection system	Population	No. of cycles	Mean gain per cycle (%)	Reference
Mass selection (M)	Hays Golden	17	3.0	Gardner, 1974
	Tropical	3	11.1	Johnson, 1963
	Krug	6	1.6	Hallauer and Sears, 1969
	Iowa Ideal	5	1.4	Hallauer and Sears, 1969
	Two Corn Belt Composites	3 & 5	3.1	Mathema, 1971
	Eight Corn Belt × Exotic or Exotic	3–10	5.5	Mathema, 1971
	Mezcla Varietales Amarillos	3–6	2.5–10.5	Arboleda and Compton, 1974
Full-sib (FS) family	$(CI21 × NC7)F_2$	7	3.0–4.0	Moll and Robinson, 1967
	Jarvis	6	3.8	Moll and Stuber, 1971
	Indian Chief	6	2.6	Moll and Stuber, 1971
	Jarvis × Indian Chief Syn 4	6	3.9	Moll and Stuber, 1971
Modified ear-to-row (ER)	Paulista Dent	3	13.6	Paterniani, 1967
	Hays Golden	4	9.4	Webel and Lonnquist, 1967
	Hays Golden	10	5.3	Bahadur, 1974

SSSS o_2	3	4.8	Dudley, Lambert, et al., 1974
SSSS fl_2	3	3.4	Dudley, Lambert, et al., 1974
D.O. o_2	3	2.4	Dudley, Lambert, et al., 1974
D.O. fl_2	3	4.6	Dudley, Lambert, et al., 1974
S_1 Family (S_1)			
BSK	4	3.9	Burton, Penny, et al., 1971
VLE	2	1.1	Genter, 1973
VCBS	2	7.2	Genter, 1973
VCBS	4	5.4	Genter and Eberhart, 1974
Fla. 767	5	4.6	Horner, Lundy, et al., 1973
S_2 Family (S_2)			
S_1 Family based on testcross family performance (S_1TC)			
Tester			
Population itself			
Mean of 5	2	5.7	Lonnquist, 1961
Fla. 767	5	3.9	Horner, Lundy, et al., 1973
Double cross or 2 single crosses			
BSK	4	1.9	Burton, Penny, et al., 1971
Two single crosses			
VLE	2	−0.7	Genter, 1973
VCBS	2	1.4	Genter, 1973
VCBS	3	7.3	Genter and Eberhart, 1974
Inbred line			
Fla. 767	5	5.4	Horner, Lundy, et al., 1973

range of environments (Genter and Eberhart, 1974). Rather limited success was realized by Genter from S_1 selection in a long-ear composite. Burton, Penny, et al. (1971) had good success with BSK (Krug Yellow Dent derivative) in Iowa.

S_2 Family Selection (S_2)

Horner, Lundy, et al. (1973) have had good success using S_2 family testing and selection in Florida (Table 5); however, to compare S_2 selection with S_1 selection, gains must be expressed on an annual gain basis because S_2 testing requires an additional year per cycle. For many agronomic traits, S_1 and S_2 family tests do provide more precise evaluations of genotypes than do other types of families; thus greater progress should be possible. If overdominance and dominance types of epistasis are nonexistent, S_1 and S_2 selection should also be effective for improving grain yield.

S_1 Family Selection Based on Testcross Performance (S_1 TC)

The S_1 family selection based on testcross family performance has been an effective procedure for the improvement of yield in populations as well as for the development of new hybrids (Table 5). The latter might be considered to be interpopulation improvement. Selection for general combining ability using the population as a tester, as was done by Lonnquist (1949, 1961) and Horner, Lundy, et al. (1973), is basically a half-sib family testing and S_1 family selection system for population improvement. This system or modifications of it have been used successfully for many years, but few consistent, long-range studies have been reported. Selection of S_1 families based on testcrosses to a double-cross hybrid or two single-cross hybrids has also generally been done to select for general combining ability and for population improvement. However, inbred lines might be developed from the S_1 lines or from the improved population to use in hybrids with the single-cross tester. Selection of S_1 families based on testcrosses to an inbred line is for specific combining ability, and the goal is generally to develop lines from the S_1 families or the selected population for use in hybrids with the tester. Nevertheless, population improvement may also occur, as indicated by Horner, Lundy, et al. (1973). Theoretically, a poor inbred line used as the tester should very effectively identify genotypes with superior genetic constitutions for selection in a population improvement program.

Reciprocal Recurrent Selection (RRS)

Of the three interpopulation selection systems listed in Table 1, the only one on which much information has been accumulated and published is the reciprocal

TABLE 6. Results reported from RRS programs

Population	Number of cycles	Gain per cycle (%)	Reference
Jarvis	6	1.8 ⎫	
Indian Chief	6	1.2 ⎬	Moll and Stuber (1971)
Jarvis × Indian Chief	6	3.2 ⎭	
BSSS (R)	5	0.4 ⎫	Eberhart, Debela, et al.
BSCB1 (R)	5	0.9 ⎬	(1973)
BSSS (R) × BSCB1 (R)	5	4.5 ⎭	

recurrent selection system proposed by Comstock, Robinson and Harvey (1949). Moll and Stuber (1971) have realized a gain of 3.2% per cycle in the interpopulation cross, and Eberhart, Debela, et al. (1973) have realized a gain of 4.5%. In both programs only modest improvements in yield levels of the populations themselves have been realized. Results are summarized in Table 6.

Moll and Stuber (1971) compared RRS with full-sib selection for the improvement of two varieties, their F_1 hybrid and the composite, which was formed by crossing the two varieties followed by four generations of random mating. Results are summarized in Table 7. Full-sib selection was more effective for improving the populations *per se*, but RRS was slightly more effective for improving the intervariety cross. Realized gains were in excellent agreement with those expected, except in the case of the intervariety composite, where realized gain was only 62% of that expected based on a genetic variance analysis and was only slightly higher than gains in the Jarvis variety.

When viewed from the standpoint of heterosis in the variety hybrid, Moll and Stuber (1971) indicate that RRS appears to be much more promising for the development of high-yielding interpopulation hybrids than the intrapopula-

TABLE 7. Comparison of RRS with FS family selection for the improvement of populations, their hybrid, and their intervariety composite (Moll and Stuber, 1971)

Population	Method	Gain per cycle (%)	
		Realized	Expected
Jarvis	FS	3.75	4.74
	RRS	1.76	
Indian Chief	FS	2.65	2.77
	RRS	1.24	
Intervariety cross	FS	2.38	
	RRS	3.19	2.69
Intervariety composite	FS	3.91	6.26

TABLE 8.   Heterosis observed in the intervariety
hybrid before and after selection (Moll and
Stuber, 1971)

	Kind of selection	
Selection cycle	Full-sib	Reciprocal
0	19.2	19.2
3	14.8	—
4	18.2	24.2
6	15.4	30.2

tion selection systems (Table 8). After six cycles of selection, heterosis from
RRS is double that realized from FS selection. Inbred lines developed from
reciprocally selected populations are likely to exhibit more heterosis than
similar lines developed from populations by intrapopulation selection methods.

Full-sib Reciprocal Recurrent Selection (FSRRS)

Full-sib reciprocal recurrent selection programs were first initiated by
Dr. J. H. Lonnquist at the University of Nebraska and A. R. Hallauer at Iowa
State University. However, to my knowledge no results have been published to
date. Empig, Gardner, et al. (1972) concluded from some computer simulation
results that FSRRS should be superior to RRS. Only half as many families are
needed with FSRRS compared to RRS to maintain the same effective population size in each population, and the long-term response using either system
depends on the effective population size maintained. The larger effective population size permitted by FSRRS offsets the higher phenotypic variance in the
prediction of response.

The major advantage of FSRRS is the natural spinoff of new lines and
hybrids for commercial use each cycle as the populations and the population
hybrid are improved.

Reciprocal Recurrent Selection Using an Inbred Tester (RRSIT)

Eberhart, Debela, et al. (1973) and Russell and Eberhart (1975) have proposed
a modification of the reciprocal recurrent selection system by using an inbred
line from the reciprocal population as the tester. Empirical results and statistical genetic theory indicate greater genetic variation among families when an
inbred line is used as the tester rather than the population. Either an elite line
or a random one might be used as the tester, and it is not clear how often the
tester line would be changed. Both theoretical and empirical information on
this system is needed.

RECURRENT SELECTION FOR OTHER TRAITS

My discussion has been focused on selection experiments designed and conducted primarily for the improvement of grain yield. However, this does not mean that other traits are unimportant. Indeed, they are very important! Selection for high-grain yields does to some extent indirectly select for disease and insect resistance and for traits correlated with yield, but many successful recurrent selection programs for traits other than yield have been reported in the literature. Excellent gains have been made in selecting for: (a) days to flower by Troyer and Brown (1972) and Hallauer and Sears (1972), (b) lower ear height by Vera and Crane (1970), (c) ear-worm resistance by Zuber, Fairchild, et al. (1971), (d) corn borer resistance by Penny, Scott, et al. (1967), (e) northern corn leaf-blight resistance by Jenkins, Robert, et al. (1954), (f) stalk rot resistance by Jinahyon and Russell (1969), (g) stalk crushing strength by Zuber (1973), (h) lysine by Dudley, Alexander, et al. (1975). Other literature could be mentioned.

For improved populations that are most useful from a commercial standpoint, simultaneous selection for higher grain yields and other desirable agronomic traits is essential. This can be accomplished in part by careful selection of individuals and families at various stages in the recurrent selection program and by useful screening techniques. However, the use of a selection index should also be given greater consideration than it has received in the past. Since the inclusion of each additional trait in the index would tend to decrease progress in improving the other traits in the index, one must carefully choose the traits to be included and the weights to be given each.

IMPROVED POPULATIONS AS SOURCES OF LINES AND HYBRIDS

One of the primary objectives of population improvement in maize is to provide germplasm pools from which new superior inbred lines can be extracted for use in hybrids. Hence we have evaluated our populations improved by mass selection in this regard.

Random sets of S_1 lines developed from the original variety and from the improved NHG(M)C9 and NHG(M)I9 mass selected populations were test-crossed to a related hybrid N6 × N6G and to an unrelated one H49 × WF9 (Harris, Gardner, et al., 1972). Hybrids of lines from improved populations yielded 17% more than comparable hybrids of lines from the original Hays Golden. In the same test, the improved populations themselves outyielded Hays Golden by 18%. Hybrid superiority was proportional to the improvement made in yield by mass selection.

Visually selected sets of S_2 lines were also developed from Hays Golden and from selected populations NHG(M)C12, NHG(M)I13, and NHG(M)P7. These lines were testcrossed to N7A × N7B and to Oh43. Two-year mean yields of hybrids involving lines selected from the three populations were 13.4%, 9.4%, and 10.4% greater than those involving Hays Golden lines crossed to the two testers. The top-yielding hybrids involving lines from each improved population exceeded the mean of the two excellent check hybrids by 4.0% with Oh43 as the tester and by 6.8% with N7A × N7B as the tester.

Russell and Eberhart (1975) crossed selected sets of S_2 lines from BSCB1(R)C5 and BSSS(R)C5 and S_3 and S_4 lines from BSSS(HT)C6 with each other. Mean yields of line crosses between any two sets were significantly higher than the population crosses. The best cross exceeded the population cross by 35%. Two BSCB1(R)C5 × BSS(R)C5 line crosses yielded significantly more than B37 × Oh43, and one of these equaled B37 × Oh43 in lodging resistance. Unfortunately, comparable lines from original populations were not available for crossing and comparison.

If overdominance or dominance types of epistasis are not important in hybrid vigor in maize, intrapopulation selection systems should be just as effective as interpopulation selection systems in developing populations to be used in hybrids. Early results with reciprocal recurrent selection showed no advantage of that system over intrapopulation selection in improving interpopulation cross performance. More recent data of Moll and Stuber (1971) and Eberhart, Debela, et al. (1973) make RRS look much more promising for the improvement of the intervariety hybrid and for the improvement of hybrids of lines selected out of the two populations.

CONCLUSIONS

We now have a considerable body of data to conclude that any of the intrapopulation selection systems used with adequate environmental and statistical controls will be effective in population improvement of yield and many other agronomic traits. Empirical results are in good agreement with expectation based on genetic variance studies and quantitative genetic theory. We also have evidence to indicate that improved populations are superior sources of inbred lines and hybrids. Improvement in hybrids of lines derived from populations is proportional to the improvement of the populations themselves.

Recent data on reciprocal recurrent selection for the improvement of the interpopulation cross is very encouraging. Heterosis increased dramatically with cycles of selection in Moll and Stuber's (1971) study. Information is needed on lines developed from both the original and the improved populations currently being used in RRS programs. Full-sib reciprocal recurrent selection looks exceptionally promising from a commercial maize-breeding standpoint, because new hybrids become available each cycle while the populations and their interpopulation cross are being improved. Reciprocal recurrent selection with inbred tester merits further consideration because of the power to discriminate among families. In the long run, the effect of the line chosen as the tester on the population to be improved, and the interpopulation cross is not as yet entirely clear.

The choice of a breeding system to use for intrapopulation improvement or interpopulation improvement will depend primarily on the expected gain per year, but also on the objectives of the breeder and the facilities available to him. Full-sib and half-sib family selection cycles can be reduced one year by making up new families during the recombination phase, but the exact effect of elimination of the recombination phase is unknown. Linkage effects are bound to be greater. More information involving precise comparisons among systems of breeding is needed.

For the improvement and maintenance of germplasm pools and the incorporation of exotic germplasm, mass selection, which involves minimum effort and permits a large effective population size minimizing inbreeding, appears to be an excellent choice. An occasional generation of inbreeding and testing of large numbers of S_1 lines that could be screened for other agronomic traits as well as yield would undoubtedly strengthen a mass selection program.

The modified ear-to-row system of Lonnquist (1964) appears to give the greatest yield increases on an annual basis; however, present data suggest that the gains realized in early generations may not be maintained over a long period of time. Full-sib family and S_1 and S_2 family selection systems as well as modified ear-to-row selection have the advantage of replication over environments, which provides a better evaluation of their broad genetic potential compared to mass selection.

The problem with almost all family selection systems reported in the literature is that the effective population size has been too small. This leads to considerable inbreeding, and many desirable genes may be lost due to genetic drift. In comparing populations improved by different systems, inbreeding must be taken into account, as was clearly pointed out by Horner, Lundy, et al. (1973). Inbreeding depression may also explain why realized response has often been reported to be slightly less than expected response. For sustained improvement over a long period of time, the effective population size must be sufficiently large to hold inbreeding to a very low level.

If improved populations are to be useful from a commercial breeding standpoint, other agronomic traits must be considered along with yield, even though yield itself is the direct result of many other traits of the plant. This is possible by careful selection at every stage in the breeding program, by use of effective screening techniques, and by the effective use of a selection index. The latter has received inadequate attention in the past.

REFERENCES

Arboleda, F. and W. A. Compton. 1974. Differential response of maize (*Zea mays* L.) to mass selection in diverse selection environments. *Theor. Appl. Genet.* **44:** 77–81.

Bahadur, K. 1974. Progress from modified ear-to-row selection in two populations of maize (*Zea mays* L.). Ph.D. thesis. Lincoln, Nebr.: University of Nebraska.

Burton, J. W., L. H. Penny, A. R. Hallauer, and S. A. Eberhart. 1971. Evaluation of synthetic populations developed from a maize population (BSK) by two methods of recurrent selection. *Crop. Sci.* **11:** 361–367.

Comstock, R. E., H. F. Robinson, and P. H. Harvey. 1949. A breeding procedure designed to make maximum use of both general and specific combining ability. *Agron. J.* **41:** 360–367.

Dudley, J. W., D. E. Alexander, and R. J. Lambert. 1975. Genetic improvement of modified protein maize. In *High-quality protein maize*. Stroudsburg, Penn.: Dowden, Hutchinson and Ross, pp. 120–135.

Dudley, J. W., R. J. Lambert, and D. E. Alexander. 1974. Seventy generations of selection for oil and protein concentration in the maize kernel. In *Seventy generations of selection for oil and protein in maize.*, J. W. Dudley (ed.). Madison, Wisc.: Crop Science Society of America, pp. 181–212.

Eberhart, S. A., S. Debela, and A. R. Hallauer. 1973. Reciprocal recurrent selection in the BSSS and BSCB1 maize populations and half-sib selection in BSSS. *Crop Sci.* **13:** 451–456.

Empig, L. T., C. O. Gardner, and W. A. Compton. 1972. Theoretical gains for different population improvement procedures. *Nebr. Agr. Exp. Sta.* No. MP26, revised.

Gardner, C. O. 1961. An evaluation of effects of mass selection and seed irradiation with thermal neutrons on yield of corn. *Crop Sci.* **1:** 241–245.

Gardner, C. O. 1968. Mutation studies involving quantitative traits. In *The present state of mutation breeding,* Gamma Field Symposium No. 7, pp. 57–77.

Gardner, C. O. 1969. The role of mass selection and mutagenic treatment in modern corn breeding. *Proc. 24th Annu. Corn and Sorghum Res. Conf.* Washington, D.C.: American Seed Trade Association, pp. 15–21.

Gardner, C. O. 1974. Evaluation of mass selection and of seed irradiation with mass selection for population improvement in maize. *Genetics* **74:** s88–s89.

Genter, C. F. 1973. Comparison of S_1 and testcross evaluation after two cycles of recurrent selection in maize. *Crop Sci.* **13:** 524–527.

Genter, C. F. and S. A. Eberhart. 1974. Performance of original and advanced maize populations and their diallel crosses. *Crop Sci.* **4:** 881–885.

Hallauer, A. R. 1967. Development of single-cross hybrids from two-eared maize populations. *Crop. Sci.* **7:** 192–195.

Hallauer, A. R. and J. H. Sears. 1969. Mass selection for yield in two varieties of maize. *Crop Sci.* **9:** 47–50.

Hallauer, A. R. and J. H. Sears. 1972. Integrating exotic germplasm into Corn Belt maize breeding programs. *Crop Sci.* **12**: 203–206.

Harris, R. E., C. O. Gardner, and W. A. Compton. 1972. Effect of mass selection and irradiation in corn measured by random S_1 lines and their testcrosses. *Crop Sci.* **2**: 594–598.

Hayes, H. K., and R. J. Garber. 1919. Synthetic production of high protein corn in relation to breeding. *J. Am. Soc. Agron.* **11**: 308–318.

Hopkins, C. G. 1899. Improvement in the chemical composition of the corn kernel. *Ill. Agr. Exp. Sta. Bull.* **55**: 205–240.

Horner, E. S., H. W. Lundy, M. C. Lutrick, and W. H. Chapman. 1973. Comparison of three methods of recurrent selection in maize. *Crop Sci.* **13**: 485–489.

Jenkins, M. T. 1940. The segregation of genes affecting yield of grain in maize. *J. Am. Soc. Agron.* **32**: 55–63.

Jenkins, M. T., A. L. Robert, and W. R. Findley, Jr. 1954. Recurrent selection as a method for concentrating genes for resistance to *Helminthosporium turcicum* leaf blight in corn. *Agron. J.* **46**: 89–94.

Jinahyon, S. and W. A. Russell. 1969. Evaluation of recurrent selection for stalk-rot resistance in an open-pollinated variety of maize. *Iowa State J. Sci.* **43**: 229–237.

Johnson, E. C. 1963. Mass selection for yield in a tropical corn variety. *Am. Soc. Agron. Abstr.,* p. 82.

Lonnquist, J. H. 1949. The development and performance of synthetic varieties of corn. *Agron. J.* **41**: 153–156.

Lonnquist, J. H. 1961. Progress from recurrent selection procedures for the improvement of corn populations. *Nebr. Agr. Exp. Sta. Res. Bull.* No. 197.

Lonnquist, J. H. 1964. Modification of the ear-to-row procedure for the improvement of maize populations. *Crop Sci.* **4**: 227–228.

Lonnquist, J. H. 1967. Mass selection for prolificacy in maize. *Der Züchter* **37**: 185–187.

Lonnquist, J. H. and N. E. Williams. 1967. Development of maize hybrids through selection among full-sib families. *Crop. Sci.* **7**: 369–370.

Lush, J. L. 1945. Animal Breeding Plans. Ames: Iowa State Univ. Press, p. 442.

Mathema, B. B. 1971. Evaluation of progress in adapted × exotic maize populations undergoing adaptive mass selection in Nebraska. M.S. thesis. Lincoln, Nebr.: Univ. of Nebraska.

Moll, R. H. and H. F. Robinson. 1967. Quantitative genetic investigations of yield in maize. *Der Züchter* **37**: 192–199.

Moll, R. H. and C. W. Stuber. 1971. Comparison of response to alternative selection procedures initiated with two populations of maize (*Zea mays* L.). *Crop Sci.* **11**: 706–711.

Paterniani, E. 1967. Selection among and within half-sib families in a Brazilian population of maize (*Zea mays* L.). *Crop Sci.* **7**: 212–216.

Penny, L. H., G. E. Scott, and W. D. Guthrie. 1967. Recurrent selection for European corn borer resistance in maize. *Crop Sci.* **7:** 407–409.

Robinson, H. F. and R. E. Comstock. 1955. Analysis of genetic variability in corn with reference to probable effects of selection. In *Population genetics: the nature and causes of genetic variability in populations. Cold Spring Harbor Symp. Quant. Biol.* **20:** 127–130.

Robinson, H. F., R. E. Comstock, and P. H. Harvey. 1949. Estimates of heritability and the degree of dominance in corn. *Agron. J.* **41:** 353–359.

Robinson, H. F., R. E. Comstock, and P. H. Harvey. 1955. Genetic variances in open-pollinated varieties of corn. *Genetics* **40:** 45–60.

Russell, W. A. and S. A. Eberhart. 1975. Hybrid performance of selected maize lines from reciprocal recurrent selection and testcross selection programs. *Crop Sci.* **15:** 1–4.

Sprague, G. F. 1966. Quantitative genetics in plant improvement. In *Plant Breeding,* K. J. Frey (ed.). Ames, Iowa: Iowa State Univ. Press, pp. 315–354.

Sprague, G. F. and S. A. Eberhart. (1977). Corn breeding. In G. F. Sprague (ed.). Madison, Wisc: American Society of Agronomy.

Troyer, A. F. and W. L. Brown. 1972. Selection for early flowering in corn. *Crop Sci.* **12:** 301–304.

Vera, G. A. and P. L. Crane. 1970. Effects of selection for lower ear height in synthetic populations of maize. *Crop Sci.* **10:** 286–288.

Webel, O. D. and J. H. Lonnquist. 1967. An evaluation of modified ear-to-row selection in a population of corn (*Zea mays* L.). *Crop Sci.* **7:** 651–655.

Zuber, M. S. 1973. Evaluation of progress in selection for stalk quality. *Proc. 28th Annu. Corn and Sorghum Res. Conf.,* Washington, D.C.: American Seed Trade Association, pp. 110–112.

Zuber, M. S., M. L. Fairchild, A. J. Keaster, V. L. Fergason, G. F. Krause, H. Hildebrand, and P. J. Loesch, Jr. 1971. Evaluation of 10 generations of mass selection for corn earworm resistance. *Crop Sci.* **11:** 16–18.

Chapter 15 ✓

POTENTIAL OF EXOTIC GERMPLASM FOR MAIZE IMPROVEMENT

Arnel R. Hallauer

Research Geneticist, Agricultural Research Service,
USDA, and Professor of Plant Breeding,
Iowa State University

Potential usefulness of exotic germplasm in maize (*Zea mays* L.) improvement programs has been emphasized for several years (Brown, 1953; Wellhausen, 1956, 1965). The Southern corn leaf blight (*Bipolaris maydis*) of 1970 was instrumental in making maize breeders aware of the possible genetic vulnerability of maize germplasm. Although the Southern corn leaf blight epidemic was related to the widespread use of Texas male sterile (T) cytoplasm in the production of hybrid seed, the possibilities that maize also may be genetically vulnerable to a pest epidemic also was theorized. Data reported by Sprague (1971) showed that a few inbred lines indeed were used extensively in hybrid production; genetic vulnerability, therefore, was a reality and not theory.

Contribution from the Agricultural Research Service, USDA, and Journal Paper No. J-8165 of the Iowa Agriculture and Home Economics Experiment Station, Ames, Iowa 50011. Project No. 1897.

Exotic germplasm can have different connotations. The term *exotic* is defined as being of foreign origin; extraneous; not native; introduced from abroad, but not fully naturalized or acclimatized. This broad definition would include all germplasm that would not be immediately adapted for a given maize-improvement program. I use exotic germplasm for all sources of germplasm that are not immediately useful or adapted for a specific maize-improvement program. In other words, physiological adjustments will be necessary to naturalize and acclimatize the germplasm to use either directly or with adapted germplasm. Relatives of maize and their contributions to maize germplasm are deferred to other chapters and are discussed by Mangelsdorf (1974).

REVIEW

Variability

Brown (1953) and Wellhausen (1956, 1965) emphasized the importance and potential of exotic germplasm for maize improvement. Both authors presented their ideas before the recent concern of genetic vulnerability. They considered exotic germplasm useful as sources for genes, and gene combinations valuable for increasing genetic variability and heterosis. Because Corn Belt maize of the United States was largely derived from a complex mixture of Southern Dents, mostly of Mexico origin, and 8–10-rowed Northern Flints, they believed greatest genetic advance could be made from complex mixtures of diverse germplasm. They recommended that different varieties and races be intermated to form a "hodge-podge" or "mess" and that selection be gradually concentrated for desirable gene combinations by use of simple selection procedures. Because of the wide diversity of maize varieties and races in Mexico, Central America, and the Caribbean Islands, they thought breeders in the United States were in an excellent position to widen the genetic base of their breeding programs by using exotic germplasm from these areas.

Goodman (1965) obtained estimates of genetic variability for a population of adapted germplasm (Corn Belt Composite) and a population that included germplasm from the West Indies (West Indian Composite). Full-sib and half-sib progenies were grown in Iowa and North Carolina to compare estimates of genetic variability and predicted gain from selection. Estimates of genetic variability and expected gains from selection were consistently greater for West Indian Composite than those for Corn Belt Composite. Mean yield of West Indian Composite was higher in North Carolina and lower in Iowa than the mean yield for Corn Belt Composite. But the upper-range of the full-sib family means for West Indian Composite exceeded those for Corn Belt Composite at both locations. Goodman's results show that genetic variability was greater for the population containing exotic germplasm and not at the expense of lower yields. At comparable yield levels and greater genetic variability, the West Indian composite would be superior to the Corn Belt Composite as germplasm in a maize-improvement program. Not only would the predicted genetic gain be greater in West Indian Composite (e.g., 15.3% vs. 9.0% for full-sib family selection), but new genes and gene combinations would be introduced in the maize improvement program. The opportunities are good that material developed from West Indian Composite would contribute significantly to heterosis of hybrids.

Chopra (1964) characterized the genetic variation in an adapted variety (Krug Yellow Dent), an exotic variety (Tabloncillo from Mexico), and in the advanced generation progeny of the cross, Krug Yellow Dent × Tabloncillo. Families derived from the advanced cross population were tested in Nebraska and Mexico, whereas the parental families were tested in their respective regions of adaptation. Mean yield of the variety-cross families was low and closer to that of Tabloncillo in Mexico than that of Krug in Nebraska. Estimates of additive genetic variance were greater for the variety cross than for either parent. Chopra concluded that hybridization with exotic germplasm was effective for increasing genetic variability. Shauman (1971) also reported greater estimates of additive genetic variance in the crossed population of Krug × Taboncillo 13 Hi Syn 3.

Romero-Franco (1965) studied composites that had incorporated Mexican and Caribbean germplasm into Corn Belt populations for the improvement of Corn Belt maize. Full-sib families developed from the advanced generations of the crosses were evaluated. Romero obtained an appreciable percentage of full-sib families whose yields equaled or exceeded the average yield of adapted check hybrids. He concluded that the yield improvement resulted because of new favorable gene combinations.

Heterosis

Heterosis among races and varieties of Central and South America was reported by Wellhausen (1965). Most of the tests were conducted in Mexico, but a few were in South America. High-parent heterosis ranged from 103% to 153% for 18 interracial crosses with an average of 136%. The interracial crosses were produced among parents of diverse origin, and all crosses yielded significantly greater than the hybrid, H-352, made from inbred lines from the race, Celaya. Wellhausen thought it was conceivable that a greater yielding variety than the race, Celaya, could be developed through a gradual concentration of the additive variance in the advanced generation of a mixture of the interracial crosses. In other tests, Antigua, Cuban Flint, Harinoso de Ocho, and Tuxpeño races would be the most immediately useful varieties for the yield improvement of U. S. Corn Belt germplasm because of their good general combining ability. Wellhausen summarized that further maize improvement through a more complete exploitation of exotic germplasm of the tropics was extremely great. He suggested the formation of four gene pools of the most promising races, crossing with Corn Belt germplasm, and backcrossing the exotic × Corn Belt crosses to the Corn Belt germplasm. To incorporate exotic germplasm into Corn Belt germplasm, Wellhausen suggested starting with small percentages ($\leq 25\%$) of the exotic germplasm.

Moll, Salhuana, et al. (1962) and Moll, Lonnquist, et al. (1965) evaluated crosses to determine heterosis as related to genetic diversity; degree of genetic diversity was inferred from ancestral relationships and geographical origin. Moll, Salhuana, et al. (1962) found greater heterosis associated with greater genetic diversity of the parental varieties when the variety crosses were tested in North Carolina. Moll, Lonnquist, et al. (1965) tested two varieties of four regions in all combinations as variety crosses (F_1s) and variety crosses selfed (F_2s). Parental, F_1, and F_2 populations were grown in the four regions of adaptation to determine the relation of heterosis to genetic diversity. Their results showed that heterosis increased with increased divergence within a restricted range of divergence, but extremely divergent crosses resulted in a decrease in heterosis. Although most studies had indicated that divergent origin of lines for expression of heterosis in hybrids was desirable, there evidently must be a limit to which this concept can be extended. Studies showing greater heterosis in hybrids among unrelated lines than among related lines, however, were on lines developed from adapted germplasm (Johnson and Hayes, 1940; Wu, 1939). But Griffing and Lindstrom (1954) showed heterosis to be greater for adapted × exotic than for adapted × adapted line crosses. Perhaps, the extremes in maize genomes had not been crossed and tested in previous studies. It can be speculated that the wide crosses tested by Moll, Lonnquist, et al. (1965) resulted in genomes that were not compatible for grain production in first generation crosses.

Eberhart (1971) evaluated two diallel series of crosses that included adapted and semiexotic varieties at three Corn Belt locations for the Corn Belt diallel and at six southern locations for the southern diallel. He identified three semiexotic Corn Belt varieties [BSTL, BS2, and PHWI(MT)C7] and two semiexotic southern varieties [PHWI(MT)C7 and Indian Chief × Diente de Caballo (S)C2] that approached or exceeded adapted varieties. Two decades of crossing and intermating were required to develop most of the semiexotic varieties, and Eberhart considered the better varieties as excellent sources of germplasm useful as breeding populations with additional genetic diversity.

Crosses among 12 well-defined races of Central and South America were tested in Brazil by Paterniani and Lonnquist (1963). Average high-parent heterosis was 14%, ranging from −19% to 84% for individual crosses. Genetic diversity among races within an endosperm type was sufficient that crosses within endosperm types were as productive as those between endosperm types. Ranking of crosses by endosperm types was dent × dent (44.4 q/ha), dent × flour (44.2 q/ha), flour × flour (42.0 q/ha), and flint × flint (36.8 q/ha). The better interracial crosses were nearly equal to the yield of the check double-cross hybrid. Itaici (a Paulistra dent), Cateto (a flint), and Carmen (a Tuxpeño from Mexico) were the races that seemed to have immediate usefulness in Brazil maize-breeding programs.

Hallauer and Malithano (1975) included three semiexotic varieties in two diallel series of crosses with Corn Belt adapted varieties tested in Iowa. The three semiexotic varieties were BS2 (formed by crossing ETO Composite with six early inbred lines and intermating) BSTL (formed by crossing Lancaster with Tuxpeño, backcrossing to Lancaster, and intermating), and Teozea (crossing adapted Corn Belt germplasm with Teosinte). Yields of the semiexotic varieties were equal to most of the adapted varieties. In variety crosses, average heterosis (12.4 q/ha) of Teozea for yield was greater than for any of the unimproved adapted varieties. Genetic diversity was evident in all variety crosses that included Teozea. Average heterosis for BSTL (11.5 q/ha) and BS2 (10.1 q/ha) in crosses with improved adapted varieties were similar to those of adapted varieties after five to seven cycles of recurrent selection for yield.

When maize breeders in the U. S. Corn Belt introduced sources of exotic germplasm, usually only germplasm from Central and South America was considered. Germplasm from other parts of the world that may have greater adaptability to the temperature regions also should be considered (Brandolini, 1969; Leng, Tavčár, et al., 1962; Lonnquist, 1974). Troyer, Forrest, and Hallauer (1968) reported data obtained for 10 early flint varieties and their diallel series of crosses. The 10 varieties were selected from 150 sources that were compared with Gaspe Flint for date of flower and other agronomic traits. Average variety heterosis (mean of parents plotted against mean of crosses) was 71%, indicating that the parent varieties were genetically diverse. Average heterosis was 43% relative to high parent and 72% relative to the midparent. Syzldecka × Motto was the highest-yielding variety cross (53.1 q/ha) and Saskachewan, the highest-yielding variety (31.2 q/ha). Crosses that included Motto as one of the parents had the highest average yields (43.1 q/ha). Early flint varieties should be valuable germplasm sources for increasing genetic variability in maize-breeding programs at the higher latitudes.

Population Improvement

The Rockefeller Foundation (1964–1965) reported that mass selection was effective for the improvement of the Chalqueno race. Eighteen varieties of the Chalqueno race were diallel crossed, and equal quantities of seed from each cross were bulked and advanced to Syn II, III, and IV by use of mass selection techniques. Selection intensity was 4% in each cycle. Average gain in yield per cycle of selection was about 10%. Wellhausen (1965) compared the effects of mass selection in a Celaya–Exotic Composite and a pure population of Celaya, formed by mixing seeds of the varieties typical of the race. The third cycle of mass selection of the Celaya–Exotic Composite yielded 70 q/ha and seemed well ahead of the pure Celaya race for comparable cycles of mass selection.

Effectiveness of adaptive mass selection for eight unadapted populations developed from Corn Belt and exotic germplasm was compared with two adapted populations (Corn Belt Composite and Early Corn Belt Composite) in a study by Mathema (1971). The initial and most advanced cycle populations were evaluated at two plant densities (34.4 and 51.7 plants/ha), and data were collected for yield, prolificacy, and six other agronomic traits. Rate of yield increase per cycle of selection was generally greater at the lower density, at which selection was practiced. Increases per cycle of selection were greater in the unadapted (3.14 q/ha) than in the adapted populations (1.72 q/ha).

Vera and Crane (1970) reported on the effectiveness of selection for lower ear height in two adapted × exotic populations, [Antigua 2D × (B10 × B14)] and (ETO × CBC)F_5, in Indiana. Selection intensity was 50% in each of two cycles for lower ear height. Selection was effective in lowering ear height 4.5% per cycle with no effect on percentage moisture. A slight reduction in yield and a pronounced reduction in lodging were not significant statistically.

Selection for early silking in two synthetic varieties including Mexican germplasm (Zapalote Synthetic and Mexican Synthetic) and one including West Indian germplasm (West Indian Composite) was reported by Troyer and Brown (1972). The Zapalote Synthetic was developed by crossing the related races, Zapalote Chico and Zapalote Grande, with 11 central Corn Belt lines. The crosses were grown in isolated plots and mass selected for recombination genotypes in southern Iowa for 10 years before selection for earliness in Minnesota. The Mexican Synthetic included several Mexican varieties that were not of Zapalote origin crossed to 10 central Corn Belt lines; mass selection also was practiced in this synthetic for 10 years in southern Iowa for desirable plant types. The West Indian Synthetic was formed by crossing high-yielding varieties from the West Indies with 10 central Corn Belt inbreds. After a generation of random mating, the synthetic was grown in isolated fields in southern Iowa for 5 years and selection practiced for recombination genotypes and West Indian phenotypes; Goodman (1965) and Eberhart (1971) also reported data on this variety. Sibmating the 5% early silking plants in each population for six cycles significantly changed every trait measured. Days to silk decreased 1.8 days per cycle accompanied by 1.00 q/ha yield increase, 1.2% grain moisture decrease, and 7.2 and 5.2 cm decrease for plant and ear height. Selection was effective for adapting the semiexotic germplasm for the more northern latitudes of the U. S. Corn Belt for semiexotic populations that had previous selection in a latitude intermediate to the origin of exotic germplasm and selection for earliness.

Hallauer and Sears (1972) compared two procedures for adapting Eto Composite for Corn Belt breeding programs: (a) crossing with six early inbred lines and intermating to form BS2 and (b) mass selection for early silking plants in isolated plantings. Heritability estimate for silking on an individual plant basis was 59% in BS2. High heritability for silking was also realized in the ETO Composite because four cycles of adaptive mass selection for earlier silking were effective in developing an earlier silking subpopulation of the ETO Composite. The fourth cycle population was 16 days earlier than the original ETO Composite. Hallauer and Sears (1972) concluded that both procedures were effective for adapting the ETO Composite to the central Corn Belt. The mass selected population, however, includes 100% exotic germplasm, whereas BS2 would include 50% exotic germplasm. The mass-selected population also may be more directly usable than BS2 because of linkage equilibrium.

Progress from 10 cycles of modified ear-to-row selection for an adapted variety (Hayes Golden) and two cycles for an Exotic Composite developed from Corn Belt and exotic germplasm for yield was compared by Bahadur (1974). Average progress was 5.26% per cycle for 10 cycles of selection in Hayes Golden and 2.68% for the Exotic Composite. The results for the Exotic Composite, however, were inconclusive because of the limited cycles of selection. Additional cycles of recombination in the Exotic Composite in future cycles of selection may release additional variability for continued progress and at a greater rate.

Evaluations of Exotic Germplasm

Reports on the use of exotic germplasm usually show positive results. Griffing and Lindstrom (1954) were one of the first to evaluate hybrids composed of inbred lines that: (a) were of Corn Belt origin, (b) included germplasm from Brazil, and (c) included 25% or 50% Mexican germplasm. They demonstrated that hybrids derived either wholly or partly from inbreds of exotic germplasm yielded more than those from adapted sources. Nelson (1972) indicated that he had been successful in using exotic germplasm in a commercial maize-improvement program. Lines that included Tuxpan germplasm were commercially useable from Florida to Texas and had contributed to wide southern U. S. adaptation, excellent combining ability, and drought and disease tolerance. Exotic inbreds also were excellent for the introgression of exotic germplasm into adapted U. S. genotypes.

Kramer and Ullstrup (1959) evaluated 1066 maize introductions from 87 countries for rust (*Puccinia sorghi*), southern rust (*Puccinia polysora*), and two leaf blights (*Helminthosporium turcicum* and *Helminthosporium maydis*).

They also crossed 572 introductions with Wf9 × Hy and collected data for yield and eight other plant and ear traits. They concluded from their evaluations that exotic germplasm provides excellent opportunities for widening the genetic base for disease resistance, but that selecting germplasm for increased yield potential would likely be less rewarding. For yield, Kramer and Ullstrup based their conclusions on the performance of the exotic germplasm relative to Wf9 × Hy, which was a widely used single-cross hybrid at the time of evaluation. At the 5% significance level, 103 (Wf9 × Hy) × introductions were not different in yield from Wf9 × Hy. It seems, therefore, that several of the exotic germplasm sources would be good gene pools for developing lines in a maize-improvement program. The (Wf9 × Hy) × introduction yields would be the average performance of the exotic variety with Wf9 × Hy and not the elite genotypes of the exotic varieties in combination with Wf9 × Hy. It seems reasonable that superior genotypes could be isolated from the exotic varieties, either by direct selection or recycling, that would be superior in crosses with Wf9 × Hy than the (Wf9 × Hy) × introduction crosses.

Thompson (1968) compared exotic varieties (primarily Tehua and Tuxpeño germplasm) with locally adapted hybrids in North Carolina for their potential in silage production. Exotic varieties yielded 13–41% more dry matter and 4–28% more estimated total digestible nutrients than the adapted full-season hybrids. Exotics had lower crude protein and more fiber than the adapted hybrids, but differences in yield of digestible protein on a land-area basis were small. Thompson concluded that exotic varieties would be useful for management regimes demanding maximum silage production but tolerating less than optimum quality.

Efron and Everett (1969) explored the potential of exotic germplasm for improving maize hybrids for the short-season temperate zones of the United States. Twelve representative short-season hybrids were crossed with bulk pollen of 10 races of Mexican maize and two sources of maize–teosinte admixtures. Progenies were developed by sibbing and selfing within the crosses and were used to develop three groups of synthetics: M—originated from crosses with multiple races of maize, C—originated from crosses with maize–teosinte admixture selected to be maize-like, and T—originated from crosses with maize–teosinte admixture selected to be teosinte-like. Group M was the most productive in green weight, stover dry matter, and total dry matter. Group M, however, was significantly lower in grain yield than either groups C or T in the silage harvest and group C in the grain harvest. In top crosses with four testers, all synthetics yielded significantly less than adapted single-cross controls, but were excellent in vegetative growth. Efron and Everett concluded that the materials observed were not so promising for grain producing, but they suggested a recurrent selection plan for concentrating the best gametes from the five best synthetics for increasing the frequency of desirable genes.

Exotic germplasm for resistance to first- and second-brood European corn borer (*Ostrinia nubilalis,* Hübner) was examined in Ithaca, New York by Sullivan, Gracen, et al. (1974). Visual plant ratings and DIMBOA analyses were made for first-brood resistance and ratings for number of live larvae, number and length of tunnels in the stalk, shank, and leaf sheath for second-brood resistance. Compared with adapted control inbred lines, the exotic selections had resistance to both broods of the corn borer. Previous research (Klun, Guthrie, et al., 1970) had shown an association between first-brood resistance and high concentrations of DIMBOA. But the exotic selections had a low concentration of DIMBOA with a high level of leaf-feeding resistance, indicating a different type of resistance. For second-brood resistance, the exotic selections had better resistance than the only previously known resistant genotype, B52.

All of the reports of crosses among races, varieties, and composites show considerable genetic diversity among germplasm sources. Estimates of genetic variability and results of selection experiments generally show increased genetic variability within populations including exotic germplasm. Maize breeders, therefore, have a wealth of germplasm to choose from, and considerable data are available to assist in the selection of germplasm that have potential in breeding programs. Except for some of the semiexotic populations (Eberhart, 1971; Hallauer and Malithano, 1975) and populations from the temperate regions (Brandolini, 1969; Leng, Tavcar, et al. 1962; Lonnquist, 1974), the germplasm is not directly usable in Corn Belt maize-breeding programs. Selection studies for adapting exotic germplasm for use in Corn Belt maize-improvement programs have been effective (Vera and Crane, 1970; Mathema, 1971; Nelson, 1972; Hallauer and Sears, 1972; and Troyer and Brown, 1972; Bahadur, 1974). For programs committed to inbred line and hybrid development, only Griffing and Lindstrom (1954) and Nelson (1972) have reported success from use of exotic germplasm; all other studies were on populations. In many programs the crucial usefulness of exotic germplasm is whether the material can be used in production of superior hybrids. Griffing and Lindstrom's (1954) results indicated that the diversity obtained from exotic germplasm enhanced crosses among adapted × unadapted lines. Although not published, experiences by maize breeders often were discouraging when inbreeding was initiated in exotic source populations. Severe inbreeding depression usually was experienced. With increased diversity this probably should be expected. Also, adequate random mating of the exotic × adapted populations was probably not sufficient to approach linkage equilibrium. Lack of coadaptive gene combinations within populations containing exotic germplasm would contribute to the disastrous effects of observed inbreeding. Recombination of the most desirable progenies from the initial samplings of adapted × exotic populations probably would have been desirable in most instances. If the favorable effects of recycling had been appreciated to allow additional cycles of recombination, future samplings of the resynthized populations probably would have been more beneficial.

EXPERIMENTAL DATA

A sample of the ETO Composite was obtained from Colombia in 1963. The ETO Composite was made up of various exotic collections, and its composition, description, and development was given by Chavarriaga (1966). In the tropics the ETO Composite has variability and good yield and has performed high in some specific combinations (Wellhausen, 1965). Because the ETO Composite had the desired characteristics needed for population improvement, it was decided to determine whether this composite could be used in Corn Belt breeding programs to increase the genetic variability of our source breeding materials.

The original sample of the ETO Composite was grown in a 0.2-ha isolation field near Ames, Iowa, in 1963. Plant (4.5 m) and ear heights (3.0 m) exceeded desired standards, and mean silking date was about August 26. The population, however, was variable for plant and ear height and silking date, and we tagged 250 plants that showed visible silk on or before August 18. Seed harvested from the 250 earliest silking plants was bulked and planted in a 0.5-ha isolation in 1964. The same procedures of mass selection for early silk emergence were repeated in each cycle, except the date of tagging was earlier in successive cycles. After six cycles of mass selection, mean flowering date was about August 1. The first four cycles of mass selection showed −3.8 days for days to silking and −15 cm for ear height decrease per cycle of selection (Hallauer and Sears, 1972). The ETO Composite population after six cycles of mass selection has been designated as BS16.

Because the original planting of the ETO Composite in 1963 was so tall and late, prospects for direct use in the Corn Belt were not promising. Crossing the ETO Composite with early germplasm seemed to be one alternative for reducing the lateness and tallness of the ETO Composite to make it useable for the central U. S. Corn Belt. In 1965 we were successful in crossing about 40 ETO Composite plants to each of six early lines (A251, A554, A575, A619, Mt42, and ND203). Five generations of intercrossing and random mating of the ETO Composite × early line crosses were completed. The resultant synthetic was designated as BS2 (Hallauer and Sears, 1972).

The BS2 and BS16 lines have been evaluated in standard yield-test plots for yield, grain moisture, lodging, and dropped ears. Table 1 presents the results for data obtained in six environments. For comparison, BSSS (Iowa Stiff Stalk Synthetic), BSK (Krug open-pollinated variety), BSTL (Lancaster × Tuxpeño, backcrossed to Lancaster), and Teozea (Corn Belt maize × teosinte) were

TABLE 1.   Agronomic data for six maize populations obtained at three locations in Iowa for 2 years

| Population | Yield (q/ha) | Grain moisture (%) | Lodging (%) | | Dropped ears (%) | Yield in crosses (q/ha) |
			Root	Stalk		
BSSS	55.3	25.5	6.5	12.6	0.7	61.5
BSKCO	50.6	23.9	8.1	28.6	0.5	59.4
BSK(S)C5	58.4	24.2	6.9	38.5	1.6	65.6
BSTL	58.1	24.3	4.6	17.5	0.6	69.0
BS2	56.2	22.2	14.2	21.2	0.0	66.7
BS16	63.9	24.7	17.7	17.5	0.2	—
Teozea	52.6	20.6	3.7	25.5	0.0	64.2
Experimental average	54.1	22.6	4.8	24.2	0.5	64.4
LSD (0.05)	7.5	0.8	5.3	8.9	1.0	3.0
Average—checks[a]	84.5	24.2	3.6	4.1	0.2	—

[a] Four single-cross checks were N28 × Mo17, B37 × B70, B37 × B45, and B37 × Oh43.

included. The BSSS and BSK lines represent adapted Corn Belt germplasm, and the proportion of exotic germplasm is 25% (BSTEL); 50% (BS2 and Teozea), and 100% (BS16) for the other populations. The BS16 line was significantly better than BSSS and BSK for yield. The BS2, BSTL, and Teozea lines were within one LSD (0.05) of the adapted populations for yield. Teozea was lower yielding than BSSS, but grain moisture also was significantly lower. Stalk lodging was greater for the populations including exotic germplasm, but except for Teozea they were within one LSD (0.05) of BSSS. The BS2, BS16, and BSTL lines were superior to BSK for stalk lodging. In all instances, the populations with exotic germplasm were superior to the open-pollinated variety, BSK. To compare progress from selection, BS2, BS16, BSTL, and BSK are undergoing S_2 recurrent selection.

Results of the S_2 recurrent selection trials for BS2, BS16, BSK, and BSTL are summarized in Table 2. The S_2 recurrent selection was initiated in BS2 and BSTL in 1971 and for the fifth cycle for BSK. The same selection was initiated in BS16 in 1974, second cycles were completed for BS2 and BSTL, and six cycles were completed for BSK. Estimates of the components of variance, repeatability, and predicted gain were similar for all populations within each year. The V_g value for BS2 and BSTL was greater than for BSK in 1971 and 1974, but all estimates were within two standard errors of one another. The V_g value for BS16 was greater than for the other populations in 1974. The lack of increased variability (V_g) in the populations including exotic germplasm may not seem encouraging. But the mean yields (Table 1) and predicted gains (Table 2) would indicate that the expected performance of selected genotypes from exotic or semiexotic germplasm would be comparable to those from adapted germplasm. Additionally, genotypes developed from exotic germplasm would alleviate the present concern of genetic vulnerability and introduce new genes in our breeding populations.

The estimates of V_g for BS2, BS16, and BSTL, however, may be conservative estimates of the genetic variability. The S_2 progenies included for yield testing were a selected segment from a larger group of S_1 progenies screened for plant and ear types, stalk rot, and first-brood European corn borer. For example, the 100 S_2 progenies of BS2 yield tested in 1971 and 1974 were selected from 339 and 492 S_1 progenies, respectively. The S_2s tested, therefore, were selected S_1 plants within selected S_1 progenies. Within BS16, 156 S_2 progenies were tested from a sample of 268 S_1 progenies. If there was an association between yield and the traits used to discard S_1 progenies, the variability may have been compressed.

The S_1 and S_2 progenies of BS16 also were evaluated for European corn-borer resistance and sorghum downy mildew [*Sclerospora sorghi* (Kulk.)]. Frequency of resistance in BS16 was high for both pests (Table 3). Frequency of resistance to the first- and second-brood European corn borer was higher than for any population previously screened. For first-brood resistance, 44% of the S_1 progenies had a rating of five or less. Forty of the S_1 lines had a visual rating less than six for second-brood resistance. Retest of 39 S_2 lines for second-brood resistance showed that 24 were within one LSD (0.05) of the resistant inbred check, B52. The B52 is the only genotype identified to have resistance to the second-brood corn borer. The BS16 seems to be an excellent source for developing genotypes having resistance to both broods. The results for BS16 for corn borer resistance agree with those of Sullivan, Gracen, et al. (1974) for exotic germplasm, but it is not known if the same DIMBOA relations found by Sullivan, Gracen, et al. (1974) are operative in BS16.

Two hundred of the BS16 S_1 progenies were evaluated for sorghum downy mildew resistance under infection in Texas in 1973. The level of natural infection was low in 1973, and 169 of the 200 S_1 progenies had 0–10% infection. The 156 S_2 progenies included for yield testing were retested for downy mildew in 1974. Level of downy mildew infection was higher in 1974, but 54 (34.6%) of the 156 entries were in the 0–10% class. These results show that recurrent selection in BS16 should be effective for developing a population having high resistance to sorghum downy mildew infection.

The BS16 line was evaluated for 2 years in southern Missouri for maize dwarf mosaic (MDM) and maize chlorotic dwarf (MCD) virus infection. Relative to the resistant and susceptible checks, BS16 had a high frequency of resistance to MDM and MCD. The BS16 line has good potential to be useful in U. S. Corn Belt breeding programs. The S_2 recurrent selection based on

TABLE 2. Pertinent portions of the analyses of variance for grain yield of selfed progenies originating from populations having 0%, 25%, 50%, and 100% exotic maize germplasm

Source	df	1971 Mean squares			1974 Mean squares			BS16(S)C0	
		BS2(S)C0	BSTL(S)C0	BSK(S)C5	BS2(S)C1	BSTL(S)C1	BSK(S)C6	df	M.S.
Entries	99	251.4	456.0	328.7	563.5	501.5	481.9	155	640.9
Entries × environments	198	70.5	57.5	71.0	92.4	126.1	126.0	310	134.4
Error	297	55.9	33.8	45.8	55.0	86.2	65.9	465	58.0
Total	594							930	
C.V. (%)		42.9	21.8	15.9	30.2	29.8	21.7		29.6
LSD (0.05)		9.7	8.6	8.4	11.1	13.0	13.0		13.4
V_g[a]		46.8 ± 11.8	66.4 ± 15.2	32.2 ± 11.0	78.5 ± 18.8	62.6 ± 16.9	59.3 ± 16.2		84.4 ± 17.2
V_{ge}[a]		7.3 ± 4.0	11.9 ± 3.0	12.6 ± 3.8	18.7 ± 4.8	19.9 ± 6.8	30.1 ± 6.5		38.2 ± 5.4
V[a]		55.9 ± 4.6	33.8 ± 2.8	45.8 ± 3.7	55.0 ± 4.5	86.2 ± 7.0	65.9 ± 5.4		58.0 ± 3.8
V_{ph}[a]		58.6	76.0	41.1	93.9	83.5	80.3		106.8
Repeatability (%)		79.9	87.4	78.4	83.6	74.8	73.8		79.0
$\bar{X}$ (q/ha)		17.4	25.8	42.1	24.5	31.1	37.4		25.7
ΔG (q/ha)[b]		8.6	10.7	7.0	11.3	9.6	9.3		11.4

[a] Components of variance estimated from the analyses of variance for entries (V_g), interaction of entries with environments (V_{ge}), pooled error (V), and phenotypic variance (V_{ph}).

[b] Gain per cycle (3 years/cycle) with a selection intensity of 20% ($k = 1.4$) and two replications in three environments.

TABLE 3. Disease and insect data obtained for BS16 S_1 and S_2 maize progenies

First-brood European corn borer—S_1 lines

Rating class[a]	Frequency
1	0
2	3
3	20
4	50
5	45
6	69
7	59
8	7
9	15
	268

CI31A[b] = 1.3
WF9[c] = 9.0

Downy mildew

	S_1 lines		S_2 lines	
	Infection (%)	Frequency	Infection (%)	Frequency
	0–10	169	0–10	54
	11–20	15	11–20	23
	21–30	7	21–30	17
	31–40	3	31–40	14
	41–50	3	41–50	13
	51–60	0	51–60	8
	61–70	2	61–70	8
	71–80	0	71–80	6
	81–90	1	81–90	6
	91–100	0	91–100	7
		200		156

Virus (Maize dwarf mosaic and maize chlorotic dwarf)

Variety	Infected plants (%)		Rating	
	1972	1973	1972	1973
BS16	0.0	28.0	1.0	2.8
Mo14W × Oh7B[b]	0.0	2.6	1.0	1.2
H55 × Mo5[c]	25.0	87.2	2.6	6.8

Second-brood European corn borer

Entries	S_1 lines		S_2 Lines	
	Rating[a]	Cavity count	Cavity mean	Range
Lines[d]	2.2	8.8	2.7	0.9–5.5
B52[b]	3.0	4.3	0.9	—
W182E[c]	9.0	39.5	15.7	—

[a] Ratings made on visual rating scale of 1 = resistant to 9 = susceptible.
[b] Resistant check.
[c] Susceptible check.
[d] 200 S_1 lines were rated 1:9 for visual damage, and only those entries (40) rating 6 or less were disected. S_2 lines, 39 of the 40 most resistant S_1 lines were retested.

yield evaluations will be continued in BS16 with selection in the S_1 progenies for disease and insect resistance. The comparisons in Tables 1 and 2 show that the semiexotic populations have good yield levels, acceptable agronomic traits, and sufficient genetic variability to ensure progress from selection. Because of the high frequency of resistance to corn borer and diseases, BS16 may have greater potential than BS2, BSTL, and BSK. Since BS16 was developed by simple mass selection for early silking in the ETO Composite, it has 100% exotic germplasm. Phenotypically, the S_1 progenies of BS16 are superior to those from BS2 and BSTL, which have 50% and 25% exotic germplasm, respectively. The generations of recombination after intercrossing certainly have not been as great in BS2 and BSTL as for BS16. Results for BS16 indicate that wide crosses of adapted and exotic germplasm require considerable intercrossing to develop productive genotypes. If the results for BS16 can be duplicated for other exotic populations, the best procedure may be to adapt productive variable exotic populations rather than crossing adapted by exotic populations.

DISCUSSION AND SUMMARY

All of the evidence reported indicates that exotic germplasm is potentially useful for increasing the genetic variability within our breeding populations, enhancing heterosis because of increased genetic diversity among breeding populations and inbred lines, and serving as sources of resistance to disease and insect pests. On a short-term basis, use of exotic germplasm for the development of inbred lines for hybrids is not so promising. It was emphasized by Brown (1953), Wellhausen (1956, 1965), and Lonnquist (1974) that adapting exotic germplasm requires a long-term program involving cyclical selection and recombination to adapt and acclimatize the exotic germplasm; devoting a lifetime (Brown, 1953) and patience (Lonnquist, 1974) are necessary. Recent evidence (Sprague, 1971) emphasizes the need for increasing the diversity of our breeding populations. Only five of the 25 most widely used public developed inbred lines were derived from open-pollinated varieties; the remaining 20 inbred lines were derived from either crosses and backcrosses of adapted lines or from Stiff Stalk Synthetic, developed by recombining 16 adapted Corn Belt lines. The B57 line, developed from Midland and released in 1963, was the most recently developed line from an open-pollinated variety. Lines developed by crossing and backcrossing adapted lines are improved for specific traits, but the genetic variability among lines has been seriously reduced. The trend has increased rather than decreased in recent years because of the trend from double-cross hybrids to single-cross hybrids. A few elite lines in single-cross hybrids can dominate a large maize acreage. To alleviate the genetic vulnerability of the hybrids, breeding populations that include exotic germplasm for the development of new lines becomes increasingly important. The time lag required to adapt the exotic germplasm and develop new lines demands prompt action by all maize breeders.

Present information suggests using populations that have undergone several years of interbreeding with only mild selection. Intense selection in early generations of crosses of adapted × exotic germplasm could be discouraging and unfeasible because of the lack of recombination of paired chromosomes of diverse germplasm. Additionally, the benefits from using exotic germplasm will not be realized if sufficient time is not permitted for a thorough mixing of adapted and exotic chromosome segments. Lonnquist (1974) indicated that there may well be barriers to recombination of paired chromosomes from crosses of diverse origin. Consequently, it would require several generations of interbreeding to develop gene pools of adaptive genotypes, that is, evolution rather than revolution to develop functional breeding populations. The example for BS2 and BS16 (Tables 1–3) indicates that BS16 is superior to BS2. The ETO Composite was developed in the 1940s and would have had greater opportunity for gene interchange than BS2 formed by recently crossing genotypes of diverse origin. Troyer and Brown (1972) allowed 5–10 generations of interbreeding before they initiated selection for adaptation in higher latitudes. Sufficient interbreeding becomes more important as the amount and diversity of germplasm increases. Most suggestions are that the populations should be developed from an array of germplasm. The rationale for this suggestion is that the high-yielding Corn Belt varieties evolved from hybridization of northern flints and southern dents. Also, some of the most productive complexes in the tropics evolved from introgression of several races of maize and teosinte. The ETO Composite also is an example of a productive variety produced by intercrossing an array of germplasm. In all instances, several generations of intermating with mild selection was permitted for gene interchange. Natural and artificial selection ultimately gave rise to gene pools of productive genotypes.

Crossing of exotic germplasm with adapted germplasm has been the usual procedure for incorporating exotic germplasm. This procedure has been successful if sufficient interbreeding with mild selection was permitted before initiation of intense selection. The West Indian Composite (Goodman, 1965; Eberhart, 1971; Troyer and Brown, 1972) is an example of adapting West Indies germplasm to the Corn Belt by crossing with Corn Belt lines; this composite seems promising using this procedure. The BS16 line, on the other hand, was developed by direct mass selection in the ETO Composite for adaptation to the U. S. Corn Belt, and it seems equally promising as a breeding population and is superior to BS2 developed by crossing the ETO Composite with six early lines. Direct selection in productive exotic gene pools would bypass the problems of recombination in adapted × exotic crosses. There is, however, no gene interchange between adapted and exotic germplasm if exotic varieties are directly adapted without crossing with adapted germplasm. This may be a serious problem, however, if the wrong choice was made in the exotic population selected for adaptive selection.

Recurrent selection seems to offer the best selection procedures needed for adapting exotic germplasm. Initially, simple mass selection may be satisfactory for highly heritable traits of flowering, plant and ear height, and disease and insect resistance. More complex procedures would be in order for increasing gene frequencies for yield and other traits (e.g., stalk quality) for modern maize production. Recurrent selection permits small, successive changes in gene frequencies of hereditary traits in successive cycles of selection, which is essential for developing functional breeding populations including exotic germplasm.

Research is still needed to determine which procedures are most effective for use of exotic maize germplasm, what proportions of exotic germplasm are needed to incorporate in adapted germplasm, how much recombination and gene flow occurs between exotic × adapted crosses, the effects of inbreeding in exotic and semiexotic populations, and the effects of direct selection in exotic varieties versus selection in exotic × adapted variety crosses. A wealth of material is available, and only the time, facilities, and imagination of maize breeders will determine how it is eventually used. Because of the demand of a hungry world for food, large acreages of maize are required, and loss of crops because of a lack of genetic variability is no excuse. Galinat (1972, 1974) has issued a warning about the genetic erosion of the sources of exotic germplasm of maize, but the maize breeder has yet to adequately sample the genetic variability available. The opportunities are great for using exotic germplasm in our breeding programs.

REFERENCES

Bahadur, Khan. 1974. Progress from modified ear-to-row selection in two populations of maize. Ph.D. thesis. Lincoln: University of Nebraska.

Brandolini, A. G. 1969. European races of maize. *Annu. Corn and Sorghum Ind. Res. Conf. Proc.* **24:** 36–48.

Brown, William L. 1953. Sources of germ plasm for hybrid corn. *Annu. Hybrid Corn Ind. Res. Conf. Proc.* **8:** 11–16.

Chavarriaga, E. 1966. Maize Eto, una varieded producida en Colombia Instituto Agropecuario. *Agr. Trop.* No. 1, 30 pp.

Chopra, R. R. 1964. Characterization of genetic variability in an adapted and an exotic variety of corn (*Zea mays* L.) and the cross derived from them. *Diss. Abstr.* **25:** order No. 64-11, 923: 2694–2695. Lincoln: University of Nebraska.

Eberhart, S. A. 1971. Regional maize diallels with U. S. and semi-exotic varieties. *Crop Sci.* **11:** 911–914.

Efron, Y. and H. L. Everett. 1969. Evaluations of exotic germ plasm for improving corn hybrids in Northern United States. *Crop Sci.* **9:** 44–47.

Galinat, W. C. 1972. Some contributions of corn's relatives to the development of its modern varieties. *Annu. Corn and Sorghum Industry Res. Conf. Proc.* **27:** 108–118.

Galinat, W. C. 1974. The domestication and genetic erosion in maize. *Econ. Bot.* **28:** 31–37.

Goodman, Major J. 1965. Estimates of genetic variance in adapted and exotic populations of maize. *Crop Sci.* **5:** 87–90.

Griffing, Bruce and E. W. Lindstrom. 1954. A study of the combining abilities of corn inbreds having varying proportions of Corn Belt and non-Corn Belt germplasm. *Agron. J.* **46:** 545–552.

Hallauer, Arnel R. and D. Malithano. 1975. Evaluations of maize varieties for their potential as breeding populations. *Euphytica* (in press).

Hallauer, Arnel R. and J. H. Sears. 1972. Integrating exotic germplasm into Corn Belt maize breeding programs. *Crop. Sci.* **12:** 203–206.

Johnson, I. J. and H. K. Hayes. 1940. The value in hybrid combinations of inbred lines of corn selected from single crosses by the pedigree method of breeding. *J. Am. Soc. Agron.* **32:** 479–485.

Kramer, H. H. and A. J. Ullstrup. 1959. Preliminary evaluations of exotic maize germplasm. *Agron. J.* **51:** 687–689.

Klun, Jerome A., W. D. Guthrie, Arnel R. Hallauer, and W. A. Russell. 1970. Genetic nature of the concentration of 2,4-dihydroxy-7-methoxy 2H-1, 4-benzoxazin 3(4H)-one and resistance to the European corn borer in a diallel set of eleven maize inbreds. *Crop Sci.* **10:** 87–90.

Leng, E. R., A. Tavcar, and V. Trifunovic. 1962. Maize of southeastern Europe and its potential value in breeding programs elsewhere. *Euphytica* **11:** 263–272.

Lonnquist, J. H. 1974. Considerations and experiences with recombinations of exotic and Corn Belt maize germplasm. *Annu. Corn and Sorghum Ind. Res. Conf. Proc.* **29:** 102–117.

Mangelsdorf, Paul C. 1974. *Corn: Its origin, evolution, and improvement.* Cambridge, Mass.: Belknap Press, Harvard Univ., 262 pp.

Mathema, Brahmaram B. 1971. Evaluation of progress in adapted × exotic maize populations undergoing mass selection in Nebraska. M.S. thesis. Lincoln: University of Nebraska.

Moll, R. H., J. H. Lonnquist, J. V. Fortuno, and E. J. Johnson. 1965. The relationship of heterosis and genetic divergence in maize. Genetics **42:** 139–144.

Moll, R. H., W. S. Salhuana, and H. F. Robinson. 1962. Heterosis and genetic diversity in variety crosses of maize. *Crop. Sci.* **2:** 197–198.

Nelson, H. G. 1972. The use of exotic germplasm in practical corn breeding programs. *Annu. Corn and Sorghum Ind. Res. Conf. Proc.* **27:** 115–118.

Paterniani, E. and J. H. Lonnquist. 1963. Heterosis in interacial crosses of corn (*Zea mays* L.). *Crop Sci.* **3:** 504–507.

Rockefeller Foundation Annual Report. 1964–1965. Program in the agriculture Science-International food crop improvement program. Maize, pp. 203–204.

Romero-Franco, J. E. 1965. The use of exotic germ plasm for the development and improvement of maize hybrids. M.S. thesis, Lincoln: University of Nebraska.

Shauman, W. L. 1971. Effect of incorporation of exotic germ plasm on the genetic variance components of an adapted, open-pollinated corn variety at two plant population densities. Ph.D. thesis. Lincoln: University of Nebraska.

Sprague, G. F. 1971. Genetic vulnerability to disease and insects in corn and sorghum. *Annu. Hybrid Corn Ind. Res. Conf. Proc.* **26:** 96–104.

Sullivan, S. L., V. E. Gracen, and Alejandro Ortega. 1974. Resistance of exotic maize varieties to the European corn borer *Ostrinia nubilalis* (Hübner). *Environ. Entomol.* **3:** 718–720.

Thompson, D. L. 1968. Silage yield of exotic corn. *Agron. J.* **60:** 579–581.

Troyer, A. Forrest, and Arnel R. Hallauer. 1968. Analysis of a diallel set of early flint varieties of maize. *Crop Sci.* **8:** 581–584.

Troyer, A. F. and W. L. Brown. 1972. Selection for early flowering in corn. *Crop Sci.* **12:** 301–304.

Vera, G. A. and P. L. Crane. 1970. Effects of selection for lower ear height in synthetic populations of maize. *Crop Sci.* **10:** 286–288.

Wellhausen, E. J. 1956. Improving American corn with exotic germ plasm. *Annu. Hybrid Corn Ind. Res. Conf. Proc.* **11:** 85–96.

Wellhausen, E. J. 1965. Exotic germ plasm for improvement of Corn Belt maize. *Annu. Hybrid Corn Ind. Res. Conf. Proc.* **20:** 31–45.

Wu, S. K. 1939. The relationship between the origin of selfed lines of corn and their value in hybrid combination. *J. Am. Soc. Agron.* **31:** 131–140.

Chapter 16

BREEDING FOR INDUSTRIAL AND NUTRITIONAL QUALITY IN MAIZE

R. G. Creech and D. E. Alexander
Mississippi State University and University of Illinois

Once again corn has demonstrated its importance. When drought hit in the Corn Belt in the midwestern United States in 1974, U. S. livestock numbers and feeding trends changed. It brought our Japanese and European Common Market friends to the USDA to discuss their adjustments. As Secretary of Agriculture Butz has said, "Two weeks of high temperatures and hot winds in the Corn Belt can make ripples in the economies around the world."

Despite an estimated 3% increase in plant corn area, worldwide corn production in 1974 was estimated at 279.2 million metric tons, which was about 10% below the 311.6 million metric tons produced in 1973, but still 7% higher than the estimated 261.1 million metric tons average of 1968–1972.[1] Most of the decline of 1974 is attributed to unfavorable weather conditions. North America's output can be blamed for most of the decline with an estimated production of 130.5 million metric tons, 17% below the 1973 level. United States production declined from 143.4 to 118.1 million metric tons, 18% below that of 1973.

[1] Source: Grain and Feed Division, Foreign Agricultural Service, USDA.

TABLE 1.   Utilization of corn for domestic use in the United States 1966–1974 (in millions of metric tons)[a]

Year	Feed	Food, industry, and seed	Total domestic	Exports	Total use[b]
1965	85.4	9.2	94.6	17.4	112.0
1966	84.5	9.4	93.9	12.4	106.3
1967	89.1	9.6	98.7	16.1	114.8
1968	90.9	9.8	100.7	13.6	114.3
1969	96.4	10.0	106.4	15.5	121.9
1970	90.9	10.1	101.0	13.1	114.1
1971	101.0	10.4	111.4	20.2	131.6
1972	109.5	10.7	120.2	31.9	152.1
1973	106.6	11.0	117.6	31.6	149.2
1974	86.6	11.4	98.0	24.8	122.8

[a] Source: USDA, Economic Research Service.

[b] Includes carryover from previous crops and imports.

In the United States corn is utilized as feed, food, seed, and in industrial raw products. Table 1 shows how corn was utilized in the United States during 1965–1974, when its use increased from 112 to about 123 million metric tons. Corn used for feed generally increased through 1972, but has subsequently decreased apparently because of an unfavorable ratio of grain to live animal prices. Corn used for food, industry, and seed has shown a generally steady increase from 9.2 to 11.4 million metric tons, which is only 9.3% of corn consumed in 1974. More than 300 million bu (7.6 million metric tons) were processed in the United States in 1974, accounting for about two-thirds of the 1973–1974 crop going into food, industrial, and seed uses. This corn provided about 4.1 billion kg (9 billion lb) of starch and 2.8 million metric tons of animal feed. Approximately 2.7 billion kg (6 billion lb) of starch was converted to corn syrup, dextrose, and maltodextrin.[2]

The corn kernel is a very valuable chemical entity, as the above paragraph confirms. The general composition of the dry dent corn kernel, although it varies somewhat according to genotype and environmental variables, is as follows:

Carbohydrates	80.0%
Protein	10.0%
Oil	4.5%
Fiber	3.5%
Minerals	2.0%

[2] Source: 1975 Corn Annual, Corn Refiners Association, Inc., Washington, D.C.

These corn products are classed as "renewable" resources as contrasted with petroleum and minerals deposited in the earth's surface. Consequently, a new supply is grown annually.

The importance of corn for industrial, food, and feed purposes has been well established. The remainder of this chapter will address the subject of genetics and breeding for industrial and nutritional quality from the standpoint of carbohydrates, protein and oil. Breeding programs are underway for other chemical constituents of the kernel; however, they will not be considered.

CARBOHYDRATES: QUANTITY AND QUALITY

The carbohydrate fraction of the corn kernel has many uses. Starch is the major component of the carbohydrate fraction in normal corn and is one of the most abundant materials in the plant world. Starch and its derivatives, such as desserts, jellies, preserves, candy, and frozen syrup and sugar, are widely used in manufacture of ice cream and other frozen fruits, soft drinks, and alcoholic beverages. Starch is also an important raw product in the manufacture of many nonfood products that have become necessities of modern life. Paper, absorbents, adhesives, and textiles, for example, require relatively large amounts of starch in their manufacture. Starch is also an important component of such varied items as medicines and explosives; and in such diverse processes as printing, ore refining and oil-well drilling. Several new industrial uses of corn starch are under development. They include water absorbents, compounds that allow the handling of rubber in a powdered form, plastics, and compounds that remove metals from water. Corn is by far the largest source of starch in the United States.

Research on the genetics and biochemistry of starch synthesis in corn has almost all been carried out in the past 40 years. Unquestionably, impetus was added by the developmental research on waxy corn that came during World War II. The potential role of high amylose starch in the production of films and fibers subsequently stimulated additional interest in the genetic regulation of starch synthesis and the breeding of corn hybrids with qualitative and quantitative differences in the starch fraction. In addition, considerable interest has been shown by sweet-corn breeders in certain mutants that affect carbohydrate properties.

The standard corn-starch granule is usually a mixture of two polysaccharides (Greenwood. 1956; Whelan. 1961). Amylopectin is the major component ranging from 75% to 85%, and is of high molecular weight (1×10^7). It consists of chains of α-D-(1-4) and α-D-(1-6)-glucosidic linkages that form a branched molecule. Amylose, the other major component, is primarily linear with α-D-(1-4)-linked glucose residues.

Amylose has special qualities that makes it particularly useful in the manufacture of certain foods and industrial products. In the late 1940s and early

1950s, when the demand first arose, it was a pressing problem to develop a practical and economical physical process that would separate amylose and amylopectin from a mixture of ordinary corn starch. Another solution was provided by the corn breeders—hybrids carrying high levels of amylose were bred. However, the new types did not carry 100% amylose; hence the physical separation of amylose and amylopectin still was required to enhance the proportion of amylose. However, the earlier success of corn breeders in developing waxy corn with 99–100% amylopectin strongly suggested that corns could be bred that possessed different levels of amylose. Bear (1958) found a simple recessive gene that changed the amylose–amylopectin ratio from 1:3 to 1:1 in mature corn endosperm without drastically reducing total starch production. Bear designated this gene *amylose-extender* (*ae*) (Kramer, Bear, *et al.* 1958). The physical and chemical properties of corn starch at an approximate 1:1 ratio was desirable for many industrial purposes and hybrid development was initiated using the *ae* gene.

Major emphasis during the last 35–40 years has been on waxy maize, comprised of starch with 100% amylopectin, and more recently on amylomaize, with amylose content 50% and higher.

Many mutants affect endosperm components in maize. Several alter the type and quantitity of carbohydrates, including starch. Collins (1909) described a waxy mutant different from normal dent corn in sugar and starch contents. Weatherwax (1922) demonstrated by iodine staining that starch in the endosperm of the Collins mutant consisted solely of a "rare" form of carbohydrate then called *erythrodextrin*, currently known as *amylopectin*. This starch stained red with iodine, in contrast to *amylodextrin*, currently termed *amylose*, which stained blue with iodine. Bates, French, et al. (1943) and Sprague, Brimhall, et al. (1943) confirmed that endosperm starch of waxy maize consisted of nearly all amylopectin. The presence of amylopectin in cereal had been demonstrated previously by Parnell (1921). Other waxy mutants have subsequently been reported by Bear (1944) and Mangelsdorf (1947). The waxy mutants produce starch that is practically all amylopectin (Bates, French, et al. 1943). Andrés and Bascialli (1941) reported on two waxy mutations in a variety of Argentine flint that possessed a small amount of amylose. They were found to be allelic to standard *wx*.

Two sugary genes, designated *sugary-1* (*su*1) and *sugary-2* (*su*2) were described by East and Hayes (1911) and Eyster (1934), respectively. Mangelsdorf (1947) reported that two gene mutations, designated *dull* (*du*) and *amylaceous sugary* (*su*am), also altered the carbohydrate properties of maize endosperm. *Amylaceous sugary* (*su*am) was later found to be allelic to the *su*1 gene. Andrew, Brink, et al. (1944) reported that the recessive gene *wx* increased sugars and water-soluble polysaccharides (WSP) in a *su*1 background. Cameron (1947) reported that the genes *su*1 and *du* interact to increase the amylose content to approximately 65%, compared to the amylose content of standard dent corn of about 25%. These two genes also increased the WSP fraction and decreased the amount of starch.

Kramer and Whistler (1949) established that the amylose content of $su1$ homozygotes was about 35%. Dvonch, Kramer, et al. (1951) and Dunn, Kramer, et al. (1953) reported that the genes du, $su1$, and $su2$ interact to increase the amylose content of starch to about 77%; however, no net increase in absolute yield of amylose was obtained because of a net decrease in amount of endosperm starch. Cameron and Cole (1959) reported that the gene combinations $su1du$ and $su1su2$ partially inhibited the accumulation of starch, compared to $su1$ alone.

Vineyard and Bear (1952) described a mutation, termed *amylose-extender* (*ae*), that substantially increased amylose content. The *ae* endosperm starch in many backgrounds contains approximately 60% amylose. Deatherage, MacMasters, et al. (1954) reported the development of an *ae* hybrid that was high in total starch yield with more than 60% amylose. Kramer, Pfahler, et al. (1958) analyzed many combinations of the genes *ae*, *du*, *su1*, *su2*, and *wx*, and reported that amylose contents ranged from none to more than 70%.

Zuber, Deatherage, et al. (1960) suggested that the range in amylose content of 54–71% in maize inbreds possessing the *ae* gene was due to modifer genes. They also reported that starch content decreased as amylose content increased.

The gene *shrunken-1* (*sh1*) is a mutant that affects endosperm by reducing starch deposition to a very significant degree, resulting in a collapsed shrunken kernel. This mutation was first described by Hutchinson (1921). Burnham (1944) described a mutant with a phenotype somewhat similar to *sh1* that was later shown by Mains (1949) to not be allelic to *sh1*. This gene was designated *shrunken-2*, *sh2*.

Laughnan (1953) studied the effects of the genes *sh2* and *su1* in the distribution of endosperm carbohydrate reserves. He found that almost 20% of the dry matter of the shrunken kernels was sugar, a 10-fold increase over standard dent kernels. Most of the sugar consisted of sucrose. A corresponding decrease in starch was observed. Total carbohydrate production was lowest in the double recessive, *sh2 su1*. In comparison with *su1* kernels the *sh2 su1* kernels showed a decrease in the accumulation of WSP. Laughnan concluded that *sh2* precedes *su1* in the synthesis of starch and WSP. His conclusion was based on the assumption that each gene controlled a specific biochemical step in starch and WSP synthesis. He pointed out that by using additional mutations in specific combinations it should be possible to gain valuable information about starch synthesis in maize.

Cameron and Teas (1953) have shown that two other mutations, *brittle-1* (*bt1*) and *brittle-2* (*bt2*), reduce endosperm starch substantially without the accumulation of WSP. Both genes apparently were responsible for an increase in both reducing sugars and sucrose and reductions in total starch. Creech, McArdle, et al. (1963) and Creech (1965) reported on the effects of the gene mutations *ae, du, sh2, su*1, *su2*, and *wx*, singly and in combinations, on qualitative and quantitative changes in carbohydrates in maize endosperm during kernel development 16–28 days after pollination. Significant differences among genotypes at different stages of kernel development were noted for all carbohydrate and dry-matter analyses. Total sugars, reducing sugars, and sucrose contents were negatively correlated with dry matter and starch contents. In general, data (Creech and McArdle, 1966) for the carbohydrates in mature kernels were similar to the findings for 28-day kernels for all the genotypes investigated, which included several genotypic combinations of *ae, du, sh2, su*1, *su2*, and *wx*.

Anderson, Uhl, et al. (1962) reported that high-amylose maize (*ae*) was higher in protein and contained relatively less endosperm starch than standard dent corn. The lower starch level in *ae* grain was suggested to be responsible for its lower test weight.

Interactions

Multiple combinations of mutants affecting kernel properties are being studied by field-corn and sweet-corn breeders in order to obtain types with more desirable chemical properties. Garwood and Creech (1972) described kernel phenotypes of genotypes possessing one to four homozygous mutant alleles.

Investigations on the role of the genes *ae* and *su*1 in phytoglycogen accumulation in maize endosperms were conducted by Ayers and Creech (1969) by quantifying and characterizing the WSP from 16 genotypes of endosperm resulting from all possible crosses among lines homozygous for normal (*Ae* and *Su*1), *ae, su*1, and *ae su*1. Only endosperms homozygous for *su*1 contained phytoglycogen. Added doses of *ae* decreased the amount of phytoglycogen in 20-day endosperms from 38.8% to 4.2% of the dry weight. The β-amylolysis-limit phytoglycogen from the double mutant suggested that it may be more loosely branched than the polysaccharide from *su*1 alone.

The data of Kramer, Pfahler, et al. (1958), Zuber, Deatherage, et al. (1960), Creech, McArdle, et al. (1963), Creech (1965) and of earlier workers, on the roles of the genes *ae, du, sh2, su*1, *su2*, and *wx* indicate that starch synthesis involves multiple systems and is indeed complex. The *ae* gene, in addition to changing the formation of amylose (Vineyard and Bear, 1952), also causes a substantial increase in sugars and reduction in total starch. In addition, *ae* in combination with *wx* and *du wx* causes dramatic increases in sugars and reductions in starch. This supports the view that *ae, du,* and *wx* may be involved in separate pathways of starch synthesis.

TABLE 2.   Amylose content of mature kernels of 24 genotypes of maize[a]

Gene combination	Amylose in starch (%)
Normal	27
ae	61
du	38
su1	29
su2	40
wx	0
ae du	57
ae su1	60
ae su2	54
ae wx	15
du su1	63
du su2	47
du wx	0
su1 su2	55
su1 wx	0
su2 wx	0
ae du su1	41
ae du su2	48
ae su1 su2	54
ae su1 wx	13
du su1 su2	73
du su1 wx	0
du su2 wx	0
su1 su2 wx	0

[a] Adapted from Kramer, Bear, et al. (1958).

The amylose content data of Kramer, Bear, et al. (1958), shown in Table 2, combined with the data on *ae* and *wx* effects on sugars and polysaccharide contents (Creech, 1965), also suggest that the normal dominant allele *Ae* may be involved with formation of branched-chain polysaccharides (amylopectin) and that the dominant normal allele *Wx* may be involved in the synthesis of straight-chain polysaccharides (amylose). Nelson and Rines (1962) reached a similar conclusion concerning *Wx*. They reported that starch granules from *wx* endosperms were deficient in uridine diphosphoglucose (UDPG) transferase activity, which was present in starch granules of *Wx* endosperms. Since *wx* endosperm lacked both amylose starch and starch granule-bound UDPG transferase activity, Nelson and Rines (1962) suggested that UDPG transferase is necessary for amylose synthesis. This enzyme catalyzes the transfer of glucose from UDPG or ADPG to an α-D-(1-4) linkage on the nonreducing end of the polysaccharide acceptor. Nelson and Tsai (1964) later reported that 17 different waxy mutants transform glucose to starch at about 10% the rate of similar preparations from kernels of nonwaxy maize. The source of the activity was indicated to be in starch granules isolated from the embryo. The endosperm, which is the site for the major amount of starch synthesis, was apparently devoid of or very low in starch granule bound transferase activity. Badenhuizen and Chandorkar (1965) suggested that the low glucosyl transferase in waxy maize starch granules may be due to the absence of straight-chain or amylose starch. Starch granules varying in amylose content were isolated and assayed for amylose content and UDPG transferase activity. Use was made of the natural increase in amylose content during the development of the kernel. There seemed to be little association between amylose content and. UDPG-transferase activity in kernels of the two high amylose strains.

Sandstedt, Strahan, et al. (1962) investigated the pancreatic digestibility of starches isolated from several genotypes. Their data are presented in Table 3. The *ae* gene appears to be associated with both high amylose (Vineyard and Bear, 1952) and high resistance to enzyme action. Kramer, Pfahler, et al. (1958) and Brown (1966) reported that *ae* starch had a very high birefringence endpoint temperature (BEPT). However, the genes *du* and *su2*, although associated with higher amylose than dent corn, seem to be associated with high digestibility. Sandstedt and co-workers concluded that amylose content was not associated with digestibility. They suggested the answer may lie in the structure of the starch granule, *ie*, in differences in the bonding of the

TABLE 3.   Comparison of the
susceptibility of various starches to
pancreatic digestion[a]

Starch source	Digestion (%)
Normal	69
ae	24
du	80
su1	71
su2	88
wx	85
ae du	42
ae su2	22
du su2	76

[a] Adapted from Sandstedt, Strahan, et al. (1962).

starch molecules and/or in possible anomalous linkages between molecules. Brown (1966) suggested that the *su2* starch was less crystalline than normal starch, but this does not account for the results with *wx*. The *su2* starch granules also had a lower BEPT than normal granules. Sandstedt suggested that modifying the digestibility of starch of commercial strains of maize may be important in improving its nutritional value, especially for nonruminants.

Four maize genotypes, normal, *su2*, *o2*, and the double mutant *o2 su2*, were evaluated for protein quality by four methods: (a) complete amino-acid analysis, (b) chemically available lysine, (c) protein efficiency ratio in the rat, and (d) protein efficiency ratio in the meadow vole (Brinkman, Shenk, et al. 1974). The four tests gave comparable results thus indicating that the improved protein characteristics imparted by the *o2* gene were retained in the double mutant. The high digestibility of the starch of *su2* did not show a benefit in the double mutant. Kernel texture is improved in the double mutant (see discussion on protein quality, pp. 258–260).

Breeding Waxy Maize

Starch produced from waxy maize differs in important respects from that of regular dent corn. Its different characteristics stem from its unique molecular structure. Starch from waxy maize consists almost solely of branched molecules.

The commercial production of waxy maize is an outstanding result of cooperative development programs spanning the fields of chemistry, agriculture, and industry. The unique type of corn from which it is produced was brought to this country from China in 1908 and was regarded as a curiosity. Its name derives from the waxy appearance of the cross-section of a kernel when cut cleanly in half.

An extensive cross-breeding program was initiated in 1937 in an effort to introduce the gene into regular high-yielding hybrid corns. The work was conducted primarily at Iowa State University by G. F. Sprague and colleagues. In 1975 an estimated 1.2–1.5 million metric tons of waxy maize was produced in the United States on an estimated 200,000–240,000 ha.

Breeding Amylomaize

Bear (1958) isolated a recessive allele (*ae*) that changed the amylose:amylopectin ratio from 1:3 to 1:1 in mature corn endosperm without materially reducing total starch production. The physiochemical properties of corn starch having a 1:1 ratio was desirable for certain industrial purposes, and hybrid development was initiated using the *ae* gene (Bear, 1958). Bear, Vineyard, et al. (1958) observed wide differences in apparent amylose content among backcross-derived inbreds. These differences suggested that a modifier complex existed and that selection in each backcross generation should be profitable (Helm, Ferguson, et al. 1967). A consistent problem has been encountered in these breeding programs. Starch yield, and hence yield per hectare, tended to be lower, and moisture and sugar content of the grain was higher, as compared to a standard dents. These problems have been largely overcome or reduced to a tolerable level, and amylomaize is becoming well established as a specialty corn. An estimated 6400 ha (16,000 acres) was planted in 1975. No single gene mutation has been discovered, to our knowledge, that will give a higher amylose content (or lower amylopectin content) without drastically reducing starch content than the standard *ae* gene. Alleles at the *ae* locus have been discovered and characterized (D. L. Garwood, J. C. Shannon, and R. G. Creech, personal communication); however, none have effected a starch superior to that of standard *ae*.

Breeding Sweet Corn with Higher Sugar Content

Most of the standard sweet-corn hybrids are single crosses involving the *su*1 gene in combination with modifier genes that increase sugar content or effect superior quality. East and Hayes (1911) described the *su*1 gene in sweet corn. Laughnan (1953) studied the effects of the genes *sh*2 and *su*1 on endosperm carbohydrate reserves. It was found that almost 20% of the dry weight of the *sh*2 kernels was sugar. The *sh*2 *su*1 kernels were lower in water-soluble polysaccharides. Laughnan concluded that *sh*2 might be useful in improving the quality of sweet corn. The hybrid, Illini X-tra Sweet, and similar hybrids, possess the gene *sh*2. The major problem with *sh*2 hybrids lies in their generally inferior seed quality. Selection for improved germination has been partially successful. Creech and McArdle (1966) suggested that *ae, du,* and *wx* might be utilized to improve the quality and quality maintenance of sweet corn. It is yet to be established whether the *ae du wx* combination will be commercially acceptable.

Genes affecting carbohydrate properties, in combination with genes affecting protein quality, show promise not only in improving the quality, but also in improving the nutritional value of sweet corn.

PROTEIN: QUANTITY AND QUALITY

Corn is usually regarded as being a carbonaceous crop, but it annually produces more protein than soybeans in the United States. This vast amount of protein, however, is of low biological quality for nonruminants, primarily because of deficiencies of lysine and tryptophan.

Proportion of protein is affected by genetic constitution and the environment, primarily nitrogen level in the soil. Environmental impact is relatively minor as compared to that of genotype. Dudley, Lambert, et al. (1974) reported that the Illinois high protein strain reached 26.6% protein in the 70th generation of selection, a value 12 standard deviations above the mean of the original population. The Illinois low-protein strain, in contrast, carried 4.4% protein. Hybrids, or even open-pollinated varieties, rarely vary more than two percentage points in protein when grown under extremes of fertility and population.

Endosperm proteins, taken as a whole, are generally quite inferior to those of the embryo. Hence improvement in quality must involve a reduction of zein, an inferior endosperm group of proteins, and an increase of the higher-quality ones as glutelins and globulins. Fortunately, a variant of this type was identified by Mertz, Bates, et al. (1964). The variant, *opaque-2* (*o2*), had existed in mutant collections for a generation but was unrecognized for its remarkable attributes. Since the 1964 report, other mutants have been found in corn with similar effects. Searches in barley and in sorghum have revealed similar mutants.

Opaque-2 corn, dubbed "high lysine" corn, appears to have some commercial promise in the United States. In eastern Europe, the Soviet Union, and in Latin America, relatively higher costs of protein supplements make its development much more attractive.

Lambert, Alexander, et al. (1969) found that *opaque-2* hybrids were on the average 8% lower in yield than their normal dent versions. However, two hybrids in a group of 10 were not found to be significantly different from their normal counterparts. Kernel damage induced by harvesting was almost double that encountered in dents in the same trial. Brown (1975) concluded that improvement in performance had occurred as a consequence of breeding, but that *o2* hybrids were inferior in yield and that their kernel texture was not well adapted to machine harvest. He suggested that these problems might be overcome by breeding types having harder endosperms. Bauman (1975) reported that some *o2* hybrids were competitive in yield with commercial dent hybrids.

Epistatic interactions involving *o2*, *fl2*, *su2*, and *wx* have been reported. Of particular interest is the report of Glover (1975) that kernel density of the double mutant *o2 su2* was nearly equal to that of ordinary dents and further, that the lysine/protein ratio of the endosperm was equivalent to that of *o2* types.

Some caution should be taken in assuming that higher kernel density is essential for maximum yield. Dudley, Lambert, et al. (1971) found that the correlation between density and yield per plant was not different from zero in an *o2* synthetic. Bauman (1975) also questioned the necessity for developing vitreous types for the American Corn Belt. However, damage to typical *o2* kernels brought about by machine harvesting and multiple handling in the grain trade is a formidable problem and possibly may be an obstacle to large scale entry into the grain trade.

Further improvements in biological value appear possible. For example, the percentage of zein in whole-kernel protein is approximately 16% in R802, a high-oil *o2* line. In contrast, ordinary *o2* stocks typically contain 20–30% zein (C. M. Wilson, personal communication). The *opaque*-2 recoveries of high-oil lines are particularly promising in that lysine as a percentage of dry matter commonly exceeds 0.5%.

OIL: QUANTITY AND QUALITY

Concentration of oil exhibits wide variation in corn—from less than 0.5% to as much as 18–20%. Embryos of ordinary corn carry 30% oil, but those of Illinois High Oil strain contain 50% oil. Since great variation exists, it seems inescapable that appropriate breeding schemes should produce hybrids with oil content substantially above the usual 4.0–4.5% level. Relatively little attention has been paid to it, however. Whether the inattention sprang from breeders' decisions to pursue more pressing problems such as yield, or from the fact that analytical services were unavailable, is uncertain. The popular view that corn was a "starch crop" unquestionably affected breeding for heightened quality— not only for oil, but for protein as well.

Impetus was given to oil breeding by the development of wide-line nuclear magnetic resonance spectroscopy (NMR) as an analytical tool (Bauman, Conway, et al. 1963; Alexander, Silvela, et al. 1967). Nondestructive analysis of bulk or single kernel samples now may be carried out in as brief a time as two seconds.

Oil content is a highly heritable trait in corn. Progress encountered by selecting the single-kernel NMR technique in Alexho Synthetic amply demonstrates this fact (Figure 1).

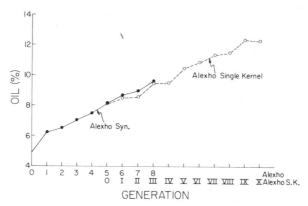

Figure 1. Progress in selecting for high oil content in Alexho Synthetic.

TABLE 4. Performance of high oil and commercial hybrids, mean of 1973–1975, Urbana[a]

	Yield (bu/A)	Oil (%)	H_2O in grain (%)	Lodged (%)
R802 A × R805	143	8.6	19	0
R802A × R806	166	8.8	21	1
R806 × H49[HO]	189	8.0	26	11
R806 × N28	173	6.4	25	1
Mo17 × N28	179	3.2	23	1
Commercial single cross	181	3.9	22	2
Highest-yielding commercial hybrid[b]	187	4.0	23	3

[a] Data for 1974 excluded because of effect of a killing freeze on performance on September 22.
[b] Mean of highest yielding commercial hybrid in 1973 and 1975 trials (not the same hybrid).

Performance of higher-oil hybrids appears to be adequate to suggest that they may be commercially feasible (Table 4). Limited commercial production of hybrids in the 6.5–8.0% range has recently been carried out in the United States. Quite aside from the requirement for adequate yield, the advantages they present to the miller and perhaps to animal feeders, particularly swine feeders, should stimulate their production. Corn oil is a premium oil and regularly is more valuable than starch. At current prices, corn oil is worth approximately four times as much as corn starch. It seems inescapable that this relationship, or one even less favorable, would encourage millers to seek corn higher in oil.

Feeding experiments with swine (Nordstrom, Behrends, et al. 1972) showed that 5–6% less high oil corn (7%) was required per unit of gain than with ordinary corn. It is perhaps even more significant to note that substantially less soybean meal was required to balance the high-oil corn diets in Nordstrom's experiments. It should be pointed out, however, that Lynch, Baker, et al. (1972) found 5.9% oil corn not to be of any advantage compared to 4% oil corn in swine feeding experiments. It seems likely that the calorie advantage of the higher-oil corn was not large enough to be detected in his experiments.

REFERENCES

Alexander, D. E., L. Silvela S., F. I. Collins, and R. C. Rodgers. 1967. Analysis of oil content of maize by wide-line NMR. *J. Am. Oil Chem. Soc.* **44**: 555–558.

Anderson, R. A., D. E. Uhl, W. L. Deatherage, and E. L. Griffin. 1962. Composition of the component parts of two hybrid high-amylose corns. *Cereal Chem.* **39**: 282–286.

Andrés. J. M. and P. C. Bascialli. 1941. Characteres hereditarios aislados en maices cultivados en la Argentina. *Univ. Buenos Aires Inst. Genet.* **2**: 1.

Andrew, R. H., R. A. Brink, and N. P. Neal. 1944. Some effects of the waxy and sugary genes on endosperm development in maize. *J. Agr. Res.* **69**: 355–371.

Ayers, J. E., and R. G. Creech. 1969. Genetic control of phytoglycogen accumulation in maize (*Zea mays* L.). *Crop Sci.* **9**: 739–741.

Badenhuizen, N. P., and K. R. Chandorkar. 1965. UDPG-alpha-Glucan glucosyl-transferase and amylose content of some starches during their development and under various external conditions. *Cereal Chem.* **42**: 44–54.

Bates, L. L., D. French, and R. E. Rundle. 1943. Amylose and amylopectin content of starches determined by their iodine complex formation. *J. Am. Chem. Soc.* **65**: 142–148.

Bauman, L. F. 1975. Germ and endosperm variability, mineral elements, oil content and modifier genes in *opaque*-2 maize. In *High quality protein maize,* E. T. Mertz (ed.). Stroudsburg, Penn.: Dowden, Hutchinson and Ross pp. 217–227.

Bauman, L. F., T. F. Conway, and S. A. Watson. 1963. Heritability of variations in oil content of individual corn kernels. *Science* **139**: 498–499.

Bear, R. P. 1944. Mutations for waxy and sugary endosperm in inbred lines of dent corn. *J. Am. Soc. Agron.* **36**: 89–91.

Bear, R. P. 1958. The story of amylomaize hybrids. *Chemurgic Digest* **17**: 5.

Bear, R. P., M. L. Vineyard, M. M. MacMasters, and W. L. Deatherage. 1958. Development of "amylomaize"—corn hybrids with high amylose starch. II. Results of breeding efforts. *Agron. J.* **50**: 598–602.

Brinkman, G. L., J. S. Shenk, R. G. Creech, and D. L. Garwood. 1974. Comparisons between rats, voles, and chemical methods for the determination of protein quality of four maize (*Zea mays* L.) genotypes. *Nutr. Rep. Internat.* **10**: 61–68.

Brown, R. P. 1966. Genetic control of starch granule development in maize endosperm. M.S. thesis. University Park, Penn.: Penn. State University.

Brown, W. L. 1975. Worldwide seed industry experience with *opaque*-2 maize. In *High quality protein maize,* E. T. Mertz (ed.). Stroudsburg, Pa.: Dowden, Hutchinson and Ross. Inc. pp. 256–264.

Burnham, C. R. 1944. Midrib color. *Maize Genet. Coop. News.* **18**: 15.

Cameron, J. W. 1947. Chemico-genetic bases for the carbohydrates in maize endosperm. *Genetics* **32**: 459–485.

Cameron, J. W. and D. Cole. 1959. Effects of the genes *su*1, *su*2, and *du* on carbohydrates in developing maize kernels. *Agron. J.* **51**: 424–427.

Cameron, J. W. and H. J. Teas. 1953. Carbohydrate relationships in developing endosperms of *brittle*-1, *brittle*-2, and related maize genotypes. *Proc. Internat. Congr. Genet.* **9**: 822–823.

Collins, G. N. 1909. A new type of Indian corn from China. *U. S. Bur. Plant Inds. Bull.* No. 161, 31 pp.

Creech, R. G. 1965. Genetic control of carbohydrate synthesis in maize endosperm. *Genetics* **52**: 1175–1186.

Creech, R. G., and F. J. McArdle. 1966. Gene interaction for quantitative changes in carbohydrates in maize kernels. *Crop Sci.* **6**: 192–194.

Creech, R. G., F. J. McArdle, and H. H. Kramer. 1963. Genetic control of carbohydrate type and quality in maize kernels. *Maize Genet. Coop. News.* **37**: 111–120.

Deatherage, W. L., M. M. MacMasters, M. L. Vineyard, and R. P. Bear. 1954. A note of starch of high amylose content from corn with high starch content. *Cereal Chem.* **31**: 50–52.

Dudley, J. W., R. J. Lambert, and D. E. Alexander. 1974. Seventy generations of selection for oil and protein concentration in the maize kernel. In *Seventy generations of selection for oil and protein concentration in the maize kernel*, J. W. Dudley (ed.). Crop Science Society of Amer. Special Publication, pp. 181–212.

Dunn, G. M., H. H. Kramer, and R. L. Whistler. 1953. Gene dosage effects on corn endosperm carbohydrates. *Agron. J.* **45**: 101–104.

Dvonch, W., H. H. Kramer, and R. L. Whistler. 1951. Polysaccharides of high-amylose corn. *Cereal Chem.* **28**: 270–280.

East, E. M. and H. K. Hayes. 1911. Inheritance in maize. *Conn. Agr. Exp. Sta. Bull.* No. 167, 142 pp.

Eyster, W. H. 1934. Genetics of *Zea mays. Bibliogr. Genet.* **11**: 187–392.

Garwood, D. L. and R. G. Creech. 1972. Kernel phenotypes of *Zea mays* L. genotypes possessing one to four mutated genes. *Crop Sci.* **12**: 119–121.

Glover, D. V., P. L. Crane, P. S. Misra, and E. T. Mertz. 1975. Genetics of endosperm mutants in maize as related to protein quality and quantity. In *High quality protein maize.* E. T. Mertz (ed.). Stroudsburg, Pa.: Dowden, Hutchinson and Ross, Inc. pp. 228–240.

Greenwood, C. T. 1956. Aspects of the physical chemistry of starch. *Adv. Carbohydrate Chem.* **11**: 335–393.

Helm, J. L., V. L. Fergason, and M. S. Zuber. 1967. Development of high-amylose corn (*Zea mays* L.) by the backcross method. *Crop Sci.* **7**: 659–662.

Hutchison, C. B. 1921. Heritable characters of maize. VII. Shrunken endosperm. *J. Hered.* **12**: 76–83.

Kramer, H. H. and R. L. Whistler. 1949. Quantitative effects of certain genes on the amylose content of corn endosperm starch. *Agron. J.* **41**: 409–411.

Kramer, H. H., R. P. Bear, and M. S. Zuber. 1958. Designation of high amylose gene loci in maize. *Agron. J.* **50**: 229.

Kramer, H. H., P. L. Pfahler, and R. L. Whistler. 1958. Gene interactions in maize affecting endosperm properties. *Agron. J.* **50**: 207–210.

Lambert, R. J., D. E. Alexander, and J. W. Dudley. 1969. Relative performance of normal and modified protein (*opaque*-2) maize hybrids. *Crop Sci.* **9**: 242–243.

Laughnan, J. R. 1953. The effect of the *sh2* factor on carbohydrate reserves in the mature endosperm of maize. *Genetics* **38**: 485–499.

Lynch, P. B., D. H. Baker, B. G. Harmon, and A. H. Jenson. 1972. Feeding value for growing-finishing swine of corns of different oil contents. *J. Anim. Sci.* **35**: 1108.

Mains, E. B. 1949. Heritable characters in maize: linkage of a factor for shrunken endosperm with the *a*1 factor for aleurone color. *J. Hered.* **40**: 21–24.

Mangelsdorf, P. C. 1947. The inheritance of amylaceous sugary endosperm and its derivatives in maize. *Genetics* **32**: 448–458.

Mertz, E. T., L. S. Bates, and O. E. Nelson. 1964. Mutant gene that changes protein composition and increases lysine content of maize endosperm. *Science* **145**: 279–280.

Nelson, O. E. and H. W. Rines. 1962. The enzymatic deficiency in the waxy mutant of maize. *Biochem. Biophys. Res. Commun.* **9**: 297–300.

Nelson, O. E. and C. Y. Tsai. 1964. Glucose transfer from adenosine diphosphate-glucose to starch in preparations of waxy seeds. *Science* **145**: 1194–1195.

Nordstrom, J. W., B. R. Behrends, R. J. Meade, and E. H. Thompson. 1972. Effects of feeding high oil corns to growing-finishing swine. *J. Anim. Sci.* **35**: 357.

Parnell, F. R. 1921. Note on the detection of segregation by examination of the pollen of rice. *J. Genet.* **11**: 209–212.

Sandstedt, R. M., D. Strahan, S. Ueda, and R. C. Abbot. 1962. The digestibility of high-amylose corn starches compared to that of other starches. The apparent effect of the *ae* gene on susceptibility to amylose action. *Cereal Chem.* **39**: 123–131.

Sprague, G. F., B. Brimhall, and R. H. Hixon. 1943. Some affects of the waxy gene in corn on properties of the endosperm starch. *J. Am. Soc. Agron.* **35**: 817–822.

Vineyard, M. L. and R. P. Bear. 1952. Amylose content. *Maize Genet. Coop. Newsl.* **26**: 5.

Weatherwax, P. 1922. A rare carbohydrate in waxy maize. *Genetics* **7**: 568–572.

Whelan, W. J. 1961. Recent advances in the biochemistry of glycogen and starch. *Nature* **190**: 954–957.

Zuber, M. S., W. L. Deatherage, C. O. Grogan, and M. M. MacMasters. 1960. Chemical composition of kernel fractions of corn samples varying in amylose content. *Agron. J.* **52**: 572–575.

Chapter 17

CURRENT AND FUTURE USE OF CYTOPLASMIC MALE STERILITY FOR HYBRID SEED PRODUCTION

D. N. Duvick and S. W. Noble
Pioneer Hi-Bred International, Inc.

Cytoplasmic male sterility (CMS) and its use for production of hybrid maize (*Zea mays* L.) can be discussed in somewhat the same way as the early Christians chose to discuss historical time. Time was described as either "B.C." or "A.D.", meaning before or after the birth of Christ. For use in maize seed production, CMS is now discussed in terms of "before 1970" or "after 1970," meaning before or after the 1970 epidemic of southern leaf blight, race T [*Helminthosporium maydis* Nisik. and Miyake (*Cochliobolus heterostrophus* Drecks.)], the epidemic that severly damaged the maize crop in most of the U. S. Corn Belt (Tatum., 1971). It is not an understatement to say that the epidemic has had extremely influential effects on thoughts of plant breeders around the world in regard to not only the use of CMS in maize and other crop plants, but also to the broader subject of genetic homogeneity and its potential for production of epidemics of disease or insect pests. "Genetic vulnerability" is the term usually used to describe this potential problem (National Academy of Sciences, 1972).

The use of CMS to produce hybrid maize has been reviewed several times prior to 1970 (Duvick, 1959, 1965, 1966) and needs no further documentation here. We propose in this review to describe the events of 1970 and their consequences, and then to review the present state of knowledge in regard to CMS and its use in production of hybrid maize seed. Emphasis is on U. S. events and research, although we discuss work from other parts of the world also. References are selective, not comprehensive.

THE EVENTS OF 1970

In 1970 CMS was being used to eliminate detasseling for most hybrid maize seed production in the United States and Canada, as well as in other parts of the world, such as Brazil, South Africa, and the Soviet Union. Two cytoplasms, S and T, were generally recognized as satisfactory (a third, "C", was known but not generally used). Most production used T cytoplasm since more inbreds were fully sterile in T than in S. It was estimated that 75–80% of the U. S. maize acreage in 1970 was in T cytoplasm (National Academy of Sciences, 1972). Many hybrids were "restored" (pollen fertility of the final hybrid was restored by introduction of $Rf1$ and/or $Rf2$ through the pollen parent) and so 100% of their plants had T cytoplasm.

Starting in the early part of the 1970 growing season, $H.$ $maydis$ moved into maize fields in Florida, Georgia, Alabama, and Mississippi, and from there, as successive crops reached susceptible stages of growth, the disease moved up the Mississippi Valley into Kentucky, Ohio, Indiana, and Illinois. Nearly all midwestern states eventually were affected by the disease. It also affected the important maize growing states in the eastern part of the country. Eventual loss of yield in the U. S. maize crop, due to the epidemic, was thought to be about 15% (Tatum, 1971). As the disease progressed it became apparent that T cytoplasm was associated with susceptibility to the disease. By October, researchers (Smith, Hooker, et al. 1970) reported that the epidemic likely was due to a new race (race T) of $H.$ $maydis$, extremely virulent on T cytoplasm, but not more virulent on normal (N), S, or C cytoplasm than the older race, now called "race O." This explained why the previous 20 years of U. S. experience with T cytoplasm, often in the presence of infections of $H.$ $maydis$, had not shown T cytoplasm to be more susceptible than N, S, or C. Exactly when or where the new race arose was not known, but it now seems possible that it was present in the Philippines in 1961 (Mercado and Lantican, 1961) and possibly was in the U. S. by at least 1969 (Hooker, Smith, et al. 1970) although at that time it was not recognized as a new race.

Also in 1969 a second disease had been identified as being more severe on T than on other cytoplasms; yellow leaf blight (*Phyllosticta maydis*) was affecting

T cytoplasm inbreds and hybrids in Pennsylvania, Minnesota, Michigan, Indiana, Wisconsin, and Ontario (Scheifele, Nelson, et al. 1969). In 1970 T cytoplasm was affected by this disease throughout much of the Corn Belt. Deleterious effects on plant health and final yield were not as clear-cut as with *H. maydis* race T, and no evidence was put forth indicating that a new race of *P. maydis*, specific for T cytoplasm, had appeared. Nevertheless, this was indeed another widespread problem associated with T cytoplasm.

The seed industry reacted quickly to the 1970 *H. maydis* epidemic. The remedy was obvious and at hand; N cytoplasm was relatively resistant to race T (i.e., race T was not more virulent than race O on N cytoplasm); N cytoplasm maintainers of all T-sterile inbreds were necessarily on hand, and the job now was to increase the N cytoplasm maintainers for use as female parents for all seed production. The females of course had to be detasseled by hand—or clipped by machine. Winter acreage was lined up; plans were made to produce some N cytoplasm hybrids in Florida, Hawaii, Argentina, and similar areas, and especially plans were made to build up supplies of N-cytoplasm foundation seed for the 1971 seed production year. Seed from the 1970 crop was sorted over; 100% T seed was generally discarded. The 100% N seed was of course at a premium, and to have enough seed to go around, blends (N and T) of hybrids with the most genetic resistance were also made available. [Genotypes varied widely in susceptibility to race T, when in T cytoplasm (Lim, Hooker, et al., 1974).]

Fortunately, the 1971 season was not as favorable for race T as that of 1970; losses were smaller even on comparable germplasm, and by 1972 N cytoplasm seed supplies were sufficient for all. Thus ended the threat from *H. maydis* race T and also from *P. maydis*

NEW DIRECTIONS

The next step had to do with new ways to overcome the expense and hazard of hand or machine detasseling. Techniques of using purely genetic sterility in combination with chromosome abnormalities were proposed. They are not included in the scope of this review, as they have been discussed in a previous review (Duvick, 1972). None is yet in commercial use, although intensive breeding work is in progress with some of them. Techniques of using chemical gametocides were—and are now—being explored by some industrial concerns. These techniques also are outside the scope of this paper. The third possibility—use of new, nonsusceptible cytoplasm—is the topic that we now consider.

Both S and C cytoplasms (the original discoveries) are resistant to *H. maydis* race T and to *P. maydis*, as noted above. Many other discoveries of sterile cytoplasm were also at hand in 1970. Many of these were examined (Nelson, Ayers, et al., 1971; Smith, Hooker, et al., 1971; Gracen, Forster, et al., 1971), and most of them were found to be resistant to race T and to *P. maydis*. However, all cytoplasms classified by fertility restoration as being in the T family of cytoplasms (Beckett, 1971) were susceptible. It soon became obvious that a second criterion for classification of new discoveries of CMS was available; any cytoplasm restored by T cytoplasm restorer genes would also be susceptible to race T. No exceptions to this rule were found; no variations in degree of susceptibility were found, provided backcrossing was sufficiently complete that variation in nuclear genotype could be excluded as a source of variability.

Both S and C cytoplasms became immediate prospects for use in producing hybrid seed without detasseling. The S cytoplasm had been used, in limited amounts, since the first work in the 1950s. Its good and bad points were known. It had fallen out of favor because many genotypes were imperfectly sterile in S cytoplasm, meaning that for these genotypes fertility was easily affected by environment, and one could not be sure of their sterility in all environments. The C cytoplasm had never been examined in detail, either with regard to sterility or its effects on yield and other performance traits. Both cytoplasms were at first suspect simply because of their category. The term "cytoplasmic male sterility" in the minds of many almost automatically signaled a warning of uniform susceptibility to some new disease or insect pest. Clearly, a period of backcrossing and testing of both cytoplasms was needed before they could prudently be used.

USE OF MULTIPLASMS

The concern that in a few years all maize might again be in S, C, or S and C cytoplasms gave rise to ideas of ways to prevent this narrowing of the cytoplasmic gene base. The term *multiplasm* was applied to a proposed technique of producing hybrids containing a blend of several kinds of male sterile cytoplasms (Grogan, 1971, Gracen and Grogan, 1974). The idea is good; the chief difficulty in implementing it is that we are not sure that very many really different forms of CMS are available. Many *discoveries* are on hand; 108 different discoveries of CMS were reported in the literature up to 1972 (Duvick, 1972), and more have been discovered since then. Unfortunately, most of these discoveries, when classified by means of restorer gene reaction turn out to be the same as S, T, or C (Beckett, 1971; Gracen and Grogan, 1974).

Some sources are of doubtful classification and may differ from S, C or T, but there are no unambiguous proofs of difference to date. All cases of possible non-S, non-T, or non-C classification have involved use of tester genotypes lacking clear-cut restorer or sterile expression in S, C, or T cytoplasm. Such testers are often referred to as *partial restorers,* or *partial steriles,* depending on degree of penetrance of their restorer genes, in any given cytoplasm. They do not work well for classifying cytoplasms because of their sensitivity to environmental effects.

If one must use such partial restorer lines to classify cytoplasms, one should use several replications, a series of well-characterized environments, and analysis of variance probably should also be used to point out significant differences, just as one would do when comparing hybrids for grain yield, another trait that is greatly affected by environment.

A much better solution of the problem is to choose testers with clear-cut reactions in S, C, and T cytoplasms. By definition, such testers give results independent of environmental fluctuations, and the testcrosses are fully fertile, fully sterile, or in the case of heterozygous restored S cytoplasm testcrosses, plants are semisterile; half of the pollen grains are fertile and half are sterile. This latter trait, incidentally, should always be scored for presence or absence in testcrosses; it is a very reliable indicator of the S cytoplasm type. A second S-cytoplasm trait, occasional extrusion of stick-like sterile anthers in nonrestored plants, should also be scored for use in identifying S cytoplasm types.

In summarizing this section, it would be advisable to attempt classification of CMS discoveries only with clear-cut, well-known tester lines; use of testers with incomplete penetrance may lead to entertaining potentially false hopes as to identification of new sterile cytoplasms.

It is theoretically possible that fertility-restoration distinctions within a family of cytoplasms can be made. But it seems likely that cytoplasms similar (even though not identical) in restorer reaction would also resemble each other in disease susceptibility. For example, to date all discoveries of T cytoplasm (classified as such by restoration) have been susceptible to *H. maydis* race T. Of course, none of these discoveries has been shown to differ even quantitatively from the original T discovery, in terms of fertility restoration.

SEARCH FOR NEW FORMS OF CMS

Continuing search may yet turn up kinds of CMS genuinely different from T, S, and C. However, 10 new discoveries of CMS, made and classified by us since 1971 in a systematic examination of maize races from around the world, have all turned out to be S, T, or C, based on restorer reactions. We also have been attempting to backcross maize into cytoplasms of the various races of teosinte with hopes of uncovering one or more types of CMS, as has been done for sorghum, wheat, and other cultivated and wild species. To date we have found nothing, but progress in hybridizing and backcrossing has not been easy.

Another way to obtain new types of CMS is to produce them by means of mutagens. Evidence is now well documented for the existence of organelle

DNA capable of coding for certain components of the organelle (Bogorad, 1975), and so in theory one should be able to cause the organelle DNA to mutate, producing mutant organelles, (However, unequivocal evidence that organelle differences are the cause of any kind of CMS in maize has not yet been put forward.)

Several reports in literature from the USSR indicate that CMS has arisen as a result of treatment with mutagens, γ-irradiation and ultrasonic waves (Lysikov, Blyandur, et al. 1969). Also from Bulgaria a report states that irradiation with γ-rays and/or neutrons resulted in lines with cytoplasmic male sterility (Stoilov, 1970). A U. S. patent has been issued to a researcher from the Soviet Union for a technique of producing CMS with streptomycin (U. S. Patent 3,594,152). No use seems to have been made of it as yet. Attempts to duplicate the technique in the U. S. have not been successful to date, nor has any use been made of the other CMS mutants, to our knowledge.

In this country also, experimenters have used chemical mutagens on T cytoplasm, attempting to change its disease susceptibility without altering its pollen-sterility attributes (Hooker, 1972), and ethyl methane–sulfonate has been investigated as a mutagen that might be able to produce new kinds of CMS (Briggs, 1975). Selection in callus culture of T-cytoplasm maize, treated with the race T toxin, has given rise to strains of callus with resistance to the toxin (Gengenbach and Green, 1975). However, plants were not regenerated from these strains, and it is not known whether cytoplasmic or nuclear genes, or both, were altered.

Thus none of these investigations have as yet given useful results, to our knowledge; however, this whole body of science—the genetics of cytoplasmic components—is so new that probably as techniques are developed and a body of knowledge is built up, we may someday be able to produce CMS at will with an array of mutagenic treatments. We may even find out the cause (or causes, more likely) of CMS!

WHAT WE KNOW ABOUT CAUSES OF CMS

This gap in our knowledge is of course one of the real problems in use of CMS; we still do not know for sure what organelles, if any, are implicated in any kind of CMS, what biochemical reactions are blocked or changed to produce the CMS phenotype, what fertility restoration does to change the biochemical reactions causing CMS, and whether certain effects sometimes associated with CMS (e.g., reduction in leaf number, plant height, disease susceptibility, and increased stress resistance) are due to separately heritable reactions.

A considerable body of information regarding cytological studies of mitochondria, chemical content, and enzyme activity of CMS strains is reported from several European countries, especially the Soviet Union (Palilova, 1972; Dmitrieva and Havkinskja, 1970; Musijko, Klujcko, et al., 1965; Borovskij, 1964).

A generalizing theory has been proposed by Turbin and Palilova (1970) as a result of some of these studies. The theory postulates a mitochondrial site of factors causing CMS; two mitochondrial genes, one structural (P_s) and one repressor (P_r^+), are proposed. The product of a nuclear controlling gene (rf), binds the mitochondrial repressor gene and allows the mitochondrial structural gene to function in N cytoplasm plants. But CMS plants make abnormally large amounts of repressor (because the mitochondrial repressor gene has mutated to a more active state: $P_r^+ \rightarrow P_r^-$) and hence a more active nuclear controlling gene (Rf) is required to control the repressor and allow the mitochondrial structural gene to operate. Thus P_r^- and rf give male sterility, whereas P_r^- and Rf give fertility restoration. Both types have "sterile" cytoplasm because of the overactive repressor, the mutant mitochondrial gene P_r^-. Note that this theory considers that rate-limiting rather than structural gene changes are the cause of CMS. No testing or application of this theory seems to have been recorded to date.

Shortly after race T was identified as specific for T-cytoplasm, mitochondria of T cytoplasm were implicated in susceptibility to the disease (Miller and Koeppe, 1971). The mitochondria showed irreversible swelling and changes in respiratory rate and oxidative phosphorylation, in presence of the toxin from race T. In 1974 and 1975 (Peterson, Flavell, et al., 1974; Peterson, Flavell, et al., 1975) studies on the electron transport chain suggested that the pathotoxin from race T has a binding site on the inner membrane of mitochondria from T cytoplasm plants and that this site is controlled by cytoplasmic DNA. Additional studies (Barratt and Flavell, 1975) indicated that restorer genes tended to partially correct the typical aberrant electron-transport reactions. This partial correction was said to suggest that products of Rf genes are mitochondrial inner membrane proteins and that they alter a defect in the mitochondrial inner membrane. However, it was acknowledged that since Rf genes were of no help in overcoming race T infection or toxin susceptibility of T cytoplasm *in vivo* (Hilty and Josephson, 1971; Gracen and Grogan, 1972), the problem must be complex.

Flavell (1974) has proposed a second general model for the mechanism of CMS based on results of these studies. In the model, anthers produce a substance capable of inhibiting altered organelles (either mitochondria or plastids) in pollen grains of CMS plants. The inhibiting substance has no action on pollen organelles of N cytoplasm plants. Restorer genes and/or environmental variations would: (a) alter amounts or availability of the "inhibiting substance" or alternatively (b) correct the altered CMS organelle structure. Any of these changes could allow the aberrant CMS organelles to operate at more or less normal levels, giving more or less normal amounts of pollen fertility. As noted above, this model is derived from observation of the action of race T toxin on mitochondria of normal and T cytoplasm plants.

Flavell's model is related to that of Turbin and Palilova (1970) in that both models postulate a restorer gene product that overcomes the effect of a repressor specific for inhibition of CMS organelle development; however, the Turbin and Palilova model postulates that the repressor is abnormal and is synthesized by CMS organelles (mitochondria), whereas the Flavell model postulates that the repressor is normal (is always present) and is synthesized by the anthers of N and also of CMS plants. Turbin and Palilova postulate normal organelle structural genes in CMS pollen; Flavell postulates abnormal structural genes.

In our opinion the Flavell model would need to be modified to allow for different restorer systems of C, S and T and also for the different race T toxin susceptibilities of C and S as opposed to T.

We have devoted considerable space in this review to studies that might explain the molecular genetic basis for CMS—in particular, Texas CMS— because it seems to us that one must know whether the association of CMS with disease susceptibility in the Texas cytoplasm is a chance occurrence peculiar to Texas CMS, or whether all changes giving rise to CMS will necessarily introduce potential susceptibility to some race of some disease, known or unknown. Corollary to elucidation of the molecular basis of CMS, (whether T, S, C, or any others to be discovered) is elucidation of the mechanism by which diseases such as *H. maydis* race T attack the plant. Such knowledge is of course very hard to obtain, but researchers are asking the questions and one can hope the answers eventually will be given.

PRESENT USE OF CMS

In the meantime, plant breeders as usual must make decisions based on information at hand; waiting for the final decisive answer would mean no action, no forward progress at all. What is being done to utilize CMS for production of hybrid maize?

We might note, first of all, that in the Soviet Union, race T and *P. maydis* have apparently never been recognized as a problem, and CMS is used in nearly all hybrid seed production. It is our understanding, based on conversational exchange, that most of the production is based on T, some on Moldavian cytoplasm. Most of the USSR maize production is in areas climatically similar to the western U. S. Corn Belt (Nebraska, South Dakota), where *H. maydis* race T also was never a serious problem, so it may be that climatic factors will continue to allow successful use of T cytoplasm in the Soviet Union.

In the United States and Canada essentially no CMS was used for seed production from 1971 through 1973. In 1974, however, a small amount of production was made in C and S cytoplasm (plus other sources, probably about the same as C or S), and it is probable that a substantial amount of seed has been made in these cytoplasms in 1975. A recent survey of 1975 seed production (Zuber, 1975) indicates that about 13% of the production was in sterile cytoplasm, with most of this being C and S cytoplasms. This survey points to the possibility that in a few years U. S. corn production may again be based on a narrow cytoplasmic base.

To be sure, extensive testing (by us and others) of C and also of S cytoplasm during and since 1970 has revealed absolutely no reason to suspect any extra susceptibility to disease or insects, or any serious defects in yielding ability or other agronomic traits. In fact, the proven advantage (in absence of cytoplasm-specific disease) of yield of sterile plants over pollen-shedding plants under yield-stress conditions, especially during drought at silking time (Vincent and Wooley, 1972) means that it is advantageous for farmers to plant hybrids in which up to 50—60% of the plants are male sterile.

CYTOPLASMIC MALE STERILITY AND THE FUTURE

It may be profitable to speculate about how widely CMS should be used in the future. But again it may be foolish to do so, since a new factor has come into the U. S. maize-breeding picture. Two class action lawsuits are pending against some of the major U. S. seed corn companies in regard to the 1970 epidemic. The outcome of these suits cannot be predicted, but they indicate the possibility that to avoid legal liability, seedsmen may have the future responsibility of correctly forecasting epidemics. A biologist knows instinctively that any decision on future performance of biological entities must be based on estimation of probabilities of occurrence of critical environmental inputs and that certainty is never possible. Thus a conflict of laws, civil and natural, seems to be in our future.

Aside from this immediate problem, a more general responsibility is placed in our hands as plant breeders—namely, the one of doing what we can to be sure that the odds are at least high that future epidemics like that of 1970 will not reoccur. What can we learn from experience?

Surely it would be advisable to stay away from an overwhelming concentration of one cytoplasm. One may argue that maize for the last 3 years has been in only one cytoplasm—N. But definition of N cytoplasm is merely "not T, C, or S" and leaves the possibility for much genetic heterogeneity that simply has not been discovered. Various studies say there is considerable, some, or very little genetic variation among "N" cytoplasms (Crane and Nyquist, 1967; Fleming, 1972; Garwood, Weber, et al., 1970; Good and Horner, 1974; Hunter and Gamble, 1968; Russell, 1972; Sprague and McGahen, 1968); they at least leave open the possibility that there is useful genetic cytoplasmic variation in regard to disease susceptibility.

We must carry forward the basic investigations into the nature of CMS. On a short-term basis U. S. funds started several good programs after 1970, but the funding periods generally are now ended and no more money is in sight. As we noted above, there is an urgent need to describe in molecular genetic terms the nature of CMS and what various types of sterile cytoplasm do or do not have in common.

An outcome of and probably a tool in making these basic investigations is creation of new kinds of CMS, using mutagens and irradiation. Present variability in kinds of CMS is too low from the point of view of making "multiplasms"; we need many more really different sterile cytoplasms. Also, if several new CMS systems can be created and compared, factors that are necessarily common to all CMS systems can perhaps be discovered and so help to explain the basics of the system.

Pest-epidemic charting and warning systems should be set up on a national and worldwide basis. This need is only incidental to the use of CMS; it is simply one of the safety factors that should be built into modern farming practice. It seems to us incomprehensible that a country such as the United States should not have such a unified system now. It is needed just as much as a national weather-forecasting and reporting service. Its reports might be incorporated into the useful *Weekly Weather and Crop Bulletin* now issued jointly by the U. S. Departments of Commerce and Agriculture.

Use of sterility systems other than CMS will have a place; chromosomal–genetic systems and chemical treatments will be used if convenient and reliable ones are developed. The chemical treatments have an extra hurdle today in the United States. They must be approved by the Environmental Protection Agency (EPA) for experimentation as well as for final farm use. This requires extensive and expensive testing, for proof of safety to humans and animals. Since total needs for chemical pollen-control agents are small compared to usage of herbicides or insecticides for general farm use, the ratio of expense of testing versus potential profit from sale of chemical pollen control agents is not very favorable. Thus there may be little incentive to development of useful pollen control chemicals. This in turn may force the seed-production industry into wider use of CMS, regardless of whether large numbers of sterile cytoplasms are ever found.

SUMMARY

Cytoplasmic male sterility for use in production of hybrid maize seed was severely curtailed after the 1970 U. S. epidemic of *H. maydis* race T. Limited use is now being made of S and C cytoplasms, with C apparently favored over S. Use of multiplasms (several sterile cytoplasms in one hybrid) is advocated, but little progress has been made in discovery or creation of sterile cytoplasms really different from T, S, or C. At least two theories have now been proposed, based on experimental data, to explain the nature of CMS on a cytological and molecular basis; neither is yet proven to be correct unequivocally. Further basic research along these lines is advocated.

REFERENCES

Barratt, D. H. P. and R. B. Flavell. 1975. Alterations in mitochondria associated with cytoplasmic and nuclear genes concerned with male sterility in maize. *Theor. App. Genet.* **45**: 315–321.

Beckett, J. B. 1971. Classification of male-sterile cytoplasms in maize (*Zea mays* L.) *Crop Sci.* **11**: 724–727.

Bogorad, L., 1975. Evolution of organelles and eukaryotic genomes. *Science* **188**: 891–898.

Borovskij, M. I. 1964. The question of transmitting characteristics of pollen sterility from certain maize forms to others by non-sexual means. *Bull. Acad. Sci. Moldav. SSR Ser. Biol. Chem.*, pp. 59–65. [From *Ref. Zh.* 1.55.12 (1966) through *Plant Breeding Abstr.* **39**: 2322.]

Briggs, R. W. 1975. Personal communication.

Crane, P. L. and W. E. Nyquist. 1967. Effects of different gene–cytoplasm systems on quantitative characters in reciprocal F_2 crosses of maize (*Zea mays* L.). *Crop Sci.* **7**: 376–378.

Dmitrieva, A. N. and O. E. Havkinskaja. 1970. Physiological and biochemical features of maize forms with cytoplasmic male sterility. In *Plant breeding using cytoplasmic male sterility.* Kiev-Urozaj (1966), pp. 458–471. [From *Ref. Zh.* 11.55.51, (1967) through *Plant Breeding Abstr.* **40**: 4532.]

Duvick, D. N. 1959. The use of cytoplasmic male-sterility in hybrid seed production. *Econ. Bot.* **13**: 167–195.

Duvick, D. N. 1965. Cytoplasmic pollen sterility in corn. *Adv. Genet.* **13**: 1–56.

Duvick, D. N. 1966. Influence of morphology and sterility on breeding methodology. In *Plant breeding.* Ames-Iowa State University Press, pp. 85–138.

Duvick, D. N. 1972. Potential usefulness of new cytoplasmic male steriles and sterility systems. *27th Annu. Corn and Sorghum Res. Conf.*, pp. 192–201.

Flavell, R. 1974. A model for the mechanism of cytoplasmic male sterility in plants with special reference to maize. *Plant Sci. Letts.* **3**: 259–263.

Fleming, A. A. 1972. Male cytoplasmic effect on reaction of maize to diseases. *Plant Dis. Rep.* **56**: 575–577.

Garwood, D. L., E. J. Weber, R. J. Lambert and D. E. Alexander. 1970. Effect of different cytoplasms on oil, fatty acids, plant height and ear height in maize (*Zea mays* L.). *Crop Sci.* **10**: 39–41.

Gengenbach, B. G. and C. E. Green. 1975. Selection of T-cytoplasm maize callus cultures resistant to *H. maydis* race T pathotoxin. *Crop Sci.* **15:** 645–649.

Good, R. L. and E. S. Horner. 1974. Effect of normal cytoplasms on resistance to southern corn leaf blight and on other traits of maize. *Crop Sci.* **14:** 368–370.

Gracen, V. E., M. J. Forster, and C. O. Grogan. 1971. Reactions of corn (*Zea mays*) genotypes and cytoplasms to *Helminthosporium maydis* toxin. *Plant Dis. Rep.* **55:** 938–941.

Gracen, V. E. and C. O. Grogan. 1972. Reactions of corn (*Zea mays*) seedlings with non-sterile, Texas male-sterile, and restored Texas male-sterile cytoplasms to *Helminthosporium maydis* toxin. *Plant Dis. Rep.* **56:** 432–433.

Gracen, V. E. and C. O. Grogan. 1974. Diversity and suitability for hybrid production of different sources of cytoplasmic male sterility in maize. *Agron. J.* **66:** 654–657.

Grogan, C. O. 1971. Multiplasm, a proposed method for the utilization of cytoplasms in pest control. Plant Dis. Rep. **55:**400–401.

Hilty, J. W. and L. M. Josephson. 1971. Reaction of corn inbreds with different cytoplasms to *Helminthosporium maydis*. *Plant Dis. Rep.* **55:** 195–198.

Hooker, A. L. 1972. Southern leaf blight of corn—present status and future prospects. *J. Envir. Qual.* **1:** 244–249.

Hooker, A. L., D. R. Smith, S. M. Lim, and J. B. Beckett. 1970. Reaction of corn seedlings with male-sterile cytoplasm to *Helminthosporium maydis*. *Plant Dis. Rep.* **54:** 708–712.

Hunter, R. B. and E. E. Gamble. 1968. Effect of cytoplasmic source on the performance of double-cross hybrids in maize, *Zea mays* L. *Crop Sci.* **8:** 278–280.

Lim, S. M., A. L. Hooker, J. G. Kinsey, and D. R. Smith. 1974. Comparative grain yields of corn hybrids in normal and Texas male-sterile cytoplasm (CMS-T) infected with *Helminthosporium maydis* race T and disease components of CMS-T corn hybrids. *Crop Sci.* **14:** 190–195.

Lysikov, V. N., O. V. Blyandur, A. I. Konotop, S. G. Byrka and Y. Orlov, 1973. A study of the possibility of obtaining maize mutants with cytoplasmic male sterility. Kishinev: Moldavian SSR, Kartja Mold. 40–47 (1969). [From Ref. *Zh.* 5.55.8 (1970) through *Plant Breeding Abstr.* **43:** 7821.

Mercado, A. D., Jr. and R. M. Lantican, 1961. The susceptibility of cytoplasm male sterile lines of corn to *Helminthosporium maydis*. *Nisik. Miy. Philippine Agr.* **45:** 235–243.

Miller, R. J. and D. E. Koeppe. 1971. Southern corn leaf blight: susceptible and resistant mitochondria. *Science* **173:** 67–69.

Musijko, A. S., P. F. Kljucko, and J. M. Sivolap. 1969. Transmission of the property of fertility restoration in maize by grafting. *Rep. All-Un Lenin Acad. Agric. Sci.* (1965), 1–4. [From *Ref. Zh.* 23.55.7 (1967) through *Plant Breeding Abstr.* **39:** 6630.]

National Academy of Sciences, Committee on Genetic Vulnerability of Major Crops. 1972. *Genetic vulnerability of major crops.* Washington, D. C.: National Academy of Science, 307 pp.

Nelson, R. R., J. E. Ayers, and J. B. Beckett. 1971. Reactions of various corn inbreds in normal and different male-sterile cytoplasms to the yellow leaf blight organism (*Phyllosticta sp.*) *Plant Dis. Rep.* **55**: 401–403.

Palilova, A. N. 1972. Biochemical and physiological characteristics of male-sterile analogues of maize. In *Geterozis v rastenievodstve*. Leningrad: Kolos (1968), pp. 215–222. [From *Ref. Zh.* 4.55.146 (1969) through *Plant Breeding Abstr.* **42**: 7596.]

Peterson, P. A., R. B. Flavell and D. H. P. Barratt, 1974. A simple biochemical assay for "Texas" cytoplasm in corn by use of *Helminthosporium maydis,* race T pathotoxin. *Plant Dis Rep.* **58**: 777–779.

Peterson, P. A., R. B. Flavell and D. H. Barratt, 1975. Altered mitochondrial membrane activities associated with cytoplasmically-inherited disease sensitivity in maize. *Theor. Appl. Genet.* **45**: 309–314.

Russell, W. A. 1972. Effects of exotic cytoplasms in agronomic characters of two maize inbred lines. *Iowa State J. Res.* **47**: 141–147.

Scheifele, G. L., R. R. Nelson, and C. Koons. 1969. Male sterility cytoplasm conditioning susceptibility of resistant inbred lines of maize to yellow leaf blight caused by *Phyllosticta zeae. Plant Dis. Rep.* **53**: 656–659.

Smith, D. R., A. L. Hooker, and S. M. Lim. 1970. Physiologic races of *Helminthosporium maydis. Plant Dis. Rep.* **54**: 819–822.

Smith, D. R., A. L. Hooker, S. M. Lim, and J. B. Beckett, 1971. Disease reaction of thirty sources of cytoplasmic male-sterile corn to *Helminthosporium maydis* race T. *Crop Sci.* **11**: 772–773.

Sprague, G. F. and J. H. McGahen. 1968. Influence of alien cytoplasms on yield of grain in *Zea mays* L. *Crop Sci.* **8**: 163–164.

Stoilov, M. 1970. Possibility of producing male sterile forms and restoring fertility in maize by means of gamma-ray and fast-neutron irradiation. In *Induced mutations in plants*. Proceedings of a symposium jointly organized by the IAEA and FAO and held at Pullman, Washington, July 14–18, 1969, pp. 509–515. Vienna: International Atomic Energy Agency, 1969, pp. 748; through *Plant Breeding Abstr.* 4446.

Tatum, L. A. 1971. The southern corn leaf blight epidemic. *Science* **171**: 1113–1116.

Turbin, N. V. and A. N. Palilova. 1970. A theoretical model for the interaction of chromosome genes and cytoplasm in the case of cytoplasmic male sterility in maize. *Rep. All-Un. Lenin Acad. Agr. Sci.* **11**: 2–4.

Vincent, G. B. and D. G. Woolley. 1972. Effect of moisture stress at different stages of growth: II. Cytoplasmic male-sterile corn. *Agron. J.* **64**: 599–602.

Zuber, M. S. 1975. Corn germplasm base in the U.S.—Is it narrowing, widening, or static? 30th Annual Corn and Sorghum Research Conference, pp. 277–286.

DISEASES AND INSECTS

Chapter 18

INTRODUCTORY REMARKS TO THE SESSION ON RESISTANCE TO DISEASE AND INSECTS

Alejandro Ortega
CIMMYT

The informative reviews presented by Drs. Ullstrup, Hooker, Dicke, and Guthrie reveal that a substantial amount of knowledge has and is being gained in breeding maize to resist pests in the U. S. Corn Belt. The pest problems that have received the most attention have been the leaf blights, rust, stalk and ear rots, viruses and their vectors, European corn borer, rootworms, and earworm.

Genetic research has revealed that often the nature of pest resistance is polygenic; however, in the case of viruses, downy mildews, blights, and rusts, the nature of resistance can also be monogenic. If both systems should be operating in the same plant, polygenic effects may be difficult to demonstrate. In the majority of the cases studied, resistance to pests is dominant to susceptibility and additive or dominant in gene action.

Unquestionably, the development of techniques to mass produce spores and insect eggs, for uniform artificial inoculation and/or infestation, has contributed to a faster and more accurate selection of truly resistant genotypes. Production of inoculum or insect eggs can now be regulated to be available at the appropriate stage of plant development. Also, the severity of the artificial inoculation and/or infestation can be regulated to permit detection of different levels of resistance. Care must be taken, however, to maintain the genetic variability of the pest through regular crossing of the laboratory strain with the feral type(s).

Only a small fraction of the total genetic diversity in maize has been used in improvement programs. More recently, and to a large extent because of susceptibility to pests, private and public improvement programs are intensifying the introgression of exotic maize germplasm into their materials. The effective utilization of such resource (genetic diversity) is very well discussed by Dr. Tsotsis. Utilization requires: (a) a logical classification to satisfy the local needs of the geographic regions to be served (maturity, texture, color, etc.), (b) an organization and structure that allows a systematic evaluation and selection, and (c) a group of scientists (breeders, entomologists, pathologists, and agronomists) who integrate harmoniously in the improvement program.

Strategy should interest leaders of maize-improvement projects, particularly in those countries where the seed industry has not yet fully developed. As an initial step, improved open-pollinated varieties with elite progenies can be developed from a "stress breeding" program. If the elite is accepted, the seed would be distributed from farmer to farmer. Improved genetically diverse populations, in which selection for pest resistance has received first priority, could be used for the production of elite lines and later of hybrids. The implementation of these programs would require the exchange and introduction of germplasm. However, quarantine regulations as implemented now in many countries can be a serious obstacle to systematic testing of new germplasm. Thus the spirit of cooperation among programs throughout the world needs to be enhanced to gradually develop a network of involved people and improved materials. As such a network develops, the monitoring of pests through strategically located, international pest nurseries could become a reality.

I would like to take this opportunity to congratulate the maize workers of the United States. Their work during the last four decades has resulted in more than a 100% increase in yield, and their scientific knowledge has permeated the entire world. Our sincere recognition to you all.

Chapter 19

EVOLUTION AND DYNAMICS OF CORN DISEASES AND INSECT PROBLEMS SINCE THE ADVENT OF HYBRID CORN

A. J. Ullstrup

Professor Emeritus, Purdue University.
Presently, Plant Pathologist, Farmers Forage Research Cooperative,
West Lafayette, Indiana

Before the development and general use of hybrid corn, much of the research on the diseases and insect pests of this crop were descriptive in nature. Diseases and insect infestations and their causal agents were described, the biological relationships were investigated, and the effects of different environmental factors on the pathogens and insects were elucidated. Differences in reaction among open-pollinated corn varieties were noted, but advances in the use of host–plant resistance were slow; what was accomplished was done by avoiding selecting seed from infested or diseased plants. Some research was done on the control of seed rots and seedling blight with different chemical seed-treating materials.

At an early stage, the basic studies on the development of inbred lines and hybrids demonstrated the relative ease of genetic manipulation of corn. Self-pollination with attendant approach to homozygosity revealed marked differences not only in agronomic characters, but also in reaction to diseases and insect pests. Inbred lines could be selected and combined into hybrids that possessed desired attributes in addition to the vigor accompanying heterosis. Hybrid vigor *per se* can, in the case of some diseases and insect infestations, reduce in part the severity of injury sustained by the less vigorous inbred parents.

With the increased use of hybrid corn over the past 40 years, employment of host–plant resistance has become important as a most effective means of "biological control" of corn insect pests (Maxwell, Jenkins, et al., 1972). This is true also in control of corn diseases where even greater emphasis is placed on genetic resistance of the host because chemical methods and other means of protection from some infections are not easily applied or are otherwise limited in their use.

SOME PROBLEM INSECTS OF CORN

The discovery in 1917 of the presence of the European corn borer, *Ostrinia nubilalis* Hübner, in the United States (Brindley and Dicke, 1963), stimulated research on ways to stem the spread of this destructive pest into the Corn Belt. Cultural practices, biological means (predators, parasites, and disease-inciting agents) and mechanical methods of restricting the advance of the borer were investigated. Open-pollinated varieties were tested for their resistance to the insect under conditions of natural infestation. Legislation was promulgated to establish quarantines restricting movement of infested materials and to make compulsory the employment of practices recommended to reduce populations of the pest. Despite these efforts the European corn borer spread to Ohio and Indiana by 1926 and to Illinois by 1939. Research centers were established by the USDA in Ohio and Indiana, and state agricultural experiment stations within the areas involved began studies to control the insect. As it advanced south and westward multivoltine forms of the insect were discovered. This complicated and added to the problems of control.

Early studies on genotype reactions to the borer were often unsatisfactory and misleading because testing under natural infestation in the field did not always reflect the type of resistance expressed (*nonpreference, antibiosis,* or *tolerance*). Furthermore, infestations could vary appreciably from year to year. Rearing of adults in cages to provide egg masses for artificial infestation was an advance in overcoming some of the disadvantages of using natural infestation to assay host materials. A milestone in this area of research, however, was the development of artificial diets used in laboratory rearing of the borer and ultimate production of egg masses. Large numbers of egg masses can now be produced when desired and, when placed on test plants, ensure relatively uniform levels of infestation (Guthrie, 1965).

The center of studies sponsored by the USDA has been located in Ankeny, Iowa, where close cooperation is maintained with the Iowa State University facilities and personnel. Several of the hybrid seed companies have begun research to develop inbred lines and hybrids resistant to the European corn borer. An appreciable number of inbred lines and hybrids are now available that combine excellent agronomic characters and high levels of resistance to leaf feeding by the first brood of the insect. Such hybrids are being grown and have effectively reduced populations of the borer in areas of severe infestation. Reduction in severity of infestation behind the periphery of advance is attributed in part to the use of such resistance. Resistance to the second brood of the European corn borer has been difficult to find; only very few inbreds (B52 and a few of only moderate resistance) have sufficient resistance to use as parental material (Pesho, Dicke, et al. 1965).

Results of studies on inheritance of genes conditioning resistance to leaf feeding by the first brood of the borer are variable depending on the parental materials used. Both monogenic and polygenic resistance have been reported (Penny and Dicke, 1956, 1957). Recurrent selection has been successful in increasing the resistance in synthetic hybrids within a comparatively short period of time (Penny, Scott, et al. 1967).

Recent basic studies on the nature of resistance to the first brood of the European corn borer have shown a glucoside to be present in resistant genotypes that, when such tissue is mechanically disrupted as in the leaf feeding of the insect, is converted by plant enzymes to the aglucone, 2,4-dihydroxy-7-methoxy-1, 4-benzoxazine-3-one (DIMBOA). This compound is chemically labile and slowly decomposes to 6-methoxy-2-benzoxazolinone (MBOA), which is chemically stable. Tissue is analyzed for MBOA which is formed in constant proportion to the concentration of DIMBOA. Only the latter is biologically active (Beck and Smissman, 1961; Klun, 1975). Studies thus far have not implicated DIMBOA in relation to resistance to the second brood of the borer. This compound has also been associated with resistance to some diseases of corn (Molot and Anglade, 1968).

Corn earworm, *Heliothus zeae* Boddie, has been an economically important pest on corn for many years. Infestations are severe in the South and in some eastern areas of the United States (Eckhardt, 1954). Sweet corn is especially susceptible but appreciable damage can occur in dent corn. Resistance was at one time believed to be due largely to husk extension but later work has shown other factors such as long, narrow, silk channels, hard kernels, and possibly antibiotic materials in the host, may be more important (Maxwell, Jenkins, et al. 1972). White corn is generally more resistant than yellow; germplasm from the South is more resistant to the insect than genotypes from the Corn Belt (Eckhardt, 1954).

General combining ability effects were highly significant in earworm damage, but genotype–environment interaction was also highly significant. Inheritance of resistance is quantitative in nature with a number of gene loci scattered over the chromosome compliment (Widstrom and Wiseman, 1973). Mass selection has been effective in increasing gene frequency for resistance to the insect in synthetic populations (Zuber, 1971).

The southwestern corn borer, *Diatrea grandiosella Dyar,* was first reported in the United States in 1913. It has spread with increasing acceleration eastward and northward and is now present in some of the Corn Belt states, where it has caused some serious damage (Fairchild, 1965). Research on host–plant resistance and results of experiments to identify sources of resistance have been variable.

The corn leaf aphid, *Rhopalosiphum maidis* Fitch, is widely distributed and has been responsible for some losses in yield (Dicke, 1969). Populations of this insect in the central and northern reaches of the Corn Belt originate from winged forms carried in on wind currents from southern areas, where they overwinter. Severe infestation may cause barrenness, especially so if corn is under a drought stress. Host–plant resistance has been observed by a number of investigators, but such resistance has not been in popular usage. Biotypes are present within the species (Everly and Miller, 1962).

Three species of rootworm attack corn. These are the northern corn rootworm, *Diabrotica longicornis* Say; the southern, *D. undecimpunctata howardi* Barber and the western, *D. virgifera* Lec. The northern and the western are the most damaging in the Corn Belt of the United States. Adults as well as larvae feed on corn, but there is no correlation in resistance of genotypes to root injury inflicted by larvae and resistance to leaf and silk feeding by adults (Painter, 1951).

The western corn rootworm, first described in Colorado in 1909 (Bryson, Wilbur, et al. 1953) has since spread eastward and is the dominant species of the three in Iowa. Resistance to silk feeding by the adults of this rootworm has been reported to be conditioned by a single gene (Sifuentes and Painter, 1964). This rootworm is an example of the development of resistance, within a species, to toxic chemicals (chlorinated hydrocarbons) that had been an effective control (Ball and Weekman, 1962). This phenomenon stimulated search for resistance. This has been successful, and high levels of resistance have been identified among inbred lines, synthetics, and populations of domestic and foreign origin (Fitzgerald and Ortman, 1964). Both tolerance, or ability to regenerate roots pruned by the rootworms, and antibiotic materials within host tissues that have a deleterious effect on the insect, are involved in resistance (Ortman and Gerloff, 1970).

Resistance to the fall armyworm has been demonstrated in a number of corn genotypes. General combining ability effect were shown to be highly significant in resistance studies among inbred lines and single crosses. Dominance and specific combining ability effects were less important. Heterosis contributed substantially to the level of resistance among F_1 progenies (Widstrom, Wiseman, et al., 1972).

The rice weevil, *Sitophilus oryzae* L., can be a serious insect pest of corn in the South (southern United States). Within southern corn germplasm an ample reservoir of resistance is available. Bird and earworm damage appear to be positively correlated with weevil infestation. A highly significant correlation has also been observed between kernel hardness and sugar content and resistance (Singh and McCain, 1963). These factors are not influenced by the extension or number of husks.

In addition to inflicting losses by direct feeding, a number of insects act as vectors for some corn diseases. Several species of leafhopper function as the only means of transmission of several virus diseases of corn. In *Cicadulina mbila* Naudé ability to transmit the virus of corn streak disease in Africa has been shown to be governed by a single dominant, sex-linked factor (Storey, 1939). Corn stunt, caused by a mycoplasma-like entity, is transmitted by five species of leafhopper. Aphids function as vectors of the so-called stylet-borne virus diseases of corn. The corn flea beetle, *Chaetocnema pulicaria* Melsh., is the primary vector of *Erwinia stewartii* (E. F. Smith) Dye, the inciting agent of bacterial wilt.

Stalk borers and earworms provide avenues of entrance into these organs for some fungus pathogens. There is some evidence that the causal agent of common smut may gain entrance through corn-borer feeding wounds. In these latter fungus diseases the insects are not obligate vectors because the pathogens can invade without the intervention of the insect pest.

CORN-DISEASE PROBLEMS

An early example of control of a corn disease through host-plant resistance is that involving bacterial wilt or Stewart's disease. An open-pollinated variety, "Golden Bantam," much in demand because of its high quality, was extremely susceptible. Because of the economic importance of the disease and demands for its control, some effort was made to find sources of resistance. Large numbers of genotypes were artificially inoculated and resistant inbred lines combined into single crosses. One of these, subsequently named "Golden Cross Bantam," combined good resistance, high yield and fine quality (Smith, 1933). It became the most popular hybrid in the sweet-corn industry. Resistance was demonstrated to be controlled by two major and one minor, dominant, supplementary, and independently inherited genes (Wellhausen, 1937).

Sweet corn is most susceptible, although a few of the popular inbred lines of dent corn now in use show unusual susceptibility to both the early, fully systemic invasion by the pathogen, and the late, or leaf-blight, phase of the disease. Inheritance of genes conditioning resistance to the leaf-blight phase has not been studied.

The first single gene to be identified conditioning resistance to a disease in corn was that controlling reaction of seedlings to a physiologic race of common rust. Later this monogenic dominant was located on chromosome 10 (Rhoades, 1935). The disease has not been of much economic importance although often widely prevalent and generally appearing after silking. One single, dominant gene, often referred to as the "Cuzco gene," has a wide spectrum of action against the many races of Puccinia sorghi Schw. (Hooker, 1969). Much basic information has been accumulated on the genetics of host–plant resistance to common rust (Saxena and Hooker, 1968). Mature plant resistance, a condition controlled by several genes, appears to play an important role in governing reaction to the many races of P. sorghi.

Southern corn rust was confined to the tropics and warm-temperate regions of the Western Hemisphere, appearing only occasionally in the northern parts of the Corn Belt in warm, wet seasons. In the late 1940s and early 1950s it became severe in West Africa (Rhind, Waterston, et al., 1952). Losses were estimated to be 50–70% in some areas. Subsequently the disease was identified in Central and East Africa (Saccas, 1955). Moving in an easterly direction the rust circled the globe. The most important losses were in West Africa, not in homogeneous hybrids, but among an array of local varieties. In 1972 and 1973 the rust became unusually prevalent in the U.S. South and Southeast. Resistance has been found among genotypes originating in the Western Hemisphere and some of these have been used to control the disease in Africa (Storey, 1958). Single, dominant genes have been found controlling specific physiologic races of Puccinia polysora Underw. and the locus of one of these has been identified on chromosome 10 (Ullstrup, 1965).

About the time hybrid corn was gaining acceptance (1939–1943), northern corn leaf blight became widespread and locally severe in some areas of the eastern Corn Belt. During this period and in this area, mild temperatures and ample moisture were general, providing a favorable environment for the disease. Hybrids available at the time were susceptible—many extremely so. The pathogen, Helminthosporium turcicum Pass., had been known in the United States for many years but had never caused a widespread epidemic.

In response to this challenge, the USDA, in cooperation with state experiment stations within the areas involved, initiated disease-resistance breeding programs. Methods of artificial inoculation were developed and large numbers of genotypes evaluated for resistance. A few inbred lines were found with sufficient resistance to be used as parental material. This resistance was quantitative in nature (Jenkins and Robert, 1959). Such material became parents of resistance inbred lines ultimately released to seed producers. They were accepted and some are still in use.

Later, an unusual type of resistance was observed in an open-pollinated variety of popcorn presumed to have originated in Peru. This variety was subsequently found to be a Ladyfinger popcorn or a variety closely similar to it. This type of reaction to *H. turcicum* is controlled by a single, dominant gene (Hooker, 1963). Symptoms are characterized by small necrotic lesions surrounded by chlorotic halos. The gene, designated at *Ht*, has been incorporated into a large number of widely used inbred lines.

Another type of resistance has been described recently (Gevers, 1975). This appears also to be a monogenic dominant in which symptoms are expressed as minute chlorotic flecks. The gene, designated as *HtN*, appears to decrease in effectiveness sometime after flowering, but it may be valuable in providing protection from the disease in early, more critical stages, of host development. Other sources of resistance are being evaluated.

Recently, what appears to be a new physiologic race of the pathogen and capable of infecting corn carrying the *Ht* gene, has been identified in Hawaii (Bergquist and Masais, 1974).

Southern corn leaf blight was not often of economic importance, except on extremely susceptible genotypes grown under warm, humid conditions. In 1961 a report from the Philippines (Mercado and Lantican, 1961) indicated that corn with Texas male-sterile cytoplasm (*Tcms*) was hypersusceptible to *Helminthosporium maydis* Nisikado & Miy., the causal agent of southern corn leaf blight. This observation was made in the "wet season"; in the "dry season" no disease-related differences were seen between normal cytoplasm and *Tcms* versions of hybrids. This difference had never been observed before and the report from the Philippines was interpreted to be that of a phenomenon associated with the warm, wet weather. In 1969 this hypersusceptibility of *Tcms* corn was seen in a few places in the United States (Scheifele; Whitehead, et al., 1970). Subsequent studies demonstrated a new physiologic race to be responsible for inciting the disease on *Tcms* corn (Hooker, 1970). This race was extremely virulent on this type of corn that constituted 80–90% of the crop in the United States. The new race was designated as race T and the old race, endemic to the United States since its description, as race O. Race T incites only a benign reaction on normal-cytoplasm corn. Reaction to race T is controlled mainly by the cytoplasm, but this is influenced to some extent by nuclear genes. Resistance to race O is polygenic in some genotypes (Pate and Harvey, 1954), monogenically conditioned in others (Smith and Hooker, 1973).

The epidemic began in the South in 1970 and moved rapidly northward as the season progressed. Most damage was sustained in the South; in the northern tier of midwestern states only moderate losses occurred. It was estimated that the monetary loss caused by the epidemic was about one billion dollars. Its ramifications and impacts were widespread (Tatum, 1971). No plant disease of any crop in the United States has caused so much loss in so short a time and yet has been so rapidly controlled as this catastrophic epidemic of southern corn leaf blight. Resistance was at hand in normal-cytoplasm corn. Supplies of such seed for the 1971 growing season were critically short. Fortunately, that season was unusually favorable for corn and supplies of the resistant, normal-cytoplasm seed corn were replenished to meet all demands for 1972.

Yellow leaf blight was first described in 1967. It is caused by *Phyllosticta maydis* Arny & Nelson (Arny and Nelson, 1971). Genotypes with *Tcms* or the "P-type" male-sterile cytoplasm are generally much more susceptible than normal-cytoplasm versions (Nelson, Ayres, et al., 1971). With the abandonment of corn with *Tcms* and other types of male-sterile cytoplasm known to confer disease susceptibility to some infectious diseases, yellow leaf blight may cease to be a hazard.

Anthracnose, gray leaf spot, and eyespot are three diseases incited by fungus pathogens that have become prevalent in the past few years when temperatures have been moderate and rainfall ample. These diseases, like some others incited by fungi, tend to become more severe where minimum tillage is practiced. All three become more severe also as the host approaches maturity. Anthracnose has caused some losses in the U. S. southeastern area and to a lesser extent in a few places in the midwestern area. Gray leaf spot has become most prevalent in Virginia and adjacent areas where minimum tillage helps prevent soil erosion in hilly fields. Eyespot, first described in Japan in 1958, appeared in the U. S. midwestern area about 10 years later (Arny, 1971). Sources of genetic resistance to each disease have been identified and could be used when warranted.

Stalk rot, long a chronic disease complex of corn, is closely associated with senescence of the host. Plants in which the onset of maturity is early and in which senescence develops rapidly, succumb to invasion by *Diplodia maydis* (Berk.) Sacc., *Gibberella zeae* (Schw.) Petch, or *Fusarium moniliforme* Sheld. These are weakly parasitic fungi that attack the host when it is under stress of aging. The particular fungus isolated from diseased stalks depends on geographical and/or climatic conditions.

Hybrid corn with its increased vigor has been much more resistant to stalk rotting and subsequent breakage than open-pollinated varieties. Resistance to breakage has become more important with the advent of mechanical harvesting. A hybrid will not be accepted by growers unless it has good "standability." Considerable progress has been made over open-pollinated varieties and the hybrids produced in early years of this era with respect to standability. Modern methods of breeding have been successful in developing high-yielding genotypes resistant to stalk breakage (Jinahyon and Russell, 1969; Sprague, 1954). Some of these advances have been somewhat negated by cultural practices such as increased plant populations per unit area and the increased use of nitrogen without consideration of nutrient balance. Destruction of functional leaf tissue by diseases, insects, or mechanical means almost invariably accentuates stalk rot.

Corn ear rots have not often been major problems. Occasionally diplodia ear rot has been acute in fields adjacent to sources of inoculum. Gibberella ear rot has been somewhat more widespread as in 1965 and again in 1972, when it was prevalent in the northeastern parts of the Corn Belt. Losses are not only from rotted ears, but more important are losses to swine growers because of the toxicity to these animals of corn infected with *Gibberella zeae*. Enteritic and endocrine disturbances follow ingestion of such feed (Caldwell, 1969; Curtin and Tuite, 1966). Although there are wide differences among inbred lines to ear rots (Ullstrup, 1970), little has been done to use genetic resistance as a means of control.

Fusarium kernel rot has caused damage to high-lysine corn. The *opaque*-2 and *floury*-2 versions, in some genetic backgrounds, are more susceptible to infection by *Fusarium moniliforme* and some other ear rotting fungi than are normal, starchy counterparts (Ullstrup, 1971).

Eight downy mildew diseases have been identified on corn under natural conditions; five of these have caused appreciable damage in Southeast Asia and India. Sorghum downy mildew, previously known only in Asia and Africa, was observed in 1961 in southeastern Texas (Reyes, 1964). Since that time it has spread in all directions and far north as Indiana (Warren, Scott, et al. 1974) and Illinois. The name of the disease derives from the crop on which it was first described over 60 years ago. In its journey northward the disease has not caused serious damage and is confined to bottomlands where "shattercane" (*Sorghum bicolor* var. *drummondii* [Steud.] Millspaugh et Chase), the host in which sexual spores of the pathogen are produced, is found. Corn genotypes have been evaluated in Texas, and the number of resistant inbreds may be sufficient to form a reservoir of material for exploitation should genetic control be required.

In 1963 the disease appeared in Thailand, where it has become a limiting factor in corn production. In Thailand the disease is rarely found on sorghum as it is in Texas. This difference, and others, in the disease as observed in Thailand and in Texas may be racial in nature. Control of the disease in Thailand is being accomplished through the development and use of composites that are far superior to the local varieties in resistance and yield. In the Philippines, where another downy mildew disease has been a serious hazard, control is being accomplished also through release to growers of resistant composites. In Taiwan sugarcane downy mildew has been controlled by resistant hybrids. The research and development of resistance to downy mildews of corn in these areas has been stimulated by the Inter-Asian Corn Program. This organization has sponsored and arranged international meetings where information and host materials are exchanged and results of cooperative experiments reported.

The two smut diseases of corn have not been of major economic importance over wide areas.

Common smut, caused by *Ustilago maydis* (DC) Cda., has on a few occasions been responsible for some losses. Selection against susceptibility has been the practice in the course of developing inbred lines, but little effort has been directed toward transferring resistance into agronomically desirable genotypes. Genes controlling reaction to the pathogen are many and distributed over several chromosomes (Saboe and Hayes, 1941).

Head smut, incited by *Sphacelotheca reiliana* (Kuhn) Clint., is found in some of the drier areas of the world. In the United States it is found in some of the western regions. In Idaho it has been troublesome in sweet-corn seed fields. Host-plant resistance has been identified (Fuentes, 1963), but this has not been used extensively in control of the disease.

Corn stunt was first described in Texas in 1945. The infectious agent was believed to be a virus, transmitted by the leafhopper *Dalbulus maidis* De L. & W. (Kunkel, 1948). The disease was subsequently found in Latin America, where it is known as "achaparramiento." In 1963 the disease caused serious damage in the delta lands of the Yazoo and Mississippi Rivers. Since it was first studied, five different leafhoppers have been identified as vectors of the etiological agent.

In 1968 evidence was presented that the causal agent was not a virus but probably a mycoplasma-like entity (Granados, Maramorosch, et al., 1968). This was based on blockage or remission of symptoms by treatment with tetracyclines that are effective against this type of infectious agent. Electron micrographs of diseased plants and infectious leafhoppers showed helical structures not present in healthy plants or noninfectious vectors (Chen and Granados, 1970; Davis, 1972).

Resistance to corn stunt is inherited in a quantitative manner, with few genes possessing strong additive effects being involved. Heterosis *per se* confers tolerance to the disease (Nelson and Scott, 1973).

In 1963 a mechanically transmitted virus was isolated from corn in the Ohio River bottomlands (Williams and Alexander, 1964). Since then the disease incited by it has been named maize dwarf mosaic (MDM) and is recognized in many parts of the United States. At least 11 species of aphid can transmit the virus (Bancroft, 1966) and at least two strains, A and B, are known (Gordon and Williams, 1970). Recent evidence suggests even more strains of the virus may be present (Louie and Knoke, 1970). The disease is similar in many respects to sugarcane mosaic (SCM) and considered by some to be a strain of the latter. Maize dwarf mosaic virus is much more virulent on corn than SCMV. Resistance is partially dominant and governed by relatively few genes (Dollinger, Findley, et al., 1970).

A more recently discovered virus disease of corn, 'maize chlorotic dwarf' (MCD), is similar to MDM, but the morphologies of the infectious particles of the two are distinct, and MCD is not mechanically transmitted. *Graminella nigrifrons* Forbes is the vector. The disease has been found in a number of places in the U. S. Midwest and South (Bradfute, Louie, et al., 1972; Pirone, 1972) and believed to be more widespread than MDM. Double infections with both viruses has a synergistic effect and more debilitating than single infections of either virus.

Corn mosaic was described in Hawaii in 1927. It has since been found in the Caribbean area, South America, and Africa. Recent evidence suggests it may be on mainland United States (Zeyen and Morrison, 1975). The disease is of some economic importance in Hawaii, where sweet corn is especially susceptible. The virus is transmitted only by *Peregrinus maidis* Ashm. Resistance is reported to be conditioned by a single gene, but dominance is not clearly evident (Brewbaker and Aquilizan, 1965).

A phanerogamic parasite, 'witchweed' (*Striga lutea* Lour.), previously known only in Asia and Africa, was discovered in the Carolinas in 1956. A small, annual plant with red, yellowish, or white flowers, bearing thousands of minute seeds, parasitizes corn, a number of other grasses, and a few dicots. Its spread appears to be contained through quarantines and the use of "catch crops" that are hosts, which when infected are plowed down, or by "trap crops," which are nonhost but secrete a stimulant that initiates germination of the seeds. Seedlings, being obligate parasites in early stages of development, soon die if no host is available (Garriss and Wells, 1956).

REFERENCES

Arny, D. C. and R. R. Nelson. 1971. *Phyllosticta maydis* sp. nova, the incitant of yellow leaf blight of maize. *Phytopathology* **61:** 1170–1172.

Arny, D. C., E. B. Smalley, A. J. Ullstrup, G. L. Worf, and R. W. Ahrens. 1971. Eyespot of maize, a new disease in North America. *Phytopathology* **61:** 54–57.

Ball, H. J. and G. T. Weekman. 1962. Insecticide resistance in adult western rootworm in Nebraska. *J. Econ. Entomol.* **55:** 439–441.

Bancroft, J. B., A. J. Ullstrup, Mimi Messieha, C. E. Bracker, and T. E. Snazelle. 1966. Some biological and physical properties of a midwestern isolate of maize dwarf mosaic virus. *Phytopathology* **56:** 474–478.

Beck, S. D., and E. Smissman. 1961. The European corn borer, *Pyrausta nubilalis* (Hbner) and its principal host plant—IX. Biological activity of chemical analogs of corn resistance factor A (6-methoxy-benzoxazolinone). *Ann. Entomol. Soc. Am.* **54:** 53–59.

Bergquist, R. R. and O. R. Masais. 1974. Physiological specialization in *Trichometasphaeria turcica* f. sp. *zeae* and *T. turcica* f. sp. *sorghi* in Hawaii. *Phytopathology* **64:** 645–649.

Bradfute, O. R., R. Louie, and J. K. Knoke. 1972. Isometric virus-like particles in maize with stunt symptoms. *Phytopathology* **62:** 748 (abstr.).

Brewbaker, J. L. and F. Aquilizan. 1965. Genetics of resistance in maize to a mosaic stripe virus transmitted by *Peregrinus maidis*. *Crop Sci.* **5:** 412–415.

Brindley, T. A. and F. F. Dicke. 1963. Significant developments in European corn borer research. *Ann. Rev. Entomol.* **8:** 155–176.

Bryson, H. R., D. A. Wilbur, and C. C. Burkhardt. 1953. The western corn rootworm, *Diabrotica virgifera* Lec. in Kansas. *J. Econ. Entomol.* **46:** 217–224.

Caldwell, R. W., J. Tuite, M. Stot, and R. Baldwin. 1969. Zearalenone production by *Fusarium* species. *Appl. Microbiol.* **20:** 31–34.

Chen, Tseh-An and R. R. Granados. 1970. Plant pathogenic mycoplasma-like organisms: maintenance in vitro and transmission to *Zea mays* L. *Science* **167:** 1633.

Curtin, T. M. and J. F. Tuite. 1966. Emesis and refusal of feed in swine associated with *G. zeae*-infected corn. *Life Sci.* **5:** 1937–1944.

Davis, R. E., R. F. Whitcomb, T. Ishijima, and R. L. Steere. 1972. Helical filaments produced by a mycoplasma-like organism associated with corn stunt disease. *Science* **176:** 521–523.

Dicke, F. F. 1969. The corn leaf aphid. *Proc. 24th Annu. Corn and Sorghum Res. Conf.* **24:** 61–73.

Dollinger, E. J., W. R. Findley, and L. E. Williams. 1970. Inheritance of resistance to maize dwarf mosaic in maize (*Zea mays* L.) *Crop Sci.* **10:** 412–415.

Eckhardt, R. C. 1954. Breeding for resistance to corn earworm. *Proc. 9th Annu. Hybrid Corn Indu. Res. Conf.* **9:** 54–60.

Everly, R. T. and M. S. Miller. 1962. Preliminary studies of the responses of the corn leaf aphid, *Rhopalosiphum maidis* (Fitch) from six geographical areas to three host crops. *Proc N.C. Branch—ESA* **17:** 25–27.

Fairchild, M. L. 1965. The Southwestern corn borer. *Proc. 20th Annu. Hybrid Corn Indu. Res. Conf.* **20:** 111–117.

Fitzgerald, P. J. and E. E. Ortman. 1964. Breeding for resistance to western corn rootworm. *Proc. 19th Annu. Hybrid Corn Indu. Res. Conf.* **19:** 46–60.

Fuentes, S. 1963. Resistance to head smut of Mexican races of corn. *Phytopathology* **53:** 24 (abstr.).

Garriss, H. R. and J. C. Wells. 1956. Parasitic herbaceous annual associated with corn disease in North Carolina. *Plant Dis. Rep.* **40:** 837–839.

Gevers, H. O. 1975. A new major gene for resistance to *Helminthosporium turcicum* leaf blight of maize. *Plant Dis. Rep.* **59:** 296–299.

Gordon, D. T. and L. E. Williams. 1970. The relationship of a maize virus isolated from southern Ohio to sugarcane mosaic virus strains and the B strain of maize dwarf mosaic virus. *Phytopathology* **60:** 1293 (abstr.).

Granados, R. R., K. Maramorosch, and E. Shikata. 1968. Mycoplasma: Suspected etiological agent of corn stunt. *Proc. Nat. Acad. Sci. (U. S. A.)* **60:** 841–844.

Guthrie, W. D., E. S. Raun, F. F. Dicke, G. R. Pesho, and S. W. Carter. 1965. Laboratory production of European corn borer egg masses. *Iowa State J. Sci.* **40:** 65–83.

Hooker, A. L. 1963. Inheritance of chlorotic lesion resistance to *Helminthosporium turcicum* in seedling corn. *Phytopathology* **53:** 660–662.

Hooker, A. L. 1969. Widely based resistance to rust in corn. In *Disease consequences of intensive and extensive culture of field crops,* J. Artie Browning (ed.) Ames: Iowa State University Special Rep. No. 64.

Hooker, A. L. and M. D. Musson. 1970. Physiologic races of *Helminthosporium maydis* and disease resistance. *Plant Dis. Rep.* **54:** 1109–1110.

Jenkins, M. T. and Alice L. Robert. 1959. Evaluating the breeding potentials of inbred lines of corn resistant to leaf blight caused by *Helminthosporium turcicum*. *Agron. J.* **51:** 93–96.

Jinahyon, S. and W. A. Russell. 1969. Evaluation of recurrent selection for stalk rot resistance in open-pollinated varieties. *Iowa State J. Sci.* **43:** 229–251.

Klun, J. A. 1975. Biochemical basis of resistance in plants to pathogens and insect hormone mimics and selected examples of other biologically active chemicals derived from plants. *Proc. Summer Inst. Biol. Control of Plant Insects and Diseases,* (F. G. Maxwell and F. A. Harris) (eds.) Jackson, Miss.: Univ. Mississippi Press.

Kunkel, L. O. 1948. Studies on a new corn virus disease. *Arkiv. für die Gesamte Virusforschung Sonderabdruck aus Band IV,* Vol. 1, pp. 24–46.

Louie, R. and J. K. Knoke. 1970. Evidence for strains of maize dwarf mosaic in southern Ohio. *Phytopathology* **60:** 1301 (abstr.).

Maxwell, F. G., J. N. Jenkins, and W. L. Parrott. 1972. Resistance of plants to insects. *Adv. Agron.* **24:** 187–265.

Mercado, A. C. and R. M. Lantican. 1961. The susceptibility of cytoplasmic male-sterile lines of corn to *Helminthosporium maydis* Nisikado & Miy. *Philippine Agr.* **45:** 235–243.

Molot, P. M. and P. Anglade. 1968. Resistance of maize inbreds to leaf blight disease and the European corn borer in relation to a compound likely identical with 6-methoxy-2(3)-benzoxazolinone. *Ann. Epiphytes* **19:** 75–95 (in French; English summary).

Nelson, R. R., J. E. Ayres, and J. B. Beckett. 1971. Reaction to various corn inbreds in normal and different male-sterile cytoplasms to the yellow leaf blight organism (*Phyllosticta* sp.). *Plant Dis. Rep.* **54:** 401–403.

Nelson, L. R. and G. E. Scott. 1973. Diallel analysis of resistance of corn (*Zea mays* L.) to corn stunt. *Crop Sci.* **13:** 162–164.

Ortman, E. E. and E. E. Gerloff. 1970. Rootworm resistance problems in measuring and its relationship to performance. *Proc. 25th Annu. Corn and Sorghum Res. Conf.* **25:** 161–174.

Painter, R. H. 1951. *Insect resistance in crop plants.* New York: Macmillan.

Pate, J. B. and P. H. Harvey. 1954. Studies on the inheritance in corn to *Helminthosporium maydis* leaf spot. *Agron. J.* **46:** 442–445.

Penny, L. H. and F. F. Dicke. 1956. Inheritance of resistance in corn to leaf feeding of the European corn borer. *Agron. J.* **48:** 200–202.

Penny, L. H. and F. F. Dicke. 1957. A single gene-pair controlling segregation for European corn borer resistance. *Agron. J.* **49:** 193–196.

Penny, L. H., G. E. Scott, and W. D. Guthrie. 1967. Recurrent selection for European corn borer resistance in maize. *Crop Sci.* **7:** 407–409.

Pesho, G. R., F. F. Dicke, and W. A. Russell. 1965. Resistance of inbred lines of corn (*Zea mays* L.) to the second brood of the European corn borer, *Ostrinia nubilalis* (Hübner). *Iowa State J. Sci.* **40:** 85–98.

Pirone, T. P., O. E. Bradfute, P. H. Freytag, M. C. Young, and C. G. Poneleit. 1972. Virus-like particles associated with a leafhopper-transmitted disease of corn in Kentucky. *Plant Dis. Rep.* **56:** 652–656.

Reyes, L. D., T. Rosenow, R. W. Berry, and M. C. Futrell. 1964. Downy mildew and head smut diseases of sorghum in Texas. *Plant Dis. Rep.* **48:** 249–253.

Rhind, D., J. M. Waterston, and F. C. Deighton. 1952. Occurrence of *Puccinia polysora* in West Africa. *Nature* (London) **169:** 631.

Rhoades, Virginia H. 1935. The location of a gene for resistance in maize. *Proc. Nat. Acad. Sci. U. S. A.* **21:** 243–246.

Saboe, L. C. and H. K. Hayes. 1941. Genetic studies of the reaction to smut and of firing in maize by means of chromosomal translocations. *J. Am. Soc. Agron.* **33:** 463–470.

Saccas, A. M. 1955. La rouille americaine du mais (*Zea mays* L.) due á *Puccinia polysora* Underw. au Cameroun et en Afrique Equatoriale Francaise. *Agron. trop. Nogent.* **10:** 499–522 (*Rev. Appl. Mycol.* **35:** 11).

Saxena, K. M. and A. L. Hooker. 1968. On the structure of a gene for disease resistance in maize. *Proc. Natl. Acad. Sci. U. S. A.* **61:** 1300–1305.

Scheifele, G. L., W. Whitehead, and C. Rowe. 1970. Increased susceptibility to southern leaf spot (*Helminthosporium maydis*) in inbred lines and hybrids of maize with Texas male-sterile cytoplasm. *Plant Dis. Rep.* **54:** 501–503.

Sifuentes, J. A. and R. H. Painter. 1964. Inheritance of resistance to western corn rootworm in field corn. *J. Econ. Entomol.* **57:** 475–477.

Singh, D. N. and F. S. McCain. 1963. Relationship of some nutritional properties of the corn kernel to weevil infestation. *Crop Sci.* **3:** 259–261.

Smith, G. M. 1933. *Golden Cross Bantam sweet corn.* USDA Circ. No. 268.

Smith, D. R. and A. L. Hooker. 1973. Monogenic chlorotic-lesion resistance in corn to *Helminthosporium maydis. Crop Sci.* **13:** 330–331.

Sprague, G. F. 1954. Breeding for resistance to stalk rot. *Proc. 9th Annu. Hybrid Corn Industry Res. Conf.* **9:** 38–43.

Storey, H. H. 1939. Transmission of plant viruses by insects. *Bot. Rev.* **5:** 240–272.

Storey, H. H., A. K. Howland, J. S. Hemingway, B. T. S. Baldwin, H. C. Thorp, and G. E. Dixon. 1958. East African work on breeding maize resistant to the tropical American rust, *Puccinia polysora. Emp. J. Exp. Agr.* **26:** 1–17.

Tatum, L. A. 1971. The southern corn leaf blight epidemic. *Science* **171:** 1113–1116.

Ullstrup, A. J. 1965. Inheritance and linkage of a gene determining resistance in maize to an American race of *Puccinia polysora. Phytopathology* **55:** 425–428.

Ullstrup, A. J. 1970. Methods of inoculating corn ears with *Gibberella zeae* and *Diplodia maydis. Plant Dis. Rep.* **54:** 658–662.

Ullstrup, A. J. 1971. Hypersusceptibility of high-lysine corn to kernel and ear rots. *Plant Dis. Rep.* **55:** 106.

Warren, H. L., D. Scott, and R. L. Nicholson. 1974. Occurrence of sorghum downy mildew on maize in Indiana. *Plant Dis. Rep.* **58:** 430–432.

Wellhausen, E. J. 1937. Genetics of resistance of bacterial wilt in maize. Ames: *Iowa Agr. Exp. Sta. Res. Bull.* **224:** 69–114.

Widstrom, N. W. and B. R. Wiseman. 1973. Locating major genes for resistance to corn earworm in maize inbreds. *J. Hered.* **64:** 83–86.

Widstrom, N. W., and B. R. Wiseman, and W. W. McMillan. 1972. Resistance among maize inbreds and single crosses to fall armyworm injury. *Crop Sci.* **12:** 290–292.

Williams, L. E. and L. J. Alexander. 1964. An unidentified virus isolated from corn in southern Ohio. *Phytopathology* **54:** 912 (abstr.).

Zeyen, R. J. and R. H. Morrison. 1975. Rhabdoviruslike particles associated with stunting of maize in Alabama. *Plant Dis. Rep.* **59:** 169–171.

Zuber, M. S., M. L. Fairchild, A. J. Keaster, V. L. Fergason, G. F. Krause, E. Hildebrand, and P. J. Loesch, Jr. 1971. Evaluation of 10 generations of mass selection for corn earworm resistance; *Crop Sci.* **11:** 16–18.

Chapter 20
GENETICS OF INSECT RESISTANCE IN MAIZE

F. F. Dicke and W. D. Guthrie

Department of Plant Breeding,
Pioneer Hi-Bred International, Inc.,
Johnston, Iowa and
Entomologist, Agricultural Research Service, USDA and
Professor of Entomology,
Iowa State University,
Ames

According to historical records, maize was well established as a cultivated crop and an important source of food among the many Indian tribes long before Columbian exploration. There is good evidence that most of the common insect pests of maize have been closely associated with the species during its evolutionary development for many hundreds of years.

The origin of maize has been a subject of investigation and speculation by many geneticists. Considerable genetic attention has been given to chromosomal comparisons among different races of maize and among the close relatives Euchlaena and Tripsacum. Mangelsdorf, 1974, has accumulated together a great deal of information on the subject.

Journal Paper No. J-8283 of the Iowa Agriculture and Home Economics Experiment Station, Ames, Iowa. Project No. 1923.

The impact of epidemics of diseases and insects on both natural and man-induced selection in the evolutionary process has had relatively little attention until recent years. It is obvious, for example, that the insect–plant relationship in the virus and spiroplasma disease complex has been present for a long time in what is believed to have been the "mother sites" of maize origin. Furthermore, maize is endowed with genetically controlled protective morphological and physiological resistance qualities that vary with the insect or disease pressures in the environment. The diverse variability in maize enables improvements in resistance to insects through modern breeding techniques. It is the purpose of this discussion to review contributions made in advancing this area of research. Both common terms, *corn* and *maize,* are used. In the approved list of common names of insects, the term *corn* is used with but a few exceptions.

BIOLOGICAL RELATIONSHIP BETWEEN INSECTS AND PLANTS

Many species of maize insects have more than one generation each season. The biological relationship between the insect and host plant may not be the same for each generation. A knowledge of the biology of an insect on the plant is imperative in host-plant resistance research. For example, during the period of egg deposition by the first-generation European corn borer, *Ostrinia nubilalis,* most dent corn in the Corn Belt is in the whorl stage of plant development. The young larvae feed primarily on the emerging spirally rolled leaves in the whorl. Factors that differentially inhibit first-generation larval establishment and survival are operative against the early larval instars. A significant rate of larval mortality is interpreted to be a high level of antibiosis against the young larvae of a first-generation infestation. Therefore, first-generation resistance actually is leaf-feeding resistance. Likewise, species such as the southwestern corn borer *Diatraea grandiosella* and *Chilo zonellus* also feed on whorl-leaf tissue during the first-generation period (Guthrie, Dicke, et al., 1960; Davis Henderson, et al., 1972; Chatterji, Bhanburkar, et al., 1971).

During the period of egg deposition by second-generation European corn borers, most corn in the central Corn Belt states is in the reproductive stage of development. On this stage of corn, the young larvae feed primarily on pollen accumulation at the axils of the leaves on ear parts and on sheath and collar tissue. Because the larvae feed extensively on sheath and collar tissue, second-generation resistance is primarily collar and sheath-feeding resistance. (Pesho, Dicke, et al., 1965; Guthrie, Russell, et al., 1971). In contrast to the European corn borer, second-generation southwestern corn borer larvae are not sheath and collar feeders, but are primarily husk feeders (Davis, Henderson, et al., 1972). Thus the site of resistance to the second-generation southwestern corn borer is mainly in the husk. In view of such differences in seasonal growth, strains of maize that are resistant during the vegetative stage of plant development may be susceptible during pollen dehiscence and later stages (Brindley, Sparks, et al., 1975).

INSECT-REARING METHODS

To evaluate plant material for resistance to insects, an adequate number of insects are required for a uniform level of infestation. The entomologist must either rely on natural field populations or rear the insect in the laboratory (Guthrie, 1975). The development of methods in recent years for mass-producing insects to infest experimental plant germplasm has materially advanced progress in host-plant resistance research.

Some species, such as the adult corn rootworms (*Diabrotica virgifera* and *D. longicornis*), are attracted to special late-planted, trap-crop maize for oviposition of the over-wintering diapausing egg population. Succeeding maize plantings on such trap-crop ground usually develop satisfactory infestations to evaluate root injury (Owens, Peters, et al., 1974).

Significant progress has been made in developing artificial diets for rearing insects. The use of wheat germ marked the advent of the modern era of practical artificial diets for rearing plant-feeding Lepidoptera. Slight modifications of the wheat-germ medium have been successful for rearing many species away from their natural host plants. The success of the artificial diets is due, in part, to the similar chemical and physical food requirements that plant-feeding Lepidoptera share. Consequently, only minor changes to existing wheat-germ diets have been necessary to make them acceptable for different species (Chippendale, 1972).

Several species of plant-feeding Lepidoptera, such as the corn earworm, *Heliothis zea*; the fall armyworm, *Spodoptera frugiperda* (Burton and Perkins, 1972); the southwestern corn borer, *Diatraea grandiosella* (Davis, Henderson, et al., 1972), *Chilo zonellus* (Dang, Anand, et al., 1970), and the European corn borer, *Ostrinia nubilalis* (Guthrie, Russell, et al., 1971), now are reared on artificial (meridic) diets to obtain eggs for infesting and evaluating plant germplasm.

Entomologists should be aware that insect cultures reared generation after generation in the laboratory or greenhouse may change genetically so that they no longer survive on plants the same as does a feral population. Data collected during the past 8 years show that the European corn borer reared continuously on a meridic diet cannot be used for resistance studies because the plant injury is too low for evaluating resistance (Guthrie, Russell, et al., 1971; Guthrie, Rathore, et al., 1974). Numerous research programs are involved in various aspects for the improvement of maize. A broad-spectrum maize-development project necessarily requires the talents of several disciplines. A general arrangement for the division of responsibility in a cooperative approach is for the entomologist to assume the labor of detailed biological relationship studies. The primary responsibility for the study of the genetics of insect resistance in plants rests with the plant breeder or geneticist. The evaluation of improvements in performance again becomes a mutual problem. In this type of approach the various disciplines can attain a common goal.

GENETICS AND BREEDING FOR INSECT RESISTANCE

European Corn Borer, *Ostrinia nubilalis*

More information on genetics and breeding for resistance is available with the European corn borer than with any other species of insect-attaching corn. Brindley and Dicke (1963) and Brindley, Sparks, et al. 1975, reviewed research on host-plant resistance through 1973. Guthrie (1974) assembled a comprehensive review on techniques, accomplishments, and future potential of breeding for resistance to European corn borer in corn. Gallun, Starks, et al. (1975) reviewed the chemical basis of resistance to first-generation borers.

Leaf-feeding (First-generation) Resistance. Resistance to leaf feeding by first-generation larvae has been easy to find (Guthrie and Dicke, 1972). At the time the European corn borer was discovered in the United States and for some 15 years thereafter only open-pollinated varieties were being grown. Although varieties of both field and sweet corn were tested and significant differences in infestation were evident, such differences resulted primarily from differential oviposition and larval survival on variable growth stages within and between varieties. There is good evidence that some European varieties possess substantial resistance to first-generation attack. Roubaud (1928) reported this type of resistance for the variety *Dent de cheval* and Horber (1961), for the varieties *Nostrano dell' isola* and *Marano*. The development of many inbred lines of corn and commercial hybrids greatly stimulated the search for resistant lines that might be utilized directly in hybrid combinations or for breeding germ plasm. The term *resistance* as used in this review refers to antibiosis as defined by Painter (1941). *Tolerance* most commonly refers to resistance to lodging and ear-holding qualities.

Extensive cooperative programs were developed during the 1930s by the USDA and many state agricultural experiment stations. Large numbers of inbred lines and crosses of both field and sweet corn were tested. During this decade the open-pollinated varieties were rapidly replaced by double-cross hybrids. Reports on these research activities were made by Meyers, Huber, et al. (1937), Patch (1937), Patch, Bottger, et al. (1938), and Patch and Everly (1945, 1948). Inbred lines with substantial resistance to larval survival were found in small numbers. The prepotency of transmitting resistance to crosses was investigated simultaneously, and it was determined that heritable qualities were involved. There was a tendency for the component inbred lines in hybrids to be susceptible, but a high percentage of inbred lines tested were intermediate in resistance. Tests of hybrid combinations indicated that multiple factors were involved in resistance. Several lines with substantial resistance were agronomically acceptable in hybrid combinations. Hybrids with improved resistance and tolerance rapidly replaced open-pollinated varieties. Prominent among inbred lines with improved resistance or tolerance were L317, R4, Oh40B, Oh07, Oh51, P8, K230, K226, MS285, Wis.CC5, and Hy. Some of the more agronomically desirable inbreds in commercial hybrids (particularly WF9, Tr, A, I153, CI.187-2, Ill.90, A334, A375, Wis.CC2, and Wis.CC6 and CC7) were susceptible. Open-pollinated varieties were the direct source material for most of the inbred lines developed from 1930 to 1940.

During the 1940s and 1950s inbred lines with a satisfactory degree of resistance were extracted from special crosses (second-cycle breeding). Representative lines from this type of breeding were: A295, A619, B49, CI.31A, Oh43, Oh45, Oh45A, Oh45B, Pa32, and Pa54. During the 1960s many experimental lines with a high level of resistance were developed by a recurrent selection technique.

In recent years farmers have largely replaced double-cross hybrids with single crosses or modified single crosses. In general, single crosses with the following combination of inbred lines are effective in reducing first-generation populations: (a) resistant × resistant, (b) intermediate × intermediate, (c) resistant × intermediate, or (d) resistant × susceptible; either dominance or incomplete dominance of resistance is necessary if the resistant × susceptible combination is to be effective.

The first systematic breeding to develop resistant inbred lines in which *Maize Amargo* (bitter maize) was used as a source was reported by Marston (1931), and Marston and Dibble (1930). Maize Amargo was crossed with Michigan varieties and selections were selfed in subsequent generations. When the frequency of plant infestation in the F_1 and F_2 generations was considered, it was concluded that resistance was recessive and susceptibility, dominant. The frequency of plant infestation in F_3 families indicated a 3:1 Mendelian ratio and a recessive resistant character. Tests of F_4 selections showed a higher level of resistance than that in the parent F_3 generation, an indication of continued segregation.

Working with first-generation, leaf-feeding resistance in F_3 and backcross progenies of susceptible × resistant (M14 × MS1), Ibrahim (1954) determined that segregation of genes at three or more loci was indicated. Similar tests with the susceptible × resistant cross (B14 × N32), indicated one or two gene pairs for resistance (Penny and Dicke, 1956). In two susceptible × resistant crosses, M14 × $gl7v17$ and WF9 × $gl7v17$, estimates in F_2 and backcross progenies showed that resistance differences were conditioned by segregation of genes at a single locus. The resistant gene was linked with the $gl7v17$ genes of the resistant parent with crossover frequencies of 31–37% (Penny and Dicke 1957).

Fleming, Singh, et al. (1958) concluded from inheritance studies that relatively few major genes control resistance to leaf feeding and overall plant damage. Segregation was explained on the basis of two pairs of major genes. Their results indicated that hybrid vigor may influence the amount of resistance.

In most cases resistance to leaf feeding by first-generation European corn borers is conditioned by genes at several loci, and effects are cumulative among loci. Ibrahim (1954), using chromosomal interchanges, concluded that leaf-feeding resistance factors differentiating the inbred line A411 from the susceptible line A344 are associated with one gene on the 3L chromosome, one gene on the 4L chromosome, and probably another on the 5L chromosome. Tests of chromosomal interchange lines also indicated resistance factors other than those carried by A411. Scott, Dicke, et al. 1966, using reciprocal translocations, reported that five chromosome arms in CI.31A and 6 arms in B49 carried genes for first-generation resistance. Guthrie and Stringfield (1961a,b) found that segregation in a 24-line synthetic variety, as measured by the net variance, diminished after each selfing but that a significant residue of segregation remained in the fifth selfed generation. If there was an average of one effectual heterozygous locus in the S_5, theoretically there should have been 2^5 or at least 32, effectual heterozygous loci five generations back in the S_0. Scott, Hallauer, et al. (1964) determined the type of gene action involved in first-generation resistance by using F_2, F_3, and selfed backcross populations of CI.31A(R) × B37(S) plus individual F_2 plants of (CI.31A × B37) × CI.31A and individual F_2 plants of (CI.31A × B37) × B37; most of the genetic variance was of the additive type.

Fleming, Singh, et al. (1958) suggested backcrossing as a method of adding resistance to otherwise desirable inbred lines. A modified backcross procedure was used to improve agronomic characteristics of Oh45 and retain its resistance to leaf feeding by the European corn borer by Guthrie and Stringfield (1961a,b). In many efforts to breed for first-generation resistance, however, the direct backcross method was not successful in transferring resistance to susceptible inbred lines. The desired genotypes could not be identified in the segregating generations; when more than two backcrosses were used, the needed level of resistance was lost. The level of resistance could be increased, however, by intermating among resistant plants in progeny of the first or second backcross (Russell, 1972). Penny, Scott, et al. (1967) showed that recurrent selection was very effective in increasing the level of resistance to first-generation corn borers in five corn populations.

Chemical Nature of Leaf-feeding Resistance Factors.   Beck (1965) discussed the fundamental concepts of plant resistance in a review of the biochemical and biophysical resistance to insect survival. A great deal of research by Beck and his associates of the University of Wisconsin was centered around identifying the chemical nature of resistance and the nutritional facets associated with it. By their incorporation of substances fractionated from corn plants in aseptic media, they obtained growth-inhibiting reactions in bioassay. They isolated three fractions that were detected to be biologically active in reducing the rate of larval growth. These substances were designated resistant factors A, B, and C (RFA, RFB, and RFC). Substances RFA and RFC were determined to be closely related and of most importance in inhibiting growth of larvae. Substance RFA was isolated in pure crystalline form and characterized as 6-methoxybenzoxazolinone (6-MBOA). Laboratory synthesis was accomplished, and a quantitative method was developed for measuring its concentration in preparations. Other investigators, particularly in Finland, characterized 6-MBOA at about the same time in connection with disease resistance (Beck, 1956, 1957; Beck and Hanec, 1958; Beck and Smissman, 1960, 1961; Loomis, Beck, et al., 1957; Smissman, Lapidus, et al., 1957). The roles of 6-MBOA and benzoxazolinone have been investigated in wheat, rye, and corn for the inhibitory effects on disease pathogens by Virtanen (1969), who found that a precursor glucoside originally was present in plants and that the formation of benzoxazolinone and 6-MBOA seemed to result from a complex enzymatic action in the crushed plant material. The precursor glucosides did not show an inhibitory effect on pathogens, but the aglucones formed showed antifungal activity. The possibility was suggested that RFA might be produced in the intestinal tract of the larvae.

Klun and Brindley (1966) provided evidence that 6-MBOA is of little importance in the phenomenon of resistance, but suggested that precursors of 6 MBOA may have an active role. Klun, Tipton, et al. (1967) showed that 2,4-dihydroxy-7-methoxy-(2H)-1,4-benzoxazine-3(4H)-one (DIMBOA) is a biochemical factor in the resistance of maize to first-generation European corn borers, occurring as a glucoside in intact maize tissue. When plant tissues are crushed, however, the glucoside is hydrolyzed by a plant glycosidase to yield free DIMBOA (Wahlroos and Virtanen, 1959), and DIMBOA decomposes stoichiometrically to 6-MBOA (Brendenbert, Honkanen, et al. 1962). Thus, for convenience, dried plant tissue is analyzed for 6-MBOA as an index of DIMBOA concentration. Klun and Robinson (1969) determined the concentration of 6-methoxy-2-benzoxazolinone (MBOA, and an indication of DIMBOA) in the whorl tissue of six inbreds at an extended leaf height of 84 cm. They showed a strong linear relationship between the logarithm of the concentration of MBOA and the leaf-feeding resistance ratings of the six inbreds. Klun, Guthrie, et al. (1970) also found a significant correlation between the concentration of DIMBOA in the leaf-whorl tissue and resistance to the first-generation borer. The correlation coefficients were −0.89 for 11 inbreds and −0.74 for 55 single crosses. Klun (1969) suggested use of chemical analysis of plant tissue for DIMBOA as a breeding technique to select for resistance to the first-generation European corn borer.

Russell, Guthrie, et al. (1975) compared the effectiveness in development of corn progenies resistant to leaf feeding in selection with the insect (artificial infestation with egg masses) and selection on the basis of DIMBOA content in leaf-whorl tissue. Selections were made in three successive inbreeding generations, starting with 198 F_3 progenies from WF9 × CI.31A (WF9 is highly susceptible; CI.31A is highly resistant). Final evaluations were of 15 F_5 families in the DIMBOA evaluation group and of 86 F_5 families in the insect infested group. All families in the DIMBOA group were resistant to leaf feeding, and eight progenies did not significantly differ from CI.31A in DIMBOA content. Sixty-five families in the insect group were resistant to leaf-feeding, but only 15 did not significantly differ from CI.31A in DIMBOA content. The data showed that chemical analysis for DIMBOA in the plant can be used to select for the DIMBOA type of resistance in segregating population. The data also showed the factors other than DIMBOA condition resistance to leaf feeding by the European corn borer. Sullivan, Gracen, et al. (1974) found that leaf-feeding resistance in some exotic corn germplasm was not conditioned by high DIMBOA content.

The DIMBOA technique may be less vulnerable to vagaries of the environment, but is too slow to be of practical value in most corn-breeding projects. The concentration of DIMBOA in the plant is not correlated with sheath-feeding (second-generation) resistance.

The technique of rearing European corn-borer larvae on a meridic diet has greatly accelerated research on second-generation resistance and its inheritance. Jennings, Russell, et al. (1974a,b) used a generation-mean analysis to determine the genetic basis of antibiosis type of second-generation resistance. Nine populations were studied: P1, P2, F_1 F_2, F_3, BC1, BC2, and selfed progenies of both backcrosses. In four different experiments, B52 was used as the resistant parent (P1), and B39, L289, Oh43, and WF9 were used as the susceptible parent (P2). The data indicated no simple genetic basis of resistance and suggested that high resistance to a second-generation infestation may be the resultant cumulative effect of an unknown number of loci. Additive genetic effects were predominant in conditioning resistance, but dominance was significant in all crosses. Scott, Guthrie, et al. 1967 showed that this high resistance of B52 also is transmitted in hybrid combinations. Jennings, Russell, et al.

(1974a,b) used a diallel analysis involving 10 inbred lines and their 45 single crosses to demonstrate further that additive type of gene action conditions resistance to a second-generation infestation. Frequencies of genes that condition second-generation resistance are low in populations of maize in the Corn Belt states. A recurrent selection program based on S_1 line evaluation is recommended to increase gene frequencies to both generations in a population that can then be used as a source for resistant lines (Russell, Guthrie, et al. 1974).

Corn Earworm, *Heliothis zea*

The corn earworm is a polyphagous species widely distributed in tropical and temperate areas. Maize is the preferred host. Varieties originating in the tropical and southern regions of the United States have long been known to have protective husk qualities and have been a source for these qualities in corn-breeding programs.

Collins and Kempton (1917), in a study of this inheritance of corn earworm resistance in crosses between sweet corn and field corn, demonstrated that characters for resistance to ear infestation was associated with the number of husk leaves and the husk extension beyond the tip of the ear. They were able to transfer these characters from field to sweet varieties. Hinds (1914) pointed out that extension and tightness of husks provided protection against earworm injury and subsequent field infestation of rice weevil. Kyle (1918) selected for protective husk qualities in a varietal development program in Georgia. With the advent of hybrid corn, many cooperative programs included corn earworm resistance as a facet in maize breeding and development of hybrids. An excellent review of the subject has been published by McMillian and Wiseman (1972).

Recurrent and continued mass selection has been an effective method for improving plant populations for corn earworm resistance (Widstrom, Wiser, et al. 1970; Zuber, Fairchild, et al. 1971b). Effects of combining ability and relative dominance were reported in several studies (Straub, Fairchild, et al. 1973; Widstrom and McMillian, 1973). Positive additive or dominance effects were indicated in these experiments. Some genetic studies using waxy-marked chromosome translocations for locating genes for resistance in inbreds varied with respect to numbers of chromosomes involved in resistance factors. Gene locations on most of the 10 chromosomes have been suggested (Robertson and Walters, 1963; Widstrom and Wiseman, 1973).

Fall Armyworm, *Spodoptera frugiperda*

The fall armyworm is one of the most important pests of maize in tropical and subtropical areas of the Americas. It infests the corn plant from early seedling to grain maturity. Ear injury through the husks is more prominent than with

Heliothis, and the succession of grain insects is perhaps more important. Tropical and southern-origin varieties or races of maize possess a higher level of resistance qualities than U. S. Corn Belt germplasm. Sources of resistance to husk infestation can be found readily in tropical races.

An increasing amount of research has been devoted to different phases of resistance to the fall armyworm. Wiseman, Painter, et al. (1967), in a test of 81 Latin-American lines of maize, reported Cuba Honduras 46-J and ETO Amarillo to be least damaged at the sheath attachment. Lines originating from Antigua tested also in other locations were considered to carry a high level of resistance. Comparative tests, in which Antigua 20-160-87 was least damaged, showed that corn was highly preferred over *Tripsacum dactyloides* by first-instar larvae of *S. frugiperda*. Differences in resistance in inbreds and single crosses were observed by Widstrom, Wiseman, et al. 1972.

Diatraea spp.

Maize is a host to several species of "stalk borers" that are injurious in tropical and subtemperate climates, particularly the southwestern corn borer *Diatraea grandiosella,* sugarcane borer *D. saccharalis,* so-called neotropical corn borer *D. lineolata,* and southern cornstalk borer *D. crambidoides*. Depending on the seasonal life history, all of these species can occur on the vegetative "whorl" or reproductive phases of maize development. The advanced instars of the larvae invade the stalk. On maize the southwestern corn borer has had more research attention than the other species. A search for sources of resistance has continued for many years. Biological host relationship studies have been pursued by Davis, Henderson, et al. (1972). In screening for sources of resistance, Davis, Scott, et al. (1973) found two lines, derived from Antigua GP02, with an intermediate level of first-brood resistance by virtue of fewer larvae per plant and smaller larvae. They reported less second-brood attack on some southern single crosses and indicated the presence of resistant genotypes in several Central American corn populations.

Corn Rootworm, *Diabrotica* spp.

The following species constitute a pest complex known as corn rootworms: (a) the western corn rootworm *Diabrotica virgifera,* (b) the northern corn root-worm *D. longicornis,* and (c) the southern corn rootworm *D. undecimpunctata howardi*. Western and northern corn rootworms overwinter as eggs in the soil and have only one generation per year.

Most of the research has been directed toward breeding for tolerance to root feeding by rootworm larvae, that is, some lines are able to develop new roots above the feeding points of the larvae almost as rapidly as the larvae destroy root tissue. Rootworm tolerance does not imply that the development of new roots is as much a response to larval feeding as it is an expression of the inherent root-development characteristics of the plant (Owens, Peters, et al., 1974). Ortman and Gerloff (1970) reported that tolerance is the most promising component of resistance and that fibrous root production is not as important as other root characteristics.

Fitzgerald, Ortman, et al. (1968) found that rootworm tolerance could be selected for in the absence of a rootworm infestation by mechanical damage of corn root systems. Ortman, Peters, et al. (1968) reported that the pounds required to pull root systems from the soil had a relatively high positive correlation with visual ratings of the root systems.

Zuber, Musick, et al. (1971a) used four single-cross hybrids and found highly significant correlations among root-pulling resistance, root volume, and root classification. They suggested that any of the three characteristics would be suitable for evaluating corn strains for tolerance to corn rootworms.

Eiben and Peters (1965) used the following criteria for evaluating inbred lines for tolerance to rootworms: (a) lodging, (b) stunting, (c) number of dead plants, (d) dry weight of roots, (e) size rating, (f) total number of crown roots, and (g) pounds of pull required for root extraction. Pounds of pull gave the best indication of rootworm related responses.

Eiben and Peters (1962) found significant differences among inbred lines for total number of crown roots, total number of root nodes, and pounds of pull required to extract roots from the ground, but they found no significant differences for the number of rootworms per plant or total damaged roots.

Wilson and Peters (1973) evaluated more than 2000 plant introductions for corn rootworm resistance and found no evidence for antibiosis, but did locate sources for tolerance.

Owens, Peters, et al. (1974) evaluated 221 random inbred lines developed from Iowa Stiff Stalk Synthetic variety for rootworm tolerance. They used root damage, lodging, root size, and secondary root growth as indicators of tolerance. Estimates of heritability indicated that selection on the basis of rootworm feeding damage alone would be ineffective. Heritability values for root size, secondary roots, and root lodging indicated that gains could be expected from selection for each of these traits. Genotypic, phenotypic, and error correlations indicated that selection for larger root systems may result in superior secondary root development, reduced feeding damage, and reduced lodging. A retest of 50 superior entries indicated that selection for rootworm tolerance resulted in advance for each component trait of tolerance. The advance for each trait reinforces the existence of interrelationships among the four root traits. Superior lines selected from the retest population have been recombined into a single synthetic population that should possess tolerance to corn rootworm feeding.

Ortman and Fitzgerald (1964) evaluated inbred lines on the basis of general appearance, firmness of anchoring, root lodging, stand reduction, adult feeding, root damage, root regeneration, and general conformity of the root system. There were differences among lines for all characters except damage. Fitzgerald and Ortman (1964) used larval counts, damage ratings, ratings of root regeneration, row ratings for general appearance, firmness of anchoring, and pounds of pull needed to extract the root system to evaluate a wide range of corn germplasm for rootworm resistance. Damage ratings, a measure of anti-biosis, did not differ among genotypes, but there were differences among genotypes for the other measurements (tolerance). These researchers suggested that tolerance was due to a naturally well-developed root system, the ability of the plant to generate new roots, the time of insect attack in relationship to the developmental stage of the plant, and environmental conditions.

Other researchers also have reported differences in corn germplasm for tolerance to corn rootworms; (Cooper, Dungan, et al. 1942; Melhus, Painter, et al. 1954). Resistance also has been shown to be heritable (Bigger, Snelling, et al. 1941). Prominent among corn germplasm with tolerance are SD10 (Shank, Beatty, et al., 1965), B54, B67, and B69 (Russell, Penny, et al. 1971). Walter (1965) observed variation among sweet-corn inbreds for lodging and development in secondary roots under a rootworm infestation.

Very little information is available on genetics and breeding for resistance to feeding by adult corn rootworms. Sifuentes and Painter (1964) obtained data from F_1, F_2, and F_3 generations of a resistant × susceptible cross and concluded that resistance to adult leaf feeding was controlled by a single recessive gene. Grandados (1967) reported differential damage by adult rootworms to silks among a wide range of materials.

Reissig and Wilde (1971) exposed water extracts and whole silks of 15 genetic sources of corn to western corn rootworm adults in the laboratory and found significant differences in beetle preference for both the water extracts and whole silks.

Luckmann, Chiang, et al. (1974) compiled a list of 426 references on the northern and western corn rootworms.

Corn-leaf Aphid, *Rhopalosiphum maidis*

The corn-leaf aphid is worldwide in distribution and periodically develop injurious populations on maize. The species has many hosts among the grasses, barley, and sorghum. Dicke (1969) discusses the biological relationship and varietal resistance for this pest. As a primary vector of MDMV, the corn-leaf aphid is involved in the incidence of this mosaic on maize and sorghum.

Host-plant resistance has been investigated for many years. Because population development is erratic, it has been difficult to obtain uniform continuity in breeding and genetic studies. Attempts are usually made to eliminate susceptible cultures in the process of inbreeding. The presence of differential resistance and heritability in crosses is well established. The more recent contributions on the subject have been made by Everly (1967) and Rhodes and Luckmann (1967).

REFERENCES

Beck, S. D. 1956. Nutrition of the European corn borer (*Pyrausta nubilalis* Hbn.). IV. Feeding reactions of first instar larvae. *Ann. Entomol. Soc. Am.* **49:** 338–405.

Beck, S. D. 1957. The European corn borer, *Pyrausta nubilalis* (Hübner), and its principal host plant: IV. Larval saccharotropism and host plant resistance. *Ann. Entomol. Soc. Am.* **50:** 247–250.

Beck, S. D. 1965. Resistance of plants to insects. *Annu. Rev. Entomol.* **10:** 207–232.

Beck, S. D. and W. Hanec. 1958. Effect of amino acids on feeding behavior of the European corn borer, *Pyrausta nubilalis* (Hübner). *J. Insect Physiol.* **2:** 85–96.

Beck, S. D. and E. Smissman. 1960. The European corn borer, *Pyrausta nubilalis* (Hübner), and its principal host plant VIII. Laboratory evaluation of host plant resistance to larval growth and survival. *Ann. Entomol. Soc. Am.* **53:** 755–762.

Beck, S. D. and E. Smissman. 1961. The European corn borer *Pyrausta nubilalis* (Hübner), and its principal host plant IX. Biological activity of chemical analogs of corn resistance factor A (6-methoxy-benzoxazolinone). *Ann. Entomol. Soc. Am.* **54:** 53–59.

Bigger, J. H., J. R. Holbert, W. P. Flint, and A. L. Lang. 1938. Resistance of certain corn hybrids to attack by southern corn rootworm. *J. Econ. Entomol.* **31:** 102–107.

Bigger, J. H., R. O. Snelling, and R. A. Blanchard, 1941. Resistance of corn strains to the southern corn rootworm *Diabrotica duodecimpunctata* (F). *J. Econ. Entomol.* **34:** 605–613.

Brendenbert, J. B., E. Honkanen, and A. I. Virtanen. 1962. The kinetics and mechanism of decomposition of 2,4-dihydroxy-1, 4-benzoxazine-3-one. *Acta Chem. Scand.* **16:** 135–141.

Brindley, T. A. and F. F. Dicke. 1963. Significant developments in European corn borer research. *Annu. Rev. Entomol.* **8:** 155–176.

Brindley, T. A., A. N. Sparks, W. B. Showers, and W. D. Guthrie. 1975. Recent research advances on the European corn borer in North America. *Annu. Rev. Entomol.* **20:** 221–239.

Burton, R. L. and W. D. Perkins. 1972. WSB, A new laboratory diet for the corn earworm and the fall armyworm. *J. Econ. Entomol.* **65:** 385–386.

Chatterji, S. M., M. W. Bhanburkar, K. K. Marwaha, V. P. S. Panwar, K. H. Siddiqui, and W. R. Young. 1971. Relative susceptibility of some promising exotic maize material to *Chilo zonellus* Swinhoe under artificial infestation. *Indian J. Entomol.* **33:** 209–213.

Chippendale, G. M. 1972. Composition of meredic diets for rearing plant feeding Lepidopterous larvae. *Proc. North Cent. Branch Entomol. Soc. Am.* **27**: 114–121.

Collins, G. N. and J. H. Kempton. 1917. Breeding sweet corn resistant to the corn earworm. *J. Agric. Res.* **11**: 549–572.

Cooper, R. R., H. G. Dungan, A. L. Lang, J. H. Bigger, B. Koehler, and O. Bolin, 1942. Illinois corn performance tests. *Ill. Agric. Exp. Sta. Bull.* No. 482, pp. 473–528.

Dang, K., M. Anand, and M. G. Jotwani. 1970. A simple improved diet for mass rearing of sorghum stem borer, *Chilo zonellus* Swinhoe. *Indian J. Entomol.* **32**: 130–133.

Davis. F. M., C. A. Henderson, and G. E. Scott. 1972. Movement and feeding of larvae of the southwestern corn borer on two stages of corn growth. *J. Econ. Entomol.* **65**: 519–521.

Davis, F. M., G. E. Scott, and C. A. Henderson, 1973. Southwestern corn borer: Preliminary screening of corn genotypes for resistance. *J. Econ. Entomol.* **66**: 503–506.

Dicke, F. F., 1969. The corn leaf aphid. *Proc. Annu. Corn and Sorghum Res. Conf.* **24**: 61–70.

Dicke, F. F. and M. T. Jenkins, 1945. Susceptibility of certain strains of field corn in hybrid combinations to damage by corn earworms. *USDA Tech. Bull.* No. 898, 36 pp.

Eiben, G. J. and D. C. Peters. 1962. Rootworms and corn root development. *Proc. North Cent. Branch Entomol. Soc. Am.* **17**: 124–126.

Eiben, C. J. and D. C. Peters. 1965. Varietal responses to corn rootworm infestation in 1964. *Proc. North Cent. Branch Entomol. Soc. Am.* **20**: 44–46.

Everly, R. T. 1967. Establishment and development of corn leaf aphid populations on inbred and single cross dent corn. *Proc. North Cent. Branch Entomol. Soc. Am.* **20**: 53–55.

Fitzgerald, P. J. and E. E. Ortman. 1964. Breeding for resistance to western corn rootworm. *Proc. Annu. Hybrid Corn Ind. Res. Conf.* **19**: 46–60.

Fitzgerald, P. J., E. E. Ortman, and T. F. Branson, 1968. Evaluation of mechanical damage to roots of commercial varieties of corn (*Zea mays* L.). *Crop Sci.* **8**: 419–421.

Fleming, A. A., R. Singh, H. K. Hayes, and E. L. Pinnell. 1958. Inheritance in maize of reaction to European corn borer and its relationship to certain agronomic characters. *Minn. Agr. Exp. Sta. Tech. Bull.* No. 266, 32 pp.

Gallun, R. L., K. E. Starks, and W. D. Guthrie. 1975. Plant resistance to insects attacking cereals. *Annu. Rev. Entomol.* **20**: 337–357.

Grandados, G. R. 1967. Differential damage to the female inflorescence of corn, *Zea mays* L. by the corn rootworms *Diabrotica virgifera* Le C., *Diabrotica longicornis* (Say) and *Diabrotica undecimpunctata howardi* Barber. Chrysomelidae: Coleoptera. M.S. thesis. Manhattan: Kansas State University.

Guthrie, W. D. 1974. Techniques, accomplishments and future potential of breeding for resistance to European corn borers in corn. In *Biological control of plants, insects and diseases.* Jackson; Mississippi Univ. Press, pp. 359–380.

Guthrie, W. D. 1975. Insect rearing and plant evaluation methods and biotype problems in host-plant resistance research. *Iowa State J. Res.,* **49:** 519–525.

Guthrie, W. D. and F. F. Dicke. 1972. Resistance of inbred lines of dent corn to leaf feeding by first-brood European corn borers. *Iowa State J. Sci.,* **46:** 339–357.

Guthrie, W. D., F. F. Dicke, and C. R. Neiswander. 1960. Leaf and sheath feeding resistance to the European corn borer in eight inbred lines of dent corn. *Ohio Agr. Exp. Sta. Res. Bull.* No. 860, 38 pp.

Guthrie, W. D., Y. S. Rathore, D. F. Cox, and G. L. Reed. 1974. European corn borer: Virulence on corn plants of larvae reared for different generations on a meridic diet. *J. Econ. Entomol.,* **67:** 605–606.

Guthrie, W. D., W. A. Russell, and C. W. Jennings. 1971. Resistance of maize to second-brood European corn borers. *Proc. Annu. Corn and Sorghum Res. Conf.* **36:** 165–179.

Guthrie, W. D. and G. H. Stringfield. 1961a. The recovery of genes controlling corn borer resistance in a backcrossing program. *J. Econ. Entomol.* **54:** 267–270.

Guthrie, W. D. and G. H. Stringfield. 1961b. Use of test crosses in breeding corn for resistance to European corn borer. *J. Econ. Entomol.* **54:** 784–787.

Hinds, W. E. 1914. Reducing insect injury in stored corn. *J. Econ. Entomol.* **7:** 203–211.

Horber, E. 1961. Versuche zur Verhinderung der vom Maikäferengerling (*Melolontha vulgaris* F.), von der Fritfliege (*Oscinella frit* L.) und vom Maizünsler *Pyrausta nubilalis* (Hübner) verursachten Schäden mittels resistenter Sorten. Landwirtsch. *Jahrb. Schweiz (75 Jahr) N. F.,* **19:** 635–669.

Ibrahim, M. A. 1954. Association tests between chromosomal interchanges in maize and resistance to the European corn borer. *Agron. J.* **46:** 293–298.

Jennings, C. W., W. A. Russell, and W. D. Guthrie. 1974a. Genetics of resistance in maize to first and second-brood European corn borer. *Crop. Sci.* **14:** 394–398.

Jennings, C. W., W. A. Russell, W. D. Guthrie, and R. L. Grindeland. 1974b. Genetics of resistance in maize to second-brood European corn borer. *Iowa State J. Res.* **48:** 267–280.

Klun, J. A. 1969. Relation of chemical analysis for DIMBOA and visual resistance ratings for first-brood corn borer. *Proc. Annu. Corn and Sorghum Res. Conf.* **24:** 55–60.

Klun, J. A. and T. A. Brindley. 1966. Role of 6-methoxybenzoxalinone in inbred resistance of host plant (maize) to first brood larvae of European corn borer. *J. Econ. Entomol.* **59:** 711–718.

Klun, J. A., W. D. Guthrie, A. R. Hallauer, and W. A. Russell. 1970. Genetic nature of the concentration of 2,4-dihydroxy-7-methoxy 2H-1, 4-benzoxazin-3 (4H)-one and resistance to the European corn borer in a diallel set of eleven maize inbreds. *Crop Sci.* **10:** 87–90.

Klun, J. A. and J. F. Robinson. 1969. Concentration of two 1,4-benzoxazinones in dent corn at various stages of development of the plant and its relation to resistance of the host plant to the European corn borer. *J. Econ. Entomol.* **62:** 214–220.

Klun, J. A., C. L. Tipton, and T. A. Brindley, 1967. 2,4-Dihydroxy-7-methoxy-1, 4-benzoxazin-3-one (DIMBOA), an active agent in the resistance of maize to the European corn borer. *J. Econ. Entomol.* **60:** 1529–1533.

Kyle, C. H. 1918. Shuck protection for ear corn. *USDA* Bull. No. 708, 7 pp.

Loomis, R. S., S. D. Beck, and J. F. Stauffer. 1957. The European corn borer [Pyrausta nubilalis (Hbn.)] and its principal host plant: V. A chemical study of host plant resistance. *Plant Physiol.* **32:** 379–385.

Luckmann, W. H., H. C. Chiang, E. E. Ortman, and M. P. Nichols. 1974. A bibliography of the northern corn rootworm, *Diabrotica longicornis* (Say), and the western corn rootworm, *Diabrotica virgifera* LeConte (Coleoplera: Chrysomelidae). *Biological notes No. 90., Ill. Nat. His. Surv., Urbana, Ill.*

Mangelsdorf, P. C. 1974. *Corn: its origin, evolution, and improvement.* Cambridge, Mass.: Harvard Univ. Press, 262 pp.

Marston, A. R. 1931. Breeding European corn borer resistant corn. *J. Am. Soc. Agron.* **23:** 960–964.

Marston, A. R. and C. B. Dibble. 1930. Investigations of corn borer control at Monroe, Michigan. *Mich. Agr. Exp. Sta. Spec. Bull.* No. 204, 47 pp.

McMillian, W. W. and B. R. Wiseman. 1972. Host plant resistance: A twentieth century look at the relationship between *Zea mays* L. and *Heliothis zea* (Boddie). *Fla. Agr. Exp. Sta. Monogr.* Ser. 2.

Melhus, I. E., R. H. Painter, and R. O. Smith. 1954. A search for resistance to the injury caused by species of Diabrotica in the corns of Guatemala. *Iowa State J. Sci.* **29:** 75–94.

Meyers, M. T., L. L. Huber, C. R. Neiswander, F. D. Richey, and G. H. Stringfield. 1937. Experiments on breeding corn resistant to the European corn borer. *USDA Tech. Bull.* No. 583, 29 pp.

Ortman, E. E. and P. J. Fitzgerald. 1964. Evaluations of corn inbreds for resistance to corn rootworms *Proc. North Cent. Branch Entomol. Soc. Am.* **19:** 92.

Ortman, E. E. and E. D. Gerloff. 1970. Rootworm resistance: problems in measuring its relationship to performance. *Proc. Annu. Corn and Sorghum Res. Conf.* **25:** 161–174.

Ortman, E. E., D. C. Peters, and P. J. Fitzgerald. 1968. Vertical-pull technique for evaluating tolerance of corn root systems to northern and western corn rootworms. *J. Econ. Entomol.* **61:** 373–375.

Owens, J. C., D. C. Peters, and A. R. Hallauer. 1974. Corn rootworm tolerance in maize. *Environ. Entomol.* **3:** 767–772.

Painter, R. H. 1941. The economic value and biological significance of plant resistance to insect attack. *J. Econ. Entomol.* **34:** 358–367.

Patch, L. H. 1937. Resistance of a single cross hybrid strain of field corn to European corn borer. *J. Econ. Entomol.* **30:** 271–278.

Patch, L. H. and R. T. Everly. 1945. Resistance of dent corn inbred lines to survival of first-generation European corn borer larvae. *USDA Tech. Bull.* No. 893, 10 pp.

Patch, L. H. and R. T. Everly. 1948. Contribution of inbred lines to the resistance of hybrid dent corn to larvae of the early summer generation of the European corn borer. *J. Agr. Res.* **76**: 257–263.

Patch, L. H., G. T. Bottger, and B. A. App. 1938. Comparative resistance to the European corn borer of two hybrid strains of field corn at Toledo, Ohio. *J. Econ. Entomol.* **31**: 337–340.

Penny, L. H. and F. F. Dicke. 1956. Inheritance of resistance in corn to leaf feeding of the European corn borer. *Agron. J.* **48**: 200–203.

Penny, L. H. and F. F. Dicke. 1957. A single gene-pair controlling segregation of European corn borer resistance. *Agron. J.* **49**: 193–196.

Penny, L. H., G. E. Scott, and W. D. Guthrie. 1967. Recurrent selection for European cornborer resistance in maize. *Crop. Sci.* **7**: 407–409.

Pesho, G. R., F. F. Dicke, and W. A. Russell. 1965. Resistance of inbred lines of corn (*Zea mays* L.) to the second brood of European corn borer, Ostrinia nubilalis (Hübner). *Iowa State J. Sci.* **40**: 85–98.

Phillips, W. J. and G. W. Barber. 1931. The value of husk protection to corn ears in limiting corn earworm injury. *Va. Agr. Exp. Sta. Tech. Bull.* No. 43, 15 pp.

Reissig, W. H. and G. E. Wilde. 1971. Feeding responses of western corn rootworm on silks of fifteen genetic sources of corn. *J. Kans. Entomol. Soc.* **44**: 479–483.

Rhodes, A. M. and W. H. Luckmann. 1967. Survival and reproduction of the corn leaf aphid on twelve maize genotypes. *J. Econ. Entomol.* **60**: 527–530.

Robertson, D. S. and E. V. Walters. 1963. Genetic studies of earworm resistance in maize. *J. Hered.* **54**: 267–272.

Roubaud, E. 1928. Biological researches on *Pyrausta nubilalis* Hb. Int. Corn Borer Invest. *Sci. Rep.* **1**: 1–40.

Russell, W. A. 1972. A breeder looks at host plant resistance for insects. *Proc. North Cent. Branch Entomol. Soc. Am.* **27**: 77–87.

Russell, W. A., W. D. Guthrie, and R. L. Grindeland. 1974. Breeding for resistance in maize to first and second broods of the European corn borer. *Crop Sci.* **14**: 725–727.

Russell, W. A., W. D. Guthrie, J. A. Klun, and R. Grindeland. 1975. Selection for resistance in maize to first-brood European corn borer leaf-feeding damage by the insect and chemical analysis for DIMBOA in the plant. *J. Econ. Entomol.* **68**: 31–34.

Russell, W. A., L. H. Penny, W. D. Guthrie, and F. F. Dicke. 1971. Registration of maize germplasm inbreds. *Crop Sci.* **11**: 140.

Scott, G. E., F. F. Dicke, and L. H. Penny. 1966. Location of genes conditioning resistance in corn to leaf feeding of the European corn borer. *Crop Sci.* **6**: 444–446.

Scott, G. E., W. D. Guthrie, and G. R. Pesho. 1967. Effect of second-brood European corn borer infestations on 45 single cross corn hybrids. *Crop Sci.* **7**: 229–230.

Scott, G. E., A. R. Hallauer, and F. F. Dicke. 1964. Types of gene action conditioning resistance to European corn borer leaf feeding. *Crop Sci.* **4:** 603–604.

Shank, D. B., D. W. Beatty, P. J. Fitzgerald, and E. E. Ortman. 1965. SD10 Inbred corn for hybrids with resistance to corn rootworm, S. D. *Farm Home Res.* **16:** 4–6.

Sifuentes, J. A. and R. H. Painter. 1964. Inheritance of resistance to western corn rootworm adults in field corn. *J. Econ. Entomol.* **57:** 475–477.

Smissman, E. E., J. B. Lapidus, and S. D. Beck. 1957. Isolation and synthesis of an insect resistant factor from corn plants. *J. Am. Chem. Soc.* **79:** 4697–4698.

Straub, R. W., M. L. Fairchild, M. S. Zuber, and A. J. Keaster. 1973. Transmission of corn earworm resistance from Zapolote Chico to topcross progenies. *Econ. Entomol.* **66:** 534–536.

Sullivan, S. L., V. E. Gracen, and A. Ortega. 1974. Resistance of exotic maize varieties to the European corn borer, *Ostrinia nubilalis* (Hübner). *Environ. Entomol.* **3:** 718–720.

Virtanen, A. I. 1969. Some aspects of factors in the maize plant with toxic effects on insect larvae. *Suom. Kemistil.* [B], **34:** 29–31.

Wahlroos, O. and A. I. Virtanen. 1959. The precursors of 6MBOA in maize and wheat plants, their isolation and some of their properties. *Acta Chem. Scand.* **13:** 1906–1908.

Walter, E. V. 1965. Northern corn rootworm resistance in sweet corn. *J. Econ. Entomol.* **58:** 1076–1078.

Widstrom, N. W. 1972. Reciprocal differences and combining ability for corn earworm injury among maize single crosses. *Crop Sci.* **12:** 245–247.

Widstrom, N. W. and J. B. Davis, 1967. Analysis of two diallel sets of sweet corn inbreds for corn earworm injury. *Crop Sci.* **7:** 50–52.

Widstrom, N. W. and J. J. Hamm. 1969. Combining abilities and relative dominance among maize inbreds for resistance to earworm injury. *Crop Sci.* **9:** 216–219.

Widstrom, N. W., W. W. McMillian. 1973. Genetic effects conditioning resistance to earworm in maize. *Crop Sci.* **13:** 459–461.

Widstrom, N. W. and B. R. Wiseman. 1973. Locating major genes for resistance to the corn earworm in maize inbreds. *J. Hered.* **64:** 83–86.

Widstrom, N. W., B. R. Wiseman, and W. W. McMillian. 1972. Resistance among some maize inbreds and single cross to fall armyworm injury. *Crop Sci.* **12:** 290–292.

Widstrom, N. W., W. J. Wiser, and L. F. Baumann. 1970. Recurrent selection in maize for earworm resistance. *Crop Sci.* **10:** 674–676.

Wilson, R. L. and D. C. Peters. 1973. Plant introduction in *Zea mays* as sources of corn rootworm tolerance. *J. Econ. Entomol.* **66:** 101–104.

Wiseman, B. R., R. H. Painter, and C. E. Wassom. 1967. Preference of first-instar fall armyworm larvae for corn compared with *Tripsacum dactyloides*. *J. Econ. Entomol.* **60:** 1738–1742.

Wiseman, B. R., C. E. Wassom, and R. H. Painter. 1967. An unusual feeding habit to measure differences in damage to 81 Latin-American lines of corn by the fall armyworm, *Spodoptera frugiperda* (J. E. Smith). *Agron. J.* **59:** 279–281.

Zuber, M. S., M. L. Fairchild, A. J. Keaster, V. L. Fergason, G. F. Krause, E. Hilderbrand, and P. J. Loesch, Jr. 1971b. Evaluation of 10 generations of mass selection for corn earworm resistance. *Crop Sci.* **11:** 16–18.

Zuber, M. S., G. J. Musick, and M. L. Fairchild. 1971a. A method of evaluating corn strains for tolerance to the western corn rootworm. *J. Econ. Entomol.* **64:** 1514–1518.

Chapter 21

GENETICS OF DISEASE RESISTANCE IN MAIZE

A. L. Hooker

Professor of Plant Pathology and Plant Genetics,
University of Illinois UC,
Urbana, Illinois

Maize is subject to numerous diseases that reduce yield and quality. Resistance is known to most, if not all, maize diseases, and breeding for resistance is the most feasible and most widely used means of disease control. It is the purpose of this chapter to describe many genetic studies of resistance that have been made. Because of page limitations it is not possible to identify and list the numerous research papers consulted.

INHERITANCE OF RESISTANCE

Resistance is the opposite of susceptibility. These plant attributes, however, are not absolute. They may be quite divergent or grade continuously from one extreme to the other. Disease reactions may vary in kind or degree. In a completely susceptible plant nothing interferes with pathogen infection and subsequent development. Resistant host plants, on the other hand, oppose the pathogen at one or more stages of its life cycle so that pathogen development is retarded. These plants are inherently less damaged by pathogens in nature whereas susceptible plants, under comparable conditions, are injured.

Disease resistance in maize embraces a wide and continuous array of interacting genetic patterns. These include Mendelian (few genes), polygenic (many genes), and cytoplasmic (extrachromosomal) systems.

Mendelian

Dominant, incompletely dominant, or recessive genes condition disease resistance in maize. The genes may be independent, loosely or closely linked, or allelic, and can occur in groups of two or three in the same plant. Sometimes a few genes act in a complementary or additive manner. Modifier genes that enhance or reduce the expression of another gene are often believed to be present in maize for disease reaction.

Polygenic

Much effective disease resistance in maize is quantitative in expression and polygenic in inheritance. Transgressive segregation sometimes occurs. Various biometrical approaches have been used to obtain estimates of genetic variation, nature of gene action, and probable number of genes conditioning resistance. The heritability of disease reaction in maize is quite high. Resistance is usually associated with additive genetic effects and partially dominant.

Cytoplasmic

Cytoplasmic control of disease reaction is of greater importance in maize than in any other crop. In both the southern and yellow leaf blight diseases, plant cytoplasm may determine the major portion of disease reaction. The two pathogens causing these diseases produce essentially the same pathotoxin that is specific in its effect on susceptible and resistant cytoplasms.

HOST–PATHOGEN GENETIC INTERACTIONS

Genes for disease reaction are conditional genes. That is, they are expressed as phenotypes only when host and pathogen genotypes interact. This may take the form of a gene-for-gene relationship.

In maize, for example, genes *Ht* and *Ht2* condition resistance to *Helminthosporium turcicum*. The most common race of *H. turcicum* is avirulent to plants with either gene. A race is known, however, that is virulent to plants with gene *Ht* but avirulent to plants with *Ht2*. Recombinants from crosses between the two races reveal that a single gene conditions avirulence or virulence in the pathogen to *Ht*. This pathogen gene interacted differentially with *Ht* but not with *Ht2*, as all fungus recombinants were avirulent to *Ht2*.

Although the difference between resistant and susceptible cytoplasms to race T of *Helminthosporium maydis* is not known to be conditioned by a single gene, recombinants from crosses between race T, which is virulent to *cms*-T cytoplasm, and race 0, which is avirulent, showed monogenic segregation for virulence and associated pathotoxin production. Gene-for-gene interactions are also suspected between maize rust pathogens and host genes that condition a hypersensitive resistance.

CHROMOSOMAL LOCATION OF RESISTANCE GENES

Several studies have been made to place genes for disease reaction on the linkage map. Chromosome translocation stocks are usually used to first determine the probable chromosomal location. Crossover frequencies in segregating populations, utilizing genetic marker stocks, are then employed to determine linkage relationships.

SPECIFIC DISEASES

Seedling Diseases

Several fungal pathogens cause seedling blights in maize. These pathogens reduce seedling stands and weaken surviving plants. Soil-borne *Pythium* spp. rot the kernels and blight seedlings where cold wet soils are present at planting. Other seed- and soil-borne pathogens cause seedling blights.

These diseases, especially those caused by *Pythium* spp., are influenced by a number of seed factors acting during production, processing, and storage. Considerable ear-to-ear variation can occur within the same genotype. In addition, the seed is comprised of pericarp, endosperm, and embryo tissues, each of which may have a different genotype.

Resistance to *Pythium* species has been studied by various workers. Differences in inbred line reaction occur, and these are associated with hybrid reaction. Differences between reciprocal crosses can be large, with hybrid reaction associated with that of the seed parent. There is no evidence for cytoplasmic inheritance of this maternal effect. When excised embryos are studied, a portion, but not all, of the seed parent effect can be removed. Resistant inbreds as either pollen or pistillate parents are more effective than susceptible ones in transmitting resistance to the F_1. The reaction of single-cross embryos is closely correlated with the average reaction of the parents. The genetic nature of inheritance seems rather complex and many genes may be involved.

Several investigators have concluded that the inheritance of resistance to *Fusarium roseum* f. sp. *cerealis* in the seedling stage is quantitative and conditioned by multiple factors. In some crosses resistance seemed to be dominant, but other crosses were not consistent. Subsequent studies showed little correlation between seedling reactions in the F_3 and F_4 generations. Both its genotype and the conditions under which the parent ear developed were believed to have an effect on seedling blight reaction. Four inbred lines and a diallel set of 12 single crosses were tested for seedling reaction to *Fusarium moniliforme*. General combining ability and maternal variances were highly significant, and additive gene action was more important than dominant gene action.

Bacterial Diseases

Bacterial leaf blight and wilt caused by *Erwinia stewartii* is an important disease in eastern North America but also occurs elsewhere. Although formerly of importance primarily on sweet corn, it is now a common disease in the United States on dent corn because several susceptible inbred types are widely used to produce hybrids.

Although low infection by *E. stewartii* can be associated with plant vigor and late maturity, highly effective resistance is also known. Resistance to seedling wilt is dominant in all crosses. In the F_2 and backcrosses of a very resistant dent inbred OSF with the susceptible flint inbred WF, Wellhausen postulated that resistance was due to two major complementary genes *Sw* and *Sw2* and a third minor gene that conditioned only a slight degree of resistance. This third gene, however, in combination with either or both *Sw* or *Sw2*, modified their expression to higher resistance. Gene *Sw* conditioned moderate resistance. Highly resistant plants contained both *Sw* and *Sw2*. Using an attenuated culture of the bacterium, single dominant gene ratios were obtained. Virulence of the pathogen is maintained or increased when it repeatedly reinfects a resistant host. Virulence can be lost in culture or by infecting susceptible plants.

We at Illinois have studied the resistant inbreds Mo5Rf, C123, and M14 and the susceptible inbreds B14A and Oh43 in various crosses and segregating populations. A single heterozygous locus was required for resistance to the wilting phase. This resistance was seen in all crosses. Inbreds C123 and Mo5Rf each have two dominant independent genes that exhibit complementary action to the late season leaf-blight phase of the disease. One of these genes is present in M14.

Leaf freckles and wilt, caused by *Corynebacterium nebraskense*, was first observed in 1969. This disease occurs in Nebraska and scattered locations in adjacent states. Inheritance studies with inoculation and with natural infection show susceptibility to be partially dominant. Segregations did not fit the classical one or two locus ratios.

Downy Mildews

Several *Sclerospora* species cause downy mildew diseases that can be very destructive in different parts of the world. Good resistance is known and breeding for resistance is effective. Resistance to one species also seems to be useful against other species.

Resistance to *S. maydis*, expressed as percentage of plants infected in the F_1, F_2, and backcross generations of 30 crosses studied in the field, segregated as a quantitative character with polygenic inheritance and additive gene action. Resistance to *S. philippinensis* also seems to be quantitative.

Resistance to *S. sorghi* in Texas appeared to be dominant, except in crosses involving one susceptible line. It was concluded that at least two and possibly three genes control resistance in the lines studied. In Thailand, Jinahyon studied one cross in the growth chamber and in the field. The F_1 and F_2 were intermediate in resistance and the backcrosses shifted toward the recurrent parent. Only additive gene effects were statistically significant.

Resistance to *S. rayssiae* var. *zeae,* expressed as percentage of plant surface with symptoms, has been studied in India. Working with natural infection on eight inbred lines and 28 F_1, F_2, and backcross populations, it was concluded that resistance was due to several genes, most of which have a minor effect but some have a major effect. Both additive and dominant gene action was indicated but additive was believed to be the most important. In some crosses there was evidence for single and in others for two dominant complementary genes. In another study three resistant and three susceptible inbreds and their 15 F_1 crosses were studied in the field following natural infection through the soil and following artificial inoculation and infection through the leaves. Resistant lines performed differently to the two methods of infection. In each test both general and specific combining ability were measured but the specific effects were much larger. Resistance was partially dominant.

Although some of the resistance to *S. sacchari* is believed to be polygenic, Chang found that resistance in A117, TW79, Ph7, and TW25 segregated as a single dominant gene in Taiwan. He concluded that this resistance gene *Dmr* is located on the short arm of chromosome 2.

Virus and Virus-like Diseases

Streak is one of the most widely distributed diseases on the coastal plains of Africa. The main vector, *Cicadulina nbila,* is favored by high humidity. Breeding for resistance has been successful. The inheritance of resistance was first reported to be complicated but more recently it has been shown in some sources to be due to a single major gene. The F_1 is intermediate in reaction between the parents. Homozygous inbreds possessing the gene differ slightly in reaction. This is believed to be due to minor genes. The resistance has been transferred to a number of inbred lines.

Mosaic stripe occurs in Hawaii and tropical countries. The virus is transmitted by the leafhopper *Peregrinus maidis*. Resistance in inbreds derived from Hawaiian Sugar is believed to be due to a single dominant gene *rM*, which may be on chromosome 1.

Maize mosaic, reported to be caused by a strain of sugarcane mosaic virus, is widespread in India. In seedling tests, resistance was monogenic, with heterozygous plants giving an intermediate reaction. In backcross populations, however, heterozygotes were classified as susceptible.

Three important virus or virus-like diseases have been recognized in the continental USA. Maize dwarf mosaic (MDM), which can be transmitted mechanically and in nature is transmitted by several aphids, was recognized in the early 1960s in river-bottom areas where johnsongrass is an important perennial host. Further south, corn stunt (CS), caused by a spiroplasma, occurs. The CS spiroplasma is transmitted by the leafhoppers *Dalbulus maidis, Graminella nigrifrons* and others. In 1972 maize chlorotic dwarf (MCD) was recognized as a distinct disease. The MCD virus is transmitted by *G. nigrifrons,* but not by *D. maidis.* The pathogens are morphologically different and can be distinguished electron microscopically. Maize dwarf mosaic and MCD occur together, and may have since the early 1960s, in Ohio river-bottom and nearby areas. Several strains of the MDM virus have been described.

In tests with plants artificially inoculated with MDM virus, resistance was controlled by one or two genes. Inbred GA209 and several other inbreds have genes for resistance on both arms of chromosome 6. The resistance in M14 is associated with the long arm of chromosome 1 or the short arm of chromosome 9. With seedlings, heterozygous genotypes are almost as susceptible as the susceptible parents of crosses. In field tests, however, where plants are naturally infected, resistance to presumably MDM is inherited as a dominant character controlled by relatively few major genes. These workers also detected minor alleles that modify MDM reaction. When crossed with C103 the resistant inbred Oh07 segregated into a two-gene ratio, but when crossed with Oh43 segregated into a single gene ratio for resistance. Inbreds differ in their susceptibility to aphid or mechanical inoculations. In a combined diallel analysis involving 10 inbred lines expressing a range of reactions, Loesch and Zuber found both general and specific combining ability effects to be significant. General combining ability was the largest. Different genes condition resistance to strains A and B of the virus.

Resistance to corn stunt has been studied in the field in Mississippi. In F_1, F_2, and backcross populations involving three resistant inbreds and one susceptible, dominance for resistance or susceptibility was lacking. There was a strong additive influence and some due to heterosis. In a diallel set of crosses involving five resistant and five susceptible inbreds, general combining ability was found highly significant in several experiments involving different locations and years. Gene action was mostly additive with resistant × susceptible hybrids expressing an intermediate reaction.

Physoderma Brown Spot

Brown spot or Physoderma disease caused by *Physoderma maydis* is generally of minor significance but can cause locally severe problems in warm, humid areas. Resistance to this obligate parasite is expressed quantitatively. Additive gene effects are most important but dominance and epistasis can be present in certain crosses. Heritability values ranged from 17–47% and a minimum of four gene loci were predicted.

Northern Leaf Blight

One of the more important diseases in the United States and European maize belts is northern leaf blight caused by *Helminthosporium turcicum*. It is also important in other temperate areas where cool weather prevails after anthesis. Several forms of resistance exist to *H. turcicum* and a number of genetic studies have been made.

Jenkins and his associates recognized a type of resistance expressed quantitatively by the number and position of lesions on plants in the field and have made several significant genetic studies of this resistance. Other studies have also been made. In general, this type of resistance is polygenic. Resistance is usually partially dominant. Many genes are involved, some of which produce major effects and others minor effects. The detection of these effects may vary with disease intensity. Additive gene action seems more important than dominance or epistatic effects in most populations. This is supported by the success of phenotypic recurrent selection for resistance. At least 12 chromosome arms carry these genes for resistance.

Several sources of resistance express chlorotic-lesions with suppressed sporulation of the pathogen. Chlorotic-lesion resistance is expressed by seedlings as well as older plants. When infected, these plants produce chemical compounds called phytoalexins that inhibit the growth of *H. turcicum*. This resistance is usually, but not always, monogenic dominant. Inbred GE440 has *Ht* on chromosome 2. Work at Illinois has shown some 34 sources of resistance from different parts of the world and representing different maize types have dominant genes at the *Ht* locus. A range of phenotypes is seen when these genes are introduced into the same inbred line. Inbred NN14 from Australia has two dominant genes for chlorotic-lesion resistance. One of these genes is at the *Ht* locus and the other is at an independent locus that can be designated *Ht2*. The *Ht* plants are more resistant in Illinois than are *Ht2* plants. In many backgrounds, the two genes interact to produce a level of resistance greater than either gene alone. Plants homozygous for *Ht* genes are more resistant than are heterozygous ones. The polygenic lesion-number and monogenic lesion-type resistances interact to reinforce the effects of each other.

Other monogenic types of resistance have been reported from South Africa. The resistant line Gto. 59-272-1-7 from Mexico was crossed with the susceptible inbred K64 by Van Schaik and Le Roux. A backcross to the resistant line yielded only resistant plants, whereas a backcross to K64 gave nearly equal numbers of resistant and susceptible plants. Resistant plants had few if any lesions but susceptible plants had many lesions. Gevers has given data on gene*HtN* from the Mexican variety Pepitilla. The resistance segregated in F_2 and backcross populations giving $3:1$ and $1:1$ ratios of lesion-free plants and plants with lesions. In both studies plants with lesions were not equally infected. Gene *HtN* seems to delay infection so that at the end of a favorable period for disease, resistant and susceptible genotypes may look the same.

Southern Leaf Blight

The disease caused by *H. maydis* is widespread in tropical and subtropical areas. In the United States it is commonly called *southern leaf blight* because the disease occurs primarily in the warmer southeastern part of the country. The 1970 epiphytotic in the United States attracted worldwide attention to this disease.

Two races of *H. maydis* have been described. Race 0 shows no specificity to plant cytoplasms and is mostly a leaf pathogen. Race T produces a specific pathotoxin and is specific in attacking plants that have *cms*T cytoplasm for male sterility. It infects the leaf, leaf sheath, husk, and ear parts. Susceptibility of *cms*T cytoplasm to *H. maydis* was first observed in the Philippines and seen in the United States in 1969. Since 1970 race T has been reported from many parts of the world.

Resistance to race 0 is of two types. One type of resistance is expressed quantitatively in the form of lesion size and percentage of leaf tissue infected. The two characters are correlated. This resistance shows polygenic inheritance. In the crosses studied by Pate and Harvey resistance was generally partially dominant to susceptibility and it was believed that relatively few genes were involved. These conclusions are supported by studies of other resistant inbred lines in Illinois. Both additive and dominance effects were significant in the analysis of generation means. In a few crosses dominance was important but most of the gene action was additive. Heritability estimates were high.

Another type of resistance to race 0 is expressed qualitatively in the form of lesion type. Resistant plants have small chlorotic lesions with inhibited fungal sporulation. In the original source of resistance, resistance was reported by Craig and Fajamisin to be controlled by two linked recessive genes having a recombination frequency of 16.83%. Resistant selections were extracted in Illinois from backcrosses involving the original resistance and Corn Belt stocks. Resistance in these selections is monogenic recessive. The gene has been designated *rhm*.

Resistance to race T is based on both cytoplasmic factors and nuclear genes. The cytoplasmic component is the most important. Mature plants with resistant cytoplasms are little affected by race T. Normal cytoplasm, and a whole series of cytoplasms for male sterility including *cms* -B, -C, -CA, -D, -EK, -F, -G, -H, -I, -IA, -J, -K, -L, -M, -ME, -MY, -PS, -R, -RB, -S, -SD, -TA, -VG, -W, and others, condition resistance to the pathogen and its pathotoxin. Genetic factors interact with *cms*T cytoplasm to condition a partial degree of resistance. General combining ability effects are high. Nuclear gene resistance has additive gene effects and is partially dominant. The same genetic factors that condition resistance to race 0 probably condition partial resistance in *cms*T cytoplasm to race T. The restorer factors *Rf* and *Rf2* that restore fertility to maize with *cms*T cytoplasm have little effect on disease reaction in the field.

Helminthosporium Leaf Spot

Race I of *Helminthosporium carbonum* is one of the more destructive pathogens of susceptible corn. However, relatively few inbreds are susceptible to this race. Susceptibility in inbred Pr is monogenic recessive. The gene has been designated *hm* and is on the long arm of chromosome 1. In later research, Nelson and Ullstrup detected another gene locus *hm2*, located on chromosome 9, for disease reaction. The interactions were such that in plants homozygous for *hm*, *Hm2/Hm2*, or *Hm2/hm2* showed intermediate susceptibility as seedlings but became more resistant as they grew older with the homozygous genotype expressing the most resistance. The *Hm* gene was fully resistant and in these plants the *hm2* locus had no effect. Two additional alleles at the *Hm* locus were detected. The *Hm*-A locus conditioned an intermediate reaction at all growth stages. The *Hm*-B homozygous plants were fully resistant at all growth stages but when heterozygous with allele *hm* were resistant as seedlings and young plants but susceptible after anthesis.

Other biotypes of *H. carbonum* also exist. In the Illinois tests resistance is frequently due to a single dominant gene.

Yellow Leaf Blight

Yellow leaf blight caused by *Phyllosticta maydis* appeared in the northern U. S. corn belt in the late 1960s. The pathogen has been reported subsequently elsewhere in the world.

The cytoplasm is important in conditioning disease reaction. The same cytoplasms that condition susceptibility or resistance to *H. maydis* condition the same reaction to *P. maydis*. Nuclear genes also modify disease reaction to *P. maydis*, with some genotypes reacting nearly the same in both resistant and susceptible cytoplasms. These genes are different from those conditioning resistance to *H. maydis*.

Common Rust

Rust caused by *Puccinia sorghi* is common on maize in temperate climates. Rust has the potential to cause large yield losses of susceptible maize, but resistance has kept the disease under control. At least two forms of resistance are known. Resistance expressed by seedlings and older plants in the form of hypersensitive infection types is specific against certain biotypes of the fungus. A maize line may be highly resistant to a certain *P. sorghi* biotype and susceptible to another. Resistance expressed by older plants in the field in the form of a low number of pustules on the leaf is of a general type and functions against all *P. sorghi* biotypes.

Many sources of specific resistance are known. Dominant genes for resistance occur at five different loci. A cytological study by Rhoades showed locus *Rp* to be located on the distal end of chromosome 10. This location has been confirmed by means of chromosome translocation stocks. Locus *Rp3* is located on chromosome 3 and *Rp4* on chromosome 4. Loci *Rp5* and *Rp6* are linked to *Rp* and hence are also on chromosome 10.

Different alleles, distinguishable by means of differential reactions to a series of *P. sorghi* cultures, occur at *Rp, Rp3*, and *Rp4*. The largest number occur at *Rp*.

The *Rp* seems to be a complex locus and is about 0.4 map units long. It is flanked by *Rp5*, 1.1 map units away on one side, and *Rp6*, 2.1 map units away on the other. Methodology used to study the *Rp* locus has been discussed in detail in papers by Saxena and Hooker. In testcross progenies of up to 19,000 plants, recombination values observed between: (a) *Rp*a and *Rp*k, (b) *Rp*g and *Rp*1, (c) *Rp*c and *Rp*k, (d) *Rp*b and *Rp*f, and (e) *Rp*a and *Rp*c were 0.27%, 0.37%, 0.16%, 0.10%, and 0.22%, respectively. No recombinations were detected between *Rp*d and other alleles tested. It is believed that the *Rp* is complex and each "allele," as first identified in conventional studies, has a functional gene and one or two nonfunctional genes closely linked together. Further evidence for this is provided by testing both reciprocal recombinants of Rpc and Rpk. The Rpc–Rpk genotype gave disease reactions to a series of rust cultures and in international disease nurseries equivalent to that of the additive effects of Rpc and Rpk. That is, if either Rpc or Rpk or both conditioned resistance, then Rpc–Rpk conditioned resistance. If neither conditioned resistance, then the recombinant conditioned no resistance.

The genes at the $Rp3$ locus exhibit a reversal of dominance to different biotypes of $P.$ $sorghi$. Six alleles at $Rp3$ have been identified. The $Rp3$ locus may also be complex and consist of dominant and recessive genes closely linked together. These putative genes, however, were not detected or if they exist they can not be more than 0.06 map units apart.

Resistance may also be due to one, two, or three recessive genes. In one resistant source, high resistance in the seedling stage is conditioned by the complementary effect of three recessive genes.

Genes that modify specific resistance have been detected in several studies. Low pustule number on mature plants is the most common type of resistance to $P.$ $sorghi$ in the United States. Inheritance studies were conducted in the field in Illinois with inbred, F_1, and F_2 populations of 15 resistant × resistant, five susceptible × susceptible, and 45 resistant × susceptible crosses. Four crosses were studied in the F_3 generation. Plants were scored in the field for percentage of leaf area infected with rust that had spread into the test plots from inoculated susceptible plants. Plant scores of the F_2 populations resembled a normal distribution. The F_2 means were near the average of the two parents, and the F_1 means were usually between the F_2 mean and the most resistant parent. Heritability values were calculated by different methods but averaged about 80%.

Southern Rust

Southern rust, caused by *Puccinia polysora*, was confined to the Western Hemisphere until 1949. It was then accidentally introduced into West Africa and within about 10 years had spread to the saturation point throughout all tropical maize-growing areas of the world. In the Western Hemisphere the pathogen causes minor losses, but in Africa, until brought under control by resistance, losses were extensive.

Like resistance to $P.$ $sorghi$, resistance to $P.$ $polysora$ is apparently of two types. Genetic variation for the adult-plant type of resistance existed in tropical varieties and this resistance has now been stabilized at a level adequate to keep losses low. The other form of resistance is specific to various races of the pathogen and has been the subject of genetic research. Genes Rpp, $Rpp2$, $Rpp9$, $Rpp10$, and $Rpp11$ have been identified and the existence of others implied. Unlike $P.$ $sorghi$, few allelic or linkage tests of genes for $P.$ $polysora$

resistance have been made. Rather, the resistance genes are identified in relation to races of *P. polysora*. The *Rpp* gene is dominant in East Africa and conditions resistance to race EA.1 but not to EA.2; *Rpp2* is an incompletely dominant gene and conditions resistance to both EA.1 and EA.2. Races PP.3, PP.4, PP.5, PP.6, PP.7, and PP.8 and resistance to them were described by Robert in 1962. Ullstrup distinguished race PP.9 and identified a single dominant gene, which he designated as *Rpp9* for resistance to it. This gene is on chromosome 10 about 1.6 crossover units from *Rp*d for resistance to *P. sorghi*. In 1961, race EA.3, virulent to *Rpp* and *Rpp2*, was recognized in Kenya. Two genes for resistance were described. The *Rpp*10 gene is fully dominant, whereas another independently inherited gene, *Rpp*11, is incompletely dominant.

Smut

Common smut or boil smut, caused by *Ustilago maydis*, is worldwide and most prevalent in warm, moderately dry areas. Because of resistance, it is no longer a significant problem in the'U. S. corn belt.

Resistance is based both on morphological and physiological characteristics of the plant and is complexly inherited. Field resistance often varies from that measured by wound inoculations of seedlings, although some maize lines are resistant to both types of infection. The location of the smut gall on the plant is an inherited feature. Resistance is neither dominant nor recessive in crosses; however, certain maize lines tend to produce predominantly resistant hybrids. Inbreds having equal levels of resistance as lines may vary greatly in the amount of resistance transmitted to hybrids. Resistance seems to be quantitative involving both additive and nonadditive gene action. Smut reaction is associated with chromosomes 1 and 2. Morphological mutants on chromosomes 4, 6, 7, and 9 are associated with susceptibility. The fungus is comprised of numerous cultural biotypes. Resistance, however, seems to be of a general type and to function against all biotypes.

Stalk Rots

Diplodia, Gibberella, and Fusarium stalk rots, caused by *Diplodia maydis, Gibberella roseum* f. sp. *cerealis,* and *Fusarium moniliforme,* respectively, cause premature plant death and contribute to stalk lodging. Any form of plant stress renders the plant more susceptible to these pathogens. Plant factors in the stalk that are associated with resistance to one of these pathogens seem to be associated with resistance to others. Resistance to the same pathogen, however, in the seedling, stalk, or ear is independent. Resistance is quantitatively expressed and a mutual balance between host genotype for resistance and pathogen isolate for virulence may determine the degree of rot spread.

Several genetic studies of stalk-rot resistance have been made. In most reports, resistance is inherited in a quantitative manner. The F_1, however, is usually more resistant than the average of its parents. Resistance to Diplodia stalk rot involves additive gene action, dominance, and in some cases, epistasis. One exception to polygenic inheritance was reported by Younis, Dahab, and Mallah for resistance to Fusarium stalk rot. These workers in Egypt obtained data that could be partitioned into ratios, expected for that of a two-gene system with resistance dominant to susceptibility.

Several chromosome arms carry genes for stalk-rot reaction. Reaction to *D. maydis* in W22 is associated with the long arm of chromosome 7 and either the short arm of chromosome 2 or both chromosomes 6 and 8. Dominant or partially dominant genes for resistance to *D. maydis* in B14 and C103 were found in the centromere region of chromosome 5 and in the long arms of chromosomes 6, 8, and 9. Genes for resistance were also found in the long arm of chromosome 1, short arm of chromosomes 3 and 9, and in the centromere region of chromosome 7. The two genes implicated in resistance to *F. moniliforme* were found to be associated with the short arm of chromosome 7 and the long arm of chromosome 10.

Resistance to Diplodia stalk rot is relatively easy to select for in segregating populations, is highly heritable, and can be improved by recurrent selection. While formerly accounting for about half of the stalk rot, *D. maydis* is now an infrequent pathogen in the U.S. corn belt.

Ear Rots

Several fungi infect maize ears. Ear rots occur in certain years when weather conditions favor them. In these seasons some hybrid pedigrees, independent of maturity, are much more damaged than are others. Therefore, differences among hybrids are genetic in nature. It would be desirable if genes for susceptibility could be removed from composite populations and from the parental inbreds of hybrids.

Ear-rot resistance depends on several factors. Koehler, in decreasing order, rated physiologic resistance, good husk coverage, and declined ears as important in Illinois. Several reports indicated that inbred line reaction per se is a poor indication of its performance in hybrids. There seems to be no correlation between inbred reaction to different ear-rot pathogens. High-lysine maize is more susceptible to *F. moniliforme* that is its normal endosperm counterpart but this reaction is modified, to a considerable extent, by other genes.

A few genetic studies of resistance have been made. Observations suggest that resistance is probably polygenic. Segregating populations derived from five inbreds were inoculated by spraying ears with a *D. maydis* spore suspension. Resistance was heritable and expressed by the ability of a plant to resist initial infection rather than subsequent pathogen development. In general, F_2 and means of both backcrosses were near the midpoint of parental reactions. Backcrosses regressed toward that of the recurrent parent. The I1190 and R4 lines were similar in terms of resistance but differed in hybrid combination. The I1190 line tended to transmit resistance to all of its crosses, where R4 was influenced by the remainder of the pedigree. In an inoculated trial, additive gene action, with some dominance, was important in ear-rot resistance to *F. moniliforme.*

Chapter 22

DEPLOYMENT OF DISEASE AND INSECT RESISTANCE IN MAIZE: BREEDING AND FIELD CONSIDERATIONS

Basil Tsotsis

DEKALB AgResearch, Inc.

Cultivated plants depend to a considerable degree on man for their survival and evolution. Maize represents an extreme case of this dependency as it is incapable of seed dispersal, is ill suited for competition with other plant species, and is only deployed in the field by man. The absence of relatives of maize growing in the wild and subjected to selection by natural evolutionary forces in most of the areas where the species is cultivated, serves to further emphasize this dependency.

The preservation and deployment of maize as a crop are functions that since the advent of genetics have been assumed by systematically organized breeding programs. The purpose of such programs is to improve, through genetics, the productivity of the species for areas where it is grown as an economic crop. In considering the genetic capacity for high productivity, it is necessary to recognize that maize genotypes resulting from breeding programs will encounter in every environment where they will be grown other biological organisms, primarily pathogens and parasitic insects that use maize as a host and are capable of coevolving with it. Failing to take into account the evolutionary significance of such organisms, definitions of productivity may become misleading as diseases and insects can drastically affect yield in maize (Stringfield, 1964). For the maintenance of the health of this crop, it is essential that protection against diseases and insects is provided through genetic mechanisms. In their absence it may become difficult, expensive, or in some cases, impossible to protect maize from their ravages by other means. Therefore, the use of genetics by maize-breeding programs is for the dual purpose of: (a) increasing the productivity of the maize plant and (b) keeping its vulnerability to insects, diseases, and other stress factors to a minimum.

In pursuing this dual objective a breeder faces two priorities that are seemingly conflicting. For the attainment of high productivity under the systems that typify intensive agriculture, extreme uniformity is desired for a given field of maize. Such uniformity can be achieved by deploying one or very few genotypes in a farmer's field. This situation can predispose that field and others like it to attacks by diseases and insects for which the deployed genotypes have no resistance. In contrast, when pursuing the objective for reduced vulnerability to pests, the breeder must recognize the need for genetic diversity in the constitution of the plants occupying a given field of maize as a means for minimizing the danger of infection or infestation by older or newer forms of parasites. The presence of an array of different genotypes in the same field, however, can be unacceptable for the attainment of high productivity.

High productivity and reduced vulnerability in maize must be attained in environments that are partly natural and partly subject to direct and repeated interventions by the farmer. Farmer inputs, such as fertilizers, herbicides, planting density, cultivation, irrigation, mechanical harvest, and rotations, are made to enhance the probability that maize can yield near the optimum of its genetic capabilities. However, practically all such inputs have a potential bearing on the incidence and the intensity of disease and insect outbreaks. As a result, breeding work needs to be conducted under conditions that reflect, not only natural evolutionary forces, but also the changes induced by technology and by consumer demands. Such conditions define the agricultural environment for which genotypes resulting from breeding programs are intended.

The preceding considerations are discussed for their breeding and field implications. The interdependence of the breeding and field considerations and the logistics for the deployment of disease and insect resistance are discussed by presenting the operational arrangements developed by the author and his colleagues at DEKALB AgResearch to implement a program for this purpose in the continental United States and Canada.

BREEDING CONSIDERATIONS

Prior to the advent of hybrids, maize was grown by farmers in the form of open-pollinated varieties that as random mating populations could change their genetic constitution in response to pressures from natural selection in the environments where they were grown. The natural selection was complimented by farmers who each season could select ears from the most desirable plants observed on their farms. The selection under local conditions was, no doubt, alternated with and influenced by crosses among different open-pollinated varieties without the direct intervention of man.

Since the advent of hybrid maize, F_1 hybrids have substituted for open-pollinated varieties in farmers' fields. The introduction of hybrids had both structural and operational consequences for the evolution of the species.

The structural change is that in deploying F_1 hybrids in farmers' fields, one or very few genotypes are used to cover a given farming area. This contrasts with the situation as it existed in open pollinates where populations of heterozygous plants representing a large array of genotypes were grown in each field of maize.

The operational change results from the fact that as open-pollinated varieties have been replaced by hybrids, it is not practical to access the genetic variability needed for making changes in the species from populations of maize evolving under natural conditions and possessing a breeding structure of their own. Plants of hybrid maize grown in a given field are essentially of the same genotype, and the farmer no longer makes provisions to save seed for his own needs. Under these conditions it has become increasingly necessary to provide the variability needed for genetic changes through formal breeding programs. In the absence of the monitoring of changes occurring in the natural environment of a given area and that in the past was provided by open-pollinated varieties, maize breeders must also design breeding methods that can substitute for this function of the open pollinates. The above changes are more pronounced in areas where open-pollinated varieties have been replaced to a large extent by hybrids and are most evident in the United States and other developed maize-producing countries.

Genetic Diversity for Yield and Resistance to Pests

The practical utilization of heterosis in maize is based on inbreeding heterozygous source materials for several generations for the development of homozygous lines. Such inbreds are subsequently combined into single, three-way, four-way, or other types of hybrids as suggested by data collected from tests designed primarily for the determination of productivity.

Inbreeding of source materials is an effective procedure for assessing their genetic endowment and allows the segregation of their diversity into streams of germplasm that can be stabilized and cataloged as inbreds. Cataloged inbred lines become definable components for the synthesis of heterotic hybrids that can eventually be deployed in the field. The ability to rapidly synthesize specific hybrids from an array of previously cataloged and evaluated inbred lines is at least as significant for the improvement of maize as the yield advantage that hybrids display over open-pollinated varieties. This is especially pertinent to the development and deployment of hybrids resistant to diseases and insects through the combination of resistant inbreds and is one of the most important means available to breeders for protecting the health of the crop.

Yield in maize is conditioned by a large and as yet undetermined number of genes. The greater portion of the genetic variance utilized for yield improvement is of the additive type. Nonadditive variance is especially important for specific combining ability. However, the determination of the general combining ability is conditioned primarily by the additive genetic variance and usually takes precedence in breeding programs. The genetic variance associated with disease and insect resistance is also considered to be primarily of the additive type in the great majority of the diseases and insects studied (Hooker, 1973). Although examples exist of simple genetic systems capable of conferring resistance to pests, it is possible in most such cases to effectively use polygenic systems in combination with simple forms of resistance. The similarity of the type of genetic variance affecting the productivity of maize and its resistance to pests suggests that the search for genetic improvement for both of these components can proceed simultaneously. Similar selection procedures can be used on the same material for the improvement of yield and resistance to pests.

Another consideration for the attainment of high productivity is that the likelihood for developing heterotic hybrids increases if their parents are developed from unrelated rather than from related source materials. This requirement encourages the maintenance of parental diversity in source materials and can help to structure breeding programs capable of developing inbreds with a low degree of relatedness. Such unrelated inbreds could be expected to also have more diversity in genetic systems for resistance to parasites. The diversity of parental inbreds and consequently that of their hybrids favorably influence both increased productivity and pest resistance. It would appear that there exist certain definable genetic and operational factors that can be jointly applied to a breeding program designed to address itself to both breeding objectives. The essential requirement for the use of common procedures for increasing productivity and reducing vulnerability in maize is the development of sufficient methods for the recognition and maintenance of the genetic diversity needed for heterosis for yield in the presence of diseases and insects.

Considerably more attention has been devoted over the years by maize breeders to methods for evaluating heterosis for yield than for the development of procedures for the detection of pest resistance in heterotic crosses of parental materials. The procedures needed for such detection are relatively simple and require the introduction of significant diseases and insects to the site where a breeding program is conducted. It is possible with appropriate techniques to generate artificial epidemics and infestations of sufficient intensity to maximize the expression of genetic resistance. Such techniques can increase heritability and allow the breeder to make rapid progress in selecting for genetic resistance to pests.

Cytoplasmic Diversity and Resistance to Diseases and Insects

Prior to the 1970 epidemic of *Helminthosporium maydis* race T and its impact on maize genotypes containing Texas Sterile Cytoplasm, it was generally assumed that host parasite interaction was conditioned by the nuclear genome of the host and the parasite. The 1970 epidemic established that in addition to genes, cytoplasm can be significant for the expression of resistance or susceptibility of the host in the presence of a specific pathogen. As a result, in the deployment of maize resistant to pests it is necessary to allow for cytoplasmic as well as nuclear genetic diversity in breeding programs. This addition compli cates the breeding work but does not alter the premise for simultaneously seeking increased productivity and reduced vulnerability. A desired form of resistance sufficient to accommodate this requirement could be nuclear genetic resistance capable of functioning regardless of the cytoplasms used as suggested in 1970 for *H. maydis* race T and also for *Phyllosticta maydis* (Hooker, 1971; Scheifele, 1970). The detection of cytoplasmic diversity is more difficult than the detection of nuclear genetic diversity. To date, a few methods for cytoplasm classification have been found useful and are based on the expression of male sterility and increased susceptibility to certain diseases. Until additional methods are developed for the recognition of differences among cytoplasms, it would appear that the evaluation of cytoplasms other than normal must be made in breeding nurseries where diseases and insects are present.

Cytoplasms other than normal in maize have only been used to induce male sterility on the seed parents used in large-scale production of hybrids. As maize hybrids can be produced by means other than cytoplasmic male sterility, its importance for the long-term development of disease and insect resistance does not appear to be as critical as it is for other hybridized species that have perfect flowers. It is conceivable that the use of cytoplasmic male sterility could act as a hindrance to maize improvement because it requires the recycling of older existing materials for the development of sterile and restorer counterparts of inbreds for which the agronomic value was already established by testing their normal counterparts in hybrids. This could contribute to the maintenance of obsolete material and occupy a portion of the facilities of a breeding program that could be better used to accommodate more genetic diversity.

Accession and Use of Genetic and Cytoplasmic Variability

One of the most critical tasks for maize breeders in the posthybrid era is the accession of source materials from which the inbred parents of improved hybrids can be obtained. Such accessions precede all other breeding operations and can define to a large degree the outcome of subsequent breeding decisions. The disappearance of open-pollinated varieties has deprived the breeder of the opportunity to use as source materials populations evolving under natural conditions. Open-pollinated varieties were used extensively as sources of inbred lines in the early days of maize hybridization; however, their use has been reduced significantly in recent years. Their place has been taken by hybrids and synthetics because of the increased probability for making greater short-term gains by using such source materials instead of open pollinates in breeding programs. At present, extensive collections of open pollinates exist in the United States and elsewhere; however, only a small fraction is used for the formation of source populations. Although such collections cannot substitute for the varieties themselves, it is important that they are maintained not as herbarium specimens but rather as potential components for the formation of source populations utilized by breeding programs. For such a use of open pollinates and other broad-based collections of materials, procedures need to be followed that can allow their evaluation for parental diversity by determining through crosses the degree of heterosis that they can display among themselves. If such evaluations were conducted in the presence of diseases and insects, it would also be possible to assess their genetic variance for resistance to parasites. Data collected by these procedures could be used for the detection of new heterotic pairs and lead through further crossing to the formation of composites that could be subjected to cyclic selection in the presence of diseases and insects for the improvement of their yield and pest resistance. Similar methods could be used for the formation of heterotic pairs of synthetic populations based on collections of inbreds and other materials.

The formation of heterotic pairs of populations and their evolution in the presence of pests could monitor changes in the disease and insect complex affecting a given area in a manner analogous to that of the old open-pollinated varieties. If the populations were to be grown under conditions approximating random mating, they would also be allowed to assume a breeding structure comparable to that of open pollinates. In addition, such populations could be grown under conditions that represent currently used or anticipated types of agricultural technology, by defining parameters such as planting densities and cultural practices, that are used now or could be adapted in the future by maize growers.

Additional protection from shifts in the disease–insect complex can be provided by conducting all the inbreeding work in the presence of pests either in selection nurseries or in areas where the pests exist naturally. As the inbreeding necessary for the development of homozygous lines is usually continued for six to seven generations, the combined effect of evolving source populations in the presence of diseases and insects, and the time required for inbred line development, may provide sufficient conditions to study the reaction of the material under development to pests and to select and catalog such reactions with relative freedom from escapes.

Once homozygous lines are evaluated for combining ability and resistance to pests they can be used for the rapid synthesis and deployment of new hybrids in the field. Maize breeders have the capability to change genotypes that can be deployed in the field more rapidly and effectively than is the case for breeders working with species not improved through hybridization. This is possible because it is not necessary to wait for genetic stabilization as the development of F_1 hybrids is the final step in the maize breeding effort and not an intermediate step in the transfer of resistance. Provisions for cytoplasmic variability can be made by using more than one cytoplasm in the formation of source populations, accompanied by methods for the maintenance of their identities during the breeding cycle.

FIELD CONSIDERATIONS

Farmer Options

Farmers using open pollinated varieties are restricted in their choices, partly because they tend to plant seed saved from the selections that they made in their fields the previous season and partly because organized seed distribution is not well developed in areas where hybrids are not used. Open-pollinated varieties under such conditions tend to encounter disease and insect problems as varietal improvement could be restricted by the frequency of genes for resistance and the unpredictability in the buildup of diseases and insects. Low-level infections and recurrent losses can become common and accepted risks.

When hybrids are used, decisions at planting time are influenced by previous knowledge on the performance of hybrids marketed in a given area and by the information provided on new genotypes by seed producers. As the work needed for hybrid development precedes the large-scale distribution of seed for specific hybrids, the degrees of resistance to pests that can be deployed at a given time in the field will depend on the attention given to the work on reduced vulnerability conducted by the originators of the hybrids.

A practice followed by many farmers in the United States is to grow several hybrids that can be distinguished as to origin and are adapted to the area where their operations are located. The use of several hybrids can allow the farmer to deploy more diversity and potentially greater resistance to pests than would be the case if only one hybrid were used over an entire farm; however, in practice this results in having individual fields planted with only one genotype. Diversity in a single field of maize could be provided by planting on it several different hybrids in strips. For a large maize-producing area or country in the temperate zone, diversity is also affected by the sensitivity of the crop to frost and the need for a north-to-south array of different hybrids to best exploit the available frost-free season. This regional diversity (Horsfall, 1972) can be helpful, but in practice it is difficult to achieve. The characteristics of hybrids available for use in a given area become rapidly known through their actual performance and resistance to pests in local fields. Locally determined performance can become a prime factor for hybrid distribution as those with the most desired attributes can be rapidly accepted in preference to other entries in the field. This acceptance can lead to situations where at a given time a small number of hybrids or even a single hybrid can dominate a large area.

A subject that received a considerable degree of discussion following the wide acceptance of single crosses in recent years has been whether single crosses are unduly contributing to the reduction of the genetic base used for hybrid deployment. Double crosses do provide more diversity than single crosses when they occupy a given field; however, from the standpoint of resistance to insects and diseases, the pertinent point is the knowledge of the resistance or susceptibility of the inbreds used and not necessarily the number of inbreds utilized for making hybrids of commercial value. A practical matter in producing a single cross hybrid is the general vigor, disease and insect resistance, and other agronomic traits of its two parental inbreds. The economics of producing single-cross hybrids are demanding and as a result, the inbreds used for such production display on the average higher levels of fitness than did the inbreds used in the past for producing four-way crosses. The distribution of single cross hybrids produced through the use of elite lines over wide areas has by far exceeded the distribution of double cross hybrids used 10–15 years ago in our experience. This, to a large extent, can be attributed to increased levels of pest resistance that permitted growing single crosses in a greater variety of environments than those occupied by double crosses in the past. Single crosses, because they are easier to synthesize and test, have also allowed for a more rapid replacement of old hybrids. Although the farmer can exercise considerable choice for genetic differences in hybrids, cytoplasmic variability in hybrids depends entirely on seed producers as cytoplasms other than normal are only used in maize as production tools. The farmer, however, can request and receive information on specific hybrids in regard to the cytoplasm used and the reaction of the specific hybrids to specific diseases and insects.

The above considerations emphasize that the deployment of disease and insect resistance depend heavily on the effectiveness of breeding programs servicing a given area and their ability to maintain a sufficient flow in the development of hybrids with high productivity. Such an effectiveness can become a deterrent against the prolonged use of the small number of genotypes that can meet the biological and the economic conditions that apply to agriculture at a given time.

Options for the Breeder and Seedsmen

Deployment of resistance to pests needs to be considered by breeders and seedsmen for two types of conditions, namely, those influenced by known diseases and insects and those caused by parasites of recent origin.

The first condition can be met by determining the reaction of hybrids to known pests while the hybrids are being developed. The earlier reference to the feasibility for simultaneously selecting for increased productivity and reduced vulnerability on the same genetic material provides an effective means for the development of this knowledge. Such knowledge can allow breeders and seedsmen to restrict the distribution of hybrids susceptible to pests existing in a given area and thus directly influence the deployment of disease and insect resistance in maize. Additionally, if breeders and seedsmen systematically attempt to produce several hybrids of diverse origin, genetic variability on a regional basis could be increased. A similar effect can be produced by following procedures for the timely introduction of new improved hybrids and the withdrawal of older hybrids from the market.

For the deployment of cytoplasmic diversity, which is completely controlled by breeders and seedsmen, it would be advisable to use more than one type of sterile cytoplasms in seed-production fields. When the choice of sterile cytoplasms is limited, it would be advisable to make only partial use of them by producing blends of normal and sterile counterparts of the same hybrid. The elimination of sterile cytoplasms from the production of hybrids can also be an effective means for deploying resistance, as was shown after the 1970 epidemic of *H. maydis* race T.

The second condition applies to parasites that exist in distant geographical areas or where recently introduced in neighboring areas and threaten to become established in additional areas through migration or dispersal. In such cases the amount of time available to first locate and then use resistance in maize to function in the presence of the new pest is of the essence. One of the first lines of defense that can be used by breeders and seedsmen in dealing with a new pest is to plant and evaluate collections of inbred lines and hybrids in areas where the new pest is occurring naturally. This should occur in advance of the natural dispersal of the parasite and with as much lead time as can be provided. The evaluation of existing hybrids can permit for the possible detection of resistance in existing genotypes. If this becomes the case, resistant hybrids could be rapidly increased and distributed. The evaluation of the reaction of existing inbred lines could also permit the rapid synthesis of new hybrids by emphasiz-ing the use of the most resistant lines. The utilization of the resistance of inbreds in new hybrid combinations can be expedited if the combining ability of the inbreds being evaluated in the presence of a new pest was previously determined. Active breeding programs capable of sustaining a flow of new materials should be well poised for their ability to generate large numbers of inbreds and have a strong background of information on combining ability and pest resistance that could be of use in such contingencies. If neither existing hybrids or inbreds show resistance to the new organisms, the problem can become acute and could require a lengthy solution, as it would be necessary to first identify and then use resistance to the new pest from unadapted sources.

LOGISTICS OF THE DEPLOYMENT OF INSECT AND DISEASE RESISTANCE

A discussion of the logistics involved in the deployment of resistance to pests based on the breeding and field considerations cited earlier is presented in the text that follows by referring to the work conducted by the maize-breeding department of DEKALB AgResearch. For purposes of organization, the needed research is aided by the recognition of the following four major areas: (a) the source populations or germplasm resources available for initiating the breeding effort, (b) the definition of the conditions chosen by breeders for the development of sources of germplasm, inbreds, and hybrids, (c) the availability of a testing program to provide a measure of the effectiveness of the breeding effort, and (d) the technology that can apply to the breeding effort and especially new concepts and methods that can expedite progress in the desired direction. The impact of each area of research on the development of resistance to pests is discussed here, as is the interdependence of the development, production, and distribution of hybrids for the deployment of reistance.

Source Populations

The procedure that has been extensively used by DEKALB maize breeders is to access and catalog broad-based genetic collections from the United States and elsewhere. The cataloging initially consists of: (a) a determination of the maturity of the accessions, based on data collected for the dates of 1%, 50%, and 100% silking, (b) an assessment of the gross morphology of accessions by collecting data on plant and ear height, leaf number, and other variables, (c) an evaluation of the general level of disease and insect resistance that the accessions display either under natural or artificial infection and infestation. Approximately 300 new accessions are evaluated in this manner each year. After the above evaluations the most promising accessions of approximately the same maturity are systematically crossed in diallel fashion, usually in groups of 10, and the resulting crosses are grown in yield trials. The objective of the trials is to determine the degree of heterosis for yield displayed among specific crosses of new germplasm accessions. If the heterosis exhibited by the cross of two new accessions equals or exceeds that shown by an established heterotic pair, for example, A × B, and can be shown to be relatively stable over seasons and environments, it is designated as a new heterotic pattern. If, for example, a new heterotic pattern is designated as M × L (for the Midland × Leaming open-pollinated varietal cross), the practice is to combine through further crossing in one pool those accessions that in crosses with Leaming behave in a manner similar to that of Midland itself and place in a second pool the varieties that in crosses with Midland behave like Leaming itself. This leads, in the example, to the formation of two composites of approximately the same maturity that display heterosis when crossed. After their formation the composites constituting a heterotic pair are grown separately and in isolation from other maize for improvement through cyclic selection. The selection scheme used most extensively in DEKALB's work for evolving the composites is mass selection. Mass selection allows one to generate conditions that approximate random mating, and it provides a breeding structure akin to that of the old open-pollinated varieties. The average populations size of each composite is not less than 8,000–10,000 plants, depending on the planting densities used. Lower planting densities are used in the earlier and higher planting densities are used in later cycles of selection. Provisions are made to use cultural practices representing those used by efficient farmers or practices that can be anticipated for the future.

In addition to the formation and evolution of synthetics constituting heterotic pairs, another consideration for source populations is the reinforcement of already existing useful heterotic patterns. New accessions containing genetic systems that can enhance the value of existing source populations are added to existing populations (that they resemble) for heterotic behavior. This allows one to extend for additional breeding cycles, the use of proven, older heterotic patterns without a reduction of their parental diversity by the addition of genetic systems not included at the time of their initial formation.

Provisions for cytoplasmic diversity are made by placing a portion of the gametes constituting a composite on several sterile cytoplasms. Plants with sterile cytoplasms are identified, and final selection at harvest are made separately for the normal and the sterile cytoplasm counterparts of the composites.

The Corn Research Department at DEKALB recognizes for the continental United States and Canada nine maturity classes arrayed from early to late in a northern–southern direction. At least three programs are assigned similar and overlapping responsibilities for the same three maturity zones, and each program is expected to develop source populations with nonduplicating parental diversity through procedures as outlined above. On the average, each of the programs utilize at any one time six to eight synthetics subjected to cyclic selection.

The work for source-population development conducted by each breeding program is coordinated for the entire breeding group by a specific research program dedicated to parental diversity and methods for obtaining it. The coordinating program maintains several field operations in the United States and Mexico for the accommodation of original accessions and their evaluation prior to their release to breeders. It also conducts collections in Mexico where existing open-pollinated varieties are obtained from farmers' fields and are gradually and over time introgressed into indigenous U. S. types. Other late maturity tropical accessions are also initially evaluated in Mexico. The introgression of late maturity accessions proceeds from south to north to bridge the maturity, photoperiodism, and other differences present in late tropical accessions and that need modification before they can be used as sources of new genes for the improvement of U. S. and Canadian Corn Belt hybrids. This arrangement allows for the introgression of materials collected at a center of origin for maize and can lead over a long time to an expansion of the genetic base used in maize improvement.

Conditions During Development

The source populations as discussed above are subjected to cyclic selection in the presence of several diseases and insects. The specific pathogens and insects used vary with the location where the synthetics are evolved. The pathogens and insects used at a given location are already established and no attempt is made to introduce new pathogens to areas where they do not already exist. For new organisms different procedures are followed, as discussed later. The introduction of the organisms applied to the composites evolved in a given area depends to a large extent on the use of artificial methods of inoculation and infestation. At present the following organisms are used as aids in selection: *Helminthosporium turcicum, H. maydis* race T and race O, *H. carbonum, Phyllosticta maydis, Kabatiella zeae, Collectotrichum graminicola, Corynebacterium nebraskense, Diplodia maydis, Gibberella zeae, Fusarium* spp.,

Ostrinia nubilalis, and *Diabrotica virgifera.* Additionally, the synthetics are planted without seed treatment early in the growing season for selection for resistance to soil-borne pathogens. Inoculations with diseases are made with virulent isolates for each pathogen. As a rule, at least two foliar diseases and one stalk-rot inciting organism are used at each of 11 DEKALB programs engaged in this work during each cycle of selection. The choice of organisms for each research site is influenced by their significance for the areas for which each breeding program develops hybrids. Artificial infestations with *Ostrinia nubilalis* egg masses are made at practically all breeding locations for both first and second brood resistance, whereas exposure of the composites to the *Diabrotica virgifera* is restricted to a few research sites because of limitations in the methods used for artificial infestation. The organisms applied at a given research location are introduced on each member of a heterotic pair. Selections made at the end of the season are based on yield and standability and although notes on each parasite could be taken, it is difficult to separate the influence of each of the specific pests on the selection criteria parameters. The intensity of selection practiced is to save 20% of the plants grown each session for each synthetic.

For organisms that do not already exist in areas where breeding programs are maintained or for organisms that cannot easily be introduced artificially with existing inoculation and infestation techniques, the practice is to plant collections of germplasm in areas where the organisms exist naturally. Observation and selection sites have been developed for virus, MDM, *Sclerospora sorghi, Cercospora zeae maydis,* and to a lesser extent for other pests existing outside the main portion of the U. S. Corn Belt. Materials exposed to organisms existing in such locations if found to be resistant are returned to the breeding sites for incorporation into composites that belong to a similar breeding group. Reconstituted composites following this introgression can be returned to the special disease and insect nurseries for additional selection.

Source populations can and have been used directly for inbred line development. In recent years a practice widely used by DEKALB maize breeders for the utilization of the genetic variability of advanced composites is to cross existing inbreds to synthetics that belong to the same breeding group. The resulting crosses are selfed with the expectation that new lines developed in this manner can be used as members of their original breeding groups. The most advanced synthetics have undergone selection for 17 cycles. Data obtained from monitoring studies indicate that the disease and insect resistance of the composites can be improved and that their parental diversity can be maintained under the conditions of selection (Tsotsis, 1972).

The development of inbred lines from composites proceeds in disease nurseries in which the same organisms applied during the evolution of the composites are also applied during each generation of inbreeding and selection.

Extensive and detailed notes on the reaction of each inbred to each of the specific organisms used are taken and become a part of the record of each inbred. The combined exposure of the source materials and their inbred progenies to diseases and insects can allow for a sufficiently long time to recognize and catalog the reactions of parents of hybrids to specific parasites. We have termed this procedure the *stress breeding concept,* as all new materials are developed under the stresses represented by diseases and insects.

Testing

Hybrids generated from existing older and newer lines are also evaluated in selection nurseries for their reaction to insects and diseases. Consequently, in assembling hybrids for combining ability and general performance tests, certain entries can be eliminated prior to their inclusion in tests if their levels of resistance to pests existing in the areas of their potential distribution were shown to be low during their evaluation in disease nurseries.

The choice of testing sites and the duration of testing prior to advancing hybrids into commercial production can be significant for determining the performance of hybrids in the presence of pests that were not applied during source population and inbred-line development. A testing program based on many diverse locations and on the use of several planting densities at each site used for the evaluation of hybrids of the same maturity can permit the collection of information not provided at the breeding sites. Our type of testing allows for up to 250 independent evaluations of a given hybrid prior to its release for production and requires a 3-year period. The advancement of a new commercial entry for a given maturity zone is based on the evaluation of approximately 1000 experimental hybrid combinations.

Technology

Breeding programs dedicated to the development of hybrids with high productivity and stable performance in the presence of diseases and insects require close cooperation among breeders, pathologists, entomologists, and geneticists. Without such cooperation it would be difficult to integrate the diverse disciplines that are needed for the operation of such a program, and tendencies might develop to unduly divide breeding problems into segments suitable for each discipline but inadequate for a timely solution of the general problems at hand. In our case, we have integrated the pathology work with breeding programs by relying on pathology expertise for the study of the dynamics of disease development, the preparation of inocula, the determination of the pathogenicity of the isolates used, and the study of the genetics of resistance to a given pathogen. In addition, we have established a disease-monitoring system in the

United States involving 22 locations of differing maturities. Specific sets of inbred lines previously classified as differential hosts for specific diseases are grown at each location and tissue from each host is systematically collected several times during the growing season. Isolates from each host are identified and evaluated with the aid of type cultures for the detection of the appearance of new biotypes of older known organisms. Isolates of previously unknown parasites are evaluated for pathogenicity with appropriate tests.

New sources of resistance as they become identified in germplasm collections are evaluated by pathologists, and procedures for their use and incorporation in breeding materials are developed jointly by breeders and pathologists. Pathology expertise is further utilized by providing diagnostic services to maize farmers, an arrangement that is also used to augment the monitoring of disease.

In the insect resistance work we are dealing with a limited number of insects, as chemical control is more widely used in maize for insects than for diseases and genetic resistance is unavailable for certain insects. We maintain laboratories for specific insects, primarily for *Ostrinia nubilalis* and the *Diabotrica virgivera*.

Another function performed through pathological and entomological work is the development of updated catalogues of the reaction of known inbreds to specific diseases and insects. This information stored in a computer becomes a part of an information retrieval system which can be used to generate lists of materials possessing resistance to specific pathogens or insects for each maturity class. This information allows the formation of small synthetic populations that can serve as specific sources of resistance and also enable the rapid generation of resistant hybrids in response to contingencies.

Interdependence of Breeding and Field Considerations. As breeding for increased productivity and reduced vulnerability requires the close cooperation and the convergence of several disciplines, so does the timely availability of resistance that can be deployed in farmers' fields. This availability can be assisted by the competence developed by the production and distribution agencies with which breeders, geneticists, pathologists, and entomologists are associated.

In the case of DEKALB AgResearch, the breeding, foundation, production, and marketing aspects of seed production are vertically integrated and their functions are coordinated by general management. For the nine maturity classes for which we develop and distribute hybrids in the United States and Canada, a total of approximately 90 hybrids are available to farmers in commercial quantities in a given season. This represents the existence of approximately 10 different hybrids of comparable maturity at a given time. Approximately 15 new hybrids are introduced in commercial volume each year, and a corresponding number are removed from production schedules. An additional 15–20 hybrids are advanced through pilot-scale productions, foundation increases, and so on, in preparation for possible future commercial-scale production. The introduction of new hybrids is made in a gradual fashion through the use of pilot produced seed which is distributed to interested farmers for field-scale evaluation. A successful new hybrid usually can reach a peak in its production in the third year following its introduction and can be maintained at high production volume for an additional 3–4-year period. After this period it is deemphasized until it is retired. The genetic basis for DEKALB's schedule as described above requires the use of in excess of 100 different inbred lines at a given time. As the commercial introduction of hybrids requires the timely availability of foundation seed, production schedules are usually developed 2 years in advance of a given production season.

The logistics for the development, testing, production, and distribution of hybrids as discussed above are cited as an example of the considerations necessary for producing hybrid seed on a national level for the United States by a large breeding and seed-production organization. Operations of this size in the United States and in most of the other maize-producing countries lead to an acute awareness of problems related to the productivity and vulnerability of the maize crop. The exposure is of a size that requires the consideration of disease and insect resistance without undue hindrance for restrictions caused by limited appreciation of either the biological or the operational parameters that can contribute to the health of the maize crop.

SYNOPSIS

Maize has a considerable degree of genetic diversity; however, to date only a small portion of it has been used for improvement in yield and the development of resistance to diseases and insects. Even with this restriction, maize yields since the advent of hybrids have increased more rapidly than in other crops that are the subject of systematic improvement through genetics. Although encounters of varying severity with diseases and insects have been experienced after the introduction of hybrids, maize breeding has become less closely associated with breeding for pest resistance alone than is the case with other field crops and more specifically, cereals. The genetic diversity available to maize as a species and the use of hybridization have allowed for the rapid and efficient synthesis of genotypes with higher yield and resistance to pests in farmers' fields. The importance of maize as a source of food, the increase in the ease of travel and the interdependence of food-producing and food-consuming nations can be expected to influence the distribution of insects and diseases that can be destructive to this species. The genetic variability that may be required for the protection of future crops of maize from pests as a result will need to be considered for the global crop as well as for the individual maize-producing areas. The need, therefore, for preserving genetic variability and incorporating it in the structure of maize-breeding programs merits strong attention. To meet this need the considerations presented in this chapter need to be expanded to reflect on the realities, both biological and technological, to be faced in the course of time. As organisms capable of parasitizing maize have the genetic capability to evolve simultaneously with it, they will undoubtedly remain as problems for breeders and farmers in the future. Cultivated plants being a part of the general biological world cannot escape entirely the outcome of their coevolution with other biological entities with which they share an environmental niche. Certain risks will always remain for agriculture despite the work of geneticists, pathologists, entomologists, breeders, seed producers, and farmers. The degree of success of an agricultural crop will depend on whether man or natural evolutionary forces will prevail at a given time. If potential problems for the health of the maize crop can be anticipated in breeding and selection nurseries and not encountered in farmers' fields, man can be expected to continue his dominant role in determining the field deployment of cultivated species such as maize.

REFERENCES

Hooker, A. L. 1971. Status of southern corn leaf blight, past and present. *Proc. 26th Annu. Corn and Sorghum Res. Conf.*, pp. 127–142.

Hooker, A. L. 1973. Maize. In *Breeding plants for disease resistance—concepts and applications.* Nelson, R. R., (ed.). University Park and London: Pennsylvania Univ. Press, pp. 133–154.

Horsfall, J. G. 1972. Genetic vulnerability of major crops. Washington, D. C.: National Academy of Sciences, 307 pp.

Scheifele, G. L. 1970. Cytoplasmically inherited susceptibility to diseases as related to cytoplasmically controlled pollen sterility in maize. *Proc. 25th Annu. Corn and Sorghum Res. Conf.*, pp. 110–138.

Stringfield, G. H. 1964. Objectives in corn improvement. *Adv. Agron.* **16:** 102–106.

Tsotsis, B. 1972. Objectives of industry breeders to make efficient and significant advances in the future. *Proc. 27th Annu. Corn and Sorghum Res. Conf.,* pp. 93–107.

Section Five

ENDOSPERM

Chapter 23

INTRODUCTORY REMARKS TO THE SESSION ON NUCLEAR–CYTOPLASMIC RELATIONS

R. I. Brawn

Funk Seed International
Bloomington, Illinois

In this session, four chapters are grouped under the general heading of nuclear–cytoplasmic interactions. We all recognize that with the advent of molecular biology it became clear that nuclear genes exercise their control through the cytoplasm. In many earlier studies, only the nucleus seemed to be involved, and in corn especially, a great number of genes were discovered and studied that conformed closely with Mendelian expectation. Then in the early 1930s, Dr. Rhoades described the first apparent exception to nuclear control in corn in which the cytoplasm was implicated as the site of control of male sterility. Subsequent studies have shown both nuclear and cytoplasmic control of a number of types of male sterility.

Later, in the 1940s, Dr. McClintock introduced the revolutionary concept of controlling elements in maize (an interpretation that preceeded the Jacob–Monod hypothesis and an idea that vigorously shook the tree of classical genetics) with two concepts: (a) that gene action was under the control of other genes that might—or might not—be closely linked with each other and (b) that under certain circumstances, these controlling elements were transported from one site in the chromosome to another—clearly anticipating the episome concept. It is another example of the great perception of Dr. McClintock that she so correctly anticipated the whole field of gene action and control now being so successfully developed by molecular biologists.

The first chapter in Section VI, is an illustration of complex interaction between episome, chromosome and cytoplasm as they relate to male sterility in maize. As Drs. Duvick and Noble have so elegantly pointed out in Chapter 17, studies on understanding cytoplasmic male sterility in corn are badly needed, yet suitable approaches to attack the problem are yet to be found. Drs. Laughnan and Gabay have discovered aberrant behavior in an S-type cytoplasmic male sterile in which apparent mutations from sterile to fertile arise as mosaic tassels. Their study of this unusual system is another contribution to extending our understanding of cytoplasmic male sterility.

The three other chapters in this section are concerned with maize endosperm. Srb, Owen, and Edgar in their book *General Genetics,* p. 92, have listed many reasons why corn is an important experimental plant, but I am surprised that they have not mentioned the endosperm. Earlier, in Chapter 30, Coe, Kermicle and Creech describe studies on endosperm that emphasizes the usefulness of this unique tissue, which is so available for chemical analysis and so visible for phenotypic classification.

Chapter 25 seems to continue the success Brink and his students have had in discovering and elucidating deviations to Mendelian inheritance patterns in corn. First, they described deviations in the waxy ratio, then important studies on controlling elements, next paramutation, now here in Chapter 29 Dr. Kermicle describes imprinting, and in Chapter 25 McWhirter and Brink report on aberrant ratios with *opaque- 7* endosperm which supports the view "that endosperm development in maize involves a canalized genetic system in the sense of Waddington's concept of canalization of developmental processes."

Nelson in Chapter 26 concludes that "although this assertation can't be proven, it appears a reasonable conclusion from the data that there are more enzyme systems participating in starch synthesis in the cereal endosperm than biochemists envisioned as essential." Nelson also points out that in relatively few instances are we able to associate a mutation with the biochemical lesion for which it is responsible. He concludes that the biochemical genetics of the endosperm is "is its infancy" and that "we do have large numbers of mutants that affect endosperm development and can serve as experimental probes to facilitate coordinated biochemical and genetical studies on the endosperm." Wilson (Chapter 27) does just this as he reports on some biochemical indicators of genetic and developmental control in endosperm. He shows that the RNase levels in developing endosperms are influenced considerably by the pollen parent, with changes in rates of synthesis and in the time period of rapid synthesis. His important conclusion is that "it must be emphasized that the endosperm is not a uniform mass of cells, but differs from top to bottom throughout development. Biochemical data on developing corn endosperm are accumulating slowly, but they must be correlated with cytological observation before they can be fully understood."

The strength of this session came from these repeated statements of how little we really know about nuclear cytoplasmic interactions and the various suggestions for future work. I look forward to another symposium such as this in another decade or two by which time a session on nuclear cytoplasmic interactions could well be the prime attraction.

Chapter 24

IMPRINTING OF GENE ACTION IN MAIZE ENDOSPERM

J. L. Kermicle

Laboratory of Genetics,
University of Wisconsin,
Madison

Equivalent gene expression following passage through male and female gametophytes is a basic, if tacit, assumption of Mendelian inheritance. Evidence collected here indicates that this assumption is unwarranted for a class of genes acting in maize endosperm. Such genes function in a manner that reflects their prior mode of sexual transmission. The term "imprinting" denotes this persisting influence of parentage.

Correns rediscovered Mendel's principles of inheritance in maize while investigating the relation between effects of pollen on kernel phenotype (xenia) and the newly discovered phenomenon of double fertilization. He associated correctly the endosperm of the mature caryopsis with one of the two products of double fertilization. So did DeVries, according to Sturtevant (1965).

Whereas double fertilization served well to explain xenia, rediscovery of the basic rules of inheritance seems scarcely to have illuminated a number of

---

Paper No. 1901 from the Laboratory of Genetics.

central questions concerning double fertilization. Some open questions arose as to the necessity for a separate fertilization for the endosperm and a triploid constitution of the tissue. We particularly want to consider here how it is that the two fertilization products, endosperm and embryo, involving the same gene combinations, follow such divergent developmental paths. Aspects of these questions remain puzzling. Our lack of understanding is generally excused on the grounds that such questions concern more directly issues of developmental, rather than transmission, genetics. I believe, however, that a significant feature of double fertilization eludes us due to a restriction we unwittingly place on the Mendelian mechanism.

It is useful in this context to classify potentially inherited information under two categories. There is, of course, the primary genetic information specifying the sequence of·bases and the termini of RNA transcripts. It is what we principally associate with Mendelian inheritance. Expression of this information during development is selective. A given gene may be active (derepressed) in one particular cell type but inactive (repressed) in another. The variation in expression is epigenetic in origin and constitutes a second category of potentially inherited information.

We want to consider here whether such epigenetic variation is transmitted from the gametophytes to the endosperm and also seek the rules governing its inheritance. But first let us consider where to look for such evidence and agree on terminology.

IMPRINTING

Inheritance of epigenetic variation presumes a potential of the genetic material to reflect some prior influence in its subsequent expression. We are concerned specifically with the effect of parental sex, that is, epigenetic variation originating during differentiation of the reproductive structures of the sporophyte, at meiosis, or during development of the two gametophytes. Any persisting influence of parentage can be termed *imprinting* of gene action, adopting the term Crouse (1960) used to denote differential elimination of maternal and paternal chromosomes in *Sciara*. It is convenient to replace the phrase "the relation between maternally and paternally imprinted forms of a gene" with a single word. I use *epiallelic* in this connection and refer to the individual gene forms as *epialleles*.

Imprinting effects comprise a subset of the general category of parental influences that are defined operationally by phenotypic inequality following reciprocal crosses. Excluded from this broad category is any variation imposed by the genotype of the maternal sporophyte and also instances of differential transmission by the two sexes involving either nuclear or extranuclear factors. Any residual effects are candidates for imprinting.

Two characteristics have been particularly useful in distinguishing imprinting from other gene action control systems: (a) observed ratios that reflect the gametic contribution of only one sex (i.e., phenotype in the immediate generation is not influenced by one parent) and (b) presence of two distinct phenotypes corresponding to the same heterozygous genotype, incomplete penetrance excepted (stated in terms of an F_2 monogenic Punnett square, the two heterozygous entries correspond to different phenotypes). Criterion (b) is self-sufficient. Unfortunately, it is cumbersome to apply to the endosperm due to unequal dosage effects. In the case of preferential paternal action, however, (a) is essentially sufficient because the phenotypic relations are contrary to expectation based on dosage.

R-MOTTLING

Dosage versus Imprinting

East and Hayes (1911) called attention to the heterozygous condition of kernels irregularly pigmented with anthocyanin in the endosperm's aleurone layer. Emerson (1918) associated this splashing of color or mottling with heterozygosity of the R factor resulting from xenia (r r/R aleurone). Heterozygous kernels resulting from the reciprocal mating were solidly colored (R R/r aleurone). Emerson discussed his findings in relation to mutable genes and to $R:r$ dosage in the triploid endosperm but noted that a full explanation was not readily formulated in these terms.

Further insight awaited Herschel Roman's derivation of a translocation between the R-bearing long arm of chromosome 10 and a B chromosome (TB-10a). Because the $B^{10(R)}$ translocation element nondisjoins in division of the generative nucleus to form sperm, one nucleus received two Rs, and the second nucleus none. The duplicate and deficient sperm result in tetrasomic and disomic endosperm R genotypes. Roman found that many r r/R R kernels are darkly mottled, thus resembling the r r/R class in this strain rather than the solidly colored R R/r class (personal communication). On the $R:r$ dosage hypothesis this finding suggested that the double dose of r in r r/R and r r/R R genotypes inhibits full color expression.

Alternatively, the mottled r r/R R and solidly colored R R/r phenotypes could relate to a difference in R expression following maternal and paternal transmission to the endosperm (Schwartz, 1965; Kermicle, 1963), that is, to imprinting. The synthesis of R R/r r kernels served to differentiate between effects of dosage and transmission. R R/r r aleurone proved to be solidly colored, like R R/r and unlike r r/R R (Kermicle, 1970.) The phenotypic dissimilarity of R R/r r and r r/R R obviously can not be reconciled with hypotheses based on $R:r$ dosage but is accommodated by imprinting. Evaluation of 12 genotypes involving combinations of a mottling R with r or r-deficient chromosomes supported the generalization that solidly colored aleurone possesses the maternal epiallele of R, whereas combinations involving only paternal R with r or r-deficient chromosomes are mottled. The maternal epiallele is dominant. No effects attributable to r dosage or difference between r and r-deficient chromosomes were observed. For this reason the particular r used (r–g; L. J. Stadler source) is treated as having a null effect, an inference also reached on the basis of paramutation tests (Brink, Styles, et al. 1968).

Allele-specific Response

The foregoing analysis was conducted utilizing R–r: *standard* and particularly its paramutant forms, since the quantitative difference between solidly colored and mottled seeds is amplified when pigmentation of the r r/R class is reduced

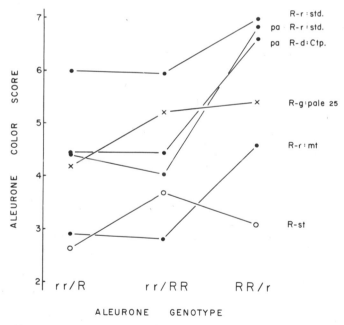

Figure 1. Aleurone pigmentation in relation to R dosage and parentage. Alleles conditioning a mottled phenotype (•) showed dosage insensitivity but were affected by parental source. Stippled (O) and pale (×) alleles were dosage-responsive but insensitive to parental source.

Figure 2. *R*-allele specific pigmentation patterns of the aleurone: (a) *R*-stippled (*r–r r–r/R-st, M–st M–st/m–st*); (b): *R*-mottled (*r–g r–g/R–g*).

by paramutation. Results of testing the *r r/R* phenotype of two other mottling alleles and of two partially pigmenting but nonmottling types is given in Figure 1.

The allele designated *R–d : Catspaw* is representative of an *R* class that pigments aleurone and various seedling parts but not anthers of mature plants. Kernel and seedling pigmentation varies both mutationally and paramutationally as though under unitary, rather than dual, genetic control (Bray and Brink, 1966; Brink, Kermicle, et al. 1970). Nonparamutant *R–d : Catspaw* (not illustrated) conditions uniform pigmentation of *r r/R* aleurone that is only slightly less intense than in *R R/r*. Mottling is shown only by the paramutant form. Paramutant *r r/R R* did not differ appreciably from *r r/R*, in contrast with the solid or nearly solid color condition of paramutant *R R/r*. This pattern of behavior parallels that of paramutant *R–r : standard*.

A third mottling allele tested for imprinting-versus-dosage effects arose as a spontaneous mutation from *R–r : standard*. Aleurone of *r r/R* genotype in this case is weakly mottled in the mutant stock and almost colorless following paramutation. Notably, the *R R/r* phenotype is mottled. The difference, therefore, between the *r r/R* and *R R/r* aleurones obtained following reciprocal crosses with *r r* in this stock is quantitative rather than qualitative. This variant (*R–mt*) was considered a likely candidate for mutational loss of function in an imprinting component of *R*. The *r r/R R* phenotype of this allele proved equivalent, however, to *r r/R*, which is appreciably lighter than *R R/r*. Thus again, the variation seen in reciprocal crosses is attributable to imprinting rather than dosage.

R-stippled is one of the two nonmottling alleles tested for imprinting. A colored spot in the case of *R-st* has a sharply defined margin of intensely pigmented cells. The pattern differs in detail from ordinary *R* mottling (Figure 2). An earlier test involving *R-st* in combination with its modifier *M-st* had indicated that dosage, rather than mottling, accounted for the *R R/r = r r/R R > r*

r/R ranking of spotting intensity (Kermicle, 1973). It is conceivable, however, that R-st reflects an imprinting response which is overridden by the dosage effect of M-st (Ashman, 1960). The results illustrated in Figure 1 are based on tests of R-st in the absence of M-st. Dosage action, rather than imprinting, is confirmed.

R–g:$pale$-25 was derived from R–sc:136 by McWhirter (1961). The aleurone is pigmented uniformly although weakly, thus giving a pale effect. As with R-st, the r r/R R class is equivalent to R R/r rather than r r/R, indicating effects of dosage rather than imprinting.

Equivalence of r r/R R and R R/r genotypes as produced by reciprocal crosses argues critically against imprinting only if other maternal influences are absent. The possibility of maternal influences that tend to offset imprinting was tested by producing R R/r and r r/R R classes on single ears using R–st as the test allele. The maternal parent was R–st r–r marked by $golden$-1; the male was homozygous TB-10a R-st. The resulting r r/R R kernels (golden, red seedlings) scored 3.84, whereas the R R/r-deficient (nongolden, green seedlings) scored 4.08 on an aleurone color scale ranging from very lightly spotted (class 2) to intensely spotted (class 6). The average difference of 0.24 that was obtained for a six-ear sample is not significant statistically. Thus the earlier indication of no detectable influence of parentage on R–st expression is upheld.

The foregoing evidence demonstrates an allele-specific pattern of R imprinting. The genetic unit of response necessarily is highly local. R-Imprinting contrasts in this respect with various cases of imprinting described in other species (Chandra and Brown, 1975) where the unit of response is a chromosome, or even a chromosome set.

Attention should be called in passing to a possible relationship between pigmentation pattern, imprinting response, and paramutational property. The three mottling alleles tested show imprinting; they are paramutable. The two nonmottling alleles gave dosage but not imprinting responses; they are paramutagenic. The possible generality of these correlations needs to be tested on a broader sample of alleles.

R-Imprinting in the Germline

A search for imprinting effects in the scutellum of embryos and in the mesocotyl and roots of seedlings proved negative (Brink, Kermicle, et al., 1970). The tests employed a class of alleles of which R–d: $Catspaw$ is a member and were chosen because of sensitivity to paramutation as reflected in embryo and seedling as well as aleurone pigmentation. Paramutation also provided a reduced grade of pigmentation especially favorable for the detection of change. The equivalent actions observed for R received via egg and sperm nuclei served to delimit the stage of R endosperm derepression to between differentation of egg from polar nuclei and fertilization.

Neither was a differential effect of parentage detected when Rs from pollen and egg were passed through a sporophytic generation and then tested for expression in the endosperm. The absence of a grandparental effect pointed again to differentiation of the female gametophyte and secondary fertilization as stages critical to R-imprinting.

Chromosomal Basis of Mottled and Solidly-colored Phenotypes

The F_2 ears descended from the parental cross $r\ r \times R\ R$ bear mottled, solidly colored, and colorless kernels in a $1:2:1$ ratio. The occurrence of both mottled and solidly colored phenotypes on single ears rules against determination of this variation by genotype of the maternal sporophyte, either nuclear or cytoplasmic. Aleurone phenotype in a given case is referable to the genotype of the two contributing gametophytes. Any information transmitted by the male is presumably nuclear. Information contributed by the embryo sac, however, could be vested either in the nucleus or in the cytoplasm. To illustrate the latter case, R but not r female gametophytes might establish a condition in the cytoplasm prerequisite for intense aleurone pigmentation. Transmission of this potential to the endosperm by the central cell of the embryo sac might account for the strongly pigmenting phenotype of $R\ R/r$ aleurone. This possibility stands in contrast to chromosomal imprinting as a mechanism of cell heredity.

Once established, a determiner or condition of the gametophyte cytoplasm might be either independent of or dependent on R for its expression. Loss of R from developing $R\ R/r$ endosperm induced by X-irradiation yielded phenotypes that argue against cytoplasmic control independent of R nuclear constitution (Kermicle, 1970). Kernels showing clonal sectors of colorless reveal a dependence of pigmentation on a nuclear factor, in this case R. Moreover, an absence of typically mottled sectors in the X-irradiated $R\ R/r$ kernel population suggested that clones of cells having a single R of maternal origin were solidly colored.

To test the possibility of interdependence between R and a gametophyte cytoplasmic factor in the determination of intense pigmentation, kernels of $R\ R/R$ endosperm constitution in which the maternal Rs were carried in unstable chromosome-10s were studied. If intense pigmentation reflected interaction between a cytoplasmic component and R, rather than a chromsomal potential of the maternal epiallele, then clones of $(-)(-)/R$ cells resulting from maternal R loss should be intensely pigmented. There would be no mottled areas. If control were chromosomal, mottled sectors on otherwise solidly colored kernels should occur. An early approach to this question utilized an unstable K10 chromosome. The incidence of chimeric kernels was low and their phenotypes often complex. The occurrence of kernels with less than fully colored sectors was suggestive nevertheless of chromosomal control (Kermicle, 1970).

Newly derived materials have made it possible to generate in large numbers chimeric kernels of a sort desired for additional tests. Two instances of heritable chromosome breakage associated with the activity of *Dissociation* (*Ds*) in chromosome 10 were established. The *Ds* is situated in these cases in the long arm between *R* and the centromere. The *R* is lost, therefore, by nonrecurrent breakage events. The mottling *R* pictured in *R R/r* genotype in Figure 3c was

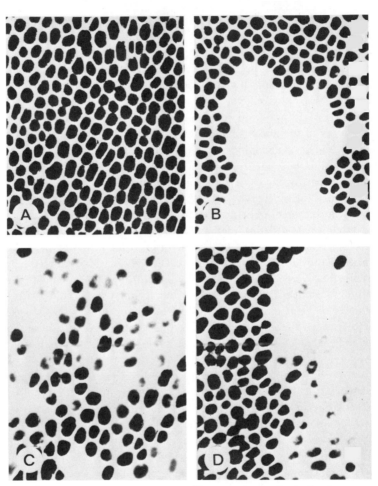

Figure 3. Aleurone pigmentation governed by paramutant *R,* indicating autonomous expression of maternal and paternal epialleles: (*A*) *R R/R* (phenotype, solidly colored). (*B*) *Ds R Ds R/r-g; P-vv P-vv/P-wr* Phenotype: solid colored with colorless sectors resulting from *Ds* induced loss of the two *R*'s. (*C*) *r-g r-g/R.* Phenotype: mottled. (*D*) *Ds R Ds R/R; P-vv P-vv/P-wr.* Phenotype: Mottled (as in *C*) superimposed on a solid color-colorless sector (as in *B*).

introduced into a *Ds* chromosome 10, combined with *P–vv*, and crossed as female with *r–g*. The *P–vv* serves to activate *Ds*, eliminating *R* in developing *R R/r* endosperm. The resulting pattern simulates that due to loss of a strong *C* allele when *Ds* is in its standard position in chromosome 9 (McClintock, 1951). Significantly, the margin of colored–colorless areas are sharp and the colored portion, expected to be of (–)*R/r* genotype, is solid color rather than mottled (Figure 3*b*).

A parallel set of crosses in which a mottling *R* replaced *r* as male parent yielded kernels of primary endosperm genotype *R R/R*. With loss of the maternal *R*s, clones of (–)*R/R* and (–)(–)/*R* cells arise. These kernels proved to be phenotypically mosaic for solid and mottled pigmentation. Boundaries between the two phenotypes often were sharp (Figure 3*d*) in spite of the highly irregular nature of the mottled phenotype. Thus there is regional, if not cellular, autonomy in expression of mottling and solid patterns in the aleurone. The principal implication, however, derives from the characteristic mottling of (–)(–)/*R* sectors and solidly colored phenotype of (–)*R/r*, both in the cytoplasm of an *R* female gametophyte. The occurrence of mottled (–)(–)/*R* sectors discounts the possibility of a nucleoplasmic as well as a cytoplasmic determiner of intense pigmentation in *R* female gametophytes. The hypothesis of differential action of maternal and paternal epialleles as a result of imprinting is thus upheld.

A REGULATOR OF *R*-MOTTLING

A Mendelian variation in the general pattern whereby a maternal epiallele of *R* is solidly colored affords some insight into the genetic control of *R*-imprinting. Three of six *R–g : 1 r–g(Nc)* progenies tested in 1973 by pollinating with *r–g* gave exclusively, or almost exclusively, solid color within the *R–g : 1* kernel offspring. The occasional occurrence of mottled kernels was not unexpected since *r–g(Nc)* belongs to a strongly paramutagenic class, isolated and described by Ashman (1965a,b). In other experiments such exceptions have proved nonheritable. The unexpected finding was the presence on ears of all 71 plants tested in the three remaining progenies of about 25% mottled and 25% solidly colored kernels rather than 50% of the latter. Plants reared from mottled kernels again produced approximately equal numbers of mottled and solid color in addition to one-half colorless seeds, following mating to *r–g* males. Solidly colored sib kernels gave 50% solid color and 50% colorless offspring in parallel matings. An additional generation of recurrent mating to *r–g* has given qualitatively similar results.

Examination of pedigree records showed that the three exceptional progenies descended from a particular *r–g(Nc)* stock; the three progenies characterized by unexceptional behavior traced to a sister $r-g(Nc)$ subline. The breeding behavior outlined in the preceding paragraph is understandable if a gene function necessary for imprinting R to its characteristic maternal level of action was missing in the exceptional $r-g(Nc)$ stock. The aleurone genotypes and phenotypes of kernels resulting from the recurrent backcrosses to $r-g$ would be as follows, using $mdr-$ and $mdr+$ to denote the variant and standard (inbred W22 source) forms, respectively, of the factor involved in *maternal derepression* of R:

Genotype	Phenotype
$R\ R/r$; $mdr-\ mdr-\ /\ mdr+$	*mottled*
$R\ R/r$; $mdr+\ mdr+\ /\ mdr+$	*solid*
$r\ r/r$; $mdr-\ mdr-\ /\ mdr+$	*colorless*
$r\ r/r$; $mdr+\ mdr+\ /\ mdr+$	*colorless*

It will be noted from the first entry that a $mdr+$ allele introduced via pollen does not override the $mdr-$ alleles introduced maternally. This outcome is believed to be due not to $mdr-$ dominance but to the presence of $mdr-$ with R in the female gametophyte at the time of R depression. An apparent reversal of dominance seen in $R\ R$; $mdr+\ mdr+ \times r-g(Nc)\ r-g(Nc)$; $mdr-\ mdr-$ crosses is explained on the basis of this view. The resulting $R\ R/r$; $mdr+\ mdr+\ /\ mdr-$ aleurone is solidly pigmented.

OTHER INSTANCES OF PREFERENTIAL MATERNAL EXPRESSION

Schwartz (1965) called attention to three cases besides R—mottling of maternal effects for which the usual criteria of cytoplasmic inheritance and maternal sporophyte effects seemed inapplicable. These include the *floury* versus *flinty* composition of endosperm, morphology of F_2 kernels borne by maize–teosinte hybrids, and the esterase-2 protein.

To this listing as candidates for imprinting can be added what appears to be a category of mutations which interrupts aleurone pigmentation in potentially solidly colored genotypes. The resulting kernel variegation is dominant when transmitted maternally but is not expressed when transmitted by pollen. Such a case (Vm) was described by Peterson (1964), one occurred in McWhirter's (1971) stocks in Australia, and at least three independent inceptions have been recovered in the course of various investigations utilizing colored aleurone stocks in my cultures. The Vm is pleiotropic for a dominant chlorophyll deficiency expressed following female or male transmission; the other cases give normal green plants.

Two among several mutations to severely defective endosperm that have occurred spontaneously in my cultures in recent years appear to involve genes activated during maternal transmission. The provisional designations are *dex-1146* and *dex-4299*. Both yield 1 : 1 distributions in F_2 rather than the three normal : one mutant characteristic of typical recessives. In reciprocal crosses with normal, heterozygous plants bear ears resembling the F_2; that is, they segregate approximately one normal to one mutant, whereas the reciprocal cross yields only normals. Again, the mutation is expressed as a dominant when introduced maternally but is not expressed when used as male.

The relation of the above cases to imprinting remains provisional since tests that equalize the dosage contributions from the two sexes have not yet been performed. Other possible bases include modified gene expressions during either megasporogenesis or female gametophyte development, which affect the endosperm only indirectly. Endosperm defectiveness associated with *elongate,* for example, or *indeterminate gametophyte* might be classified mistakenly as resulting from imprinting were their respective modifications of megasporogenesis and embryo sac development not known (Rhoades and Dempsey, 1966; Kermicle, 1971).

PREFERENTIAL PATERNAL ACTIVATION

Randolph (1935) reported a marked crossing barrier between his newly synthesized tetraploid maize lines and diploid strains, including direct progenitors of the tetraploids. Fertilization occurred, but seed development failed at an early stage following 2X × 4X crosses in both directions. The morphological abnormalities have been described in detail (Cooper, 1951). Seed failure in this instance as in a variety of other circumstances is attributable to dysfunction of the endosperm (Brink and Cooper, 1947). The basis for failure has been ascribed variously to ploidy imbalance between embryo and endosperm, between maternal tissue and endosperm, or between all three components.

Rather than consider these hypotheses in relation to seed failure, the converse approach may be considered. Are there examples of deviation from the ploidy ratio of 2X embryo : 3X endosperm : 2X maternal tissue where seed development nevertheless is normal? Table 1 lists five such instances in maize. In the final case, the endosperm is imbalanced with respect to both other two components. It deserves closer consideration.

When diploid plants carrying *indeterminate gametophyte* (*ig*) are pollinated by tetraploid, a majority of the seeds fail at a stage typical of diploid × tetraploid crosses. About 10–20% develop normally, however, yielding a plump class of mature kernels. By embryological examination Dr. Bor-Yaw Lin in our laboratory has found that *ig* embryo sacs frequently differentiate more than two polar nuclei. These can be involved jointly in fusion with a sperm nucleus, thus elevating endosperm ploidy level. By correlating the frequency of various ploidy

TABLE 1.   Instances of normal endosperm development in maize that do not conform to the standard ploidy balance between maternal (2X) : endosperm (3X) : embryo (2X) components of the kernel

Exceptional circumstance	Ploidy level			Reference
	Maternal	Endosperm	Embryo	
Unreduced egg, 2X(e1) × 4X	2	6	4	Rhoades and Dempsey (1966)
Parthenogenesis, 2X × 2X	2	3	1	Chase (1964)
Meiotic restitution, 1X × 2X	1	[3]	2	Chase (1964)
Recurrent partheno-genesis, 1X × 2X	1	[3]	1	Ford (1952)
Supernumerary polar nuclei, 2X (ig) × 4X	2	6	3	Lin (1975)

levels of 5–7-day endosperms with the frequency of various mature seed classes, Lin has related the occurrence of plump seed to the 6X condition.

What essential condition does the 6X class satisfy that is not met by endosperm of other ploidy levels following this cross? Evidently, a condition intrinsic to the endosperm is involved since the maternal parent is common to all kernel classes and the embryos are triploid throughout. We interpret success of the 6X endosperm class as reflecting the optimum balance of functionally differentiated genomes: two maternal to one paternal. Seed failure on this view is a matter of epigenetic imbalance rather than intertissue ploidy imbalance.

By similar reasoning, Lin has associated a miniature seed class from *ig ig* × *Ig Ig* crosses with endosperm tetraploidy. The embryo in this case is diploid. The miniature kernels average about half the normal size and regularly are germinable. Lin (1975) has contrasted this class with the underdeveloped, abortive class resulting from 2X × 4X crosses. Both have tetraploid endosperm. He notes that the 2 : 2 composition of the latter is a larger deviation quantitatively from the standard 2 : 1 ratio than is the 3 : 1 makeup resulting from *ig ig* × *Ig Ig* crosses.

RELATED PHENOMENA

Phenomena in maize involving gene action control include paramutation and so-called mutable alleles. The source of altered expression appears not to be the same as for imprinting as indicated as follows:

Phenomenon	Source of variation
Paramutation	Allelic interaction
Mutable alleles	Transposable controlling element
Imprinting	Epigenetic differentiation of reproductive structures

A feature common to these systems is the inheritance in somatic cells of an altered gene potential. In the case of paramutation, the variation regularly is transmitted to progeny (Brink, 1956, 1960). A sequence of gene action programming by controlling elements, involving presetting and erasure, has been described by McClintock (1968). Exceptionally, the preset condition is transmitted to offspring.

DOUBLE FERTILIZATION RECONSIDERED

Let us look again at double fertilization, granting the possibility of differences in gene action potentials between parental genomes. The embryo and endosperm play unlike roles in seed formation. The embryo provides for genetic continuity; the endosperm supports this function by nourishing the embryo and sustaining early seedling growth. The female progenitor of the embryo is a gamete, and the female progenitor of the endosperm is a differentiated cell. Divergence between embryo and endosperm in developmental rate and path (e.g., cellular vs. free nuclear) ensues from the zygote and primary endosperm cell stages. This differentiation clearly is prestaged in the egg and central cells of the female gametophyte, assuming the two sperm in given pollen grains to be equivalent.

To say that the central cell is differentiated only adds to the puzzle as to why double fertilization should be adaptively significant. Why shouldn't the central cell differentiate and itself serve the function of an endosperm, as in gymnosperms? Or, why shouldn't the endosperm in sexually reproducing species differentiate directly from the sporophyte as in some autonomous apomicts? Brink (1952) has considered the possibility of a unique significance of biparental origin of the endosperm due to heterosis.

A novel form of hybrid vigor may derive from epigenetic differentiation between polar and sperm nuclei. Differences in the combination of derepressed genes could establish a condition of epistatic hybridity within the endosperm, contributing to the rapid mitotic activity and aggressive early growth so frequently noted by plant embryologists. It is of interest to note further that the functional complementation resulting from nuclear differentiation would not be limited to cross-pollinated species. The same sort of interaction, or "epihybridity," could occur in self-pollinated species. It perhaps is not mere chance that the self-pollinated cereal crops number among man's most valuable cultivars and that the endosperm in these instances represents a major concentration of stored materials of nutritional significance to him.

ACKNOWLEDGMENTS

These studies were conducted at the Wisconsin Agricultural Experimental Station with funds supplied by ERDA Contract AT(11-1)-1300 and NIH Grant GM15422. I thank Beverly Oashgar for technical assistance, Dr. Bor-Yaw Lin for preparing Figures 2 and 3, and Professor R. A. Brink for sharpening my interest in this subject.

REFERENCES

Ashman, R. B. 1960. Stippled aleurone in maize. *Genetics* **45**: 19.

Ashman, R. B. 1965a. Mutants from maize plants heterozygous $R^r R^{st}$ and their association with crossing over. *Genetics* **51**: 305.

Ashman, R. B. 1965b. Paramutagenic action of mutants from maize plants heterozygous $R^r R^{st}$. *Genetics* **52**: 835.

Bray, R. A. and R. A. Brink. 1966. Mutation and paramutation at the R locus in maize. *Genetics* **54**: 137.

Brink, R. A. 1952. Inbreeding and crossbreeding in seed development. In *Heterosis*. Ames: Iowa State College Press, p. 81.

Brink, R. A. 1956. A genetic change associated with the R locus in maize which is directed and potentially reversible. *Genetics* **41**: 872.

Brink, R. A. 1960. Paramutation and chromosome organization. *Quart. Rev. Biol.* **35**: 120.

Brink, R. A. and D. C. Cooper. 1947. The endosperm in seed development. *Bot. Rev.* **13**: 423.

Brink, R. A., J. L. Kermicle, and N. K. Ziebur. 1970. Derepression in the female gametophyte in relation to paramutant R expression in maize endosperms, embryos and seedlings. *Genetics* **66**: 87.

Brink, R. A., E. D. Styles and J. D. Axtell. 1968. Paramutation: directed genetic change. *Science* **159**: 161.

Chandra, H. S. and S. W. Brown. 1975. Chromosome imprinting and the mammalian X chromosome. *Nature* **253**: 165.

Chase, S. S. 1964. Monoploids and diploids of maize: a comparison of genotypic equivalents. *Am. J. Bot.* **51**: 928.

Cooper, D. C. 1951. Caryopis development following matings between diploid and tetraploid strains of maize. *Am. J. Bot.* **38**: 702.

Crouse, H. V. 1960. The controlling element in sex chromosome behavior in *Sciara*. *Gentics* **45**: 1429.

East, E. M. and H. K. Hayes. 1911. Inheritance in maize. *Conn. Agr. Exp. Sta. Bull.* **167**: 1–142.

Emerson, R. A. 1918. A fifth pair of factors, *A a*, for aleurone color in maize, and its relation to the *C c* and *R r* pairs. *Cornell Univ. Agr. Exp. Sta. Memoir* **16**: 226.

Ford, L. E. 1952. Some cytogenetic aspects of maize monoploids and monoploid derivatives. Ph.D. thesis. Ames: Iowa State College.

Kermicle, J. L. 1963. Metastability of paramutant forms of the *R* gene in maize. Ph.D. thesis. Madison: University of Wisconsin.

Kermicle, J. L. 1970. Dependence of the *R*-mottled aleurone phenotype in maize on mode of sexual transmission. *Genetics* **66**: 69.

Kermicle, J. L. 1971. Pleiotropic effects on seed development of the *indeterminate gametophyte* gene in maize. *Am. J. Bot.* **58**: 1.

Kermicle, J. L. 1973. Allele specific response of the *R* gene in maize to parental imprinting. *Proc. 13th Internat. Congr. Genet.* (abstr.), p. S134.

Lin, B. Y. 1975. Parental effect on gene expression in maize endosperm development. Ph.D. thesis. Madison: University of Wisconsin.

McClintock, B. 1951. Chromosome organization and genic expression. *Cold Spring Harbor Symp. Quant. Biol.* **16**: 13.

McClintock, B. 1968. Genetic systems regulating gene expression during development. In *Control mechanisms in developmental processes*. New York: Academic, pp. 84–112.

McWhirter, K. S. 1961. Paramutation studies with self-colored mutants from the stippled allele in maize. Ph.D. thesis. Madison: University of Wisconsin.

McWhirter, K. S. 1971. Aleurone color variegation involving the *R* locus. *Maize Genet. Coop. News.* **45**: 179.

Peterson, P. A. 1964. The dominant mutable *V^m mp-1817*. *Maize Genet. Coop. Newsl.* **38**: 80.

Randolph, L. F. 1935. Cytogenetics of tetraploid maize. *J. Agr. Res.* **50**: 591.

Rhoades, M. M. and E. Dempsey. 1966. Induction of chromosome doubling at meiosis by the *elongate* gene in maize. *Genetics* **54**: 505.

Schwartz, D. 1965. Regulation of gene action in maize. *Genetics Today*, Oxford: Pergamon, p. 131.

Sturtevant, A. H. 1965. *A History of genetics*. New York: Harper and Row, p. 28.

Chapter 25

CANALIZATION OF ENDOSPERM DEVELOPMENT IN OPAQUE-7 MAIZE

K. S. McWhirter and R. Alexander Brink

Department of Agricultural Botany,
The University of Sydney,
Sydney, Australia and
Laboratory of Genetics,
The University of Wisconsin,
Madison

The significance of the *opaque*-2 and *floury*-2 mutations for increasing the lysine content of maize endosperm proteins is widely recognized (Mertz, Bates, et al., 1964; Nelson, Mertz, et al., 1965). Another mutant, designated (*o7*), also increases the lysine content of endosperm proteins (Misra, Jambunathan, et al., 1972) and appears to be similar to *opaque*-2 in potential for improving the nutritional value of maize. The *opaque*-7 mutant assorts as a simple Mendelian recessive in W22, the inbred dent corn line in which it was discovered (McWhirter, 1971). However, following outcrosses of W22 *o7* to numerous

Paper No. 1880 from the Laboratory of Genetics, University of Wisconsin, Madison, Wisconsin, 53706 other normal endosperm strains, opaque kernels are recovered with much lower than conventional Mendelian frequencies. The present report is concerned with the genetic basis of the irregular expression of the o7 endosperm phenotype. It will be shown that a rational explanation of the experimental results is provided by assuming that; (a) the *opaque*-7 endosperm phenotype is a threshold character (Falconer, 1960) and (b) normal endosperm development in maize is a canalized, or genetically buffered, process.

THE *opaque*-7 MUTANT

This mutant was discovered as a spontaneous mutation in a W22 inbred subline (McWhirter, 1971). The o7 locus is approximately 23 map units distal to *R* on the long arm of chromosome 10 (McWhirter, 1973). The o7 allele has been characterized as a high lysine mutant (Misra, Jambunathan et al., 1972) that increases the biological value of the endosperm proteins (Buttenshaw, McWhirter, et al. 1973). It also reduces endosperm size and protein content of the whole seed (McWhirter, unpublished).

GENETIC MODIFICATION OF *opaque*-7 EXPRESSION

In contrast to regular Mendelian assortment of O7 and o7 alleles in matings within the W22 inbred line, marked deviations were observed in crosses of W22 *opaque*-7 with various other normal endosperm lines. Data from two crosses observed, which illustrate the direction and magnitude of the observed deviations, are given in Table 1. Less than the expected proportion of opaque endosperm kernels was observed in all of these matings. Following self-pollinations, the frequencies of opaque kernels on individual ears ranged within 3.1–14.8%, whereas 25% opaque kernels was expected. In backcrosses the frequency of opaque kernels also was reduced, and more extreme deviations from the expected 50% of opaque kernels occurred in $O7/o7$ ♀♀ × W22 $o7/o7$ ♂♂ crosses than in reciprocal matings.

Following the observation of deviations similar to these in outcrosses of W22 *opaque*-7 to 11 commonly grown inbred dent strains, a survey was undertaken to establish the frequency and the distribution among diverse kinds of maize of genetic modifiers of *opaque*-7 expression. Crosses of W22 *opaque*-7 with a broad sample of maize genotypes, including commercial inbred strains, Latin American strains (from Bolivia, Colombia, Guatemala, Mexico, Peru, and Venezuela), maize × teosinte derivatives, and also flint, pop, and sugary endosperm types, were studied. In all these groups a high proportion of genotypes

TABLE 1.   Modified distributions of translucent and *opaque*-7 kernels from two different O7/o7 heterozygotes

Matings	Number of kernels			Total kernels	Percent opaque and semiopaque kernels
	Translucent	Semiopaque	Opaque		
	F_1 from W22 inbred × B37 inbred				
Selfed[a]	492	9	10	511	3.7
Backcrossed[b]	1130	109	194	1433	21.1
Testcrossed[c]	1361	99	669	2119	36.2
	Heterozygous plants of M14/W23 × W22 origin				
Selfed[a]	7342	417	365	8124	9.6
Backcrossed[b].	3787	396	813	4996	24.2
Testcrossed[c]	1966	87	607	2660	26.1

[a] Mating O7/o7 selfed.

[b] Mating O7/o7 ♀♀ × W22 o7/o7 ♂♂.

[c] Mating W22 o7/o7 ♀♀ × O7/o7 ♂♂.

were associated with reduced frequencies of opaque kernels in backcross or test-cross matings of $O7/o7$ heterozygotes (Table 2).

Frequencies of opaque kernels ranged within 0–57%, with a modal value at the 20–30% frequency. Only 21 of 139 genotypes tested, gave nonsignificant deviations from a one normal : one opaque kernel ratio. Also the frequencies of opaque kernels for different heterozygotes varied continuously, and were not clustered about the proportions expected on the assumption of simultaneous segregation at one, two or three suppressor loci. Evidently genetic modifiers of $o7$ expression are very widespread in the species, and the varying proportions of $o7$ segregates, cannot be explained by assuming a wide distribution of dominant suppressor genes. Transformation of observed deviations to multiples of the standard deviation from a hypothetical 1 : 1 ratio (Table 2), showed that a high frequency of the genotypes tested produced a deviation lying between 6 and 10 standard deviations below the 1 : 1 ratio.

No simple Mendelian hypothesis has been found capable of a consistent explanation of these distributions. A feature of the results in Table 1 is that the observed F_2 ratios do not agree with those predicted from the corresponding backcrosses and testcrosses. This discrepancy eliminates most simple Mendelian hypotheses. Moreover, in the same matings that produced reduced frequencies of opaque kernels, alleles at the linked R locus were observed to assort in the expected Mendelian proportions. Therefore, explanation of the opaque versus normal endosperm distributions must be sought in genetic modification of

TABLE 2. Distributions of maize genotypes that produce modification of expression of opaque-7 in O7/o7 ♀ × W22 o7/o7 ♂ or W22 o7/o7 ♀ × O7/o7 ♂ matings

Source of genotypes tested	Deviation from a one translucent: one opaque kernel ratio (× standard deviation)							
	<2×	>2× <4×	>4× <6×	>6× <8×	>8× <10×	>10× <12×	>12× <14×	>14×
Dent maize inbreds	6	8	5	12	14	1	—	—
Flint populations	2	—	2	8	6	—	1	1
Sugary inbreds	1	1	—	4	4	—	—	1
Popcorn inbreds	4	1	1	7	3	5	—	2
Maize × teosinte derivatives	1	2	4	7	7	—	—	—
Latin American populations	11	4	1	3	5	1	—	—
Virginia Blue Ridge Dent	—	—	—	2	—	—	—	—
Totals	16	16	13	43	39	7	1	4

expression of the *o7* allele rather than in aberrant assortment of alleles at the *O7* locus. The observed distributions usually were discrete, that is, comprised relatively distinct normal and opaque kernel classes. Collectively, these observations of a reduced, but variable, frequency of opaque kernels, with a marked discontinuity of the two phenotypes, are suggestive of the inheritance pattern expected for a threshold character (Falconer, 1960). For such characters a polygene complex producing continuous variation underlies a visible discontinuity of phenotype expression. The distributions in question can be interpreted by assuming that in the presence of the *o7* allele, the assortment and expression of polygenes results in continuous variation in a developmental activity leading to a normal endosperm phenotype. Discontinuity of phenotypes and widely varying proportions of the alternative endosperm types result from the existence of a developmental threshold. Only kernels with a genotypic value resulting in an activity level below the threshold, express an opaque phenotype.

The above reasoning leads to the view that endosperm development in maize is a canalized developmental process (Waddington, 1957, 1966; McWhirter, 1973). Normal endosperm development is homeostatic, or buffered, and genetic control has so evolved that the process resists modification by extraneous genetic or nongenetic causes. Stability is achieved by natural selection of components of a polygene complex, which tends to maintain normal development in endosperms homozygous for the *o7* allele. The canalization is responsible for the observed reduction in frequency of opaque kernels in populations heterozygous at the *O7* locus.

Two methods are available, using present data, for examining the validity of this hypothesis. The first is to show that the hypothesis provides a self-consistent explanation of the evidence on inheritance of opaque endosperm in *O7/o7* heterozygotes. The second is to test for polygenically determined variability about the postulated developmental threshold.

A DEVELOPMENTAL THRESHOLD INTERPRETATION OF *opaque-7* INHERITANCE

A threshold hypothesis permits potentially any frequency of opaque kernels within the limits of 0–50% to be produced in *O7/W22o7* ♀ × W22 *o7/o7* ♂ matings. The observed frequencies of opaque kernels were within these limits, allowing for sampling variation. Assuming a normal distribution of genetic effects, very low percentages of opaque kernels are expected in populations with a mean level of the critical developmental activity in *o7* kernels with 1–3 standard deviations above a constant threshold. Percentages of opaque kernels near 50% are expected for populations with a mean below the threshold. Within the W22 inbred line, the mean developmental activity in *o7* kernels must be below the threshold, and a low variance is expected. The majority of all populations studied had frequencies of opaque kernels in the range 20–30% (Table 2). When expressed as a proportion of the total number of homozygous opaque kernels expected, these frequencies imply mean developmental activities in the first backcross populations, as estimated by probit analysis (Falconer, 1965; Rendel, 1967), of between +0.253 and −0.253 standard deviations from the threshold point. The threshold hypothesis makes no definite prediction on frequency of opaque kernels. Nevertheless two regularities are expected in the data from related families. Both are evident in the results at hand.

First, assuming only additive gene action, it is expected that the population means estimated for F_2 and backcross populations, are seriated in direct proportion to the contribution of the W22 genotype to the population. In Table 3 the data on inheritance of $o7$, in crosses with inbred B37, are summarized in terms of the threshold hypothesis. The observed frequencies of opaque kernels are expressed as a percentage of the expected total of homozygous $o7$ segregates. Means of the F_2 and reciprocal backcross populations are then estimated from these values by probit transformation. The mean values are expressed in multiples of the standard deviation characteristic of the population, with the threshold taken as zero. The three populations are expected to differ in variance, with $\sigma^2 F_2 = 1.25$, $\sigma^2_{BC} = 1.0$, and $\sigma^2_{(reciprocal\ BC)} = 0.25$[1]. Calculation of weighted means represents an attempt to place the observed mean values on a common scale. The results of this calculation (see column 6, Table 3) show that the population means are in the relation expected on the threshold hypothesis.

Second, reciprocal backcrosses differ in the relative contribution of the W22 inbred genotype. If a simple linear relationship between population means and the proportion of W22 genotype in the population is assumed, there should be a consistently higher frequency of opaque kernels in backcross matings in which W22 $o7/o7$ serves as the female parent, compared with the exact reciprocal matings.

A total of 28 reciprocal matings of individual plants of $O7/o7$ genotype were available for comparison. In 23 of these the W22 $o7/o7$ ♀ × $O7/o7$ ♂ mating produced frequencies of opaque kernels, ranging from 2% to 27.5%, that were higher than the frequency in the reciprocal mating. Data from two families of plants were analyzed for statistical significance of the difference in paired mean values for the populations of kernels produced by reciprocal matings. In both families the mean differences between populations were significant (family W22 × B37: mean difference = 0.66σ; t6 = 9.10, P < 0.001; family M14/W23 × W22: mean difference = 0.48σ; t7 = 7.35, P < 0.001). These data, therefore, are in accord with the expectation of a consistent difference between reciprocal crosses.

[1] We are indebted to Dr. Sewall Wright for calculation of these variances.

TABLE 3. Distributions of progenies of O7/o7 plants of W22 × B37 origin interpreted according to the threshold hypothesis

Population	Observed opaque kernels (%)	Opaque kernels (%) among o7/o7/o7 homozygotes	Normal deviate	Population standard deviation	Estimated population mean (xs) = 1.0	Proportion of W22 genotype
(1)	(2)	(3)	(4)	(5)	(6)	(7)
O7/o7 Selfed	3.7	14.8	+1.045	1.118	+1.168	0.5
O7/o7 Backcrossed	21.1	42.2	+0.197	1.0	+0.197	0.66
O7/o7 Testcrossed	36.2	72.4	−0.625	0.5	−0.313	0.83

The single greatest difference between reciprocal matings was observed for plants of pearl pop inbred Ames 4894 × W22 origin. The difference between reciprocal matings should be greatest when parental mean values are markedly different. The near absence (0.6%) of opaque kernels in the $O7/o7$ ♀♀ × W22 $o7/o7$ ♂♂ matings, in this case, is evidence that the parents are extremely divergent. In parallel, the reciprocal matings produced a 27.5% higher frequency of opaque kernels, corresponding to a difference of 1.658σ in the respective population means. This result is consistent with a threshold interpretation, but is disconcerting on almost any other basis.

Progeny test data, obtained by growing on samples of translucent, and in some cases opaque, kernels from a number of backcross matings, effectively dispose of any simple Mendelian hypothesis for opaque endosperm inheritance. The particular progenies chosen for test represented a seriated sample over the observed range of frequencies of occurrence of opaque kernels. When pollinated with the W22 $o7$ homozygous stock, these plants produced frequencies of opaque kernels ranging from low values (10–20%) to values significantly greater than 50%, and occasionally near 100%.

Occurrence of plants with greater that 50% opaque kernels is unexpected on any simple Mendelian hypothesis but is compatible with the threshold hypothesis, because these progenies are expected to contain both homozygous ($o7/o7$) and heterozygous ($O7/o7$) plants. Also, in 24 of the 26 progenies studied, the distributions for individual plants were in obvious disagreement with simple Mendelian assortment, but gave some evidence of quantitative inheritance of the genetic effects involved in inheritance of endosperm type. There was a crude parent–progeny relation, in that parent ears with a significant deviation toward a low frequency of opaque kernels tended to produce progenies with an excess of plants deviating in the same direction. These deviations, however, were usually less extreme than that characterizing the respective parent ear. This evidence is suggestive of a quantitative shift of progeny means in the direction of the recurrent parent value, as is expected for a quantitative trait in a backcross.

A major prediction of the threshold hypothesis is that a proportion of homozygous $o7$ kernels ($o7/o7/o7$ genotype) will manifest a translucent kernel phenotype. In 16 of the progenies studied there was assortment also for linked R alleles affecting anther color. The expected distributions for anther color within the translucent kernel class were calculated for each progeny on the basis of regular assortment for R alleles, a recombination value between the two loci of 0.23, and the observed incidence of opaque kernels. In 15 of the progenies the observed numbers of plants with red or green anthers did not deviate significantly from the number expected. Yet in eight of these same progenies, containing a total of 221 plants, there were only two individuals that were classified as homozygous *opaque*-7. Assuming a regular assortment of $o7$

alleles along with linked R alleles, approximately 74 plants homozygous $o7$ were expected. Thus in this particular group of progenies a large number of plants, although homozygous $o7$, gave distributions of both translucent and opaque kernels when backcrossed again with the W22 $o7$ stock. Such progenies regularly consisted of two classes of plants, with the frequency of opaque kernels in one class being approximately twice the frequency in the alternative class, thus paralleling the expected distribution of homozygous ($o7/o7$) and heterozygous $O7/o7$) plants.

QUANTITATIVELY GENETIC MODIFICATION OF *opaque-7* EXPRESSION

Populations that differ in frequency of opaque kernels may be expected to do so, according to the threshold hypothesis, because of differences in population means, in variances, or in both means and variances for the developmental activity leading to normal endosperm development. Support for the threshold hypothesis has been obtained from two experiments that detect genetic differences of a quantitative nature among related individual plants. These differences may be interpreted as indicative of quantitative variation in mean developmental effects, because in each case the experimental design permits the assumption that variances in the populations being compared do not greatly differ. In the first experiment, significant heterogeneity for distribution of normal and opaque kernels was found among plants of $O7/o7$ genotype produced by the mating M14/W23 $O7/O7$ ♀ × W22 $o7/o7$ ♂ (Table 4).

Quantitative genetic differences among homozygous $o7$ plants recovered from backcrosses of diverse $O7/o7$ heterozygotes with W22 $o7/o7$ were detected in

TABLE 4.   Evidence of quantitative genetic variation among related $O7/o7$ heterozygous plants

Mating	Number of plants	Percentage of opaque kernels			Heterogeneity[a] χ^2	Probability value
		Low	Mean	High		
O7/o7 Plants from mating M14/W23 O7/o7 ♀ × W22 o7/o7 ♂						
Selfed	20	3.8	9.6	14.8	94.3	<0.001
Backcrossed	13	11.8	24.2	29.7	51.8	<0.001
O7/o7 Plants from mating W22 o7/o7 ♀ × B37 O7/O7 ♂						
Selfed	3	2.8	3.7	4.6	0.25	0.8–0.9
Backcrossed	7	18.1	21.1	24.2	4.53	0.5–0.7

[a] Calculated by the Brandt–Snedecor method (Mather, 1951).

TABLE 5. Genetic variation among recovered homzogyous opaque-7 (o7/o7) plants

	Testcross female parents						
	W22 o7/o7			Recovered o7/o7			
O7/o7 Heterozygous pollen parent	Opaque kernels (%)	Heterogeneity[a] χ^2	P Value		Opaque kernels (%)	Heterogeneity[a] χ^2	P Value
1 F1783	24.9	20.4	<0.01[b]		26.7	8.4	<0.01
2 F1811	23.3	2.7	>0.5		27.0	10.3	<0.05
3 Ven 309	51.9	1.5	>0.8		49.5	1.8	>0.8
4 Ven 336	51.8	3.8	>0.5		49.6	1.7	>0.2
5 Black Beauty	19.3	8.4	>0.05		17.8	5.8	>0.2
6 Pearl Pop	51.5	1.4	>0.8		58.3	12.6	<0.02
7 Pointed Pop	18.3	1.8	>0.8		20.1	11.6	<0.05

[a] All χ^2 values associated with four degrees of freedom.

[b] Significant heterogeneity due to one highly deviant cross.

the second experiment. A single pollen collection from an *O7/o7* plant was used to pollinate five homozygous *o7/o7* plants within a recovered family and also five plants of inbred W22 *o7/o7*. Seven heterozygous *O7/o7* plants of diverse parentage were used as pollen parents. The data in Table 5 show that significant heterogeneity was found in three families of matings on recovered *o7/o7* plants, whereas there was no significant heterogeneity among control matings on W22 *o7/o7* plants. This is the result of interest in Table 5. It establishes the existence of genetic variation for percentage of opaque kernels among at least some recovered *o7* homozygous plants.

ENDOSPERM PHENOTYPE AND LYSINE CONTENT

Opaque endosperm and high lysine content of endosperm proteins are related phenotypic effects in *o7* strains. Analyses by the method of Beckwith, Paulis, et al. (1975) of translucent, semiopaque, and fully opaque kernels from five different *O7/o7* × *o7/o7* backcross ears, selected from larger populations because they contained significant numbers of semi-opaque as well as the other two classes of seeds, showed that the order of lysine level invariably was opaque > semiopaque > translucent. The values for crude protein content were in reverse order, namely, translucent > semiopaque > opaque (Table 6). Reductions in protein content ranged within 11–30%, and increases in lysine content (g/100 g protein) ranged within 30–54%, in relative values. The materials currently available are not suitable for measuring the degree of independence of lysine content and endosperm texture. Therefore, the extent to which a high lysine content can be conserved by selection in cryptic opaques (i.e., *o7* homozygotes of translucent phenotype) remains an open question.

DISCUSSION

The pattern of inheritance observed for *opaque-7* endosperm supports the view that endosperm development in maize involves a canalized genetic system, in the sense of Waddington's (1957, 1966) concept of canalization of developmental processes. When viewed in this framework control of endosperm development involves three components: (a) a major gene, the dominant *O7* allele, which establishes a sufficient level of activity for normal endosperm development, (b) a polygene complex, similar in physiological importance to the major gene, that produces variation in the level of developmental activity, and (c) a threshold of endosperm developmental activity that results in discrete levels of phenotype expression. The critical developmental process subject to

TABLE 6.   Lysine and crude protein (Nx6.25) content of translucent, semiopaque, and opaque whole kernels borne on single ears resulting from five different O7/W22 o7 × W22 o7 matings[a]

Source of O7 gene	Kernel class	Crude protein (%)	Relative protein content	Lysine mg/g meal	Lysine g/16 g of N	Relative lysine content
Hickory	translucent	12.65	100	3.614	2.857	100
King	semi-opaque	11.73	93	4.017	3.424	120
	opaque	10.54	83	4.962	4.706	130
Shoe	translucent	11.20	100	3.614	3.227	100
Peg	semi-opaque	10.81	96	3.642	3.370	104
	opaque	10.01	89	4.198	4.192	130
Pearl	translucent	13.58	100	4.198	3.093	100
Pop	semi-opaque	11.86	87	5.199	4.382	109
	opaque	11.07	81	5.518	4.985	129
Maiz	translucent	12.78	100	3.461	2.708	100
Amargo	semi-opaque	10.94	86	4.990	4.562	168
	opaque	10.68	84	5.838	5.466	202
Pointed	translucent	10.16	100	3.183	3.134	100
Pop	semi-opaque	9.49	93	3.948	4.161	133
	opaque	7.12	70	4.031	4.817	154

[a] Data provided by O. E. Nelson, Jr. and R. Huang.

threshold action has not been identified. An interaction might be involved at the histological level between the young endosperm and the multicellular antipodals that lie athwart the nutrient stream flowing from the parent sporophyte to the endosperm. Randolph (1936) showed that persistence of the initially prominent and highly active antipodals varied widely from one seed to another within a given variety.

These three components can be considered a consequence of natural selection for a favorable endosperm type. Rendel (1967) has pointed out that the evolutionary reason for a threshold in a canalized developmental process is that a particular level of expression is advantageous to the individual, and natural selection operates to stabilize the phenotype at this level. Stability is promoted when the polygene complex, assembled as a result of natural selection, can substitute physiologically for the major gene. Evidence for this effect in the o7 material is provided by progeny test data showing that some homozygous *o7* endosperms regularly manifest a translucent phenotype. Rendel (1967) also has argued that a common, essential, property of canalized processes will be both a lower and an upper threshold. The region between the thresholds, the canalized plateau, is the region in which control of development is complete. Natural selection should result in the accumulation of polygenes that maintain the level of developmental activity at about the middle of the canalized plateau. Many maize genotypes, therefore, are expected to have a polygene complex of similar phenotypic effect. This is the explanation for the clustering of genotypes (Table 2) when categorized by the extent of the deviation observed from a one translucent : one opaque ratio in backcrosses of *O7/o7* heterozygotes.

The view that endosperm development is a canalized process has implications for an understanding of the evolution of endosperm types in maize. The evidence points to the canalized phenotype as being that of normal dent maize. Opaque endosperm represents a deviation below the lower threshold, whereas flint and/or popcorn types represent phenotypes above the upper threshold. Evidence for the latter idea comes from the inheritance of *o7* in crosses with the pearl pop inbred Ames 4894.

The normal process of endosperm development is disrupted (decanalized) by mutation at the *o7* locus. It is not necessary, however, for qualitative changes in the process to be directly associated with the *o7* mutant allele. The observed inheritance pattern is consistent with the mutant allele being associated with a marked reduction in mean level of a developmental activity that follows essentially the same course in both normal and mutant genotypes.

The canalization hypothesis developed for *o7* can be extended to *opaque-2*, by assuming that the *o2* mutant reduces the developmental activity well below a common threshold. Few genotypes produced by polygene combinations will cross the threshold and modify *opaque-2* to a normal phenotype. Such genotypes occasionally occur (Vasal, 1975; Pollacsek, Caenen, et al., 1972), however, and the results of studies of them appear to be compatible with a threshold interpretation.

The *opaque-7* may provide favorable material for the selection of "modified opaque" endosperm strains high in lysine (Vasal, 1975; Mertz, 1974). It evidently is more sensitive to genetic modification than is *o2*. The present study shows that modified *o7* genotypes are numerous among the progeny of crosses with many normal endosperm strains. Provided there is some independence of endosperm phenotype and lysine content, significant opportunity should be presented for selection of modified opaque, high lysine combinations. Also, recognition of the presence of quantitatively varying genetic effects in decanalized populations should stimulate attempts to select more extreme levels of protein and amino-acid concentration in opaque maize.

ACKNOWLEDGMENTS

Several individuals, R. I. Brawn, W. Galinat, M. M. Goodman, C. J. Hartman, A. Manwiller, E. C. Rossman, and E. Thorne, kindly permitted access to a wide variety of pollen sources at Homestead, Florida. The Regional Plant Introduction Station at Ames, Iowa furnished seed of numerous noncommercial strains. We are also much indebted to Funk Seeds International and also to the Mexican-based CIMMYT for off-season growing facilities and to Professor J. R. Laughnan, for field and laboratory facilities to the senior author at the University of Illinois in 1974.

REFERENCES

Beckwith, A. C., J. W. Paulis, and J. S. Wall. 1975. Direct estimation of lysine in corn meals by the ninhydrin color reaction. *J. Agr. Food Chem.* **23:** 194–196.

Buttenshaw, R., K. S. McWhirter, and D. M. Walker. 1973. Animal response to diets containing *opaque*-7 endosperm proteins. *Maize Genet. Coop Newsl.* **47:** 174–175.

Falconer, D. S. 1960. *Introduction to quantitative genetics.* Edinburgh: Oliver and Boyd.

Falconer, D. S. 1965. The inheritance of liability to certain diseases, estimated from the incidence among relatives. *Ann. Hum. Genet.* (London) **29:** 51–76.

Mather, K. 1951. *The measurement of linkage in heredity.* London: Methuen, p. 22.

McWhirter, K. S. 1971. A floury endosperm, high lysine locus on chromosome 10. *Maize Genet. Coop. Newsl.* **45:** 184.

McWhirter, K. S. 1973. Linkage relations of *opaque*-7 with marker loci in linkage group 10. *Maize Genet. Coop. Newsl.* **47:** 171–173.

McWhirter, K. S., unpublished data.

Mertz, E. T. 1974. Genetic improvement of cereals. *Nutr. Rev.* **32:** 129–131.

Mertz, E. T., L. S. Bates, and O. E. Nelson. 1964. Mutant gene that changes protein composition and increases lysine content of maize endosperm. *Science* **145:** 279–280.

Misra, P. S., R. Jambunathan, E. T. Mertz, D. V. Glover, H. M. Barbosa, and K. S. McWhirter. 1972. Endosperm protein synthesis in maize mutants with increased lysine content. *Science* **176:** 1425–1427.

Nelson, O. E., E. T. Mertz, and L. S. Bates. 1965. Second mutant gene affecting the amino acid pattern of maize endosperm proteins. *Science* **150:** 1469–1470.

Pollacsek, M., M. Caenen, and M. Rousset. 1972. Mise en évidence d'un deuxième gène suppresseur du gène *opaque*-2 chez le maïs. *An. Genet.* **15:** 173–176.

Randolph, L. F. 1936. Developmental morphology of the caryopsis in maize. *J. Agr. Res.* **53:** 881–916.

Rendel, J. M. 1967. *Canalisation and gene control.* London: Logos Press.

Vasal, S. K. 1975. Use of genetic modifiers to obtain normal-type kernels with the *opaque*-2 gene. In *High quality protein maize.* Stroudsburg, Penn.: Dowden, Hutchinson and Ross, pp. 197–216.

Waddington, C. H. 1957. *The strategy of the genes.* London: Allen and Unwin.

Waddington, C. H. 1966. *Principles of development and differentiation.* New York: Macmillan.

Chapter 26

GENE ACTION AND ENDOSPERM DEVELOPMENT IN MAIZE

Oliver E. Nelson, Jr.
Department of Genetics,
University of Wisconsin,
Madison

GENE ACTION AND ENDOSPERM DEVELOPMENT IN MAIZE

One of the gratifying aspects of the present volume from my viewpoint has been the emphasis in both invited and demonstration papers on endosperm development and its genetic control. This is a subject of considerable practical importance and one on which I intend to focus, particularly with regard to the poorly understood pattern of physiological specialization of the endosperm in relation to the sporophytic and gametophytic generations and with regard to our limited understanding of starch and protein synthesis in that tissue.

The endosperm, which forms the site of depositions of large quantities of starch during endosperm development in maize, nourishes the embryo during the early stages of development and the plantlet after germination (Brink and Cooper, 1947; Maheshwari, 1954). However, in seeds homozygous for mutations (*brittle*-1, *brittle*-2, and *shrunken*-2) that are markedly disruptive of

Laboratory of Genetics Paper No. 1916.

starch synthesis and endosperm development, the embryo may be even larger than in normal seeds. This does not negate conclusions concerning the vital role of the endosperm in nourishing the embryo since the effect of these mutations does not become apparent until 14–16 days postfertilization (Tsai, Salamini, et al., 1970).

With our present knowledge of developmental genetics and the existence of nonallelic isoenzymes, it is not surprising that many genes essential for normal endosperm development can mutate to inactivity without affecting either gametophytic or sporophytic tissue. For the sporophyte, it is known that nonallelic isoenzymes catalyze the same reactions as does Wx in the endosperm (a starch granule-bound nucleoside diphosphate sugar–starch glucosyl transferase) (Akatsuka and Nelson 1966) and $Sh2$ and possibly $Bt2$ in the endosperm (adenosine diphosphate glucose pyrophosphorylase) (Preiss, Hammel, et al., 1971; Hannah and Nelson, 1975) and Sh in the endosperm (sucrose synthetase) (Chourey, personal communication). It has also been shown by biochemical methods that the glucose phosphate isomerase of the sporophyte is a different enzyme than the major enzyme of the endosperm (Salamini, Tsai, et al., 1972) (see Table 1). Thus the genome of maize contains numerous genes that are active only in endosperm development. This is not invariably so, as Mangelsdorf (1926) reported in the first extensive study of mutants affecting endosperm development. He found in studies of *defective seed*-1 through *defective seed*-14 that the disruption of development for both embryo and endosperm was approximately comparable and that if the homozygous mutant seeds were viable, the plants produced lacked vigor. A number of mutants that are possibly not allelic to the defectives reported by Mangelsdorf affect both endosperm and sporophyte development. I suspect that many of these mutants will be shown to affect steps in protein synthesis.

Several mutant genes that affect starch synthesis and thus endosperm development have the same effect on the gametophytic generation. The best-known case is that of the *waxy* gene, whose effect was shown to be extended to the microspore by Brink and MacGillivray (1924) and Demerec (1924). It was

TABLE 1.   The reactions catalyzed by the enzymes to be discussed.

Enzyme	Reaction
Nucleoside diphosphate sugar–starch glucosyl transferases	$Gn + XDPG \rightarrow G_{n+1} + XDP$
Adenosine diphosphate glucose pyrophosphorylase	$G\text{-}1\text{-}P + ATP \leftrightarrows ADPG + PP$
Sucrose synthetase	$Sucrose + UDP \rightleftarrows UDPG + fructose$
Phosphorylase	$Gn + G\text{-}1\text{-}P \rightleftarrows G_{n+1} + Pi$
Glucose phosphate isomerase	$F\text{-}6\text{-}P \rightleftarrows G\text{-}6\text{-}P$

later shown by Brink (1925) that the mutant gene had the same effect on the starch of the female gametophyte. The gene *amylose-extender* (*ae*), which increases the proportion of the starch that is amylose for the endosperm, has the same effect in the pollen (Banks, Greenwood, et al., 1971). The *waxy* gene, in addition to its effect on the type of starch produced, also conditions a small but significant deviation from an expected Mendelian ratio in self-pollinations of $\pm/wx$ plants. This deviation is much exaggerated in plants homozygous for the *sugary* (*su*) mutant (Brink and Burnham, 1927). The *sugary* mutant also conditions a change in the carbohydrate storage products of the endosperm since the principal storage product is the water-soluble polysaccharide, phytoglycogen (Sumner and Somers, 1944), but the *su* mutation does not increase the amount of water-soluble polysaccharide in the microspores (Pfahler and Linskens, 1974). Thus two of three mutations that effect the greatest change in the nature of polysaccharide storage products exert a similar effect on the microspore and, in the case of *waxy*, on the megaspore as well.

Several mutations are known that drastically curtail the total quantity of polysaccharides synthesized and apparently do so by direct effects on polysaccharide synthesis. One might cite *brittle*-1, *brittle*-2, *shrunken*-1, and *shrunken*-2 as examples of this class with *bt1*, *bt2*, and *sh2* being particularly disruptive of starch synthesis. Yet the male gametophytes carrying these mutations are equally competitive with gametophytes carrying their nonmutant alleles, as shown by the expected Mendelian ratios in backcross and F_2 generations. This result would not be expected if the microspores were as deficient in starch content, as are endosperms homozygous for the mutant genes. In fact, it is known that *shrunken*-2 pollen has a starch content equal to that of nonshrunken pollen (Pfahler and Linskens, 1974) although the gene apparently conditions a change in the content of free amino acids in the pollen, proline being particularly enhanced in the mutant pollen grains (Pfahler and Linskens, 1970).

The effect of the *waxy* mutation on the male gametophyte has been useful in genetic investigations (Nelson, 1962, 1968, 1975), and its example has tended to obscure the fact that this is a rare situation. Mutations that condition more serious lesions in starch synthesis have no apparent effect on starch synthesis in the gametophytic generation. The explanation may be either that the pathway(s) of starch synthesis is different in the pollen than in the endosperm or that if the same pathway(s) is followed, nonallelic isozymes catalyze the same steps in the two tissues. Since several genes that have major effects on the type of polysaccharides synthesized in the endosperm have similar effects on the microspore, some common steps must be presumed. There is evidence accumulating, however, of considerable difference between the microspore and the endosperm in regard to starch synthesis (Bryce, personal communication). Sucrose synthetase, which is believed to catalyze the first step in the conversion of sucrose to starch in most tissues (Delmar and Albersheim, 1970), is not detectable in mature pollen grains, whereas invertase, which hydrolyzes sucrose, is very active. This is in marked contrast to the endosperm where invertase activity is very low (Tsai, Salamini, et al., 1970). Two other enzymes, adenosine diphosphate glucose pyrophosphorylase and the soluble adenosine diphosphate glucose–starch glucosyl transferases (Ozbun, Hawker, et al., 1971), which may be essential for rapid starch synthesis in the endosperm, are detectable but display little activity in the pollen when compared, for example, to phosphorylase, which is active in both tissues.

In addition to the necessity of nonmutant alleles at a number of loci for normal development of the endosperm, there is evidence of a necessary contribution from maternal tissue for normal endosperm development. Mangelsdorf (1926) in his survey of de mutants reported that heterozygotes of de–pl ($+/de$–pl) when selfed did not produce mutant seeds. In the F_2 generation 25% of the plants produced only mutant seed regardless of the pollen parent, so the presence of a normal allele at the de–pl locus in the embryo and the endosperm does not suffice to condition normal development. The lesion in such an instance might be either a disruption of translocation of photosynthetate, nitrogenous compounds, or salts to the developing seed or else a specific compound, a hormone, essential for development. If the latter case pertained, it would be of interest since mutagenesis experiments in higher plants do not result in the isolation of auxotrophic mutants other than thiamin-requiring mutants (Nelson and Burr, 1973). Such experiments are conducted on the assumption that competent maternal tissue cross-feeds the defective embryo and allows development to the dormant stage, permitting the investigator then to rescue the mutant plantlet by experimental manipulation. One of the reasons for repeated failures might be the inadequacy of the assumption on which the experimental plans are based. To find an obligate contribution (other than carbohydrates, simple nitrogenous compounds, and salts) by maternal tissue to the developing seed might stimulate reexamination by physiologists of the exact contribution by the sporophyte.

We know a number of loci at which mutations affect the quality or quantity of starch synthesized and whose primary effect is apparently on starch synthesis. These loci are *amylose-extender* (ae), *brittle-1* (bt), *brittle-2* ($bt2$), *dull* (du), *shrunken-1* (sh), *shrunken-2* ($sh2$), *sugary-1* (su), *sugary-2* ($su2$), and *waxy* (wx). Creech (1968) has shown the net effect on carbohydrate synthesis of substituting the mutant alleles at many of these loci either singly or in double or triple combinations. His conclusion in agreement with that of the other geneticists who have considered the effect of these mutations on starch synthesis (Kramer, Pfahler, et al., 1958; Nelson and Rines, 1962) was that multiple pathways were involved, giving rise to a most complex pattern. Although this assertion cannot be proven, it appears a reasonable conclusion from the data that there are more enzyme systems participating in starch synthesis in the cereal endosperm than most biochemists envision as essential. The true pattern of complexity will not be known until we understand the primary biochemical lesion associated with each mutant affecting starch synthesis.

As an aid to a further discussion of starch synthesizing mutants, it might be useful to review very briefly the enzymes in the endosperm capable of starch synthesis, that is, capable of catalyzing the formation of α-1,4 linkages between glucose molecules disregarding the branching enzymes involved in α-1,6 bond formation. At one time it was considered that phosphorylase was responsible for starch synthesis in plants (Porter, 1962). However, Leloir, De Fekete, et al. (1961) demonstrated the presence in starch granules isolated from developing bean cotyledons or from developing maize endosperms of a starch granule-bound enzyme that transferred glucose from the nucleoside diphosphate sugar, uridine diphosphate glucose, to the nonreducing end of starch molecules, and later with Recondo (1961), Leloir showed that adenosine diphosphate glucose was a preferred substrate for the enzyme. Frydman and Cardini (1965) subsequently reported the existence of a soluble adenosine diphosphate glucose-starch glucosyl transferase that could utilize only adenosine diphosphate glucose as a substrate, and Ozbun, Hawker, et al. (1971) reported two soluble ADPG–starch glucosyl transferases, one requiring a primer and the other capable of synthesis without a primer provided the concentrations of citrate (or acetate) and bovine serum albumin were sufficiently high in the reaction mixture. There is thus a variety of starch-synthesizing enzymes in the developing maize endosperm that utilize several different substrates. Lavintman, Tandecarz, et al. (1974) have reported two other enzymes that were detected in proplastid preparations and may well be implicated in the initiation of starch synthesis. It is certain that the enzymes utilizing nucleoside diphosphate sugars are responsible for a major portion of starch synthesis since the reactions are energetically more favorable than that catalyzed by phosphorylase for starch synthesis (Hassid, 1962) and since the mutations that reduce the synthesis of adenosine diphosphate glucose are highly disruptive of starch synthesis (Tsai and Nelson, 1966). Nevertheless, a role in starch synthesis for phosphorylase cannot be ruled out in most starch-synthesizing tissues (Badenhuizen, 1973), and DeFekete and Vieweg (1973) have ascribed a principal role to phosphorylase in starch synthesis in the bundle sheath cells in maize leaves. The relative role of phosphorylases versus glucosyl transferases or one glucosyl transferase in preference to another is obscure and probably will remain so until we detect mutants that have lost a particular activity.

In relatively few instances are we able to associate a mutation with the biochemical lesion for which it is responsible. Nelson and Rines (1962) showed that starch granules from $wx/wx/wx$ endosperms have very low starch granule-bound glucosyl transferase activity. The data on bound glucosyl transferase activity as a function of Wx dosage in diploid and tetraploid maize show a linear proportionality between gene dosage and activity such that a preparation of $Wx/Wx/Wx/Wx/Wx/Wx$ starch granules has twice the activity that is associated with a preparation of $Wx/Wx/Wx$ starch granules, but the amylose content of the starch in both types of starch granules is the same (Akatsuka and Nelson, 1969; Tsai, 1965). This makes it unlikely that the bound glucosyl transferase activity is an artifactual consequence of a soluble glucosyl transferase being bound to the amylose produced as the real consequence of action of the wx locus, as suggested by Akazawa and Murata (1965), after finding that the soluble glucosyl transferase(s) of developing rice endosperms would bind firmly to amylose but not to amylopectin. More recently Tanaka and Akazawa (1968) have concluded that the soluble and bound glucosyl transferases are different enzymes since the binding of the soluble enzyme to amylose does not alter substrate specificity so that UDPG is utilized effectively as a substrate, as is characteristic of the glucosyl transferases bound to starch granules. The soluble glucosyl transferases utilize only ADPG as a substrate.

The mutant alleles at the *shrunken*-2 and *brittle*-2 loci condition major losses in adenosine diphosphate glucose pyrophosphorylase (G–1–P + ATP $\rightleftarrows$ ADPG + PP) activity as shown by Tsai and Nelson (1966). Dickinson and Preiss (1969) reported that endosperms homozygous for either mutant gene still retained 5–10% of normal activity, and this is characteristic of all mutant alleles at these loci. This residual activity may be due to a pyrophosphorylase not coded by either of these loci (Hannah and Nelson, 1975), although it is possible that null mutations at these loci are lethal so that a selected sample of mutants has been examined.

One other mutation has recently been associated with an enzymatic lesion. Chourey (1975) has reported that *shrunken*-1 mutants have only 10% of the normal activity of sucrose synthetase. This is of interest since the Sh allele specifies a major protein band in the developing endosperm (Schwartz, 1960), and this band on elution from starch gels has sucrose synthetase activity (Chourey, 1975). The residual activity observed in mutant endosperms is associated with a minor band that migrates slightly faster than the Sh protein, and since it is observed in all sh mutants examined it is probably coded by a locus other than *shrunken*-1.

We know only this at the present time concerning the primary lesions in the various starch-synthesizing mutants, so the limited extent of our knowledge concerning the mutants affecting starch synthesis is apparent. This is particularly true for mutations at loci structural for enzymes that catalyze the formation of α-1,4 glucosidic linkages. Only *waxy* (if my interpretation is correct) actually belongs on this category, and no other mutant has been identified as coding for the presence of a phosphorylase or one of the glucosyl transferases.

To this group of α-1,4 glucan-forming enzymes can be added another starch-granule bound enzyme whose existence we have only recently detected.

Preparations of starch granules from $wx/wx/wx$ endosperms have about 10% of the activity of similar preparations from $Wx/Wx/Wx$ endosperms (Nelson and Tsai, 1964). At one time we believed that this activity could be accounted for by contaminating starch granules from the $waxy$ embryos (which are as active as those from normal embryos), but this is clearly not the case. Investigations of $waxy$ starch granules in comparison with normal starch granules have demonstrated the presence in the $waxy$ starch of a bound glucosyl transferase with a much lower K_m than the enzyme assayed in nonwaxy starch granules. The lower the substrate (ADPG) concentration at which the mutant and normal starch granules are assayed, the less the difference in activity observed between the two genotypes until at the lowest substrate levels employed (0.1 μM and 1 μM) the $waxy$ preparation appears to be as active as normal. Then incorporation relative to the normal preparation decreases rapidly with increasing substrate concentration and reaches a plateau at a much lower level than does normal (Table 2). The data in this table indicate a K_m (ADPG) of 6×10^{-5} M for the $waxy$ starch granules and 3.5×10^{-3} M for nonwaxy starch granules. This latter figure is in reasonable agreement with the

TABLE 2.   Incorporation of glucose from ^{14}C-ADPG[a] into $Wx/Wx/Wx$ and $wx/wx/wx$ endosperm starch granule preparations in 30 min at various substrate concentrations

ADPG Concentration	Glucose incorporated (pmol)	
	wx/wx/wx	Wx/Wx/Wx
0.1 μM	1.7	1.6
1.0 μM	16.7	16.8
10.0 μM	106.7	163.0
100.0 μM	429.0	1,520.0
1,000.0 μM	881.0	10,622.0
10,000.0 μM	818.0	51,460.0

[a] The reaction mixture contained 5 mg of starch granules, glycyl-glycine buffer 4 μmol, pH 8.5; 0.08 μmol EDTA; 4 μmol KCl, and the indicated concentrations of ADPG. Incubation was at 37°C for 30 min, after which the reaction was terminated by the addition of chilled methanol (70% V/V) containing 1% KCl. The starch pellet after centrifugation was resuspended in PO_4 buffer, trapped on a glass fiber filter, and counted in a Beckmen scintillation counter. All counts were corrected for the counts observed in reaction mixtures stopped immediately after addition of substrate.

previous estimate of a K_m (ADPG) for nonwaxy starch granules derived from the endosperm of 2.5×10^{-3} M (Akatsuka and Nelson, 1966). The amount of incorporation observed at these low substrate concentrations is a function of incubation time, although with a steadily decreasing velocity (Table 3). Note that at 60 min over 50% of the substrate had been incorporated.

The same amount and pattern of activity has been found for three other *waxy* mutants, B7, m1, and m8, for which this activity has been investigated. This renders unlikely the hypothesis that the activity seen in preparations of the reference *waxy* allele is due to an enzyme altered by mutation since the probability is slight that all four mutational events would have produced enzymes with greatly decreased K_m values. The *waxy* mutations have revealed a second starch granule-bound glucosyl transferase not affected by the *waxy* locus. Presumably then, $Wx/Wx/Wx$ starch granules contain this system in addition to the bound enzyme system that is dependent on the normal alleles at the *waxy* locus. It may be difficult to detect the absence of this system in the presence of the bound system associated with the *waxy* locus and thus necessary to assay double mutant stocks—homozygous for *waxy* and for the mutant one desires to investigate.

As with other aspects of starch synthesis, the role of this enzyme system won't be understood until we identify a mutant that does not possess the activity. We do not find on the basis of double-mutant assays that the *amylose-extender* affects the activity of this enzyme. Its great affinity for the substrate might indicate a role in starch synthesis early in the endosperm development when the quantity of ADPG is known to be very small (Tsai, Salamin, et al., 1970). It could possibly act to synthesize primer molecules which other starch-synthesizing enzymes would elongate although other candidates have been proposed for this role such as phosphorylase (Tsai and Nelson, 1968) and the

TABLE 3.   The incorporation of glucose from ^{14}C-ADPG into *Wx/Wx/Wx* and *wx/wx/wx* endosperm starch granule preparations over 90 min in a reaction mixture containing 40 pmol ($/\mu M$) of ADPG[a]

Time of incubation (min)	Glucose incorporated (pmol)	
	wx/wx/wx	Wx/Wx/Wx
5	6.1	6.9
10	8.2	10.9
20	13.3	14.3
30	15.2	15.4
60	21.6	21.6
90	23.4	24.7

[a] The conditions of the assay are as stated in Table 1.

soluble ADPG–starch glucosyl transferase, which can initiate starch synthesis under certain stated conditions (Ozbun, Hawker, et al., 1971). Even more promising candidates for the initiation of starch synthesis are the proplasted enzymes reported by Lavintman, Tandecarz, et al. (1974) that transfer glucose from UDP glucose to protein acceptor molecules. These enzymes have very low K_m values (UDPG) as does the enzyme reported here, and all may have a role early in the development of the starch granule.

It is interesting and probably significant that with three enzymes concerned with starch synthesis, mutations that affect enzymatic activity have in two instances allowed the detection of less active nonallelic isoenzymes (*waxy* and *shrunken*-1). This may well be the case also with ADPG pyrophosphorylase (*brittle*-2 and *shrunken*-2), but this is not certain. These instances illustrate the complexities of attempting to identify all of the reactions contributing to starch synthesis in nonmutant endosperms, not to mention their relative importance.

Although starch is the greatest component quantitatively of the endosperm, the proteins collectively may constitute 7–15% of the endosperm weight, depending on genotype and environmental circumstances. Attention has recently been focused on the proteins and genes that affect their relative proportions since some of these mutants (*opaque*-2, *floury*-2, and *opaque*-7) markedly change the amino-acid profile of the collective endosperm proteins in ways that greatly enhance the nutritive value of the corn for humans and monogastic animals (Mertz, Bates, et al., 1964; Nelson, Mertz, et al., 1965; McWhirter, 1971). Specifically, these changes are increases in the lysine, tryptophan, aspartate, and glycine content and decreases in the content of leucine, alanine, and glutamate. The net effect of gene substitutions at these loci is understood as consisting of a partial repression of synthesis of the prolamine (alcohol-soluble) fraction and, apparently as a secondary response, a large increase in free amino-acid content as well as in the water- and salt-soluble proteins (Jimenez, 1968; Misra, Jambunathan, et al., 1972). Identical patterns of changes in the solubility fractions of the endosperm proteins and in amino-acid content are found in the high lysine mutants of barley and sorghum (Ingversen, Køie, et al., 1973; Nelson, 1976; Axtell, in press). The single possible exception to the generalization that substantial changes in amino-acid patterns of cereal grains arise only by partial suppression of prolamine synthesis and a compensatory oversynthesis of other proteins of quite different overall amino-acid composition is the *hiproly* mutant of barley (Munck, 1972). The apparent basis of the higher lysine content is greater synthesis in the mutant of several lysine-rich proteins that are produced in lesser quantities in nonmutant endosperms. The results of analysis of this mutant, the analyses from the other barley mutant, *Risø 1508* (Ingversen, Koie, et al., 1973), the sorghum mutant (*hl*) and the maize mutants, support the hypothesis that the most probable route to altered amino-acid composition of cereal seeds will be changes in the proportions in which proteins are normally produced. At the same time, further attention is being given to the possibility that increases in the contents of limiting essential amino acids could be achieved by freeing the biosynthetic pathway for an amino acid from the regulating mechanism(s) that limit production of that amino acid to the quantities used in protein synthesis and other reactions that may utilize that particular amino acid (Nelson, 1969). Recently Green and Phillips (1974) have devised a selection system designed to detect mutants in the cereals that are overproducers of lysine, threonine, or methionine, and it should be possible in the near future to test the feasibility of this approach.

In spite of reasonably complete data covering the net effect of genes such as *opaque*-2, *opaque*-7, and *floury*-2 on the endosperm proteins, we do not yet understand in any instance the primary biochemical lesion responsible for the effect. There are, in fact, relatively few investigations of protein synthesis in cereal endosperms. In this respect the findings of Ben and Francis Burr (1976) constitute an important advance in our understanding of the system. The protein bodies that are the site of zein deposition of the developing endosperm have polyribosomes associated with them as electron micrographs reveal. The polyribosomes can be stripped from the isolated protein bodies by detergent treatment. Such polyribosomes actively synthesize protein if supplied with a soluble fraction that contains the transfer RNAs and amino–acyl synthetases. The product of the reaction is not separable from zein synthesized *in vivo* by SDS gel electrophoresis. This is, to my knowledge, the first isolation of an apparently pure messenger RNA from a higher plant.

There is thus a unique apparatus allowing synthesis of zein at the site of its deposition, and as the Burrs (1976) note, a number of genes may well be necessary for this to take place, considering that several genes disrupt the process. A multigenic supporting system for zein synthesis would also account for the observations that double mutants (*o2/o2*; *o7/o7* or *o2/o2*; *fl2/fl2*) do not give increased lysine contents over the single mutants (Misra, Mertz, et al., 1975).

An intriguing aspect of some mutants that affect the synthesis of the zein component is the extension of an effect to the sporophytic generation. The mutant, *opaque*-6, is an example, as are several other nonallelic mutants that have not been as well investigated. The effect of *opaque*-6 on endosperm protein synthesis is very similar to the effect of *opaque*-2—for example, pronounced reduction of the prolamine fraction, an increase in the free amino acids, albumins, and globulins, and either no change or a slight decrease in the glutelin content (Ma and Nelson, 1975). Consequently, the amino-acid pattern of *o6/o6/o6* endosperms is similar to those of *o2/o2/o2* or *o7/o7/o7* endosperms. Yet *o6/o6* plants die as seedlings. Since the storage proteins presumably serve only as a readily available source of reduced nitrogen for the young plant once germination is initiated, the lethality of a mutant affecting only storage protein synthesis is difficult to explain. We could be observing the effects of a mutation affecting a function vital for sporophytic development, but not endosperm development. However, the lack of this function in the endosperm might condition the alteration in the proportions of endosperm proteins synthesized. It is also possible that by changing the proportions of the solubility fractions (and the proportions of various proteins within each solubility class), the mutation may depress the synthesis of an albumin or globulin essential during early stages of plant growth but not synthesized *de novo* during germination. A number of high-lysine mutants in sorghum have also proven to be seedling lethals (Axtell, personal communication).

Mutants are also known that reduce the synthesis of all classes of endosperm proteins. The *shrunken*-4 mutant has a low content of water- and salt-soluble proteins. Although this mutant has much reduced phosphorylase activity, this is not the primary lesion in the mutant tissue as initially believed (Tsai and Nelson, 1969). Burr and Nelson (1973) have shown that the mutant endosperms contain about $\frac{1}{8}$ as much B6, pyridoxal phosphate, as do normal endosperms. Since pyridoxal phosphate is a cofactor for a number of the other enzymes in addition to phosphorylase, the reduced quantity in the mutant could well account for the lower protein content. Foard, Ma, et al. (1974) have also reported two other mutants, *de**-91 and *de**-92, in which both the soluble proteins and the storage proteins are much reduced. No basis is known for these abnormally low protein contents. The plants grown from mutant seeds are normal.

In discussing our knowledge of genic control of endosperm development, I have endeavored to emphasize two points. The first is our relatively limited knowledge of both starch and protein synthesis in the endosperm although these are areas of considerable practical importance. Biochemical genetics of the endosperm, which is a unique and highly specialized tissue, is in its infancy. We do have large numbers of mutants that affect endosperm development and can serve as experimental probes to facilitate coordinated biochemical and genetical studies of the endosperm. The second point is our inability to make a physiologically coherent interpretation of the mutations affecting endosperm growth in which gametophyte or sporophyte development is or is not simultaneously affected. Apparently, no mutation affecting the quantity of starch synthesized in the endosperm has a parallel effect on either the gametophyte or the sporophyte, whereas the two genes influencing the relative proportions of amylose to amylopectin (*wx* and *ae*) have the same effect on the gametophyte, but not the sporophyte. The mutations that we know to affect protein synthesis in the endosperm apparently do not have an effect on the gametophyte and may or may not affect the sporophyte. The noncorrespondence can be explained formally on the basis of either different pathways of biosynthesis in the endosperm as contrasted to the gametophyte or sporophyte or nonallelic isoenzymes catalyzing the same reaction in the endosperm and the gametophyte or sporophyte generation. There is some evidence that both explanations are valid in different cases. More detailed information concerning the reactions governed by loci whose mutant effects either are or are not extended to the gametophyte or sporophyte might well prove illuminating to our understanding of how the genic control of specialized functions evolves and functions.

REFERENCES

Akatsuka, T. and O. E. Nelson. 1966. Starch granule-bound adenosine diphosphate glucose-starch glucosyl transferases of maize seeds. *J. Biol. Chem.* **241:** 2280–2286.

Akatsuka, T. and O. E. Nelson. 1969. Studies on starch synthesis in maize mutants. *J. Jap. Soc. Starch Sci.* **17:** 99–115.

Akazawa, T. and T. Murata. 1965. Adsorption of ADPG-starch transglucosylase by amylose. *Biochem. Biophys. Res. Commun.* **19:** 21–26.

Axtell, J. D. (In press). Naturally occurring and induced genotypes of high lysine sorghum. Paper presented at the Third Research Coordinated Meeting of the FAO/IAEA/GSF Seed Protein Improvement Program. Hahnenklee, German Federal Republic, Vienna: Internat. Atomic Energy Agency.

Badenhuizen, N. P. 1973. Fundamental problems in the biosynthesis of starch granules. *Ann. N. Y. Acad. Sci.* **210:** 11–16.

Banks, W., C. T. Greenwood, D. D. Muir, and J. T. Walker. 1971. Pollen starch from amylomaize. *Die Stärke* **23:** 380–382.

Brink, R. A. 1925. Mendelian ratios and the gametophyte generation in angiosperms. *Genetics* **10:** 359–394.

Brink, R. A. and C. R. Burnham. 1927. Differential action of the sugary gene in maize on two alternative classes of male gametophytes. *Genetics* **12:** 348–378.

Brink, R. A. and D. C. Cooper. 1947. The endosperm in seed development. *Bot. Rev.* **13:** 423–541.

Brink, R. A. and J. H. MacGillivray. 1924. Segregation for the waxy character in maize pollen and differential development of the male gametophyte. *Am. J. Bot.* **11:** 465–469.

Burr, B. and F. A. Burr. 1976. Zein synthesis in maize endosperm by polyribosomes attached to protein bodies. *Proc. Nat. Acad. Sci. U. S. A.* **73:**515–519.

Burr, B. and O. E. Nelson. 1973. The phosphorylases of developing maize seeds. *Ann. N. Y. Acad. Sci.* **210:** 129–138.

Chourey, P. S. 1975, Sucrose synthetase in *Sh* and *sh* endosperms. *Maize Genet. Coop. Newsl.* **49:** 161–162.

Creech, R. G. 1968. Carbohydrate synthesis in maize. *Adv. Agron.* **20:** 275–322.

De Fekete, M. A. R. and G. Vieweg. 1973. The role of phosphorylase in the metabolism of starch. *Ann. N. Y. Acad. Sci.* **210:** 170–180.

Delmar, D. and P. Albersheim. 1970. The biosynthesis of sucrose and nucleoside diphosphate glucoses in *Phaseolus aureus. Plant Physiol.* **45:** 782–786.

Demerec, M. 1924. A case of pollen dimorphism in maize. *Am. J. Bot.* **11:** 461–464.

Dickinson, D. B. and J. Preiss. 1969. Presence of ADP-glucose pyrophosphorylase activity in *shrunken-2* and *brittle-2* mutants of maize endosperm. *Plant Physiol.* **44:** 1058–1062.

Foard, D., Y. Ma, and O. E. Nelson. 1974. The mutations *de**-91 and *de**-92. *Maize Genet. Coop. Newsl.* **48:** 169–172.

Frydman, R. B. and C. E. Cardini. 1964. Soluble enzymes related to starch synthesis. *Biochem. Biophys. Res. Commun.* **17:** 407–411.

Green, C. E. and R. L. Phillips. 1974. Potential selection system for mutants with increased lysine, threonine, and methionine in cereal crops. *Crop Sci.* **14:** 827–830.

Hannah, L. C. and O. E. Nelson. 1975. Characterization of adenosine diphosphate glucose pyrophosphorylases from developing maize seeds. *Plant Physiol.* **55:** 297–302.

Hassid, W. Z. 1969. Biosynthesis of oligo-saccharides and polysaccharides in plants. *Science* **165:**137–144.

Ingversen, J., B. Koie, and H. Doll. 1973. Induced seed protein mutant of barley. *Experentia* **29:** 1151–1152.

Jimenez, J. R. 1968. The effect of the *opaque-2* and *floury-2* genes on the production of protein in maize endosperm. Ph.D. thesis. West Lafayette, Ind.: Purdue University.

Kramer, H. H., P. L. Pfahler, and R. L. Whistler. 1958. Gene interactions in maize affecting endosperm properties. *Agron. J.* **50:** 207–210.

Lavintman, N., J. Tandecarz, M. Carceller, and C. E. Cardini. 1974. Role of UDP-glucose in the biosynthesis of starch. Mechanism of formation of a glucoproteic acceptor. *Eur. J. Biochem.* **50:** 145–155.

Leloir, L. F., M. A. R. DeFekete, and C. E. Cardini. 1961. Starch and oligosaccharide synthesis from uridine diphosphate glucose. *J. Biol. Chem.* **236:** 636–641.

Ma, Y. and O. E. Nelson. 1975. Amino acid composition and storage proteins in two new high-lysine mutants of maize. *Cereal Chem.* **52:** 412–419.

Maheshwari, P. 1954. *An introduction to the embryology of angiosperms.* New York: McGraw Hill, pp. 221–267.

Mangelsdorf, P. C. 1926. The genetics and morphology of some endosperm characters in maize. *Conn. Agr. Exp. Sta. Bull.* No. 279, p. 513.

McWhirter, K. S. 1971. A floury endosperm, high lysine locus on chromosome 10. *Maize Genet. Coop. Newsl.* **45:** 184.

Mertz, E. T., L. S. Bates, and O. E. Nelson. 1964. Mutant gene that changes protein composition and increases lysine content of maize endosperm. *Science* **145:** 279–280.

Misra, P. S., R. Jambunathan, E. T. Mertz, D. V. Glover, H. M. Barbosa, K. S. McWhirter. 1972. Endosperm protein synthesis in high-lysine maize mutants. *Science* **176:** 1425–1427.

Misra, P. S., E. T. Mertz, B. V. Glover. 1975. Characteristics of proteins in single and double endosperm mutants of maize. In *High quality protein maize. Proc. CIMMYT–Purdue Symp. on Protein Quality in Maize,* El Batan, Mexico, 1975. Stroudsburg, Penn.: Dowden, Hutchinson and Ross.

Munck, L. 1972. Improvement of nutritional values in cereals. *Hereditas* **72:** 1–128.

Nelson, O. E. 1962. The *waxy* locus in maize. I. Intralocus recombination frequency estimates by pollen and by conventional analysis. *Genetics* **47:** 737–742.

Nelson, O. E. 1968. The *waxy* locus in maize. II. The location of the controlling element alleles. *Genetics* **60:** 507–524.

Nelson, O. E. 1969. Genetic modification of protein quality in plants. *Adv. Agron.* **21:** 171–194.

Nelson, O. E. 1976. Interpretive summary and review. In *Genetic improvement of seed proteins.* Proceedings of a workshop, March, 1974. Washington, D. C.: Board of Agriculture and Renewable Resources, National Research Council.

Nelson, O. E. 1975. The *waxy* locus in maize. III. Effect of structural heterozygosity on intragenic recombination and flanking marker assortment. *Genetics* **79:** 31–44.

Nelson, O. E. and B. Burr. 1973. Biochemical genetics of higher plants. *Annu. Rev. Plant Physiol.* **24:** 493–518.

Nelson, O. E., E. T. Mertz, and L. S. Bates. 1965. Second mutant gene affecting the amino acid pattern of maize endosperm proteins. *Science* **150:** 1469–1470.

Nelson, O. E. and H. W. Rines. 1962. The enzymatic deficiency in the *waxy* mutant of maize. *Biochem. Biophys. Res. Commun.* **9:** 297–300.

Nelson, O. E. and C. Y. Tsai. 1964. Glucose transfer from adenosine diphosphate glucose to starch in preparations of waxy seeds. *Science* **145:** 1194–1195.

Ozbun, J. L., J. S. Hawker, and J. Preiss. 1971. Adenosine diphosphoglucose–starch glucosyl transferases from developing kernels of waxy maize. *Plant Physiol.* **48:** 765–769.

Pfahler, P. L. and H. F. Linskens. 1970. Biochemical composition of maize (*Zea mays* L.) pollen. I. Effects of the endosperm mutants, waxy (*wx*), shrunken (*sh2*) and sugary (*su*) on the amino acid content and fatty acid distribution. *Theor. Appl. Genet.* **40:** 6–10.

Pfahler, P. L. and H. F. Linskens. 1974. Biochemical composition of maize (*Zea mays* L.) pollen. II. Effects of the endosperm mutants, waxy (*wx*), shrunken (*sh2*) and sugary (*su*) on the carbohydrate and lipid percentage. *Theor. Appl. Genet.* **41:** 2–4.

Porter, H. K. 1962. Synthesis of polysaccharides of higher plants. *Ann. Rev. Plant Physiol.* **13:** 303–328.

Preiss, J., C. Hammel, and A. Sabraw. 1971. A unique adenosine diphosphoglucose pyrophosphorylase associated with maize embryo tissue. *Plant Physiol.* **47:** 104–108.

Recondo, E. and L. F. Leloir. 1961. Adenosine diphosphate glucose and starch synthesis. *Biochem. Biophys. Res. Commun.* **6:** 85–88.

Salamini, F., C. Y. Tsai, and O. E. Nelson. 1972. Multiple forms of glucose–phosphate isomerase in maize. *Plant Physiol.* **50:** 256–261.

Schwartz, D. 1960. Electrophoretic and immunochemical studies with endosperm proteins of maize mutants. *Genetics* **45:** 1419–1427.

Sumner, J. B. and G. F. Somers. 1944. The water-soluble polysaccharides of sweet corn. *Arch. Biochem.* **4:** 7–9.

Tanaka, Y. and T. Akazawa. 1968. Substrate specificity of the granule-bound chloroplastic starch synthetase. *Plant Cell Physiol.* **9:** 405–410.

Tsai, C. Y. 1965. Correlation of enzymatic activity with *Wx* dosage. *Maize Genet. Coop. Newsl.* **39:** 153–156.

Tsai, C. Y. and O. E. Nelson. 1966. Starch-deficient maize mutant lacking adenosine diphosphate glucose pyrophosphorylase activity. *Science* **151:** 341–343.

Tsai, C. Y. and O. E. Nelson. 1968. Phosphorylases I and II of maize endosperm. *Plant Physiol.* **43:** 103–112.

Tsai, C. Y. and O. E. Nelson. 1969. Mutations at the *shrunken*-4 locus in maize that produce three altered phosphorylases. *Genetics* **61:** 813–821.

Tsai, C. Y., F. Salamini, and O. E. Nelson. 1970. Enzyme of carbohydrate metabolism in the developing endosperm of maize. *Plant Physiol.* **46:** 299–306.

Chapter 27

SOME BIOCHEMICAL INDICATORS OF GENETIC AND DEVELOPMENTAL CONTROLS IN ENDOSPERM

Curtis M. Wilson

Research Chemist, North Central Region,
USDA Agricultural Research Service and
Department of Agronomy,
University of Illinois,
Urbana

The developing seed is a great natural storehouse for energy from the sun. The corn plant (*Zea mays* L.), with its C4 photosynthetic system, is one of the most efficient converters of sunlight into products with stored energy. Increased emphasis on biochemical studies of developing seeds should lead to identification of genetic and developmental traits which will produce improved nutritional quality, larger seeds, and higher yields.

Many developmental studies include the measurement of the starch, protein, or enzyme contents of the whole endosperm. Researchers have long known that the endosperm of corn is not homogeneous at harvest, and that it does not

Cooperative investigations of the Agricultural Research Service, United States Department of Agriculture, and the Department of Agronomy, University of Illinois, Urbana–Champaign.

develop uniformly. Although the first stages of differentiation seem to begin in the basal region (Duvick, 1955), subsequent differentiation processes—such as starch synthesis (Shannon, 1974, and references therein), protein body formation (Duvick, 1961), and storage of ^{14}C from photosynthetic products (Shannon, 1974)—begin near the apex of the endosperm and proceed downward. Storage of starch and protein is maximized when these processes are coordinated such that the endosperm cells are successively filled to capacity from top to bottom. The activity stops when the maternal plant has ceased photosynthesis and the process of senescence has disassembled most of the leaf cellular contents for transport into the seed. The entire process must be accomplished during the available growing period. If the timing is not coordinated, the seed may be filled before the maternal plant has lost its capacity to supply nutrients; alternatively, the kernels may be immature at frost. These controls are ultimately determined by the genetic nature of the plant, but the operation of the controls depends on enzymes and hormones. We enter the area of uncertainty with these subjects.

Brown and Robinson (1955) presented a provocative outline of developmental metabolic changes that I have adapted and enlarged to illustrate the development of corn endosperm (Figure 1). The four, somewhat arbitrarily selected, time periods or metabolic states are identified by certain characteristic reactions. Through the first 15 days the endosperm becomes coenocytic; then cell walls form and growth is largely by nuclear and cell division. The genes controlling these processes have been turned on by some as yet unknown reaction. These active genes (G_1) specify that a cell may synthesize only certain of

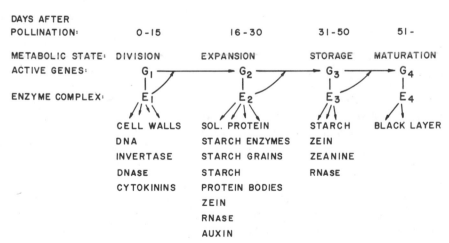

Figure 1. Sequential metabolic states during corn-endosperm development (see text for details).

its many potential enzymes, which make up the measurable enzyme complex E_1. For example, Tsai, Salamini, et al. (1970) reported that invertase activity peaks at 12 days after pollination (DAP). Another enzyme found at this time is deoxyribonuclease (DNase) (C. Wilson unpublished). E_1 produces DNA, cell walls, and the many membranes that go into the cell structures. At this time the endosperm also contains maximum levels of zeatin (Miller, 1967), which may control cell division.

The metabolic state of a cell is not fixed; with time it is changed by the products of certain enzymes and by the turning on or off of selected genes. Thus, the endosperm reaches the state characterized as a period of cell and tissue expansion. The active genes (G_2) now begin to specify a rapid increase in the many enzymes (E_2) involved in the synthesis of storage products. The enzymes involved in starch synthesis are the best known (Ozbun, Hawker, et al., 1973; Tsai, Salamini, et al., 1970); they apparently reach a peak during this time. However, the usual techniques inefficiently extract ribonuclease (RNase) from older endosperms (C. Wilson, unpublished), and the same is probably true for other enzymes. The endosperm is also changed by the appearance of bodies specialized for the storage of starch and protein, and these bodies enlarge as starch and zein synthesis begin (Duvick, 1961; Shannon, 1974). RNase also increases rapidly while DNase decreases (C. Wilson, unpublished). Auxin content is high at this time (Hinsvark, Hourff, et al., 1954).

The period of continued rapid synthesis of the major storage products, starch, zein, and certain other proteins such as zeanine, is reached next (G_3). The enzymes that synthesize these products plus appropriate metabolic respiratory enzymes that provide energy are present. At the final or maturation stage genes (G_4) are turned on which specify the enzyme complex, E_4, which produces dark substances—possibly polyphenols—that seal off the base of the endosperm. The seed dries and retains viability for some years until conditions are suitable for germination.

The timing of the above events varies among varieties, and also from crown to base in each endosperm. The enzyme complex of each metabolic state must initiate the reactions that change that state to the next. We may think that the way to increase the amount of a certain compound is to increase the activity of the enzyme that synthesizes it. This scheme suggests that the rate of change from one metabolic state to another is also important in determining the final size of a tissue or seed or the total amount of an enzyme product. If we wished to increase a substance synthesized most rapidly in metabolic state 2, we could select for an increased level of the appropriate enzymes, or for a reaction which inhibits the change from G_2 to G_3. We need to remember that timing controls are applied to individual cells, and our gross analyses, usually made on large masses of cells representing several metabolic states, do not differentiate at that level. Most importantly, then, controls must be exerted not just on levels of enzymes, but on the time period over which key enzymes are functioning at an optimum level.

Moureaux and Landry (1972a) examined the rates at which various proteins were synthesized in normal (O2) and *opaque-2* (o2) (1972b) kernels (Figure 2). Soluble proteins were synthesized most rapidly during the early stages of growth, the zeanine fraction of the glutelins at a later stage, and zein for a long period overlapping the periods for both of them. The *opaque-2* mutation delayed the appearance of zein, lowered its rate of synthesis, and shortened its production period. *Opaque-2* showed increases over normal for salt-soluble protein and zeanine during the later stages of development.

Dalby and Tsai (1974) found that zein was accumulated rapidly between 16 and 40 DAP in normal kernels of four inbreds. In the *opaque-2* versions of these inbreds, the increase was slower and zein synthesis ceased at about 30 DAP. A modified opaque-2 version of B14 showed an intermediate rate of zein accumulation that took as long to reach a final level as it did in the normal kernel. Zein accumulation was much higher in the modified *opaque-2* than in

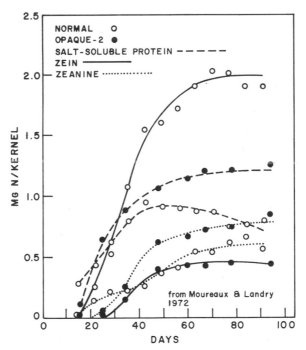

Figure 2. Content per grain of three protein fractions during kernel development (Moureaux and Landry, 1972a,b).

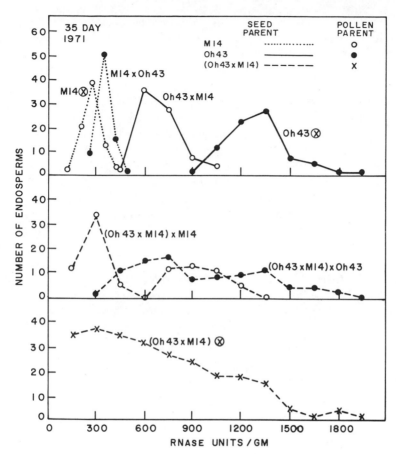

Figure 3. RNase Distribution among endosperms of two inbreds, their reciprocal crosses, and the selfed and backcrossed F_2 generation (Apel, 1972).

the unmodified opaque-2. When the *opaque*-2 inbreds were crossed, the importance of timing was illustrated. The rate of zein accumulation was similar to that in the inbred parents, but the shut-off point was not reached until about 40 DAP; thus, the total zein content was higher in the hybrid. Tsai and Dalby (1974) reported that the zein percentage declined in an *opaque*-2 inbred when the dry weight continued to increase after zein synthesis stopped.

In studies on RNase in corn, I have assumed that this enzyme is involved in the control of protein synthesis, although this function has not been demonstrated for RNase in any organism. Three major types of RNA are known (messenger RNA, transfer RNA, and ribosomal RNA), and different RNases may have different specificities for these classes of substrates. Although corn roots contain three types of enzymes that degrade RNA (Wilson, 1968), only RNase I is found in the later stages of endosperm development (Wilson, 1973). RNase I is a stable enzyme that is readily extracted from mature corn endosperms in the presence of high salt concentrations, so it is a convenient subject for the study of genetic and developmental controls on enzyme activity.

Table 1 shows the wide range of RNase levels found in the endosperms of several inbreds and their *opaque*-2 versions (Apel, 1972). The *opaque*-2 endosperms contained from 2 to 6 times more RNase than did the normals.

Apel (1972) determined the frequency distribution of RNase concentrations in individual endosperms of the corn inbreds M14 and Oh43, the reciprocal hybrids, and the selfed and backcrossed F_2 generation at 35 DAP (Figure 3). He tested the hypothesis that the difference in RNase levels between the two inbreds might be controlled by two alleles at one locus, but the statistical analyses were inconclusive. Modifier genes or the environment may affect the

TABLE 1.   RNase Contents in corn endosperm 50 days after pollination. The data are averages of individual analyses on 20 endosperms from each of three ears (Apel, 1972)

Inbred	RNase, units/g		o2/N
	Normal	Opaque-2	
M14	218	1065	4.9
Oh45B	280	1430	5.1
N28	295	680	2.3
R803	392		
W64A	419	1509	3.6
R801	461	1768	3.8
Oh7A	602	2925	4.9
WF9	792		
B37	796		
WF9[HO]	797	1287	1.6
B57	839	2333	2.8
Oh43	840	3708	4.4
R802	862	5075	5.9
B14A	993		
Oh7N	1337	3569	2.7
High/low	6.1	7.5	

$$\frac{\text{High } opaque\text{-}2}{\text{Low normal}} = 23.2$$

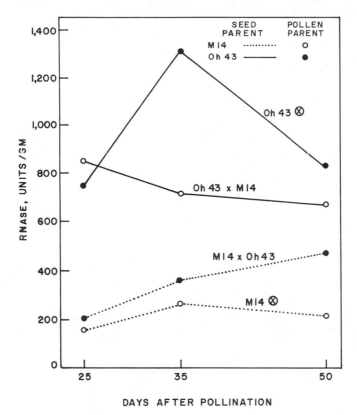

Figure 4. Changing RNase concentrations in developing endosperms of two corn inbreds and their reciprocal crosses in 1971 (Apel, 1972).

RNase content. One obvious difference is that the F_2 and the backcross seeds are borne on vigorous hybrid plants. Figures 4 and 5 show that the relative RNase levels change during development; at 18–25 DAP the Oh43 endosperm pollinated by M14 actually had a higher level of RNase than did the selfed Oh43, though the latter reached a higher level at 35–38 DAP. This pattern has occurred repeatedly. The reciprocal cross, M14xOh43, had low RNase at 20 and 25 DAP but almost equalled Oh43xM14 at 50 DAP (Figures 4 and 5). The rates of RNase synthesis and the times of maximum RNase varied so widely among the inbreds and hybrids that a simple genetic analysis was not possible.

Crosses between M14 and N28, two low-RNase inbreds, produced no surprises, and both inbreds and higher RNase levels when pollinated by the high

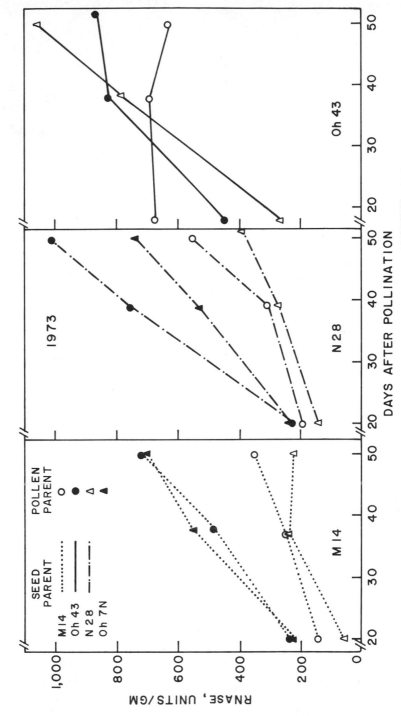

Figure 5. Effect of pollen parent on the RNase concentration in developing endosperms of three inbreds in 1973.

RNase inbreds Oh43 and Oh7N (Figure 5). The influence of N28 pollen on Oh43 endosperm RNase levels appeared to be opposite to that of M14 pollen.

As mentioned earlier, the processes of development start in the crown and proceed toward the base. When endosperms were divided into three parts, RNase levels were high in the endosperm crown at 20 days and only 10% as high in the base (Table 2). The content in the middle and basal regions increased with age and approached that seen earlier in the crown. Distribution was similar whether the overall RNase level was high or low and whether it was in *opaque*-2 or normal endosperms. It seems likely that similar differences in activity will be found for other enzymes, such as those involved in starch synthesis, although the timing may be different.

Ribonuclease is only one of a number of buffer-soluble proteins (albumins

TABLE 2.   RNase Distribution in endosperm as a function of kernel age

(1972)[a]	Oh43			
Age (days)	20	30	40	50
Crown	936	813	650	1065
Middle	398	623	485	670
Base	141	284	595	1170
Whole	466	594	580	910

(1973)[a]	Oh43		Oh43xM14		Oh43xN28	
Age (days)	18	38	18	38	18	38
Crown	756	1174	1033	793	504	1117
Middle	461	797	442	738	254	1026
Base	66	505	179	534	51	647
Whole	450	847	590	651	281	929

(1974)[b]	R802 +/+/+		R802 o2/o2/o2	
Age (days)	15	20	15	20
Crown half	688	1096	1085	1600
Basal half	89	360	202	344
Whole	331	726	540	920

[a] Units/g fresh weight, average of individual assays on parts from three endosperms.

[b] Units/g fresh weight, eight endosperm halves assayed together at 15 days, four halves at 20 days.

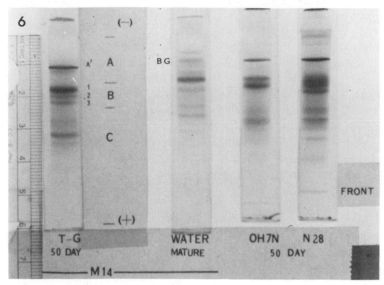

Figure 6. Polyacrylamide gel electrophoresis of buffer soluble proteins from endosperms of three inbreds. Single endosperms, 50 DAP, were homogenized in pH 8.3 tris–glycine (T–G) buffer, centrifuged, and applied to the gel. A mature M14 endosperm was homogenized with water. Symbol BG marks a blue–green band. The proteins moved from top to bottom.

and globulins) present in corn endosperms. Polyacrylamide gel electrophoresis (PAGE) is a powerful, relatively simple procedure for rapid qualitative analysis of protein mixtures. Details of my methods are presented elsewhere, but the basic procedure is that of the pH 8.9 disc gel system of Davis (1964). The proteins shown in Figure 6 are albumins and globulins. The relative intensities of three protein bands, termed B1, B2, and B3, are under genetic control in developing corn endosperms. B1 is a prominent band in all inbreds. The M14 pattern, (B1 > B2 = B3) is also found in Oh43, R803, Oh45B, B14A, and R801. In a second group of inbreds, including W64A and B57 and represented here by Oh7N, bands B1 and B2 have nearly equal intensity (B1 = B2). The inbred N28 shows equal B bands (B1 = B2 = B3), but visual inspection suggests that B1 and B2 are actually multiple bands. Some inbreds have a pattern of B1 > B3 > B2, but this may be just a variation of the first pattern.

The A region of 50-day extracts contains a unique band labelled A′, which is very thin and intense. In mature extracts the A′ band is usually not seen, but a new band that stains blue-green with Amido black appears near its position.

The C-region patterns differ slightly among inbreds, but the bands are less intense and the relationships have not been studied.

When two inbreds are crossed, the B-band pattern resembles that of the maternal parent much more than would be expected from the gene dosage level. Usually PAGE cannot distinguish the hybrid from its maternal parent (Figure 7). However, PAGE technique is not quantitative, and we cannot yet conclude that maternal dominance is complete. The inbred patterns are less distinct in mature seeds (Figure 8), though the B1 = B2 pattern remains distinct. Most of the extracts studied were from endosperms that had been harvested at 50 days and stored at -15°C.

Characteristically, the endosperms of *opaque*-2 varieties contain a high level of soluble proteins that can be analyzed by PAGE. Figure 9 shows split-gel comparisons of nearly equal amounts of the soluble proteins of single phenotypically normal and *opaque*-2 endosperms taken from a single ear of each of several inbreds that had been pollinated with a mixture of normal and *opaque*-2 pollen. (WF9 samples were taken from separate ears.) The patterns for the heterozygous normal endosperms were indistinguishable from those for the homozygous normal. In most *opaque*-2 inbreds, one or both of the B2 and B3 bands were greatly reduced or eliminated. Dalby, Cagampang, et al. (1972) noted the loss of what I call the B2 band from W64A-o2. However, the loss of bands is not the only response, for Oh45B responded to the *opaque*-2 gene by

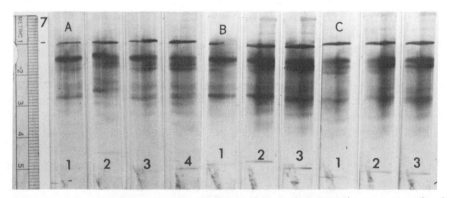

Figure 7. Split-gel polyacrylamide gel electrophoresis (two samples on one gel) of soluble proteins from inbreds and reciprocal crosses, with pollen parent listed second: A: 1—M14 × M14 and M14 × N28, 2—M14 × M14 and N28 × N28, 3—M14 × N28 and N28 × M14, 4—N28 × M14 and N28 × N28; B: 1—Oh43 × Oh43 and Oh43 × Oh7N, 2—Oh43 × Oh43 and Oh7N × Oh7N, 3—Oh7N × Oh43 and Oh7N × Oh7N; C: 1—N28 × N28 and N28 × Oh7N, 2—N28 × N28 and Oh7N × Oh7N, 3—Oh7N × N28 and Oh7N × Oh7N. All samples were 50-day endosperms except Oh43 × Oh7N on gel B1, which was a mature sample.

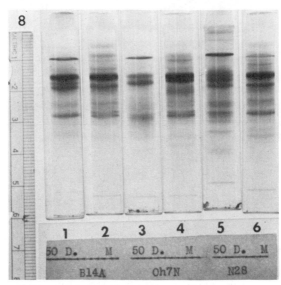

Figure 8. Comparative PAGE of 50-day (50 D.) and mature (M) endosperm proteins from three inbreds.

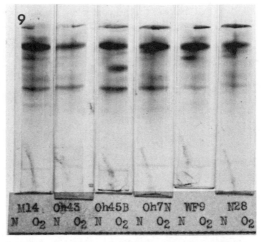

Figure 9. Split-gel PAGE of soluble endosperm proteins from normal ($o_2/o_2/+$) and opaque-2 ($o_2/o_2/o_2$) versions of six inbreds.

producing a new protein in the C region and increasing the relative amounts of the B2 protein and another C protein. We might conclude that the *opaque*-2 gene does not directly affect any of the soluble proteins studied here, but that each inbred may respond by increasing or decreasing the relative proportions of certain proteins.

The protein patterns of the B bands are not constant throughout development. Figure 10 shows a sequence of M14 endosperms sampled from 20 days through maturity. The top half was extracted separately from the base. At 37 days the expected B1 > B2 = B3 pattern showed for the top, and the base resembled N28 with B1 = B2 = B3. With increasing maturity, the B2 and B3 bands became less prominent. The B1 protein may be a storage protein, whereas B2 and B3 may be metabolically functional during development and either become insoluble or are digested during the final stages of maturation.

Dalby, Cagampang, et al. (1970, 1972) proposed, from their data on the rates of total endosperm RNase synthesis and zein synthesis, that the *opaque*-2 gene becomes inactive during the third week after pollination. Later work of Dalby and Tsai (1974), as well as the data of Moureaux and Landry (1972a,b, and Figure 2), show that zein synthesis may be shut off at 30 DAP in opaque-2 endosperms rather than at 40 DAP, as in the normals. Studies on the incorporation of ^{14}C-leucine and ^{14}C-lysine into endosperm protein showed that these processes and the reaction by which lysine was converted into other amino acids differed between normals and opaques throughout development

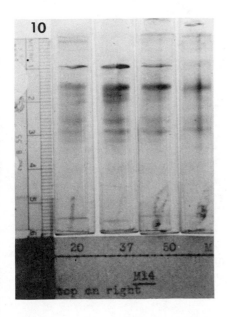

Figure 10. Differences in PAGE patterns of inbred M14 throughout development. Endosperms were harvested at 20, 37, and 50 days postpollination and at maturity. Each endosperm was cut in half; extract from top half was placed on the right side of a split gel, and extract from bottom half was placed on left side.

(Sodek and Wilson, 1970). The nonuniform distribution of protein bodies (Duvick, 1961) and RNase (Table 2) at early stages of development shows that the regions of high zein and RNase levels in normal endosperms are not always the regions of most rapid synthesis of these two proteins. Thus rate equations cannot be derived from the data. I conclude that we have no compelling evidence yet that the *opaque*-2 gene becomes inactive early in endosperm development.

The data presented here have provided a few signs of developmental controls in the corn endosperm, but they also have provided a few contradictions. The RNase levels in developing endosperms were influenced considerably by the pollent parent, with changes in rates of synthesis and in the time period of rapid synthesis. Conversely, the soluble proteins showed almost no changes when foreign pollen was used, except in the case of the *opaque*-2 endosperm, which showed the usual recessive nature of the *opaque*-2 gene. It must be emphasized that the endosperm is not a uniform mass of cells, but differs from top to bottom throughout development. Biochemical data on developing corn endosperms are slowly accumulating, but they must be correlated with cytological observations before they can be fully understood.

REFERENCES

Apel, G. A. 1972. Studies on Nucleases in *Zea mays* L. M.S. thesis. University of Illinois, Urbana-Champaign.

Brown, R. and E. Robinson. 1955. Cellular differentiation and the development of enzyme proteins in plants. *Symp. Soc. Study Devel. Growth* **14**: 93–118.

Dalby, A. and G. B. Cagampang. 1970. Ribonuclease activity in normal, *opaque*-2, and *floury*-2 maize endosperm during development. *Plant Physiol.* **46**: 142–144.

Dalby, A., G. B. Cagampang, I. Davies, and J. J. Murphy. 1972. Biosynthesis of proteins in cereals. In *Symposium: seed proteins,* G. E. Inglett (ed.). Westport, Conn.: Avi Publ. Co., p. 39–51.

Dalby, A. and C.-Y. Tsai. 1974. Zein accumulation in phenotypically modified lines of opaque-2 maize. *Cereal Chem.* **51**: 821–825.

Davis, Baruch J. 1964. Disc electrophoresis. II. Method and application to human serum protein. *N. Y. Acad. Sci.* **121**: 404–427.

Duvick, D. N. 1955. Cytoplasmic inclusions of the developing and mature maize endosperm. *Am. J. Bot.* **42**: 717–725.

Duvick, D. N. 1961. Protein granules of maize endosperm cells. *Cereal Chem.* **38**: 374–385.

Hinsvark, O. N., W. H. Hourff, S. H. Wittwer, and H. M. Sell. 1954. The extraction and colorimetric estimation of indole-3-acetic acid and its esters in developing corn kernels. *Plant Physiol.* **29**: 107–108.

Miller, C. O. 1967. Cytokinins in *Zea mays*. *Ann. N. Y. Acad. Sci.* **144:** 251–257.

Moureaux, T. and J. Landry. 1972a. La maturation du grain de maïs. Evolution qualitative et quantitative des différentes formes azotées. *Physiol. Veg.* **10:** 1–18.

Moureaux, T. and J. Landry. 1972b. Effets du gène *opaque*-2 sur la protéogenèse du grain de mais au cours de la maturation. *C. R. Acad. Sci. D.* **274:** 3309–3312.

Ozbun, J. L, J. S. Hawker, E. Greenberg, C. Lammel, J. Preiss, and E. Y. C. Lee. 1973. Starch synthetase, phosphorylase, ADP glucose pyrophosphorylase, and UDPglucose pyrophosphorylase in developing maize kernels. *Plant Physiol.* **51:** 1–5.

Shannon, J. C. 1974. *In vivo* incorporation of Carbon-14 into *Zea mays* L. starch granules. *Cereal Chem.* **51:** 798–809.

Sodek, L. and C. M. Wilson. 1970. Incorporation of leucine-^{14}C and lysine-^{14}C into protein in the developing endosperm of normal and *opaque*-2 corn. *Arch. Biochem. Biophys.* **140:** 29–38.

Tsai, C.-Y. and A. Dalby. 1974. Comparison of the effect of *shrunken*-4, *opaque*-2, *opaque*-7, and *floury*-2 genes on the zein content of maize during endosperm development. *Cereal Chem.* **51:** 825–829.

Tsai, C.-Y., F. Salamini, and O. E. Nelson. 1970. Enzymes of carbohydrate metabolism in the developing endosperm of maize. *Plant Physiol.* **46:** 299–306.

Wilson, C. 1968. Plant Nucleases. II. Properties of corn ribonucleases I and II and corn nuclease I. *Plant Physiol.* **43:** 1339–1346.

Wilson, C. M. 1973. Plant Nucleases. IV. Genetic control of Ribonuclease activity in corn endosperm. *Biochem. Genet.* **9:** 53–62.

Chapter 28

INTRODUCTORY REMARKS TO THE SESSION ON TISSUE BIONOMICS

James L. Brewbaker
University of Hawaii, Honolulu

To most of us here, *Zea mays* is an object of immediate research interest and we herald the advantages it presents for basic research inquiries. Not the least of these is the potential practical significance of our inquiries. Maize is indeed an incredible economic crop, one that appears in the agriculture of every nation and whose products appear reputedly in 2400 of the 10,000 different items marketed in the average American supermarket!

We are justly proud of the precocious role played by maize in basic research in many fields, as in the science of cytogenetics. We are often rather amused to find a human biologist, for example, "discovering" basic cytological aberrations in man four decades after their full elaboration in maize. We unhesitatingly accord primacy to maize in the development of plant mutation genetics and quantitative genetics.

It is then somewhat disarming to appraise the minor role maize has played in the growth of physiological, biochemical, and developmental biological research. Chloroplasts emerge from spinach leaves, cell cultures from a carrot,

423

peroxidases out of a lowly horseradish plant, whereas maize sits on the shelf—its choroplasts, peroxidases, and cells unused. And it is less than amusing when we find the physiologist employing mere weeds to gratify his interest in basic events like the hormonal control of flowering, when we reflect on the impressive value to the world were all corn bred with daylength insensitivity.

The speakers in this session generally represent a reversal of this trend, and have placed maize—with its several disadvantages—at the core of their studies of tissue bionomics. Let me try first to define "tissue bionomics" and then reflect on some of the disadvantages as well as the history of use of maize in such studies, as I introduce the topics in Chapters 29–33 to you.

"Tissue bionomics" is a somewhat academic term for studies of differentiation and development. It literally refers to the ecology of cells in tissues and organs. Like ecology generally, it can be used to cover a multitude of disciplines, of which the following chapters focus principally on developmental genetics.

The genetic control of plant development appears at times to be at about the level of ignorance that cytogenetics was in the early days of R. A. Emerson. In that·era, Weatherwax, Keisselbach, Randolph, Edgar Anderson, and others provided elegant morphological descriptions of the corn plant and its parts, and these authors often remain the best recourse when questions of maize developmental biology arise. Dr. Coe discusses (Chapter 30) the kind of evidence, painstakingly gained with specific genetic markers, that is advancing this frontier. He is assembling evidence for the precise sequence of events in the development of surely the most significant tissue of maize, its endosperm. We may hope in the future to see his evidence and techniques applied to understanding gene action triggering biosynthetic and morphological changes during this development, as of starch and carotene synthesis, and of high-sucrose and defective-endosperm mutants.

Many lines of evidence, isozymic in particular, confirm the fact that genes are more often "turned off" than "on." This repression of gene action evidently occurs at the level of Messenger RNA (mRNA) transcription, and derepression calls for positive, often tissue-specific action. Some of us feel that to understand derepression fully is to understand differentiation. Dr. Kermicle (Chapter 24) brings for us a refreshing view of gene action in maize endosperm, with evidence for parental "imprinting" that persists from parent to offspring to affect action at the R locus in maize. He shows that alleles from one parent fail to derepress fully in the endosperm, whereas those from the other parent "turn on," apparently bearing an "imprint" of the parental source of the gene. Consistent to an extent with literature from insects and animals, Kermicle's data encourage a reexamination of many genetic phenomena that depart from simple expectations. Imprinting may seem an awesome addition to those who have attempted vainly to explain paramutation to fledgling geneticists, but it is clearly on the road to a full explanation of gene action and tissue bionomics.

Few tissues are as simple and enigmatic as the pollen grain. The transitory life of maize pollen—a few hours at best in the corn field—is an eternal source of frustration to the corn breeder. Annoyingly, we know no more about the nature of this precocious demise of pollen than when Andronescu and Knowlton studied it in the early 1920s. Pollen from 70% of the flowering plants, however, can be frozen or dried and stored for years, including grasses like Pennisetun. The bright young graduate student who achieves this feat with maize pollen will be instantly famous. He might also save the seed industry millions annually—a point seedsmen might wisely consider when distributing grants! Dr. Pfahler (Chapter 33) discusses corn pollen for us, reflecting on the remarkably few genes whose action is known in this tissue, and on some future possibilities for genetic research.

The annualism of maize is as frustrating to the biologist as the ephemerism of pollen. From the day of planting, maize progresses with singular determination to its death. No one has offered to review for this conference the genetics of annualism and perennialism (e.g., teosinte), probably because we remain so ignorant of it.

However, we begin to see the dawn of a new day in the *in vitro* studies of Dr. Green and his colleagues (Chapter 32). Green has developed technics that permit ready differentiation of maize plants from scutellar callus, allowing the completion of the maize life cycle through tissue culture. Dr. Green reviews culture methods with maize endosperm and diploid tissues, describing in detail his research with amino-acid media supplementation. Preliminary studies of cell-suspension cultures and the race T pathotoxin indicate some exciting avenues for future research with maize tissues in culture.

Dr. Berlyn and colleagues (Chapter 36) review for us the photosynthetic biochemistry of maize. Their elegant studies of tobacco tissue culture provide models for both the establishment of biosynthetic pathways and the potential selection of mutants with enhanced photosynthetic and controlled respiratory activities. We find that maize excels in its self-control of photorespiration, a model that may help improve other less efficient species. At the same time, it is intimated that maize genotypes with more efficient dark respiration, especially during hot nights, may be forthcoming to help raise the annoying yield plateau of tropical maize production.

Basic research funding by federal agencies ultimately drives much of our endeavor in developmental biology and in physiological and biochemical genetics. We may express the hope that the chapters in this section herald a return of attention to maize by basic researchers in these fields. The kinds of disadvantages of corn that have been discussed here—its annualism, size, ephemeral pollen, and obstinate tissue cultures—are basically petty, and surmountable through research. Many of us here feel, however, that a peculiar and crippling disadvantage of corn has been the tremendous potential applicability of research on it. Like many other crops, corn sounds a bit too agricultural to compete for funding with the glamor queens of basic research like the colon bacilli and Rhesus monkeys. A hungry world must encourage us to forge stronger links between basic biological inquiry and the needs of man.

Chapter 29

NUCLEAR AND CYTOPLASMIC MUTATIONS TO FERTILITY IN S MALE-STERILE MAIZE

John R. Laughnan and Susan J. Gabay

Professor of Genetics and
Research Associate, School of Life Sciences,
University of Illinois,
Urbana

Although cytoplasmic male sterility (CMS) in maize and other plant species has been used to advantage in the production of hybrid seed, the underlying cause of this type of male sterility is not well understood. It is still possible to entertain widely divergent views concerning the mechanism of action involved in the inheritance of the several known types of CMS in maize and other species, and there is no evidence to suggest that only a single mechanism is involved. Perhaps the major reason for this lack of understanding is that male-fertile cytoplasms have proved to be highly stable in the hands of the breeder and geneticist. There have been numerous independent discoveries of CMS in

Research supported by the Department of Agronomy and the Department of Genetics and Development, University of Illinois, Urbana; by CSRS grants 177-15-04, and PL 89-106, by NSF grant BMS75-20115; and by Illinois Foundation Seeds Company, Inc., Tolono, Illinois.

strains of maize from a variety of geographic locales, but as far as we are aware, in all such cases the CMS condition existed in these strains before discovery and either appeared as natural segregants or became apparent only after the use of specific crossing procedures. There appear to be no documented instances of change from the normal cytoplasmic condition to the male-sterile form in the hands of the experimenter. Moreover, although presumptive changes or reversions from CMS to normal phenotype have been observed on occasion in maize (based on personal communications with R. P. Bear, J. B. Beckett, D. N. Duvick, and H. T. Stinson), rigorous proof of heritable change at the cytoplasmic level is not available for these cases, and such changes, in any event, have been regarded as extremely rare occurrences. One such instance of reversion (Jones, 1956), occurring in the inbred line WF9, however, was found to be heritable and, on the basis of analyses carried through several generations, was interpreted as a case of change at the cytoplasmic level. Reversions from male-sterile to male-fertile phenotype have been reported in CMS strains of pearl millet, *Pennisetum typhoides* (Burton, 1972, 1977; Clement, 1975), where it appears that these reversions may involve changes at either the cytoplasmic or nuclear level.

In a preceding paper on this program (Duvick and Noble, 1978) attention was called to the current and future practical applications of cytoplasmic male sterility in the production of hybrid seed. We do not intend to expand on this subject; instead, we discuss the significance of genetic investigations of a number of reversions to fertility in certain strains of maize carrying the S-type cytoplasmic male sterility (*cms*S). These reversions have been shown to involve changes or mutations at the nuclear level in some instances and at the cytoplasmic level in others, and are highly strain dependent. Whereas the studies to be discussed do not provide a basis on which to distinguish between several alternative mechanisms for S-type cytoplasmic male sterility, they do suggest hypothetical models that are subject to test.

CYTOPLASMIC MALE-STERILE SYSTEMS IN MAIZE

The male-sterile condition involves a breakdown in normal development of pollen grains during the postmeiotic stages of microsporogenesis, and this is later manifested in its extreme form as a barren tassel in which anthers fail to exsert. Female fertility is unaffected, however, so that propagations are made by employing the male-sterile individual as the pistillate parent in crosses. In maize, as in numbers of other species of plants, two kinds of inherited male sterility have been described, genic (nuclear) and cytoplasmic (extranuclear). In the first case the male-sterile phenotype is determined strictly by nuclear genotype, usually a homozygous recessive condition. Many nuclear gene loci of this type, designated *ms*, have been identified in maize. In the second case, although the primary basis for male sterility is extranuclear, certain nuclear genes, called restorers of fertility *(Rf)*, can countermand this effect of the cytoplasm. Although an exception to the rule has been described in *Vicia faba* (Bond, Fyfe, et al., 1966), restorers produce no heritable change in the cytoplasm and may, therefore, be regarded as suppressors of the male-sterile phenotype.

The first case of cytoplasmic male sterility in maize was described by Rhoades (1931, 1933). In this instance the data suggested that only extranuclear inheritance was involved in determination of the male-sterile phenotype. Since this strain was lost it was not possible to expand the studies to determine whether restorer genes were also involved.

Since this first discovery, maize workers have encountered numerous cases of cytoplasmic male sterility, but it appears that many of these represent repeated encounters with relatively few cytoplasmically determined male-sterile phenotypes. Thus two types of *CMS* have been recognized for some time in maize (Duvick, 1965; Edwardson, 1970); one of these is designated T (Texas) and the other S (USDA). A third type designated C (Charrua) was identified recently by Beckett (1971).

Although T and S male-sterile plants are phenotypically similar, the two types of male-sterile cytoplasm, designated hereafter as *cms*T and *cms*S, are nevertheless distinguishable on the bases of: (a) the nuclear restorer genes that function in their restoration, (b) the mode of action of these restorer genes, and (c) susceptibility to *Helminthosporium maydis*, race T. In the case of *cms*T there are two known gene loci for restoration, designated *Rf* and *Rf2*; they act as complementary genes in restoring fertility such that plants with T cytoplasm and the genotype *Rf–Rf2–* are restored male fertiles, whereas plants with nuclear genotypes *rf rf Rf2–*, *Rf–rf2 rf2*, or *rf rf rf2 rf2* are male sterile (nonrestored). Restoration of *cms*T is sporophytic in nature since all the pollen of a plant of genotype *Rf rf Rf2 rf2* is functional even though only one-fourth of the pollen grains from such a plant carry both restoring alleles *Rf* and *Rf2*. In the case of *cms*S only one major restorer gene, designated *Rf3* (Buchert, 1961; Duvick, 1965), has so far been described and, in contrast with *cms*T, restoration is gametophytic in character; that is, the phenotype is determined by the genotype of the pollen grain, not of the sporophyte. Thus *cms*S plants with the genotype *rf3 rf3* are male-sterile, those with genotype *Rf3 Rf3* have all normal pollen, whereas *Rf3 rf3* heterozygotes have one-half normal and one-half aborted pollen grains, the former having the genotype *Rf3* and the latter, the genotype *rf3*. Like *cms*T, *cms*C has a sporophytic mode of restoration, but the restorer genes in this case have not been identified. These relationships are summarized in Table 1.

Because it has only recently been identified, *cms*C has not been intensively

TABLE 1. General characteristics of T, C, and S cytoplasms in maize

Male-sterile cytoplasm	Nuclear restorer genes	Mode of restoration
T	Rf and Rf2	sporophytic
C	not known	sporophytic
S	Rf3	gametophytic

investigated even though it, as well as cmsS, is now being used extensively in the production of hybrid corn (Duvick and Noble, 1978). On the other hand, the characteristics and behavior of S and T sterile cytoplasms, including their responses to restorer genes, are well documented. Both cmsS and cmsT, in particular the latter, have been employed along with appropriate restorer genes in the commercial production of hybrid seed corn, and in the course of this enterprise there has been the opportunity to observe large numbers of male-sterile progenies. From these and other observations, Duvick (1965) concluded that cmsS and cmsT are surprisingly stable to genetic mutation. As noted earlier, there are no recorded instances of spontaneous mutation in the cytoplasm from normal to male-sterile condition in maize.

Since the studies reported here deal with cmsS, it will be helpful to review the crossing procedures and inheritance involved in this system. Table 2 provides a summary of the pollen phenotypes associated with various rf3 nuclear restorer genotypes in the presence of normal and S cytoplasm. When male-sterile plants of genotype (1) (Table 2) are crossed as females with males of genotype (2), all F_1 plants are phenotypically male-sterile and are type (1) in genotype, that is, cmsS rf3 rf3. In the seed trade a plant of genotype (2) is

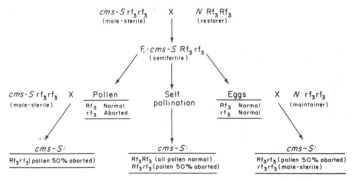

Figure 1. Crossing scheme illustrating the gametophytic mode of restoration of cmsS (N refers to normal cytoplasm).

TABLE 2.   Pollen phenotypes associated with restorer genotypes in the cms-S system

Type	Cytoplasm	Restorer genotype	Male-fertility characteristic
(1)	cmsS	rf3 rf3 (nonrestoring)	sterile
(2)	Normal	rf3 rf3 (nonrestoring)	fertile (maintainer)
(3)	Normal	Rf3 Rf3 (restoring)	fertile (restorer)
(4)	cmsS	Rf3 rf3 (restoring)	semifertile
(5)	cmsS	Rf3 Rf3 (restoring)	fertile

called a *maintainer* because when it is used as a pollen parent in crosses with the type (1) female parent, it maintains, or recapitulates, the male-sterile phenotype among the progeny. Plants of genotype (3) have normal cytoplasm but are homozygous for the nuclear restorer genotype; when crossed as male parents with male-sterile plants of genotype (1), all F_1 offspring are cmsS Rf3 rf3 or genotype (4). These plants shed pollen abundantly, but when pollen from individual anthers is examined it is apparent that the plants are semifertile, having about 50% normal and 50% aborted grains (Buchert, 1961). Therefore, when plants of genotype (4) are again crossed with male-sterile plants of genotype (1), only Rf3 pollen functions and the progeny are again all semifertile and of genotype (4). If plants of genotype (4) are self-pollinated, all F_2 offspring, because of uniparental inheritance of the cytoplasmic condition, have S cytoplasm, but half of these will be semifertile, having the nuclear restorer genotype Rf3 rf3, and the other half will be homozygous for the restorer (Rf3 Rf3) and fully fertile despite the sterile cytoplasm; the latter correspond to genotype (5) in Table 2. The various crosses discussed above are illustrated in Figure 1.

The aborted pollen grains of cmsS Rf3 rf3 plants are distinctive; examination of fresh pollen samples from such plants reveals that the aborted grains are customarily highly shriveled, yellow in appearance, and frequently adherent to the walls of normal pollen grains in the same sample. They can thus be distinguished readily from aborted pollen grains produced by most translocations and inversions. We have used this distinguishing feature to advantage in the test-cross analyses of cmsS Rf3 rf3 plants heterozygous for *Inversion 2a* or for one of a number of translocations.

BACKGROUND INFORMATION

Source materials for the studies reported here consist of seven sweet-corn inbred lines carrying the *shrunken-2* allele. These lines were converted to CMS versions through a program of backcrossing. The male-sterile cytoplasm incorporated into these lines comes from a Vg (dominant *vestigial glume*) source whose pattern of restoration places it in the *cms*S group. As a result of this conversion program, S male-sterile and male-fertile maintainer versions [genotypes (1) and (2), respectively, in Table 2] of the seven inbred lines are available. The inbred lines involved are R839, R851, R853, R853N, R825, M825, and El.

The crossing procedure employed in the search for male-fertile exceptions in the M825 pedigree is illustrated in Figure 2. Male-sterile *cms*S plants are crossed with maintainer pollen parents to produce F_1 progeny which, except for exceptional instances of male fertility, are expected to be male-sterile. The latter are crossed in turn with M825 maintainer pollen parents to produce another generation of progeny that may be searched for male-fertile exceptions. Exceptional male-fertile individuals may have entirely fertile tassels or may exhibit fertile–sterile tassel sectors. In either case they are routinely crossed as pollen parents with *cms*S male-sterile testers and are also either self-pollinated or crossed with pollen from maintainer plants. As indicated in Figure 3 this procedure is designed to determine whether the male-fertile condition is assignable to a change at the cytoplasmic level from S to normal, or alternatively, to a chromosomal mutation at a restorer locus.

The first instance of male-fertile exceptions occurred among the progeny of M825 *cms*S plants. When these exceptions were analyzed by the procedures described above it was apparent (Singh and Laughnan, 1972) that the newly acquired male-fertile trait was propagated through the female in crosses by maintainer plants, but was not transmitted through the male germ cells to test-cross progeny. Since it was not possible to explain the newly arisen male fertility on the basis of dominant or recessive nuclear restorer genes, it was con-

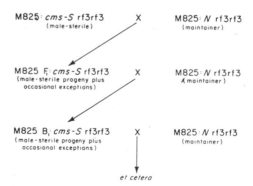

Figure 2. Crossing procedure employed in the search for male-fertile exceptions that appear either as plants with entirely fertile tassels or plants with fertile–sterile tassel sectors.

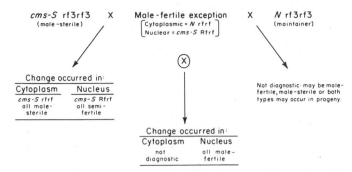

Figure 3. Crossing procedure employed to determine whether the male–fertile exception is based on a change in cytoplasm (*cms*S to N) or in nucleus (*rf* to *Rf*).

cluded that the exceptional plants resulted from a change from male-sterile to male-fertile condition in the cytoplasm. The fertile condition continued to be inherited through the female germ cells but was not transmitted through the pollen. Over 300 additional and independently occurring instances of changes from S male-sterile to male-fertile condition have now been identified and analyzed according to these same test procedures. Most of these occurred in inbred line M825 which appears to be particularly prone to the event, but male-fertile exceptions have been observed in each of the other six inbred lines as well. In the vast majority of these cases the progenies of testcrosses with S male-sterile plants were male-sterile, indicating that the newly arisen male fertility derived from a change in the cytoplasm.

There have been a number of male-fertile exceptions, however, whose test-cross analyses indicate that the causal event involves a change at the nuclear level. When these plants were crossed with *cms*S male-sterile testers, the progeny were all male semifertile (Figure 1) and these semifertile plants, on further testing exhibited the behavior expected of nuclear restorers of *cms*S (Figure 3). Results of studies dealing with four of these strains have been published (Laughnan and Gabay, 1973). These investigations confirmed that the basis for male fertility in these strains resides in the nucleus and provided the first indication that we might be dealing with a male-fertility element having the characteristics of an episome; it appeared that the fertility element could be "fixed" in the cytoplasm (autonomous), or in the nuclear genome (integrated). Soon thereafter six additional cases of mutations leading to nuclear restoration in S cytoplasm were identified.

The 10 new restorer strains have been assigned Roman-numeral designations until such time as their relationships with the already described restorer loci *Rf, Rf2*, and *Rf3* have been defined. Seven of the ten cases arose in inbred line backgrounds, two in F_1 populations and one in an F_2 progeny. In all but one the female parent of the male-fertile exception was a male-sterile *cms*S plant; restorer IV, the single exception, arose in a maintainer plant of inbred line R853N. The genetic behavior of the ten newly arisen restorers has been considered elsewhere (Laughnan and Gabay, 1973, 1975).

NATURALLY OCCURRING RESTORERS OF CMS-S

A number of established inbred lines of corn carry a nuclear gene capable of restoring fertility in plants carrying S cytoplasm (Duvick, 1965; Beckett, 1971). The gametophytic basis for restoration of *cms*S was first identified by Buchert (1961), whose study involved the restorer gene carried by inbred line Ky21. This restorer of fertility was initially designated "Rf^2" but, since "Rf_1" and "Rf_2" have been used to designate the gene loci that function in *cms*T restoration, it was suggested (Duvick, 1965) that "Rf_3" ($Rf3$, according to nomenclature recently adopted for maize) be used to symbolize the *cms*S restorer.

The finding that the male fertility conferred by the ten newly identified *cms*S restorers referred to in the previous section is inherited on a Mendelian basis, and that, as we shall later see, these restorers occupy different chromosomal sites, indicated a need for additional information concerning the so-called naturally occurring *cms*S restorers. Are the natural restorers that are found in different strains of maize allelic, or do they represent genes at different loci, each capable of restoration? The information provided in the following paragraphs suggests that in these naturally occurring restorer strains only a single locus is involved in *cms*S restoration.

Duvick (1957) made all possible crosses between inbred lines Ky21, BH2, CE1, JG3, and JG5, each of which carries the ability to restore pollen fertility in plants carrying *cms*S. The resulting F_1 plants were then crossed with *cms*S male-sterile plants of inbred line WF9. Each of the ten F_1 combinations gave testcross progenies that were all fertile, indicating that the restorers carried by the five lines are allelic.

We recently employed another procedure to test for allelism between five *cms*S restorer lines: Ky21, CE1, Tr, C103, and CI21E. The *cms*S homozygous restorer version of one of the lines to be tested was crossed as female parent with a second line, and pollen samples of F_1 offspring were examined for the frequency of aborted grains. If the restorer genes in the tested lines are not allelic, the F_1 pollen sample should exhibit approximately 25% of pollen grains of the shriveled type characteristic of abortion found in the *cms*S system. If, on the other hand, the restorers are allelic, all or nearly all pollen grains produced by F_1 plants should be normal. Eight of the 10 possible F_1 combinations were tested in this way, and all five were involved in one or more of these combinations; in each case the F_1 plants produced all-normal pollen records, indicating that the restorer genes in these five lines are allelic. Since two of these lines, Ky21 and CE1, were among those analyzed by Duvick (1957), it may be concluded that these eight lines in all likelihood carry the same restorer, although the possibility cannot be excluded that different but closely linked restorer loci are involved.

The Rf and $Rf2$ genes that restore cmsT have, by the use of translocation testers, been assigned to chromosomes 3 and 9, respectively (Duvick, Snyder, et. al., 1961; Snyder and Duvick, 1969). Several different sources of evidence indicate that $Rf3$, the restorer of cmsS, is in the chromosome 2 linkage group in maize. Preliminary evidence (Singh, 1969), based on a pollen technique that is described in detail later, suggested that the restorer gene carried in inbred line CE1 might be assigned to either chromosome 2 or 3, probably the former. Unpublished studies (S. W. Noble, personal communication) involving the use of translocation testers, indicate that the cmsS restorer carried by inbred line CE1 is located in chromosome 2, probably in the long arm.

We have recently undertaken studies to confirm the assignment of cmsS restorers of a number of inbred lines to the chromosome 2 linkage group. One of these concerns a cmsS restorer ($Rf3$–$In2$a) in an inbred tester strain carrying $inversion$-2a (breakpoints = 2S.75-2L.80), and the linked plant color allele B. This $inversion$-2a strain, here designated Rf In B, was crossed as a pollen parent with cmsS male-sterile plants that carried neither the inversion nor the plant color allele. The F_1 offspring from these crosses have the genotype cmsS : Rf In B/rf N b, in which N designates the noninversion genotype, and exhibit 75–80% pollen abortion including both the shriveled yellow grains typically associated with the rf abortion genotype in S cytoplasm, and the non-shriveled, empty grains found in pollen samples from plants heterozygous for the inversion. These F_1 plants were crossed as female parents with pollen from green maintainer plants not carrying the inversion (rf N b/rf N b). Plants in 10 testcross progenies were scored for the male-fertile versus male-sterile trait, for B versus b plant color phenotype, and for presence versus absence of the inversion, the latter distinction being made on the basis of ear sterility; it was also possible to score pollen samples of all but two of the male-fertile offspring to confirm presence or absence of the inversion. The data for these ten testcross progenies (Table 3) indicate that the restorer factor ($Rf3$–$In2$a) carried in the inversion stock is closely linked with both B and $In2$a, the recombination values being 5.7% and 3.0%, respectively. These data indicate that $Rf3$–$In2$a is located in chromosome 2, but they do not reveal whether the restorer locus is included in the inversion or whether it lies outside the inversion but close to one or the other of its breakpoints. Additional data on recombination between $Rf3$–$In2$a and the inversion come from two testcross progenies not involving segregation for B plant color alleles; among 73 offspring there were four recombinants. The combined data for these two sets of experiments indicate that *In*2a and the *Rf*3 factor carried in the inversion chromosome are recombined with a frequency of 3.4%.

The *In*2a restorer strain referred to above was crossed as a pollen parent with *cms*S plants of inbred line Tr, which carries a natural restorer (*Rf*3–*Tr*) of S. The male-fertile F_1 plants were crossed as female parents with maintainer plants. If the restorer genes involved in this cross are in different linkage groups, about one-quarter of the testcross offspring are expected to be male-sterile. In two such progenies there was a total of 119 plants, and all were male-fertile, indicating that the natural restorer carried by inbred line Tr is also located in chromosome 2. Since Tr is one of the eight restoring lines for which allelism was earlier indicated, it appears that all these restorers are assignable to the chromosome-2 linkage group.

From linkage studies involving translocation heterozygotes there is direct independent evidence for the assignment of three naturally occurring *cms*S restorers to chromosome 2. The restorers involved are those carried by inbred lines CE1 and Tr, as well as one, (*Rf*3–*Vg*) not previously mentioned, carried by the *Vg* (*vestigial glume*) strain in which *cms*–*Vg*, a member of the S group of sterile cytoplasms, was identified. Table 4 presents summarized data on recombination between the *Rf* allele in each of these strains and the breakpoint in one or more of four chromosome 2 translocation testers. The tested plants carried the S-type male-sterile cytoplasm and were heterozygous for the indicated translocations and restorers. They were either crossed as female parents by maintainer testers or were employed as male parents in crosses with *cms*S : *rf rf* (male-sterile) plants to produce the testcross progenies. In the first case progeny plants were scored for male fertility or sterility, and for presence or absence of the translocation; in the second case, since only *Rf* pollen grains are expected to function, all offspring are expected to be male-fertile, and it is only necessary to score for presence or absence of the translocation. The data in Table 4 indicate a loose linkage between each of the three restorers and the T2-9b breakpoint in chromosome 2, which is located at 0.2 in the short arm. Each of the three recombination values, however, represents a significant departure from expectations on the assumption of independent segregation. The data also indicate linkage between the restorer of two inbred lines (CE1 and Tr) and three translocations with breakpoints well out in the long arm of chromosome 2. On the assumption that a single restorer locus is involved, the combined information on linkage and recombination from both inversion and translocation heterozygotes (Tables 3 and 4) suggests that *Rf*3 is located in the proximal region of the long arm of chromosome 2. It is not likely that it lies beyond 2L.80 since it rarely recombines with *In*2a (Table 3), one of whose breakpoints is at 2L.80, but shows significant recombination (Table 4) with T2-6(4717) and T2-4b, whose chromosome 2 breakpoints are 2L.75 and 2L.80, respectively. While the restorer shows relatively high recombination with T2-9b

TABLE 3. Distribution of progeny from the testcross cmsS: *Rf In B/rf N b* × *rf N b/rf N b*

Family	Total progeny	Phenotypic distribution of testcross progeny								
		Parentals		Recombinants[a]						
		Rf In B	rf N b	Rf N b	rf In B	Rf In b	rf N B	Rf N B	rf In b
1	57	24	28	1	1	2	1	0	0
2	59	26	28	1	2	0	2	0	0
3	60	31	29	0	0	0	0	0	0
4	22	10	12	0	0	0	0	0	0
5	21	13	8	0	0	0	0	0	0
6	24	14	7	0	1	1	1	0	0
7	22	11	9	0	1	1	0	0	1
8	23	8	14	0	0	0	0	0	1
9	23	6	14	0	0	1	1	0	1
10	24	7	15	0	1	1	0	0	0
Totals	335	150	164	2	6	6	5	0	2

[a] Recombination: *Rf–In* = 10/335 (2.98%); *Rf–B* = 19/335 (5.67%); *In–B* = 13/335 (3.88%).

437

TABLE 4. Percent recombination between naturally occurring S restorers and breakpoints in four chromosome-2 translocations

Source of natural restorer	Heterozygous parent	Chromosome 2 translocations and their breakpoints							
		T2-9b (2S.20)		T2-7(4519) (2L.65)		T2-6(4717) (2L.75)		T2-4b (2L.80)	
		No. plants	Recombination (%)	No. plants	Recombination (%)	No. plants	Recombination (%)	No. plants	Recombination (%)
Vestigial stock	♀	78	30.8[b]						
Inbred line CE1	♀	192	41.7[a]	275	24.4[b]	237	31.2[b]	92	40.2
	♂			91	34.1[a]	79	22.8[b]		
Inbred line Tr	♀	278	36.3[b]			95	23.2[b]		

[a] Significant at $P = 0.05$.

[b] Significant at $P = 0.01$, based on expectation of independent segregation.

(Table 4), whose chromosome-2 breakpoint is at 2S.20, it probably is not located distal to this site since independent evidence indicates that the restorer segregates independently of the $ws3$, lg, and $gl2$ markers that lie beyond B in the short arm.

In summary, the naturally occurring cmsS restorers carried by four inbred strains of maize ($Rf3-In2a$, $Rf3-CE1$, $Rf3-Tr$, and $Rf3-Vg$), exhibit linkage with various chromosome-2 marker sites. Moreover, tests for allelism involving three of these four restorers and restorers carried in other strains of maize indicate that all 10 natural cmsS restorers so far tested are in the chromosome-2 linkage group and may be considered allelic.

LINKAGE RELATIONS OF THE NEWLY ARISEN RESTORERS

If nuclear restoration of fertility to plants carrying S male-sterile cytoplasm is brought about by the presence of the restoring allele $Rf3$ in chromosome 2, it is reasonable to anticipate that the restorers carried in the 10 newly arisen exceptional strains occurred by mutation of the $rf3$ nonrestoring allele to $Rf3$; if so, these restorers should be allelic with the naturally occurring restorer, and should be located in chromosome 2. On the other hand, if the exceptional event that has occurred in these strains involves a male-fertility element (F) with the characteristics of an episome (Laughnan and Gabay, 1973, 1975), there is no *a priori* reason to anticipate that its integration would be restricted to the $rf3$ site in chromosome 2. In fact, the studies on linkage relationships of the naturally occurring S restorers reported in the previous section were stimulated by an earlier finding indicating that at least some of the 10 new restorers that arose as fertile exceptions in our experiments are not allelic with the S restorer of inbred line CE1 (Laughnan and Gabay, 1975). These experiments indicated that none of the restorers I through VI is allelic with $Rf3$ carried by CE1; moreover, similar tests indicated that the restorers in strains I through VI are not allelic with each other.

Following this indication that the new restorers may occupy unique chromosomal sites, we undertook to establish their linkage relationships. The conventional technique, employing *waxy* translocation testers for this purpose, would be laborious since it requires that each of the 10 restorer strains be crossed with each of the translocation testers, following which it is necessary to analyze each of the testcross progenies for *waxy* and for the mature plant trait, male fertility. A new procedure (Gabay and Laughnan, 1974) takes advantage of the fact that genic restoration of S cytoplasm occurs at the gametophytic level. Since plants with S cytoplasm that are heterozygous for a restorer gene produce equal numbers of normal (Rf) and aborted (rf) pollen grains, it is possible to obtain a preliminary indication of the linkage group for a particular restorer through analysis of iodine-stained pollen samples from plants heterozygous for both the restorer and a particular *wx*-linked reciprocal translocation. The procedure involves an initial cross of a plant with S cytoplasm that is heterozygous for a genic restorer (*Rf rf*), as female parent, with a nonrestoring tester plant that is homozygous for one or another *wx*-linked translocation. All F_1 offspring should be heterozygous for the *wx* alleles and for the translocation, and approximately half of these, having received the *rf* allele from the female parent, should be male-sterile. The remaining half, those carrying the *Rf* restoring allele from the female parent, should be semifertile, with about 25% normal pollen grains (Figure 4). If the restorer gene being tested is located on a chromosome other than the two that are involved in the *waxy* translocation carried by the male parent, blue- and red-staining normal pollen grains should occur with equal frequencies. If linkage is encountered, however, more than 50% of normal pollen grains should stain blue, the proportion of blue and red grains being a function of the genetic distance or recombination between the *Rf* and *wx* loci.

Using the pollen method described above, linkage groups have been identified for six of the 10 new restorers (Table 5). Only pollen data from translocation heterozygotes in which a linkage with the respective restorer was evident are presented here. Since each of the restorers was involved in crosses with other chromosome 9 translocation heterozygotes that did not provide evidence of linkage, it is apparent that none of the six restorers treated in Table 5 is in chromosome 9. The pollen data suggest that restorers I and VIII are located in chromosome 8, restorers IV and VII in chromosome 3, and restorers

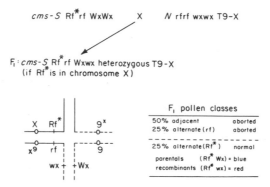

Figure 4. Pollen method for determination of linkage group of a new restorer, designated *Rf**. The new restorer strain carrying S cytoplasm is crossed as female parent with the *waxy* translocation tester series and iodine-stained pollen samples from F_1 plants exhibiting 75% abortion are analyzed. Linkage of the new *Rf** restorer with one of the chromosomes involved in the interchange is detected on the basis of a discrepant ratio for blue : red-staining pollen grains.

TABLE 5. Pollen and testcross data from *cmsS: Rf N Wx/rf T wx* translocation heterozygotes indicating linkage group assignments for six of the 10 new restorers

New restorer	wx Translocation tester	Pollen grains scored				Testcross data from cms-S: rfrf wxwx females × cmsS: Rf N Wx/rf T wx[a]			
		Total	Wx (blue)	wx (red)	Recombination (%)	Total	Wx	wx	Recombination (%)
I	8-9d	937	819	118	12.6				
	8-9(6673)	634	604	30	4.7	191	181	10	5.2
IV	3-9c	716	645	71	9.9	934	873	61	6.5
VII	3-9c	858	554	304	35.4				
VIII	8-9d	722	487	235	32.5				
	8-9(6673)	911	628	283	31.1	197	119	78	39.6
IX	1-9c	778	675	103	13.2				
	1-9(4995)	761	591	170	22.3	260	177	83	31.9
X	1-9c	1062	843	219	20.6				
	1-9(4995)	796	515	281	35.3	809	540	269	33.3

[a] Data scored on testcross ears; since *rf* pollen does not function, the frequency of waxy kernels provides a direct estimate of *Rf-Wx* recombination.

441

IX and X in chromosome 1. Testcross data, available for five of the six restorers (Table 5) support these chromosome assignments. The data do not rule
out the possibility that restorers IX and X may be located at the same site in
chromosome 1, but they do suggest that restorers I and VIII and restorers IV and
VII are at different locations in chromosomes 8 and 3, respectively. Restorers II,
III, V, and VI have not yet been located, but it is evident from the earlier mentioned negative tests for allelism involving these restorers and the restorer carried
by inbred line CE1, that if they are in chromosome 2 they are not at the standard
$Rf3$ site. In addition, as previously indicated, restorers I through VI are not
allelic with each other. Restorers II, III, V, and VI, therefore, are neither at the
restorer I site in chromosome 8 nor at the restorer IV site in chromosome 3.

DISCUSSION

What is the origin of the new restorers? One possibility is that they arose by
mutational events at a variety of different restorer (suppressor) loci in the maize
nuclear genome. None of the six new restorers assigned to linkage groups is in
chromosome 2, which has been identified as the location for the 10 $cmsS$ restorers
encountered in natural restorer strains. If there exists a number of gene sites for
$cmsS$ restoration in the maize genome, it is not surprising that we should have
encountered new restorer mutations at five or six different locations, or that none
of these is in chromosome 2. Given the same assumption, however, why is it that
all ten natural restorers carried in the established inbred lines that have been
analyzed, turn out to be located in chromosome 2? It would be reasonable to
expect that if the mutational events encountered in our experiments are
representative of those that have occurred in maize populations that provided the
source for the natural restorer strains, the restorer genes carried by the latter
would be found in a variety of chromosomal locations. A consideration of the
genetic behavior of the new restorers appears to provide an answer to this question; as noted elsewhere (Laughnan and Gabay, 1975), most, but not all, of the
new restorers are associated with deleterious phenotypic effects not related to
male fertility and sterility. Among these are reduced transmission through the
female gametophyte, reduction in kernel size, and lethality of the restorer
homozygote, all thus far observed in S cytoplasm. Of the 10 new restorers, only
restorer IV, the one that originated in a maintainer plant and has been placed in
chromosome 3, appears to be without these adverse effects. If it is argued that
most restorer mutations occurring in nature, like the ones we have encountered
in our studies, are deleterious and at a disadvantage in survival, and that the
standard $Rf3$ restorer is unique in that it does not have these properties, this
might account for the finding that all of the 10 natural restorers thus far analyzed
were found to be located in chromosome 2. For obvious reasons, we have

intensified efforts to identify larger numbers of naturally occurring cmsS restorer stains and to determine the linkage groups of the restorers that they carry.

Despite the above arguments in support of the mutation hypothesis, we feel it is an insufficient model to account for the origin of the new restorers. For one thing, the fact that most of the new restorers share similar deleterious side effects is difficult to reconcile with the idea that they originate by independent mutational events at different gene loci. Moreover, there is evidence that cyto-plasmic and nuclear male-fertile exceptions have a common origin; these two kinds of male-fertile exceptions have arisen in the same strains, and both may be expressed initially as either entirely male-fertile plants or as fertile–sterile tassel sectors. In this connection restorer VIII is of special interest. It arose as an exceptional male-fertile plant among the progeny of a cross involving an M825 cmsS male-sterile plant and an M825 maintainer pollen parent. When crossed as a female parent with a maintainer, the exceptional plant gave only male-sterile progeny, but when crossed as pollen parent with cmsS male-sterile testers it gave both male-fertile and male-sterile offspring, the former pre-dominating. Restorer VI, which arose in an exceptional plant with a fertile tassel sector, gave mixed testcross progenies similar to those just described for restorer VIII. The appearance of both male-fertile and male-sterile offspring in these instances suggests that both cytoplasmic and nuclear fixations of the male-fertile element occurred in pollen produced by the original exceptional plants and hence that these two kinds of events are associated.

Most of the male-fertile exceptions we have observed occurred in progeny of crosses involving inbred line M825. In this particular background the vast majority of fertile exceptions are cases of cytoplasmic change, instances of nuclear "intergrations" being relatively infrequent. We have recently identified two genetic strains in which there is both a high frequency of initial fertile events and a relatively high frequency among these of nuclear exceptions. A number of the latter have been found to exhibit a unique behavior that we must conclude is a function of the genetic backgrounds from which they were isolated. In our previous experience with the original restorer strains, when pollen from the male-fertile exception was used in crosses with male-sterile testers, all, or almost all, of the offspring were semifertile (50% pollen abor-tion), and when these F_1 plants were again testcrossed they produced all semifertile offspring, thus indicating a nuclear basis for male fertility. The newly encountered cases referred to above, however, are characterized by what appears to be delayed "fixation" of nuclear fertility; thus the first testcross of a fertile sector in these instances produces a progeny with a substantial number of male-sterile plants along with the familiar semifertile plants whose pollen record (50% abortion) is indicative of nuclear restoration. When these F_1 semifertile plants are testcrossed they again produce both male-sterile and semifertile progeny. In some cases this unusual pattern of persistent segregation has now been maintained through three generations of testcrosses. Although our current understanding of this phenomenon does not provide a basis for sound conclusions, it seems likely that in these cases the fertility element is carried neither in the cytoplasm nor in the chromosome-integrated state, but that it resides in the nucleoplasm or in a tenuous state of association with a chromosome. As such, it is transmitted through the male germ cell but is subject to loss or to chromosomal integration in the succeeding generations. This would account for the occurrence in these testcross progenies of: (a) male-sterile offspring, (b) semifertile offspring whose testcrosses again yield both male-sterile and semifertile offspring, and (c) semifertile offspring whose testcross progenies carry only semifertile individuals.

A model that proposes to account for the spontaneous male-fertile exceptions must account for the observation that the underlying change may occur in either the cytoplasm or the nucleus; it should also deal with the ultimate origin of male fertility in these instances. Male-fertile exceptions of either class first make their appearance among the progeny of crosses of *cms*S male-sterile with maintainer plants (Figure 2). In view of this it may be considered that they arise either as a result of a qualitative change, or mutation, in a preexisting but nonfunctional fertility element in the *cms*S male-sterile female parent, or, alternatively, as the result of occasional transfer of a normal fertility element of cytoplasmic origin through the male germ cells of the maintainer pollen parent. Although evidence currently available does not permit a clear distinction between these two alternatives, the preliminary studies cited above that deal with semifertile exceptions that exhibit persistent segregation appear to support the hypothesis of male germ cell transmission. These plants appear to carry a fertility element that is labile in the sense that it can be transmitted through the male germ cells to offspring in which it may appear in one of several states. Given this capability in transmission, it is only required that the fertility element in a maintainer plant move from a fixed condition, perhaps in a cytoplasmic organelle, to a free or labile state rendering it capable of transmission through the nucleus of a pollen grain.

According to the model we currently favor to account for the male-fertile exceptions, the initiating event occurs in the male-fertile maintainer parent and involves the movement of a male-fertility episome (F) from the cytoplasm, or cytoplasmic organelle, into the nucleus. The frequency of this event, and thus the frequency of sperm nuclei carrying the F episome varies with the genotype, some inbred lines, for example M825, being much more prone to this event than others. When such an F-carrying sperm fertilizes an egg contributed by the *cms*S male-sterile plant, the zygote carries the F episome in the diploid nucleus. In the course of the vegetative development of this plant, the F episome may continue to be propagated in the nucleoplasm free of the chromosomes, it may be lost altogether from a particular cell or cell line, move into the cytoplasm and become established (propagate) there, or become integrated into one or another chromosome. This model makes a strong commitment in regard to the ultimate origin of the fertile event but is consistent with the appearance of the several kinds of fertile sectors we have observed, that is, those in which the F factor appears to have reestablished in the cytoplasm, those in which it has become integrated into a chromosome, those in which both events appear to have occurred, and those in which it appears that the F episome continues to be propagated in the nucleus, as an extrachromosomal entity, through one or more generations before chromosomal integration is achieved, or F is lost from the nucleus.

ACKNOWLEDGMENT

We are indebted to Dr. E. B. Patterson, Dr. D. B. Walden, and Mr. Carl E. Hall for many helpful discussions and for direct assistance in carrying out field and laboratory studies.

REFERENCES

Beckett, J. B. 1971. Classification of male-sterile cytoplasms in maize (*Zea mays* L.). *Crop Sci.* **11**: 724–727.

Bond, D. A., J. L. Fyfe, and G. Toynbee-Clarke. 1966. Male sterility in field beans (*Vicia faba* L.) III. Male sterility with a cytoplasmic type of inheritance. *J. Agr. Sci. Camb.* **66**: 359–367.

Buchert, J. G. 1961. The stage of genome–plasmon interaction in the restoration of fertility to cytoplasmically pollen-sterile maize. *Proc. Nat. Acad. Sci. U. S. A.* **47**: 1436–1440.

Burton, G. W. 1972. Natural sterility maintainer and fertility restorer mutants in Tift 23A$_1$ cytoplasmic male-sterile pearl millet, *Pennisetum typhoides,* (Burm.) Stapf and Hubb. *Crop Sci.* **12**: 280–282.

Burton, G. W. 1977. Fertile sterility maintainer mutants in cytoplasmic male sterile pearl millet. *Crop Sci.* **17**: 635–637.

Clement, W. M., Jr. 1975. Plasmon mutations in cytoplasmic male-sterile pearl millet, *Pennisetum typhoides. Genetics* **79**: 583–588.

Duvick, D. N. 1957. Allelism of FR genes of inbreds which restore pollen fertility to WF9^s. *Maize Genet. Coop. Newsl.* **31**: 114.

Duvick, D. N. 1965. Cytoplasmic pollen sterility in corn. *Adv. Genet.* **13**: 1–56.

Duvick, D. N., and S. W. Noble. 1978. In *Maize breeding and genetics,* D. B. Walden (ed.). New York: John Wiley & Sons, Chap. 17.

Duvick, D. N., R. J. Snyder, and E. G. Anderson. 1961. The chromosomal location of Rf_1, a restorer gene for cytoplasmic pollen sterile maize. *Genetics* **46:** 1245–1252.

Edwardson, J. R. 1970. Cytoplasmic male sterility. *Bot. Rev.* **36:** 341–420.

Gabay, S. J. and J. R. Laughnan. 1974. Linkage analysis in the male gametophyte. *Maize Genet. Coop. Newsl.* **48:** 44–45.

Jones, D. F. 1956. Genic and cytoplasmic control of pollen abortion in maize. Genetics and Plant Breeding. *Brookhaven Symp. Biol.* **9:** 101–112.

Laughnan, J. R. and S. J. Gabay. 1973. Mutations leading to nuclear restoration of fertility in S male-sterile cytoplasm in maize. *Theor. Appl. Genet.* **43:** 109–116.

Laughnan, J. R. and S. J. Gabay. 1975. An episomal basis for instability of S male sterility in maize and some implications for plant breeding. In *Genetics and the biogenesis of cell organelles,* C. W. Birky, Jr., P. S. Perlman, and T. J. Byers (eds.). Columbus: Ohio State Univ. Press.

Rhoades, M. M. 1931. Cytoplasmic inheritance of male sterility in *Zea mays. Science* **73:** 340–341.

Rhoades, M. M. 1933. The cytoplasmic inheritance of male sterility in *Zea mays. J. Genet.* **27:** 71–93.

Singh, A. 1969. Mutation of the S cytoplasmic element for male sterility in maize. Ph.D. thesis. Urbana: University of Illinois.

Singh, A. and J. R. Laughnan. 1972. Instability of S male-sterile cytoplasm in maize. *Genetics* **71:** 607–620.

Snyder, R. J. and D. N. Duvick. 1969. Chromosomal location of Rf_2, a restorer gene for cytoplasmic pollen sterile maize. *Crop Sci.* **9:** 156–157.

Chapter 30

THE ALEURONE TISSUE OF MAIZE
AS A GENETIC TOOL

E. H. Coe, Jr.*

USDA Agricultural Research Service and
University of Missouri,
Columbia

Many significant studies in genetics have depended on the aleurone tissue of corn as the "petri dish" or the "stage" on which the observations were made. This chapter focuses on several examples of research on the special characteristics of this tissue and its development.

Mendel and Correns in their work with maize no doubt were attracted to it as research material in part by the *bulk* and *visibility* of the aleurone tissue of the kernel. Aleurone color factors were the materials with which East and Hayes (1911) demonstrated one- and two-factor F_2 ratios, including the complementary factor $9:7$ ratio and the $9:3:4$ and $13:3$; the study of these

Cooperative Investigations of the Agricultural Research Service, USDA, and the Missouri Agricultural Experiment Station. Journal Series No. 7402.

* Geneticist, Agricultural Research Service, USDA, and Professor of Agronomy, University of Missouri, Columbia, Mo. 65201. Sheila McCormick and David Chen contributed skillfully to the data analysis.

interactions was enhanced by the ease with which the aleurone tissue could be viewed, classified, and counted on or off the ear.

East and Hayes also described mosaic kernels, which were later shown by Emerson (1921) to be attributable to spontaneous chromosomal losses in heterozygously marked individuals during development. Emerson's observations were based on events within a *field* (or "lawn") of tissue whose *development is systematic* in clonal hierarchies, and this field permitted the events to be recognized. Among applications of this knowledge is the early radiation research of Stadler (1928), in which he employed mosaics to test whether X-rays would induce losses of chromosomes. In addition to showing that X-rays do induce high frequencies of mosaics, he recognized and characterized the recovery ("abortive loss") phenomenon, in which segments of chromosome, broken spontaneously or by X-irradiation, are lost from the majority of the endosperm or plant but are recovered in limited areas as stable, functional segments (Stadler, 1930). We return to further consideration of the systematic development of clonal hierarchies later.

In his discovery of the dominant factor, *Dt,* which causes dots of color in the aleurone tissue of *a* (colorless) kernels, Rhoades (1936) also employed the aleurone tissue as a field on which the events could be observed. In a classic study, the dots were soon shown to be due to induced mutations from *a* to *A,* elaborated in clones (Rhoades, 1938). Rhoades found that numbers of events occurring with various dosages for *a* and *Dt* could be counted in this field, and he identified a linear relationship for *a* dosage and an exponential one for *Dt* dosage. Nuffer (1955) used the counting procedure in this tissue field to characterize two additional factors, *Dt2* and *Dt3,* and to extend the exponential relationship to dosages higher than three.

A number of further instances in which the field character and visibility of the aleurone tissue have been employed for genetic purposes might be mentioned. Subdividing the mosaic field, Stadler (1939) categorized classes of events induced in pollen by X-rays and ultraviolet light (UV) according to their extent (whole-kernel losses, 15/16ths, etc.) and established the striking fact that UV differs from X-rays in "strand resolution"—in other words, that UV-induced losses following pollen irradiation are mostly half-fractional events while X-ray-induced losses are mostly whole-kernel events (Figure 1). Seeking events of the reverse type, Stadler (1944) tested for X-ray induction of mutations from *a* to *A* (in the absence of *Dt*) by irradiating at a stage calculated to lead to dot-sized mutational events; he found no such events in a field of 16 million potential sites. McClintock (1951b) also conducted tests with X-rays similar to those of Stadler, as did Neuffer (1966) with X-rays, UV, and ethyl methanesulfonate (EMS). Although these additional tests showed no induced mutations of the *a* to *A* type, they did yield new occurrences of mutator activity for *Dt* and *Spm,* a phenomenon that McClintock (1951b, 1965) later correlated

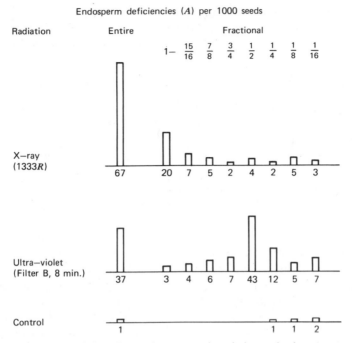

Figure 1. Comparison of the effects of X-rays and UV light on the frequencies of entire and fractional losses in the endosperm [from Stadler (1939)].

with chromosome breakage. By scanning the tissue field in thousands of kernels, Doerschug (1973) succeeded in recovering germinally two new occurrences of *Dt* induced by chromosome breakage. Tests showing that the "resolving power" of X-rays is less fine than that of UV and EMS (Nuffer, 1957; Neuffer and Ficsor, 1963) were conducted through observations on the aleurone field. The occurrence of different states of mutable factors and of their regulators, *Ac, Dt, En,* and *Spm* (McClintock, 1951a, 1965; Nuffer, 1957; Peterson, 1961), was experimentally recognized by their mutational expressions on the extended field of aleurone cells.

Analyses by Stadler and Uber (1942) designed to test whether different wavelengths of UV differently affected fractional losses showed that types of fractional losses were disappointingly comparable at all wavelengths. However, this paper deserves a place among the readings of aspiring geneticists for other reasons. First, it is a model example of precise designing of physical experimentation on biological materials; second, it showed that the effectiveness of different wavelengths of UV coincided with the absorption spectrum of nucleic acids in 1942, well before the Watson–Crick era (an example of a clue whose importance was little appreciated in its time); and third, it contains in one paragraph a modest but revealing observation on a further characteristic of the aleurone tissue that has been useful in genetic analysis—in addition to its bulk, visibility, and systematic development, this tissue has an extended *active biosynthetic state.* Stadler and Uber were concerned with the potential confounding between the phenotypic effects of losses of *A* and losses of *Pr.* They observed that endosperms that had a fractional loss of *A* and also had a complementary fractional loss of *Pr* (e.g., half colorless, *a,* and half red, *pr*) displayed a margin of purple cells, which they interpreted as a result of interaction of *A pr* cells with adjacent *a Pr* cells in kernels arising from two coincidental losses in complementary halves. The similar occurrence of purple margins between colorless *C-I Bz* and bronze *C bz* areas, reported matter-of-factly by McClintock (1951a) with breathtaking photographs (Figure 2), is a result of exchange of unknown intermediates from one cell type into the other, followed by biosynthetic activity in the receiving cell type. Rhoades (1952) determined that the deposit of pigment was greater on one side of the border, suggesting that the diffusion was more effective in one direction. With this activity defined, later observations (Coe, 1957) that indicated diffusion of materials from the aleurone tissue into the pericarp in the presence of intensifier, *in,* could easily be accommodated and led to postulation of an order of action for two groups of factors—those recessives that do and those that do not confer gold pericarp color over *in* aleurone tissue. A provisional sequence of action for the aleurone color factors was soon defined by testing artificially for exchange of intermediates between living pieces of aleurone tissue (Reddy and Coe, 1962), a test that has been difficult to enlarge on or even conduct repeatedly but that seems consistent with each new piece of information. With the help of the bulky character of the aleurone tissue, extraction and identification of some of the anthocyanin pigments and related compounds in various genotypes has been

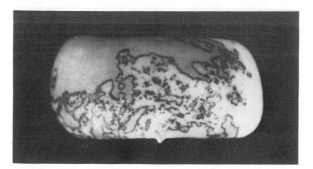

Figure 2. Borders of anthocyanin formed by complementary interaction of products diffusing between aleurone cells carrying the inhibitor *C-I* with *Bz* (colorless background) and aleurone cells carrying *C* with *bz* (lightly colored regions) [from McClintock (1951a)].

carried out (Coe, 1955; Harborne and Gavazzi, 1969; Kirby and Styles, 1970), and in one instance the biosynthetic events have been related to genetic control of enzymes (Larson and Coe, manuscript in preparation).

The biosynthetic activity of the aleurone tissue that is perhaps most intriguing genetically is the synthesis of anthocyanin in colorless aleurone (c) strains during germination (Kirby and Styles, 1970), which Chen (1973) showed not only to be specific to the c locus under germinating conditions with light but specific to certain recessive-colorless alleles, $c-p(W22)$ and $c-p(K55)$, and not to others, $c-n(W23)$ and $c-Ref$. Considering this information along with the existence of alleles that condition pigment synthesis while the kernel is maturing, and of alleles with dominant inhibitory effect, we see that different alleles regulate synthesis specifically in different periods of development in precisely the same cells. Chen also tested whether the $c-p$ response is determined at the cell, tissue, or whole-kernel level by restricting light passage to a tiny window cut in an aluminum foil wrapper; she found that the margin of response was very sharply defined, which indicates that the response is determined at the individual-cell level. If light is given before maturation, the effect of the light exposure will be stored without visible pigment synthesis until germination, when pigment will be synthesized in the dark. In relation to unresolved questions about the synthesis of α-amylase and proteases during germination (Dure, 1960; Harvey and Oaks, 1974), this inducible pigment property may be a tool that will help unravel just how specific enzymes are induced in particular cells under particular conditions. Corn aleurone tissue is clearly unlike that of barley at germination in that it does not appear to be the site of synthesis of α-amylase or, at least, it is not responsive to gibberellin induction.

One further property of the aleurone tissue that has proved interesting, although it is not yet understood, is its *phasic constitution,* recognized first in the $R-nj$ (Navajo) type, in which the crown of the aleurone tissue becomes pigmented while the sides remain colorless. McClintock (1965), Peterson (1966), and Doerschug (1973) have described and pictured similar phasic expression and the inverse, in which only the sides are pigmented, for mutability factors.

Returning to the characteristic of systematic development in clonal hierarchies, it is worth noting that not only the mosaics studied by Emerson (1921) but also many other phenomena, such as the bridge–breakage–fusion (B–B–F) events reported by McClintock (1939, 1941), are recognizable because sister cells generate sister clones in a systematic, hierarchial relationship to one another. If these events or others, such as the fractional losses studied by Stadler (1939), developed in a field from broken-up, irregular clonal relationships, they would be exceedingly difficult to classify. The same applies to sectoring and variegation for markers on the short arm of chromosome 9 following breakage at Ds and to the somatic segregation of the Ac factor that originally demonstrated its mutability (McClintock, 1951a). At the other extreme of development, counting dots down to the two-celled size to identify relative frequencies of events at different stages (Nuffer, 1961) presupposes that pairs of adjacent cells are systematically related as sisters. It has, in fact, been generally recognized in all of these studies that the developmental process in this tissue is systematic; it might be desirable also to identify some parameters of the process, a consideration greatly stimulated by photographs (Figure 3) and observations of McClintock (1965), in which genetic marking identified distinct clones radiating from the center of the endosperm to the surface (like conical fans expanding outward into a sphere). We might, then, attempt to relate the morphological sequence of events with the clonal proliferation of the endosperm that culminates with the aleurone tissue surrounding the inner endosperm.

The development of the endosperm, which Randolph (1936) has described in morphological detail, begins with the triple fusion of the two polar nuclei and one sperm at around 24 h postpollination. The first nuclear division occurs within 2–4 h, and the daughter nuclei migrate to positions opposite each other, flanking the zygote. Within 12 h after the triple fusion a second division occurs, often a third, and the four or eight still-free nuclei distribute around the

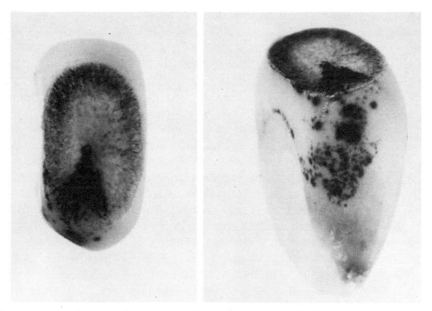

Figure 3. Developmental patterns in the endosperm revealed by mutational changes proliferated into clones. The kernel (shown in two views) underwent an early change in state of *Ac,* expressed in the wedge-like clone as internal dark-staining (*Wx*) subclones terminated by colored (*A*) aleurone clones [from McClintock (1965)].

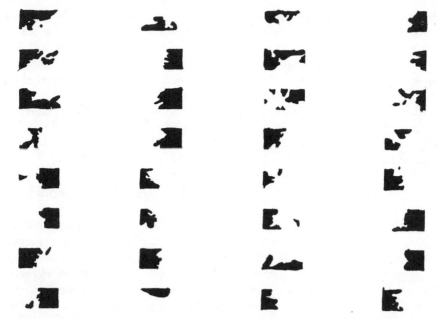

Figure 4. Half-fractional losses of C induced by UV-light treatment of pollen, with the aleurone tissue of the kernel (embryo toward top) represented diagrammatically as a rectangle within which losses are shown as white blank areas.

periphery of the embryo sac. The free-nucleate condition continues through the seventh to eighth division (128–256 nuclei) at 3 days postpollination; walls begin to form only after that stage. According to Randolph, there is striking synchrony in the early divisions, which continues into the cellular stage.

To identify clones of cells and relate them to this morphological description, we can use radiation to induce losses of dominant markers. Events induced at definite times before or during the developmental process allow us to deduce a more detailed picture; we can then specify division planes and the degree to which the patterns are systematic.

Treatment of pollen with UV will yield losses that are mostly half-fractionals in the endosperm (Stadler, 1939), in which we can ask whether the first division is oriented more often in certain planes than in others. Figure 4 shows diagramatically a sample from a group of 214 half-fractionals for loss of C induced by UV on pollen. That vertically divided halves are frequent is obvious; this is familiar to anyone who has conducted similar experiments. To define this numerically each pattern was judged, with the aid of a transparent straightedge, to be symmetrical or asymmetrical across lines dividing the rectangle vertically, horizontally, diagonally to the upper left, and diagonally to the upper right. Among 214 kernels, 182 were generally asymmetrical across a vertical line—that is, 85% were consistent with a vertical first division plane, but only 29% were consistent with a horizontal; 66% were consistent with left diagonal and 74% with right diagonal planes. These data show that the first division plane must be oriented vertically in the majority of the kernels. Since the embryo is directed toward the tip of the ear, the division plane is truly vertical, in planes radiating from the axis of the upright ear. With respect to the morphological observation that the two daughter nuclei of the first division migrate to positions flanking the zygote, this flanking must be to the left and right of the fertilized egg viewed from outside the ear.

These division events can be investigated more fully by X-irradiation during the early divisions, since fractional events of half-kernel size (similar to those from UV treatment of pollen) are induced by X-rays applied around 24 h postpollination at about the time of the triple fusion; smaller events can be anticipated with later treatments as the succeeding cell divisions progress through the sequence described by Randolph (1936). Ears from the cross of $a \times A$, X-irradiated with 1000 rad at 21 h after pollination and at 2-h intervals during the next 12 h, show a decrease in average area of sectors and an increase in frequency of loss (Figure 5) consistent with the occurrence of the first division and the S phase of the second cycle during this 12-h period. The earliest sectors are largely asymmetrical across the vertical plane (Figure 6) rather than the horizontal, but this difference disappears by the second division, when most are asymmetrical diagonally. Thus it is clear that the first-division plane is vertical and the second diagonal.

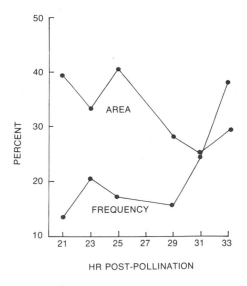

Figure 5. Percent of endosperm area affected and frequency of sectorial losses of A induced by 1000 rad of X-rays over 12-h period.

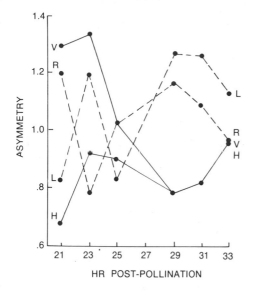

Figure 6. Relative asymmetry—across vertical (V), horizontal (H), left diagonal (L), and right diagonal (R) planes in the endosperm—of sectorial losses of A induced by X-rays over 12-h period.

We might suppose from these observations that the process continues, each division occurring at an angle to the preceding one in cycles. But to form conical fans like those pictured by McClintock (1965) requires a change in the process. According to Randolph (1936), instead of a block of cells dividing in unison there is a cambium-like proliferation at the surface of the block and cells are laid down internally by periclinal divisions at the periphery. Cell enlargement is primarily responsible for growth in the internal region, and anticlinal divisions at the surface must compensate for this growth. Considering the aleurone tissue as a sheet of cells that is the end product of this anticlinal division pattern, we can deduce the patterns of the final divisions from study of sectors that arise during the final divisions. In kernels of *C–I C C* constitution spontaneous losses of *C–I* occur frequently, resulting in clones of colored cells in a colorless ground. When the number of cells is determined in each such clone, the distribution (Figure 7) shows a distinctly binary pattern, in which clones with even numbers of cells predominate and clones with power-of-two cell numbers are particularly frequent. The power-of-two pattern shows that even these late divisions occur in unison, continuing the tendency that Randolph noted in the early divisions. When the individual sizes are accumulated into classes at powers of two, the frequency of the loss events is seen to be constant (per division) for the four, eight, and 16-cell sizes as expected, but lower for the one- and two-cell sizes, an effect for which I have found no satisfactory explanation. With regard to the division planes, we can ask what the clones may look like in form, and the relative frequencies of the various forms.

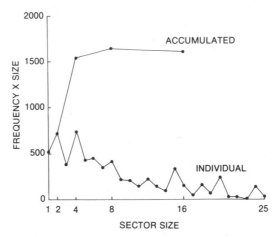

Figure 7. Spontaneous loss frequencies for *C–I* in the endosperm according to size (number of cells), distributed individually and by accumulation around powers of 2.

The clonal forms and frequencies (Figure 8) show that three, four, five-cell clusters are more often compact than linear, while six, seven, and eight-cell clusters are more often elongated than compact. Around 16, the clusters are frequently compact (square or triangular). Finally, clones that arise following mutation of *R–st* to self-color (Figures 9 and 10) commonly show groupings in squares (four and 16 cells) and rectangles (eight and 32 cells). These are precisely the sorts of distributions expected from division in alternating perpendicular planes.

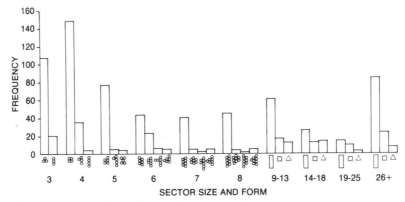

Figure 8. Frequencies of spontaneous sectorial losses of *C–I* according to sizes and forms of sectors.

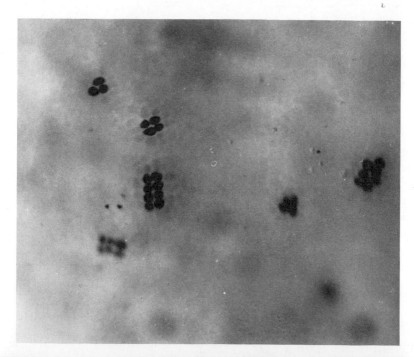

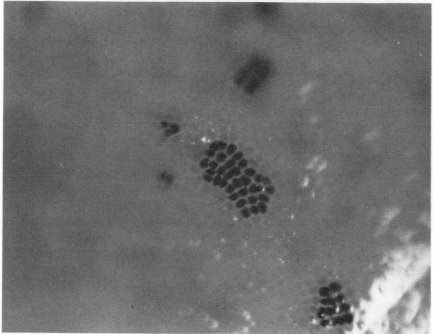

Figures 9 and 10. Clones of aleurone cells elaborated following mutations from *R–st* to self-color.

In summary, the characteristics of the aleurone tissue that have been useful for genetic studies are its bulk and visibility, its extended field display that is formed by systematic developmental processes, its active biosynthetic state, and its phasic constitution. Some parameters of the systematic developmental process by which the tissue is elaborated can be specified. In particular, the first-division plane is vertically oriented, whereas the second is diagonal, and late-division planes alternate perpendicularly, elaborating rectangular-square-rectangular-square forms. Because these events are systematic and are generally consistent, it has been possible to observe the effects of mutational changes in this field, follow the sequence of chromosomal breakage events as in the B–B–F cycle, and identify at least some aspects of the genetic control of biosynthetic processes in this tissue.

REFERENCES

Chen, Shu-Mei Hsu. 1973. Anthocyanins and their control by the C locus in maize. Ph.D. thesis. Columbia: University of Missouri.

Coe, E. H. 1955. Anthocyanin synthesis in maize, the interaction of A_2 and Pr in leucoanthocyanin accumulation. *Genetics* **40:** 568 (abstr.).

Coe, E. H., Jr. 1957. Anthocyanin synthesis in maize—A gene sequence construction. *Am. Natur.* **91:** 381–385.

Doerschug, E. B. 1973. Studies of dotted, a regulatory element in maize. I. Inductions of dotted by chromatid breaks. II. Phase variation of dotted. *Theor. Appl. Genet.* **43:** 182–189.

Dure, L. S. 1960. Site of origin and extent of activity of amylases in maize germination. *Plant Physiol.* **35:** 925–934.

East, E. M. and H. K. Hayes. 1911. Inheritance in maize. *Conn. Agr. Exp. Sta. Bull.* No. 167.

Emerson, R. A. 1921. Genetic evidence of aberrant chromosome behavior in maize endosperm. *Am. J. Bot.* **8:** 411–424.

Harborne, J. B. and G. Gavazzi. 1969. Effect of Pr and pr alleles on anthocyanin biosynthesis in *Zea mays*. *Phytochemistry* **8:** 999–1001.

Harvey, B. M. R. and A. Oaks. 1974. The hydrolysis of endosperm protein in *Zea mays*. *Plant Physiol.* **53:** 453–457.

Kirby, L. T. and E. D. Styles. 1970. Flavonoids associated with specific gene action in maize aleurone, and the role of light in substituting for the action of a gene. *Can. J. Genet. Cytol.* **13:** 934–940.

McClintock, B. 1939. The behavior in successive nuclear divisions of a chromosome broken at meiosis. *Proc. Nat. Acad. Sci. U. S. A.* **25:** 405–416.

McClintock, B. 1941. The association of mutants with homozygous deficiencies in *Zea mays*. *Genetics* **26:** 542–571.

McClintock, B. 1951a. Chromosome organization and genic expression. *Cold Spring Harbor Symp. Quant. Biol.* **16:** 13–47.

McClintock, B. 1951b. Mutable loci in maize. *Carnegie Inst. Wash. Yearbook* **50:** 174–181.

McClintock, B. 1965. The control of gene action in maize. *Brookhaven Symp. Biol.* **18:** 162–184.

Neuffer, M. G. 1966. Stability of the suppressor element in two mutator systems at the A_1 locus in maize. *Genetics* **53:** 541–549.

Neuffer, M. G. and G. Ficsor. 1963. Mutagenic action of ethyl methanesulfonate in maize. *Science* **139:** 1296–1297.

Nuffer, M. G. 1955. Dosage effect of multiple *Dt* loci on mutation of *a* in the maize endosperm. *Science* **121:** 399–400.

Nuffer, M. G. 1957. Additional evidence on the effect of X-ray and ultraviolet radiation on mutation in maize. *Genetics* **42:** 273–282.

Nuffer, M. G. 1961. Mutation studies at the A_1 locus in maize. I. A mutable allele controlled by *Dt. Genetics* **46:** 625–640.

Peterson, P. A. 1961. Mutable a_1 of the *En* system in maize. *Genetics* **46:** 759–771.

Peterson, P. A. 1966. Phase variation of regulatory elements in maize. *Genetics* **54:** 249–266.

Randolph, L. F. 1936. Developmental morphology of the caryopsis in maize. *J. Agr. Res.* **53:** 881–916.

Reddy, G. M. and E. H. Coe, Jr. 1962. Inter-tissue complementation: A simple technique for direct analysis of gene-action sequence. *Science* **138:** 149–150.

Rhoades, M. M. 1936. The effect of varying gene dosage on aleurone colour in maize. *J. Genet.* **33:** 347–354.

Rhoades, M. M. 1938. Effect of the *Dt* gene on the mutability of the a_1 allele in maize. *Genetics* **23:** 377–397.

Rhoades, M. M. 1952. The effect of the bronze locus on anthocyanin formation in maize. *Am. Natur.* **86:** 105–108.

Stadler, L. J. 1928. Genetic effects of x-rays in maize. *Proc. Nat. Acad. Sci. U. S. A.* **14:** 69–75.

Stadler, L. J. 1930. Recovery following genetic deficiency in maize. *Proc. Nat. Acad. Sci. U. S. A.* **16:** 714–720.

Stadler, L. J. 1939. Genetic studies with ultraviolet radiation. *Proc. VII. Internat. Congr. Genet.* pp. 269–276.

Stadler, L. J. 1944. The effect of X-rays upon dominant mutation in maize. *Proc. Nat. Acad. Sci. U. S. A.* **30:** 123–128.

Stadler, L. J. and F. M. Uber. 1942. Genetic effects of ultraviolet radiation in maize. IV. Comparison of monochromatic radiations. *Genetics* **27:** 84–118.

Chapter 31

PIGMENT-DEFICIENT MUTANTS: GENETIC, BIOCHEMICAL, AND DEVELOPMENTAL STUDIES

D. S. Robertson, I. C. Anderson, and M. D. Bachmann
*Departments of Genetics, Agronomy, and Zoology,
Iowa State University,
Ames*

Mutants can be useful in answering questions about plant biochemistry and development. We are not the first workers to realize this, and informative studies utilizing mutants of many species have been reported in the literature. Because of space limitations we are not able to review these earlier reports.

The intent of this chapter is to demonstrate how an interdisciplinary study, utilizing some selected plastid pigment-deficient mutants of maize, has provided information on the biosynthesis of plastid pigments, their interrelationships, and how pigment alterations might be related to the development of plastid structures.

Journal Paper No. J-8292 of the Iowa Agriculture and Home Economics Experiment Station, Ames. Project No. 2035, supported in part by grants G-19395, GB-1584, and GB-5538 from the National Science Foundation. We gratefully acknowledge this support and the expert assistance of Nancy E. Thompson in carrying out some of the electron-microscopic studies.

GENETIC MATERIAL

The white endosperm-albino seedling (white-albinos) mutants used in our studies have been investigated by several workers [see Robertson (1975) for a review]. Mutants of 14 different loci have been characterized genetically. In addition, there are several (viz., *lc, ly,* and *z*) that appear in the literature in which little or no genetic information has been reported.

At some loci, alleles are known that yield pale green (pastel) seedlings, and at other loci, green seedling alleles have been found. The seedling phenotype of the *cl* mutant can vary from albino to pale green to green, depending on which of several known alleles are present at an independent modifier locus. Mutable alleles have been found at several of the white-albino loci. None of these has been characterized in detail, but preliminary investigations suggest that independent loci are involved in the control of mutability in each instance.

In recent years we also began an investigation of the luteus mutants. Although these mutants are characterized by yellow seedlings, they are far from a homogeneous group. They can vary in the intensity of yellow pigmentation from very pale (off-white) to deep yellow. They also are quite variable with respect to the presence of chlorophyll. Some seem completely devoid of this pigment, but others have various amounts of chlorophyll. Different mutants will vary with respect to amount of chlorophyll, when it develops, and how it is distributed in the leaf (e.g., tip greening or uniformly distributed in the leaf). Preliminary observations suggest that the amount of greening observed in some mutants may vary depending on environmental factors (e.g., heat and/or light).

Although luteus mutants have been known for some time, not much is known about their genetics. Some have been placed to chromosomes, and a very few have been mapped. The most extensive mapping has been done for six nonallelic mutants located on chromosome 6 (Robertson, 1973).

METHODS

Biochemical Studies

Procedures for the isolating and determining the concentration of carotenoids, chlorophyll, and protochlorophyllide have been described in previous papers (Bachmann, Robertson, et al., 1973; Richardson, Robertson, et al., 1961; Robertson, Bachmann, et al., 1966).

For carotenoid precursor determinations, dim-light grown (0.5–1.0 ft-c) seedlings were used, and pigments extracted by the same procedure as those for chlorophyll and carotenoids. The optical density of the hexane extract was read on a Spectronic 505 at 285 nm, 348 nm, 400 nm, and 472 nm for phytoene, phytofluene, ζ-carotene, and lycopene, respectively. The following formulas were used to determine concentration of the four precursors: (a) phytoene (O.D./140 × vol./wt. = mg/g), (b) phytofluene (O.D./150 × vol./wt. = mg/g), (c) ζ-carotene (O.D./220 × vol./wt. = mg/g), and (d) lycopene (O.D./320 × vol./wt. = mg/g) (Porter and Lincoln, 1950).

Cytological Studies

The steps in preparing material for cytological studies have been described by Bachmann, Robertson, et al. (1967, 1969, 1973).

BIOCHEMICAL STUDIES

White Albinos

Most of these mutants, when grown in a seedling bench in the light, have paper-white seedlings. Early in the morning, some, however, will be pale green at the very base where the seedling is just emerging from the sand. Seven of these mutants, besides being albino, are also viviparous (germinate prematurely). In such mutants the prematurely germinating seedlings on the ear frequently will be pale green when the husks are opened at harvest time. These observations suggest that the albinos have the ability to synthesize chlorophyll under conditions of low light intensities. To test the ability of these mutants to synthesize chlorophyll, some dark-grown mutants were analyzed for the presence of protochlorophyllide. While some dark-grown seedlings were exposed to light for 1 min and extracted to determine chlorophyllide formation, other dark-grown seedlings were exposed to 1 min of light and then returned to the dark for 1 h, after which the pigments were extracted to determine the amount of chlorophyll synthesized. All mutants tested had the ability to synthesize protochlorphyllide and to convert it to chlorophyll (Table 1). Because these mutants were not in an isogenic background, it was not possible to determine how much of the observed variation was due to differences in genetic background and how much was due to differences in the synthetic potential of the various mutants.

Dim-light-grown mutant seedlings also have the ability to synthesize chlorophyll (Table 2). Again, variation is considerable, due in part to differences in genetic background of the stocks and environmental differences. The foregoing experiments establish that all the white-albino mutants tested have the ability to synthesize chlorophyll. These results confirm the observations of Smith, Durham, et al. (1959) on chlorophyll synthesis in the white-albino mutants.

When grown in the light, most of these mutants also lack colored carotenoids. Grown under dim-light conditions, however, many produce

TABLE 1. Protochlorophyllide, chlorophyllide, and chlorophyll content of a group of white-albino mutants

Mutant	Protochlorophyllide dark grown (mg/g fr. wt.)	Chlorophyllide dark grown + 1 min of light (mg/g fr. wt.)	Chlorophyll dark-grown + 1 min of light followed by 1 h of dark (mg/g fr. wt.)
Normal	0.0045	0.0040	0.0032
y10-8624	0.0027	0.0030	0.0019
lw	0.0048	0.0042	0.0036
lw2	0.0062	0.0042	0.0044
cl-7716	0.0058	0.0031	0.0031
vp5-Mumm #1	0.0022	0.0020	0.0008
y7-Wisconsin #2	0.0039	0.0018	0.0014
w3	0.0028	0.0020	0.0014
al	0.0028	0.0020	0.0008

TABLE 2. Chlorophyll synthesis in dim-light-grown normal and white-albino seedling

Mutant	Light intensity	Temperature	Chlorophyll (mg/g fr. wt.)
Normal	5	22°C	0.355
Normal	0.5-1	22°C	0.226
Normal	0.5-1	27°C	0.238
cl	5	22°C	0.132
cl	0.5-1	27°C	0.032
w3	0.5-1	22°C	0.173
w3	0.5-1	27°C	0.146
lw	0.5-1	27°C	0.077
lw-6474	0.5-1	22°C	0.031
vp5	0.5-1	22°C	0.093
vp5-Mumm #1	0.5-1	22°C	0.376
vp5-Mumm #2	0.5-1	22°C	0.248
vp5	0.5-1	27°C	0.123
al	0.5-1	27°C	0.175
y7-Wisconsin #2	0.5-1	22°C	0.610
y7	0.5-1	27°C	0.055
ps	0.5-1	27°C	0.037
y9	0.5-1	27°C	0.179
y10-8624	0.5-1	22°C	0.022

precursors of carotene that can be characterized by their absorption spectra. Of the 14 known mutants of the white endosperm–albino class, seven accumulate two or more of these precursors (Table 3). Other workers have reported similar carotene-precursor accumulations in some of these mutants. Sandler, Laber, et al. (1968) and Troxler, Lester, et al. (1969) found phytoene and phytofluene in the albescent mutant. They did not report the presence of ζ-carotene, although Troxler, Lester, et al. (1969) indicated that a "unidentified substance" was present. Treharne, Mercer, et al. (1966) confirmed our observation that $w3$ accumulated phytoene, phytofluene, and ζ-carotene, but they also found indications of neurosporene in some samples of $w3$. Faludi-Daniel's research group (Garay, Demeter, et al., 1972; Horvath, Kissimon, et al. 1972) have reported on the accumulation of carotenoid precursors in two mutants; one (z) accumulates phytoene, phytofluene, and ζ-carotene, and the other (ly) phytoene, phytofluene, and lycopene as well as α and β-carotene, Δ ,and γ-carotene, antheraxanthin, and other xanthophylls. The relationship of these mutants to the ones in our studies is unknown. The mutant that accumulates ζ-carotene could be any one of several of ours or a completely new mutant. The lycopene mutant could be the same as our ps. The lc mutant of Andres (1951), which he reports accumulates lycopene, probably is allelic to ps.

A proposed biosynthetic pathway from phytoene to β-carotene is given in Figure 1. It seems that $w3$ exerts its affect at the point of conversion of neurosporene to lycopene in as much as Treharne, Mercer, et al. (1966) have reported the accumulation of neurosporene in this mutant. The accumulation of lycopene by ps suggests that, in corn, β-carotene synthesis passes through this precursor and that ps might be blocking the conversion of lycopene into γ-carotene. How the four mutants, al, $vp5$, $vp9$, and $y9$, fit into the scheme of

TABLE 3. Accumulation patterns of carotene precursors for seven mutants grown at 27°C in dim light (0.5–1 ft-c)

Mutant	Phytoene (mg/g fr. wt.)	Phytofluene (mg/g fr. wt.)	ζ-Carotene (mg/g fr. wt.)	Lycopene (mg/g fr. wt.)
vp2	0.1153	0.0092	0	0
vp5	0.0562	0.0039	0.0019	0
al	0.0346	0.0135	0.0043	0
y7	0.0301	0.0156	0.0123	0
w3	0.0419	0.0153	0.0089	0
ps[a]	0.0876	0.0288	0	0.0635
y9	0.0437	0.0340	0.0350	0

[a] Determination made on extract of scutella. This mutant is strongly viviparous and vigorous seedlings are difficult to obtain.

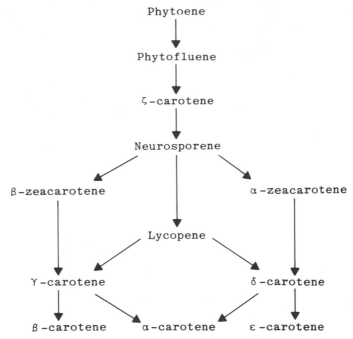

Figure 1. Proposed biosynthesis of carotenes (Goodwin and Mercer, 1972).

things is not obvious because these four nonallelic mutants accumulate the same three precursors.

Each step in the pathway from phytoene to lycopene involves the creation of a single double bond (a stepwise desaturation). Attempts at using mutants to dissect the carotene biosynthetic pathway between phytoene and lycopene generally have not been very successful. Jensen, Cohen-Bazire, et al. (1961), as a result of studies with purple bacteria, concluded that a single enzyme was responsible for the dehydrogenations from phytoene to neurosporene. De La Guardia, Aragon, et al. (1971), concluded from studies of mutants in *Phycomyces* that the enzymes involved in this pathway are organized as a complex that operates as an assembly line. Each enzyme receives the substrate from the previous one and can pass it to the next. Once attached to the complex, the substrate stays attached until the end product is produced. Defective members of the complex will result in the liberation of the substrate from the complex and the accumulation of precursor molecules. Goldie and Subden (1973) found a *Neurospora* mutant that accumulates phytoene, but failed in several "mutant hunts" to find strains with genetic lesions for steps after phytoene dehydrogenase. As did Jensen, Cohen-Bazire, et al. (1961), they concluded that a single multifunctional enzyme is responsible for the dehydrogenations involved. Garber et al. (1975) working with *Ustilago violacea,* propose a multienzyme aggregate composed of four identical dehydrogenases and one or two identical cyclases to accomplish carotenogenesis in this fungus. Kushwaha, Suzue, et al. (1970) were able to demonstrate *in vitro* the steps involved in the conversion of phytoene to lycopene by using enzyme systems extracted from fruits of different tomato varieties.

The results of studies on the maize mutants are not without their difficulties of interpretation. They suggest that more than one enzyme might be involved in the stepwide dehydrogenations. Because *vp2* accumulates phytoene and phytofluene, but none of the more desaturated precursors, this mutant may represent a genetic lesion for the phytofluene to ς-carotene step. But how are the four mutants that accumulate phytoene, phytofluene, and ς-carotene related to the carotenoid biosynthetic pathway? These all seem involved in the dehydrogenation responsible for converting ς-carotene to neurosporene. Four mutants for such a simple biochemical step would seem to strain the classical "one gene, one enzyme" hypothesis or even the more modern "one gene, one polypeptide" hypothesis. Perhaps our tests were not sensitive enough to pick up small amounts of the precursors later than ς-carotene that might accumulate in these four mutants. Such was the case with *w3*, which Treharne, Mercer, et al. (1966) found accumulates some neurosporene. Our tests did not reveal the presence of this compound. At this time, however, there is no evidence for the presence of any precursors later than ς-carotene for any of these four mutants. The identical accumulation pattern of these four mutants might be expected if, as has been proposed (Garber et al., 1975) an aggregate of four identical dehydrogenases is responsible for the steps between phytoene and lycopene. The dehydrogenase could be a multimeric enzyme consisting of four or more different monomers, each under the control of a different structural gene. If so, it is possible that the four mutants are the structural genes for different monomers, each of which is essential for the dehydrogenase to function in the ς-carotene to neurosporene step. It may be that none of the mutants is acting directly in the biosynthetic pathway controlling the production of altered enzymes that function in carotenoid synthesis. It could be that these genes alter membrane structure or some other element in the milieu of the chloroplasts that is necessary for the normal functioning of the enzyme controlling the step from ς-carotene to neurosporene. For this enzyme there may be several genetically controlled properties of the chloroplast, which, if altered, will interfere with its normal functioning. Hence there are four mutants that seem to affect the same biosynthetic step. Obviously, much more research is needed on the relationship of these four mutants to carotenoid biosynthesis. For that matter, indirect effects, such as alteration of membrane attachment sites rather than direct effects on enzyme structure, might be involved in all of these mutants.

A difference was observed in the absorption spectra for lycopene in the scutella and leaves of *ps* (Figure 2). The scutellar lycopene has an absorption spectrum characteristic of translycopene while the spectrum of the lycopene from the leaves seems to be that of a *cis* isomer, possibly neo-D-lycopene (Zechmeister, 1962). The cause for the difference in isomerism in the two tissues is not certain. The scutella were obtained directly from freshly harvested ears that had been protected from light by the husks, whereas the leaves were from dim-light-grown plants. Perhaps the presence of light resulted in the conversion of the trans isomer into the *cis* form in the seedlings.

Seven of the white-albino mutants do not seem to accumulate any carotenoid

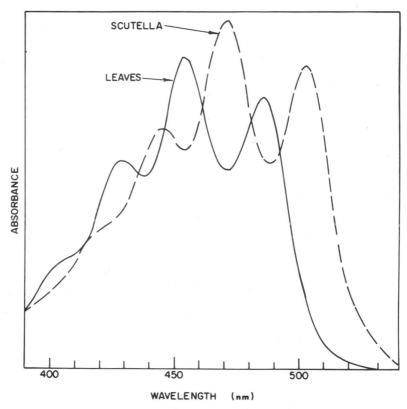

Figure 2. Absorption spectra for lycopene obtained from scutella and leaves of *ps*.

precursors that absorb in the UV portion of the spectrum. Presumably carotenoid synthesis in these is blocked before phytoene.

In summary, the studies on the pigment development in the white-albino mutants have shown that they retain the ability to synthesize chlorophyll but have defective carotenoid synthesis.

If these mutants can make chlorophyll, why is this pigment not observed in light-grown material? Work on photosynthetic bacteria by Cohen-Bazire and Stainer (1958) and Fuller and Anderson (1958) have shown that colored carotenoids are required to protect bacteriochlorophyll from photodestruction. Stainer (1959) proposed that carotenoids played this protective role in all photoautotrophs. The white-albino mutants provided the opportunity to test this hypothesis for higher plants because it has been proved that they can make chlorophyll, yet lack colored carotenoids. Anderson and Robertson (1960, 1961), Richardson, Robertson, et al. (1961), and Robertson, Bachmann, et al. (1966) were able to demonstrate that colored carotenoids probably were involved in protecting chlorophyll from photoautooxidation in corn.

Biochemical studies of the luteus (yellow seedling) mutants are just in their infancy. Among these mutants should be those that are the reciprocal of the white-albinos (i.e., defective in their chlorophyll synthesis, but normal with respect to carotenoid synthesis). As indicated in the Genetic material section, many of these mutants vary with respect to the intensity of yellow pigmentation and the amount of chlorophyll present. Table 4 lists the chlorophyll, carotene, and xanthophyll concentrations for 20 mutants that could be grouped in the luteus class grown under 2000 ft-c of light and at 27°C. Most of these mutants synthesis some chlorophyll. If any or all of these are chlorophyll mutants, they are at best leaky mutants. None comes close to possessing normal levels of the carotenoid pigments. This is true even for the mutants classified as dark yellow. Limited tests of dim-light grown material (Table 5) suggest that chlorophyll production is much reduced. Normal seedlings produce less carotene under dim light, whereas three of the mutants tested produced more carotene at the lower light intensity. The one mutant that produced less carotene at dim light (*l*Blandy #4*) also produced the most under bright light. No consistent pattern of xanthophyll accumulation was associated with difference in light intensities.

Several luteus mutants have been tested for protochlorophyllide synthesis when grown in the dark, with and without feeding of δ-aminolevulinic acid. These tests are just in the preliminary stages, but of the mutants tested all but one unfed mutant have shown some protochlorophyllide synthesis, but not near the extent of normals, again indicating that the mutants are leaky. Most of these mutants show an increase in protochlorophyllide synthesis when supplied δ-aminolevulinic acid and indications of the accumulation of one or more of the intermediates in chlorophyll synthesis. One mutant, *l*Blandy #4*, produces

TABLE 4. Phenotype and chromosome location and chlorophyll, carotene, and xanthophyll concentrations for 23 chlorophyll deficient mutants grown at 27°C and under 2000 ft-c of light

Mutant	Phenotype	Chromosome	Chlorophyll (mg/fr. wt.)	Carotene (mg/fr. wt.)	Xanthophyll (mg/fr. wt.)
Normal	green	—	1.674	0.0980	0.1140
l3	almost albino	—	0	0.0015	0.0055
l4	very slight yellow	10	trace	trace	0.0066
l7	pale yellow–green	9	0.3220	0.0077	0.0395
l10	dark yellow	6	trace	0.0029	0.0140
l²1106	yellow, leaf tips slightly green	4	0.0992	0.0019	0.0187
l²4106	light yellow–green	—	0.0781	0.0028	0.0103
l²4117	yellow–green	—	0.1824	0.0051	0.0335
l11 (l²4120)	yellow, leaf tips green	6	0.0935	0.0021	0.0175
l12 (l²4920)	pale yellow, leaf tips green	—	0.0486	0.0021	0.0117
l²4923	dark yellow, some green	—	0.2667	0.0205	0.0350
py²PI 177593	pale yellow, leaf tips green	—	0.1914	0.0059	0.0165
yg²PI 183367	yellow with some green	—	0.2380	0.0070	0.0375
yel nec²PI 217486	yellow, necrotic leaf tips	8	0.1639	0.0128	0.0369
l²Blandy #3	yellow with some green	6	0.3115	0.0176	0.0432
l²Blandy #4	dark yellow	—	0.0055	0.0240	0.0195
l²Brawn #1	yellow–green	6	0.5939	0.0222	0.0800
yd	yellow–dwarf	3	0.0140	0.0015	0.0070
w	very pale yellow	6	0.0180	0.0008	0.0091
w²Italy #1	very pale yellow	—	0.0140	0.0000	0.0030
w15 (w²8896)	pale yellow	6	trace	0.0006	0.0090

TABLE 5. Chlorophyll, carotene, and xanthophyll concentrations in four luteus mutants grown under 0.5-1 ft-c at 27°C (selected from Table 2, Bachmann, Robertson, et al. (1967)]

Mutant	Chlorophyll (mg/g fr. wt.)	Carotene (mg/g fr. wt.)	Xanthophyll (mg/g fr. wt.)
Normal	0.0245	0.0212	0.0182
w15 (w²-8896)	none	0.0016	0.0045
l²-Blandy #4	none	0.0142	0.0180
l3	none	0.0042	0.0030
l11 (l²-4120)	0.0580	0.0075	0.0130

little or no protochlorophyllide in the dark, but when fed δ-aminolevulinic acid it accumulates an as yet undetermined intermediate of chlorophyll synthesis.

CHLOROPLAST DEVELOPMENT IN MUTANTS

As is true for other C4 plants, the structure of mesophyll chloroplasts in maize differs from those found in the bundle sheath cells. The mesophyll chloroplasts (Figure 3) have an intricate membrane system consisting of stacks of closely packed disk-shaped grana thylakoids connected by more widely spaced stroma lamellae. This complex membrane structure is imbedded in the stroma, and the whole structure is enclosed by a double-unit membrane. The chloroplasts of the bundle sheath cells usually are more elongated than those of mesophyll cells and have long lamellae, with only an occasional small two- or three-disc grana. Our studies have been concerned primarily with the mesophyll chloroplasts of the classes of pigment-deficient mutants already described in this chapter. To obtain the maximum information about chloroplast development from mutant material, plants have to be studied under a variety of environmental conditions. Thus far our studies have been with bright-light-grown, dark-grown, and dim-light-grown plants, and in a couple of selected cases, development has been followed in dark-grown material that has been exposed to bright light for up to 24 h. Temperature also influences development, but this environment factor has been involved in only a few of our investigations.

Light-grown Material

Bright light is responsible for the bleaching of chlorophyll in the albino mutants. Thus one might expect to find the greatest departure from normal development under such conditions.

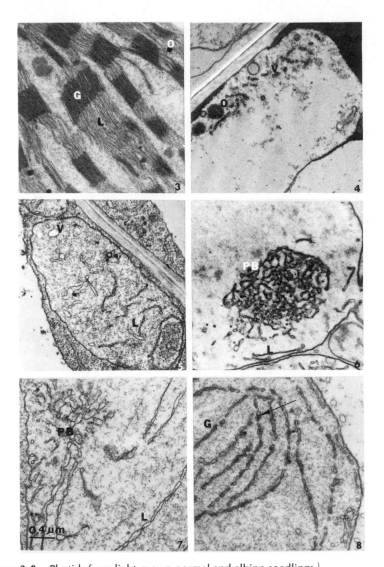

Figures 3–8. Plastids from light-grown normal and albino seedlings.[1]

Figure 3. Normal mesophyll: 12 h of light (2000 ft-c), 12 h of dark, 22°C. Magnification × 16,500

Figure 4. Albino (w3): 12 h of light (2000 ft-c), 12 h of dark, 27°C (Bachmann, Robertson, et al., 1967). Magnification × 16,500

Figure 5. Albino (vp5): 12 h of light (2000 ft-c), 12 h of dark, 27°C. Magnification × 16,500

Figure 6. Albino (cl): 12 h of light (2000 ft-c), 12 h of dark, 27°C. Magnification × 16,500
Figure 7. Albino (cl): 12 h of light (2000 ft-c), 12 h of dark, 22°C. Magnification × 16,500
Figure 8. Albino (cl): 12 h of light (2000 ft-c), 12 h of dark, 22°C. Magnification × 16,500

472

Albino Mutants.   In a limited sampling of our material we have found a difference between the chloroplasts of the mutants that accumulate later carotenoid precursors and those that do not. Chloroplasts of the accumulators, $w3$ (Figure 4) and $vp5$-*Mumm #1* (Figure 5) frequently are irregular in shape and often almost devoid of internal structures. Occasionally vesicles are present and also short segments of lamellae, some of which appear to have swollen into irregular vesicles. In contrast, the nonaccumulating mutants usually show more internal structure. One of these, *cl,* frequently has chloroplasts with a loosely organized prolamellar body (Figures 6, 7) and occasional scattered, somewhat swollen, and perforated lamellae (Figure 7). Occasional chloroplasts are found with lamellae that show foldbacks or small grana stacks (Figure 8). Small grana stacks also were observed in the plastids of *lw,* another nonaccumulator.

Luteus Mutants.   The chloroplasts of $w15$ ($w\dsuppress{}8896$) seedlings grown in the light are irregular in shape, often with loosely organized prolamellar bodies consisting of swollen structural elements (Figure 9). Occasional plastids are observed that also have scattered lamellae, and rarely, groups of parallel, but nonappressed, lamellae are observed. Larger-than-normal osmiophilic globules also are observed. This mutant possessed little or no chlorophyll under these conditions, but did accumulate some carotene and xanthophyll (Table 4). Like *cl,* it is capable of making abnormal elements of the prolamellar body in the absence of protochlorophyllide.

The plastids of *l*Blandy #4,* which also does not have chlorophyll but accumulates more carotenoids than $w15$ (Table 4), has quite a different structure. A few nearly empty plastids are found that have only an occasional vesicle. Most plastids had parallel lamellae and large dense osmiophilic globules (Figure 10). Some, however, had more irregularly arranged lamellae. No hint of a prolamellar body was observed. Perhaps this is related to the fact that this mutant does not make protochlorophyllide.

Pale-green Mutants.   The pale green $w3$-8686 allele of $w3$, when grown in the light, produces plastids that usually have the normal "football" shape. Most have many lamellae running the length of the plastid. Several small grana, of five to six or fewer disks are found associated with the lamellae, usually, although not universally near the ends of the lamellae (Figure 11).

---

[1] Key to symbols used in Figures 3–35: C—cytoplasmic crystal, CE—common center from which grana radiate, D—two disc thylakoid units, G—grana, L—lamellae, LG—large and/or long grana, O—osmiophilic globules, PB—prolamellar body, PL—parallel groups of lamellae, S—starch, TS—thylakoid stack, V—vesicles (see text for discussion of structural details of micrographs).

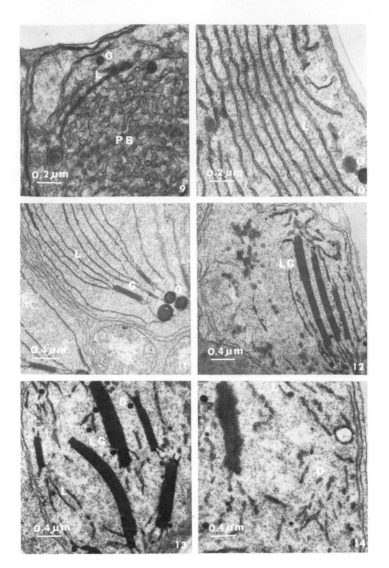

Figures 9–14. Plastids from light-grown mutant seedlings.

Figure 9. Luteus [*w*15 (*w*-8896)]: 12 h of light (2000 ft-c), 12 h of dark, 27°C. Magnification × 33,500

Figure 10. Luteus (*l*-*Blandy* #4): 12 h of light (2000 ft-c), 12 h of dark, 27°C. Magnification × 33,500

Figure 11. Pale green (*w*3-8686): 12 h of light (2000 ft-c), 12 h of dark, 27°C (Bachmann, Robertson, et al., 1967). Magnification × 16,500

Figure 12. Pale green suppressed albino (*cl cl Clm2 Clm2*): 12 h of light (2000 ft-c), 12 h of dark, 22°C. Magnification × 16,500

Figure 13. Pale green suppressed albino (*cl cl Clm5 clm*): 12 h of light (2000 ft-c), 12 h of dark, 22°C. Magnification × 16,500

Figure 14. Pale green suppressed albino (*cl cl Clm5 clm*): 12 h of light (2000 ft-c), 12 h of dark, 22°C. Magnification × 16,500

Osmiophilic globules are frequently found. It seems that this mutant, which has both carotenoids and chlorophyll, but markedly reduced amounts compared with normals [i.e., chlorophyll 11.1%, carotene 7.9, and xanthophyll 45.0% of normal at 22°C; Richardson, Robertson, et al. (1961)], is very restricted in its ability to form grana.

Suppressed Albino Mutants.   The albino phenotype of *cl* can be partly or completely suppressed, depending on the combinations of alleles present at the independent *Clm* locus (Robertson, Bachmann, et al., 1966). Homozygosity at this locus for *Clm2* results in seedlings with 55.2%, 51.1%, and 47.5% of the normal levels of chlorophyll, carotene, and xanthophyll, respectively, at 22°C. Plants heterozygous for *Clm5* have 39.5%, 22.9%, and 46.9% of the normal levels of chlorophyll, carotene, and xanthophyll, whereas plants homozygous for *Clm5* produce 93.6%, 93.2%, and 84.9% of the normal levels of chlorophyll carotene and xanthophyll, respectively. Homozygous *Clm3* plants have 80.4%, 67.4%, and 99.7% of normal for the same pigments. The chloroplasts of the *cl cl Clm2 Clm2* seedlings were very diversified. Some approached normal in appearance while others had a few lamellae with a few small grana stacks. Still others had a limited membrane system with a few very long grana made up of closely appressed thylakoids (Figure 12), and others were of irregular shape and seemed completely disorganized with scattered vesicles and fragments of membranes. It may be that the reduced levels of plastid pigments observed in this material are not distributed uniformly among the chloroplasts and that the different plastid morphologies are the result of variations in pigment present in any given chloroplast.

The *cl cl Clm5 clm* seedlings also had a variety of chloroplasts. Most were more round than normal. Some had a few grana that were normal in size and appearance, but with few connecting stroma lamellae and scattered lamellar fragments elsewhere in the plastids. Others (Figure 13) had a few very long grana consisting of closely packed thylakoids with greatly reduced loculi. Also in these plastids, fragments of lamellae and osmiophilic globules were scattered throughout the stroma in no organized fashion. In other chloroplasts, there were long grana in which the thylakoid stacking was more normal. Rarely, chloroplasts were observed in which bits of lamellae, often associated into two-disc units, were scattered at random throughout the stroma (Figure 14). Perhaps the diversity in types of plastids of this mutant material may again be due to differences in pigment concentrations among plastids.

Plants homozygous for the *Clm5* modifier gene (*cl cl Clm5 Clm5*), which have near normal levels of chlorophyll, carotene, and xanthophyll, produce a nearly normal appearing chloroplast (Figure 15). The *cl cl Clm3 Clm3* plastids all were very uniform in appearance. They all approached normal in shape and internal structure (Figure 16). The observations made on *cl cl Clm5 Clm5* and

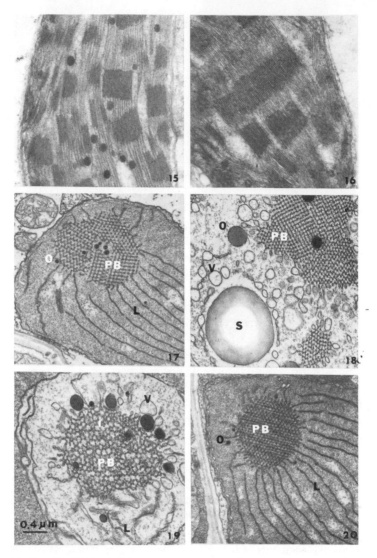

Figures 15–16. Plastids from light-grown mutant seedlings.

Figures 17–20. Plastids from dark-grown normal and albino seedlings.

Figure 15. Green suppressed albino (*cl cl Clm5 Clm5*): 12 h of light (2000 ft-c), 12 h of dark, 22°C. Magnification × 16,500

Figure 16. Green suppressed albino (*cl cl Clm3 Clm3*): 12 h of light (2000 ft-c), 12 h of dark, 22°C. Magnification × 16,500

Figure 17. Normal: 27°C. Magnification × 16,500

Figure 18. Albino (*w3*) plastid with a crystalline-like prolamellar body: 27°C (Bachmann, Robertson, et al., 1967). Magnification × 16,500

Figure 19. Albino, *w3*, plastid with a nonordered prolamellar body: 27°C (Bachmann, Robertson, et al., 1967). Magnification × 16,500

Figure 20. Albino (*cl*): 22°C. Magnification × 16,500

476

cl cl Clm3 Clm3 suggest that, when seedlings approach normals in the level of their plastid pigments, so does their plastid morphology.

Dark-grown Material

Bright-light conditions may not be the most conducive to the expression of the optimum plastid development of mutants. We have seen from the chemical studies that many of the mutants, under conditions of reduced light, retain the ability to synthesize chlorophyll, which is rapidly destroyed under bright light. When grown in the dark the wide divergence in structure found between mutant and normal chloroplasts of light-grown material is not observed. Because both normal and white-albino mutants can produce protochlorophyllide in the dark, and because it has been shown that the accumulation of protochlorophyllide in dark-grown material is associated with the formation of the crystalline-like prolamellar body in etioplasts (Henningsen and Boynton, 1969, 1970; Kahn, 1968a), prolamellar bodies might be expected in both normals and white albinos. Figure 17 shows a typical plastid of a dark-grown normal seedling. Dominating the structure is a crystalline-like prolamellar body with long, continuous lamellae that radiate from it. Also present are small, dense osmiophilic globules.

Albino Mutants. The plastids of dark-grown *w3* seedlings were of two types. About one-third had organized (crystalline-like) prolamellar (Figure 18) bodies. These plastids lacked the lamellae observed in normal seedlings. Numerous rounded vesicles were observed as were numerous globules of various sizes and density. Starch was observed in some. About two-thirds of the chloroplasts had a more random, more vesiculate system of tubules making up the prolamellar body (Figure 19). Such plastids had irregular and flattened vesicles and isolated segments of lamellae.

Most of the chloroplasts of *cl* seedlings (Figure 20) have crystalline-like prolamellar bodies with a well-developed system of lamellae radiating from them. In most respects, they are very similar to normal plastids. An occasional plastid is observed with only a few short lamellae. Most have small dense osmiophilic globules.

Pale-green Mutants. The plastids of the pale-green allele of *w3*, *w3-8686*, have essentially normal-looking plastids (Figure 21) with crystalline-like prolamellar bodies and an extensive system of lamellae. The globules appear to be larger and less electron dense than those of normals or *cl*.

Plastid Development in Dim-light-grown Plants

Dim-light growing conditions often are favorable for studying the full potential for the development in plastid structure of mutants because these conditions often permit the accumulation of plastid pigments and their precursors. Such accumulation does not occur in bright light because of the photooxidation that takes place under these conditions. Bachmann, Robertson, et al. (1973) described the development of plastids in albino, pale-green, yellow-green, and luteus (yellow) seedlings grown at 0.5–1 ft-c. Under this light intensity at 27°C, normals make about one-eighth of the chlorophyll and one-fifth of the carotenoids of their light-grown counterparts (2000 ft-c). The plastids of dim-light-grown normals are more spindle-shaped and have continuous, long lamellae and grana stacks of two or three thylakoids. Grana vary in length from very short ones to those that extend about a third the length of the plastids (Figure 22). Normal plastids from material grown at 5 ft-c and 22°C accumulate more chlorophyll than those grown at 0.5–1 ft-c (27°C) (0.355 mg/mg and 0.238 mg/gm, respectively; Table 2). At this greater light intensity, there is a much more extensive development of grana, some consisting of a dozen or more discs. The stroma lamellae frequently look swollen. In some plastids, the membrane systems run the full length of the plastid; in others, they seem to radiate from one or more centers (Figure 23). The more extensive grana development in these plastids may be partly the result of the greater chlorophyll level.

Albino Mutants.   Bachmann, Robertson, et al. (1973) found that dim-light-grown seedlings of $w3$ grown at 27°C contained numerous lamellae, which were frequently arranged in parallel groups of 10 or more, with occasional grana consisting of two disks.

The plastids of $vp5$ (Figure 24) never seem to develop even the most rudimentary grana. Plastids of this mutant, unlike those of $w3$, frequently had well-organized prolamellar bodies, in many instances, more than one per plastid. Plastids with as many as four were observed. Mutants $w3$, and $vp5$ were alike in that the plastids frequently had groups of parallel lamellae and large electron-transparent globules were present. Plastids of $vp5$ differed from $w3$ in the presence of starch granules in many $vp5$ plastids.

Although $w3$ and $vp5$ do not approach normal in their ability to initiate grana in dim light, one albino, $y7$ ($vp9$), had a more extensive grana formation than normals (Figure 25). Like $w3$ and $vp5$, the chloroplasts of $y7$ have large electron-transparent globules, and like $vp5$ (but not $w3$), $y7$ plastids also have prolamellar bodies.

Chloroplasts of ps, which accumulates lycopene, phytoene, and phytofluene, but only about 10% as much chlorophyll as normals, frequently are irregular in

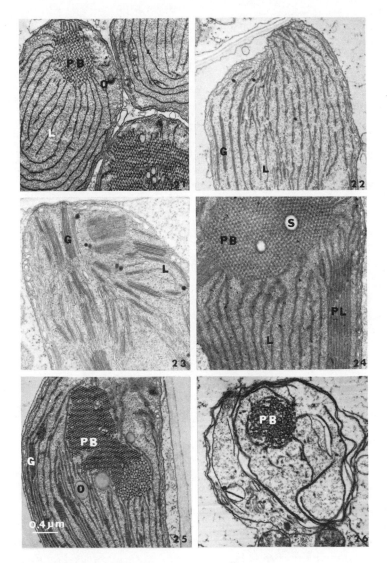

Figure 21. Plastid from a dark-grown mutant seedling.

Figures 22-26. Plastids from dim-light-grown normal and mutant seedlings.

Figure 21. Pale green (w3-8686): 27°C (Bachmann, Robertson, et al., 1967). Magnification × 16,500

Figure 22. Normal: 0.5–1 ft-c, 27°C (Bachmann, Robertson, et al., 1967). Magnification × 16,500

Figure 23. Normal: 5 ft-c, 22°C. Magnification × 16,500

Figure 24. Albino (vp5): 0.5–1 ft-c, 27°C. Magnification × 16,500

Figure 25. Albino [y7(vp9)]: 0.5–1 ft-c, 27°C. Magnification × 16,500

Figure 26. Lycopene accumulating mutant (ps): 0.5–1 ft-c, 27°C. Magnification × 16,500

shape with disorganized prolamellar bodies and undulating lamellae. Sometimes two or three lamellae can stack together but do not form true grana (Figure 26). True grana were not found.

Bachmann, Robertson, et al. (1973) found that the plastids of the albinos *cl* and *lw,* neither of which produces carotenoid precursors when grown at 0.5–1 ft-c at 27°C, formed crystalline-like prolamellar bodies but few lamellae and no grana stacks. Seedlings of *cl* grown at 5 ft-c and 22°C had little lamellar formation. Only occasional chloroplasts were found with lamellae, and these had only a few or no grana stacks (Figure 27). Most plastids were irregular in outline and had irregularly arranged, swollen lamellae and large irregular vesicles (Figure 28). An occasional rudimentary, disorganized prolamellar body was observed.

Luteus Mutants.   Bachmann, Robertson, et al. (1973) described in detail the plastids of dim-light-grown mutants *w*15 (*w*-8896), 13, and *l*∗*Blandy* #4, all of which make little or no chlorophyll, but various amounts of carotenoids. The plastids of *w*15 and *l*3 were quite similar, with large loosely organized prolamellar bodies made up of elongated, membranes-limited vesicles of various dimensions. Lamellae were few, and osmiophilic globules were numerous and larger than those of normal plastids. Although the vesicular structures observed in *w*15 and *l*3 are far from crystalline prolamellar bodies, they nonetheless probably represent the unstabilized protein skeleton of the prolamellar body made in the absence of protochlorophyllide. Similar structures are retained in the light-grown *w*15 (Figure 9). In contrast to *w*15 and *l*3, *l*∗*Blandy* #4 has two types of plastids, neither of which has the loosely organized prolamellar bodies. One type of plastid is similar to those of light-grown albinos. It is irregular in shape and has only a few lamellae, vesicles, and osmiophilic globules. The other type, often found in the same cells as the first type, is elongated, with numerous lamellae extending the length of the plastid. These occasionally are folded into structures that look like small grana, but probably are not. Dense osmiophilic globules of normal size were found. The seedlings of *l*∗*Blandy* #4 have about 5 times the carotenoid pigments of *w*15 and *l*3 and just under the concentration found in normals. The greater carotenoid content may result in the more normal lamellar development of some *l-Blandy* #4 plastids. The plastids *l*11 (*l*-4120), which have 24% chlorophyll, 71% xanthophyll, and 35% carotene of normal under dim-light conditions, occasionally have the loosely organized prolamellar body similar to *w*15 and *l*3, with a few lamellae and osmiophilic globules of normal size (Figure 29). Others, however, have a more extensive system of lamellae and no prolamellar body-like structure. These occasionally have closely packed groups of thylakoids in which the lumens of the thylakoids are much more restricted than is usually observed in normal grana. These may not be true grana, but may

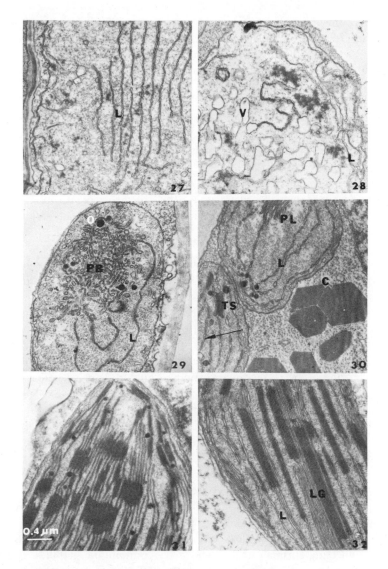

Figures 27–32. Plastids from dim-light-grown mutant seedlings.

Figure 27. Albino (*cl*): 5 ft-c, 22°C. Magnification × 16,500

Figure 28. Albino (*cl*): 5 ft-c, 22°C. Magnification × 16,500

Figure 29. Luteus [*l*11 (*l**4120)]: 0.5–1 ft-c, 27°C. Magnification × 16,500

Figure 30. [*l*11 (*l**4120)] with small grana (arrow): 0.5–1 ft-c, 27°C. Magnification × 16,500

Figure 31. Suppressed albino (*cl cl Clm2 Clm2*): 5 ft-c, 22°C. Magnification × 16,500

Figure 32. Suppressed albino (*cl cl Clm2 Clm2*): 5 ft-c, 22°C. Magnification × 16,500

481

represent groups of parallel lamellae that have become fused. Figure 30 shows such a stack in one plastid and what might be a precursor stage of parallel lamellae in another. Plastids of these mutants frequently have small two-disk true grana. Thus this mutant, which possesses some chlorophyll, can initiate grana formation.

Pale-green Mutants. Bachmann, Robertson, et al. (1973) reported that plastids of the pale-green allele of $w3$, $w3$-8686, were very similar to normals except that they occasionally had groups of parallel lamellae. Plastids of the pale-green F_1 between $w3$ and w-3-8686 ($w3$-8686/$w3$) also were very similar to $w3$-8686, whereas the reciprocal F_1 ($w3$/$w3$-8686 differed from $w3$-8686 in having prolamellar bodies besides the other structural configurations typical of the pale-green allele (i.e., parallel lamellae, grana, and osmiophilic globules).

Suppressed Albinos. Bachmann, Robertson, et al. (1973) found that the pale-green partially suppressed cl albino seedling cl cl $Clm2$ $Clm2$, grown under 0.5–1 ft-c at 27°C, frequently had one or more ordered prolamellar body to which were attached long parallel lamellae. Grana consisting of two or three thylakoid, were numerous. A few small, electron-dense osmiophilic globules were present. The same mutant material grown at 5 ft-c and 22°C produced various types of plastids. A very few were completely disorganized, with only some vesicles and a few membranes. Some had numerous more or less normal grana of various thicknesses, consisting of two to 30 or more thylakods (Figure 31), while others had a few very long grana, consisting of up to 8 thylakoids (Figure 32), similar to those in bright-light-grown cl cl $Clm2$ $Cm2$ seedlings (Figure 12). Sometimes the grana would seem to radiate out from a center that resembled the remnant of a prolamellar body. The number of osmiophilic globules varied. The more normal appearing the chloroplast (Figure 31), the more globules were present. Also, in the more abnormal plastid (Figure 32), the stroma lamellae were swollen. The plastids of the pale-green partially suppressed albino, cl cl $Clm5$ clm (heterozygous for the $Clm5$ modifier), were of several types. Some had a few large grana of longer-than-average length, with very few connecting stroma lamellae and a few osmiophilic globules. Rarely, some were found that were practically empty, with bits of randomly arranged lamellae. Some were observed with one to several prolamellar bodies connected by relatively long grana of up to 15 thylakoids (Figure 33). These also had a few small osmiophilic globules. Seedlings from the homozygous suppressed mutant, cl cl $Cl5$ $Clm5$, have chloroplasts similar to those of the heterozygote. Generally, however, the homozygous $Clm5$ seedlings have more grana and smaller and less crystalline prolamellar body centers (Figure 34). The plastids of the green suppressed seedling cl cl $Clm3$ $Clm3$ appear normal in structure.

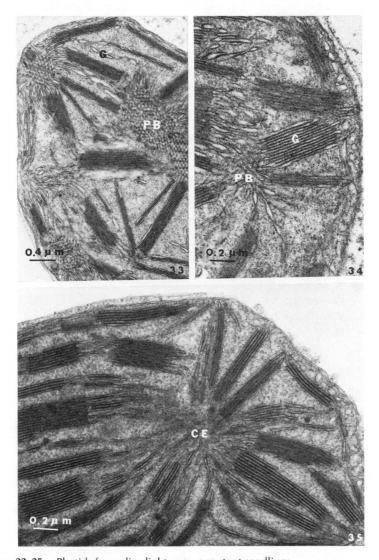

Figures 33–35. Plastids from dim-light-grown mutant seedlings.
Figure 33. Suppressed albino (*cl cl Clm5 clm*): 5 ft-c, 22°C. Magnification × 16,500
Figure 34. Suppressed albino (*cl cl Clm5 Clm5*): 5 ft-c, 22°C. Magnification × 33,500
Figure 35. Suppressed albino (*cl cl Clm3 Clm3*): 5 ft-c, 22°C. Magnification × 33,500

Many, but not all, however, have one or more centers from which the grana seem to radiate (rudimentary prolamellar bodies?) (Figure 35). Some seem to have few, if any, osmiophilic globules, and others have the normal number.

OBSERVATIONS AND CONCLUSIONS

Studies such as we have been considering generally raise more questions than they answer. There are, however, some observations that can be made and some tentative conclusions that can be drawn.

Since we have been concerned here only with mutants with altered plastid pigments (i.e., carotenoids and chlorophylls), our conclusions center around what can be learned about the relationship between these pigments and plastid membrane structure. These membranes are indeed complex, consisting of various substances, such as proteins and lipids, besides the chloroplast pigments. Models have been proposed as to how the chemical components of the membranes may be related Anderson, 1975; Benson, 1971; Weier and Benson, 1967). Most models incorporate the chlorophyll and carotenoid pigments. Thus it is not surprising that changes in these pigments are accompanied by changes in membrane configuration. Although the pigments undoubtedly are important elements of the membranes, other components are involved, and these "other components" by themselves may be sufficient for what seems to be normal membrane formation, especially under low-stress conditions (e.g., low light intensities).

Light-grown Seedlings

Generally, the absence of plastid pigments, chlorophylls, and colored carotenoids is very detrimental to the development of normal plastid structure under conditions of high light intensity, as was seen in this study with $w3$ (Figure 4) and $vp5$-*Mumm* #1 (Figure 5) and by Troxler, Lester, et al. (1969) and Walles (1971) with ly. Walles (1965), Shumway and Weier (1967), and Von Wettstein (1958, 1961) also observed plastids of albino tissue that were deficient in membrane structure. Our work on mutants such as $w3$ and $vp5$ suggests that their lack of internal membrane structure is not due to some block in the steps necessary to assemble membranes. Plastids from dark-grown plants, or from the early stages after dark-grown seedlings, are exposed to light (Bachmann, Robertson, et al. 1967) or from dim-light-grown seedlings of these mutants, have seemingly normal membrane development. Thus the deficiency in membranes in light-grown seedlings is probably due to membrane instability under bright-light conditions in the absence of colored carotenoids and (or) chlorophyll. The mutants $w3$, $vp5$, and al accumulate later carotenoid precursors. Two mutants, cl and lw, which are not accumulators, retain membrane structures (e.g., rudimentary prolamellar bodies, lamellae, and even small grana) (Figures 6–8). Von Wettstein, Henningsen, et al. (1971) describe an albino mutant, $alb17$, of barley that does not disperse its prolamellar body. Why these nonaccumulating mutants of maize should show more membrane stability is not known. Like the "accumulators," they lack both colored carotenoids and chlorophyll. The presence of what seem to be normal, albeit rudimentary, grana in some cl plastids indicates that grana initiation can occur in the absence of chlorophyll.

The two luteus mutants, $w15$ and $l^*Blandy$ #4, both of which produce colored carotenoids but little or no chlorophyll, showed more extensive membrane production than the albinos. Like cl, $w15$ (Figure 9) sometimes had disorganized prolamellar bodies with swollen structural elements. As mentioned earlier, the presence of crystalline-like prolamellar bodies generally is thought to be the result of accumulation of protochlorophyllide (Henningsen and Boynton, 1969, 1970; Kahn, 1968a). Prolamellar body disorganization is usually thought to follow exposure to light and the conversion of protochlorophyllide to chlorophyll (Henningsen and Boynton, 1969, 1970; Kahn, 1968b; Treffry, 1970). On exposure to light, the crystalline-like structure of the prolamellar body in normal plant material begins to disorganize, and the membranous elements seem to contribute to the formation of an extensive lamellar system which, usually after a lag phase, goes on to develop the extensive grana system of fully developed chloroplasts (Bogorad, 1967; Gunning, 1965; Gunning and Jagoe, 1967). The initiation of grana corresponds approximately to the time that Virgin, Kahn, et al. (1963) and Bogorad (1967) observed that dark-grown seedlings first begin to synthesize large amounts of chlorophyll after exposure to light. Seedlings of cl and $w15$ seem defective with regard to the mechanism responsible for the dispersion of the elements of the prolamellar body. These mutants grown under bright-light conditions are devoid of chlorophyll and its precursors, yet they still retain what seem to be modified elements of the prolamellar body. What is responsible for blocking the complete dispersal of the prolamellar body membranes is not obvious. It is not the lack of chlorophyll $per se$ inasmuch as $w3$ and $vp5$ in the light also lack this pigment and yet do not have any remnant of the prolamellar body. Nor does the presence or absence of colored carotenoid pigments seem to be involved because cl lacks them, whereas $w15$ possesses them. One general conclusion that can be drawn from the observation of light-grown luteus mutants is that their plastids frequently show more extensive membrane production (Figure 10) than that observed in the albinos. This would imply that the presence of colored carotenoids imparts some stability to membranes under bright-light conditions. This conclusion is supported by the observation that *l*Blandy* #4, which has more colored carotenoids than *w*15, has a more extensive lamellar development.

Mutants with reduced pigment concentrations (pale greens) (viz., *w*3-8686, *cl cl Clm*2 *Clm*2, *cl cl Clm*5 *clm*) exhibit several different structural patterns (Figures 12, 14). It might be that these plants with less than normal concentrations of chlorophylls and carotenoids have plastids that vary in the amounts of these pigments that they contain. Those with lesser amounts may not be able to support very extensive membrane synthesis, whereas others possessing more of these pigments may develop a more normal lamellar and grana configuration. Large grana such as were observed in *cl cl Clm*2 *Clm*2 (Figure 12) have been reported in other pale-green mutant plants. Bachmann, Robertson, et al. (1967, 1969) reported similar grana in the pale-green mutant *w*3-8686, and Chollet and Paolillo (1972) described large grana in *vl*8. Similar, but not identical, arrays of thylakoids have been reported by Von Wettstein (1960) in xantha mutants of barley. In the barley mutants the groups of parallel thylakoids were not closely appressed as in true grana. Bachmann, Robertson, et al. (1969) observed such parallel arrangements of thylakoids as described by Von Wettstein in several of their mutants. The extremely electron-dense thylakoid stacks observed in *cl cl Clm*5 *clm* (Figure 13) are in reality very closely packed thylakoids, with very reduced loculi. Similar structures were observed in the pale-green *w*3-8686 mutant of corn by Bachmann, Robertson, et al. (1969) and in barley mutants by Nielsen (1970). Bachmann, Robertson, et al. (1969) interpreted such configuration as representative of degenerating plastids, but Nielsen (1970) interpreted them as crystalloids.

Suppressed mutants with near normal concentrations of plastid pigments (viz. *cl cl Clm*3 *Clm*3 and *cl cl Clm*5 *Clm*5) have normal plastid morphology (Figures 15 and 16).

Dark-grown Seedlings

In the dark-grown material studied, normals, albinos, and pale greens (Figures 17–21) all had essentially the typical crystalline-like prolamellar body observed by many other workers in dark-grown plants. Seedlings of the albino, *w*3, however, were observed sometimes to have less crystalline, more randomly arranged units in the prolamellar body (Figure 19). Because both *w*3 and *cl* albino seedlings lack colored carotenoids and yet *cl* has a normal appearing prolamellar body and lamellae system, it is unlikely that carotenoid deficiency accounts for the poor lamellar system of *w*3 plastids. The two mutants differ, however, in the amount of protochlorophyllide each produces. The *w*3 seedlings produce about half as much as *cl* (Table 1), which produces an amount equal to that of normals.

The observations of this dark-grown material support the suggestion that protochlorophyllide production is essential to normal prolamellar and lamellar structure. It seems that colored carotenoids do not play an obvious structural role. These observations are in agreement with those of Troxler, Lester, et al. (1969). Seedlings of *al* grown in the dark accumulated carotenoid precursors (Table 3) but no colored carotenoids, and they had a level of protochlorophyllide equal to that of normal seedlings. They found plastids of dark-grown *al* plants to have crystalline-like prolamellar bodies with accompanying lamellae. They appeared in all regards similar to normal, except that the *al* plastids did not possess any osmiophilic globules.

Dim-light-grown Seedlings

Plants of differing genotypes respond differently to dim light with regard to plastid pigment synthesis. Under dim-light and 27°C, normal seedlings produce about one-eighth the chlorophyll and one-fifth the carotenoid pigments as normal grown at the same temperature and 2000 ft-c. Not all mutants necessarily respond to high and low light intensities in the same manner with respect to pigment production. The albinos used in our study usually will accumulate more chlorophyll under dim-light conditions, varying from 17% to 79% of the normals grown under the same conditions (Bachmann, Robertson, et al. 1973; Tables 1 and 2). Some mutants, such as *cl cl Clm2 Clm2*, possess less chlorophyll under dim light than under bright, as do normals; yet, under bright-light conditions, the final chlorophyll concentration reaches only 27% of normal. At low-light intensities, however, *cl cl Clm2 Clm2* produces as much chlorophyll as normals do under these conditions. Similar variation in carotenoid pigments has been found. Normal plants have more of these pigments under bright-light conditions, and the albinos usually little or none under either light intensity, whereas the luteus and pale-green mutants will vary in their responses (Bachmann, Robertson, et al., 1973; Tables 1 and 2).

Normals grown at 27°C, 0.5–1 ft-c have plastids with numerous lamellae and grana stacks consisting of two or three thylakoids (Figure 22). Thus the lower plastid pigment concentrations do not permit the extensive grana formation found under bright-light conditions that permit the accumulation of substantially more plastid pigments. If the light intensity is increased to 5.0 ft-c (22°C) there is marked increase in grana formation, with normal-sized grana being very common (Figure 23). At this light intensity 0.355 mg/g (fresh weight) of chlorophyll is present, which is a 1.5-fold increase over that present at 0.5–1 ft-c. From observations on normals it seems that the amount of grana formation is directly correlated to the level of plastid pigment production. The albinos grown under bright-light conditions do not possess chlorophyll, and most do not exhibit any grana formation or extensive membrane synthesis (a possible exception being *cl;* see Figure 8). We have shown, however, that all albinos used in this study can make the chlorophyll precursor protochlorophyllide (Table 1) and also possess the ability to convert this to chlorophyll [Table 1; Smith, Durham, et al. (1959)]. It has been demonstrated for some albinos [Table 2; Bachmann, Robertson, et al. (1973)] and probably holds for all albinos that we have studied that plants grown under dim-light conditions accumulate some chlorophyll. Some of these albinos possess the ability to initiate grana under dim-light conditions, and some do not. The mutants *w*3 (Bachmann, Robertson, et al., 1973) and *y*7 (*vp*9) (Figure 25) have this ability. Seedlings of *w*3 under dim light can make 79% of the chlorophyll of normals. If the presence of chlorophyll is one of the prerequisites for grana initiation, the presence of a relatively high concentration in *w*3 may account for grana initiation in this mutant under these conditions. If this relationship between chlorophyll and grana initiation holds, then *y*7 (*vp*9) would be expected to have greater chlorophyll concentration than *w*3, because the *y*7 grana initiation is considerably more extensive (Figure 25). Unfortunately, the chlorophyll values for the seedlings from which the cytological samples were obtained is not known. The observations on *vp*5 seem to support the foregoing assumption inasmuch as this mutant, which accumulated less chlorophyll than did *w*3 (i.e., 0.123 mg/g fr.wt. for *vp*5 vs. 0.146 mg/g fr.wt. for *w*3), shows no grana initiation. Bachmann, Robertson, et al. (1973) found that *cl* and *lw* had considerably less chlorophyll when grown at 0.5–1 ft-c than did *w*3 and that neither *cl* nor *lw* showed any grana initiation. Seedlings of *cl* grown at 5 ft-c and 22°C produced considerably more chlorophyll than did *cl* seedlings at 0.5-1 ft-c (0.0416 mg/g fr.wt. at 0.5–1 ft-c vs. 0.132 mg/g fr.wt. at 5 ft-c) and yet no grana were observed in the 5 ft-c material. Again, we see that seedlings with less chlorophyll than that found in the dim-light grown *w*3 do not initiate grana. Perhaps there is some critical level of chlorophyll near that of *w*3 (i.e., 0.146 mg/g fr.wt.) that has to be reached before grana can be initiated. If this is true, the concentration of 0.0370 mg/g fr.wt. of chlorophyll in *ps* seedlings may account for why no grana are observed (Figure 26). Walles (1971), however, in a lycopenic mutant (unplaced and not allele-tested with *ps*), found grana stacks in seedlings grown at 2–3 ft-c. His material produced 30% the chlorophyll of normal −3 times that observed for *ps*. Perhaps the greater chlorophyll concentration in Walle's preparation accounts for grana initiation in these plastids. Gyurjan, Rakovan, et al. (1969) also studied the same lycopene mutant as did Walles and at about 0.5 ft-c and 9 ft-c. They found that chloroplasts of this mutant frequently had prolamellar bodies and grana. Occasional chloroplasts would have just lamellae, frequently arranged in parallel groups of concentric thylakoids. At 0.5 ft-c of illumination this mutant had chlorophyll in the amounts of 0.5 mg/g dry weight (d.w.) (half the concentration of normal). In their material, however, if the light was increased to 93 ft-c, which resulted in 10.4 mg/g d.w. of chlorophyll, the chloroplast did not continue to show grana, but rather there were groups of concentrically arranged, parallel lamellae, closely associated but not forming true grana stacks. Thus although high chlorophyll concentration is sometimes associated with grana formation, it does not by itself ensure that it will occur. The observations of Gyurjan, Rakovan, et al. (1969) on an albino (ζ-carotenic mutant), that accumulated phytoene, phytofluene, and ζ-carotene support this conclusion. At 0.5 ft-c this mutant accumulated slightly more chlorophyll than did the *ly* mutant, yet the ζ-carotenic mutant did not form grana, although some contained extensive lamellae formation. Many contained disorganizing prolamellar bodies, but most had more irregular internal structure with numerous vesicles. Although the chlorophyll increased slightly at 9 ft-c, less lamellar structure was observed and more vesicles, whereas at 93 ft-c the chlorophyll level was below that of 0.5 ft-c, and the plastids showed signs of deteriorating. Thus we see again that the presence of chlorophyll does not ensure grana initiation.

Troxler, Lester, et al. (1969) observed plastids of *al* at low-light intensity (ca. 95 ft-c). Under these conditions *al*, unlike other mutants that we have considered, makes normal amounts of carotenoids and chlorophyll. The plastid structure of the mutant seedlings grown under these condition was essentially comparable to normals. Both normal and mutant plastids showed grana initiation, and some of both had portions of the prolamellar body present.

Because the albino mutants that initiate grana in dim light lacked colored carotenoids, it is obvious that these pigments are not essential to the initial stages of grana formation. Also, most albinos show extensive lamellae formation; thus these structures also do not require carotenoids for their production. Because several of these same mutants grown under bright-light conditions are essentially devoid of lamellae and it is known that under such conditions chlorophyll is destroyed, it would seem that lamellar structure cannot be retained in the absence of chlorophyll in some mutants. Complicating the picture, however, is the observation of small grana in some plastids of light-grown *cl* seedlings (Figure 8), which are completely devoid of chlorophyll.

Of the luteus mutants studied under dim light (Bachmann, Robertson, et al., 1973) and this study (Figures 29, 30), only one produced any chlorophyll, *l*11 (*l*⁻4120) (0.0580 mg/g fr.wt.). The other three (*w*15, *l*3, and *l*⁻*Blandy* #4) did not. Various amounts of carotenoids were produced. Plastids of *w*15, *l*13, and *l*11 (*l*⁻1420) all had disorganized prolamellar bodies with large vesicular elements, but very little lamellar structure. Plastids of *l*⁻*Blandy* #4 did not have prolamellar bodies. Biochemical studies have shown that this mutant does not make protochlorophyllide. The lack of a prolamellar body in these seedlings supports the hypotheses that protochlorophyllide accumulation is essential for the formation of crystalline-like prolamellar bodies. Also, *l*⁻*Blandy* #4, which has the highest concentration of carotenoids of the three luteus mutants (Bachmann, Robertson, et al., 1973), has plastids with the most extensive lamellar system, suggesting that the presence of colored carotenoids may be sufficient for some lamellar formation even in the absence of chlorophyll.

The pale-green mutants studied (Bachmann, Robertson, et al., 1973) all produced near-normal amounts of chlorophyll and carotenoids at 0.5–1 ft-c. Under these conditions all produced lamellae and initiated grana, usually with no more than two or three thylakoids. Some mutants retained their prolamellar bodies, and some did not. One of the pale-green mutants that retain pro-lamellar bodies was the partially suppressed albino *cl cl Clm2 Clm2*. In this study some plastids of this genotype from plants grown at 5 ft-c had very exten-sive grana stacks, comparable to those of normals grown under these conditions (Figure 31). Other plastids differed from normal in having a few very long grana (Figure 32). In neither case did the plastids of plants grown at 5 ft-c have prolamellar bodies. In this respect they differ from the material grown at 0.5–1 ft-c. This would suggest that, in this mutant at high light intensities, the prolamellar body disappears as grana development becomes more extensive. This was not the case, however, in the *cl cl Clm5 clm* mutant, which had extensive long grana stacks consisting of many thylakoids as well as several prolamellar bodies (Figure 33). Under bright-light conditions this mutant is known to possess lower concentrations of chloroplast pigments than does *cl cl Clm2 Clm2* (Robertson, Bachmann, et al. 1966). A similar differential in plastid pigments may exist between seedlings of these two genotypes at 5 ft-c, thus accounting for the presence of prolamellar bodies in *cl cl Clm5 clm*, but not in *cl cl Clm2 Clm2*. If this is the case, the *cl cl Clm5 Clm5* seedling would be expected to have reduced prolamellar bodies and more normal grana con-figuration, because Robertson, Bachmann, et al., (1966) have shown that it possesses a higher concentration of chloroplast pigments than does *cl cl Clm5 clm*. This is indeed the case (Figure 34). Seedlings heterozygous for *Clm5* and those homozygous for *Clm5* grown under dim light, however, are much less divergent in structure then seedlings of these same two genotypes grown in bright light (see Figures 13, 14, and 15). Dim-light growing conditions, by pre-venting the destruction of plastid pigments in the heterozygous *Clm5* seedlings, result in homzygotes and heterozygotes with very similar pigment concentra-tions (i.e., *Clm5 Clm5*-chlorophyll 0.0977 mg/g fr.wt., carotene 0.0071 mg/g fr.wt., and xanthophyll 0.0082 mg/g fr.wt., *Clm5 clm*-chlorophyll 0.0916 mg/g fr.wt., carotene 0.0080 mg/g fr.wt., and xanthophyll 0.0080 mg/g fr.wt.). This similarity in pigment accumulation is probably responsible for the structural similarity. Under bright-light conditions it is known that the heterozygote has about half the plastid pigments of the homozygote. This marked deficiency in plastid pigments is reflected in the structural anomalies of the heterozygote. As would be expected, *cl cl Clm3 Clm3*, which also produced high concentrations of plastid pigments, has near-normal grana formation (Figure 35). In the last three genotypes the lamellae and grana structures seem to radiate from prolamellar bodies or their remnants, which confirms the suggestion that elements of the prolamellar body contribute to the formation of membrane structure in lamellae and grana.

The presence of crystalline-like prolamellar bodies in some of the plastids of dim-light-grown mutants suggests that the plastids have photochlorophyllide present. Henningsen and Boynton (1970) found in barley seedlings grown in the dark and then exposed to dim light (1.9 ft-c) that the prolamellar bodies disperse and that after 4 h they reformed as crystalline-like prolamellar bodies. In many cases, several large and small prolamellar bodies per plastid were found. This reforming of the prolamellar bodies was accomplished by the reaccumulation of protochlorophyllide. It is very probable that the dim-light-grown maize mutants that have crystalline-like prolamellar bodies also accumulate protochlorophyllide. This undoubtedly is true for *cl cl Clm5 clm*, which has plastids that are nearly identical to those seen in barley.

REFERENCES

Anderson, I. C. and D. S. Robertson. 1960. Role of carotenoids in protecting chlorophyll from photodestruction. *Plant Physiol.* **35:** 531–534.

Anderson, I. C. and D. S. Robertson. 1961. Carotenoid protection of porphyrins from photodestruction. *Proc. 3rd Internat. Congr. Photobiol. Copenhagen.* Amsterdam: Elsevier, pp. 477–479.

Anderson, J. M. 1975. Possible location of chlorophyll within chloroplast membranes. *Nature* **253:** 536–537.

Andres, J. M. 1951. Une neuva mutacion que goviernca la sintesis de la licopina y carotinas en el maiz. *Rev. Argent. Agron.* **18:** 136–142.

Bachmann, M. D., D. S. Robertson, C. C. Bowen, and I. C. Anderson. 1967. Chloroplast development in pigment deficient mutants of maize. I. Structural anomalies in plastids of allelic mutants at the *w3* locus. *J. Ultrastruct. Res.* **21:** 41–60.

Bachmann, M. D., D. S. Robertson, C. C. Bowen, and I. C. Anderson. 1973. Chloroplast ultrastructure in pigment-deficient mutants of *Zea mays* under reduced light. *J. Ultrastructure. Res.* **45:** 384–406.

Bachmann, M. D., D. S. Robertson, and C. C. Bowen. 1969. Thylakoid anomalies in relation to grana structure in pigment-deficient mutants of *Zea mays. J. Ultrastruct. Res.* **28:** 435–451.

Benson, A. A. 1971. Lipids of chloroplasts. In *Structure and function of chloroplasts,* M. Gibbs (ed.). New York-Heidelberg: Springer-Verlag, pp. 130–148.

Bogorad, L. 1967. Biosynthesis and morphogenesis in plastids. In *Bio-chemistry of chloroplasts,* Vol. 2, T. W. Goodwin (ed.). London and New York: Academic, pp. 615–631.

Chollet, R. and D. J. Paolillo, Jr. 1972. Greening in a virescent mutant of maize. I. Pigment, ultrastructural, and gas exchange studies. *Z. Pflanzenphysiol.* **68:** 30–40.

Cohen-Bazire, G. and R. Y. Stanier. 1958. Specific inhibition of carotenoid synthesis in a photosynthetic bacterium and its physiological consequences. *Nature* **181:** 250–252.

De La Guardia, M. D., G. M. G. Aragon, F. J. Murillo, and E. Cerda-Olmedo. 1971. A cartenogenic enzyme aggregate in *Phycomyces:* Evidence from quantitative complementation. *Proc. Nat. Acad. Sci. U. S. A.* **68:** 2012–2015.

Fuller, R. C. and I. C. Anderson. 1958. Suppression of carotenoid synthesis and its effect on the activity of photosynthetic bacterial chromatophores. *Nature* **181:** 252–254.

Garay, A., S. Demeter, K. Kovacs, G. Horvath, and A. Faludi-Daniel. 1972. Circular dichroism spectra of system I particles from normal chloroplasts and carotenoid deficient mutants of maize. *Photochem. Photobiol.* **16:** 139–144.

Garber, E. D., M. L. Baird, and D. J. Chapman. 1975. Genetics of *Ustilago violacea.* I. Carotenoid mutants and carotenoids. *Bot. Gaz.* **136:** 341–346.

Goldie, A. H. and R. E. Subden. 1973. The neutral carotenoids of wild-type and mutant strains of *Neurospora crassa. Biochem. Genet.* **10:** 275–284.

Goodwin, T. W. and E. I. Mercer. 1972. Introduction to plant biochemistry. New York: Pergamon, 278 pp.

Gunning, B. E. S. 1965. The greening process in plastids. I. The structure of the prolamellar body. *Protoplasma* **60:** 111–130.

Gunning, B. E. S. and M. P. Jagoe. 1967. The prolamellar body. Pages 655–676 In *Biochemistry of chloroplasts,* T. W. Goodwin, (ed.). London: Academic.

Gyurjan, I., J. N. Rakovan, and A. Faludi-Daniel. 1969. Chloroplast differentiationa and $^{14}CO_2$ fixation in normal and mutant maize leaves. Pages 63–73 In *Progress in photosynthesis research,* Vol. I, H. Metzner (ed.). Tubingen: International Union of Biological Sciences.

Henningsen, K. W. and J. E. Boynton. 1969. Macromolecular physiology plastids VII. The effect of brief illumination on plastids of dark-grown barley leaves. *J. Cell Sci.* **5:** 757–793.

Henningsen, K. W. and J. E. Boynton. 1970. Macromolecular physiology of plastids. VIII. Pigment and membrane formation in plastids of barley low-light intensity. *J. Cell Biol.* **44:** 290–304.

Horvath, G., J. Kissimon, and A. Faludi-Daniel. 1972. Effect of light intensity on the formation of carotenoids in normal and mutant maize leaves. *Photochemistry* **11:** 183–187.

Jensen, S. L., G. Cohen-Bazire, and R. Y. Stanier. 1961. Biosynthesis of carotenoids in purple bacteria: A re-evaluation based on considerations of chemical structure. *Nature* **192:** 1168–1172.

Kahn, A. 1968a. Developmental physiology of bean leaf plastids. II. Negative contrast electron microscopy of tubular membranes in prolamellar bodies. *Plant Physiol.* **43:** 1769–1780.

Kahn, A. 1968b. Developmental physiology of bean leaf plastids III. Tube transformation and protochlorophyll(ide) photoconversion by flash irradiation. *Plant Physiol.* **43:** 1781–1785.

Kushwaha, S. C., G. Suzue, C. Subbarayan, and J. W. Porter. 1970. The conversion of phytoene–^{14}C into acyclic, monocyclic and dicyclic carotenes and the conversion of lycopene-15, 15-^{3}H to mon- and dicyclic carotenes by soluble enzyme systems obtained from plastids of tomato fruits. *J. Biol. Chem.* **245:** 4704–4717.

Nielsen, O. F. 1970. A gene controlling the association of chlorophyll with chloroplast membranes in barley. In *Microscopie electronique 1970,* P. Favard, (ed.), pp. 179–180; Comm. Sept. Congr. Internat. Grenoble, Vol. 3. Soc. Fr. Microsc. Electr., Paris.

Porter, J. W. and R. E. Lincoln. 1950. I. *Lycopericon* selection containing a high content of carotenes and colorless polyenes. II. The mechanism of carotene biosynthesis. *Arch. Biochem.* **27:** 390–403.

Richardson, L. B., D. S. Robertson, and I. C. Anderson. 1961. Genetic and environmental variation: Effect on pigments of selected maize mutants. Science **138:** 1333–1334.

Robertson, D. S. 1973. Linkage relationships of chlorophyll defective mutants on chromosome 6. *Maize Genet. Coop. Newsl.* **47:** 82–87.

Robertson, D. S. 1975. A survey of the albino and white endosperm mutants of maize: Their phenotypes and gene symbols. *J. Hered.* **66:** 67–74.

Robertson, D. S., M. D. Bachmann, and I. C. Anderson. 1966. Role of carotenoids in protecting chlorophyll from photodestruction-II. Studies on the effect of four modifiers of the albino cl_1 mutant of maize. *Photochem. Photobiol.* **5:** 797–805.

Sandler, C. L., J. Laber, W. D. Bell, and K. H. Hamilton. 1968. Light sensitivity of plastids and plastid pigments present in the albescent maize mutant. *Plant Physiol.* **43:** 693–697.

Shumway, L. K., and T. E. Weier. 1967. The chloroplast structure of iojap maize. *Am. J. Bot.* **54:** 773–780.

Smith, J. H. C., L. J. Durham, and C. F. Wurster. 1959. Formation and bleaching of chlorophyll in albino corn seedlings. *Plant Physiol.* **34:** 340–345.

Stainer, R. Y. 1959. Formation and function of photosynthetic pigment system in purple bacteria. In *The photochemical apparatus: Its structure and function.* Brookhaven Symposia in Biology, No. 11. Upton, N. Y.: Brookhaven National Laboratory, pp. 43–53.

Treffry, T. 1970. Phytylation of chlorophyllide and prolamellar body transformation in etiolated peas. *Planta* **91:** 279–284.

Treharne, K. J., E. I. Mercer, and T. W. Goodwin. 1966. Carotenoid biosynthesis in some maize mutants. *Phytochemistry* **5:** 581–587.

Troxler, R. F. F., R. Lester, F. O. Craft, and J. T. Albright. 1969. Plastid development in albescent maize. *Plant Physiol.* **44:** 1609–1618.

Virgin, H. I., A. Kahn, and D. Von Wettstein. 1963. The physiology of chlorophyll formation in relation to structural changes in chloroplasts. *Photochem. Photobiol.* **2:** 83–91.

Von Wettstein, D. 1958. The formation of pastid structures. In *The photochemical apparatus: Its structure and function.* Brookhaven Symposia in Biology, No. 11. Upton, N. Y.: Brookhaven National Laboratory. pp. 138–159.

Von Wettstein, D. 1960. Multiple allelism in induced chlorophyll mutants. II. Error in the aggregation of the lamellar discs in the chloroplasts. *Hereditas* **46:** 700–708.

Von Wettstein, D. 1961. Nuclear and cytoplasmic factors in development of chloroplast structure and function. *Can J. Bot.* **39:** 1537–1545.

Von Wettstein, D., K. W. Henningsen, J. E. Boynton, G. C. Kannangara, and O. F. Nielsen. 1971. The genic control of chloroplasts development in barley. Pages 205–223 In *Autonomy and biogenesis of mitochondria and chloroplasts,* N. K. Boardman, A. W. Linanne, and R. W. Smillie (eds.). Amsterdam–London–New York: Elsevier-North-Holland Publishing Co., pp. 205–223.

Walles, B. 1965. Plastid structures of carotenoid-deficient mutants of sunflower (*Helianthus annuus* L.). The white mutant. *Hereditas* **53:** 247–256.

Walles, B. 1971. Chromoplast development in a carotenoid mutant of maize. *Protoplasma* **73:** 159–175.

Weier, T. E. and A. A. Benson. 1967. The molecular organization of chloroplast membranes. *Am. J. Bot.* **54:** 389–402.

Zechmeister, L. 1962. *Cis–trans* Isomeric carotenoids vitamins A and arylpolyenes. New York: Academic.

Chapter 32

CELL .AND TISSUE CULTURES OF MAIZE

C. E. Green
Department of Agronomy and Plant Genetics,
University of Minnesota,
St. Paul, Minnesota

Recent advances in tissue and cell culture methodology have resulted in increased interest among plant geneticists and breeders who anticipate using these techniques to study numerous research areas. These advances include the ability to obtain large numbers of plants with the gametic chromosome number (haploid) by inducing mitotic cell division in the microspores within cultured anthers (Sunderland, 1973) or in isolated immature pollen grains (Nitsch, 1974). As a result, large populations of cultured haploid cells have become available for mutagenesis and the selection of defined mutants (Carlson, 1973). Fertile homozygous plants have been regenerated from haploid cell cultures by spontaneous (Kasperbauer and Collins, 1972) or chemically induced (Burk, Gwynn, et al., 1972) chromosome doubling. Protoplasts produced by enzymatic removal of the cell wall have been cultured successfully, and subsequently, plants have been regenerated from amphidiploid (Nagata and Takebe, 1971), diploid

Paper no. 1609, Miscellaneous Journal Series, Minnesota Agricultural Experiment Station.

(Grambow, Kao, et al., 1972) and haploid cells (Ohyama and Nitsch, 1972). Protoplasts have been utilized to investigate interspecific (Carlson, Smith, et al., 1972) and intraspecific (Kao, Constabel, et al., 1974; Kartha, Gamborg, et al., 1975) cell fusions, nuclear (Potrykus and Hoffman, 1973) and chloroplast (Bonnett and Ericksson, 1974) transplantation, and DNA uptake in plant cells (Ohyama, Gamborg, et al., 1972).

The objective of developing these *in vitro* techniques is to achieve major simplifications in the genetic and organizational complexities of whole plants. Such simplification would allow the investigation of fundamental genetical and developmental processes at or near the cellular level. Currently, these procedures are primarily available in a few model systems such as tobacco (*Nicotiana* sp.), carrot (*Daucus carota*), and petunia (*Petunia hybrida*). Steady progress, however, has been achieved in the formation of plants with the gametic chromosome number from cultured anthers of rice [*Oryza sativa* L.; Niizeki and Oono (1968)], barley (*Hordeum vulgare* L.; Clapham, 1973), wheat (*Triticum aestivum* L.; Picard and de Buyser, 1973), *Triticale* (Wang, Sun, et al., 1973), and rye (*Secale cereale* L.; Thomas and Wenzel, 1975). Recent advances in maize cell and tissue culture, which are discussed in this chapter, should promote the investigation of metabolic, physiological, and developmental processes at the cellular level. The availability of efficient *in vitro* culture systems coupled with the well-documented knowledge on the genetics and physiology of maize would offer intriguing genetic and breeding research opportunities.

ENDOSPERM CALLUS CULTURES

The first tissue cultures of any grass species were established from the endosperm of immature maize kernels of an unidentified genotype in 1949 (LaRue, 1949). The successful initiation and maintenance of these cultures required the addition of a tomato juice supplement to White's medium (White, 1943). Exogenous hormones were not required. In subsequent years endosperm cultures of maize have been successfully established in many additional investigations.

In the methods generally used for endosperm callus initiation, young ears 7–12 days after controlled pollination are husked and surface sterilized in 0.5–2.5% sodium hypochlorite. Then the crown, or upper portion, of the immature kernel is sliced off, exposing the developing endosperm, which is then easily removed for culture. The basal media formulations of White (1943), Nitsch (1951), and Straus (1960) have been utilized successfully for growing endosperm callus. Endosperm cultures frequently have been incubated in the dark at 25°C, but optimal growth rates also have been observed up to 32°C (Sun and Ullstrup, 1971).

Endosperm Age and Genotypic Variability

Two extremely important parameters for callus initiation and growth are the age and genotype of the endosperm source. Various genotypes and kernel ages from which successful endosperm cultures have been maintained are summarized in Table 1. Although several endosperm types are listed in this table, callus cultures were initially derived from a few *sugary* lines. Among 15 maize genotypes, including *wx* (*waxy*), *Su* (*starchy*), and *su* (*sugary*) types, Straus and LaRue (1954) observed that 12-day postpollination endosperm tissue from only two sugary lines, Surprise Sweet and Black Mexican Sweet, could be cultured consistently and the callus maintained by subculture. Sternheimer (1954) also initiated callus from 10–12-day postpollination Black Mexican Sweet endosperm. Among 25 *su, Su, wx, bt* (*brittle*), *fl* (*floury*), and popcorn inbreds and 32 singlecrosses, Tamaoki and Ullstrup (1958) observed that 8–11-day postpollination endosperm from four *sugary* types could be maintained as callus for at least 4–8 months. In their study endosperm callus was also initiated from several *Su* and *wx* genotypes, but the cultures died after 2–5 months of growth. Sun and Ullstrup (1971) established vigorous callus from 7–10-day postpollination endosperm from the inbred P14 *susu*. They reported no or limited callus initiation with 24 additional *su* and two *Su* inbreds.

Tabata and Motoyoshi (1965) established callus cultures from 9-day postpollination endosperm from the *Su* inbreds S41 and S42. When these inbreds were reciprocally crossed, endosperm from the resulting F_1 kernels produced callus cultures that grew more rapidly *in vitro* than either inbred parent. Endosperm callus from S41 × S42 was maintained for at least 6 years. Attempts to culture endosperm from the *wx* inbred S43 and the *su* inbred S44 were unsuccessful. These results indicated that callus could be initiated and maintained from *starchy* endosperm from some inbreds and singlecrosses. Further, the type of endosperm such as *starchy* or *sugary*, was not as important for callus growth as were other genetic factors. Among 23 *Su* inbred and hybrid lines Shannon and Batey (1973) successfully initiated and maintained callus from 7–10-day postpollination endosperm from the inbreds A636 and R168 and the hybrids P3369A and Px610.

A number of endosperm mutants in addition to *su* have been successfully cultured. Tsai (1970) established endosperm callus from the inbred W64A, which was homozygous for the *ae* (*amylose extender*) gene. Callus also has been established from endosperm of *sh2* (*shrunken*-2) W64A (Shannon and Batey, 1973). The singlecross hybrid Oh43 × B37 was used recently as the common

TABLE 1. Summary of maize genotypes and kernel ages from which endosperm callus cultures have been successfully established and maintained

Maize line	Endosperm type	Endosperm age (days postpollination)	Reference
Surprise Sweet	sugary	12	(Straus and LaRue, 1954)
Black Mexican Sweet	"	10–12	(Straus and LaRue, 1954)
Country Gentleman	"	8–11	(Tamaoki and Ullstrup, 1958)
Yellow Sugary su2su2	"	8–11	(Tamaoki and Ullstrup, 1958)
Late Season Yellow Sugary	"	8–11	(Tamaoki and Ullstrup, 1958)
Oh43 × B37 su2su2	"	7–10	(Shannon and Batey, 1973)
P14 susu	"	7–10	(Sun and Ullstrup, 1971)
S41 × S42	starchy	9	(Tabata and Motoyoshi, 1965)
S42	"	9	(Tabata and Motoyoshi, 1965)
A636	"	7–10	(Shannon and Batey, 1973)
R168	"	7–10	(Shannon and Batey, 1973)
P3369A	"	7–10	(Shannon and Batey, 1973)
PX610	"	7–10	(Shannon and Batey, 1973)
Oh43 × B37 aeae	amylose extender	7–10	(Shannon and Batey, 1973)
W64A aeae	"	7–10	(Tsai, 1970)
W64A sh2sh2	shrunken-2	7–10	(Shannon and Batey, 1973)

background to study various homozygous endosperm mutants, including *ae, fl, wx, du (dull), sh (shrunken), su2 (sugary-2), h (soft starch),* and *o2 (opaque-2)* (Shannon and Batey, 1973). Among these, actively growing endosperm cultures were established from *ae* and *su2*.

In summary, callus has been established successfully only from 7–12-day postpollination endosperm tissues and from a limited number of genotypes. This variation in optimum age is probably to be expected considering the maturity differences between various genotypes and the environmental changes among geographical locations and years.

Growth Studies

Growth studies with clone 1-C endosperm cultures by Straus and LaRue (1954) revealed that erratic callus growth was eliminated when filtered yeast extract was substituted for the tomato juice in White's medium. The effectiveness of yeast extract over tomato juice was also observed by Tamaoki and Ullstrup (1958). Subsequently, a defined medium in which 15 mM L-asparagine effectively replaced yeast extract was developed for the growth of Black Mexican Sweet endosperm callus (Straus 1960). Sun and Ullstrup (1971), however, concluded that asparagine was only partially effective in replacing yeast extract for the growth of P14 *susu* endosperm callus. Other additives, such as coconut milk and casein hydrolysate, have been less effective than yeast extract or asparagine in supporting endosperm callus growth (Straus and LaRue, 1954; Tamaoki and Ullstrup, 1958; Straus, 1960).

Examination of the ability of ten carbon sources to support endosperm growth revealed that lactose, galactose, arabinose, rhamnose, and glycerol were ineffective (Straus and LaRue, 1954). Addition of sucrose, fructose, and glucose to the culture medium, however, supported excellent callus growth. Maximum growth, as measured by fresh weight increase, occurred at 2% sucrose. Sun and Ullstrup (1971) observed that sucrose was a better carbon source than glucose and that increasing the sucrose concentration from 2 to 4% resulted in slightly improved callus growth.

Growth regulators such as indoleacetic acid (IAA), α-naphthalene acetic acid (NAA), 2,4-dichlorophenoxyacetic acid (2,4-D), kinetin, or gibberellic acid have not been useful for the initiation or maintenance of endosperm callus. On the contrary, 0.5 mg of NAA or 2,4-D/liter caused the death of endosperm cultures after two to three transfers on medium containing these hormones (Tamaoki and Ullstrup, 1958). Straus (1960) observed that IAA concentrations above 0.25 mg/liter severely inhibited the growth of Black Mexican Sweet endosperm callus. Stowe, Thimann, et al. (1956) demonstrated that LaRue's (1949) original Clone 1-C endosperm callus contained about 1 mg of IAA/kg fresh weight. This concentration was very high and suggested that these endosperm cultures synthesized sufficient IAA and perhaps other growth regulators to allow their continued growth as callus.

Large quantities of maize endosperm for biochemical studies have been grown in semicontinuous suspension culture (Graebe and Novelli, 1966a). With this procedure one culture produced 5 kg of tissue over a 5-month period. Ribosomes were isolated from these tissues and used in a cell-free amino-acid incorporation system (Graebe and Novelli, 1966b).

Morphology and Cytology

The morphological and cytological characteristics of developing and established endosperm callus has been studied by Straus (1954). At 12 days postpollination the endosperm of Surprise Sweet kernels was a relatively homogenous tissue with a clearly differentiated peripheral meristematic layer two to three cells thick. After two days *in vitro* the peripheral layer was four cells thick in many places, and starch granules and leucoplasts were clustered around the nuclei. Four days after culture initiation the outermost layer was many cells thick, and localized areas of meristematic activity in this layer had resulted in raised nodules of tissue on the endosperm surface. At this stage regions of starch deposition were observed in the interior cells of the endosperm. After 6 days *in vitro* the nodular meristematic regions on the endosperm surface had greatly increased in size, and starch deposition was particularly evident in the interior cells of the endosperm. Endosperm callus after 8 days growth *in vitro* had a characteristic nodular appearance. Continuous callus growth occurred by cell enlargement and rapid division in the meristematic regions of the nodules.

The chromosome complement of maize endosperm is normally triploid ($3n = 30$). When Straus (1954) examined division figures in clone 1-C callus, he found that polyploid, hypoploid, or other aneuploid chromosome complements were almost as frequent as the normal chromosome number. Hyperploid cells having as many as 105 chromosomes and aneuploid cells with 41 and 48 chromosomes were observed. Hypoploid mitotic figures with 10 chromosomes were not as common as cells with polyploid figures. In addition to these mitotic disturbances, he observed multinucleate cells, chromosome bridges, and lagging chromosomes. Approximately 30% of the examined anaphase figures showed single, double, or four-chromosome bridges as well as lagging chromosomes.

Genetic Control

Genetic control of endosperm callus formation was demonstrated with four unrelated inbred lines, S41, S42, S43, and S44 by Tabata and Motoyoshi (1965). Callus was successfully initiated with a slow growth rate from S41 and a moderate growth rate from S42. Endosperm tissues from S43 and S44 did not produce callus. Among all possible crosses involving these four inbreds, endosperm callus only developed from S41 × S42 and the reciprocal cross S42 × S41. These hybrid cultures grew much more rapidly than those from either inbred parent and exhibited a maternal effect during callus initiation. That is, callus was initiated from 73% of S41 and 77% of S41 × S42 endosperm explants. This was compared to callus initiation from 22% of S42 and 35% of S42 × S41 endosperm explants. Further genetic tests were conducted on endosperm tissues of self-pollinated F_1 (S41 × S42) plants and the F_1 × S41 and F_1 × S42 backcrosses. Endosperm from the progeny of these crosses segregated for S41 (slow), S42 (moderate), and F_1 (rapid)-type callus growth as well as a "no callus" class. These progeny tests suggested that the two inbreds, S41 and S42, differed in at least two genetic factors controlling callus formation from endosperm tissues. Furthermore, studies with the inbred S43 suggested that it possessed a genetic factor inhibiting callus formation. The S41 × S42 endosperm cultures were subsequently used to isolate protoplasts (Motoyoshi, 1971). Many of these protoplasts were multinucleated, suggesting to the authors that numerous fusions occurred during protoplast isolation. Nuclear division was observed in some protoplasts, but no evidence of cell-wall regeneration or cell division was obtained.

Sun and Ullstrup (1971) have also shown that endosperm callus initiation was under genetic control. They utilized the inbred P14 *susu*, which produced a rapidly growing callus, and W64A and four of its near-isogenic conversions that did not produce callus. When W64A and its isogenic conversions were selfed and reciprocally crossed with P14 *susu*, significant effects on *in vitro* endosperm growth in the hybrid combinations were observed. With one exception, the amount of callus growth was directly proportional to the dosage of P14 genome in the hybrid.

DIPLOID CALLUS CULTURES

In contrast to endosperm tissues, callus cultures originating from diploid tissues were not successfully established until more recently. Exemplifying the early unsuccessful studies are the results of Tamaoki and Ullstrup (1958), who attempted to initiate callus from node, internode, and intercalary meristem tissues on a variety of culture media.

Maize callus cultures of diploid origin were first successfully initiated from Golden Bantam sweet corn (Mascarenhas, Sayagover, et al., 1965). In this study sterilized seedlings were germinated and grown aseptically for 7–8 days, followed by the isolation of 5-mm-long root and stem sections. Callus was successfully initiated only from the explanted root tissues and was maintained for at least 15 months on White's medium containing NAA and S-diphenyl urea

(DPU) or NAA, coconut milk, and corn steep liquor. A mixture of callus and roots was always obtained during subculture. During repeated transfers in the presence of DPU, the growth rate of these cultures was progressively stimulated until it was about fivefold higher than the original rate. Sheridan (personal communication), however, observed that repeated transfer in the presence of DPU had no effect on the growth of callus derived from an advanced generation of the singlecross W23 × M14. Gresshoff and Doy (1973) have also initiated maize callus with NAA and kinetin on B5 medium (Gamborg and Eveleigh, 1968). Their tissue source was macerated embryos from inbred B48 kernels that had been sterilized and then germinated for 48 h.

Tassel meristems isolated from 56-day-old Minhybrid 519 plants formed callus on MS medium (Murashige and Skoog, 1962) containing 10–20 μM 2,4-D (Linsmaier-Bednar and Bednar, 1972). These cultures were maintained for at least 6 years, during which time they were used to study the influence of light and exogenous cytokinin/auxin levels on root initiation and growth from maize callus. This study also demonstrated that 2,4-D was more effective than NAA or IAA for maize callus propagation. Similar conclusions were obtained in maize callus investigations by Sheridan (1975) and Green, Phillips, et al. (1974). Burr and Nelson (1972) have initiated maize callus from the mesocotyl region of germinating whole kernels on MS medium containing high concentrations of 2,4-D. This method was particularly successful with the inbred A632.

Ultrastructure Studies

Johnson and Holden (1974) investigated the ultrastructure of maize callus cells by transmission electron microscopy. Parenchymal cells from the first coleoptilar node region of germinating inbred B146 seedlings were compared with callus cells developed from this parenchymal tissue on MS medium containing 2,4-D and coconut milk. The callus cells had a meristematic appearance characterized by a dense cytoplasm filled with numerous free and bound ribosomes and a greater number of mitochondria and plastids than the nodal parenchyma cells. The most striking difference was the small scattered vacuoles in the callus cells and the large central vacuole in the parenchymal cells. The cytoplasm in the parenchymal cells of the coleoptile was found in a thin parietal layer. Long segments of endoplasmic reticulum, mitochondria with well-developed cristae, and plastids with limited lamellar development and starch grains were found in both cell types.

Genetic Variability

Genetic variability has been observed for both the frequency of callus initiation and its growth rate among 40 field and sweet-corn inbreds and single crosses (Green, Phillips, et al., 1974). Callus was initiated from 10 of 17 inbreds

(40%) and 17 of 23 singlecrosses (70%) when isolated mature embryos were grown on MS medium containing 2,4-D. Under these conditions the normal morphological development of the embryos ceased in a few days, and further incubation resulted in the appearance of callus on the coleoptile at the first node region. Wide variations in growth rate were observed for callus cultures established from different genotypes. An example of a slow-growing callus was the single-cross Oh43 × Oh5, which had a 3.7 fold fresh-weight increase during 30 days incubation at 30°C. Under similar conditions the more rapidly growing callus from Oh51A × Os420 had a 26.5 fold fresh weight increase. This callus has been maintained for 3.5 years by periodic transfer to fresh medium.

Growth Studies

Maize callus has been successfully grown on unsupplemented medium in several investigations (Linsmaier-Bednar and Bednar, 1972; Gresshoff and Doy, 1973; Green, Phillips, et al., 1974). However, attempts have been made to improve callus initiation or growth by supplementation of the culture medium with yeast extract, casamino acids, malt extract, or coconut milk. These attempts have yielded mixed results. Addition of 800 mg of yeast extract/liter and 200 mg casamino acids/liter stimulated Oh51A × Os420 callus growth by 40 and 45%, respectively (Green, Phillips, et al. 1974). In the same investigation coconut milk at 1–20% inhibited callus growth, whereas malt extract from 100–1600 mg/liter had no effect. Johnson and Holden (1974), however, successfully used 1% coconut milk to initiate callus from the inbred B146. Sheridan (1975) found that 1 g of casamino acids/liter was required for successful subculture of newly initiated callus but, after subsequent growth this supplementation was no longer necessary.

A promising approach to achieve more rapid growth rates for maize callus may be the supplementation of the culture medium with specific stimulatory amino acids or other cellular metabolites. The individual effects of 1 mM concentrations of each of the 20 common amino acids on Oh51A × Os420 callus growth are shown in Table 2. Twelve of the amino acids were found to at least slightly stimulate callus growth. Four of these, L-lysine, L-glutamine, L-serine, and L-asparagine stimulated growth by 68%, 51%, 37%, and 30%, respectively. Eight amino acids were inhibitory with L-histidine, L-tyrosine, L-threonine, and L-methionine retarding callus growth by 43%, 44%, 47%, and 57%, respectively. In these studies the growth stimulation or inhibition caused by an individual amino acid was usually clearly expressed at a 1 mM concentration (Table 3).

The growth rate of maize callus was accelerated further by various combinations of the stimulatory amino acids (Table 4). A nearly additive growth response was observed for two of these combinations. For example, the increased callus growth from lysine was 72% and from glutamine was 49%, for a total of 121%. The lysine–glutamine combination stimulated growth by 117%. Similarly, the totaled stimulation for lysine, glutamine, and serine individually was 154%, whereas the combination of these three amino acids increased callus growth by 138%. Of the two reduced nitrogen sources, glutamine was more effective individually and in combination with other amino acids than was asparagine. One cautionary note is that these amino-acid supplementation studies were conducted on only one callus genotype, Oh51A ×

TABLE 2.   Influence of 20 common amino acids on Oh51A × Os420 callus growth. Each amino acid was tested at a 1 m*M* concentration in MS medium containing 2 mg of 2,4-D/1. All amino acids were sterilized by autoclaving, except L-glutamine, which was filter sterilized. Each treatment was inoculated with callus weighing approximately 20 mg fresh-weight and incubated at 28–30°C for 30 days. Callus growth was expressed as the ratio of the final to the initial fresh weight and was termed "fold fresh-weight increase." The data represent the mean of at least 12 replications per treatment.

	Callus growth	
Treatment	Fold fr. wt. increase	Percent stimulation or inhibition (−)
Control	18.6	0
L-Lysine	31.3	68
L-Glutamine	28.1	51
L-Serine	25.4	37
L-Asparagine	24.2	30
L-Alanine	23.8	28
L-Aspartic acid	23.3	25
L-Proline	22.9	23
L-Glutamic acid	21.9	18
L-Arginine	20.2	9
Glycine	19.7	6
L-Phenylalanine	19.2	3
L-Cysteine	19.1	3
L-Isoleucine	16.3	(−12)
L-Valine	12.5	(−33)
L-Leucine	12.4	(−33)
L-Tryptophan	12.3	(−34)
L-Histidine	10.6	(−43)
L-Tyrosine	10.4	(−44)
L-Threonine	9.8	(−47)
L-Methionine	8.1	(−57)
LSD (5%)	2.1	

TABLE 3. Influence of various concentrations of L-lysine (Lys), L-glutamic acid (Glu), and L-methionine (Met) on Oh51A × Os420 callus growth. Methods were the same as described in Table 2. The data represent the mean of at least 12 replications per treatment

	Callus growth	
Treatment	Fold fr. wt. increase	Percent stimulation or inhibition (−)
Control	17.4	0
0.5 mM Lys	23.8	37
1.0 mM Lys	27.7	60
2.5 mM Lys	18.5	6
5.0 mM Lys	14.2	(−12)
0.5 mM Glu	20.1	15
1.0 mM Glu	19.1	10
2.5 mM Glu	19.1	10
5.0 mM Glu	16.3	(−6)
0.1 mM Met	17.9	3
0.5 mM Met	11.1	(−36)
1.0 mM Met	7.8	(−55)
LSD (5%)	1.9	

TABLE 4. Influence of combinations of the stimulatory amino acids, L-lysine (Lys), L-glutamine (GluNH$_2$), L-asparagine (AspNH$_2$), and L-serine (Ser) on Oh51A × Os420 callus growth. Methods were the same as described in Table 2. The data represent the mean of at least 16 replications per treatment

	Callus growth	
Treatment	Fold fr. wt. increase	Percent stimulation
Control	17.8	0
1 mM Lys	30.6	72
1 mM GluNH$_2$	26.5	49
1 mM Ser	23.7	33
1 mM AspNH$_2$	23.1	30
1 mM Lys + 1 mM GluNH$_2$	38.6	117
1 mM Lys + 1 mM GluNH$_2$ + 1 mM Ser	42.5	138
1 mM Lys + 1 mM GluNH$_2$ + 1 mM AspNH$_2$	35.6	100
LSD (5%)	2.3	

Os420. Whether lysine, glutamine, serine, and asparagine or their combinations stimulate callus growth from other genotypes is unknown.

Plant Regeneration

The regeneration of maize plants from callus cultures has been an elusive goal until recently. In studies previously discussed in this chapter callus was initiated from a diverse variety of maize tissues. Although root development was observed in all of these investigations, leaves or complete plants were never obtained. Recently, callus was initiated from the scutellum of immature maize embryos from which large numbers of plants have been regenerated (Green and Phillips, 1975). These embryos were obtained initially from 18- and 20-day postpollination ears of inbred A188 or A188 × W22 *RnjRnj*. After the husked ears were sterilized, the embryos were excised from the kernels and transferred onto MS medium containing 15 mM L-asparagine and 1 mg, 2 mg, 4 mg, or 6 mg of 2,4-D/liter.

Orientation of the embryos on the medium had a marked influence on the formation of scutellar callus. When embryos were oriented with the flat plumule-radicle axis in contact with the medium (Fig. 1a), callus was sub-

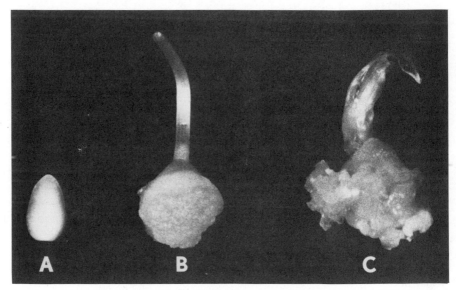

Figure 1. Scutellar callus initiation from A188 × W22 *RnjRnj* embryos: (a) 2.5-mm long embryo with scutellar tissues oriented upward on MS medium containing 15 mM L-asparagine and 2 mg 2,4-D/liter; (b) similarly oriented embryo after 10 days of incubation; (c) scutellar callus development after 18 days of incubation. Embryos and scutellar callus were incubated at 28–30°C in a 16/8-h photoperiod of 1200 lux cool-white fluorescent light.

TABLE 5.Genotypes from which scutellar callus capable of plant regeneration has been isolated. The ear parent is listed first in the crosses

Genotype	Current Age of Culture (months)
A632T[b]/W22 *Rnj Rnj* × A188[2c]	5[a]
Wf9T[b]/W22 *Rnj Rnj* × A188[2c]	6
Black Mexican *wxwx susu*	15
Black Mexican *wxwx SuSu*	15
A188 × W22 *Rnj Rnj*	25
A188 × W23	3[a]
W23 × A188	12
A188	19
W64A	13
A619	13
B9A	4[a]

[a] Indicates the time in months after callus initiation when these cultures died.

[b] The T indicates Texas male-sterile cytoplasm.

[c] A188[2] indicates the second backcross to A188.

sequently initiated from the exposed scutellum. When embryo orientation was reversed by placing the scutellum downward, the exposed plumule–radicle axis germinated, producing roots and leaves, but no scutellar callus was formed.

Callus initiation began with a generalized enlargement of the scutellum, resulting in the formation of a dome shaped tissue within a few days. After 10 days of incubation the shape of the scutellum had become irregular due to outgrowths on its surface (Figure 1b). A nodular white or pale-yellow callus was visible in localized regions of the scutellum after 18 days. After 30 days these primary cultures developed callus masses of 1–2 cm diameter, which frequently exhibited visible organization such as localized chlorophyll development or small 1–2-mm-long leaves. A 2,4-D concentration of 2 mg/liter has been most effective for scutellar callus initiation under these conditions. Historically, it is interesting to note that LaRue (1952), who cultured scutellar tissues from immature embryos of the maize lines, Evergreen and Golden Bantam, observed that some scutella developed "papillate or even callus-like outgrowths on their surfaces." Unfortunately, this work predated the widespread use of auxins and cytokinins to initiate callus growth. Scutellar callus capable of plant regeneration has been initiated from immature embryos of 11 of 31 tested genotypes (Table 5). The cultures from three of these genotypes, A632T/W22 *RnjRnj* × A188[2], A188 × W23, and B9A died within a few months of callus initiation.

Cultures from the remaining eight lines have remained viable and are currently being maintained by periodic subculture to fresh MS medium.

Plant regeneration from scutellar callus of A188 × W22 *RnjRnj* and other genotypes was accomplished by initially lowering and then completely removing the 2,4-D from the culture medium (Green and Phillips, 1975). The differentiation and growth of plants from A188 × W22 *RnjRnj* scutellar callus is illustrated in Figure 2. Figure 2*a* shows a piece of maize callus at the time of inoculation onto MS medium containing 15 m*M* L-asparagine and 0.25 mg of

Figure 2.   Plant differentiation from A188 × W22 *RnjRnj* callus: (a) callus on MS medium containing 15 m*M* L-asparagine and 0.25 mg 2,4-D/liter at 0 days of incubation; (b) same culture after 30 days of incubation; (c) multiple plant differentiation after 30 days of growth on MS medium without 2,4-D; (d) six-leaf-stage regenerated plant established in soil. Incubation conditions were the same as Figure 1.

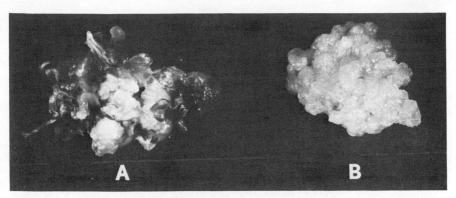

Figure 3. Variation in scutellar callus derived from different genotypes: (a) W64A callus growing on MS medium containing 2 mg of 2,4-D/liter; (b) Black Mexican *wxwx susu* callus growing on MS medium containing 1 mg of 2,4-D/liter.

2,4-D/liter. This callus exhibited no chlorophyll or leaves and was obtained from cultures grown on MS medium containing 2 mg of 2,4-D/liter. After 30 days of incubation this tissue mass had increased greatly in size and was covered with numerous leaves and occasional short roots (Figure 2*b*). When this organized culture was transferred to hormone-free medium, six plants were recovered (Figure 2*c*). In total, many hundreds of complete plants have been regenerated from various A188 × W22 *RnjRnj* callus cultures.

The regeneration of plants from callus cultures established from other genotypes was in principle similar to that just described. Differences between genotypes, however, were observed for culture characteristics at various 2,4-D concentrations. An extreme example was the inbred W64A from which cultures were initiated that contained many leaves when maintained at 2 mg of 2,4-D/liter (Figure 3*a*). This complex mixture of callus and leaves would not lend these cultures to future studies that depend on the availability of unorganized callus. The other extreme was represented by cultures established from the line Black Mexican *wxwx susu* (Figure 3*b*). These cultures were originally initiated at 2 mg of 2,4-D/liter but subsequently were maintained at 1 mg of 2,4-D/liter, where little chlorophyll developed and leaves were rarely differentiated. This Black Mexican callus readily differentiated leaves at 0.25 mg of 2,4-D/liter. Plants were then regenerated from these cultures when they were transferred to hormone-free medium.

When two-leaf-stage regenerated plants from A188 × W22 *RnjRnj* or A188 callus were transplanted directly to steam sterilized soil, 10–15% of the plants became established and grew (Figure 2*d*). Regenerated plants grown in soil had normal appearing leaves, ears, and tassels. Although many abnormal leaf shapes were observed in cultures during plant differentiation, no albino, striated, or chimeral leaves or plants were found. Chromosome counts in root-tip cells from five regenerated plants indicated that each had 20 chromosomes. Meiotic tissue has not yet been examined cytologically, nor have crosses been attempted to recover progeny from regenerated plants.

Embryo Age and Size

The influence of embryo age on the formation of scutellar callus and plant regeneration was studied by isolating embryos from the inbreds A188, A632, B9A, and W64A at 2-day intervals from 14 through 24 days after self-pollination (Table 6; Green and Phillips, 1975). This study showed that variability existed between genotypes in the ability to initiate differentiating cultures. For example, scutellar callus capable of differentiation was initiated from A188 at all embryo ages except 24 days, whereas none were recovered from A632 at any embryo age. In this experiment 18-day embryos were approximately 3 mm long and were grown on MS medium containing 15 mM L-asparagine and 2 mg 2,4-D/liter. Further studies have shown that plants grown at different temperatures and photoperiods had embryos that were markedly different in size. For example, embryos of 3 mm length have been isolated from ears whose age range was 13–20 days after pollination. These results indicated that embryo age was not as important as embryo size. Embryos can be smaller but generally not longer than 2.5–3 mm for scutellar callus initiation.

Future Directions

Effective utilization of maize-cell cultures to investigate metabolic, physio-logical, and developmental processes at the cellular level will depend on the

TABLE 6.   Influence of embryo age and genotype on scutellar callus formation (Green and Phillips, 1975)

	Postpollination embryo age (days)					
Inbred	14	16	18	20	22	24
	Percent embryos forming differentiating cultures[a]					
A188	18	15	21	14	9	0
A619	0	5	6	5	0	0
A632	0	0	0	0	0	0
B9A[b]	0	5	8	0	0	0
W64A	0	0	11	0	0	0

[a] Each value represents the percent of 60 embryos.
[b] Subsequently, these B9A cultures died during the fourth month of growth (fourth subculture).

continued development of *in vitro* techniques. An important objective will be the growth of well-dispersed or single haploid and diploid cells and subsequent plant regeneration from these cultures. This technique could open productive avenues for mutant selection in large, relatively homogenous cell populations grown on defined culture media. Plant regeneration would provide the necessary link between *in vitro* cellular manipulations and the investigation of their developmental and genetical manifestations in the whole plant. This also provides for the evaluation and utilization of new genotypes with agricultural potential.

Maize callus has grown satisfactorily in liquid medium, but usually as clumps of tissue with a size range of 1–10 mm in diameter [Figure 4a; Sheridan (1975). Suspension cultures with greatly improved cell-dispersion have been obtained recently from Black Mexican Sweet callus (Figure 4b). The cell aggregates in these cultures ranged in size from a few cells to 1–2 mm in diameter. Black Mexican callus was initiated from seedling stem section explants on MS medium containing 4 mg of 2,4-D/liter. Both the callus and suspension cultures were subsequently maintained on the same culture medium.

As cell and tissue culture techniques continue to improve, increased attention will focus on the utilization of these cultures to investigate fundamental problems in biology and agriculture. One early attempt to utilize diploid maize tissue cultures was an extension of the amino-acid supplementation studies previously discussed in this chapter. When all possible combinations of the amino acids lysine, threonine, and methionine were tested by adding them to culture medium, two had dramatic effects on callus growth (Green and Phillips, 1974). In this study it was observed that coaddition of lysine and threonine severely inhibited callus growth and that addition of low concentrations of methionine with the lysine and threonine prevented this growth inhibition. One interpretation of these results was that lysine and threonine feedback inhibited the activity of early enzymes in the aspartate family biosynthetic pathway, thus preventing normal methionine synthesis. This lysine–threonine inhibition has provided a potential selection system to isolate feedback mutants that might overproduce one or more of the amino acids lysine, threonine, or methionine.

Maize tissue cultures have also been used to study the effects of *Helminthosporium maydis* race T pathotoxin on callus growth (Gengenbach and Green, 1975). Texas male-sterile cytoplasm (*cms*T) maize is susceptible to this fungus and its pathotoxin, whereas nonsterile (N) cytoplasm maize is resistant. In the presence of toxin, callus from A619N maize grew normally, whereas the growth of callus from A619 *cms*T maize was inhibited up to 100% by increasing toxin concentrations in the culture medium. Selection for resistant A619 *cms*T callus was initiated with cultures grown on sublethal toxin concentrations. The fastest-growing callus or callus sectors were subcultured for addi-

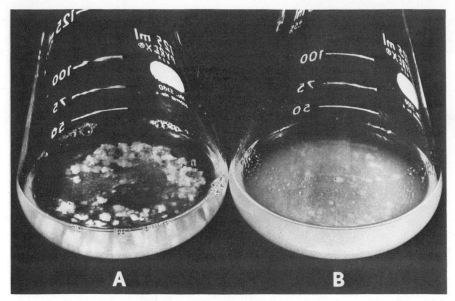

Figure 4. Maize suspension cultures grown in MS medium containing 4 mg of 2,4-D/liter: (a) Tom Thumb popcorn; (b) Black Mexican sweet corn. Cultures courtesy of Dr. W. F. Sheridan.

tional selection cycles. After four cycles, 14 callus lines were selected that were resistant based on their rapid growth rates in the presence of high toxin concentrations. This resistance has remained stable after 293 days of growth in the absence of toxin. Mitochondria isolated from these resistant callus lines differed from their unselected A619 (*cms*T) counterparts. They exhibited NADH oxidation, oxidative phosphorylation, malate-driven DCPIP reduction, and mitochondrial swelling, which was characteristic of A619N seedling or callus mitochondria.

NOTE. Since preparation of this manuscript several significant maize tissue culture papers have appeared. Freeling, Woodman et al. (1976) discussed developmental differences of totipotent, friable, and nonfriable maize callus types initiated from immature embryos. Harms, Lörz et al. (1976) reported plant regeneration from mesocotyl callus initiated from the variety "Prior." Green (1977) discussed factors influencing plant regeneration from scutellar callus, the characterization of regenerated plants, and the initiation and characterization of suspension cell cultures. Gengenbach, Green et al. (1977) regenerated plants from *H. maydis* race T pathotoxin resistant cell lines selected from *cms T* callus. Genetic tests showed that the resistance to pathotoxin and race T spores was maternally inherited.

REFERENCES

Bonnett, H. and T. Ericksson. 1974. Transfer of algal chloroplasts into protoplasts of higher plants. *Planta* **120**: 71–79.

Burk, L. G., G. R. Gwynn, and J. F. Chaplin. 1972. Diploidized haploids from aseptically cultured anthers of *Nicotiana tabacum*. *J. Hered.* **63**: 355–360.

Burr, B. and O. E. Nelson. 1972. Induction and maintenance of maize callus tissue. *Maize Genet. Newsl.* **41**: 202–203.

Carlson, P. S. 1973. Methionine sulfoximine-resistant mutants of tobacco. *Science* **180**: 1366–1368.

Carlson, P. S., H. H. Smith, and R. D. Dearing. 1972. Parasexual interspecific plant hybridization. *Proc. Nat. Acad. Sci. U. S. A.* **69**: 2292–2294.

Clapham, D. 1973. Haploid *Hordeum* plants from anthers *in vitro*. *Zeit. Pflanzenzucht.* **69**: 142–155.

Freeling, M., J. C. Woodman, and D. S. K. Cheng. 1976. Developmental potentials of maize tissue cultures. *Maydica* **21**: 97–112.

Gamborg, O. L. and D. E. Eveleigh. 1968. Culture methods and detection of glucanases in suspension cultures of wheat and barley. *Can. J. Biochem.* **46**: 417–421.

Gengenbach, B. G. and C. E. Green. 1975. Selection of T-cytoplasm maize callus cultures resistant to *Helminthosporium maydis* race T pathotoxin. *Crop Sci.* **15**: 645–649.

Gengenbach, B. G., C. E. Green, and C. M. Donovan. 1977. Inheritance of selected pathotoxin resistance in maize plants regenerated from cell cultures. *Proc. Natl. Acad. Sci. U. S. A.* **74**: 5113–5117.

Graebe, J. E. and G. D. Novelli. 1966a. A practical method for large-scale plant tissue culture. *Exp. Cell Res.* **41**: 509–520.

Graebe, J. E. and G. D. Novelli. 1966b. Amino acid incorporation in a cell-free system from submerged tissue cultures of *Zea mays* L. *Exp. Cell Res.* **41**: 521–534.

Grambow, H. J., K. N. Kao, R. A. Miller, and O. L. Gamborg. 1972. Cell division and plant development from protoplasts of carrot cell suspension cultures. *Planta* **103**: 348–355.

Green, C. E. 1977. Prospects for crop improvement in the field of cell culture. *Hort Sci.* **12**: 131–134.

Green, C. E. and R. L. Phillips. 1974. Potential selection system for mutants with increased lysine, threonine, and methionine in cereal crops. *Crop Sci.* **14**: 827–830.

Green, C. E. and R. L. Phillips. 1975. Plant regeneration from tissue cultures of maize. *Crop Sci.* **15**: 417–421.

Green, C. E., R. L. Phillips, and R. A. Kleese. 1974. Tissue cultures of maize (*Zea mays* L.): Initiation, maintenance, and organic growth factors. *Crop Sci.* **14**: 54–58.

Gresshoff, P. M. and C. H. Doy. 1973. *Zea mays*: Methods for diploid callus culture and the subsequent differentiation of various plant structures. *Aust. J. Biol. Sci.* **26**: 505–508.

Harms, C. T., H. Lörz, and I. Potrykus. 1965. Regeneration of plantlets from callus cultures of *Zea mays* L. *Z. Pflanzenzuchtg.* **77**: 347–351.

Johnson, J. W. and D. J. Holden. 1974. Ultrastructure of callus tissue of *Zea mays. Can. J. Bot.* **52**: 251–254.

Kao, K. N., F. Constabel, M. R. Michayluk, and O. L. Gamborg. 1974. Plant protoplast fusion and growth of intergeneric hybrid cells. *Planta* **120**: 215–227.

Kartha, K. K., O. L. Gamborg, F. Constabel, and K. N. Kao. 1975. Fusion of rapeseed and soybean protoplasts and subsequent division of heterokaryocytes. *Can. J. Bot.* **52**: 2435–2436.

Kasperbauer, M. J. and G. B. Collins. 1972. Reconstitution of diploids from leaf tissue of anther-derived haploids in tobacco. *Crop Sci.* **12**: 98–101.

LaRue, C. D. 1949. Cultures of the endosperm of maize. *Am. J. Bot.* **36**: 798.

LaRue, C. D. 1952. Growth of the scutellum of maize in culture. *Science* **115**: 315–316.

Linsmaier-Bednar, E. M. and T. W. Bednar. 1972. Light and hormonal control of root formation in Zea mays callus cultures. *Devel. Growth Differ.* **14**: 165–174.

Mascarenhas, A. F., B. M. Sayagover, and V. Jagannathan. 1965. Studies on the growth of callus cultures of *Zea mays. In Tissue culture,* C. V. Ramakrishnan (ed.). The Hague: Dr. W. Junk, pp. 283–292.

Motoyoshi, F. 1971. Protoplasts isolated from callus cells of maize endosperm. *Exp. Cell Res.* **68**: 452–456.

Murashige, T. and F. Skoog. 1962. A revised medium for rapid growth and bioassays with tobacco tissue cultures. *Physiol. Plant.* **15**: 473–497.

Nagata, T. and I. Takebe. 1971. Plating of isolated tobacco mesophyll protoplasts on agar medium. *Plants* **99**: 12–20.

Niizeki, H. and K. Oono. 1968. Induction of haploid rice plant from anther culture. *Proc. Jap. Acad.* **44**: 554–557.

Nitsch, C. 1974. Pollen culture—a new technique for mass production of haploid and homozygous lines. In *Haploids in higher plants,* K. J. Kasha (ed.). Guelph: University of Guelph, pp. 123–135.

Nitsch, J. P. 1951. Growth and development *in vitro* of excised ovaries. *Am. J. Bot.* **38**: 566–577.

Ohyama, K., O. L. Gamborg, and R. A. Miller. 1972. Uptake of exogenous DNA by plant protoplasts. *Can. J. Bot.* **50**: 2077–2080.

Ohyama, K. and J. P. Nitsch. 1972. Flowering haploid plants obtained from protoplasts of tobacco leaves. *Plant Cell Physiol.* **13**: 229–236.

Picard, E. and J. de Buyser. 1973. Obtaining haploid plants of *Triticum aestivum* L. by the culture of anthers *in vitro. C. R. Acad. Sci. Paris Ser.* D **277**: 1463–1466.

Potrykus, I. and F. Hoffman. 1973. Transplantation of nuclei into protoplasts of higher plants. *Zeit. Pflanzenphysiol.* **69**: 287–289.

Shannon, J. C. and J. W. Batey. 1973. Inbred and hybrid effects on establishment of *in vitro* cultures of *Zea mays* L. endosperm. *Crop. Sci.* **13**: 491–493.

Sheridan, W. F. 1975. Tissue cultures of maize. I. Callus induction and growth. *Physiol. Plant.* **33:** 151–156.

Sternheimer, E. P. 1954. Method of culture and growth of maize endosperm *in vitro. Torrey Bot. Club Bull.* **81:** 111–113.

Stowe, B. B., K. V. Thimann, and N. P. Kefford. 1956. Further studies of some plant indoles and auxins by paper chromatography. *Plant Physiol.* **31:** 162–165.

Straus, J. 1954. Maize endosperm tissue grown *in vitro.* II. Morphology and cytology. *Am. J. Bot.* **41:** 833–839.

Straus, J. 1960. Maize endosperm tissue grown *in vitro.* III. Development of a synthetic medium. *Am. J. Bot.* **47:** 641–647.

Straus, J. and C. D. LaRue. 1954. Maize endosperm tissue grown *in vitro.* I. Cultural requirements. *Am. J. Bot.* **41:** 687–694.

Sun, M. H. and A. J. Ullstrup. 1971. *In vitro* growth of corn endosperm. *Torrey Bot. Club Bull.* **98:** 251–258.

Sunderland, N. 1973. Pollen and anther culture. In *Plant tissue and cell culture (botanical monographs,* Vol. 2) H. E. Street (ed.). Berkeley–Los Angeles: University of California Press, pp. 205–239.

Tabata, M. and F. Motoyoshi. 1965. Hereditary control of callus formation in maize endosperm cultured *in vitro. Jap. J. Genet.* **40:** 343–355.

Tamaoki, T. and A. J. Ullstrup. 1958. Cultivation *in vitro* of excised endosperm and meristem tissues of corn. *Torrey Bot. Club Bull.* **85:** 260–272.

Thomas, E. and G. Wenzel. 1975. Embryogenesis from microspores of rye. *Naturwissenschaften* **62:** 40–41.

Tsai, S. 1970. Carbohydrate and amylase enzymes of maize endosperm tissue cultures. Ph.D. thesis. University Park: Pennsylvania State University, Dissertation Abstracts 11–16, 680, Univ. Microfilms, Ann Arbor.

Wang, Y.-y., C.-s. Sun, C.-c. Wang, and N.-f. Chien. 1973. The induction of the pollen plantlets of *Triticale* and *Capsicum annum* from anther culture. *Scientia Sincia* **16:** 147–151.

White, P. R. 1943. *A handbook of plant tissue culture.* Lancaster: J. Cattel Press.

Chapter 33

BIOLOGY OF THE MAIZE MALE GAMETOPHYTE

Paul L. Pfahler

Department of Agronomy,
University of Florida,
Gainesville

The primary function of the gametophytic generation in higher plants is to transmit genetic information to the next sporophytic generation. The male gametophyte has a more active and demanding role in the fertilization process than the female gametophyte and a specialized structure and unique properties to perform this role. The pollen grain is independent of the sporophyte on which it is produced for a period of its life cycle and must retain germinability during this period. Also, to effect fertilization, the grain must compete with other grains in germination and growth through the style. In contrast, the female gametophyte remains surrounded and nourished by the sporophyte during its complete life cycle, passively awaiting the arrival of the pollen tube.

This chapter reviews the information concerning the maize male gametophyte from the time of its release from the sporophyte through fertilization. The importance and future possibilities of male gametophyte research is explored in the concluding section.

517

MATURE POLLEN-GRAIN CHARACTERISTICS

Morphology

When expanded to maximum volume, the pollen grains are spheroidal in shape and have a mean diameter of 90–100 μ. When released, they are dry and powdery and their shape is somewhat irregular.

Many factors are recognized that alter pollen diameter. These include the location in the tassel where the grain is produced, the environmental conditions to which the sporophyte is exposed, ploidy level, and genetic factors (Banerjee and Barghoorn, 1971; Pfahler, 1965; Rumbaugh and Whalen, 1972, 1973). Genetic variation is produced both by complex quantitative factors with extremely low heritability (Rumbaugh and Whalen, 1972, 1973) and simply-inherited qualitative factors (Neuffer, Jones, et al. 1968).

The single pore in the exine is circular in shape with a diameter of 7–9 μ (Irwin and Barghoorn, 1965). Apparently, pore diameter is under genetic control and has been suggested as a method for cultivar identification (Nair, 1962).

Pollen-wall architecture, which is associated with the ektexine or outer layer of the extine, appears at low magnifications to be minutely roughened with a spinulate exine pattern. At higher magnifications a strong pattern with spinules arranged in a regular order can be recognized (Erdtman, 1969; Irwin and Barghoorn, 1965). Minute holes are distributed between and around the spinules.

Chemical Composition

The yellow color is produced by a flavonol pigment, quercetin, which is localized in the exine (Redemann, Wittwer, et al., 1950). No definite information is available about the site of quercetin production, but since the walls of empty grains are yellow, quercetin production is probably associated with the sporophyte rather than the microscope protoplast.

Numerous studies reporting the mineral element content and ash percentage are available (Anderson and Kulp, 1922; Knight, Crooke, et al., 1973; Pfahler and Linskens, 1974; Todd and Bretherick, 1942). Comparisons among results are difficult because of differences in genotypes tested, collection techniques, locations and years in which the plants were grown, and the methods of analysis employed. In one study (Pfahler and Linskens, 1974), a significant genotypic effect was present with an indication that heterozygosity level was a factor. The mechanism associated with mineral uptake by the plant and absorption by the pollen grain is not known. Also, the specific role of mineral elements in the pollen grain is not understood.

A number of carbohydrates have been recognized and measured (Anderson and Kulp, 1922; Banks, Greenwood, et al., 1971; Nielson, Grommer, et al.,

1955; Nilsson, Ryhage, et al., 1957; Pfahler and Linskens, 1971; Todd and Bretherick, 1942; Zuber, Deatherage, et al., 1960). Starch as small granules is abundant and probably acts as an energy source. It is usually composed of two fractions, amylose and amylopectin. Many qualitative genetic loci are known to alter the amylose content of endosperm starch (Creech, 1965; Nelson, 1967). The absence of amylose in wx pollen grains produced on heterozygous (Wx wx) plants was the first clear indication that the genotype of the pollen grain could alter the contents of the grain independent of the sporophyte on which it was produced. However, with other loci governing amylose content of the endosperm starch, no correlation between the amylose content of the pollen and endosperm starch was found (Zuber, Deatherage, et al., 1960).

The starch and sugar content of pollen have been reported in a number of studies (Anderson and Kulp, 1922; Haber and Gaessler, 1942; Nielson, Grommer, et al., 1955; Pfahler and Linskens, 1972; Todd and Bretherick, 1942). The starch and sugar content of the pollen grains was not affected by the alleles at the waxy (wx), sugary ($su1$), and shrunken ($sh2$) loci (Pfahler and Linskens, 1971).

Many studies have been conducted on the protein and amino-acid content in an effort to elucidate mechanisms of protein breakdown and synthesis (Anderson and Kulp, 1922; Linskens and Pfahler, 1973; Pfahler and Linskens, 1970; Ray Sarkar, Wittwer, et al., 1949; Sawada, 1960; Todd and Bretherick, 1942). Evidence is available to suggest that the alleles at the wx, $su1$, and $sh2$ loci interact during storage in altering the content of some free amino acids (Linskens and Pfahler, 1973; Pfahler and Linskens, 1970). In general, however, the complex processes associated with protein breakdown and synthesis are not understood especially in such a specialized cell as the pollen grain.

Many other biologically-active substances have been recognized (Brewbaker, 1971; Fathipour, Schlender, et al., 1967; Fukui, Teubner, et al., 1958; Knox, 1971; Knox and Heslop-Harrison, 1969; Mitchell and Whitehead, 1941; Scandlios, 1969; Walden, 1959). The primary purpose of these studies was to determine how these substances function and interact to maintain the pollen grain while it is independent and to initiate germination. Genetic control of a large number of isozymes has been established, but no concept of their function has been developed (Brewbaker, 1971). Since the isozyme composition of the pollen grain is rarely identical to the surrounding tapetal tissue, most enzymes and isozymes in the grain are probably synthesized after meiosis, under the control of the genotype of the maturing microspore.

Fatty-acid and lipid contents were also reported (Anderson, 1923; Anderson and Kulp, 1922; Fathipour, Schlender, et al., 1967; Fukui, Teubner, et al., 1958; Pfahler and Linskens, 1970, 1971). The alleles at the wx, $su1$, and $sh2$ loci did not alter the contents of either (Pfahler and Linskens, 1970, 1971).

Viability

When applied to pollen grains, *viability* refers to the ability to germinate on the silk, participate in fertilization and produce a normal kernel. It involves the reaction of the pollen grains to the environmental conditions encountered during the period in which the grain is independent of the sporophyte.

The ultimate measure of viability is *in vivo* or fertilization ability tests in which one grain is placed on one silk to determine if a normal kernel is produced. Technically, this procedure would be possible in maize but very difficult on a scale necessary to obtain statistically meaningful results. However, some quantitative methods involving controlled pollinations have been developed (Jones and Newell, 1948; Knowlton, 1922; Pfahler and Linskens, 1972; Walden and Everett, 1961). In general, the results indicated that maximum viability was obtained when pollen grains were stored at temperatures just above 0°C under high humidity. Some viability was found in pollen grains stored for 292 h at 35°F (1.6°C) under 95–98% relative humidity, but in general only a small proportion remained viable for more than 6 days at any storage temperature (Walden, 1967). The results on humidity effects are quite inconsistent, probably because the moisture content of the grains at the time of release vary considerably depending on air humidity, water stress conditions of the plant, and collection methods. The moisture content of the grain when placed in storage alters the humidity effects during storage presumably because of the rate of desiccation or moisture absorption. In general, humidity over 70% produced maximum viability in most studies. Limited evidence exists of a genotype effect on viability as measured by *in vivo* tests (Pfahler and Linskens, 1972). Comparisons between the alleles at the *wx, su*1, and *sh*2 loci indicated that pollen grains containing the recessive alleles maintained viability slightly longer than their dominant counterparts.

Limited use of *in vitro* germination tests has been made to estimate viability. The major difficulty associated with this method is that the relationship between *in vitro* fertilization ability and *in vitro* germination is not clear, but a parallel obviously exists. Numerous studies (Bair and Loomis, 1941; Cook and Walden, 1965, 1967; Knowlton, 1922; Pfahler, 1967) have been conducted to determine the requirements for optimum *in vitro* germination. In general, a medium containing sucrose, agar, calcium, and boron consistently produced the best germination percentages and pollen-tube growth in most studies. However, considerable variability, the source of which could not be determined, was found among studies. One major source of variation was found to be pollen or pollen-source genotype (Pfahler, 1968, 1970, 1971; Pfahler and Linskens, 1973). This effect was so pronounced that the value of calcium and boron in increasing germination depended, in some cases, on the pollen or pollen source genotype. Also, pollen grains containing the alleles at the *wx, su*1, and *sh*2 loci differed after storage at 2°C in *in vivo* fertilization ability and *in vitro* germination capacity (Pfahler and Linskens, 1972).

Another test for viability is the use of vital stains. Many stains, including tetrazolium salts and benzidine–hydrogen peroxide, have been tested, although no direct comparison between the staining results and fertilization ability has been conducted. In theory, vital stains indicate the presence and activity of certain enzyme systems considered essential to sustain vital life processes. In practice, it has been shown that many of the enzyme systems measured by these stains are present in an active form long after the viability in terms of fertiliza tion ability is lost (Brewbaker, 1971).

TRANSMISSION

Male transmission of genetic elements by mature pollen grains includes two interrelated aspects—ability and opportunity. Transmission "ability" concerns the ability of pollen grains containing different genetic elements to compete in fertilization on the same silk. Thus ability is associated with the relationship between fertilization ability of the pollen grain and its genotype. Transmission "opportunity" involves the factors that influence the pollen grains containing different genetic elements landing on the same silk and competing in fertilization. Many factors alter both transmission ability and transmission opportunity (Pfahler, 1975).

GERMINATION AND POLLEN-TUBE GROWTH

The silk is divided at the tip with two shallow grooves running its full length (Kiesselbach, 1949). Growth is rapid from a meristematic region near the ovary. Silk hairs, which cover the entire silk surface develop from a single cell on that surface. Fully developed hairs consist of four columns of cells loosely fitted together in a circular pattern. An intercellular canal, which develops at the point where the cells of each column meet, runs the entire length of the hair.

Germination processes are initiated when the dry pollen grains contacts and adheres to the silk. Generally, the grain is intercepted by the silk hair, but it can function if it lands on the silk body (Kiesselbach, 1949; Weatherwax, 1955). Liquid is immediately absorbed from the silk, and the pollen tube emerges from the pore within 3–5 min (Kihara and Hori, 1966). The pollen tube then grows (frequently outside the silk exposed to the atmosphere until it enters the silk), usually between the cells of the silk hair, but sometimes by penetrating the body of the silk directly. Once inside the silk body the tube soon reaches and follows the sheath cells surrounding the vascular tissue to the base of the silk. The tissue through which the tube grows is distinct from the surrounding tissue and has been termed "transmitting tissue" (Rosen, 1971). This tissue is morphologically similar and presumed functionally related to the cells on the silk surface. At the base of the silk the tube enters the ovarian cavity and follows the inner surface of the ovary wall to the micropyle. The tube grows between the cells of the nucleur tissue until the embryo sac is reached. On entering the embryo sac the end of the tube ruptures releasing its two sperm nuclei. Fertilization ensues.

Although subject to much variation because of differences in genotypes, environments, pollination hour, and silk lengths, the time required between pollination and fertilization is estimated at about 25 h, with about 5–10 h required for 3 cm of tube growth in the distal end of the silk (Walden, 1967; Weatherwax, 1955). Also, the germination of grains on the silk is not uniform. After 24 h grains at five different stages of germination have been reported (Kihara and Hori, 1966).

After pollen-tube formation, migration of the nuclei in the pollen grain begins. In the ungerminated grain, the three nuclei lie on the side of the grain opposite the pore (Kihara and Hori, 1966; Korobova, 1974). During the initiation of germination the tube nucleus experiences an intense period of metabolic activity and exceeds both sperm nuclei in volume. When the tube develops the sperm nuclei migrate into the tube, followed by the tube nucleus, which by this time has changed shape and volume. After a period of tube growth, the tube nucleus joins the sperm nuclei, and all three form a tight group behind the tip of the tube.

Once inside the silk an enlargement of the tube occurs a short distance from the tip, and this form is maintained until tip-rupturing occurs (Kiesselbach, 1949; Weatherwax, 1955). Only a short segment of the tube extending from the actively-growing tip contains the protoplast. As the tube grows, callose plugs are formed between the enlarged area and the older portions of the pollen tube. The tube shows interrupted rather than continuous growth, with a period of active growth followed by an inactive period (Korobova, 1974). Callose plugs are formed during the inactive phase. After the older segment of the tube is separated from the tip with a callose plug, the silk tissue expands and collapses the empty tube segment.

The interaction of the pollen grain and silk during the initiation of germination and silk penetration is not understood. The silk produces a small amount of exudate from which various chemical substances, including enzymes capable of digesting starch to maltose and glucose have been isolated (Iwanami, 1957; Martin and Brewbaker, 1971). The silk is also covered by a lipid or cutin coating to increase resistance to wetting and desiccation (Martin and Brewbaker, 1971). The entrance to the silk is usually between the silk hairs, and whether cutin is present in this area is not known. However, since penetration through the silk body has been observed, a cutin-degrading enzyme is probably released from the tube tip.

The growth of the pollen tube once inside the silk requires large amounts of energy and nutrients. This is especially true in maize since the tube must grow a considerable distance. Quite likely, the pollen grain reserves are exhausted early in the process, and continued growth must depend on energy and nutrients supplied by the silk. Limited information on the chemical composition of silks is available (Britikov, Musatova, et al., 1964; Hoerner and Snelling, 1940; Knight, Crooke, et al., 1973; Pfahler and Linskens, 1974), and most aspects of the pollen grain-silk interaction are not well understood.

Another intriguing aspect of this process is how the pollen tube grows between the cells of the silk. The growth of the tube tip appears entirely intercellular with no silk cells being destroyed (Kiesselbach, 1949; Weatherwax, 1955). Rather the cells appear to be pushed aside or compressed. This conclusion is supported by the fact that after the tip of the tube containing the protoplast passes, the silk cells expand or close in and collapse the empty tube. Essentially, no information is available on this growth aspect.

Related to the mechanism of how pollen tubes grow through the silk is the question of why tubes grow in the direction of the ovary. Possibly, the "transmitting tissue" physically guides and directs the growing tube. Apparently, the tubes must grow through this tissue, which forms a continuous pathway to the ovary. Another possibility is chemotropism. A calcium gradient with a high concentration at the ovary grading to a low concentration at the site of pollen germination was suggested from a number of studies with other species (Mascarenhas and Machlis, 1962; Rosen, 1964). A comparison of the upper and lower halves of maize silk suggested that no distinct calcium gradient was present (Knight, Crooke, et al., 1973). Results using *in vitro* techniques indicated that a calcium gradient did not effect tube orientation (Cook and Walden, 1967).

FERTILIZATION

When the tip of the tube ruptures, releasing the two sperm nuclei into the embryo sac, double fertilization occurs (one sperm nucleus fuses with the egg nucleus, forming a $2n$ zygote, and the other sperm nucleus unites with the two polar nuclei producing a $3n$ endosperm nucleus). This process provides for the "xenia" effect in which the male parent directly influences the endosperm and kernel characters. Many loci associated with xenia effects are known (Curtis, Brunson, et al., 1956; Leng, 1949; Neuffer, Jones, et al., 1968).

Normally, the sperm nuclei from only one pollen tube are involved.

However, a low frequency of heterofertilization occurs in which the sperm nucleus from one tube unites with the egg nucleus while the sperm nucleus from another tube fuses with the polar nuclei (Sprague, 1932).

Generally, the sperm nuclei from the same tube are genetically identical. With B chromosomes, nondisjunction at the second pollen mitosis produces different numbers of B chromosomes in the embryo and endosperm (Roman, 1948).

Normally, the female contributes a haploid set of chromosomes to the embryo. With the *ig* (indeterminate gametophyte) locus, a low frequency of androgenetic monoploids that originate from one sperm nucleus developing into the embryo without fusion with the egg nucleus are produced (Kermicle, 1969).

Radioactive tracer studies and the occurrence of heterofertilization have suggested that some tubes whose sperm nuclei are not involved in fertilization discharge their contents into the embryo sac (Linskens, 1969; Polyakov, 1964; Sprague, 1932). Numerous studies have reported that progeny originating from pollination with a mixture of pollen from diverse genetic sources differed from progeny obtained from pollination with pollen from one genetic source (Polyakov, 1964; Raicu and Popovici, 1964). These differences could not be explained using known genetic theory and mechanisms. To add to the interpretation problems, these differences were reported to be maintained over a number of generations. Their explanation was that normal double fertilization involving only one tube occurs in both pollination types. The differences result from those tubes that discharge their contents in the same embryo sac but are not involved in fertilization *per se*. Therefore, the cytoplasm of progeny from mixed pollinations contain the contents of tubes from a diverse genetic background, whereas the cytoplasm of progeny from single source pollinations contain tube contents from only one genetic source. According to this explanation, kernels with identical embryo and endosperm genotypes could differ in their cytoplasm, and these differences could be maintained for a number of generations. With present knowledge this process is difficult to visualize. However, the phenomenon of *dauermodifikation* has been described in which cytoplasmic changes are induced by environmental conditions and transmitted but eventually lost after a number of generations (Goldschmidt, 1955). Also, mitochondrial complementation in maize has been related to hybrid vigor (McDaniel and Sarkissian, 1968). Such complementation requires the contribution of mitochondria from the pollen grains.

IMPORTANCE AND FUTURE POSSIBILITIES

The primary function of the male gametophyte is to transmit genetic material to the next sporophytic generation. Depending on the mating system of the species, new genetic combinations, which are necessary if the species is to adapt to new environments, are possible. Many predictive models in population and quantitative genetics assume random mating or the independence of pollen genotype from fertilization ability. The application of these models to maize whose mating system appears random would lead to incorrect conclusions are many loci where differential pollen transmission has been detected. The presence of differential pollen transmission may be advantageous, especially if the factors associated with competitive ability are known and can be readily manipulated. As an example, control of sex ratio is very desirable in animal breeding (Church, 1974). The separation of X-bearing and Y-bearing sperm from semen has been partially successful using ultracentrifugation, electrophoresis, and chromatography. At present the most effective treatment has produced only a 60:40 ratio, but at least this suggests that shifts by semen treatment are possible. In plant species, some type of pollen treatment before pollination might result in superior progeny and populations. Thus if manipulated correctly, differential pollen transmission might be valuable in plant-improvement programs.

The effect of multipollination on the phenotype of the progeny in later generations has not been adequately explored. The interpretation of the results contained in earlier reports was vague, and thus the presence and importance of this effect was generally discounted. However, more recent results were interpreted on the basis of known fertilization events and processes and hence the possibility of a multipollination effect increased considerably. More detailed research with better controls should be conducted in this research area to determine the presence, nature, and significance of a multipollination effect.

The factors limiting interspecific and intergeneric hybridization are still largely unknown. The most obvious factor is that the pollen grain from one species does not germinate and grow through the style of another species. The process associated with germination and growth include many physical and chemical factors in both the pollen grain and style that interact to make fertilization possible. A better understanding of these complex pollen grain–style interactions is necessary if elimination of this major barrier to fertilization between species and genera is successful.

Because of its specialized structure and unique properties, the male gametophyte of higher plants is ideally suited for many studies that would be difficult or impossible using the female gametophyte or sporophyte. Genetic fine-structure studies of specific loci have been largely restricted to lower plants and animals since a critical analysis requires the classification of large numbers of meiotic products. In higher plants, pollen grains can be used since they are produced in large numbers and each contains one meiotic product. Pioneering studies of the fine structure of the waxy locus in maize have been reported (Nelson, 1959, 1962, 1975). The possibility of analyzing the fine structure of the amylose extender (ae) locus in maize was also suggested (Creech and Kramer, 1961). The use of this method is restricted to those loci whose alleles alter the pollen grain in such a manner that the genotype of each individual grain can be recognized.

Our understanding of differentiation in higher plants might be enhanced by studying the male gametophyte. Recent reports using *Nicotiana* have indicated that functional, flowering haploid plants can be produced from microspores using artificial medium (Nitsch, 1971; Nitsch and Nitsch, 1969). No success with maize has as yet been reported, but the production of pollen plantlets from species (*Triticum aestivum* and *Triticale*) within the grass family has been indicated (Ouyang, 1973; Wang, 1973). Using plants produced under these circumstances, a more critical and extensive analysis of differentiation processes at all stages of development would be possible.

Study of the male gametophyte would contribute to our limited knowledge of biochemical and physiological genetics in higher plants. The protoplast of the pollen grain is generally developed under the control of the nucleus of the grain. Thus, the physiological effect of certain alleles could be closely examined without the confounding effects of dominance, epistasis, and differentiation. Also, the effects of various agents such as mutagens and environment on this genetic–physiologic relationship could be more closely identified using the male gametophyte as the test material. The capacity of the pollen grain to be germinated under artificial conditions allows for a critical analysis of the relationship between various alleles and the biochemistry, physiology, and growth processes associated with the cell.

REFERENCES

Anderson, R. J. 1923. Composition of corn pollen. II. Concerning certain lipoids, a hydrocarbon and phytosterol occurring in the pollen of white flint corn. *J. Biol. Chem.* **55**: 611–628.

Anderson, R. J. and W. L. Kulp. 1922. Analysis and composition of corn pollen. Preliminary report. *J. Biol. Chem.* **50**: 433–453.

Bair, R. A. and W. E. Loomis. 1941. The germination of maize pollen. *Science* **94**: 168–169.

Banerjee, U. C. and E. S. Barghoorn. 1971. Factors controlling pollen grain size in maize. *Maize Genet. Coop. Newsl.* **45**: 244–245.

Banks, W., C. T. Greenwood, D. D. Muir, and J. T. Walker. 1971. Pollen starch from amylomaize. *Die Stärke* **23**: 380–382.

Brewbaker, J. L. 1971. Pollen enzymes and isozymes. In *Pollen: development and physiology*, J. Heslop-Harrison (ed.). London: Butterworth, pp. 156–170.

Britikov, E. A., N. A. Musatova, S. V. Vladimirtseva, and M. A. Protsenko. 1964. Proline in the reproductive system in plants, In *Pollen physiology and fertilization*, H. F. Linskens (ed.). Amsterdam: North-Holland, pp. 77–85.

Church, R. B. 1974. Molecular and reproductive biology in animal genetics. *Genetics* **78:** 511–524.

Cook, F. S. and D. B. Walden. 1965. The male gametophyte of *Zea mays* L. II. *In vitro* germination. *Can. J. Bot.* **43:** 779–786.

Cook, F. S. and D. B. Walden. 1967. The male gametophyte of *Zea mays* L. III. The influence of temperature and calcium on pollen germination and growth. *Can. J. Bot.* **45:** 605–613.

Creech, R. G. 1965. Genetic control of carbohydrate synthesis in maize endosperm. *Genetics* **52:** 1175–1186.

Creech, R. G. and H. H. Kramer. 1961. A second region in maize for genetic fine structure studies. *Am. Natur.* **95:** 326–328.

Curtis, J. J., A. M. Brunson, J. E. Hubbard, and F. R. Earle. 1956. Effect of the pollen parent on oil content of the corn kernel. *Agron. J.* **48:** 551–555.

Erdtman, G. 1969. *Handbook of palynology.* New York: Hafner, 486 p.

Fathipour, A., K. K. Schlender, and H. M. Sell. 1967. The occurrence of fatty acid methyl-esters in the pollen of *Zea mays*. *Biochem. Biophys. Acta* **144:** 476–478.

Fukui, H. H., F. G. Teubner, S. H. Wittwer, and H. M. Sell. 1958. Growth substances in pollen. *Plant Physiol.* **33:** 144–146.

Goldschmidt, R. B. 1955. *Theoretical genetics.* Berkeley: Univ. California Press, 563 pp.

Haber, E. S. and W. G. Gaessler. 1942. Sugar content of sweet corn pollen and kernels of inbred and hybrid strains susceptible to tassel infestation by aphis. *Am. Soc. Hort. Sci.* **40:** 429–431.

Hoerner, I. R. and R. O. Snelling. 1940. Effect of pollination upon chemical composition of silks of certain inbred lines of maize. *J. Am. Soc. Agron.* **32:** 213–215.

Irwin, H. and E. S. Barghoorn. 1965. Identification of the pollen of maize, teosinte and *Tripsacum* by phase contrast microscopy. *Bot. Mus. Leafl., Harvard Univ.* **21:** 37 57.

Iwanami, Y. 1957. Physiological researches on pollen. XI. Starch grains and sugars in stigma and pollen. *Bot. Mag. (Tokyo)* **70:** 38–43.

Jones, M. D. and L. C. Newell. 1948. Longevity of pollen and stigmas of grasses: Buffalograss, *Buchloe dactyloides* (nutt.) Engelm., and corn, *Zea mays* L. *J. Am. Soc. Agron.* **40:** 195–204.

Kermicle, J. L. 1969. Androgenesis conditioned by a mutation in maize. *Science* **166:** 1422–1424.

Kiesselbach, T. A. 1949. The structure and reproduction of corn. *Univ. Nebr. Res. Bull.* No. 161, 96 pp.

Kihara, H. and T. Hori. 1966. The behavior of nuclei in germinating pollen grains of wheat, rice and maize. *Der Zuchter* **36:** 145–150.

Knight, A. H., W. M. Crooke, and J. C. Burridge. 1973. Cation exchange capacity, chemical composition and the balance of carboxylic acids in the floral parts of various plant species. *Ann. Bot.* **37:** 159–166.

Knowlton, H. E. 1922. Studies in pollen, with special reference to longevity. *Cornell Univ. Agr. Exp. Sta. Mem.* **52:** 751–793.

Knox, R. B. 1971. Pollen wall enzymes: taxonomic distribution and physical localization. In *Pollen: development and physiology,* J. Heslop-Harrison (ed.). London: Butterworth, pp. 171–173.

Knox, R. B. and J. Heslop-Harrison. 1969. Cytochemical localization of enzymes in the wall of the pollen grain. *Nature (London)* **223:** 92–94.

Korobova, S. N. 1974. On the behavior of sperms in the process of fertilization of higher plants. In *Fertilization in higher plants,* H. F. Linskens (ed.). Amsterdam: North-Holland, pp. 261–270.

Leng, E. R. 1949. Direct effects of pollen parent on kernel size in dent corn. *Agron. J.* **41:** 555–558.

Linskens, H. F. 1969. Fertilization mechanisms in higher plants, p. 189–253. In *Fertilization,* Vol. 2, C. B. Metz and A. Monroy (eds.). New York: Academic, pp. 189–253.

Linskens, H. F. and P. L. Pfahler. 1973. Biochemical composition of maize (*Zea mays* L.) pollen. III. Effects of allele × storage interactions at the waxy (*wx*), sugary (*su*-1) and shrunken (*sh*-2) loci on the amino acid content. *Theor. Appl. Genet.* **43:** 49–53.

Martin, F. W. and J. L. Brewbaker. 1971. The nature of the stigmatic exudate and its role in pollen germination. In *Pollen: development and physiology,* J. Heslop-Harrison (ed.). London: Butterworth, pp. 262–266.

Mascarenhas, J. P. and L. Machlis. 1962. Chemotropic response of *Antirrhinum majus* pollen to calcium. *Nature (London)* **196:** 292–293.

McDaniel, R. G. and I. V. Sarkissian. 1968. Mitochondrial heterosis in maize. *Genetics* **59:** 465–475.

Mitchell, J. W. and M. R. Whitehead. 1941. Responses of vegetative parts of plants following application of extract of pollen from *Zea mays. Bot. Gaz.* **102:** 770–791.

Nair, P. K. K. 1962. Pollen grains of cultivated plants. III. Great millet and maize. *Indian J. Agr. Sci.* **32:** 196–200.

Nelson, O. E. 1959. Intracistron recombination in the *Wx/wx* region of maize. *Science* **130:** 794–795.

Nelson, O. E. 1962. The waxy locus in maize. I. Intralocus recombination frequency estimates by pollen and conventional analyses. *Genetics* **47:** 737–742.

Nelson, O. E., Jr. 1967. Biochemical genetics of higher plants. *Annu. Rev. Genet.* **1:** 245–268.

Nelson, O. E. 1975. The waxy locus in maize. III. Effect of structural heterozygosity on intragenic recombination and flanking marker assortment. *Genetics* **79:** 31–44.

Neuffer, M. G., L. Jones, and M. S. Zuber. 1968. The mutants of maize. *Crop. Sci. Soc. Am. Publ.,* Madison, Wisconsin. 74 pp.

Nielson, N., J. Grommer, and R. L. Lunden. 1955. Investigations on the chemical composition of pollen from some plants. *Acta Chem. Scand.* **9:** 1100–1106.

Nilsson, M., R. Ryhage, and E. von Sydow. 1957. Constituents of pollen. II. Long chain hydrocarbons and alcohols. *Acta Chem. Scand.* **11:** 634–639.

Nitsch, J. P. 1971. The production of haploid embryos from pollen grains. In *Pollen: development and physiology,* J. Heslop-Harrison (ed.). London: Butterworth, pp. 234–236.

Nitsch, J. P. and C. Nitsch. 1969. Haploid plants from pollen grains. *Science* **163:** 85–87.

Ouyang, T. 1973. Induction of pollen plants from anthers of *Triticum aestivum* L. cultured *in vitro. Scientia Sincia* **16:** 79–95.

Pfahler, P. L. 1965. Fertilization ability of maize pollen grains. I. Pollen sources. *Genetics* **52:** 513–520.

Pfahler, P. L. 1967. *In vitro* germination and pollen tube growth of maize (*Zea mays* L.) pollen. I. Calcium and boron effects. *Can. J. Bot.* **45:** 839–845.

Pfahler, P. L. 1968. *In vitro* germination and pollen tube growth of maize (*Zea mays*) pollen. II. Pollen source, calcium, and boron interactions. *Can. J. Bot.* **46:** 235–240.

Pfahler, P. L. 1970. *In vitro* germination and pollen tube growth of maize (*Zea mays*) pollen. III. The effect of pollen genotype and pollen source vigor. *Can. J. Bot.* **48:** 111–115.

Pfahler, P. L. 1971. *In vitro* germination and pollen tube growth of maize (*Zea mays*) pollen. IV. Effects of the fertility restoring Rf_1 locus. *Can. J. Bot.* **49:** 55–57.

Pfahler, P. L. 1975. Factors affecting male transmission in maize (*Zea mays* L.) In *Gametic competition in plants and animals,* D. L. Mulcahy (ed.). Amsterdam: North-Holland.

Pfahler, P. L. and H. F. Linskens. 1970. Biochemical composition of maize (*Zea mays* L.) pollen. I. Effects of the endosperm mutants, waxy (*wx*), shrunken (*sh*-2) and sugary (*su*-1) on the amino acid content and fatty acid distribution. *Theor. Appl. Genet.* **40:** 6–10.

Pfahler, P. L. and H. F. Linskens. 1971. Biochemical composition of maize (*Zea mays* L.) pollen. II. Effects of the endosperm mutants, waxy (*wx*), shrunken (*sh*-2) and sugary (*su*-1) on the carbohydrate and lipid percentage. *Theor. Appl. Genet.* **41:** 2–4.

Pfahler, P. L. and H. F. Linskens. 1972. *In vitro* germination and pollen tube growth of maize (*Zea mays* L.) pollen. VI. Combined effects of storage and the alleles at the waxy (*wx*), sugary (*su*-1) and shrunken (*sh*-2) loci. *Theor. Appl. Genet.* **42:** 136–140.

Pfahler, P. L. and H. F. Linskens. 1973. *In vitro* germination and pollen tube growth of maize (*Zea mays* L.) pollen. VIII. Storage temperature and pollen source effects. *Planta (Berlin)* **111:** 253–259.

Pfahler, P. L. and H. F. Linskens. 1974. Ash percentage and mineral content of maize (*Zea mays* L.) pollen and style. I. Genotypic effects. *Theor. Appl. Genet.* **45:** 32–36.

Polyakov, I. M. 1964. New data on use of radioactive isotopes in studying fertilization of plants. In *Pollen physiology and fertilization,* H. F. Linskens (ed.). Amsterdam: North-Holland, pp. 194–199.

Raicu P. and I. Popovici. 1964. Contribution à l'étude de la fécondation polyspermique chez *Zea mays* L. In *Pollen physiology and fertilization,* H. F. Linskens (ed.). Amsterdam: North-Holland, pp. 208–218.

Ray Sarkar, B. C., S. H. Wittwer, R. W. Lueke, and H. M. Sell. 1949. Quantitative estimation of some amino acids in sweet corn pollen. *Arch. Biochem.* **22:** 353–356.

Redemann, C. T., S. H. Wittwer, C. D. Ball, and H. M. Sell. 1950. The occurrence of quercetin in the pollen of *Zea mays. Arch. Biochem.* **25:** 277–279.

Roman, H. 1948. Directed fertilization in maize. *Proc. Nat. Acad. Sci. U. S. A.* **34:** 36–42.

Rosen, W. G. 1964. Chemotropism and fine structure of pollen tubes. In *Pollen physiology and fertilization,* H. F. Linskens (ed.). Amsterdam: North-Holland, pp. 159–166.

Rosen, W. G. 1971. Pistil–pollen interactions in *Lilium.* In *Pollen: development and physiology,* J. Heslop-Harrison (ed.). London: Butterworth, pp. 239–254.

Rumbaugh, M. D. and R. H. Whalen. 1972. Variation in pollen grain size of inbred maize lines. *Maize Genet. Coop. Newsl.***46:** 171–172.

Rumbaugh, M. D. and R. H. Whalen. 1973. Measurement of pollen grain size in the diallel. *Maize Genet. Coop. Newsl.* **47:** 170–171.

Sawada, Y. 1960. Physiological and morphological studies on pollen grains. XVIII. Nitrogen metabolism of maize pollen. *Bot. Mag. (Tokyo)* **73:** 252–257.

Scandlios, J. G. 1969. Alcohol dehydrogenase in maize: genetic basis for isoenzymes. *Science* **166:** 623–624.

Sprague, G. F. 1932. The nature and extent of hetero-fertilization in maize. *Genetics* **17:** 358–368.

Todd, F. E. and O. Bretherick. 1942. The composition of pollens. *J. Econ. Entomol.* **35:** 312–317.

Walden, D. B. 1959. Preliminary studies on longevity of corn pollen and related physiological factors. Ph.D. thesis. Ithaca, N. Y.: Cornell University, 232 pp. University Microfilms, Ann Arbor, Mich. (Diss. Abstr. **20:** 3488–3489; (Libr. Congr. Card. No. Mic. 60-622).

Walden, D. B. 1967. Male gametophyte of *Zea mays* L. I. Some factors influencing fertilization. *Crop Sci.* **7:** 441–444.

Walden, D. B. and H. L. Everett. 1961. A quantitative method for the *in vivo* measurement of the viability of corn pollen. *Crop Sci.* **1:** 21–25.

Wang, Y. 1973. Induction of pollen plantlets of *Triticale* and *Capsicum annuum* from an anther culture. *Scientia Sincia* **16:** 147–151.

Weatherwax, P. 1955. The structure and development of reproductive organs. In *Corn and corn improvement,* G. F. Sprague (ed.). New York: Academic, pp. 89–121.

Zuber, M. S., W. L. Deatherage, M. M. MacMasters, and V. L. Fergason. 1960. Lack of correlation between amylose content of pollen and endosperm in maize. *Agron. J.* **52:** 411–412.

GENE ACTION

Chapter 34

GLOSSY MUTANTS: LEVEL OF ACTION AND LEVEL OF ANALYSIS

A. Bianchi

Experimental Institute for Cereal Research,
Rome, Italy

The cuticles of leaves, fruits, and other plant organs are often covered with a wax layer (Martin and Juniper, 1970).

Frequently the cotyledonary leaves are without waxes, but the mature leaves bear wax extrusions. In *Zea mays* the first five or six leaves are endowed with typical wax extrusions, whereas the other leaves are waxless (Bianchi and Marchesi, 1960). The production of the wax layer is under genetic control; there are species in which the cuticle is typically waxless, and species that possess wax extrusions. Of the latter, there are many mutants, usually recessive, interfering with the waxiness of the leaf surface. Such mutants, whose leaf surface is consequently characteristically glossy, have been described in *Brassica oleracea* (Anstey and Moore, 1949; Hall, Matus, et al., 1965; Thompson, 1963) in various *Eucalyptus* species (Barber and Jackson, 1957), in *Hordeum vulgare* (Lundquist and von Wettstein, 1962) in *Pisum sativum*

With the important cooperation of F. Salamini for the genetical aspects, C. Lorenzoni, for the electron microscopy analyses, and G. Bianchi for the chemical studies.

533

(Hall, Matus, et al., 1965) and in *Zea mays* [for the literature of this species, see Borghi and Salamini (1966)].

This chapter is a condensed description of our work on glossy mutants in maize that has been conducted during a period of more than 15 years making use of several techniques that permit us to analyze leaf wax extrusions at various levels of physiogenetic action with levels of analysis of different resolving power.

THE GENETICAL ANALYSIS

An analysis of about 500 samples of open-pollinated Italian maize populations was started in 1958. Out of a total of more than 4500 self-pollinated ears, 246 segregated glossy types. About one half of these ears were discarded because they were recognized to possess the same inherited factor present within the progeny studied; seven types have been abandoned because of poor viability. Additional genetic analysis has been restricted to 127 stocks, five of which were homozygous for two independent glossy mutants. Table 1 reports the results of the genetical tests carried out to identify the different mutants.

A possible explanation of the findings is that *gl*1 and *gl*3 correspond to highly mutable loci and possess relatively less disadvantageous function in the heterozygotes. Similar interpretations are supported by experimental results; Sprague and Schuler (1961), after elimination of all morphological mutants, have found, in a rebuilt isolated population that underwent four years of open pollination, nine cases of *gl*1, one case of *gl*3, and three other glossy mutants in a total of 1137 progenies examined. Lorenzoni, Pozzi, et al., (1965) have recognized a higher rate of recessive mutants in samples of populations from areas less favorable to maize cultivation. Table 2 reproduces the distribution of

TABLE 1.   Classification of 125 glossy mutants isolated in Italian open-pollinated populations, on the basis of appropriate test crossing (Borghi and Salamini, 1966)

Locus	No. of glossy mutants	Percent of total
*gl*1	52	41.6
*gl*2	14	11.2
*gl*3	40	32.0
*gl*4	2	1.6
*gl*6	5	4.0
*gl*7	3	2.4
gl unknown	9	7.2

TABLE 2.   Number of ears segregating for glossy types and their percentage in the main three Italian districts (Borghi and Salamini, 1966)

Climatic areas	Segregating ears	Examined ears	Percent segregated
Northern Italy	97	1719	5.64
Central Italy	87	1261	5.90
Southern Italy	45	512	8.77

glossy mutants in the three main areas of Italian maize cultivation. Going from northern Italy, the best region for maize growing, to southern Italy, where maize meets the greatest difficulties, the glossy mutants appear to increase in frequency.

A possibility for analyzing the nature and the size of the $gl1$ locus may be realized by intracistron recombination. This had been demonstrated to be feasible in maize by Nelson (1959, 1962) in the case of wx locus, whose action affecting the pollen grains provided material with which it became possible to carry out experiments with resolving power similar to that of microbial genetics. Borghi and Salamini (1966) attempted a similar analysis on the basis of the numerous independently originated $gl1$ factors mentioned above. Some of them were selected and crossed to a stock possessing a standard $gl1$ allele. The F_1 thus obtained were self-pollinated, and the same was done for the homozygous $gl1$ selected. The seed was produced on a fairly large scale, with appropriate crossing and self-pollinating techniques that assured control of undesired contamination (isolation, marker genes, etc.) and was germinated in the greenhouse and scrutinized for the appearance of Gl normal phenotype. In such a way 11 alleles have been studied in the homozygous condition, together with 17 heteroallelic crosses. The results of this analysis are summarized in Table 3.

TABLE 3.   Rate of appearance of Gl phenotype in homozygotes for 11 alleles of the $gl1$ locus and in 17 progenies of heteroallelic compounds (Salamini and Borghi, 1966)

Type of progeny	No. of seedlings	No. of gametes	No. of Gl recovered	$Gl \times 10^{-4}$	Fiducial limits for $F = 0.05$[a]
Homoallelic	121,095	242,190	4	0.17	0.05–0.42
Heteroallelic	750,091	1,500,182	358	2.39	2.32–2.48

[a] From Stevens (1942).

It is evident that a higher rate of the normal *Gl* phenotype was encountered among the seedlings grown from the crossed populations as compared to that derived from stocks homozygous for given *gl*1 alleles (2.39×10^{-4} vs. 0.17×10^{-4}). Moreover, within the homoallelic crosses, the variability was remarkable, as illustrated in Figure 1. The order of magnitude with which the normal *Gl* phenotype occurred in certain combinations was quite similar to the recombination values calculated by Nelson (1959, 1962, 1968) for pseudoalleles at the *wx* locus.

In a second experiment the *gl*1 alleles studied were transferred into the common background of the inbred WF9. After three backcrosses the alleles (indicated in this chapter as *gl*1a, *gl*1b, etc.) were crossed with a stock, provided by the Maize Genetics Cooperative, Illinois University, possessing the chromosome-7 markers *o*2 (opaque endosperm), *gl*1 (glossy seedling), and *sl* (slashed leaf). The map distance between *o*2 and *gl*1 is 20, and between *gl*1 and *sl*, 14. The *gl*1 allele of the above stock is referred to as *gl1-st* (standard).

The F_1 plants thus obtained, cultivated in isolation, were fertilized with *o*2 *gl*1-st *sl* pollen. The kernels obtained, after classification as normal or opaque, were planted in the greenhouse, The seedlings having normal (nonglossy) phenotypes were classified with regard to the slashed (*sl*) character. It appears

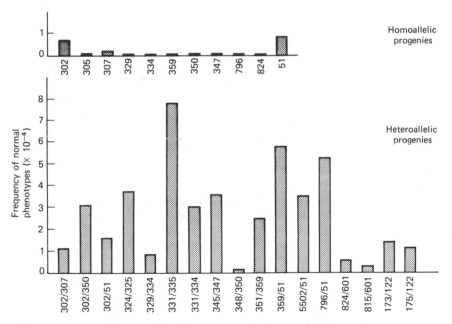

Figure 1. Reversion rate *gl*1 ⟶ *Gl*1 in the progeny of self-pollinated plants homozygous for *gl*1 alleles and their various heteroallelic combinations.

TABLE 4.   Genotype of the *Gl* reversions in backcrosses (*O2 gl* 1-x *Sl*/*o2 gl* 1-st *sl*) × *o2 gl* 1-st *sl* in relation to the flanking markers *o2* and *sl* (Salamini and Lorenzoni, 1970)

Allele (x) crossed with gl 1-st	Number of reversions and their genotype with respect to outside markers				
	O2 Sl	O2 sl	o2 Sl	o2 sl	Total
i	1			3	4
a	3		1		4
f	1				1
m			1	2	3
d	3				3
c	2	2		1	5
b	2	2	1	1	6
l	5	5	3	5	18
g	2	6	3	3	14
h	2	4	3	4	13
Total	21	19	12	19	71
Percent	29.5	26.8	16.9	26.8	100

from the data in Table 4 that about 60% of the *Gl* gametes have parental combinations (*O2 Sl* and *o2 sl*). In fact, out of 71 *Gl* seedlings, 40 (21 + 19) are nonrecombinant for outside markers, whereas 31 (19 + 12) have recombinant outside markers.

If we assume that recombination as traditionally considered gives rise to the observed revertants, the parental classes should be scarce. As the data give a different result, it may be suggested that "gene conversion" is active when within-locus recombination is studied. A similar interpretation may be obtained from Nelson's (1959, 1962) data at the *waxy* locus in maize and from Green's (1960) data for an example from *Drosophila*.

Our data make it possible to obtain some indication of the occurrence of polarized crossing over by grouping together the reversions derived from crosses between *gl*1-st and the b, l, g, and h alleles. These mutants are probably located in a distal position with respect to *gl*1-st. From the backcrosses studied, 24 reversions originated without recombinant outside markers. Out of these 11 show the *O2sl* genotype, whereas 13 possess the *o2sl* genotype. As the two classes are numerically similar, it seems that at the *gl*1 locus, polarized crossingover is not occurring.

In conclusion, the formal analysis of glossy mutants in maize has been extended to a level that is unusual for higher organisms that are unwieldy on a numerically large scale. However, our analysis has not revealed really new aspects of genetical behavior, facts, or phenomena. So let us consider another approach to the experimental research on glossy mutants.

THE ULTRASTRUCTURE RESEARCH

In various species when the leaf surface of the glossy mutations are studied with the electron microscope, the normal phenotype shows cuticular waxes, whereas the mutant plants are characterized by waxless surfaces or by a reduction in the number or size of the wax crystals (Bianchi and Marchesi, 1960; Bianchi and Salamini, 1975a; Lundquist, von Wettstein-Knowles, et al., 1968; Hall, Matus, et al., 1965; Hallam, 1967; Salamini, 1963).

The use of the transmission electron microscope and the carbon replica-plating technique provides important details on wax extrusions. In Figure 2 we reproduce the normal maize phenotype as contrasted to that of gl3. The former is characterized by a high degree of crystallization on the whole leaf surface. In the mutant the wax extrusions are reduced in size, their degree of crystallization is poor, and a large fraction of the leaf cuticle is waxless. The introduction of the scanning electron microscope has enabled us recently to obtain quantitative data on the distribution of the wax extrusions in various mutants, even though this technique provides few details about minute structures of the wax.

The mutants gl1, gl2, gl3, gl4, gl5, gl6, gl7, gl8, gl9, gl14, gl15, gl16, and glg, previously transferred by four or five backcrosses to WF9, have been studied with the scanning electron microscope by Lorenzoni and Salamini (1975) and are illustrated in Figure 3. For each mutant the number of wax bodies for surface unit, the percent of wax-covered surface, and the mean surface of the wax crystal have been calculated.

When compared with their normal counterpart, the epicuticular waxes present on the surface of the mutant phenotypes are reduced in size and/or frequency. Some mutations, as gl1 and gl9, are very "drastic" since they eliminate the wax layer almost completely; other mutants severely affect the surface waxiness but retain an organized wax layer; gl15 appears to affect the wax distribution only slightly.

According to the shape of their wax crystals the normal and the 13 mutant phenotypes may be assigned to four groups. The *star* shape is typical of the normal and gl15 waxes. A single wax extrusion appears branched; the lateral crystals join the elementary structures altogether. The waxes are clearly crystalline with sharp borders.

The *reduced star* shape is typical of the mutations gl4, gl7, gl14, and gl16. The structure of the wax is still multibranched, but the lateral projections are shorter than the normal. The crystalline structure is maintained.

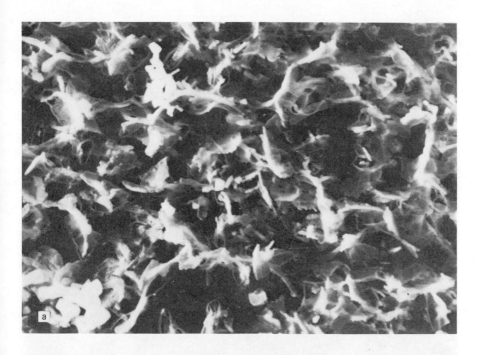

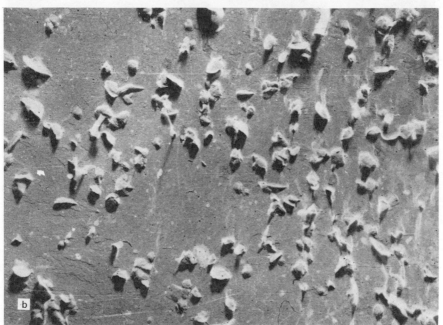

Figure 2. Wax extrusions of leaf cuticle: (a) *Gl*1 seedlings × 14,000 magnification; (b) *gl*3 seedlings × 8000 magnification as seen at the transmission electron microscope (Salamini, 1963).

539

The mutant *gl5* shows two distinct types of wax extrusions. The first is abnormally long and deprived of lateral branches, and the second lies flat on the cuticle, covering a high percentage of the entire surface.

The mutants *gl1*, *gl2*, *gl3*, *gl6*, *gl8*, *gl9*, and *glg* possess an elementary form of the wax that is slightly crystalline with a round border and with the shape of a rice grain.

Biometrical data on the wax bodies obtained from magnified micrographs of the wax layer are presented in Table 5. The normal (*Gl1*) wax layer covers

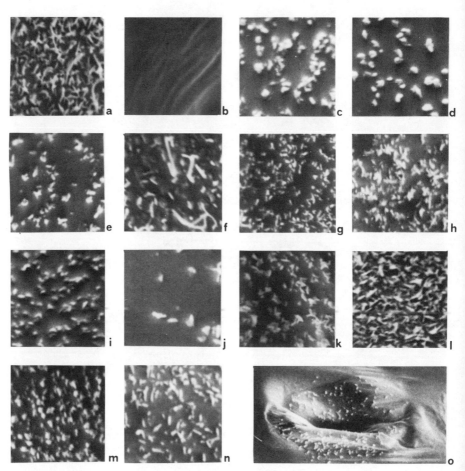

Figure 3. Wax on the adaxial leaf cuticle of seedling: *G1* (a), *gl1* (b), *gl2* (c), *gl3* (d), *gl4* (e), *gl5* (f), *gl6* (g), *gl7* (h), *gl8* (i), *gl9* (j), *gl14* (k), *gl15* (l), *gl16* (m), *glg* (n), at the scanning electron microscope, magnification × 3750. In the (o) stoma, magnification × 1250 of seedling *gl1*.

TABLE 5. Morphological traits of the wax layer present on the surface of normal and mutant maize seedlings (Lorenzoni and Salamini, 1975)

Locus	Wax-covered surface		Wax extrusions on a surface of 1×10^{-4} mm²		Single wax body surface	
	Percent	Percent of the normal	Number	Percent of the normal	1×10^{-7} mm²	Percent of the normal
Gl	48.1	100	138	100	3,48	100
gl1	0–6.4	13[a]	variable	—	variable	—
gl2	10.9	23	114	83	0,96	28
gl3	11.5	24	103	75	1,12	32
gl4	18.3	38	61	43	3,00	86
gl5	36.1[b]	75	87	63	4,15	119
gl6	22.2	46	203	147	1,09	31
gl7	21.9	45	117	85	1,87	54
gl8	12.1	25	118	85	1,03	29
gl9	0–11.9	25[a]	variable	—	variable	—
gl14	32.3	67	100	72	3,23	93
gl15	37.9	79	86	62	4,41	127
gl16	15.9	33	140	101	1,14	33
glg	variable[c]	—	variable	—	variable	—

[a] When present, the wax layer bears particles of variable size; frequently the wax layer is absent (0% of covered surface).

[b] Data take into consideration the long particles characterizing this phenotype and the flatter particles layered under the former (see Figure 3f).

[c] The presence of variable phenotypes does not permit a clear quantification of the wax traits for this mutation.

48% of the leaf surface. The mutations $gl5$, $gl14$, and $gl15$ are similar in this respect to the normal situation; $gl1$ and $gl9$ reduce drastically the amount of wax extruded. Furthermore, the leaf surfaces of these two mutants are discontinuous with respect to the waxiness; whereas some epidermial cells bear small wax grains, other cells are waxless. The normal (Gl) frequency of the wax extrusions is about 140 crystals on a surface of 1×10^{-4} mm²; each element has a mean surface of 3.48×10^{-7} mm². Mutations such as $gl16$ are normally endowed with wax extrusions, but the single crystals are reduced in size. Mutants $gl4$, $gl5$, $gl14$, and $gl15$ show an almost normal particle size, but the frequency of crystals for unit area is abnormally low. The remaining mutations ($gl1$, $gl2$, $gl3$, $gl6$, $gl7$, $gl8$, $gl9$, and glg) are characterized by a negative variation in either number or size of the wax extrusions. The leaf surfaces of $gl6$ bear a higher-than-normal number of particles, yet the particles are clearly reduced in size. This mutation affects the morphology of a single branched wax in such a way that the lateral projections are small and separated from one another. As a result, the mutant surface is crowded with little sized waxes.

Mutation $gl1$ shows interesting differences at the cellular level. As previously reported by Bianchi and Salamini (1967), in normal maize the guard cells of the stoma are waxless, whereas the accessory cells show a normal wax layer. When seedling homozygous for $gl1$ are examined, the accessory cells that bear *rice grain* waxes can be differentiated from the surrounding waxless epidermial cells (Figure 3).

Reduced wax layers have been reported to characterize the guard and accessory cells in *Pisum sativum* and in *Brassica oleracea* (Martin and Juniper, 1970). This pattern appears on $gl1$ leaves. In this respect $gl1$ resembles the glabrous *Eucalyptus* sp. described by Hallam (1967). The same author describes in *Eucalyptus polyanthemos* the presence of different types of waxes. This observation can also be made in the maize mutations $gl1$ and $gl9$, whose wax extrusions are variable in size, distribution, and even shape. These mutations show waxed leaf borders. An opposite distribution of the wax layer (less abundant on the border and more concentrate around the midrid) has been reported by Kravkina (1972) in *Agropyron repens*. Among the maize mutations, only $gl15$ shows an almost normal phenotype on the second or third leaves. This mutant, however, has the fourth leaf without waxes, whereas in normal seedlings the glossiness becomes evident starting from the sixth leaf (Bianchi and Marchesi, 1960). A similar behavior has been reported in *Hordeum vulgare* for some *cer* mutations (Lundquist and von Wettstein, 1962).

THE CHEMICAL APPROACH

The *glossy* mutants modify the size and/or distribution of the epicuticular waxes. It has been suggested several times that the morphology of the waxes is strictly correlated with their chemical composition. Hallam (1967) collected and recrystallized *Eucalyptus* sp. waxes and obtained wax crystals similar to the ones observed on untreated leaf surfaces. A relationship between the chemical composition of the wax and its crystal structure is suggested, for example, if one considers that whereas the tube-like waxes have a high proportion of β-diketones, the percentage of these compounds in the plate-like waxes is negligible. Moreover, the waxes from mutated plants appear chemically different from normal waxes when analyzed (Hall, Matus, et al., 1965; Lundquist, von Wettstein-Knowles, et al., 1968; Macey, 1974; Macey and Barber, 1969a,b).

Biochemistry of the Epicuticular Waxes

The very long chains of the wax organic compounds are synthetized in the epidermis starting from a long chain C-16–C-18 fatty acid substrate and with successive additions of C-2 units (Kolattukudy, 1968, 1970). If the elongation is followed by a decarboxylation of the fatty acids, hydrocarbons with an odd number of carbon atoms are generated (Kolattukudy, Buckner, et al., 1972; Kolattukudy and Walton, 1972). If the decarboxylation does not take place, the fatty acid may be again elongated, exuded as free acid into the wax, or esterified or reduced to aldehyde or reduced to free alcohol via aldehyde (Kolattukudy, 1971; Kolattukudy, Buckner, et al., 1972). In Figure 4 a general pathway for the synthesis of the components of the epicuticular waxes is reported.

This pathway has been derived from data obtained in various species. When the wax metabolism of a given plant is considered, a compound or a metabolic step of the Figure 4 may be altered. For example, the species *Brassica oleracea*,

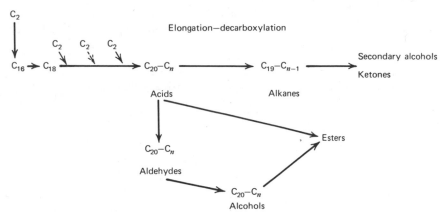

Figure 4. Outline of the biosynthetic mechanism for the formation of the epicuticular wax compounds.

Pisum sativum, and *Allium porrum,* whose waxes have been extensively studied, differ clearly one from another. In *Brassica* the C-29 predominates within the hydrocarbons (>90%); the more abundant fatty acids have 26, 28, and 30 carbon atoms, but acids with a lower number are also present; the primary alcohols are represented by C-12–C-16 and C-26–C-28 compounds, accompanied by the C-26, C-28, C-29, and C-30 aldehydes (Purdy and Truter, 1963; Macey and Barber, 1969b). In *Pisum* 98.4% of the alkanes contain the C-31 compound; the alcohols and aldehydes contain 26, 27, 28, 29, 30, 31, and 32 carbon atoms with the even members predominating; the free acids extend from C-16 to C-32, the majority of them being C-26, C-28, and C-30 (Macey and Barber, 1969a). In *Allium* C-8–C-32 fatty acids are present, together with C-21—C-33 akanes (Cassagne, 1972; Cassagne and Lessire, 1974). In normal maize (Table 6) the waxes present on the leaf surface of normal seedlings are composed by alkanes (9.57%), aldehydes (9%), primary alcohols (62%), and esters (19.43%), whereas the free acids are absent (Bianchi, Avato, et al., 1977). The alkanes are represented by compounds with a number of carbon atoms in the range C-19–C-33, with a predominance of C-27 (10%), C-29 (23%), and C-31 (36%); the C-32 compounds 1-dotriacontanal and 1-dotriacontanol represent 98% and 100% of the aldehydes and alcohols, respectively (Table 7). Maize differs from *Pisum sativum* and *Brassica oleracea* because it synthesizes at least 14 alkanes (with C-31 predominating), whereas the other two synthesize one alkane. On the other hand, in normal maize only one aldehyde and one alcohol are present, whereas in *Pisum* and *Brassica* these two classes of wax components are represented by a large number of compounds. The maize waxes are also different from those of *Allium,* in which numerous alkanes and numerous fatty acids are present. On the basis of our data the general wax biosynthetic pathway of Figure 4 has been tentatively adapted to the wax synthesis for maize (Figure 5). Maize offers a clear demonstration of the biosynthetic relationship existing between fatty acids, fatty aldehydes, alcohols, and the predominating alkanes. This relationship was previously pointed out by Kolattukudy and collaborators (Kolattukudy, Buckner, et al., 1972).

Effect of the allele substitutions on the chemical wax composition

Tables 6 and 7 show the chemical composition of the waxes in normal and mutant (*gl*1, *gl*2, and *gl*7) maize seedlings. All mutants cause a large reduction of wax production. Moreover, significant changes in wax composition are noticeable, involving mainly alkanes, aldehydes, and alcohols. If percent composition of the waxes is considered, esters are higher in the mutants, alcohols are lower, and alkanes and aldehydes are irregular. However, when their absolute quantities per 1000 g of fresh weight or per plant are compared, the quantity of esters of the mutants is decreased. The *gl*7 seedlings synthesize

TABLE 6. Composition of leaf-surface wax (preliminary data of G. Bianchi and colleagues)

Components	Percent of the waxes				Fresh weight, mg/100 g				mg/plant			
	Gl	gl7	gl2	gl1	Gl	gl7	gl2	gl1	Gl	gl7	gl2	gl1
Alkanes	9.6	4.51	12.07	15.12	56.46	11.27	14.48	10.58	0.134	0.019	0.022	0.018
Aldehydes	9.0	31.55	6.12	4.95	53.10	78.87	7.34	3.46	0.126	0.133	0.011	0.006
Alcohols	62.0	41.41	22.46	26.05	365.80	103.52	26.95	18.25	0.868	0.174	0.040	0.032
Esters	19.4	22.53	54.25	50.38	114.63	56.32	65.10	35.26	0.272	0.094	0.982	0.061
Unknown	—	—	5.10	3.50	—	—	6.12	2.46	—	—	0.009	0.004
Total	100	100	100	100	590	250	120	70	1.400	0.421	0.181	0.122

TABLE 7. Percent composition of normal and mutants wax (preliminary data of G. Bianchi and colleagues)

No. of carbon atoms	Genotype	Hydrocarbons				Free primary alcohols				Aldehydes			
		GI	gl7	gl1	gl2	GI	gl7	gl1	gl2	GI	gl7	gl1	gl2
18													
19		0.21	1.44	—	2.27								
20		0.42	—	1.05	0.57								
21		0.42	1.43	1.05	—								
22		1.05	0.48	1.08	—			2.00	1.07				
23		1.26	0.95	1.60	—								
24		1.26	0.72	1.60	—			6.00	—				
25		3.88	1.91	3.04	7.50								
26		1.26	0.95	2.18	0.57			12.50	1.34				
27		10.06	6.68	17.65	17.05						1.07	0.58	
28		1.89	1.43	2.89	2.05			17.90	17.52		5.00	2.61	20.15
29		23.27	20.05	45.23	64.55						1.15	1.09	4.03
30		3.67	2.86	2.58	—			27.93	74.35		12.51	10.45	72.16
31		36.65	61.10	14.25	5.40						1.08	2.54	1.18
32		1.26	—	—	—	−100	≥98	32.8	5.72	≥98	79.00	82.73	1.48
33		8.39	—	—	—								
34		—	—	5.80	—								

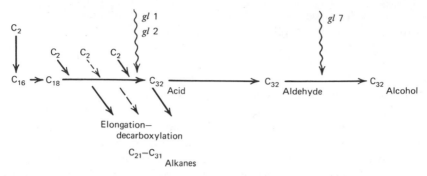

Figure 5.   Hypothetical scheme for the synthesis of epicuticular wax in *Zea mays*.

reduced amounts of alkanes and alcohols, whereas the amount of aldehydes is higher than normal. The mutations *gl*1 and *gl*2 are similar to each other, they reduce alkanes, aldehydes, and alcohols more so in *gl*1 than in *gl*2. The mutations *gl*1 and *gl*2 also affect the carbon-number distribution within alkanes, aldehydes, and alcohols. The alkanes of the two mutants have much less of the C-31 and more of the C-29 component, whereas in the normal and in the mutant *gl*7, C-31 is the predominant component. In the aldehydes and alcohols of the *gl*1, *gl*2, and *gl*7 mutants, C-22–C-30 (mainly those with an even number of carbon atoms) are present together with the C-32 component.

In *gl*1 and *gl*2 a genetic block in the step C-30 → C-32 is suggested; as a result, the C-29 alkane predominates, whereas the C-32 components 1-dotriacontanal and 1-dotriacontanol decrease. Mutations whose effects are similar to those of *gl*1 and *gl*2 of maize are: (a) *gl*5 of *Brassica oleracea* and (b) *wa*, *wb*, *was* of *Pisum sativum* (Macey and Barber, 1969a,b). The maize mutation *gl*7 seems to induce less dramatic effects on the quantity and quality of the epicuticular waxes. The plants homozygous for this mutation bear waxes with reduced amount of alkanes and alcohols and with a greater amount of aldehydes than normal. This mutation might either block or inhibit the biosynthetic reductive step relating the fatty aldehydes to the fatty alcohols. This step is normally catalyzed by a TPNH (reduced triphosphopyridine nucleotide)-dependent aldehyde reductase (Kolattukudy, Buckner, et al., 1972).

CONCLUSIONS

The experimental attack on glossy mutants has utilized genetic, microscopic, and chemical approaches. The genetic analysis has confirmed that in maize the use of seedling material such as the glossy mutants can lead to results comparable to those that we obtain with more advanced material. The ultrastructural research has clearly demonstrated that at the level of resolution of the electron microscope we can detect differential effects of different alleles and loci, suggesting the need for further experimental work at the chemical level.

In the not-too-distant future, once the different glossy mutants have been understood in chemical terms, it may also be possible to study the chemistry of the different mutational sites of a given glossy mutant, such as *gl1*, that is genetically complex. As already mentioned, at this locus numerous alleles and pseudoalleles have been encountered; they may confirm that in maize, as in other higher organisms, the genetic unit may be a much larger entity than its counterpart in prokaryotes. Support for this statement has emerged from a comparison of genetic and chemical observations on *Drosophila melanogaster* and *Escherichia coli* (Gellbart, McCarron, et al., 1974). In maize a similar experimental approach is under way, as reported in this chapter for different genes, for alleles of the same *locus*, and on combinations of alleles at the same locus and at different loci, in future.

SUMMARY

In maize open-pollinated populations the plants heterozygous for glossy (*gl*) mutants are relatively frequent. Such types are conditioned by many different genetic loci, among which *gl1* may represent more than 40% of the glossy cases; other frequent mutations refer to *gl3* (>30%) and *gl2* (>11%). Several *gl1* mutants of independent origin have been genetically studied and assigned to different mutation sites in the *gl1* cistron.

The wax extrusions on the second and third leaf surfaces of *gl1*, *gl2*, *gl3*, *gl4*, *gl5*, *gl6*, *gl7*, *gl8*, *gl9*, *gl14*, *gl15*, *gl16*, and *glg* plants have been studied with electron microscope techniques. The results demonstrate that in maize, glossy mutations affect the leaf waxiness in specific and unique ways.

Chemical analysis of *gl1*, *gl2*, *gl7*, and the normal phenotype demonstrated that there was a large reduction of wax production, which is decreasing from *Gl* to *gl7* and *gl2*, with extreme reduction in *gl1*. The main constituents of epicuticular wax were alkanes, aldehydes, alcohols, and esters. Within the alkane the C31 predominates, whereas only 1-dotriacontanal and 1-dotriacontanol, respectively, constitute the aldehydic and alcoholic fractions.

In *gl1* and *gl2* a genetic block in the elongation–decarboxilation step C-30 → C-32 is recognized; as a result, the C29 alkane is preferentially synthesized, together with aldehydes and alcohols containing 22, 24, 26, 28, 30, or 32 carbon atoms. It is proposed that the mutation *gl7* either blocks or inhibits the reductive step that leads specifically to fatty alcohols starting from fatty aldehydes.

REFERENCES

Anstey, T. H. and J. F. Moore. 1949. Inheritance of glossy foliage and cream petals in green sprouting broccoli. *J. Hered.* **45:** 39–41.

Barber, H. N. and W. D. Jackson. 1957. Natural selection in action in *Eucalyptus*. *Nature* **179:** 1267.

Bianchi, A. and G. Marchesi. 1960. The surfaces of the leaf in normal and glossy maize seedlings. *Z. Vererbungls.* **91:** 214–219.

Bianchi, A. and F. Salamini. 1967. Ultrastrutture di superfici fogliari di mutanti ratio-indotti in *Triticum durum*. *Genet. Agr.* **18:** 183–194.

Bianchi, G. and F. Salamini. 1975. Glossy mutants of maize. IV. Chemical composition of normal epicuticular waxes. *Maydica* **20:** 1–3.

Bianchi, G., P. Avato, and F. Salamini. 1975. Glossy mutants of maize. VI. Chemical composition of *gl2* waxes. *Maydica* **20:** 165–173.

Bianchi, G., P. Avato, and F. Salamini. 1977. Glossy mutants of maize. VII. Chemistry of *glossy* 1, *glossy* 3 and *glossy* 7 epicuticular waxes. *Maydica* **22:** 9–17.

Borghi, B. and F. Salamini. 1966. Analisi genetica di mutanti glossy di mais. I. Frequenze per i diversi loci. *Maydica* **11:** 45–57.

Cassagne, C. 1972. Etude de la biosynthèse des alcanes dans l'épiderme des feuilles d'*Allium porrum* L. *Qual. Plant. Mater. Veg.* **21:** 257–290.

Cassagne, C. and R. Lessire. 1974. Étude de la biosynthèse des alcanes dans l'épiderme des feuilles d'*Allium porrum*. II. Synthèse des acides gras et leur rélation avec les alcanes. *Physiol. Veg.* **12:** 149–163.

Gellbart, W. M., M. McCarron, J. Pandey, and A. Chovniok. 1974. Genetic limits of the xanthine dehydrogenase structural element within the rosy locus in *Drosophila melanogaster*. *Genetics* **78:** 869–886.

Green, M. M. 1960. Double crossing-over or gene conversion with the white loci in *Drosophila melanogaster*. *Genetics* **45:** 15–18.

Hall, D. M., A. I. Matus, J. A. Lamberton, and H. N. Barber. 1965. Infraspecific variation in wax on leaf surfaces. *Aust. J. Biol. Sci.* **18:** 323–332.

Hallam, N. D. 1967. An electron microscope study of the leaf waxes of the genus *Eucalyptus*, L'Heritier. Ph.D. thesis. Melbourne, Australia: University of Melbourne.

Kolattukudy, P. E. 1968. Biosynthesis of surface lipids. *Science* **159:** 498–505.

Kolattukudy, P. E. 1970. Biosynthesis of cuticular lipids. *Ann. Rev. Plant Physiol.* **21:** 165–192.

Kolattukudy, P. E. 1971. Enzymatic synthesis of fatty alcohols in *Brassica oleracea*. *Arch. Biochem. Biophys.* **142:** 701–709.

Kolattukudy, P. E., J. S. Buckner, and L. Brown. 1972. Direct evidence for a decarboxylation mechanism in the biosynthesis of alkanes in *B. oleracea*. *Biochem. Biophys. Res. Commun.* **47:** 1306–1313.

Kolattukudy, P. E. and T. J. Walton. 1972. Metabolism of alkyl glyceryl ethers and their noninvolvement in alkane biosynthesis in plants. *Arch. Biochem. Biophys.* **150**: 310–317.

Kravkina, I. M. 1972. Morfologija i raspredelenie voskovyh otlozenij na poverhnosti kutikuly *Agropyron repens* (L.) Beauv. *Bot. Z.* **57**: 519–524.

Lorenzoni, C., M. Pozzi, and F. Salamini. 1965. Frequenza di mutanti in popolazioni italiane di mais. *Genet. Agr.* **19**: 146–158.

Lorenzoni, C. and F. Salamini. 1975. Glossy mutants of maize. V. Morphology of the epicuticular waxes. *Maydica* **20**: 5–19.

Lundquist, U. and D. von Wettstein. 1962. Induction of *eceriferum* mutants in barley by ionizing radiations and chemical mutagens. *Hereditas* **48**: 342–362.

Lundquist, U., P. von Wettstein-Knowles, and D. von Wettstein. 1968. Induction of *eceriferum* mutants in barley by ionizing radiations and chemical mutagens, II. *Heriditas* **59**: 473–504.

Macey, M. J. K. 1974. Wax synthesis in *Brassica oleracea* as modified by trichloroacetic acid and glossy mutations. *Phytochemistry* **13**: 1353–1758.

Macey, M. J. K. and H. N. Barber. 1969a. Chemical genetics of wax formation on leaves of *Pisum sativum. Phytochemistry* **9**: 5–12.

Macey, M. J. K. and H. N. Barber. 1969b. Chemical genetics of wax formation on leaves of *Brassica oleracea. Phytochemistry* **9**: 13–23.

Martin, J. T. and B. E. Juniper. 1970. *The cuticles of plants.* London, Edward Arnold Press.

Nelson, O. E. 1959. Intracistron recombination in the *Wx/wx* region in maize. *Science* **130**: 794–795.

Nelson, O. E. 1962. The waxy locus in maize. I. Intralocus recombination frequency estimates by pollen and by conventional analyses. *Genetics* **47**: 737–742.

Nelson, O. E. 1968. The waxy locus in maize. II. The location of the controlling element alleles. *Genetics* **60**: 507–524.

Purdy, S. J. and E. V. Truter. 1963. Constitution of the surface lipid from the leaves of *Brassica oleracea var. capitata. Proc. Roy. Soc. B.* **158**: 536–565.

Salamini, F. 1963. La superficie fogliare del mais. *Maydica* **8**: 67–72.

Salamini, F., and B. Borghi. 1966. Analisi genetica dei mutanti *glossy* di mais. II. Frequenze di reversione al locus *gl* 1. *Genet. Agr.* **20**: 239–248.

Salamini, F. and C. Lorenzoni. 1970. Genetical analysis of glossy mutants of maize. III. Intracistron recombination and high negative interference at the *gl*1 locus. *Molec. Gen. Genet.* **108**: 225–232.

Sprague, G. F. and J. F. Schuler. 1961. The frequencies of seed and seedling abnormalities in maize. *Genetics* **46**: 1713–1720.

Stevens, W. L. 1942. Accuracy of mutation rate. *J. Genet.* **43**: 301–307.

Thompson, V. T. 1963. Resistance to the cabbage aphid. (*Brevicoryne brassicas* in Brassica plants. *Nature* **198**: 209.

Chapter 35

A MODEL FOR CONTROL OF THE LEVEL OF GENE EXPRESSION IN MAIZE

Drew Schwartz

Department of Plant Sciences,
Indiana University,
Bloomington

One of the most interesting and important problems in genetics today is the regulation and control of gene action during differentiation and development. Studies on the control of gene expression in bacteria have provided considerable information as to how genes can be "turned on" and "off." These studies have indicated the existence of both negative and positive control of gene action. In negative control a regulatory gene makes a product that represses or "turns off" a structural gene. In positive control the regulatory gene produces a product that renders a structural gene active. The operon consists of an operator, promoter, and structural genes. In the case of the much studied lactose operon in *E. coli,* negative control involves repression of transcription by the binding of a repressor protein to the operator (Pardee, Jacob, et al., 1959), whereas positive control is at the promoter site that is activated by CAP

This research was supported by a National Science Foundation grant GB 35462X.

(catabolite activating protein), the cAMP (cyclic adenosine monophosphate) receptor protein (Zubay, Schwartz, et al., 1970).

These model systems readily account for an "on–off" switch type of control in which a gene or operon is in one condition or the other. Such all-or-none differences in which a protein is either present or almost completely absent are, of course, found between tissues. For example, the *Sh*1 protein specified by the *Sh*1 gene on chromosome 9 in maize is formed in high concentration in the endosperm (Schwartz, 1960). However, it is found only in this tissue, and no trace of this protein can be detected in any other plant part, not even in the scutellum of the *Sh*1 kernels. The immunological test for the presence of this protein is extremely sensitive. In addition to presence or absence differences in proteins, tissues and organs show variation in enzyme and protein level. The relative concentrations of specific proteins can vary strikingly from tissue to tissue. The early electrophoretic studies of Markert and Hunter (1959) on the mouse esterases provide a clear example of such tissue-specific differences. A large number of esterase bands are observed after electrophoretic separation. In addition to tissue-specific variation in the number of bands, there are striking differences in the relative concentrations of the various isozymes. The level of enzyme activity and the balance between the activities of a number of enzymes may be just as important to normal development and cellular differentiation as tissue-specific differences in the presence or absence of an enzyme. This chapter is concerned with the problem of regulation of the level of gene expression during development. One mechanism proposed for the regulation of gene activity is by a feedback repression-type control. In such cases the concentration of a product is maintained at an optimum level by product repression of a gene that specifies an enzyme in its biosynthetic pathway. A good example of this is the control of histidine synthesis in *Salmonella* (Brenner and Ames, 1971). The level of expression of the genes in the histidine operon is controlled by the concentration of histidine in the cell. Histidine is converted to histidyl transfer RNA (tRNA), which is the corepressor of the histidine operon (Lewis and Ames, 1972). At high histidine levels in the cell, the corepressor level is high, and the operon activity is reduced. This causes a drop in the concentration of histidine in the cell and thus less conversion to the histidyl tRNA. The decrease in the level of the corepressor results in derepression of the operon and an increase in expression of its structural genes. Such a regulatory scheme is well suited to maintain an optimum level of product but cannot readily serve to maintain different levels of gene expression in the diverse tissues of multicellular organisms. To do so would necessitate tissue-specific differences in the affinity for binding between the corepressor and the operator site, or in rate of conversion of product to corepressor.

The level of gene expression can also depend on the concentration of effector, repressor, or activating factor in a cell. However, this simply pushes the problem back one step to the mechanism for regulation of level of expression of those genes which specify or produce the effector, repressor, or activator.

Based on the investigations in our laboratory on the role of gene competition in the regulation of the activity of the alcohol dehydrogenase ($Adh1$) gene in maize, (Schwartz, 1971, 1973) I have formulated a mechanism for the regulation of the level of gene expression that could operate during differentiation of higher organisms to produce tissue- or organ-specific differences in the balance of gene activities. This scheme does not involve any novel regulatory systems, but uses only those control mechanisms for which evidence already exists, such as repression and derepression, for rendering genes inactive and active.

I propose that a given gene, X, is under positive control and its transcription requires the binding of a specific activating protein to its promoter. The activating protein is produced in low but equal concentration in all cells. The low concentration of activating protein limits the transcription of gene X such that it is functional at a level somewhat below its maximum capacity. A change in the level of expression of gene X can be accomplished by simply derepressing another gene or genes that possess a promoter that binds the same activating protein. These regulating genes may, but need not have any function other than the regulation of the level of expression of gene X. The regulating genes thus compete with X for the limited activating protein. If the promoters of the regulating gene and X compete equally for the activating protein, then the activity of gene X will be reduced 50%. If the regulating gene cannot compete as well as X for the activating protein, X activity will be set at a higher level. Reduction of the level of expression of gene X below 50% will result from derepression of a regulating gene that competes better than X for the limiting activating protein or from an increase in competition by the derepression of more regulating genes. Thus by the use of an all-or-none mechanism whereby genes are either "turned on" or "off," it is possible to set the limit of expression of a given gene at a particular level. Different levels of activity for the gene can be established during development by derepressing different regulating genes. According to this scheme, the regulatory feature of the regulating gene lies solely in its transcription, and it is not necessary that it specify a functional protein. In fact, these genes need not be translated at all. In view of the great excess of DNA in higher organisms, over and above that accounted for by the estimated number of functional genes, (Judd, Shen, et al., 1972) this is not an unreasonable proposition.

The proposed competition model for the quantitative regulation of gene expression sets the level of activity of a particular gene by derepressing other genes that compete with it for a limited activating factor. The model requires: (a) the existence of activating factors, (b) activation of multiple genes by the same factor, (c) competition between genes for the activating factor, and (d) that in repressed state a gene cannot compete for the factor. Evidence is available for all of these points from studies of gene regulation in bacteria and maize. The existence of factors that activate a gene, as in positive control of gene expression, are well documented in bacteria. The first system described was by Englesberg (1971) for the arabinose operon. In 1970 we presented evidence that the amount of alcohol dehydrogenase synthesized in corn kernels is independent of gene dosage and set by the limited concentration of a factor required for gene activation (Schwartz, 1971).

Point (b), namely, multiple genes activated by the same factor, is also well documented in bacteria. Cyclic AMP and CAP are catabolite activating molecules that serve to activate a number of different genes for diverse catabolic pathways (Lee, Wilcox, et al., 1974; Nissley, Anderson, et al., 1971; Zubay, Schwartz, et al., 1970). In the maize ADH system we have presented evidence that two genes, $Adh1$ and $Adh2$, compete for the same limited activating factor (Schwartz, 1971). Furthermore, analysis of the phenomenon of dosage compensation in *Drosophila* strongly suggests that groups of genes on the X chromosome are each activated by a single species of activating molecule that is present in limited concentration (Maroni and Plaut, 1973; Schwartz, 1973).

Point (c) concerns differences in the ability of genes to compete for the activating factor. The work on maize ADH has shown that two naturally occurring alleles of the $Adh1$ gene ($Adh1F$ and $Adh1S$) have a twofold difference in competitive ability at certain stages in development. If alleles of the same gene can differ in competitive ability, such differences must also exist between loci.

Point (d) is the most critical to the proposed regulatory model. Since the genotype does not change during differentiation and development, the number of potentially competing regulating genes is the same in all cells. The model proposes, however, that a potentially competing gene can compete only when it is in an active or derepressed state. No direct evidence on this point is as yet available from the bacterial studies, although some results are suggestive. *In vitro* studies on the mechanism of the repression of operon transcription by the repressor molecule indicates that binding of repressor to the operator site of an operon renders the promoter incapable of binding RNA polymerase (Squires, Lee, et al., 1975). This repressor-induced change in the promoter may similarly prevent the attachment of activator molecules, thus eliminating the ability of such genes to compete for the activating molecule. The studies on the maize ADH system provide direct evidence for point (d). The alcohol dehydrogenase gene $Adh2$ specifies a relatively inactive enzyme (Schwartz, 1966). The genes $Adh1$ and $Adh2$ are not linked (Freeling and Schwartz, 1973). In the maturing embryo and uninduced young seedling, where $Adh2$ is not active, this gene does not compete with $Adh1$. However, when the seedling is subjected to anaerobic conditions as by immersion in water, the $Adh2$ gene is derepressed and now competes with $Adh1$ for the limited activating factor. In the absence of competition with any other genes, plants homozygous for a strong or weakly competing *Adh*1 allele produce the same amount of enzyme. However, this is not the case when *Adh*1 is competing with another gene for the limited factor. Direct evidence that it is the *Adh*2 gene that competes with *Adh*1 for the activating factor, rather than some other gene that is similarly regulated, comes from our recent studies on the interaction between *Adh*1 and *Adh*2 alleles. We have worked with a strain of maize that shows high ADH synthesis in the scutella of kernels submerged 2–3 days underwater. The F_2 population of kernels segregating for the electrophoretic variant *Adh*1F and *Adh*1S alleles, and *Adh*2N and *Adh*2P alleles were individually analyzed. The *Adh*2 gene is derepressed in the scutella of these soaked seeds, and the relative activities of the *Adh*1 and *Adh*2 alleles can be determined by comparing the intensities of the newly synthesized intergenic heterodimer (set II) isozyme bands. It was observed that under these conditions the relative activities of the alleles of one gene vary with the allelic constitution at the second locus. For example, the *Adh*1F allele is much more active than the *Adh*1S allele in *Adh*2N/*Adh*2N kernels than in *Adh*2P/*Adh*2P homozygotes. The difference in intensities between the IF·2N and 1S·2N intergenic heterodimer bands is much greater than the IF·2P and 1S·2P heterodimers. Similarly, the *Adh*2N allele is much more active than the *Adh*2P allele in *Adh*1F/*Adh*1F kernels than in *Adh*1S/*Adh*1S homozygotes. The difference in intensities of the 1F·2N and 1F·2P intergenic heterodimer bands is much more striking than the difference between 1S·2N and 1S·2P. This interaction between the unlinked *Adh* loci can best be accounted for by postulating that both genes compete for an activating factor that is just barely limiting in the induced scutella. The *Adh*1F allele competes better than *Adh*1S, and *Adh*2N competes better than *Adh*2P. Competition can be observed only when the factor being competed for is limiting. Thus the differential expression of two alleles that differ in competitive ability will be greater when the second locus competes strongly since there will be less factor available for their activation. Similarly, the amount of activating factor available for the same two alleles will be higher and hence less limiting in genotypes where the gene at the second locus competes poorly.

Recent studies in our laboratory suggest that the difference in competitive ability is a reflection of the number of functioning promoter-like sites at the locus of the gene. The alleles *Adh*1F and *Adh*1S compete equally well for the limited factor in the maturing embryo, but in the seedling and at all other stages in development *Adh*1F competes about twice as well as *Adh*1S. We have isolated an EMS-induced mutant in which the alteration is in the regulatory portion of the *Adh*1F locus. This mutant competes only half as well as *Adh*1S in the maturing embryo but equally as well as *Adh*1S in the seedling, and so on. To account for this result we have proposed that the *Adh*1F has two promoter sites that are functional at all times, whereas *Adh*1S also has two promoters, but one of these is functional only in the maturing kernel and not in the seedling. This could account for the difference in competitive ability of the same two alleles, $Adh1F$ and $Adh1S$, in the maturing embryo and in the seedling. If the EMS-induced mutation inactivated one of the promoters, the mutant would have only half the number of functional promoters as $Adh1S$ in the maturing embryo, but both the mutant and $Adh1S$ would have the same number of functioning promoters in the seedling. Analysis of the alcohol dehydrogenase content of F_2 kernels segregating for the mutant and wild-type $Adh1S$ alleles reveals that although the wild type allele is twice as active as the mutant in heterozygotes, the same amount of enzyme is synthesized in all three F_2 genotypes. This evidence establishes that the mutation is in the regulatory portion of the $Adh1$ locus and not in the structural gene.

REFERENCES

Brenner, M. and B. N. Ames. 1971. The histidine operon and its regulation. In *Metabolic pathways*, Vol. 5, H. J. Bogel (ed.). New York: Academic, pp. 349–387.

Englesberg, E. 1971. Regulation in the L-arabinose system. In *Metabolic Pathways*, Vol. 5, H. J. Vogel (ed.). New York: Academic, pp. 257–296.

Freeling, M. and D. Schwartz. 1973. Genetic relationships between the multiple alcohol dehydrogenases of maize. *Biochem. Genet.* **8:** 27–36.

Judd, B. H., M. W. Shen, and T. C. Kaufman. 1972. The anatomy and function of a segment of the X chromosome of *Drosophila melanogaster. Genetics* **71:** 139–156.

Lee, N., G. Wilcox, W. Gielow, J. Arnold, P. Cleary, and E. Englesberg. 1974. *In vitro* activation of the transcription of araBAD operon by araC activator. *Proc. Nat. Acad. Sci. U. S. A.* **71:** 634–638.

Lewis, J. A. and B. N. Ames. 1972. Histidine regulation in *Salmonella typhimurium*. XI. The percentage of transfer RNA[His] charged *in vivo* and its relation to the repression of the histidine operon. *J. Molec. Biol.* **66:** 131–142.

Markert, C. L. and R. L. Hunter. 1959. The distribution of esterases in mouse tissues. *J. Histochem. Cytochem.* **7:** 42–49.

Maroni, G. and W. Plaut. 1973. Dosage compensation in *Drosophila melanogaster* triploids. II. Glucose-6-phosphate dehydrogenase activity. *Genetics* **74:** 331–342.

Nissley, S. P., W. B. Anderson, M. E. Gottesman, R. L. Perlman and I. Pastan. 1971. *In vitro* transcription of the gal operon requires cyclic adenosine monophosphate and cyclic adenosine monophosphate receptor protein. *J. Biol. Chem.* **246:** 4671–4678.

Pardee, A. B., F. Jacob, and J. Monod. 1959. The genetic control and cytoplasmic expression of "inducibility" in the synthesis of B-galactosidase by *E. coli. J. Molec. Biol.* **1:** 165–178.

Schwartz, D. 1960. Electrophoretic and immunochemical studies with endosperm proteins of maize mutants. *Genetics* **45:** 1419–1427.

Schwartz, D. 1966. The genetic control of alcohol dehydrogenase in maize: Gene duplication and repression. *Proc. Nat. Acad. Aci. U. S. A.* **56:** 1431–1436.

Schwartz, D. 1971. Genetic control of alcohol dehydrogenase—a competition model for regulation of gene action. *Genetics* **67:** 411–425.

Schwartz, D. 1973. Comparisons of relative activities of maize Adh_1 alleles in heterozygotes-analyses at the protein (CRM) level. *Genetics* **74:** 615–617.

Schwartz, D. 1973. The application of the maize-derived gene competition model to the problem of dosage compensation in *Drosophila. Genetics* **75:** 639–641.

Squires, C. L., F. D. Lee, and C. Yanofsky. 1975. Interaction of the *trp* repressor and RNA polymerase with the *trp* operon. *J. Molec. Biol.* **92:** 93–112.

Zubay, G., D. Schwartz, and J. Beckwith. 1970. Mechanism of activation of catobolite-sensitive genes: a positive control system. *Proc. Nat. Acad. Sci. U. S. A.* **66:** 104–110.

Chapter 36

USE OF TISSUE CULTURE FOR GENETIC MODIFICATION OF PHOTOSYNTHETIC BIOCHEMISTRY

M. B. Berlyn, I. Zelitch, J. Polacco, and P. R. Day·
Departments of Biochemistry and Genetics,
The Connecticut Agricultural Experiment Station,
New Haven

It is well known that maize is a highly efficient species with respect to photosynthesis. We discuss the biochemical and cellular characteristics that contribute to this efficiency and our efforts to assess and manipulate some of these characteristics in cell culture. We include not only carbon-fixing reactions, but also respiration reactions responsible for utilization or loss of fixed carbon. The .techniques currently becoming available in somatic cell genetics offer new possibilities for a genetic approach to the study of photosynthesis and light and dark respiration in higher plants. These processes are of obvious and unique importance in plant cells, and their control would have very great practical significance in terms of future development of superior crop plants.

The classic methodology of microbial genetics has often been invoked as a model for work in somatic cell genetics. There are several prerequisites for the use of cell cultures in this way. Those most relevant to our studies are:

1. We must know or be able to elucidate the specific pathway(s) of interest.

559

2. The phenotype under consideration must be expressed and distinguishable in culture.

3. Selective pressures must be devised that are related to specific parts of the pathway and are capable of differentiating between wild-type and mutant cells.

4. Variants with the desired phenotype must be subjected to genetic analysis. Since genetic analysis cannot at present be done in cell culture, it is necessary to derive whole plants before formal genetics can begin.

Straightforward and basic procedures such as mutagenesis and efficient plating of cells are necessary for the use of cell cultures but at present are fraught with technical difficulties. Regeneration of intact plants from cells has been accomplished in some, but certainly not all, species examined. Efforts to improve these techniques are intensifying in many laboratories, but for the most part are recent and fairly preliminary. Nevertheless, several examples of mutant plants derived from carrot and tobacco cells selected in culture do exist (Carlson, 1973; Maliga, Breznovits, et al., 1973; Widholm, 1974). Our approach is to use a very amenable species, *Nicotiana tabacum,* to learn how to make important kinds of selections. We hope that future improvements in basic technical procedures will facilitate more widespread genetic applications of the study of pathways and selective systems which we will describe. In Chapter 32, which is concerned with recent progress in tissue culture of corn, it is suggested that many of these techniques may soon be applicable to corn.

A genetic approach to the study of photosynthesis and respiration in a model species, using powerful selective techniques currently possible with cell cultures of that species, may be of interest to plant breeders because of the correlation between net photosynthesis and productivity.

THE PATHWAYS OF CO_2 ASSIMILATION AND RESPIRATION

The existence of an alternate or auxiliary pathway for each of these necessary functions may allow alteration and possibly elimination of one pathway without total loss of plant viability. We are investigating the possibility that mutants lacking inefficient types of respiration may in fact have improved rates of net photosynthesis. Mutants can be used to study the relative contributions and interrelationships of the various pathways.

Ribulose Diphosphate Carboxylase and the C-3 Pathway

The carboxylation of ribulose diphosphate (RudP) catalyzed by RudP carboxylase is of course the primary reaction of CO_2 fixation by photosynthesis. The first detectable product is the 3-carbon compound, 3-phosphoglyceric acid, hence the name C-3 pathway. In addition to the carboxylation reaction, RudP carboxylase can catalyze the oxygenation of RudP in a reaction of significance to the respiration of immediate products of photosynthesis, as discussed in a later section (cf. Figure 4).

This important enzyme has unusual structural properties and an unusual mode of inheritance. Ribulose diphosphate carboxylase may comprise 40–50% of the total soluble protein of the chloroplast. It is a large protein with a molecular weight of approximately 500,000, consisting of large subunits with a molecular weight of about 60,000 and small subunits with a molecular weight of 12,000–15,000 (Kawashima, and Wildman 1970), which have been further resolved by isoelectric focusing on polyacrylamide gels (Kung, Sakano, et al., 1974; Sakano, Kung, et al., 1974). The large subunits are believed to contain the catalytic sites, and the small subunits are postulated to have regulatory functions (Sugiyama, Ito, et al., 1971). Evidence from several species indicates that the pattern of inheritance and the sites of synthesis are different for the two types of subunits. The large polypeptides are synthesized on 70S chloroplast ribosomes (Criddle, Dau, et al., 1970) and by *in vitro* chloroplast RNA (Hartley, Wheeler, et al., 1975) and show a maternal pattern of inheritance (Sakano, Kung, et al., 1974; Chan and Wildman, 1972). The small subunits are synthesized on 80S ribosomes (Criddle, Dau, et al., 1970), and information for marker peptides of these subunits is transmitted through both pollen and egg (Sakano, Kung, et al., 1974; Kawashima and Wildman, 1972). It is generally assumed, therefore, that the large subunits are encoded by chloroplast DNA, whereas the small subunits are encoded by chromosomal DNA.

Somatic cell-culture methods have been used in a novel approach to the study of the nuclear–cytoplasmic interactions involved in carboxylase synthesis. Protoplast-fusion techniques were used to obtain a parasexual hybrid plant of *Nicotiana glauca* and *N. langsdorffii*; the banding pattern after isoelectric focusing of RudP carboxylase on polyacrylamide gels indicated the presence of the small subunits of both species and the large subunits of *N. glauca* only, although both chloroplast types would be expected in the fusion product (Kung, Gray, et al., 1975). In the same study the incorporation of *N. suaveolens* chloroplasts into albino protoplasts of *N. tabacum* resulted in green callus cells and subsequently a regenerated variegated plant that had both parental types of RudP carboxylase large polypeptides; but surprisingly, both types of RudP carboxylase small polypeptides were present as well.

Mutations affecting RudP carboxylase would be of interest from the standpoint of carbon fixation, respiration of substrates derived from early photosynthetic products, and nuclear–cytoplasmic relationships in plants. An examination of RudP carboxylase from leaves of a large number of mor-phological mutants of tomato revealed altered enzymatic and electrophoretic properties of the enzyme in four mutant plants but did not establish gene–enzyme correlations (Anderson, Wildner, et al., 1970). Cell cultures may provide an opportunity for obtaining a wider variety of carboxylase mutants. Mutations resulting in total loss of carboxylation activity would obviously be lethal to intact plants, but they could be maintained in culture. However, the absence of regenerated viable plants (or of alternate, parasexual means for analysis) greatly diminishes their potential interest. On the other hand, it is conceivable that mutations affecting the oxygenase activity could be recovered (e.g., by use of inhibitors or conditions specifically affecting this reaction), and perhaps such mutant cells could be converted to viable plants. Conditionally lethal (e.g., temperature-sensitive) mutants of RudP carboxylase could also be maintained.

Sugar phosphate intermediates of the Calvin cycle regulate ribulose diphosphate carboxylase activity in a complex way (Zelitch, 1975; Robinson and Gibbs, 1974). Further clarification of the observed regulatory effects are necessary before they become useful as tools for genetic analysis of regulation of photosynthetic processes.

Phosphoenolpyruvate Carboxylase and the C-4 Pathway

An auxiliary pathway for CO_2 assimilation involving the carboxylation of phosphoenolpyruvate is very important in photosynthesis in maize and other species with characteristically high rates of photosynthesis. The product of the carboxylation reaction, oxaloacetate, is converted to aspartate and malate and decarboxylation of these 4-carbon compounds is followed by refixation via the C-3 carbon reduction pathway. The C-4 pathway is predominant in sugarcane (Kortschak, Hartt, et al., 1965; Hatch and Slack, 1966), maize, sorghum, and other species, especially tropical grasses.

There is evidence for biochemical specialization of tissues within the leaf of a C-4 plant. Leaves of maize and other C-4 species are distinguished by usually having well-developed chloroplast-containing bundle sheath cells surrounding minor veins. The enzymes leading to oxaloacetate synthesis are present primarily in the cytoplasm of mesophyll cells, whereas ribulose diphosphate carboxylase and the carbon-reduction cycle reactions occur primarily in the bundle-sheath chloroplasts. It appears that malate or other C-4 compounds are transported from mesophyll to bundle sheath, where they are decarboxylated and refixed in the C-3 pathway.

The genus *Atriplex* and the genus *Molluga* include both C-3 and C-4 species. Crosses between a C-3 and C-4 species of *Atriplex* resulted in hybrids intermediate in anatomy, phosphoenolpyruvate carboxylase levels, and other biochemical characteristics, but resembling C-3 plants in photosynthetic capacity (Björkman, 1973). *Molluga verticillata* is a naturally occurring species that also is intermediate between C-3 and C-4 species in leaf anatomy, early photosynthetic products, and perhaps respiratory rates, but not photosynthetic

TABLE 1.   Typical rates of net photosynthesis in single leaves of various species at high illuminance and 300 ppm of CO_2 in air

Species	Net photosynthesis $(mg\ CO_2\ dm^{-2}\ h^{-1})$
Maize	46–63
Sugarcane	42–49
Sorghum	55
Bermuda grass, *Cyandon dactylon*	35–43
Pigweed, *Amaranthus edulis*	58
Sunflower	37–44
Cattail, *Typha latifolia*	44–69
Tobacco	16–21
Sugar beet	24–48
Orchard grass	13–24
Wheat	17–31
Bean	12–17
Oak	10
Maple	6
Dogwood, *Cornus florida*	7

rates (Kennedy and Laetsch, 1974). Björkman (1973) suggests that a small number of genes determine C-4 characteristics, but hybrids lack coordination between the biochemical and anatomical characteristics.

The C-4 reaction sequence itself does not appear to be highly efficient. In fact, it requires two more ATP molecules per molecule of CO_2 fixed than C-3 photosynthesis. Yet leaves of the C-4 plants maize, sugarcane, and sorghum assimilate CO_2 at high irradiance in air at rates much faster than most other crop species (Table 1). This efficiency is correlated with higher productivity of C-4 crops. The differences in net photosynthesis are not significantly attributable to differences in maximal photosynthetic rates (at saturating CO_2 and high irradiance) but result in large part from differences in respiration reactions that occur in the light and are associated with photosynthetic substrates. This light respiration, photorespiration, is discussed in a later section.

With further development of tissue-culture techniques for C-4 plants it may be possible by mutational studies to assess the importance of specific reactions of this pathway to the photosynthetic economy of these species.

Dark Respiration via the Cytochromes

The oxidation of photosynthetically formed sugars, yielding ATP to drive cellular processes, is most efficiently accomplished when molecular oxygen is the electron acceptor and the cytochromes are the terminal electron carriers.

Respiratory mutants lacking specific sites of oxidative phosphorylation in the electron-transport chain are well-known in eukaryotes, for example, *poky* and *mi* mutants of *Neurospora* and *petite* mutants of yeast (Gillham, 1974). The alternate or incomplete means of oxidation used by these mutants results in slow rates of growth.

Dark Respiration via the Alternate Pathway

Mitochondria from tissue of several plant species were found to possess, in addition to the cytochrome pathway, an alternate pathway of respiration, insensitive to the cytochrome inhibitors antimycin and cyanide and inhibited by hydroxamic acids, such as salicylhydroxamic acid (SHAM) (Schonbaum, Bonner, et al., 1971). Figure 1 shows this pathway as it relates to the path of electron flow through the cytochromes in plant mitochondria. The P:O ratio for oxidation via this alternate pathway is one, in contrast to a ratio of 3 for the cytochrome oxidase pathway. The proportion of respiration accounted for by this less efficient pathway in mitochondria of a number of plant species and tissues ranged within 1–100%, with a value of 15–20% most often encountered (Bahr and Bonner, 1973). This is the main pathway of electron flow utilized in *poky* cytoplasmic mutants of *Neurospora crassa* that lack cytochromes a and b and have a low P:O ratio (Lambowitz, Slayman, et al., 1972). Table 2 shows respiration of leaf disks of maize and tobacco in the presence and absence of salicylhydroxamic acid and cyanide, indicating that these tissues also have a high proportion of respiration through the alternate pathway. Although SHAM alone does not inhibit total respiration, SHAM in the presence of cyanide eliminates a large portion of the cyanide-resistant respiration. We are

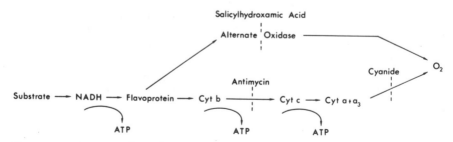

Figure 1. The normal (antimycin- and cyanide-sensitive) and alternate (salicylhydroxamic acid-sensitive) pathways of electron flow and ATP formation in plant mitochondria [reproduced from Zelitch (1975); copyright 1975, American Association for the Advancement of Science]. Abbreviations: NADH—reduced nicotinamide adenine dinucleotide; cyt—cytochrome.

TABLE 2.   Effect of cyanide and salicylhydroxamic acid (SHAM) on respiration of leaf disks of maize and tobacco. Oxygen consumption of leaf disks was assayed manometrically in Warburg vessels with 10 mM potassium phosphate buffer, pH 7.0, and the inhibitor. Final pH of the fluid was 7.1–7.2. In the cyanide-containing vessels, potassium cyanide was also added to the 1 N NaOH contained in the centerwells

	Inhibitor	O_2 Consumed	Inhibition
		μmol g^{-1} fr. wt. h^{-1}	%
Maize	0	11.8	—
	1.0 mM CN$^-$	3.6	69
	1.3 mM SHAM	12.8	−8
	1.0 mM CN$^-$ + 1.3 mM SHAM	1.5	87
Tobacco	0	13.9	—
	1.0 mM CN$^-$	12.1	13
	1.3 mM SHAM	13.8	1
	1.0 mM CN$^-$ + 1.3 mM SHAM	3.2	77

interested in eliminating by genetic means the inefficient SHAM-sensitive pathway to see if plants with cytochrome oxidase respiration only are more efficient.

Photorespiration

In 1959 Decker and Tió observed that leaves of many species exhibited a post-illumination outburst of CO_2, which they attributed to a fast form of respiration that occurred only in the light. An example of differences in respiration observed in light and dark is shown in Figure 2. This increased CO_2 evolution correlated with biochemical evidence that glycolic acid, an early product of photosynthesis (Kortschak, Hartt, et al., 1965), was rapidly oxidized to CO_2 and that glycolate was produced at rates sufficient to account for photorespiration in inefficient species, but it is not typically rapid in maize and other C-4 plants (Zelitch, 1958; 1959). It is now well established that glycolic acid is the primary substrate of photorespiration (Zelitch, 1971). The pathways of glycolate metabolism are shown in Figure 3. The reactions leading to synthesis of glycolate itself are not well understood. Several reactions known to produce glycolic acid are shown in Figure 4. One of these reactions, the oxygenation of ribulose diphosphate, is catalyzed by the enzyme RudP carboxylase. There are several lines of evidence supporting and limiting the effective functioning of each of these mechanisms, and it seems likely that multiple pathways occur in the same tissue. Microbial genetics has been used extensively to dissect and elu-

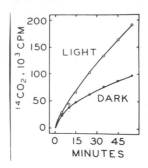

Figure 2. The effect of light (○) and darkness (●) on the rate of $^{14}CO_2$ released from tobacco leaf disks [from Zelitch, (1968)]. The experiment was carried out as in Figure 5, and 2.12×10^6 cpm of $^{14}CO_2$ was fixed. At zero time, one of the duplicate flasks was covered with aluminum foil.

cidate complex pathways and we hope that somatic cell genetics can be equally productive in clarifying this complex metabolic pathway in plants.

Photorespiration is observed in many plants but is not typically rapid in maize and other C-4 plants, as shown in Figure 5. How does maize avoid what appears to be wasteful loss of CO_2? Can we mimic maize-like efficiency by blocking photorespiration in other species? These are questions of prime interest in our consideration of the regulation of CO_2 fixation and respiration.

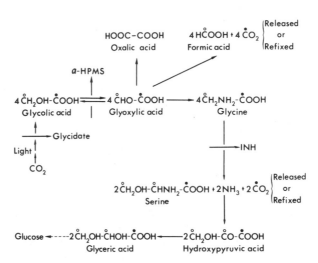

Figure 3. The pathways of carbohydrate synthesis and CO_2 evolution from glycolic acid. The symbols indicate the fate of the carbon atoms of glycolate. The inhibitors, blocking specific reactions are glycidate, α hydroxy-2-pyridinemethanesulfonic acid (α-HPMS), and isonicotinic acid hydrazide (INH).

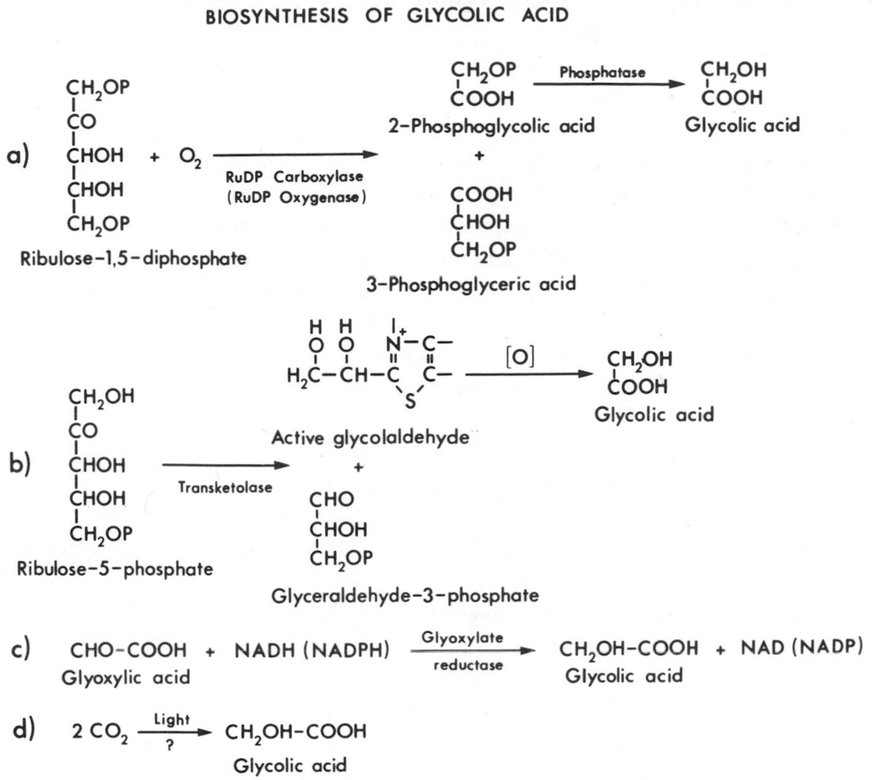

BIOSYNTHESIS OF GLYCOLIC ACID

a)

$$
\begin{array}{l}
CH_2OP \\
CO \\
CHOH \\
CHOH \\
CH_2OP
\end{array}
+ O_2 \xrightarrow[\text{(RuDP Oxygenase)}]{\text{RuDP Carboxylase}}
$$

Ribulose-1,5-diphosphate

$$
\begin{array}{l}
CH_2OP \\
COOH
\end{array}
\xrightarrow{\text{Phosphatase}}
\begin{array}{l}
CH_2OH \\
COOH
\end{array}
$$

2-Phosphoglycolic acid   Glycolic acid

+

$$
\begin{array}{l}
COOH \\
CHOH \\
CH_2OP
\end{array}
$$

3-Phosphoglyceric acid

b)

$$
\begin{array}{l}
CH_2OH \\
CO \\
CHOH \\
CHOH \\
CH_2OP
\end{array}
\xrightarrow{\text{Transketolase}}
$$

Ribulose-5-phosphate

Active glycolaldehyde

$$
H_2C-CH-C \quad C- \xrightarrow{[O]}
\begin{array}{l}
CH_2OH \\
COOH
\end{array}
$$

Glycolic acid

+

$$
\begin{array}{l}
CHO \\
CHOH \\
CH_2OP
\end{array}
$$

Glyceraldehyde-3-phosphate

c)   CHO-COOH + NADH (NADPH) $\xrightarrow[\text{reductase}]{\text{Glyoxylate}}$ CH$_2$OH-COOH + NAD (NADP)

Glyoxylic acid   Glycolic acid

d)   2 CO$_2$ $\xrightarrow[?]{\text{Light}}$ CH$_2$OH-COOH

Glycolic acid

Figure 4. Four biochemical pathways leading to the synthesis of glycolic acid [from Zelitch (1975); copyright, 1975, American Association for the Advancement of Science].

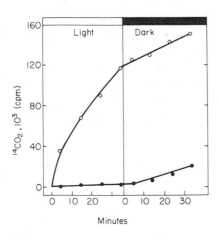

Figure 5. Comparison of $^{14}CO_2$ released in the light and dark by tobacco (O) and maize (●) leaf disks at 30° [from Zelitch, (1968)]. After 45 min in air at 2000 ft-c the flasks (75 ml) were closed and 5 μmol of $^{14}CO_2$ (1.08 × 10^6 cpm) were released. After another 45 min, at zero time CO$_2$-free air was passed through the vessels at a rate of 3 vols/min, and then released $^{14}CO_2$ was collected and the radioactivity determined.

567

PHOTOSYNTHESIS AND RESPIRATION IN CELL CULTURES

Photosynthesis

If we are to alter photosynthetic biochemistry initially in cells and ultimately in plants, the phenotype that we most need to maintain in culture is a dependence on photosynthesis for cell growth. Although there have been a few reports of photoautotrophic growth in plant-cell cultures, it is widely believed that undifferentiated cell cultures are incapable of prolonged autotrophic growth. We have sought to establish and document continuous autotrophy in cell cultures. Our model system has been *Nicotiana tabacum*. We have used haploid callus cultures from lines previously studied with respect to rates of photorespiration and net photosynthesis of leaves. As reported in detail elsewhere (Berlyn and Zelitch, 1975), by modifying a standard medium and providing a CO_2-enriched atmosphere, we have obtained autotrophic growth for many consecutive passages for all of the lines tested, despite their differences in origin and in photosynthetic efficiency as determined in intact leaves. Approximately threefold increases in dry weight were observed in each 3-week passage as cultures were subcultured continuously for three to eight passages. Typical photosynthetic rates for cultured cells were 3–4 μmol CO_2/g fwt h^{-1} or 200–300 μmol CO_2 per mg chlorophyll h^{-1}. All of the cell lines tested responded similarly to this regime; however, strain differences were observed for autotrophic growth on plates of solid medium in air. The culture JWB*su* grew best under these conditions, whereas the culture 918.11 could not be successfully maintained on such plates in air. These cultures were derived from plants of quite different varieties and with different rates of photorespiration and net photosynthesis. The line JWB *su* originated from a green haploid plantlet among the progeny obtained by anther culture (Nakata and Tanaka, 1968) of the yellow *Sulfur* (aurea) mutant (Burk and Menser, 1964) of variety John Williams Broadleaf. The *aurea* mutant is heterozygous for the dominant mutation *Su* (*Sulfur*) and anthers from *Su/su* plants give rise to white (*Su*) and green (*su*) haploid plantlets (Burk, 1970). Zelitch and Day (1968) have shown that *Su/su* plants have high photorespiration, whereas *su/su* siblings have low photorespiration and high net photosynthesis. The culture 918.11 is derived from a haploid plant obtained by anther culture of a plant of Connecticut Shade Variety 918, which has high photorespiration and low net photosynthesis. The JWB*su* cells were grown continuously under autotrophic conditions in air for 6 months with dry-weight increases two- to threefold per month.

As in previous studies by others (Davey, Fowler, et al., 1971; Hanson and Edelman, 1972), we have observed low levels of CO_2 assimilation in heterotrophically grown plant cell cultures. We find that the rates of assimilation are higher in cells grown on low concentrations of naphthaleneacetic acid and are often higher in autotrophically grown cells. Each of these observations is in agreement with one or more of the earlier studies of temporary or sustained photoautotrophic growth of plant cultures (Bergmann, 1967; Corduan, 1970; Hanson and Edelman, 1972. Chandler, de Marsac, et al., 1972. Koth (1974) observed parallel increases in chlorophyll and ribulose diphosphate carboxylase synthesis after transfer of heterotrophic dark-grown tobacco cultures to light. The requirement for 2–4,D in the growth medium of maize is antagonistic to chlorophyll synthesis.

The photosynthetic rates that we have observed in photoautotrophic cell culture are comparable to leaves on a chlorophyll basis but very low on a fresh-weight basis. Growth rates are much slower than heterotrophic rates. Nevertheless, we are hopeful that the rates of growth and CO_2 assimilation achieved by these cultures will be adequate to allow selection of photosynthetic and photorespiratory mutants in culture.

Respiration

Our attempts to assess quantitative levels of photorespiration in callus cells have been hampered by the low assimilation of $^{14}CO_2$ per gram of tissue. Nevertheless both photorespiration and glycolate synthesis are detectable in autotrophically grown callus cells.

Dark respiration has been assayed in cell suspensions of *N. tabacum*. These results have been reported in detail by Polacco and Polacco (in press). Both cyanide-sensitive and cyanide-insensitive, salicylhydroxamate (SHAM)-sensitive (alternate pathway) components of respiration were observed in these cultures. Antimycin alone inhibited 25% of tobacco respiration in a well-dispersed suspension culture, and the residual (antimycin-resistant) respiration was virtually eliminated by addition of 80 μg/ml (0.52 mM) SHAM. Salicylhydroxamate alone inhibited 77% of total respiration. Less disperse cultures had lower levels of alternate oxidase, usually about 25% of total respiration. Cyanide, but not antimycin, effectively inhibited the cytochrome system in these cultures. Alternate oxidase levels are sufficient to maintain cell viability for at least 48 h when the cytochrome oxidases are inhibited by antimycin A. When cells exposed to SHAM or antimycin for 48 h are washed, plated, and incubated for 12 days, growth is apparent for cells exposed to one of the inhibitors but not for cells exposed to both inhibitors.

Cyanide-resistant respiration has also been observed in mitochondria from cell cultures of sycamore, and the levels increased during lag phase (Wilson, 1971). Although we have demonstrated cyanide-insensitive respiration in maize leaves and in mitochondria isolated from etiolated shoots of maize, we have not yet extended these observations to callus cells of this species.

MAIZE: A NATURALLY OCCURRING MODEL FOR REGULATION OF PHOTORESPIRATION ENZYMES

Rates of Glycolate Synthesis

The enzymes for glycolate synthesis and oxidation are not lacking in maize leaves, although the reaction rates are typically very slow. The suggestion that photorespiration rates are actually high in maize but are masked by a superior ability to recycle CO_2 within the tissues has been refuted by several lines of evidence (Goldsworthy and Day, 1970; Troughton, 1971; Osmund and Harris, 1971; Volk and Jackson, 1972). Experiments with labeled CO_2, pyruvate, and acetate showed that the low level of glycolate in maize was derived mainly from organic acids, whereas tobacco had an additional rapid pathway for glycolate synthesis from early products of photosynthetic CO_2 fixation (Zelitch, 1973).

We consider it very important to ascertain the exact mechanism involved in "shutting off" the photorespiratory system in maize. The regulation of the enzymes of glycolate metabolism in maize leaves may offer clues for making less efficient plants more maize-like. We are interested in blocking steps in the glycolate pathway in inefficient species (e.g., tobacco) and ascertaining the effects of these alterations on photorespiration and photosynthesis.

Conditions that Control Photorespiration

A variety of atmospheric factors affect photorespiration rates. These include CO_2 concentration, O_2 concentration, and temperature. Chemical inhibitors of the glycolate pathway (Figure 3) also block photorespiration. A comparison of net photosynthesis in maize, as a model low photorespiration plant, and tobacco, as a typical high photorespiration plant, in experiments that vary these regulating factors provide evidence for the importance of light respiration as a factor in net photosynthesis. An assessment of the ways in which maize responds to these conditions may help us to formulate hypothetical expectations for a low photorespiration mutant of a high photorespiration species.

Inhibitors of glycolate synthesis or oxidation (see Figure 3) inhibit photorespiration and enhance net photosynthesis in tobacco but do not increase net photosynthesis in maize (Zelitch, 1966, 1973, 1974). Glycidate inhibits glycolate synthesis but does not affect the RudP carboxylase (oxygenase) reaction (Zelitch, 1976), indicating that other reactions are responsible for a large portion of glycolate synthesis in inefficient species.

At saturating light intensities, increases in the ambient CO_2 concentration increase net photosynthesis. The differences in net photosynthesis between C-3 and C-4 species are obscured at saturating CO_2 concentrations and high irradiation. For example, at 150 ppm CO_2, the CO_2 assimilation rate of maize was 2.7 times greater than that of tobacco but at 1000 ppm, was only 1.3 times greater (Zelitch, 1971; Hesketh and Moss, 1963). We have argued against efficient CO_2 recycling systems as the direct mechanism which prevents photorespiratory CO_2 loss in maize. However, the recycling of CO_2 between the C-4 and C-3 pathway, from the malic enzyme reaction to RudP carboxylase, may have an indirect effect on photorespiration. The transport and subsequent decarboxylation of malate in the bundle sheath cells may raise the CO_2 concentration within these cells to levels that inhibit glycolate synthesis and photorespiration. The CO_2 concentration in the bundle-sheath cells has been calculated to be 5 times greater than the concentration in other photosynthetic cells (Hatch, 1971).

When oxygen concentration is decreased from 21% to 0–3%, with normal CO_2 and high irradiance, net CO_2 fixation in less efficient species increases (Forrester, Krotkov, et al., 1966a,b). Maize leaves are unaffected by lowered oxygen concentrations and oxygen inhibition is observed only at concentrations greater than 50% (Zelitch, 1966; Laing and Forde, 1971). The observation that oxygen inhibits photosynthesis was first made by Warburg and is often termed the *Warburg effect.*

The increases in net photosynthesis that occur at high CO_2 or low O_2 in inefficient species are paralleled by decreases in glycolic-acid synthesis and in photorespiration under these conditions (Zelitch, 1971; Forrester, Krotkov, et al., 1966a,b; Laing and Forde, 1971). Oxygen effects on photosynthesis and glycolate synthesis can be reversed by high levels of CO_2. The RudP (carboxylase) oxygenase reaction, producing phosphoglycolate and phosphoglycerate (see Figure 4), has been proposed as the mechanism mainly responsible for the Warburg (oxygen) effect (Laing, Ogren, et al., 1974). Evidence against this view has also been presented (Zelitch, 1975). Regardless of the mechanism, this oxygen effect may be useful for mutant selection, and mutants may in turn clarify the role of the oxygenase and other reactions in the glycolate pathway.

In inefficient species, a 10°C temperature rise approximately doubles gross photosynthesis and dark respiration but increases photorespiration by a factor of 3 or more (Zelitch, 1971). Consequently, net photosynthesis does not increase. Blocking photorespiration with a glycolate pathway inhibitor does not change net photosynthesis at 25°C but increases it 2–5 times at 35° (Zelitch, 1966). The oxygen effect on net photosynthesis is also greater at higher temperatures.

In contrast, maize and other efficient species, unhampered by high photorespiration, show large increases in net photosynthesis at high temperatures. This comparison is shown in Figure 6. This figure also illustrates that at low temperatures maize is less efficient than the so-called inefficient species. It appears that evolution of corn and tropical grasses resulted in adaptation for effi-

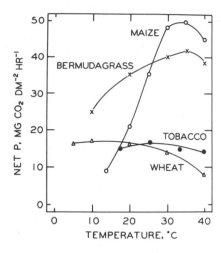

Figure 6. Effect of temperature on net photosynthesis in single leaves of maize (O), Bermuda grass (X), tobacco (●), and wheat (A) at 0.03% CO_2 in air at high illuminance [from Zelitch (1971)].

cient photosynthesis and rapid growth at high temperatures in spite of poorer efficiency at low temperatures. This adaptation involved complex morphology as well as regulatory phenomena, which are not completely understood at present.

Genetic Control of Respiration and CO_2 Assimilation

The differences in photorespiration and net photosynthesis between species are striking. Differences within species can also be observed. Low-photorespiration, high-photosynthesis lines within different varieties of tobacco have been studied by Zelitch and Day (1968, 1973). Different varieties of *Zea mays* have also been shown to have differences in net photosynthesis (Fousova and Avratovscukova, 1967; Heichel and Musgrave, 1969) and respiration (Heichel, 1971). Heichel (1971) examined respiration rates of two inbred varieties of maize previously shown to have different rates of net photosynthesis (Heichel and Musgrave, 1969). He observed faster respiration in the variety with lower photosynthetic capacity and slower dry-weight accumulation. This presumably reflects differences in some form of dark respiration in these varieties.

The differential effects of CO_2 and O_2 concentrations on high and low photorespiration plants have been used in attempts to select efficient individuals from a population of inefficient plants. Widholm and Ogren (1969) and Cannell, Brun, et al. (1969) combined maize and soybean plants in a closed system and observed senescence of the soybean plants as CO_2 became limiting after approximately 1 week. The corn plants survive longer, and presumably soybean mutants with reduced photorespiration would also survive. However, among several hundred thousand mutagenized soybean seedlings and approximately 2500 cultivars, survivors have not been recovered. Heichel (1973) combined tobacco seedlings with genotypes Su/su (high photorespiration) and su/su (low photorespiration) in a closed system with a high oxygen–low CO_2 atmosphere and observed prolonged survival of the low-photorespiration line. This system also can be used for screening mutagenized seedlings for low-photorespiration variants.

The selective methods used thus far do not appear to be sufficiently powerful. That is our primary motivation for turning our attention to somatic cell genetics.

SELECTIVE SYSTEMS FOR RECOVERING MUTANT. PLANT CELLS

We are attempting many different types of selections for mutants affected in CO_2 fixation or respiration. The beneficial effects of high CO_2 on autotrophic growth of callus cells and the preliminary correlation between autotrophic growth of cell cultures in air and low photorespiration rates in leaf disks of the parent plants suggest the hypothesis that low photorespiration is important for photoautotrophic survival of cell cultures. Oxygen concentration may be a useful selective agent, since rapidly photorespiring plants are more sensitive to a high oxygen atmosphere than are plants with slow rates of photorespiration. A 1–3-week exposure to oxygen at levels that inhibit CO_2 fixation and increase glycolate synthesis in leaves of C-3 plants is quite toxic to autotrophically grown tobacco callus. We are investigating this effect and attempting to recover cells that survive under these conditions.

Inhibitors of specific reactions in the glycolate pathway are shown in Figure 3. All of these compounds are toxic to tobacco callus cells. Intermediates in the pathway such as glycolate and oxalate also inhibit growth. Small foci of growth occur among mutagenized cells on media containing one of these inhibitors and are being tested as possible resistant mutants. The adaptation of the low photorespiring, photosynthetically efficient species to high temperature suggests that high temperatures may apply selective pressures against high photorespiration.

Although maize can hardly be improved with respect to photorespiration, improvements in the efficiency of dark respiration may be possible. We have shown that only about 70% of the respiration in leaf disks of maize is cyanide-sensitive (Table 2). Polacco is working on negative selection systems in cell cultures that will allow recovery of mutants totally sensitive to antimycin or cyanide. A selective system utilizing arsenate killing of growing cells appears promising, since the presence of both SHAM and antimycin has a sparing effect on wild-type cells exposed to 5mM arsenate for 24 h, whereas cells in the presence of antimycin alone are arsenate-sensitive. Mutants selected for resistance to arsenate in the presence of antimycin or cyanide alone may be incapable of the less efficient, cyanide-insensitive respiration and thus be forced to rely on the efficient pathway only. There is evidence that in the fungus *Ustilago maydis* mutants resistant to the succinate dehydrogenase inhibitor carboxin are also sensitive to antimycin (Georgopoulos and Sisler, 1970; Georgopoulos, Alexandri, et al., 1972) suggesting possible loss of alternate pathway respiration. A carboxin-resistant line has also been obtained in tobacco. Future work will characterize the properties of isolated mitochondria from the putative mutants and examine the hypothesis that such mutants may have more efficient respiration. Application of selective systems affecting dark respiration to maize may allow recovery of mutants with streamlined utilization of the substrates produced by its extremely efficient photosynthetic process.

REFERENCES

Anderson, W. R., G. F. Wildner, and R. S. Criddle. 1970. Ribulose diphosphate carboxylase III. Altered forms of ribulose diphosphate carboxylase from mutant tomato plants. *Arch. Biochem. Biophys.* **137**: 84–90.

Bahr, J. T. and W. D. Bonner. 1973. Cyanide-insensitive respiration. I. The steady states of skunk cabbage spadix and bean hypocotyl mitochondria. *J. Biol. Chem.* **248**: 3441–3445.

Bergmann, L. 1967. Wachstum grüner Suspensionskultures von *Nicotiana tabacum var.* "Samsun" mit CO$_2$ als Kohlenstoffquelle. *Planta* **74**: 243–249.

Berlyn, M. and I. Zelitch. 1975. Photoautotrophic growth and photosynthesis in tobacco callus cells. *Plant Physiol.* **56**: 752–756

Björkman, O. 1973. Comparative studies on photosynthesis in higher plants. *Photophysiology* **8**: 1–63.

Burk, L. G. 1970. Green and light-yellow haploid seedlings from anthers of sulfur tobacco. *J. Hered.* **61**: 279.

Burk, L. G. and H. A. Menser. 1964. A dominant aurea mutation in tobacco. *Tobacco Sci.* **8**: 101–104.

Cannell, R. O., W. A. Brun, and D. N. Moss. 1969. A search for high net photosynthetic rates among soybean genotypes. *Crop Sci.* **9**: 840–841.

Carlson, P. S. 1973. Methionine-sulfoximine resistant mutants of tobacco. *Science* **180**: 1366–1368.

Chan, P. H. and S. G. Wildman. 1972. Chloroplast DNA codes for the primary structure of the large subunit of Fraction I protein. *Biochim. Biophys. Acta* **277**: 677–680.

Chandler, M. T., N. T. de Marsac, and Y. de Kouchkovsky. 1972. Photosynthetic growth of tobacco cells in liquid suspension. *Can. J. Bot.* **50**: 2265–2270.

Corduan, G. 1970. Autotrophe Gewebekulturen von Ruta graveolens und deren $^{14}CO_2$-markierungsprodukte. *Planta* 91: 291–301.

Criddle, R. S., B. Dau, G. E. Kleinkopf, and R. C. Huffaker. 1970. Differential synthesis of ribulose-diphosphate carboxylase subunits. *Biochem. Biophys. Res. Commun.* 41: 621–627.

Davey, M. R., M. W. Fowler, and H. E. Street. 1971. Cell clones contrasted in growth, morphology and pigmentation isolated from a callus culture of *Atropa belladonna* var. *lutea*. *Phytochemistry* 10: 2559–2575.

Decker, J. P. and M. A. Tió. 1959. Photosynthetic surges in coffee seedlings. *J. Agr. Univ. Puerto Rico* 43: 50–55.

Forrester, M. L., G. Krotkov, and C. D. Nelson. 1966a. Effect of oxygen on photosynthesis, photorespiration, and respiration in detached leaves. I. Soybean. *Plant Physiol.* 41: 422–427.

Forrester, M. L., G. Krotkov, and C. D. Nelson. 1966b. Effect of oxygen on photosynthesis, photorespiration, and respiration in detached leaves. II. Corn and other monocotyledons. *Plant Physiol.* 41: 428–431.

Fousova, S. and N. Avratovscukova. 1967. Hybrid vigor and photosynthetic rate of leaf discs in *Zea mays* L. *Photosynthetica* 1: 3–12.

Georgopoulos, S. G., E. Alexandri, and M. Chrysayi. 1972. Genetic evidence for the action of oxathiin and thiazole derivatives of the succinic dehydrogenase system of *Ustilago maydis* mitochondria. *J. Bacteriol.* 110: 809–817.

Georgopoulos, S. G. and H. D. Sisler. 1970. Gene mutation eliminating antimycin A-tolerant electron transport in *Ustilago* maydis. *J. Bacteriol.* 103: 745–750.

Gillham, N. W. 1974. Genetic analysis of the chloroplast and mitochondrial genomes. *Ann. Rev. Genet.* 8: 347.

Goldsworthy, A. and P. R. Day. 1970. Further evidence for reduced role of photorespiration in low compensation point species. *Nature* 228: 687–688.

Hanson, A. D. and J. Edelman. 1972. Photosynthesis by carrot tissue cultures. *Planta* 102: 11–25.

Hartley, M., A. Wheeler, and J. Ellis. 1975. Protein synthesis in chloroplasts. V. Translation of messenger RNA for the large subunit of Fraction I protein in a heterologous cell-free system. *J. Molec. Biol.* 91: 67–77.

Hatch, M. D. 1971. The C_4-Pathway of photosynthesis. Evidence for an intermediate pool of carbon dioxide and the identity of the donor C_4-dicarboxylic acid. *Biochem. J.* 125: 425–432.

Hatch, M. D. and C. R. Slack. 1966. Photosynthesis by sugarcane leaves. A new carboxylation reaction in the pathway of sugar formation. *Biochem. J.* 101: 103–111.

Heichel, G. H. 1971. Confirming measurements of respiration and photosynthesis with dry matter accumulation. *Photosynthetica* 5: 93–95.

Heichel, G. H. 1973. Screening for slow photorespiration in *Nicotiana tabacum* L. *Plant Physiol.* 51 (suppl.): 42.

Heichel, G. H. and R. B. Musgrave. 1969. Varietal differences in net photosynthesis of *Zea mays* L. *Crop Sci.* **9**: 483–486.

Hesketh, J. D. and D. N. Moss. 1963. Variation in the response of photosynthesis to light. *Crop Sci.* **3**: 107–110.

Kawashima, N. and S. G. Wildman. 1970. Fraction I Protein. *Annu. Rev. Plant Physiol.* **21**: 325–358.

Kawashima, N. and S. G. Wildman. 1972. Studies on Fraction I Protein. IV. Mode of inheritance of primary structure in relation to whether chloroplast or nuclear DNA contains the code for a chloroplast protein. *Biochim. Biophys. Acta* **262**: 42–49.

Kennedy, R. A. and W. M. Laetsch. 1974. Plant species intermediate for C_3, C_4 photosynthesis. *Science* **184**: 1087–1089.

Kortschak, H. P., C. E. Hartt, and G. O. Burr. 1965. Carbon dioxide fixation in sugarcane leaves. *Plant Physiol.* **40**: 209.

Koth, P. 1974. Changes in the enzyme pattern of greening tissue cultures of *Nicotiana tabacum* var. "Samsun." *Planta* **120**: 207–211.

Kung, S. D., J. C. Gray, S. G. Wildman, P. S. Carlson. 1975. Polypeptide composition of Fraction 1 protein from parasexual hybrid plants in the genus *Nicotiana. Science* **187**: 353–355.

Kung, S. D., K. Sakano, and S. G. Wildman. 1974. Multiple peptide composition of the large and small subunits of *Nicotiana tabacum* Fraction I protein ascertained by fingerprinting and electrofocusing. *Biochim. Biophys. Acta* **365**: 138–147.

Laing, W. and B. J. Forde. 1971. Comparative photorespiration in *Amaranthus* soybean, and corn. *Planta* **98**: 221–231.

Laing, W., W. L. Ogren, and R. H. Hagemann. 1974. Regulation of soybean net photosynthetic CO_2 fixation by the interaction of CO_2, O_2, and ribulose 1,5-diphosphate carboxylase. *Plant Physiol.* **54**: 678–685.

Lambowitz, A. M., C. W. Slayman, C. L. Slayman, and W. D. Bonner. 1972. The electron transport components of wild type and *poky* strains of *Neurospora crassa. J. Biol. Chem.* **247**: 1536–1545.

Maliga, P., A. Sz.-Breznovits, and L. Marton. 1973. Streptomycin-resistant plants from callus culture of haploid tobacco. *Nature New Biol.* **244**: 29–30.

Nakata, K. and M. Tanaka. 1968. Differentiation of embryoids from developing germ cells in anther culture of tobacco. *Jap. J. Genet.* **43**: 65–71. Also described by Nitsch, J. P. and C. Nitsch, *Science* **163**: 85 (1969) and Sunderland, N. and F. M. Wicks, *Nature* **224**: 1227 (1969). We used the modification described by Anagnostakis, S. in *Planta* **115**: 281 (1974).

Osmund, C. B. and B. Harris. 1971. Photorespiration during C_4 photosynthesis. *Biochim. Biophys. Acta* **234**: 270–282.

Polacco, J. C., and M. L. Polacco. (In press). Inducing and selecting valuable mutation in plant cell culture: a tobacco mutant resistant to carboxin. *Annals. N.Y. Acad. Sci.*

Robinson, J. M. and M. Gibbs. 1974. Photosynthetic intermediates, the Warburg effect; and glycolate synthesis in isolated spinach chloroplasts. *Plant Physiol.* **53:** 790–797.

Sakano, K., S. D. Kung, and S. G. Wildman. 1974. Identification of several chloroplast DNA genes which code for the large subunit of *Nicotiana* Fraction I proteins. *Molec. Gen. Genet.* **130:** 91–97.

Schonbaum, G. R., W. D. Bonner, B. T. Storey, and J. T. Bahr. 1971. Specific inhibition of the cyanide-insensitive respiratory pathway in plant mitochondria by hydroxamic acids. *Plant Physiol.* **47:** 124–128.

Sugiyama, T., T. Ito, and T. Akazawa. 1971. Subunit structure of ribulose-1,5-diphosphate carboxylase for *Chlorella ellipsoidea. Biochemistry* **10:** 3407–3411.

Troughton, J. H. 1971. The lack of carbon dioxide evolution in maize leaves in the light. *Planta* **100:** 87–92.

Volk, R. J. and Jackson, W. A. 1972. Photorespiratory phenomena in maize. Oxygen uptake isotope discrimination and carbon dioxide efflux. *Plant Physiol.* **49:** 218–232.

Widholm, J. M. 1974. Cultured carrot mutants: 5-methyltryptophan-resistance trait carried from cell to plant and back. *Plant Sci. Lett.* **3:** 323–330.

Widholm, J. M. and W. L. Ogren. 1969. Photorespiratory-induced senescence of plants under conditions of low carbon dioxide. *Proc. Nat. Acad. Sci. U. S. A.* **63:** 668–675.

Wilson, S. B. 1971. Studies of the growth in culture of plant cells XIII. Properties of mitochondria isolated from batch cultures of *Acer pseudoplatanus* cells. *J. Exp. Bot.* **22:** 725–734.

Zelitch, I. 1958. The role of glycolic acid oxidase in the respiration of leaves. *J. Biol. Chem.* **233:** 1299–1303.

Zelitch, I. 1959. The relationship of glycolic acid to respiration and photosynthesis in tobacco leaves. *J. Biol. Chem.* **234:** 3077–3081.

Zelitch, I. 1966. Increased rate of net photosynthetic carbon dioxide uptake caused by the inhibition of glycolate oxidase. *Plant Physiol.* **41:** 1623–1631.

Zelitch, I. 1968. Investigations of photorespiration with a sensitive ^{14}C assay. *Plant Physiol.* **43:** 1829–1837 (1968).

Zelitch, I. 1971. *Photosynthesis, photorespiration and plant productivity.* New York: Academic Press.

Zelitch, I. 1973. Alternate pathways of glycolate synthesis in tobacco and maize leaves in relation to rates of photorespiration. *Plant Physiol.* **51:** 299–305 (1973).

Zelitch, I. 1974. The effect of glycidate, an inhibitor of glycolate synthesis, on photorespiration and net photosynthesis. *Arch. Biochem. Biophys.* **163:** 367–377.

Zelitch, I. 1975. Improving the efficiency of photosynthesis. *Science* **188:** 626–633.

Zelitch, I. (In press.) Biochemical and genetic control of photorespiration. In *CO₂ Metabolism and productivity of plants,* R. H. Burris and C. C. Black (eds.). College Park, Md.: University Park Press.

Zelitch, I. and P. R. Day. 1968. Variation in photorespiration. The effect of genetic differences in photorespiration on net photosynthesis in tobacco. *Plant Physiol.* **43:** 1838–1844.

Zelitch, I. and P. R. Day. 1973. The effect on net photosynthesis of pedigree selection for low and high rates of photorespiration in tobacco. *Plant Physiol.* **52:** 33–37.

Chapter 37

INDUCTION OF
GENETIC VARIABILITY

M. G. Neuffer

Department of Agronomy,
The University of Missouri,
Columbia

The reasons for conducting experiments designed to produce large numbers of new mutants in corn have differed according to the interests of the individual investigators. Various mutation experiments have been carried out to determine the nature of the gene (Stadler, 1946), to define the biological consequences of atomic radiation (Anderson, 1948; Anderson, Longley, et al., 1949), to describe the attributes of controlling elements (McClintock, 1951), to define the genetic code and the mechanism of information transfer, and to develop new types of mutants for use as tools in genetic and physiological studies and as a source of genetic variability for plant improvement. (It should be pointed out that corn is especially rich in natural genetic variability, so this was not an urgent reason for such studies.) The experiments conducted to study specific loci were successful, whereas those designed to find new mutants often were not.

Contribution from the Missouri Agricultural Experiment Station. Journal Series No. 7404.

Through the years three major means of inducing mutation have been successfully employed, namely radiation, controlling elements and chemical
mutagens. Each has had its advantages and problems. This chapter deals
chiefly with chemical mutagens, but the problems dealt with relate to mutation
experiments in general. A lot of work in mutation experiments, especially those
designed to produce new mutants with corn, has gone unreported, and some of
the published reports have not been entirely satisfactory. Generally, dormant
seed was treated because it is easy to handle, and the dormant embryo can
tolerate considerable abuse. The selfed seed was grown for an M_1 that was outcrossed or selfed to produce an M_2, which in turn was grown and selfed for
mutant segregation. Insufficient attention was given to the number of germ-cell
primordia, the ontogeny of the germ line, and the redundancy that occurs in
sampling gametes from a mutant sector. The following is a consideration of
these factors as they relate to the efficiency of sampling in mutation experiments.

Corn, with separate male and female flowers (geitonogamy), may be compared with single-flowered (autogamous) species such as barley and
Arabidopsis, in which more successful mutation experiments have been
conducted. If one treats dormant seeds with a mutagen and produces M_1
plants, the proportion of mutant gametes from each plant will depend on the
intensity of treatment and the number of primordial cells in the seed that
contribute to the germ line.

As can be seen from Table 1, the combined seed progeny from a selfed
autogamous M_1 plant receiving one mutational event may give different proportions of normal and mutant progeny depending on the primordial germ-cell
number (GCN). For a GCN of 2, it will be $7:1$ ($4:0 + 3:1$), for 4, $15:1$ ($3 \times
4:0 + 3:1$), and for 8, $31:1$ ($7 \times 4:0 + 3:1$). For a geitonogamous plant with
concordant germ cells, the rates would be $15:1$, $63:1$, and $255:1$ for the same
GCN numbers.

TABLE 1.   The frequency of mutant individuals in the M_2 from an M_1 in which
a mutant has been induced in one primordial cell for autogamous versus
geitonogamous plants with different primordial germ-cell numbers. The same
primordia (concordance) for male and female germ cells are assumed for the
latter; for nonconcordance, no mutants would appear in the M_2

	Germ-cell number					
	1	2	4	6	8	10
M_2 Ratio (autogamous)	3:1	7:1	15:1	23:1	31:1	39:1
M_2 Ratio (geitonogamous)	3:1	15:1	63:1	143:1	255:1	399:1

TABLE 2. Comparison of the efficiency of different methods of treatment and handling for mutagenesis

Treatment and handling	M_1 Plants	Treated genomes	Mutants produced at 25% rate	M_2 Selfed	Mutants detected	Plants grown $M_1 + M_2$	Mutants per plant
Seed:							
M_1 Crossed; large M_2	100	1600[a]	400	4700[b]	380	4800	0.08
M_1 Selfed; large M_2							
Concordance	100	1600	400	2300	380	2400	0.16
Nonconcordance	100	3200[a]	800	4700	760	4800	0.16
M_1 Selfed; minimal M_2							
Concordance	1000	16,000	4000	1000	500	2000	0.25
Nonconcordance	1000	32,000[a]	8000	1000	500	2000	0.25
Pollen:							
M_1 Selfed	2000	2000	500	—	500	2000	0.25

[a] Assuming eight nonconcordant primordial cells each for the tassel and ear.
[b] Required to detect 95% of mutants; 7200 would be required to detect 99%.

According to Anderson, Longley, et al., (1949) and Steffensen (1968), the primordial cell number for the corn tassel is between four and eight, probably nearer eight. It has not been clearly established whether the same primordial cells participate in the production of both male and female gametes (concordance) or whether different primordial cells are involved. Preliminary unpublished data from Coe and Neuffer indicates that different cells participate, and nonconcordance is the rule.

Treatment effective at the seed stage may affect any one of the primordial cells independently and produce a sector generating half normal and half mutant gametes. Assuming eight primordial cells, the population treated will be 16 genomes times the number of seeds treated. For those mutants having no selective disadvantage, $1/16$ of the gametes from an M_1 plant undergoing a mutational event will carry the mutant. The M_1 plants may be outcrossed or selfed with the results indicated in Table 2.

If one treats 100 kernels, each carrying eight diploid primordial germ cells, the resulting M_1 plants will carry 1600 treated genomes. If the treatment is 25% effective, then 400 mutants will have been produced. Crossing each M_1 plant with an untreated plant will produce 100 ears that will preserve all the mutants. A planting of these seeds with a subsequent self will be required to express any recessive mutants.

A probability table will tell us that a sample of 47 individuals is required to give a 95% certainty of obtaining one individual occurring at a frequency of one in 16. Therefore, to detect 95% (380) of the 400 mutants in the 100 ears will require planting 47 kernels from each for a total of 4700 plants to be selfed. Thus an input of 4800 plants (100 M_1 + 4700 M_2) will produce 380 mutants with an efficiency of 0.08 mutants per cultured plant.

Selfing is more efficient than outcrossing because each self samples two sets of eight-cell primordia (tassel and ear) for a total of 3200 treated genomes and 800 mutants. This will require 4800 individuals to detect 760 mutants (an efficiency of 0.16).

The tendency is to try to save all mutants, but this is counterproductive. Note that a sample of 52 kernels (5300 plants) would be required to save 99.9% of the mutants (efficiency: outcross, 0.071; self, 0.013).

Maximal sampling raises the additional problem of duplication and the confounding of mutants. A single mutational event in a primordial cell will be duplicated many times through succeeding cell divisions to gamete formation; this is of some consequence in that there are many loci whose recessive alleles are similar, especially if viewed in an F_2 background with unknown numbers of modifiers segregating. Only one mutant of a particular type can be accepted in the progeny of each M_1 plant.

A more efficient sampling method, developed by Redei (1974a,b) for *Arabidopsis* but generalized in application, is the treatment of a large number of seeds for the M_1 followed by a minimum sampling for the M_2. For corn the minimum sample would consist of one kernel from each M_1 ear. Following previous assumptions, the treatment of 1000 kernels will affect 32,000 genomes and produce 1000 M_1 plants carrying 8000 mutants. Planting one from each progeny will test 2000 gametes and save 500 ($\frac{1}{16}$ of 8000) of the mutants. In terms of total input, an investment of 2000 selfed plants (1000 M_1 + 1000 M_2) will yield 500 mutants for an efficiency of 0.25; of course, 7500 mutants will be lost. Furthermore, since only one sample of two gametes is taken from each M_1 plant, the problem of duplicate copies is eliminated; if corn has nonconcordant male and female germ cell primordia, there is no chance of both gametes carrying the same mutant.

An equally efficient approach with some additional advantages is the treatment of pollen, which has been technically difficult because corn pollen is so fragile; the paraffin-oil technique developed by Coe (1966) and perfected by Neuffer (1972) has been used successfully.

PARAFFIN-OIL METHOD FOR POLLEN TREATMENT

The procedure is as follows:

1. Suspend 0.11% by volume of ethyl methanesulfonate (Eastman #7830) in light domestic paraffin-oil (Fisher) by vigorous stirring for an hour or more.

2. Mix EMS suspension with fresh pollen in a plastic vial with a cap, using at least 15 volumes of oil per volume of pollen.

3. Shake the pollen-oil mixture periodically to prevent the pollen from clumping in the bottom of the vial.

4. After 50 minutes apply the pollen-oil mixture to fresh silks using a #10 camel-hair brush; apply only enough to get a good distribution of pollen. Stirring with the brush is necessary between every application.

   (*Extreme precautions—mixing under a hood, protective clothing, sanitary disposal—should be taken to protect all persons associated with the work. Chemical mutagens are potentially carcinogenic and are particularly penetrating in oil.*)

The resulting kernels will produce M_1 plants that can be screened for dominant mutants and then selfed to produce an M_2 that will segregate for recessive mutants. Each pollen grain represents one treated genome, and each kernel has one treated and one untreated genome in its embryo.

Using the same calculations as for seed treatment, an investment of less than 100 plants for treatment will produce several thousand kernels. If 2000 of these are grown as M_1 plants and selfed, they will carry 500 mutants (25%), all of which may be expressed in the M_2. The efficiency will be slightly less than 0.25 (counting treated plants).

Pollen treatment has other advantages: (a) each mutant is an independent event not complicated by sectorial redundancy, (b) all mutants are saved, (c) variations in germ cell primordial number, and sex differences do not enter into comparisons of mutation frequencies, and (d) dominant mutants are easily recognized and ready for testing.

From several experiments using the paraffin-oil technique, 3113 transmissible mutants have been obtained. The total includes a large number of new types as well as most of the known variations from normal that fall within the limits of the observations made; mutants affecting anthocyanin pigmentation and mature plant characteristics, for example, were not always observable. The distribution of mutant types is roughly similar to the distribution of spontaneous mutants, as shown in Table 3, where types and frequencies are compared with the established collection of mutants listed in *The mutants of maize* (Neuffer, Jones, et al., 1968). Of course, such a comparison is somewhat inaccurate because the distribution of these mutants reflects the interests of investigators in particular types.

The data in Table 4 are from one experiment in which the most types were accurately recorded. The 3919 kernels obtained were planted and produced 3461 M_1 seedlings, which were examined periodically for altered phenotypes resulting from possible dominant mutations. The majority of the plants were normal, but some abnormal types were observed. Some were damaged by insects or disease; others were haploids or aneuploids, and 18 appeared to be dominant mutants. Twelve of these died before flowering and were tentatively recorded as dominant lethals. Six plants survived to maturity; of these, two produced only normal progeny on outcrossing, while four transmitted the mutant phenotype and proved to be legitimate dominant cases.

The 3172 selfed ears were examined for seed set and segregating kernel mutants. Of these, 851 showed partial sterility indicative of aborted gametes. A 3:1 ratio for an altered kernel phenotype was noted on 859 ears, which were classified as carrying a recessive kernel mutant. Four ears segregated nearer a 1:3 ratio for what are probably dominant kernel mutants. A 40-kernel sample was planted in a sand bench and the seedlings scored for segregating mutants; 751 were found. Of the 3919 kernels, 458 failed to grow and therefore received no test.

The seedling mutants included the commonly seen variations in chlorophyll such as white, yellow, yellow–green, pale green, and virescent; within each of these groupings there were some individual variants associated with the altered chlorophyll phenotype. Such things as vitality and leaf size, shape, and texture

TABLE 3.   Distribution of the types of seedling mutants induced by EMS treatment compared to the frequency of the same types among the 262 known mutants listed in Neuffer, Jones, et al. (1968)

	EMS	Percent	Control	Mutant of maize	Percent
Chlorophyll mutants	389	45	2	50	45
Albino	113	13	0	18	16
Luteus	32	3.7	2	12	11
Yellow–green	14	1.6	0	2	1.8
Pale green	131	15	0	5	4.3
Virescent	99	11	0	13	12
Variable mutants					
(Light- and/or temperature-sensitive)	104	12	1	16	14
Piebald and crossbanded	28	3.2	0	5	4.5
Streaked and speckled	46	5.3	0	1	0.9
Striped like sr	15	1.7	0	4	3.6
Striped like gs, ys, etc.	15	1.7	1	6	5.3
Small plant	181	21	0	13	12
Seedling lethal	102	12	0	1	0.9
Adherent	21	2.4	0	1	0.9
Glossy	23	2.6	0	13	12
Other	50	5.7	0	18	16
Total mutants	870	100	3	112	100

make it possible to recognize different loci in many cases. Other variants included such things as white or pale-yellow seedlings with yellow or green streaks, seedlings with green and white stripes, seedlings with green and white or otherwise altered crossbands, and seedlings with contrasting spots of light or dark green tissue. The mutants also included variations in morphology, such as dwarfing, adherent and split leaves, differences in leaf size, shape, and texture, and variations in viability, such as necrotic lethals. The kernel mutants included variations in: (a) size and shape (e.g., wrinkled, shrunken, dented, or small), (b) surface characteristics (e.g., rough and pitted); and (c) texture and color (e.g., opaque, translucent, brown, white, speckled); some of these were germless or had a weak embryo, and some had normal-appearing embryos but failed to germinate. Scattered among these mutants were rare cases that were dominant rather than recessive.

TABLE 4. Total heritable changes induced by treatment of pollen with ethyl methanesulfonate in paraffin oil. Frequencies are expressed as percent of M_1 seed tested, M_1 plants grown or selfed ears tested, as they apply in each case

Treatment	M_1 Seed	M_1 Plants	Dominant lethal plants	Dominant viable plants	Selfed ears	Semi- sterile ears	Dominant endosperm mutants	Recessive seedling mutants	Recessive kernel mutants	Total heritable changes
EMS	3919	3461	12	4	3172	851	4	751	855	2477
Percent		88	0.3	0.1	81	27	0.1	24	27	63
Control	637	575	0	0	451[a]	1	0	3	0	4
Percent		90	<0.16	<0.16	—	0.16	<0.16	0.52	0	0.7

[a] Some control ears were used for other purposes.

586

For the purpose of this presentation a few of the most interesting new types of mutants are described in more detail.

DIURNAL CROSSBANDED SEEDLING MUTANTS

Several mutants were found that expressed alternate crossbands of normal and mutant tissue, especially on the earlier leaves, in patterns corresponding to a 24-h growth period. Figure 1 shows a mutant with alternate bands of normal green and albino tissue with intermediate pale-green regions in between, suggesting a transitory period of modest activity between states of normal activity

Figure 1. First three leaves of a crossbanded mutant seedling showing alternating bands of normal green and white tissue separated by intermediate regions of pale green.

and complete inactivity for some agent necessary for chlorophyll production. Figure 2 is an example of alternating bands of normal and necrotic lethal tissue. In Figure 3 the basic mutant phenotype is a pale-green lethal with superimposed bands of tissue where the chlorophyll bleaches and produces necrotic crossbands. Figure 4 shows a similar expression, except that the basic phenotype is a tan color. The mutant designated *nec*3 is brown when it first emerges, and the leaves are rolled tightly together; when the leaves are unrolled manually, they have a distinctive tan color with crossbands of dark-brown pigment spaced at regular intervals. No chlorophyll is observed. Each of these phenotypes (Figures 1–4) can be interpreted as being mutants that are sensitive to light and/or temperature or some other variable condition that fluctuates between favorable and unfavorable in a 24-h period.

Figure 2. First two leaves of a crossbanded seedling (left) with normal and necrotic bands.

Figure 3. First two leaves of a pale-green mutant seedling (left) with crossbands of bleached tissue.

SEEDLING LETHALS

In addition to the mutants that were lethal because of the absence of or reduced amounts of chlorophyll, there were several types that grew normally to the three- or four-leaf stage and then died rather abruptly. In a classic case (Figure 5), the mutant seedlings are normal until the three-leaf stage when, one after another, the tips of the leaves suddenly change from green to a watery brown and then wilt; drops of a dark-brown exudate appear on the tips and margins of the affected leaves. This effect moves rapidly down the plant until, within 48 h of onset, the whole plant is affected and soon dies. There are numerous variants from this type, most of which show the mutant phenotype at about the time the nutrient supply from the endosperm is exhausted. However,

Figure 4. First two leaves of two tan necrotic (*nec3*) mutant seedlings with dark-brown crossbands in contrast to a normal seedling. The leaves of the mutant seedling on the left have been unrolled manually to show the color pattern.

one case grew normally until the six- or eight-leaf stage, when abrupt deterioration and death occurred.

SPOTTED SEEDLINGS

In this type seedling leaves are uniformly pale green with dark-green spots (Figure 6) or, alternatively, normal green with pale-green spots. The spots are not distributed according to cell lineages but appear to be scattered at random. Some groupings that indicate response to favorable or unfavorable conditions do occur. The spots are small, round, and fuzzy in outline and suggest the diffusion of a gene product.

PHOTOSYNTHETIC ELECTRON-TRANSPORT MUTANTS (*hcf*)

Among the seedling mutants was a unique class that demonstrates the usefulness of these materials. Miles (1975) and his students have been studying the pathways of photosynthetic electron transport and phosphorylation and looking for mutants that are defective in some part of the transport system. Using a technique employing UV light and a red filter, they found a total of 57 mutants among the 3172 M_2 progenies tested, including 18 defective in photosystem I, 25 defective in photosystem II, and 14 defective in some other photosystems.

KERNEL MUTANTS

The 3172 M_1 plants tested by selfing produced 855 ears that were segregating for at least one mutant kernel type, including those resembling known

Figure 5. The three leaves on the left from a necrotic lethal mutant seedling show necrosis moving from the tip to the base of each leaf.

Figure 6.   A pale green spotted mutant (left); the leaves are pale green with many small round fuzzy spots of normal-green tissue.

mutants as well as many new ones. Some of these mutants were viable, but most were defective in either endosperm or embryo or both and were lethal.

Figures 7–12 show examples of speckled aleurone, rough or pitted endosperm (like *et*), colorless floury-defective endosperm (*clf*), shrunken opaque endosperm, discolored defective endosperm and collapsed crown, respectively. All of these except the *et*-like mutants are nonviable. Over 228 were recorded as opaque; of these 101 were viable, but only 18 produced reasonably normal seedlings.

DOMINANT MUTANTS

Mutation from normal to dominant, whether spontaneous or induced, is a rare occurrence. From a population of 7738 M_1 plants produced by the treat-

Figure 7.   The full-colored ear from a self, segregating for a speckled kernel mutant.

ment of pollen with EMS, 10 viable transmissible dominant mutants were obtained. They included one striped virescent, one tiny dwarf, one yellow-green plant, two different disease lesion mimics, one yellow-streaked type, one shrunken endosperm, one collapsed endosperm, and two etched endosperm types. The 3461-plant subset of the population (Table 4) produced eight of the above and 12 probable dominant lethal cases. From other treatments with

Figure 8.   A colored ear segregating for a pitted (*et*-like) endosperm.

Figure 9.   A colored ear segregating for colorless floury-defective (*clf*).

EMS, five additional viable dominants—one yellow–green, two different intermediate dwarfs, one speckled leaf, and one lazy mutant—were obtained, bringing the total of the induced viable dominant mutants to 15.

SOME CONSIDERATIONS REGARDING FREQUENCIES

The total mutation rate from this experiment was quite high. If semisterile ears are counted, 63% of the treated pollen grains receiving a test carried a mutant of some type; the figure is 41% if semisteriles are excluded.

Figure 10.   A colorless ear segregating for a shrunken opaque mutant.

Figure 11. A colorless ear segregating for a discolored defective endosperm mutant.

The frequency per locus can be estimated in a rough way by selecting a few easily recognizable loci and determining their rates. Four such loci are *su* on chromosome 4, *wx* on 9, *clf* on 1, and *nec3* on 5. Their rates among the 3172 were 4, 3, 2, and 3, respectively, for an average rate of 3 and an average frequency of 0.9 per thousand. Since, on the average, each locus would have mutated three times in this population, the 1616 recessive mutants represent 539 loci.

By similar reasoning the 57 electron-transport mutants found by Miles represent 19 loci—six in photosystem I, eight in photosystem II, and five in unknown systems.

The eight viable dominant mutants represent a rate of 2×10^{-3}, as compared to 412×15^{-3} for the recessive mutants.

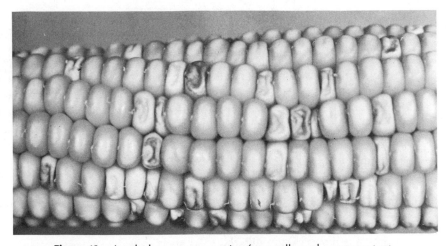

Figure 12. A colorless ear segregating for a collapsed crown mutant.

THE ABSENCE OF AUXOTROPHIC MUTANTS

The failure to obtain obligate auxotrophic mutants in corn and other eukaryotes has been a puzzle, especially in view of the remarkable success with fungi and bacteria. Numerous attempts by the author and others to grow various types of lethal and sublethal mutants on supplemental media have been mostly unsuccessful. Only a few mutants have been found in higher plants (Gavazzi, Nava-Racchi, et al., 1975; Li and Redei, 1968; Nelson and Burr, 1973; Neuffer, 1974). The large collection of mutants produced in these experiments affords an excellent opportunity to reexamine this problem. Among the mutants already isolated may be some that respond to supplemental feeding. If we have yet to recognize the phenotype of true auxotrophs in higher plants, then the segregating M_2 should be a rich source of such mutants once a phenotype has been identified.

One possible explanation may lie in intercellular transport of gene products. It is possible that the mutants that are commonly observed and studied involve genes whose product is not highly transportable (nondiffusible, or cell limited or unstable). Failure of transport beyond one or two neighboring cells is suggested by the fact that chimeras for most of the known recessive mutants in corn have distinct borders. Distinct borders are generally found for chimeras resulting from chromosome loss, from reversions arising through the action of controlling elements and from spontaneous mutations.

If transport is an important factor, what would be the phenotypic expression of various types of mutants with and without effective intercellular transport of gene product? A series of predictions can be framed for the alternatives of transportable versus nontransportable product and for universal versus stage-specific vital functions, such as chloroplast assembly, chlorophyll synthesis, or starch storage in the endosperm.

1. If one treats mature pollen with a mutagenic agent and produces in the proembryo nucleus a recessive mutant for a gene controlling a universally vital function involving a nontrasportable product, one may expect the following consequences. The mutant will potentially be a cell lethal but will survive through the gametophyte generation because of the covering of the sperm by the tube nucleus and will form a viable zygote. The lethality of the mutant will not be tested until sporogenesis when, for the first time, cells will arise with only a mutant allele in the nucleus. These will lack the vital function and will abort. The phenotype will, therefore, be a normal-appearing F_1 plant with 50% aborted pollen and ovules (semisterile pollen and ear); the mutant will not be transmitted to the next generation. A number of examples of this have been found.

2. A mutant that controls a vital function for which the product can be transported will survive through the F_1 as above, but the gametes produced

will have the advantage of gene product from normal diploid cells of the supporting tissue. Thus the gametes of the F_1 plant may survive to achieve fertilization. (This may or may not be true for the microspores since the male gamete does go through a short period of independent existence.) Assuming both male and female gametes succeed, the selfed F_1 will produce 25% homozygous mutant kernels. The mutant embryos may not be lethal at first, however, because the endosperm may carry stored gene product supplied by the ear parent. Lethality will occur when this endo-sperm supply is exhausted and the seedling must make its own product. The mutant would hence be expressed in the F_2 as a normal seedling that dies when the endosperm nutrients are exhausted (about the three-leaf stage). The necrotic lethals appear to be of this kind. If the male gamete does not survive, then only normal pollen grains will affect fertilization, and the F_2 will include only normal plants, half of which will be heterozygous. The mutant would escape detection unless special techniques such as differential transmission of linked markers are used to detect it.

3. A mutant that controls a stage-specific vital function such as chloroplast assembly and a product that is confined to the cell where it is produced should form viable mutant gametes, because chloroplasts are not necessary for gametogenesis. The F_1 selfed will produce 25% homozygous mutant embryos, which will grow into chlorophyll-less seedlings that survive only as long as the endosperm nutrients last. This type of mutant would be expressed in the F_2 as white or yellow seedlings that die at endosperm depletion. The commonly occurring w, wl and l mutants would fit this category.

4. A mutant that controls chlorophyll synthesis through a product that is transportable and stored may have F_2 seedlings that initially are normal or nearly normal green (depending on the efficiency of transport), if the ear parent supplies the gene product and if it is present in the embryo and endosperm until depleted by the seedling. At depletion, the green seedling should deteriorate in one of two ways. If intact chlorophyll can continue to function without the product, the first two or three leaves will be normal green, and subsequent leaves will be more white or yellow. The plant will live for some time on the photosynthesis of the first leaves but will eventually die. If, however, the manufactured chlorophyll in the first leaves requires continued gene product to function, at depletion the existing leaves will begin to fade or discolor, and the seedling will die rather abruptly. Two alternative phenotypes would appear for the F_2 of this type of mutant: (a) initially green seedlings that produce newer leaves that are white or yellow and survive to the fifth- or sixth-leaf stage and (b) initially green seedlings that begin to fade rather abruptly and die. Mutants representing the first of these types have not been seen, but a number of the

latter have been obtained as a result of pollen treatment with EMS. Most likely the transport situation will not be all or none but a matter of degree, the amount of transported product being somewhat less than that used in normal cell function.

5. A mutant controlling a vital function for which transport occurs to a less than adequate degree will likely be expressed as an altered kernel type, since the endosperm will have an inherited lack that is only partially satisfied by maternal feeding. The embryo will also be somewhat defective but will probably germinate to produce a weak seedling that dies as soon as the limited endosperm reserves supplied by maternal feeding are exhausted. The degree of defectiveness of the endosperm and the viability of the embryo will depend on how critical the missing constituent is and how effectively it can be transported and stored. Mutants at different loci should exhibit a wide range of variation in these aspects. A large number of the new endosperm mutants produced by EMS treatment fit very well into this class.

6. A mutant controlling a function concerned only with starch storage in the endosperm and having a partially transportable constituent should be expressed as a somewhat defective endosperm and a nearly normal seedling that eventually recovers full vigor, whereas a mutant controlling starch storage and having a nontransportable constituent should produce a defective endosperm with a normal embryo that germinates poorly due to an initial lack of needed nutrients. Mutants of both types appear frequently among the EMS-induced cases.

From these considerations one can logically conclude that auxotrophic mutants should be found among those necrotic seedling mutants [item (2) above] that grow normally and die abruptly at the three- to four-leaf stage and among the partially defective endosperm mutants that are associated with a weak embryo and a seedling that dies at an early stage [item (5)]. The preliminary testing of some mutants in the first class has produced no definitive results as yet, but the proline-requiring mutant reported by Gavazzi, Nava-Racchi, et al., (1975) appears to be of the latter type.

The alternative conclusions are that auxotrophic mutants do not have phenotypes that are presently recognizable or that they are not expressed in higher plants because of the complexity of cell relationships in tissues. The first possibility requires further investigation, and the second is not so surprising when one considers observations of chimeras of adjacent tissues that differ in two mutant phenotypes (Coe, Chapter 30, this volume; McClintock, 1951). The boundary between them may be a line of normal tissue, showing that there is a diffusion of precursors from one sector into the other; however, the fact that this border is never more than a few cells in width indicates that precursors move only a very short distance before they are used up. Herein may lie an important difference between higher plants and microorganisms in their response to supplemental feeding: in microorganisms, replaced nutrients need only travel through the wall and membranes of one cell, whereas in higher organisms the substance must move across sizable groups of cells. Since normal cells in higher organisms do not require supplementary feedings of amino acids, no provision exists for transporting them within the plant. Therefore, it may be impossible in most cases to feed a whole organism except with those substances, such as carbohydrates and growth hormones, that are normally transported.

SUMMARY

In a comparison of treatment and sampling methods in chemical mutagenesis experiments, it was established that pollen treatment and seed treatment with minimum sampling were more efficient than seed treatment with statistically significant sampling.

Material from the treatment of pollen with ethyl methanesulfonate produced exceptionally large numbers of new mutants representing most known types and many new ones, including 15 dominant mutants.

A comparison of frequencies shows that: (a) 63% of the treated pollen grains were mutated (41% excluding semisteriles) for a frequency of 412×10^{-3}, (b) frequency of viable dominant mutants was 2×10^{-3}, (c) the rate for four specific loci was $3/3172$, or 0.9×10^{-3}, and (d) it was estimated that the 1616 recorded mutants represent approximately 539 separate loci.

Screening for *hcf* mutants demonstrates the usefulness of this material for finding special types of mutants. The segregating M_2 material is already on hand; only the proper screening technique for each special type needs to be developed.

The large collection of mutants obtained provides a means for reexamining the question of auxotrophic mutants in higher plants.

ACKNOWLEDGMENT

Technical assistance from Karen A. Sheridan and Evelyn Bendbow is gratefully acknowledged.

REFERENCES

Anderson, E. G. 1948. On the frequency and transmitted chromosome alterations and gene mutations induced by atomic bomb radiations in maize. *Proc. Nat. Acad. Sci. U. S. A.* **34**: 386–390.

Anderson, E. G., A. E. Longley, C. H. Li, and K. L. Retherford. 1949. Heriditary effects produced in maize by radiations from the Bikini atomic bomb I. Studies on seedlings and pollen of the exposed generation. *Genetics* **34:** 639–646.

Coe, E. H., Jr. 1966. Liquid media suitable for suspending maize pollen before pollination. *Proc. Missouri Acad. Sci.* **3:** 7–8.

Gavazzi, G., Milvia Nava-Racchi, and Chiava Tonelli. 1975. A mutation causing proline requirement in *Zea mays. Theor. Appl. Genet.* **46:** 339–345.

Li, S. L., and G. P. Redei. 1968. Thiamine mutants of the crucifer, *Arabidopsis. Biochem. Genet.* **3:** 163–176.

McClintock, B. 1951. Chromosome organization and genic expression. *Cold Spring Harbor Symp. Quant. Biol.* **16:** 13–47.

Miles, C. D. 1975. Genetic analysis of photosynthesis. *Stadler Symp.* **7:** 135–154.

Nelson, O. E., Jr. and B. Burr. 1973. Biochemical genetics of higher plants. *Annu. Rev. Plant Physiol.* **24:** 493–518.

Neuffer, M. G. 1972. *In vitro* germination of pollen as a measure of effectiveness of chemical mutagens in maize. *Agron. Abstr.* (1972): 36.

Neuffer, M. G. 1974. Absence of auxotrophic mutants in corn and other eukaryotes. *Maize Genet. Coop. Newsl.* **48:** 118–120.

Neuffer, M. G., L. Jones, and M. S. Zuber. 1968. *The mutants of maize.* Madison, Wisc.: Crop Science Society of America.

Redei, G. P. 1974a. Economy in mutation experiments. *Z. Pflanzenzuchtg.* **73:** 87–96.

Redei, G. P. 1974b. Induction of auxotrophic mutations in plants. In *Genetic manipulations with plant materials,* L. Ledous (ed.). London: Plenum, pp. 329–350.

Stadler, L. J. 1946. Spontaneous mutation of the *R* locus in maize. I: The aleurone and plant color effects. *Genetics* **31:** 377–394.

Steffensen, D. M. 1968. A reconstruction of cell development in the shoot apex of maize. *Am. J. Bot.* **55:** 354–369.

Chapter 38

CONTROLLING ELEMENTS: THE INDUCTION OF MUTABILITY AT THE *A*2 AND C LOCI IN MAIZE

Peter A. Peterson

Agronomy Department,
Iowa State University,
Ames

Controlling elements in maize have focused attention on a series of phenomena that have elicited interest from diverse biological disciplines. That a locus can be suppressed and regulated in activity by genetically identifiable elements has been well documented in several laboratories by independent investigators working with different materials (McClintock, 1951a; Brink and Nilan, 1952; Peterson, 1953, 1960; Nuffer, 1961; Doerschug, 1973).

Controlling elements apply to units, often detected in changes in their chromosome position (transposition), that can be identified and observed only by their control of functioning genes. Generally, and there are exceptions (Peterson, 1964), the controlled gene is in a functionless condition, and the regular or irregular change in activity represents the excision of an element

Journal Paper No. J-8297 of the Iowa Agriculture and Home Economics Experiment Station, Ames, Iowa 50011. Project No. 1884. Support from NSF GB38328.

from its locus-suppression position to a position apart from the locus, which fully restores the function of the controlled gene. The gene events to be discussed in this chapter are generally observed in the endosperm, primarily due to the ease of observation of the expression of distinct genes: for example, from the phenotype colorless to colored that relates to the change involved in $a \rightarrow A$. The same events take place in the germ line; however, only early events that affect the germ cells have a chance of being transmitted for additional tests. The critical experiments that led to the identification of these events are discussed in the following paragraphs.

ESTABLISHMENT OF THE CONCEPT OF CONTROLLING ELEMENTS

In McClintock's (1951a) early studies of controlling-element phenomena that followed an unexpected appearance of large numbers of mutable loci, several features of controlling elements were established. Some of the key crosses and analyses of progeny that led to the identity of these features are outlined here.

Her first studies with the *Ac–Ds* (see Table 1 for definition of terms) system were accompanied by chromosome-breakage events observed as variegated sec-

TABLE 1.   Designation of symbols and terms (applicable to this chapter)

Allele or element	Definition and function
I	Inhibitor: controlling element associated with the locus; suppresses locus activity until changed or removed by *En* (not to be confused with *I* of *CI*, a dominant color inhibiting allele at the c locus) (Peterson, 1960).
En	Enhancer: a regulatory element necessary for mutability: can be located at the *a* locus or at a position independent of *a*; exists in several states, *En*-flow and *En*-crown, (Peterson, 1966a) (with *a–m*1, *En* acts similarly to *Spm* as a bifunctional unit; *Sp* suppresses the action of the *a–m*1 allele until *m* causes a mutation event to occur) (McClintock, 1956b); *En-sw* allele shows strong suppression, weak *m*utability with a^{m-1} (Figure 12*b*).
Ds	Dissociator: a controlling element that acts like *I* but is triggered only by *Ac* (Activator).
a–m	A mutable allele, in this case, for *a* (as contrasted at *c–m*, *a2–m*, etc.).

TABLE 1.   (*Continued*)

Allele or element	Definition and function
*a–m*1	Functional, suppressible allele: one of the alleles at the *a* locus that is colored (nonvariegated) in the absence of *Spm* and variegated (colored spots on a colorless background) in the presence of *Spm* (McClintock, 1956b), represents a tester for the presence of *Spm* and *En*.
a–m(r)	A colorless allele that responds to *En*, with *En*, gives purple colored sectors on a colorless background in the aleurone tissue of the kernel.
a2–m(r)	Same as *a–m(r)* except at the *a2* locus.
c–m(r)	Same as *a–m(r)* except at the *c* locus.
a–m(nr)	The non-responsive (*nr*) type; these are found at each of the mutable sites and are identified as colorless or pale derivatives of mutable alleles.
Chromosome 9	loci affecting the endosperm: *C* versus *c*-color versus noncolor; *Sh* versus *sh*-round versus shrunken; *Wx* versus *wx*-purple versus red staining *Bz* versus *bz*-purple versus bronze aleurone color.
Chromosome 5	loci affecting the endosperm: *A2* versus *a2*—color versus noncolor; *Bt* versus *bt*-round versus brittle-shrunken.
Controlling elements:	elements such as *Ds* and *I* that suppress gene function when in *cis* position to the locus; also *receptor elements* (Fincham and Sastry, 1974).
Regulatory elements:	elements such as *Ac*, *Spm*, and *En* that trigger controlling elements, such as *Ds* and *I*, so that the locus under control becomes functional or, as in the case of *a–m*1, is activated to suppress locus activity.

tors in the endosperm (Figure 1) and were unlike previous cases of bridge-breakage–fusion (B–B–F) cycles initiated by crossing over at meiosis between the short arms of chromosome 9, one of which had an inverted terminal duplication. (*Variegation* refers to a mosaic of phenotypes in a tissue and appears following the loss of the various dominant genes, which allows expression of the other alleles. For example, the intermittent loss of the dominant *C*I by a breakage event allows expression of the color allele, *C*, which leads to the variegated pattern.) These newly arisen B–B–F type events exposed a pattern of events expressed as a variegated endosperm that McClintock (1938, 1941) had previously described. In this instance, however, by following the pattern of events of the several genes under observation, she

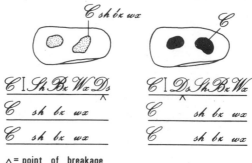

∧ = point of breakage

Figure 1. Diagramatic representation of observed phenotypes on kernels following *Ac*-induced breaks at *Ds* at the positions indicated. The break in the marked chromosomes of the kernel on the left results in losses that lead to the phenotypic expression of the C *sh bz wx* phenotypes. On the right, the change in the position of *Ds* adjacent to *Cl* results in a different expression, namely, that of C. Each of the kernels arose from the cross of C *sh bz wx* with no *Ds* in chromosome 9 by plants (male) having *Cl Sh Bz Wx* and *Ds* at the positions indicated and *Ac* (McClintock, 1951a).

could determine that there was a definite time and frequency of breakage events and that the site of chromosome breakage was not constant, as would be expected with B–B–F cycles initiated by a cross-over event in a terminal inverted duplication. Two results were recorded. There is a preferred site of chromosome breakage *apart* from the inverted terminal duplication that induced the breaks (McClintock, 1949, 1950). This was identified as the site where the chromosome preferentially dissociated. The unit responsible was called *Dissociator (Ds)*. Further, and quite surprisingly, *Ds* could move from one site on the chromosome to another and its location could be mapped. Detection of the position and movement (transposition) of *Ds* by using a well-marked chromosome is illustrated in Figure 1. The change of the position of *Ds* from a point proximal to *Wx* to a point proximal to *Cl* could be detected by the different phenotypes. Finally, it was shown that *Ds* breaks occurred at definite times and in definite frequencies based on the heritability of the observed variegated patterns.

   The *Ds*-induced breakage events were associated with marked loci *Cl*, *Sh*, *Bz*, and *Wx* on chromosome 9. The *Cl* is a dominant color-suppressor, and in crosses with *C,* breakage events leading to the loss of *Cl* would produce a varied series of color patterns (for typical examples, some of the *En* related patterns are illustrated in Figure 8).

   In some crosses, such as *C sh bz wx/C sh bz wx* × *Cl Sh Bz Wx/C sh bz wx,* half the progeny showed the variegated pattern, and the other half were fully colorless, without the expected colored patterns on a colorless background (Figure 2, progeny type 2). When these colorless kernels, *Cl Sh Bz Wx/C sh*

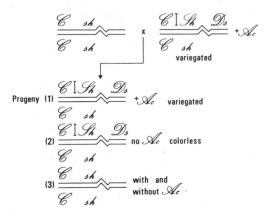

Figure 2. A type of cross that demonstrates a two-unit system of controlling–regulatory elements. From the intercross of the progeny type shown (described in text), it could be demonstrated that a separable factor Activator, *Ac,* triggers *Ds* breaks.

bz wx, were crossed by the colored, shrunken sibs (Figure 2,.progeny type-3), some of the crosses [e.g., *CI Sh Ds/C sh* no *Ac* (colorless) × *C sh/C sh, Ac*] revealed progeny with colored spots on a colorless background, the result of breakage events (Figure 3). Now it could be shown that *Ds,* quiescent as revealed by the selected colorless kernels among the progeny of this cross, could be "activated" by an independently assorting dominant factor to reveal breaks. The factor that "activated" the breaks at *Ds* was called *Activator* (*Ac*) (McClintock, 1947). Thus a two-unit interacting system was established. Within the context of this discussion, a locus-situated element such as *Ds* is termed the *controlling element,* and the element such as *Ac* that induces the breaks at the controlling element, the *regulatory element* (see Figure 5). It

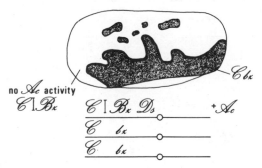

Figure 3. Diagramatic representation of the phenotypes expressed following *Ds* breaks. The female parent was *C bz,* and the male parent, *CI Bz Ds* and *Ac.* The *C bz* phenotype appears following *Ds* break induced by *Ac.*

should be recalled that *Ds* breaks are induced by *Ac*. This results in two free chromatid ends that fuse to give an acentric and a dicentric fragment. The bridge that appears in the subsequent division again results in fused ends, repeating the cycle. If appropriate genetic markers on the *Ds*-containing chromosome are present, a mosaic of cells (variegation) will appear following these events.

The next revealing observation concerned the origin of an unstable allele from the transposition of *Ds*. It originated in the following way. When *yg c sh wx/yg c sh wx* is used as a female in crosses with the male *Yg C Sh wx/Yg C Sh wx Ds Ac/+*, the expected progeny if *Ac* is present should include kernels with colorless areas on a colored background. (In *Yg C Sh Wx Ds/yg c sh wx*, the *Ds* breakage and subsequent loss of the arm leads to colorless, *c*, spots on a colored background, *C*, Figure 4). An exceptional kernel had the reverse pattern, with colored spots on a colorless background (Figure 4). This indicates a change from *c* to *C*. Reconstruction of the events that lead to the origin of *c-m*-1 indicates that *Ds* moved from its original position and became inserted close to the *C* allele. Associated with this change in position of *Ds* is a change in the functioning of the *C* allele, which now expresses itself as recessive *c*. This allele, designated *c-m*-1, is colorless until an *Ac* is introduced which causes mutability, colored areas on a colorless background. Thus the new *c-m*-1 allele is a *C* allele whose activity is inhibited by the adjacent *Ds* locus. The loss of *Ds* from the *C Ds* complex restores *C* activity. Dissociator activity could not be separated from the *c* → *C* behavior, which suggests that these new events at the *c* locus can be identified with the presence of *Ds* and respond to *Ac* in the same manner as the *Ds* that induced chromosome breaks. Thus in the isolation of a changed *c* associated with *Ds* behavior and subsequent control of *Ac*, it was established that the component associated with breaks at preferred sites

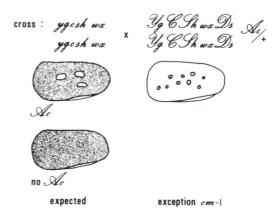

Figure 4. Diagramatic representation of the expected kernel types and of the origin of the exceptional c-*m*1 allele from the cross indicated.

(namely, associated with *Ds*) is now associated with the control of a locus that responds to signals from the second unit, the *Activator* (*Ac*). In this way, McClintock related the mechanism responsible for mutations at mutable loci as identical to the *Ds–Ac* interactions occurring during endosperm divisions. Although *Ds* was involved with breaks during mitotic divisions and the activities associated with mutable loci, *Ds* at the mutable loci did not exhibit breaks due to a change in the "state" of *Ds*. In addition, other independent instances of *Ds–Ac* type of mutability (such as that responsible for the origin of *bz m*-1 and *bz m*-2) appeared at loci that were originally dominant and functional. This lends further support to evidence for transposition of these elements to new locations. Furthermore, regulation by *Ac* of events at *c* and at *bz* is strikingly *nonrandom*. The uniformity of the events is very revealing, and McClintock (1951a,b) suggested that this represented a form of control of development on which she elaborated in later studies (McClintock, 1956a–c, 1965a,b, 1967a,b). It was in the context of these observations that controlling elements were identified and established in the control and regulation of gene activity.

If it were not for the discovery of *Ds* chromosome breakage, the transposition of the elements, and the association of the same breakage events with *c–m,* and if the McClintock studies had begun with the discovery of *c–m,* the conclusions probably would not have been very different from Rhoades's studies (1938, 1941) with *dotted* (*Dt*) elements, where a colorless allele at the *a* locus responds specifically to an independently located *Dt* element. In the Rhoades case, a specific allele is activated by *Dt* situated elsewhere in the genome. It was the control of a battery of genes by a single regulatory element that was significant and the identity of a receptor element, namely *Ds,* as well as the "burst" of new mutable genes among the progeny of crosses involving B–B–F events. The regularity of this control and the universality of the control of gene function encouraged McClintock to relate these observations to aspects of development.

In the 25 years since her early reports (McClintock, 1948, 1949, 1950a,b), additions have been made to the information about controlling elements. These are reviewed. Phenomena associated with the general topic include problems of origin, transposition, induction of new sites of mutability, relation of the elements to chromosome linearity, the state of the element(s), and the origin of novel variation.

ORIGIN OF CONTROLLING ELEMENTS[1]

Although the initial observations came from studies of the induction of controlling-element activity by B–B–F cycles (McClintock, 1947, 1948, 1951a,b)

[1] See Table 1 for definition of terms.

and although additional studies (Bianchi, Salamini, et al., 1969; Doerschug, 1973; Orton and Brink, 1966) have supported the induction mechanism, genetic materials subsequently identified as controlling elements (Emerson, 1917; Rhoades, 1936) were observed in materials from natural populations. The element *En*, although found in a treated population of maize exposed during the Bikini A-Bomb tests (Peterson, 1953), as well as in a genetic nursery population after B–B–F cycles [as *Spm* (McClintock, 1953, 1956a–c)—*En* can be considered similar to *Spm* (Peterson, 1965)], has been found by Gonella and Peterson (1975a) in numerous races of maize from naturally occurring populations. Thus the major controlling element systems such as *Mp* that can be considered as identical to *Ac, En–Spm,* and *Dt* (Nuffer, 1955) have been found in natural indigenous populations. An additional system has been identified by Gonella and Peterson (1975b) in a maize race associated with the Cuna Indian tribe of Colombia. Thus it can be said that controlling elements are pervasive among a wide assortment of cultivated and natural populations from diverse geographical origins, even though the *Ac–Ds* system was uncovered in events following the B–B–F cycle. Whether B–B–F can cause the *de novo* induction of elements cannot be definitely established despite the induction of *Dt*-like elements in stocks originally devoid of such elements (McClintock, 1950a; Doerschug, 1973). Fincham and Sastry (1974) suggested that the elements are present, but in a quiescent state or hidden in heterochromatin and thus unable to function before the stresses and relocation of materials associated with B–B–F or mutagenic agents (Neuffer, 1966). Once activated in a strain of maize, however, the transposition of elements provides the possibility of control of every gene that has been tested.

SYSTEMS OF CONTROLLING–REGULATORY-ELEMENT INTERACTIONS

A controlling-element system is defined by a specific two-unit interaction between the *cis*-located controlling element, the receptor of signals, and a regulatory element where signals originate. Thus far, three controlling-element systems have been described, and the controlling and regulatory elements, respectively, are described as *Ds–Ac, a–dt–Dt,* and *I–En* or *Spm* (Peterson 1965). It is possible to establish the identity and separateness of the elements of a system only when a responding receptor (the controlling element) is associated with a locus whose function is phenotypically impaired (Fig. 5). Thus the availability of a *Ds* stock without *Ac* provides a test for regulatory elements. This type of test established that *Mp* is indistinguishable from *Ac* in its effect on *Ds* in causing chromosome breakage (Barclay and Brink 1954), (Figure 5). Furthermore, in the reciprocal test, *Ac* can simulate the inde-

	Receptor elements			
Regulatory elements	Ds	I	a–m1	a–dt
Ac	+	−	−	−
Mp	+	−	−	−
En	−	+	+	−
Spm	−	(?)	+	−
Dt	−	−	−	+
F–cu	−	−	−	−
Unknown	( )	( )	( )	( )

Figure 5. A scheme of controlling–regulatory element interaction that describes the specificity between controlling and regulatory elements. In the case of the a–m1 response to *En* or *Spm,* the pigmentation in kernels is suppressed until a mutation event takes place. The *F–cu* system (Gonella and Peterson, 1978).

pendently located *Mp* effect in causing the light-variegated phenotype. Curiously, however, a colorless, nonpericarp-striping *P* allele that is responsive to *Mp,* just as *Ds* is responsive to *Ac,* has not been isolated. Brink and Barclay (1954) have ascribed to *Mp* the properties of both *Ds* and *Ac*; it is a *cis*-located element rendering the *P* locus unstable and acting as *Ac* on breakage events at *Ds.* It shows dosage effects as well.

This disparity between the functions of the controlling and regulatory elements has been discussed previously (Peterson, 1968, 1970a,b). In the case of the origin of *a2–m* 1 1511, the unstable condition arises with the insertion of *En* at the locus, in this instance, *a2.* This can be considered as an autonomous, one-element system. It will be considered as a one-element system until a derivative allele, receptive to a regulatory element has been isolated. In some cases, the insertion of *En* at the locus results in only autonomous behavior (Table 2) on *a2–m* 4 1596. No responsive allele has been isolated thus far among many tested colorless derivatives (Table 2, column 4) of *a2–m* 4 1596. In every case, in evidence to be presented within this report, the instability was initiated by the insertion of the regulatory element, *En,* at the locus in question which produced a varied series of derivative alleles. In the case of mutable pericarp, no such responsive allele has been reported and thus it is considered an autonomous allele. Smith's (1960) description of mutable luteus also fits this autonomous allele pattern. With a responsive allele, however, tests can be made to show whether a relationship exists among the systems such as was done with *Mp* in triggering *Ds* chromosome breakage.

When new mutable alleles arise, it is necessary to test them against the known responding alleles specific for certain regulatory elements (Peterson,

TABLE 2. The origin of an unstable allele at the *a2* locus *a2–m* (4 1596) following the insertion of *En*. The progeny frequency of a number of tests following the cross $\dfrac{a2m(coarse)\ Bt}{a2\ bt} \times \dfrac{a2\ bt}{a2\ bt}$

Ear #	Round				Brittle		Percent colorless total round
	Colored (1)	Coarse (2)	Fine (3)	Colorless (4)	Coarse (5)	Colorless (6)	(7)
−1	8	78	1	66	6	148	43.1
−2	3	116	8	35	6	166	21.6
−3	1	82	0	22	4	125	21.0
−4	2	37	0	24	0	60	38.1
−5	5	31	4	44	2	70	52.4

w m A NEWLY-ORIGINATED UNSTABLE GENE
GREEN STRIPES ON WHITE LEAVES

ORIGINAL CROSS: *w m* X AN *a En* TESTER: PALE-COLORED, ROUND

$$w\ m\ w\ m,\ A\ A \times W\ W,\ a\ m_{-1}\ sh_2/a\text{-}m(l)\ Sh_2$$

F1 X RECESSIVE TESTER

GREEN-PLANTS, X GREEN-PLANTS,
COLORED-SEED COLORLESS, SHRUNKEN-SEED

$$W/w\text{-}m,\ A/a\text{-}m_{-1}\ sh_2\ \ X\ \ W/W,\ a\ sh_2/a\ sh_2$$
$$a\text{-}m(l)\ Sh_2$$

IF MUTABILITY APPEARS, *En* IS PRESENT IN THE *w-m* STOCK BUT A RELATIONSHIP IS NOT ESTABLISHED

SELECT *a* PROGENY   1. MUTABLE ROUND AND SHRUNKEN
    2. NON-MUTABLE ROUND,-COLORLESS AND PALE COLORED SHRUNKEN,-PALE COLORED

SELF THE *a* SELECTIONS

HYPOTHESIS: IF *w-m-En*, *w m* SHOULD APPEAR ONLY IN 1.
IF *w-m* IS NOT RELATED TO *En* , *w m* SHOULD APPEAR IN 1. AND 2. EQUALLY.

Figure 6. An idealized scheme for establishing the relationship of a newly originated mutable allele to an established controlling-element system.

1966a,b) (Table 5). Thus instances such as Rhoades's *a–m* and Sprague's *a–m* (personal communication) are cases in point. Ideally, tests should be made utilizing tester loci unrelated to the unknown, backcrossing followed by selection of the known, and subsequent testcrossing for the unknown (see Figure 6 for the idealized scheme). A correspondence of the tester loci mutability with the mutability of the known provides necessary proof for establishing the relationship of the unknown to a system.

RELATIONSHIP BETWEEN THE CONTROLLING AND REGULATORY ELEMENTS OF A MUTABLE SYSTEM

In strains of maize containing regulatory elements or autonomous unstable alleles, new mutables arise (McClintock, 1951a,b, 1953, 1958; Peterson 1956, 1963). Usually, in analysis of the newly arisen mutable allele, the regulatory element, in the case of colorless forms, reveals the origin of a responsive allele; for example, $a2–m(r)$ responds to the regulatory element, *En*. The response is in the form of spots representing phenotypic changes from colorless to colored. To recapitulate, what was originally isolated as an autonomous allele or a one-element system subsequently gave rise to a two-element system. Therefore, *A2* could be described as $(A2–En)$ (Peterson, 1968, 1970a). The responsive derivative is colorless and stable in the absence of *En* and can be described as $(A2I)$ (*I*, the responding element). The origin of *I* is now considered. There are two possibilities. The newly arisen unstable allele could have arisen by the insertion of *I* and *En*. Arguments against this possibility have previously been presented (Peterson, 1970a). Briefly, one would expect stable, colorless, yet responding alleles at the *a2* and *c* loci to arise. Of more than 27 colorless forms isolated, no responding alleles could be identified. In every instance, the regulatory element is identified with the newly arisen mutable allele.

The second possibility is that *En* alone could instigate the observed mutability, with the resultant transpositions leaving a residue that results in the responsive condition. The resultant allele responds only to the regulatory element originally located at the locus. It is as if the original "infection" by the regulatory element provided the locus with the responsiveness identified with the new allele. In a series of arguments, Peterson (1970a) favored this second alternative. Of course, it should be noted that *Ds* alone has been responsible for the induction of controlled loci in the presence of *Ac* (McClintock, 1951a).

TRANSPOSITION

Transposition of controlling and regulatory elements in a chromosomal complex involves identification of an element at one location and evidence of relocation to a new position. This was demonstrated for both *Ds* and *Ac* by following the changes in linkage relations of each of the elements with marked loci (McClintock, 1951b). The *Ac* element changed position from one in which it was linked to *C wx* to one in which it was unlinked.

In autonomous one-element systems such as the original *a2–m* 1 1511 (Peterson, 1968) colored spot appearing on a colorless background arises from a transposition event. A transposition event requires excision of the element adjacent to the gene, a process that may leave the locus unimpaired in function, modified in function, or functionless (Fowler and Peterson, 1974). One of the most frequent events at the mutable loci associated with color is the very frequent change to colorless. This occurs with both the *a2* and *a* loci and represents changes to nonfunctioning, nonresponding forms of the locus. These changes are accompanied by transposition of *En* away from the locus (Peterson, 1961, 1968, 1970b).

The most convincing evidence for transposition comes from observation of reciprocal events in twin sectors because one of the pair is the necessary alternative to the other member. This has been developed most thoroughly in the mutable pericarp, *P–rr*, studies, originally described by Emerson (1914, 1917) and subsequently investigated by Brink and his students (Brink, 1954; Brink and Nilan, 1952; Greenblatt, 1974). The element *P–vv* expresses a medium variegated phenotype and mutates to a full red (*P–rr*) and to a lighter variegated form. In subsequent tests of the lighter variegated kernels, medium variegated was recovered. This, along with the correspondence between the appearance of red and light variegated progeny from medium variegated, indicated that the medium variegated, *P–vv*, was caused by suppression of *P–rr* by a conjoined *Mp* (*P–rr–Mp*). Subsequent loss of *Mp* from (*P–rr–Mp*) would release the pigment-producing potential of *P–rr* (the red part of a twin sector), and this, by the assortment of *trMp* with the sister strand *P–rr–Mp*, leads to *P–rr–Mp* + *trMp* (yielding the light sector). Thus the appearance of twin sectors of red and light variegated amidst a background of medium variegated kernels, as well as the nearly equal numbers of red and light derivatives from medium-variegated testcrosses, is interpreted as loss at one position and gain at another position. This evidence provides clear support for the physical movement or transposition of an element from one position to another and negates the possible alternative of deactivation followed by reactivation of elements at diverse chromosome positions.

Analysis of twin sectors (Greenblatt, 1968, 1974) provides evidence to support the hypothesis that *Mp* transposition occurs during chromosome replication (McClintock, 1949, 1956) and that the chromosome strand that contributed the *trMp* is the one newly replicated. Moreover, the site of deposition of *trMp* is the *old strand,* not the replicating one that contributed *trMp*. Greenblatt (1968, 1974) has presented persuasive arguments that support the conclusion that the excision of *Mp* from the replicating strand occurs during the period of replication, yet *Mp* deposition is restricted to a nonreplicating strand.

VARIATION IN PHENOTYPIC DERIVATIVES ARISING FROM THE ACTION OF CONTROLLING ELEMENTS AT VARIOUS LOCI

The most common expressions arising from the action of controlling elements on mutable loci are the changes to full color and to colorless that appear as germinal changes as well as kernel sectors (Peterson, 1970b, 1975). In addition, a series of alleles differentiated by quantitative grades of pigment formation has been identified at the *C* locus (McClintock, 1951a) and at the *a2* locus (Fowler and Peterson, 1974), as well as at the *a* locus (Rhoades, 1941; Peterson, 1970b). Even the full-color derivatives can be differentiated. Some of the colored spots on a colorless background give dark rims at the interface of the colored and colorless areas, whereas others do not (McClintock, 1951a, 1952; Peterson, 1966a,b). An approach to the mechanism that generates these varied alleles may lie in intracistron analyses of a number of these alleles arising from diverse mutables, for the purpose of locating the intracistron site of controlling element deposition.

MATERIALS AND METHODS

Gene symbols and terms are listed and described in Table 1.

Induction of Controlling–regulatory–element Instability

Development of Seed Stocks for Use in Isolation Plots.   In isolation plot experiments, plants that develop uniformly and do not tiller provide the best insurance against pollen contamination. Stocks known to contain *En* were crossed to B-chromosome and Abnormal-10 bearing chromosome stocks and then selfed to increase the *En* content of the stocks. Subsequently, these *En*-containing stocks were crossed 2–3 times to a W22-color converted line that provided excellent vigor, increased uniformity, and depressed tiller formation.

Several *En* sources were used in this study: (a) *En* from stocks with various *a–m* alleles (varied states of an autonomous mutable) (Peterson, 1961) or (b) from stocks with *En* on a chromosome independent of *A*. Genotypes of stocks used in isolation plots with respect to the key alleles were *A2A2 CC RR A A* or *A2A2 CC RR A a–*. In addition, each of the stocks possessed the dominant *Bt, Sh Wx* alleles in homozygous condition.

Isolation Plot Procedure. Isolation plots small enough to accommodate 6000–7000 plants were grown with two ear-bearing rows alternating with one pollinator row. In the *c* plots the pollen parent was homozygous recessive for *c sh wx y.* In each of the plots, female rows were hand-detasseled, and open silks were pollinated with pollen from adjacent pollinator rows. Detasseling was continued daily during the tasseling period.

Detection of Mutable Alleles. Each of the harvested ears was examined, and exceptional variegated or colorless kernels on otherwise fully purple ears were selected. Remnant seed of these exceptional bearing ears was saved for further tests.

First, in a test for *En,* the presence of *En* or its involvement in certain reactions is identified by the use of *responsive* alleles of the tester locus such as *a–m(r), a2–m(r),* and *c–m(r).* Each of these alleles when used as a tester expresses a pattern(s) of spots on kernels with a colorless background. Certain *En* states exist that elicit a different phenotype pattern against a standard receptive element (Figure 9*a*). The *a–m1* and *a2–m1* alleles also are testers for *En.* In this case, however, the kernel is colored (various grades of paleness to full color—"A" 2 state), and the response to *En* is expressed by suppression of color and mutability, which results in spots of color on a colorless background (Figure 9*b*).

Second, in a test for *a–m(r)* allele, colorless or pale alleles frequently appear among the progeny of crosses. Confirmation of their composition can be obtained by crossing *a2–m(r)* or *a–m(r),* depending on the locus being investigated. If *En* is present, the original isolates are considered to be nonresponding. If *En* is not present, an additional test is needed. If crossing to an *En* tester, *a2 bt En,* also gives negative results, it can be concluded that nonresponding (*nr*) types have arisen.

RESULTS

Induction of Controlling-element Instability at Selected Loci

Two loci, *a2* and *c,* were selected as targets for the induction of controlling-element instability in isolation plot tests. These tests were conducted over a 7-year period beginning in 1960. In the examination of each of the harvested ears, each kernel showing exceptional behavior was isolated, the ear source was identified, and the sib seed was saved for confirmation tests of each of the isolates. Each isolate selected was tested in numerous ways. From such tests 72 exceptions were confirmed in an examination of approximately 14,908,154 gametes. The confirmed *a2* and *c* mutables, their parental source, the number of original isolates, and the rate of occurrence are given in Table 3.

TABLE 3. Listing of newly originated a2 and c mutable alleles in isolation plots between 1960 and 1966, the number of original isolates, the number of confirmed mutables, and the rate of occurrence

a2–m or c–m		No. of gametes tested	No. of isolates	No. of confirmed mutants	Rate × 10⁻⁶
a2–m	1 1511[b]	646,760	102	2	3.1
a2–m	1 1524				
c–m	1 1702	672,980	123	1	1.5
c–m	2 3132[a]		122	1	
c–m	3 1529	1,233,490	512	4	3.2
c–m	3 1766				
c–m	3 2126				
c–m	3 2129				
a2–m	4 1533	1,664,880	237	7	4.2
a2–m	4 1596				
a2–m	4 1626				
a2–m	4 1629				
a2–m	4 1643				
a2–m	4 1667				
a2–m	4 1676				
c–m	4 1763	2,158,800	357	11	5.1
c–m	4 1779				
c–m	4 1851				
c–m	4 1863				
c–m	4 1899				
c–m	4 1905				
c–m	4 1936				
c–m	4 1963				
c–m	4 2097				
c–m	4 2132				
c–m	4 2132				
c–m	4 2134				
a2–m	5 5040	2,452,800	194	8	3.3
a2–m	5 5059				
a2–m	5 5064				
a2–m	5 5160				
a2–m	5 5169				
a2–m	5 5173				
a2–m	5 5179				
a2–m	5 5194				
c–m	5 5196	2,455,824	410	23	9.4
c–m	5 5207				
c–m	5 5208				
c–m	5 5232				

TABLE 3. (*Continued*)

a2-m or c-m		No. of gametes tested	No. of isolates	No. of confirmed mutants	Rate × 10⁻⁶
c-m	5 5277				
c-m	5 5292				
c-m	5 5301				
c-m	5 5305				
c-m	5 5320				
c-m	5 5351				
c-m	5 5364				
c-m	5 5368				
c-m	5 5370				
c-m	5 5374				
c-m	5 5378				
c-m	5 5386				
c-m	5 5398				
c-m	5 5423				
c-m	5 5437				
c-m	5 5453				
c-m	5 5482				
c-m	5 5483				
c-m	5 5607				
a2-m	6 8080	1,527,384	421	6	3.9
a2-m	6 8140				
a2-m	6 8144				
a2-m	6 8291				
a2-m	6 8379				
a2-m	6 8410				
a2-m	6 8487				
c-m	6 8602	1,647,512	279	7	4.3
c-m	6 8620				
c-m	6 8648				
c-m	6 8655				
c-m	6 8663				
a2-m	7 8018	447,724	230	2	4.5
a2-m	7 8141				
Total		14,908,154		72	

^{a} Count not available.

^{b} The numbers identifying each mutable allele refer to the originating family where confirmed.

The rates of origin of mutable alleles for the different years varied from 1.5 to 9.4×10^{-6}. Variation in rates of origin is discussed in a later section.

The several tests conducted for each of the 72 newly arisen mutables include the assays described in the following paragraphs.

Confirmation of the Newly Arisen Mutable Loci.   Confirmation of a new mutable is established by a testcross with the pollen tester parent *a2 bt/a2 bt* or *c sh wx/c sh wx*. The appearance of these markers among the testcross progeny confirms the origin of the exception. Further confirmation is established by testing the sib seed (remnant) appearing on the ear bearing the exception. For each of the exceptions, colored sibs from remnant seed are tested. For the *a2* tests the appearance of *A2 Bt/a2 bt* and, for the *c* tests, the appearance of *C Sh Wx/c sh wx* among the testcross progeny, provide unequivocal proof that each plant giving rise to ears bearing exceptions is

$$\frac{A2\ Bt}{A2\ Bt} \quad \text{or} \quad \frac{C\ Sh\ Wx}{C\ Sh\ Wx}$$

Each of the 72 exceptions isolated arose from an ear that was *A2A2* or *CC*. This strongly supports the supposition that each of the exceptions arose from an allele that changed from *A2* to *a2–m* or *C* to *c–m*.

The *a* Genotype.   For each confirmed *a2* or *c* new mutable a sample of the remnant seed, as well as progeny of the exception, was tested for the *a* content by crossing with *a–dt sh2*, an *a* tester. Among the progeny, either all the resulting ears were fully colored or half were fully-colored. The former would indicate that the originating genotype was *A A* and, the latter, *A a*. Among the latter, the *a* allele was *a–m* or *a–s*. Among the 72 exceptions, 54 were *A A* and the remainder, *A/a–m* or *A/a–s*. This indicates that a large majority of the newly arisen *a2* and *c* mutable alleles arose in the absence of an *a–m* or any other known mutable allele.

Tests for *En*.   The *En* from the ear bearing the exception was tested by crossing both the remnant sibs and the individual selected exceptions by a specific *a–m(r)*, an allele responsive to *En* that serves as a standard for determining the mutability pattern induced by *En*. In the cases where the originating ear was *A A*, the resulting heterozygote *A/a–m(r)* was testcrossed by *a–dt sh2/a–dt sh2*. The resulting segregation of mutable versus nonmutable (spotted versus colorless; Table 4) among the noncolored progeny provided information on the number of *En* in the originating stock. Examples of supporting data on the determination of the number of *En* in the remnant sibs of newly arisen mutables are given in Table 4.

TABLE 4.   The progeny arising from several crosses of the F_1s of remnant sibs of originating mutable alleles with an *En* tester, *a–m(r)/a–m(r)*, testcrossed by *a sh2/a sh2* and summaries of the progenies indicating the number of *En* in the ears in which the exceptions arose

A.   *Progeny of F_1s A/a–m(r) (En?)* × *a sh2/a sh2:*

	F_1 ear #	Colored	(1) Spotted	(2) Colorless	(3) Total	% 2/3	No. En
c–m 5 5207	−1	no count	112	13	125	10.4	3
	−2	no count	78	15	93	16.1	3
	−3	no count	84	112	196	57.1	1
a2–m 5 5059	−1	no count	77	31	108	28.7	2
	−2	no count	95	12	107	11.2	3
	−3	no count	95	107	202	53.0	1
	−4	no count	41	50	91	54.9	1
a2–m 5 5040	−1	no count	126	63	189	33.3	2
	−2	no count	87	73	160	45.6	1
	−3	no count	74	65	139	46.8	1
	−4	no count	138	18	156	11.5	3

B.   Summary of tests for the number of *En* among segregating progeny of remnant sibs of newly originating mutable alleles.

	# of En			
	0	1	2	3
a2–m 5 5059	3	6	12	2
a2–m 5 5040	1	10	12	1
c–m 5 5207	1	3	8	4

Relationship to the *En* System.   Simultaneously with the tests for the presence of *En* in the remnant sibs, crosses were made between each of the isolated, newly arisen, mutables, *a2–m* or *c–m*, and the *En* tester stocks, *a–m(r)/a–m(r)* or *a–m(r)/a–m1*. Where the originating ear was *A A*, the colored F_1s were testcrossed by *a sh2/a sh2*. The resulting *a* mutable and nonmutable progeny were collected and tested separately by *a2 bt* or *c sh wx*, depending on the allele being tested. If, for example, in the case of *c–m* a correspondence exists between the mutability of the newly originated *c*-mutable allele and the *En* system, the testcrosses by *c sh wx/c sh wx* would yield *c*-mutability when *a*-mutability is tested and *c*-nonmutability when *a*-nonmutability is tested. Examples of these crosses and the resulting progeny types are outlined in Table 5 for *c–m* 5 5277 and *c–m* 5 5292. The same series of crosses was made to confirm the *a2–m* relationship to *En*.

When mutable *a* selections are testcrossed by *c sh wx/c sh wx*, the progeny are mutable or indeterminate. The indeterminate class includes the *CC* types that would be associated with an independently segregating *En* from the cross *A/a-m(r) C/c-m* × *c sh wx/c sh wx*. In contrast, when nonmutable kernels are selected, the *c* tests indicate nonmutability. When exceptions to this correla-

TABLE 5.   Relationship to the *En* system: genotypes, crosses, and progeny of a newly derived mutable allele crossed to appropriate testers

The originating mutable allele c-m or a2-m × *an En—responding tester CC a-m(r)/a-m1*

   example   *c-m/c sh wx A/A* × *C/C a-m(r)/a-m1 sh2*
                  colored round × pale round

$$\frac{C/c\text{-}m^a}{c\ sh\ wx}\ \frac{A/a\text{-}m(r)}{a\text{-}m1\ sh2} \times a\ sh2/a\ sh2\ CC$$

                                    a selections

Select and separate   ⎧ mutable; variegated (round and shrunken)
                                    ⎨ not mutable:
Test *a* selections × *c sh wx*   ⎩ pale (round and shrunken) colorless-round

The following typical results are obtained. (In each case mutable vs. nonmutable kernel separation from 10–12 ears was made).

	Ear	*a* Selection	Mutable	No mutability	Indeterminate
				× *c sh wx/c sh wx*	
c-m 5 5277					
	−1	mutable	12	0	4
		not mutable	0	9	5
	−2	mutable	12	0	15
		not mutable	0	0	20
	−3	mutable	19	0	24
		not mutable	0	10	12
	−4	mutable	12	0	16
		not mutable	0	12	16
c-m 5 5292					
	−1	mutable	22	5	6
		not mutable	0	3	16
	−2	mutable	28	0	3
		not mutable	0	1	13
	−3	mutable	7	0	12
		not mutable	1	0	6
	−4	mutable	27	0	9
		not mutable	0	0	10

[a] Identified and used the *C/c(m)* heterozygote by a simultaneous cross of each plant to *c sh wx/c sh wx*.

TABLE 6. Rate of *En* controlled mutable alleles arising at the *a2* and *c* locus in separate tests from a single *En*-containing seed source. Parental source, derivative alleles, and rate of occurrence (see Materials and methods section for development of the populations)

1965 Mutants:

Parental source[a]	a2-m	No. of gametes tested	No. of confirmed mutants	Rates $\times 10^{-6}$	c-m	No. of gametes tested	No. of confirmed mutants	Rates $\times 10^{-6}$
3 1098 × 1095		247,632	0	0	5 5196	270,144	3	10
$\frac{A/a\text{-}m\ sh}{A}$ × W22-Cl					5 5207 5 5208			
3 1102 × 828		229,152	0	0	5 5232	284,928	1	3.5
$\frac{a\text{-}m/a\text{-}m}{A}$ × W22-Cl								
3 1104 × 828	5 5040	344,736	1	2.9	5 5277	358,176	4	11.1
$\frac{a\text{-}m(pa\text{-}pu)/a\ sh}{A}$ × W22-Cl					5 5292 5 5301 5 5305			
3 1106 × 828	5 5059	384,720	1	2.6	5 5320	415,968	4	9.6
$\frac{p\text{-}m/a\ sh}{A}$ × W22-Cl					5 5351 5 5364 5 5607			
3 1118 × 828	5 5064	274,848	1	3.6	5 5368	367,584	4	10.8
$\frac{a\ sh/a\ sh\ 3\text{-}4\ En}{A}$ × W22-Cl					5 5370 5 5374 5 5378			

Cross	Culture	Kernels	Mutants	Rate	Culture	Kernels	Mutants	Rate
3 1101 × 1095 $a\text{-}m/A \times \dfrac{A/a\text{-}m\ sh}{A}$		402,192	0	0	5 5386 5 5398 5 5423 5 5437	353,136	4	11.3
3 1100 × 1095 $a\text{-}m/A \times \dfrac{A/a\text{-}m\ sh}{A}$	5 5160 5 5169 5 5173 5 5179	450,576	4	8.9	5 5453 5 5482 5 5483	310,464	3	9.7
3 1111 × 1095 $\dfrac{A/a\ sh}{a\text{-}m}\,2B \times \dfrac{A/a\text{-}m\ sh}{A}$	5 5194	118,944	1	8.4		89,040	0	0
1966 Mutants:								
4 2214 × 1474	6 8080	298,080	3	10	6 8487	413,292	1	2.4
$\dfrac{A}{A} \times \dfrac{A}{a\ sh}\,2\text{-}3\ En$	6 8140 6 8144							
4 1456-57 × 828 $A/a\text{-}m \times$ W22-Cl		457,056	0			502,172	0	
4 1472 × 828	6 8291	361,560	1	2.8	6 8602	502,172	6	11.9
$\dfrac{a\text{-}m(pa\text{-}pu)}{a\text{-}m(pa\text{-}pu)}/A \times$ W22-Cl					6 8613 6 8620 6 8648 6 8655 6 8663			
4 1464,65 × 828 $a\text{-}m/A \times$ W22-Cl	6 8379	410,688	2	4.9		229,876	0	

[a] sh refers to shrunken-2, chromosome 3.

tion occur, such as the single nonmutable *a* selection that showed *En* in the *c–m* 5 5292 tests (-3 of Table 5), and are further tested, it becomes evident that the original *a* selection was misclassified. A number of the *a* mutable selections that did not show mutability in the *c* tests such as −1 of *c–m* 5 5292 are confirmed to carry *En* and these are considered to be *c–m(nr)* types that arise frequently among some of the pattern types (Peterson, 1970b).

This correlation between the presence of *En* as determined by the standard *a–m(r)* selection and the appearance of the *a2* or *c* mutability establishes the relationship between the new mutable alleles and the *En* system. These results, coupled with data presented in previous sections, establish that *En* became inserted at or near the *A2* or *C* loci and gave rise to new mutable alleles. These tests are not exhaustive enough to establish the site of *En* insertion (within the structural locus or adjacent to it); however, one can presume on the basis of precedence (Peterson, 1956, 1968, 1970a) that *En* is closely allied to the locus. These results support the conclusions that *En* was transposed from some position in the genome to the *A2* or *C* locus. The differential rates of the origin of *a2–m* or *c–m* are now examined.

Rate of Occurrence of *a2* and *c* Mutable Alleles

In tests for the induction of *a2–m* and *c–m,* in 1963 and 1964 comparable seed sources were used in the isolation plots, which would provide a judicious comparison of rate differences in the origin of mutables at the two loci. These tests and the seed sources are given in Table 6.

It is clearly evident that, with the same source population, the incidence of *c*-mutable alleles in most instances is higher than that of *a2*–m mutables. In many of the direct comparisons the rate is 2–3 times higher. In two of the 1966 comparisons the frequency of *a2–m* is higher. There is no clear explanation for these differences because the location of *En* in the originating stocks, which influences its eventual site deposition after transposition (Van Schaik and Brink, 1959; Peterson, 1970b), was not known. If, in the stocks favoring the appearance of *c–m,* the *En* was located in chromosome 9, the higher rate could be due to *En* proximity to the locus involved. In some cases where the rates between *a2* and *c* are similar ($\sim 9/10^6$), such as in tests with the cross 3 1100 × 1095, it could be presumed that each of the loci, *A2* and *C,* was an equal target for *En* insertion. Under such circumstances *En* would have approximately one chance in 100,000 (Table 6) of being inserted in a target locus. Because an average of two *En* were present in a number of the plants of the sample population, the rate would be halved to 1 per 200,000.

It is unfortunate that a precise control of *En* chromosome location was not made in these experiments, but, the primary aim was to "load" the sampled population with *En*. Nevertheless, even if the *En* location were initially known, subsequent attempts to develop sufficient quantities of seed stock for tests would have altered the initial location by transpositions, as has been reported in a number of studies. If these *En*-originated *c* or *a*2-mutables represent origins from an *En* that is likely to be transposed from one chromosome to another chromosome rather than to sites on the same chromosome, then these rates may reveal an approximation of the rate of *En* insertion at any chromosome site, or they may mean that a locus such as *c* or *a* 2 represents $1/100,000$ to $1/200,000$ of the possible sites of a genome at which insertions can occur.

DISCUSSION

General Features of Controlling and Regulatory Elements

In the composite diagram of controlling and regulatory element activity shown In Figure 10A, an *autonomously controlled allele* is being freed from the suppressive activity of the regulatory element. (Although a pachytene figure is shown, the event occurs during both mitotic and meiotic divisions). Phenotypically, this results in the expression of color and is accompanied by transposition of the regulatory element from the locus to another position on the same chromosome or on another chromosome. This is true for *P–vv* (Brink and Nilan, 1952) and evident for the *Ac–Ds* (McClintock, 1951a) and *En–Spm* systems (Peterson, 1961, 1968, 1970a). An additional and, for some alleles, very frequent occurrence is the permanent allelic change to a nonresponsive, colorless, or pale derivative; this is common to most systems studied: (a) *Ac–Ds* (McClintock, 1948), (b) *En* (Peterson, 1961, 1968, 1970a), and (c) *Dt* (Neuffer, 1961; Rhoades, 1938). In the case of *P–vv*, however, it is a rare change (Brink, 1958). The permanency of the change to the null-form, as well as the nonresponsive nature of the event, suggests that in the transposition event, emission of the element might have caused a small deletion at the locus (d, in Figure 10A) (Peterson, 1970a).

In Figure 10B regulatory element activity from an independent position, expressed as *En* activity, is causing the loss or change of *I* by a signal (*s*) that leaves the controlled locus fully restored in function. That *s* is a constant and not an intermittent signal is supported by observations with the *a–m*1 type alleles where color is produced in the absence of the regulatory element and *a–m*1 is colorless in the presence of *En* except when a mutation event occurs (Figure 11B). Behavior of the *a–m*1 alleles provides evidence that the regulatory element signal (*s*) in Figure 10 is continuously on. Nevertheless, the mutation event, although occurring uniformly in time and frequency for a given regulatory element (Figure 9B), occurs in only a few cells (Figure 8A, parts a–c, and Figure 9A).

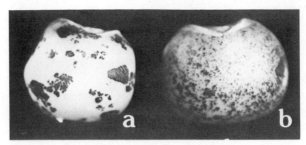

Figure 7. The *c–m* 5 5292 allele, which illustrates the effect of Restrainer, *Rst,* on the coarse pattern (a) resulting in a fine pattern (b). The cross is:

$$\frac{c\text{-}m(coarse)Sh\ Wx\ Rst/+}{c\ sh\ wx} \quad \frac{c\ sh\ wx}{c\ sh\ wx}$$

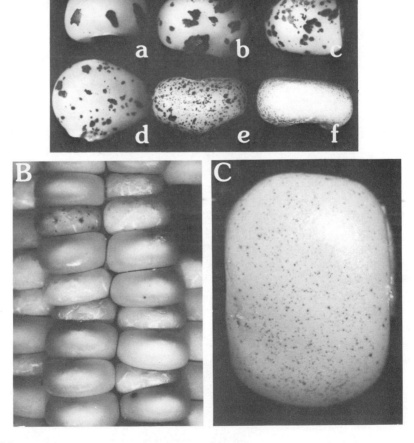

624

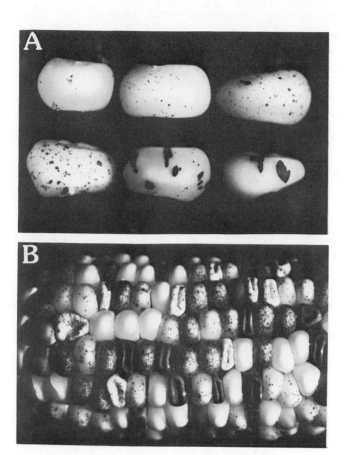

Figure 9. Consistency and diversity of pattern types: (A) diversity of pattern types triggered by an assortment of regulatory element types (*En*) on a common receptor-controlled a2–m(r) allele. Each individually isolated *En* activates a distinguishable heritable pattern; (B) consistency of response on an ear containing two different controlled alleles (a–m(r)) and (a–m1) with a common *En*.

Figure 8. A series of states of controlling-element alleles: (A) (a–f) a series of a2–m alleles' patterns ranging from early occurring mutation events (large spots) to very late occurring (small spots); (B) the expression of *En–sw* on the a–m1 allele illustrating the strong suppression but weak mutability with occasional exceptions as evident in upper left; (C) a2–m 4 1629 showing the very late, largely single-celled spots that are identified with this controlling-element allele.

What then determines the mutation event leading to changes from a to A? The time and frequency of mutation events as determined by pattern types (Figures 7 and 8) vary widely, which indicate variability in excision of the element at the locus. In addition, the s signal can be modified by environment (Rhoades, 1941; Peterson, 1958), and there are unique regulatory elements with s signals that are modified by particular tissues that cause an extinction of the signal as evident in phase variation (Peterson, 1966a; Doerschug, 1973; Schwartz, 1960) or cyclical changes (McClintock, 1948, 1964, 1965a, 1967b). Phase changes are caused by extinction of the signal in specific areas of tissue (Figure 11A), not changes in the time and frequency of regulatory element activity. There is, therefore, a time in the development of tissues when regulatory element effects are not compatible with the cellular environment.

Another feature of regulatory element signals is the lateness of the mutation event associated with increased dosage; this is especially evident in the Ac–Ds

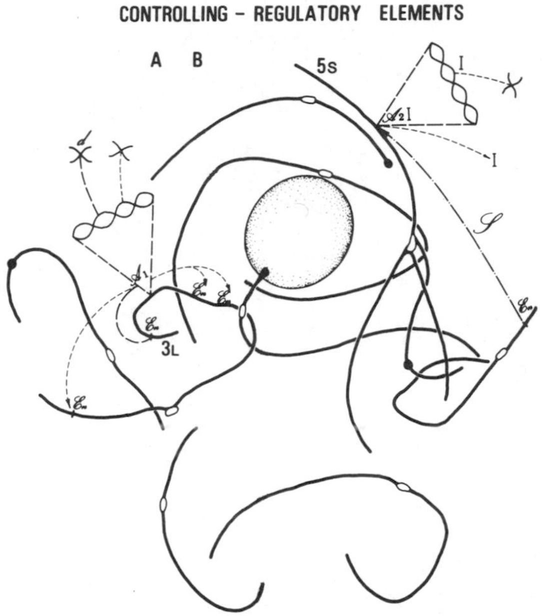

Figure 10. A model of controlling–regulatory-element activity. Diagramatic representation of the action of regulatory elements on a controlled allele. In (A), *En* is being transposed from an autonomous mutable allele at the *a* locus to an assortment of receptor sites. Dark arrow (heavy line) illustrates a transposition event, and other arrows (lighter lines) illustrate alternative sites (Peterson, 1970b). The magnified view attempts to show excision of regulatory element and depicts a deletion as is hypothesized for the a-m(nr) types. In (B), a two-element system is shown. A signal, *S*, originates from *En* that excises the *I* element from a controlled allele leading to the restoration of gene activity at the controlled locus. The magnified view illustrates the excision of an *I* element from the chromosome.

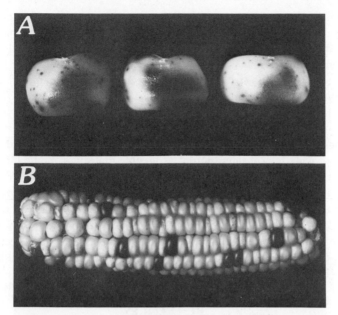

Figure 11. Variation in *En* activity: (A) phase change of *En* illustrating the extinction of *En* activity and reversion to a–*m*1 coloration in the kernel of a–*m*1 *sh2/a Sh2 a Sh2/a Sh2 En*; (B) an ear illustrating the uniformity of the expression of the *En–sw* allele showing the strong suppression and weak mutability.

system (McClintock, 1948, 1951a) and with mutable pericarp (Brink and Nilan, 1952; Brink, 1954). Although the presence of several regulatory elements should result in more *En* signals (*s* in Figure 10B), under such circumstances, the excision event is delayed. The later the event occurs in the development of endosperm tissue, the greater is the number of cells potentially present to mutate. Thus the light types among *P–vv* may be the result of interference with the *En* signals. It may mean that saturating cells with regulatory element signals acts to repress the excision of elements from controlled loci.

Although the composite diagram (Figure 10) attempts to show a universal picture of controlling and regulatory elements, it is plainly evident that differences in behavior of elements exist. Whereas with *P–vv*, transposition of an element leads to obligate reinsertion (Greenblatt, 1974), this is not true with *En*. The *Mp* emissions from *P–vv* reinsert 65% of the time within a linkage (Van Schaik and Brink, 1959). In studies of transposition from the a–*m*2 allele, *Spm* was found on the same chromosome 38% of the time (McClintock, 1962). In the *Ac* system with *bz–m*2, changes to stable, colorless types were accompanied by *Ac* changes of slightly more than 50% that were linked to the original site (McClintock, 1956a,b, 1962). With *En* and with one particular mutable allele, a–*m*–*papu*, *En* transposition from the original site at the *a* locus and deposition at a linked distance occurs approximately 25% of the time (Peterson, 1970b). Differences between controlling and regulatory element systems and even within systems indicate that elements probably differ in reinsertion capabilities.

The maize controlling elements have previously been compared to similar phenomena in prokaryotes (Peterson, 1967, 1970a), specifically the *IS* elements and the bacteriophage, *Mu* (Starlinger and Saedler, 1972; Taylor, 1963; Bukhari and Zipser, 1972; Howe and Bade, 1975). Similarities between these three types of elements do not imply that the elements are identical. They do, however, provide for consideration of a possible molecular model for the maize controlling-element system (McClintock, 1961; Peterson, 1970a). With current advances in plant DNA isolation and identification (Flavell and Smith, 1975) and with the development of precise analytical methods in maize (Freeling and Brown, 1975), it should not be long before molecular probes begin to provide an explanation of controlling and regulatory elements that have been a source of speculation for over a half a century—from Emerson's studies with mutable pericarp (Emerson, 1914) to the International Maize Symposium in 1975.

REFERENCES

Barclay, P. C. and R. A. Brink. 1954. The relation between modulator and activator in maize. *Proc. Nat. Acad. Sci. U. S. A.* **40:** 1118–1126.

Bianchi, A., F. Salamini, and R. Parlavecchio. 1969. On the origin of controlling elements in maize. *Genet. Agrar.* **22:** 335–344.

Brink, R. A. 1954. Very light variegated pericarp in maize. *Genetics* **39:** 724–740.

Brink, R. A. 1958. A stable somatic mutation to colorless from variegated pericarp in maize. *Genetics* **43:** 435–447.

Brink, R. A. and R. A. Nilan. 1952. The relation between light variegated and medium variegated pericarp in maize. *Genetics* **37:** 519–544.

Bukhari, A. I. and D. Zipser. 1972. Random-insertion of Mu-1-DNA within a single gene. *Nature New Biol.* **236:** 240–243.

Doerschug, E. B. 1973. Studies of *Dotted*, a regulatory element in maize. I. Inductions of *Dotted* by chromatid breaks. II. Phase variation of *Dotted*. *Theor. Appl. Genet.* **43:** 182–189.

Emerson, R. A. 1914. The inheritance of a recurring somatic variation in variegated ears of maize. *Am. Natur.* **48:** 87–115.

Emerson, R. A., 1917. Genetical studies of variegated pericarp in maize. *Genetics* **2:** 1–35.

Fincham, J. R. S. and G. R. K. Sastry. 1974. Controlling elements in maize. *Annu. Rev. Genet.* **8:** 12–50.

Flavell, R. B. and D. B. Smith. 1975. Genome organization in higher plants. *Stadler Gent. Symp.* **7:** 47–70.

Fowler, R. and P. A. Peterson, 1974. The a2m(r-pa-pu) allele of the En-controlling element system in maize. *Genetics* **76**: 433–446.

Freeling, M. and E. Brown. 1975. *In situ* staining of pollen grains for alcohol dehydrogenase activity. *Maize Genet. Coop. Newsl.* **49**: 19–21.

Gonella, J. and P. A. Peterson. 1975a. The presence of *En* among some maize lines from Mexico, Colombia, Bolivia and Venezuela. *Maize Genet. Coop. Newsl.* **49**: 73.

Gonella, J. and P. A. Peterson 1978. In *Maize breeding and Genetics,* D. B. Walden (ed.). New York: John Wiley & Sons, p. 774.

Greenblatt, I. M. 1968. The mechanism of modulator transposition in maize. *Genetics* **58**: 585–597.

Greenblatt, I. M. 1974. Movement of modulator in maize: A test of an hypothesis. *Genetics* **77**: 671–678;

Howe, M. M. and E. G. Bade. 1975. Molecular biology of bacteriophage *Mu. Science* **190**: 624–632.

McClintock, B. 1938. The fusion of broken ends of sister half-chromatids following chromatid breakage at meiotic anaphases. *Mo. Agr. Exp. Sta. Res. Bull.* **290**: 1–48.

McClintock, B. 1941. The stability of broken ends of chromosomes in *Zea mays. Genetics* **26**: 234–282.

McClintock, B. 1947. Cytogenetic studies of maize and neurospora. *Carnegie Inst. Wash. Year Book* **46**: 146–152.

McClintock, B. 1948. Mutable loci in maize. *Carnegie Inst. Wash. Year Book* **47**: 155–169.

McClintock, B. 1949. Mutable loci in maize. *Carnegie Inst. Wash. Year Book* **48**: 142–154.

McClintock, B. 1950a. Mutable loci in maize. *Carnegie Inst. Wash. Year Book* **49**: 157–167.

McClintock, B. 1950b. The origin and behavior of mutable loci in maize. *Proc. Nat. Acad. Sci. U. S. A.* **36**: 344–355.

McClintock, B. 1951a. Chromosome organization and genic expression. *Cold Spring Harbor Symp. Quant. Biol.* **16**: 13–47.

McClintock, B. 1951b. Mutable loci in maize. *Carnegie Inst. Wash. Year Book* **50**: 174–181.

McClintock, B. 1952. Mutable loci in maize. *Carnegie Inst. Wash. Year Book* **51**: 212–219.

McClintock, B. 1953. Induction of instability of selected loci in maize. *Genetics* **38**: 579–599.

McClintock, B. 1956a. Controlling elements and the gene. *Cold Spring Harbor Symp. Quant. Biol.* **21**: 197–216.

McClintock, B. 1956b. Intranuclear systems controlling gene action and mutation. *Brookhaven Symp. Biol.* **8**: 58–74.

McClintock, B. 1956c. Mutation in maize. *Carnegie Inst. Wash. Year Book* **55**: 323–332.

McClintock, B. 1958. The suppressor-mutator system of control of gene action in maize. *Carnegie Inst. Wash. Year Book* **57**: 414–429.

McClintock, B. 1961. Some parallels between gene control systems in maize and in bacteria. *Am. Natur.* **95**: 265–277.

McClintock, B. 1962. Topographical relations between elements of control system in maize. *Carnegie Inst. Wash. Year Book* **61**: 448–461.

McClintock, B. 1964. Aspects of gene regulation in maize. *Carnegie Inst. Wash. Year Book* **63**: 592–602.

McClintock, B. 1965a. The control of gene action in maize. *Brookhaven Symp. Biol.* **18**: 162–184.

McClintock, B. 1965b. Components of action of the regulators *Spm* and *Ac*. *Carnegie Inst. Wash. Year Book* **64**: 527–537.

McClintock, B. 1967a. Genetic systems regulating gene expression during development. *Develop. Biol.* (suppl.) **1**: 84–112.

McClintock, B. 1967b. Regulation of pattern of gene expression by controlling elements in maize. *Carnegie Inst. Wash. Year Book* **65**: 568–578.

Neuffer, M. G. 1966. Stability of the suppressor element in two mutator systems at the A_1 locus in maize. *Genetics* **53**: 541–549.

Nuffer, M. G. 1955. Dosage effect of multiple *Dt* loci on mutation of a in maize endosperm. *Science* **121**: 399–400.

Nuffer, M. G. 1961. Mutation studies in the A_1 locus in maize. I. A mutable allele controlled by *Dt*. *Genetics* **46**: 625–640.

Orton, E. R. and R. A. Brink. 1966. Reconstitution of the variegated pericarp allele in maize by transposition of modulator back to the *P* locus. *Genetics* **53**: 7–16.

Peterson, P. A. 1953. A mutable pale green locus in maize. *Genetics* **38**: 682.

Peterson, P. A. 1956. An a_1 mutable arising in pg^m stocks. *Maize Genet. Coop. Newsl.* **30**: 82.

Peterson, P. A. 1958. The effect of temperature on the mutation rate of a mutable locus in maize. *J. Hered.* **49**: 120–124.

Peterson, P. A. 1960. The pale green mutable system in maize. *Genetics* **45**: 115–118.

Peterson, P. A. 1961. Mutable a_1 of the *En* system in maize. *Genetics* **46**: 759–771.

Peterson, P. A. 1963. Influence of mutable genes on induction of instability in maize. *Proc. Iowa Acad. Sci.* **70**: 129–134.

Peterson, P. A. 1964. The dominant mutable *V–m–mp–1817*. *Maize Genet. Coop. Newsl.* **38**: 80–81.

Peterson, P. A. 1965. A relationship between the *Spm* and *En* control systems in maize. *Am. Natur.* **99**: 391–398.

Peterson, P. A. 1966a. Phase variation of regulatory elements in maize. *Genetics* **54**: 249–266.

Peterson, P. A. 1966b. Linkage and control of mutability of *w–m 13*—a white seedling mutable. *Maize Genet. Coop. Newsl.* **40:** 64–65.

Peterson, P. A. 1967. A comparison of the action of regulatory systems in maize and lysogenic bacteria. *Genetics* **56:** 581.

Peterson, P. A. 1968. The origin of an unstable locus in maize. *Genetics* **59:** 391–398.

Peterson, P. A. 1970a. Controlling elements and mutable loci in maize: their relationship to bacterial episomes. *Genetics* **41:** 33–56.

Peterson, P. A. 1970b. The *En* mutable system in maize. III. Transposition associated with mutational events. *Theor. Appl. Genet.* **40:** 367–377.

Peterson, P. A. 1975. Controlling elements in maize and the timing of mutability events. *Genetics* **80:** s63–s64.

Rhoades, M. M. 1936. The effect of varying gene dosage on aleurone colour in maize. *J. Genet.* **33:** 347–354.

Rhoades, M. M. 1938. Effect of the *Dt* gene in the mutability of the a_1 allele in maize. *Genetics* **23:** 377–397.

Rhoades, M. M. 1941. The genetic control of mutability in maize. *Cold Spring Harbor Symp. Quant. Biol.* **9:** 138–144.

Schwartz, D. 1960. Analysis of a highly mutable gene in maize. A molecular model for gene instability. *Genetics* **45:** 1141–1152.

Smith, J. D. 1960. A mutable locus in maize. Ph.D. thesis. Ames: Iowa State University.

Starlinger, P. and H. Saedler. 1972. Insertion mutations in microorganisms. *Biochimie* **54:** 177–185.

Taylor, A. L. 1963. Bacteriophage-induced mutation in *Escherichia coli. Proc. Nat. Acad. Sci. U. S. A.* **50:** 1043–1051.

Van Schaik, N. W. and R. A. Brink. 1959. Transpositions of modulator, a component of the variegated pericarp allele in maize. *Genetics* **44:** 725–738.

Section Eight

CYTOGENETICS

Chapter 39

INTRODUCTORY REMARKS TO THE SESSION ON CYTOGENETICS

E. Dempsey

Department of Plant Sciences,
Indiana University
Bloomington

The methodology of cytogenetics has been utilized in many fundamental biological problems of theoretical interest, such as the nature of the gene, the mechanism of chromosome movement, the genetic effects of chromosome aberrations, and the role of heterochromatin in the cell economy. Cytogenetic analysis has also been of great importance in taxonomic and evolutionary studies, where the relationships between various taxa or the ancestral origins of certain groups were under study. Pairing between homologous or homoeologous chromosomes in hybrids and similarities in chromosome morphology of different species are criteria by which the degree of relationship has been deduced.

The dual approach of cytogenetics, combining the analysis of chromosome structure with studies of the transmission of marker genes, has vastly enriched the science of genetics. The impact of cytogenetics on *Drosophila* investigations has long been recognized and continues to be significant. For example, the recent studies on the fine structure of the white-zeste region of the X chromosome (Judd, et al. 1972) could not have been accomplished without the aid of the banding pattern in the giant chromosomes. The chapters on

635

cytogenetics in this volume further confirm the importance of the combined attack; in all cases, the study of the maize chromosomes at the pachytene stages was an indispensable part of the analysis.

Considerable information has been amassed on the nature and function of heterochromatin since it was first described in 1928 by Heitz. Its essential inertness has been repeatedly demonstrated in studies with *Drosophila,* the mealy bug, maize, and a variety of mammals, including man. Very few typical genes are located in heterochromatin, and individuals possessing heterochromatic chromosomes or segments of chromosomes usually resemble their sibs without the heterochromatin or show only subtle quantitative differences. Nevertheless, as documented in Chapter 40, a surprising number of genetic effects can now be attributed to heterochromatin. Rhoades (Chapter 40) discusses the various activities of maize heterochromatin at some length, reviewing the earlier discoveries and describing a new instance where heterochromatin causes sex differences in recombination. Most of the identifiable "phenotypes" involve some aspect of chromosome behavior, and the activity of the heterochromatin is restricted to very specific periods in the life cycle. In this regard maize heterochromatin may be compared with the hetrochromatic Y of *Drosophila,* which is functional only during the prophase stage of meiosis (Hess and Meyer, 1968).

The phenomena induced by heterochromatin in maize include preferential segregation, formation of neocentromeres, enhancement of recombination, and loss of chromosome segments. The addition of heterochromatin, in the form of chromosome knobs, the extra segment on abnormal chromosome 10, or B chromosomes, leads to unexpected chromosomal events, either at meiosis or at the second spore mitosis, and these aberrations are detected genetically by a change in segregation ratios, differences in recombination frequencies, or noncorrespondence of endosperm and embryo phenotypes. Rhoades points out that certain of these phenomena require interaction of two types of heterochromatin, whereas others involve only a single heterochromatic inducer. Included in the discussion of preferential segregation are previously unpublished photographs showing the poleward migration of fragments possessing neocentromeres but no true centromere. It is obvious from these studies that heterochromatin may on occasion have dramatic effects, which vary with the specific kind of heterochromatin, and that it can no longer be classified as genetically inert.

In the second paper of this session, see Chapter 11, McClintock has made use of studies of chromosome morphology to reconstruct patterns of migration of maize races in Central and South America. A vast amount of information is available on the pachytene chromosomes of representatives of geographic races of the Americas. McClintock has based her analysis on her own observations reported in 1959 and 1960, as well as on data recorded more recently by Drs. Longley, KatoY, and Blumenschein. Differences among the races had been noted with respect to knob number, size, and position, number of B chromosomes, type of nucleolus organizing region, and other less obvious characteristics of the genome. Unlike the races and species of *Drosophila,* there have been few reports of chromosomal rearrangements in the races of maize, with the exception of a single paracentric inversion in chromosome 8, which has appeared in both maize and teosinte. McClintock concentrates her attention on the chromosome knobs, which are recognized to be stable, heritable modifications of chromosome structure, and on the heterochromatic supernumerary B chromosomes. Specific knob constellations, such as the Andean or Venezuelan complexes, have been identified, and the geographic distribution of plants possessing the entire complex or elements of a complex has been determined. Following distribution patterns of both entire knob complexes and individual knobs, McClintock has discerned paths of migration and has been able to specify the probable site of origin of many of the complexes. A surprising number of these have been traced to Guatemala and southwest Mexico. In addition, McClintock has found evidence for introgressions from maize strains along the path of the migration. Her studies provide a satisfying confirmation of earlier work by Wellhausen and associates, who based their conclusions about the past history of maize races on the blending of morphological characteristics.

In Chapter 41, Burnham briefly summarizes the standard meiotic behavior of interchange heterozygotes and then presents a number of ingenious studies on translocations that have been conducted in his laboratory. Prominent among these is the analysis of crosses between translocation stocks possessing interchanges involving the same two chromosomes. The resulting hybrids have been classified into four groups depending on the position of the breakpoints in the two translocations. The complex pachytene configurations found in the different F_1s are described and the genetic consequences of their meiotic segregation are detailed. Crossovers in the "differential" segment, between the two points where exchange of pairing partner occurs, leads in some cases to new chromosomes containing duplications or inserted chromatin from the nonhomologue and in other cases reconstructs the original untranslocated chromosome. Burnham and colleagues have made use of the intercrossed translocation stocks to study the initiation of pairing and have concluded, on the basis of cytological observations, that synapsis begins near the ends of chromosome arms. Other applications include construction of specific duplications of short chromosome segments, such as portions of the nucleolar organizer region. Duplications may also be generated from translocation heterozygotes by $3:1$ segregation of the ring of 4. Tertiary trisomics arising from a number of translocations are being tested in schemes designed to give all male-sterile progenies, as a means of avoiding detasseling in hybrid corn production.

Burnham has devoted several pages to a discussion of future research on translocations. He lists a number of areas where investigations are needed, mentioning problems of a theoretical nature as well as those of direct interest to the breeder. This section of his chapter will prove of interest to students beginning their own research on maize cytogenetics.

The nucleolus organizing region (NOR) has been under intensive study in a number of laboratories, since this organelle was shown to be responsible for synthesis of ribosomal RNA (Ritossa and Spiegelman, 1965). In addition to frequently cited work on *Drosophila, Xenopus,* and *Triturus,* numerous important studies have been made on the NOR of maize, beginning with the pioneering papers of McClintock (1934) and Lin (1955) and culminating in recent investigations by Phillips. In Chapter 43, Dr. Phillips presents a comprehensive review of the earlier work, including observations on the structure of the nucleolus and NOR at both light- and electron-microscopic levels, analyses of their chemical composition, and cytogenetic studies on aberrations possessing a breakpoint in the NOR. The majority of the chapter is concerned with the function of the NOR; after summarizing evidence for precise localization of the ribosomal RNA genes (rDNA) in the organizer of maize by DNA–RNA hybridization, Phillips presents a functional map of the NOR based on his study of 19 translocations with breakpoints at varying positions in the NOR. He concludes that the secondary constriction is the active site of rRNA synthesis, but two other sites exist in the heterochromatic portion that may become active when separated from the secondary constriction. Phillips believes the NOR has both a synthetic and organizing function, but the two kinds of activities have not yet been localized in separable entities within the NOR. Maize differs from *Drosophila* and *Xenopus* in that very high multiplicities of rRNA genes are present in the NOR; moreover, neither compensation nor magnification have been detected in maize. Phillips speculates that the extra copies of rDNA are needed at certain times in the life cycle or under conditions of peak demand and suggests that the high content of potentially functional rRNA genes constitutes a reservoir of genetic material, which obviates the necessity for compensation. As a result of the work of Phillips and his associates, more is known in maize than in any other organism about the structure and function of the nucleolus organizing region.

REFERENCES

Heitz, E. 1928. Das Heterochromatin der Moose. I. *Jahrb. wiss. Bot.* **69**: 762–818.

Hess, O. and G. F. Meyer. 1968. Genetic activities of the Y chromosome in *Drosophila* during spermatogenesis. *Adv. Genet.* **14**: 171–223.

Judd, B. H., M. W. Shen, and T. C. Kaufman. 1972. The anatomy and function of a segment of the X chromosome of *Drosophila melanogaster*. *Genetics* **71**: 139–156.

Lin, M. 1955. Chromosomal control of nucleolar composition in maize. *Chromosoma* 7: 340–370.

McClintock, B. 1934. The relation of a particular chromosomal element to the development of the nucleolus in *Zea mays*. *Zellforsch. w. mikr. Anat.* 21: 294–328.

Ritossa, F. M. and S. Spiegelman. 1965. Localization of DNA complementary to ribosomal RNA in the nucleolus organizer region of *Drosophila melanogaster*. *Proc. Nat. Acad. Sci. U. S. A.* 53: 737–745.

Chapter 40

GENETIC EFFECTS OF HETEROCHROMATIN IN MAIZE

M. M. Rhoades

Department of Plant Sciences,
Indiana University
Bloomington

INTRODUCTION

According to Heitz's (1928, 1929) definition, heterochromatin differs from euchromatin in that it remains in a condensed state throughout the nuclear cycle, whereas euchromatin is lightly staining during the metabolically active interphase and is in a contracted condition only from prophase through telophase of mitosis. Both types contain DNA. Heterochromatin was originally defined by the cytological criterion of heteropycnosis. It later became apparent that the condensed, deeply staining bodies of chromatin, although they may be large in bulk, had no or few active genes; they were genetically inert. Gross deficiencies of heterochromatin are tolerated with no or little visible phenotypic effect, whereas small deficiencies of euchromatin are usually lethal. Time of replication is another characteristic by which heterochromatin differs from euchromatin. Heterochromatin replicates later in the cell cycle than euchromatin and in certain insect cells, it fails to do so at all. Synthesis of RNA is suppressed in heterochromatin; this accounts for its genetical inertness since the

641

genes are repressed. Constitutive heterochromatin is permanently in a tightly coiled state, whereas facultative heterochromatin is capable of converting into euchromatin at certain times and under certain conditions. Heterochromatin and euchromatin are not necessarily different substances but are different states of the same substance, differing in degree of coiling and association with chromosomal proteins.

Judged by some of the above criteria, the following structures in the maize nucleus contain heterochromatin: (a) chromosomal knobs, (b) B chromosomes, (c) abnormal chromosome 10, and (d) chromatin adjacent to the centromeres (centric heterochromatin). The functioning of the heterochromatic nucleolus organizing region of chromosome 6 is considered in detail by Phillips (Chapter 43) and is not discussed here.

In this survey of the genetic effects of heterochromatin in maize, we do not consider its postulated role in gene regulation at specific loci because there is no cytological evidence of the putative heterochromatic elements. This is not to say that insertion of minute bits of heterochromatin into or adjacent to a structural gene could not alter its activity; that this occurs seems likely from the work on controlling elements. However, we prefer to confine our attention to those cases where the participation of heterochromatin in controlling a given phenomenon is based on cytological evidence of its presence and not on ingenious speculation.

For many years no specific activity had been ascribed to knobs, B chromosomes, or the centric heterochromatin of maize. It was reasonable to assume that they were inert. However, we now know that they are responsible for or participate in a surprising number of unusual phenomena and that they differ in their genetic effects.

B CHROMOSOMES AND ENHANCED RECOMBINATION

Among the genetic effects in maize caused by the B chromosome is that of increased crossing over. Hanson (1965, 1969), working with regions in the long arm of chromosome 3 and the short arm of chromosome 9, found no or little change in crossing over with fewer than four Bs and a slight increase with four or more. Nel (1973) reported that Bs significantly increased recombination in regions adjacent to the centromere of chromosome 5 and to a lesser degree in chromosome 9. Higher crossover values were found in the chromosome 5 data as the number of Bs increased. However, these increases are relatively modest compared to an exceptional case where a single B chromosome more than doubled the amount of crossing over (Rhoades, 1968). An internal piece of chromosome 3 constituting $\frac{1}{10}$ of the length of the long arm was inserted into the short arm of chromosome 9 between the *bz* and *wx* loci to give the Tp9 chromosome. Crossing over was studied in plants homozygous for Tp9 and heterozygous for the closely linked *C, Sh,* and *Bz* genes to the left of the transposed piece and *Wx* to the right. The inserted piece of 3L increased the length of 9S by 20% and approximately doubled the physical distance in the pachytene chromosome between the flanking markers to the left and right. Closely related plants with two normal chromosomes 9 had 17.7% recombination between *C* and *Wx*. Inasmuch as these two loci in Tp9 Tp9 individuals were physically further separated by the length of the inserted piece of 3L, a marked increase in *C–Wx* crossing over was expected on the logical assumption that a longer segment of chromatin provided more opportunity for genetic exchange than did a shorter one. Surprisingly, little or no increase in recombination was found for the *C–Wx* or *Sh–Wx* intervals in Tp9 Tp9 plants over that occurring in N9 N9 sibs. It was concluded that no exchanges occurred in the transposed piece of 3L in Tp9 Tp9 sporocytes, since the *C–Wx* and *Sh–Wx* values were not sensibly different from those in N9 N9 plants. However, the presence of a single B chromosome dramatically increased crossing over in Tp9 Tp9 meiocytes from 17.7% to 37.0%. As is evident in the data given in Table 1, there was a small but consistent dosage effect for increased numbers of Bs.

The B chromosome consists chiefly of heterochromatic segments that are not uniformly distributed over the length of the chromosome (see Figure 1). Adjacent to the nearly terminal centromere is a knob-like region of centric heterochromatin followed by a more euchromatic segment, then four large blocks of heterochromatin and finally, a small relatively euchromatic tip of two chromomeres. To determine which portion of the B contains the factor(s) responsible for enhanced crossing over in Tp9 Tp9 sporocytes, Ward (1973b)

TABLE 1. Percentage of C–*Wx* recombination in C Tp *wx/c* Tp *Wx* plants differing in numbers of accessory B chromosomes (from Rhoades, 1968)

0B	1B	2Bs	3Bs
13.3	40.7	42.6	42.5
15.1	36.0	38.1	43.1
15.0	32.5	44.4	44.6
18.0	40.4	43.7	38.6
20.0		38.4	
21.4		36.4	
21.3			
14.4			
16.6			
M = 17.7	*M* = 37.0	*M* = 40.4	*M* = 42.0

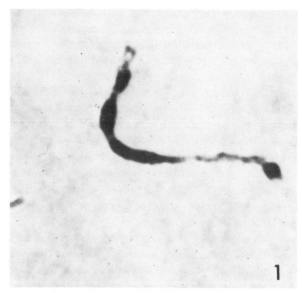

Figure 1. Photomicrograph of a B chromosome bivalent at pachynema showing the large heterochromatic blocks.

employed a series of B–A translocations with progressively larger segments of the B in the B^A chromosome. All tested segments of the B gave increased recombination values; there was no localization of a factor in the B for increased recombination in Tp9 Tp9 plants. Ward (1975) studied crossing over in Tp9 Tp9 plants of different genetic backgrounds. The C–Wx value in plants without B chromatin ranged within 16–27%. The addition of B chromatin, achieved by using balanced TB–A translocations, resulted in increased C–Wx values but the increase was greatest in those plants with the lowest C–Wx values.

In Tp9 Tp9 plants without B chromosomes apparently no exchanges take place in the transposed piece of 3L, whereas crossing over occurs in this segment in the presence of one or more B chromosomes (Rhoades, 1968). This statement is based on the absence of chiasma interference in the C–Sh and Sh–Wx regions in 0B plants and its presence in plants with Bs. It is argued that most of the crossovers between Sh and Wx in a 0B plant are to the right of the inserted segment of 3L. An exchange in this region is physically removed from the C–Sh interval by the considerable length of the piece of 3L, in which there is no crossing over. The decrease in interference is comparable to that produced by increasing the map distance and the probability of a second exchange occurring in the adjacent C–Sh region would not be adversely affected. However, if Bs induce exchanges in the transposed piece of 3L, those crossovers taking place near its distal end would be expected to decrease the probability of a second crossover in the neighboring *C–Sh* interval. Unfortunately, the inserted segment of 3L was not marked by mutant genes, which would permit an unequivocal demonstration of crossing over within the transposed piece.

Although some of the genetic effects ascribable to heterochromatin involve the interaction of different kinds of heterochromatin (e.g., K10 with knobs and B chromosomes with knobs), B chromosomes alone are able to induce a striking increase in crossing over in Tp9 Tp9 homozygotes. Ward (1975) reported that K10 also enhanced *C–Wx* crossover values in Tp9 Tp9 plants but to a lesser extent than did B chromatin, and here too, the increase was achieved without interaction with B chromatin or with knobs.

Additional information on the effect of B chromosomes on recombination comes from Chang and Kikudome (1974), who reported that maize plants with odd numbers of Bs had higher crossover values in megasporocytes than did those with even numbers; this odd–even effect is comparable to that found earlier in rye by Jones and Rees (1969) and Kirk and Jones (1970).

Another well-known genetic effect of B chromosomes is the induction of nondisjunction of the B centromere at the second microspore division. This phenomenon, which has been exhaustively studied by Roman (1947) and Carlson (1970, 1973a,b), is not discussed in this chapter, since the topic of B chromosome nondisjunction is reviewed by Carlson in Chapter 44.

There have been many investigations with various plants on the genetic activities of B chromosomes. Most of the effects have been of a regulatory or quantitative nature. An exception is the highly specific effect reported by Evans and Macefield (1973), where Bs controlled chromosome pairing in species hybrids of *Lolium*. This situation involves a striking qualitative change and in this regard is comparable to the effect of Bs on crossing over in Tp9 homozygotes.

CHROMOSOME LOSS INDUCED BY THE INTERACTION OF B CHROMOSOMES AND KNOBS

Before presenting the data on chromosome loss induced by the interaction of two heterochromatic elements, B chromosomes and knobs, let us first review the course of male gametogenesis in maize. Two somatic mitoses take place in the development of a haploid microspore into the male gametophyte, the pollen grain. The first of these gives rise to the vegetative and generative cells. The vegetative nucleus does not redivide, but the generative cell undergoes a second mitosis to form two identical sperm cells. On germinating, the pollen grain sends forth a pollen tube that grows down the style to discharge its two sperm cells into the embryo sac. One of them fertilizes the egg to form the zygote, and the other unites with the two polar nuclei to produce the triploid endosperm of the maize kernel. Presumably it is a matter of chance as to which sperm fuses with the egg and which unites with the polar nuclei.

However, under certain circumstances chromosome behavior at the second microspore mitosis is frequently abnormal. The two sperm do not receive the same chromosomal endowment; one has the normal complement, but the other is deficient for parts of certain chromosomes. Dissimilar sperm are a consequence of some event occurring at and only at the second microspore division. The circumstance responsible for this event is the combination of two heterochromatic elements, B chromosomes and knobs, in a specific genetic background. In a series of three papers (Rhoades and Dempsey, 1972, 1973a; Rhoades, et al., 1967), we have described the unusual genetic situation where the interaction of Bs and knobs leads to chromatin loss. Here we briefly review the pertinent data and then present some unpublished observations. The facts are as follows.

Loss at the second microspore mitosis involves only knobbed chromosomes and occurs only in spores with two or more B chromosomes. Knobless chromosomes undergo no loss, irrespective of the presence or absence of Bs. Little or no elimination takes place in microspores with 0 or 1 B chromosome. Two or more Bs are required, and there is no increase in frequency of loss in spores having more than two Bs. B chromosomes from various sources are equally effective in inducing loss of segments from knobbed chromosomes. The frequency of loss depends on knob size; the larger the knob, the greater the rate of loss. All knobbed chromosomes are subject to loss and the rate of concomitant loss involving two different knobbed chromosomes is higher than that expected from independent events. Thus joint loss appears to take place in competent cells. Modifying genes exist that are able to alter the rate of loss. We have found no loss coming from some plants having both knobbed chromosomes and the requisite number of Bs but lacking the necessary modifiers. The assortment at anaphase of two heterologous chromosomes, each of which has undergone loss, is at random to the two poles. Moreover, nondisjoining B chromosomes and deficient A chromosomes assort independently at the second microspore anaphase. It is a remarkable and significant fact that B chromosomes undergo a high rate of nondisjunction in the same mitosis where they induce chromatin loss in knobbed members of the complement. Indeed, it is this discovery that led to the development of the following hypothesis.

According to current dogma, B chromosomes undergo nondisjunction because the delayed replication of the heterochromatic knob adjacent to the centromere prevents the passing of the two chromatids to opposite poles at anaphase (Rhoades and Dempsey, 1972, 1973a). It is hypothesized that B chromosome nondisjunction and elimination of knobbed chromatin both result from delayed replication of heterochromatic segments—the proximal knob of the B and the more distal knobs of the A chromosomes. The juxtaposition to the centromere of the unreplicated proximal knob of the B results in nondisjunction, whereas the delayed replication of the more distally placed knobs in the A chromosomes leads to formation of a dicentric bridge as the two sister centromeres move to opposite poles. Anaphase stretching causes the bridge to break. The great majority of breaks are close or adjacent to the centromere, but breaks at other positions in the bridge have been detected. Following bridge breakage, one of the two sperm cells receives an intact and the other a deficient chromosome with a freshly broken end. Healing of the broken end occurs in zygotes and the deficient rod chromosome is stable throughout subsequent divisions of the developing sporophyte. There is, however, no healing of freshly broken ends when introduced into the primary endosperm nucleus (McClintock, 1938, 1941b). Fusion of the two broken ends occurs when the chromosome replicates. A dicentric chromosome is formed that will undergo the chromatid type of breakage–fusion–bridge (B–F–B) cycle during endosperm development. Breakage of the bridge at different locations gives rise to diverse kinds of deficient chromosomes.

When a high-loss male parent with a knobbed chromosome possessing dominant alleles is used in testcrosses, all F_1 endosperms and sporophytes will have the dominant phenotype of the male parent if there has been no elimination. However, if loss occurs, the progeny will contain some individuals exhibiting one or more of the recessive phenotypes, depending upon the extent of the lost segment. Figure 2 illustrates how the appearance of recessive phenotypes is diagnostic of the position of bridge breakage. Data from high-loss plants with a knobbed chromosome 9 are particularly instructive because 9S has the C and Wx markers that affect the endosperm and are some distance apart. When pollen from a high-loss plant with a large terminal knob on 9S and the C and Wx alleles was used in crosses to c wx testers, several distinctive kernel phenotypes were produced as a consequence of breaks in 9S. Let region (1) be

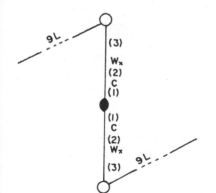

Figure 2. Diagram of a dicentric bridge at anaphase caused by the postulated failure of the terminal knob on 9S to replicate at the second microspore division. Breakage of the bridge in specific regions produces diagnostic endosperm phenotypes.

the knob–C interval, region (2) the C–Wx, and region (3) the Wx–centromere interval. Breaks in (1) followed by a B–F–B cycle in the endosperm mitoses give kernels variegated for C–c and Wx–wx, whereas a break in (2) produces a kernel with colorless aleurone and an endosperm mosaic for Wx and wx sectors. Rupture in (3) yields a colorless c kernel with all wx starch. The kernel phenotypes are diagnostic for positions of bridge breakage. Data from such a cross are given in Table 2.

An unexpected feature of Table 2 is the occurrence of the 20 stable c Wx kernels. We postulated (Rhoades and Dempsey 1973a) that these c Wx kernels could arise from: (a) a break between C and Wx followed by healing to form stable ends, as occurs in the embryo, (b) a cryptic B–F–B cycle in chromosome 9 throughout endosperm development, where bridge breakage occurs at the same position at every anaphase owing to weak or incomplete fusion of the broken ends, (c) a mutation from C to c, or (d) an interstitial deletion of the C gene. We have no evidence which proves or disproves any of these hypotheses, but recent observations (Rhoades and Dempsey, 1975) have revealed 'a new type of structural change that we feel is responsible for the stabilization of freshly broken ends in the endosperm.

Two translocation chromosomes isolated from high-loss progenies proved to be deficient for the tip of 9S including the knob and the C locus but possessed the Wx allele and were capped by a terminal piece from a heterologous chromosome. One of these arose in a microspore where bridge breakage produced a chromosome 9 lacking the terminal tip, which formed an acentric fragment. Chromosome 4 with a knob in the long arm likewise formed a dicentric bridge in the same microspore. Breakage of the second bridge yielded an acentric 4L fragment with a freshly broken end. The joining of the broken end of the deficient 9 with the broken end of the acentric 4L fragment produced a translocated 9^4 chromosome deficient for a terminal segment of 9S, possessing the Wx allele and capped by a terminal segment of 4L including its distal knob. If such a chromosome were introduced into the primary endosperm nucleus, it would not undergo a B–F–B cycle. The mature kernel would be colorless and have no Wx–wx mosaicism. Such a mechanism, we suggest, is the correct explanation of the stable c Wx kernels.

Nondisjunction of the maize B chromosome does not occur during the divisions of the megagametophyte; it is a phenomenon restricted to the second division of the male gametophyte. Chromatin elimination of knobbed A chromatin takes place in the same cell division where B chromosomes nondisjoin. Our conclusion that elimination of knob-bearing chromatin did not occur during development of the embryo sac was based on the failure to find kernels with colorless aleurone when individuals of the high-loss line with knobbed chromosomes 3 carrying the A allele were used as the female parent in crosses with recessive a pollen parents. The reciprocal crosses yielded as high as 20% color-

TABLE 2. Distribution and frequency of breaks in the short arm of chromosome 9-based on number of kernels showing whole loss and variegation for the C and Wx genes

Kernel phenotype	C, Wx Kernels, no break, colored aleurone, blue-staining starch	C-c, Wx-wx Kernels from breaks in (1); mosaic colored-colorless aleurone, blue- and red-staining starch	c, Wx-wx Kernels from breaks in (2); colorless aleurone, blue- and red-staining starch	c, Wx Kernels from breaks in (2); colorless aleurone, blue-staining starch	c, wx Kernels from breaks in (3); colorless aleurone, red-staining starch	Sum
Number	5342	72	52	20	1865	7351

less kernels. However, the basis for this conclusion should be carefully examined. The lack of colorless kernels in the female cross cannot be taken as conclusive evidence that no loss occurred. Development of the female gametophyte is more complex than that of the male. Three mitoses give rise to an eight-nucleated embryo sac. One of the polar nuclei and the egg nucleus are sisters lying in the micropylar half of the embryo sac, whereas the other polar nucleus comes from the chalazal portion. The triploid endosperm is the result of the fusion of a sperm with the two polar nuclei, one coming from the micropylar region and the other from the chalazal. The expected frequencies of embryos and endosperms deficient for the *A* gene following chromatin loss at the first, second, and third mitosis are given in Table 3.

As Table 3 shows, chromatin loss during embryo sac development leads to deficient endosperms much less frequently than to deficient embryos. In fact, if elimination of the knobbed *A* segment occurred only at the first megaspore mitosis or in either the micropylar or chalazal nucleus at the second division, no kernels with endosperms deficient for *A* would be found. If loss were restricted to these times, we would erroneously conclude from the endosperm phenotypes that there was no chromatin loss during embryo sac development. To determine whether there is loss, testcrosses were made with high-loss plants having knobbed chromosomes 3 and the dominant *Gl6*, *Lg2*, and *A* genes as the female parent and embryo, rather than endosperm, loss was determined. The recessive glossy and liguleless phenotypes would be expressed in the F_1 seedlings lacking the dominant *Gl* and *Lg* alleles. High-loss plants giving 10–12% loss of *A* in the endosperm, when used as the pollen parent in testcrosses, produced no recessive phenotypes in female testcrosses. In a seedling popula-

TABLE 3. Deficiency for the A allele in embryos and endosperms following chromatin loss at the three mitoses in embryo sac development

Time of postulated A loss in embryo sac development	Frequency of embryos deficient for A (%)	Frequency of endosperms deficient for A (%)
First megaspore division	50	0
Second megaspore division		
Loss in micropylar nucleus	50	0
Loss in chalazal nucleus	0	0
Coincident loss in both nuclei	50	25
Third megaspore division		
Loss in any one of the four nuclei	12.5	0
Loss in any two nuclei	25.0	4.17
Loss in any three nuclei	37.5	12.5
Loss in all four nuclei	50.0	25.0

tion in excess of 4000 from female testcrosses, no plants were found with the glossy or liguleless phenotypes expected if there was elimination during the embryo sac mitoses. The only division in female gametogenesis that would not produce recessive seedling phenotypes is the second mitosis, with loss restricted to a chalazal nucleus. The data strongly suggest that B-chromosome-induced loss of knobbed A chromatin is restricted to the second microspore division and that it does not take place in the embryo sac.

Another aspect of the high-loss phenomenon merits attention. As mentioned above, two dissimilar sperm cells are produced after chromatin loss at the second microspore division, and it may be asked whether preferential fertilization of the egg and polar nuclei by the differing sperm cells occurs. Preferential fertilization was first discovered in maize carrying a TB–A translocation where nondisjunction of the B^A chromosome at the division of the generative nucleus gives a hyperploid sperm with two B^A chromosomes and a hypoploid with none (Roman, 1947, 1948). The hyperploid sperm preferentially fertilizes the egg and the hypoploid sperm unites with the two polar nuclei. Carlson (1969) demonstrated that preferential fertilization of the egg by the hyperploid sperm with two B^9 chromosomes did not occur when several intact Bs were present in both sperm cells (designated the *swamping effect*). Dissimilar sperm are also formed at the second microspore division in high-loss plants; one sperm is hypoploid, and the other is usually euploid. Chromosomal elimination in high-loss plants occurs only when several Bs are present. Thus both sperm cells would carry Bs. Any potential that the euploid sperm might have for preferential fertilization would be suppressed if the swamping effect found by Carlson were operating. Random fertilization of egg and polars by dissimilar sperm should give equal numbers of deficient embryos and endosperms. These can be recognized by genetic markers. In our 1967 paper (Rhoades et al., 1967), we reported a slight excess of deficient endosperms in a limited sample. The data suggested that the two dissimilar sperm probably were randomly involved in double fertilization of egg and polar nuclei. This tentative conclusion was in accord with Carlson's prediction.

However, extensive data subsequently acquired were at variance. As a rule, deficient endosperms were found more frequently than deficient embryos, but there was a considerable range in different crosses. The matter called for further investigation. We now conclude that fertilization is truly random and that the variation in rates of loss in endosperm versus embryo is caused by the tendency of deficient embryos to abort, producing either germless kernels or those with embryos so abnormal that they are unable to germinate. The triploid endosperm with its diploid and euploid genome from the female parent is buffered against the deleterious effects of aneuploidy or genic unbalance. It is able to develop to maturity even though it has a deficient chromosome. Deficient embryos are not buffered by polyploidy and with varying frequencies, pre-

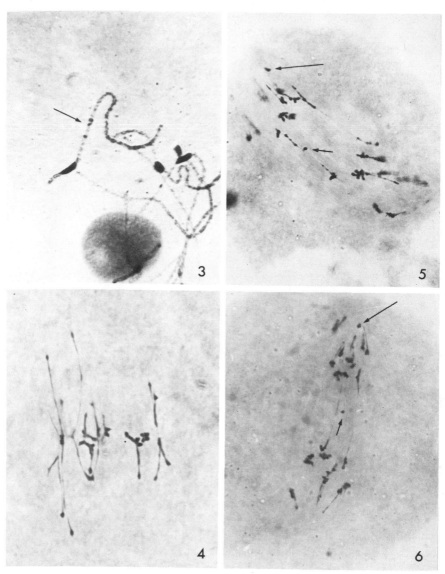

Figure 3. Homozygous K10 bivalent at pachynema illustrating the large knob and the three small knobs (indicated by arrow) that lie proximal to the extra segment of K10.

Figure 4. Neocentric activity at MII in a K10 K10 plant. One knobless dyad (chromosome 9) has no neocentromeres; the seven dyads with knobbed long arms have a single arm stretched toward each pole; the two dyads (chromosomes 1 and 3) with knobs in the long and short arms have neocentromeres in both arms. The rightmost dyad has neocentromeres in both arms, the knobless chromosome 9 dyad is second from the right, and one of the seven dyads with one knobbed arm is third from the right.

sumably dependent on modifying genes, fail to develop during embryogenesis. In short, preferential fertilization is not believed to occur. For a fuller account of these experiments, see our report in the 1973 Maize Genetics Cooperation Newsletter (Rhoades and Dempsey, 1973b).

GENETIC EFFECTS OF THE K10 HETEROCHROMATIN

Races of maize have two kinds of chromosome 10, the shortest member of the complement. The usual type, or N10, with an arm ratio of $1:2.8$ has deeply staining chromomeres in both arms adjacent to the centromere (centric heterochromatin) and lightly staining chromomeres in the distal ends. The abnormal type, or K10, differs from N10 by having an extra segment of chromatin at the end of the long arm. Races with this extra piece have been reported from the Southwestern United States and from Latin America (Longley, 1937, 1938). Its length at pachynema is approximately equal to 10S. The proximal portion is euchromatic as is a smaller distal segment. A large conspicuous knob lies between the two euchromatic pieces (Figure 3). The N10 differs from K10 not only in lacking the extra chromatin, but also in the distal one-sixth of 10L, where K10 has three small knobs or prominent chromomeres not present in N10. Although intimate pachytene pairing of the dissimilar distal ends of the two long arms is observed at pachynema in heterozygotes, they must differ structurally other than by the three small knobs because no crossing over takes place in this segment (Kikudome, 1959).

The large heteropycnotic knob on the K10 chromosome is superficially similar to the large knobs on other chromosomes, but it clearly differs from them by its ability to induce a number of unusual phenomena. Whether the euchromatic chromomeres spanning the K10 knob play a role, however minor, in these phenomena has not been unequivocally determined, but the evidence at hand from experiments in which the extra segment has been dissected for activity suggests that the knob heterochromatin is chiefly responsible (Miles, 1970). The K10 chromosome produces the following effects: (a) preferential segregation during megasporogenesis, (b) neocentromere formation, and (c) enhanced recombination in centric heterochromatin and in the euchromatin of structural heterozygotes.

Preferential segregation and neocentromere formation involve the interaction of K10 with knobs on other chromosomes. Preferential segregation is

---

Figures 5 and 6.   Anaphase I with dicentric bridges and acentric fragments arising from crossing over in K10 K10 plants heterozygous for *In*3a and *In*7a. Both chromosomes 3 were knobless, and the two chromosomes 7 had a large knob in the long arm. The knobless acentric *In*3a fragment (short arrow) lies passively at the side of the *In*3a bridge, whereas the large knobbed acentric *In*7a fragment (long arrow) has been moved to the pole by neocentric activity.

an example of meiotic drive, which has been defined as the unequal production of two alternative classes of gametes where equality is expected. Rhoades (1942) demonstrated that the K10 chromosome was preferentially recovered at megasporogenesis in K10 N10 plants. Instead of the expected 1 : 1 ratio, 70% of the megaspores received the K10 chromosome. The salient results of subsequent investigations (Rhoades and Vilkomerson, 1942; Longley, 1945; Rhoades, 1952; Emmerling, 1959; Kikudome, 1959; Rhoades and Dempsey, 1966; Miles, 1970) are briefly discussed.

The K10 chromosome is preferentially recovered in K10 N10 heterozygotes, but random segregation of genetically marked chromosomes 10 occurs in N10 N10 and K10 K10 homozygotes. However, in K10 N10 and K10 K10 plants, preferential segregation occurs for other chromosomes of the set when they are heterozygous for knobs. Segregation is random for homozygous knobbed or knobless chromosomes irrespective of the K10 constitution. It is the knobbed chromatid of heteromorphic dyads (with one knobbed and one knobless chromatid) that is preferentially recovered. Formation of a heteromorphic dyad by crossing over between the knob and the centromere is an essential prerequisite for preferential segregation. Segregation is random in the homomorphic KK and kk dyads also formed in Kk heterozygotes. The dependence of preferential segregation on crossing over to give heteromorphic dyads was demonstrated by a parallel reduction of preferential segregation and of crossover values in structural heterozygotes (Rhoades and Dempsey, 1966).

Neocentromeres were observed by Rhoades and Vilkomerson (1942) and by Rhoades (1952). A fuller account is found in Rhoades and Dempsey (1966). They occur in K10 K10 and K10 N10 plants, but not in N10 N10. Meiosis proceeds normally in K10 plants until MI and the bivalents are normally cooriented on the spindle. Before the onset of anaphase I disjunction, spindle fibers arise from distal portions of the arms when knobs are present. The distal neocentric regions undergo a precocious anaphase movement, pulling the ends of the chromosomes to the pole in advance of the true centric region and the remainder of the separating dyads. Neocentromeres are also found at the second meiotic metaphase and anaphase. Again there is a precocious poleward movement of neocentromeres, and the knobbed member of a heteromorphic dyad is the one that first reaches the pole. Because only knobbed arms form neocentromeres, the maximum number of stretched arms at anaphase can be predicted if the knob constitution of the ten bivalents at pachynema has been determined (see Figure 4).

The genetic data demonstrated that preferential segregation occurs in K10 plants when one member of a pair is knobbed and the other knobless, that heteromorphic dyads coming from crossing over are essential, and that it is a knobbed chromatid that is preferentially recovered. The concordance between the cytological observations on neocentromeres and the genetic data on preferential segregation led Rhoades (1952) to suggest that neocentromere formation accounted for preferential segregation. He hypothesized that the maintenance of the AI orientation of heteromorphic dyads, with the knobbed chromatid nearest the pole, into the second meiotic division of megasporogenesis would result in the basal spore preferentially receiving a knobbed chromatid [see Rhoades and Dempsey, (1966) for a detailed account]. It is a plausible hypothesis for which considerable evidence has been adduced.

According to the Rhoades hypothesis, the K10 chromosome causes the overproduction by the true centromere of a substance concerned with spindle-fiber formation. In N10 N10 plants, this putative spindle substance remains within the confines of the centric region but in K10 plants the excess moves distally along the arms. Interaction between knob and spindle substance from the centromere gives rise to neocentric regions that form precociously acting chromosomal fibers. The involvement of the true centromere in neocentric activity is indicated by the behavior of the acentric fragments produced by crossing over in heterozygous paracentric inversions. The knobless acentric fragment from In3a heterozygotes displayed no neocentric activity at AI, whereas the knobbed acentric fragment from In7a heterozygotes had neocentromeres which moved poleward in advance of the disjoining dyad. Figures 5 and 6 are from a K10 plant heterozygous for In3a and In7a. Both the normal and inverted chromosomes 3 are knobless, whereas the normal and inverted chromosomes 7 carry a large knob in the loop region. The smaller knobless acentric fragment from In3a lies passively by the side of its dicentric bridge at AI, whereas the larger knobbed acentric fragment from In7a has already progressed to the pole. Neither fragment shows any indication of neocentric activity at the second meiotic division. The inability of the knobbed fragment to congress on the MII spindle or to undergo active movement on the spindle indicates that no fiber-forming material is available. That which it received from the centromere prior to pachynema had been depleted in AI and was not replenished in the second division because the fragment was no longer physically connected to a centric region. The behavior of acentric fragments illustrates two points. First, it substantiates the conclusion that knobs are essential for neocentric activity and second, it reveals the true centromere as the source of the material needed for the formation of neocentromeres.

The effect of K10 on recombination was discussed in Rhoades and Dempsey (1966) and in Rhoades (1968). Briefly, K10 enhances recombination in proximal heterochromatic, but not in distal euchromatic, regions of either homozygous knobbed or knobless chromosomes. The K10 partially compensates for the localized reduction in crossing over caused by knob heterozygosity and brings about a marked increase in crossing over in euchromatic regions of plants heterozygous for structural changes, such as inversions and translocations. The enhanced crossover values are correlated with the more intimate synapsis observed in structural heterozygotes (Rhoades, 1968). Although K10 alone is able to produce this crossover enhancement in the absence of other kinds of heterochromatin, a synergistic effect is found when K10 and knobs are combined.

Emmerling (1959) and Miles (1970) studied rod derivatives of the K10 chromosome to see if the diverse genetic effects could be assigned to particular regions. These modified K10 chromosomes owe their inception to a ring chromosome obtained by Emmerling from X-irradiation of a K10 chromosome. The ring arose by fusion of broken ends where one break was very near the tip of 10S and a second break approximately in the middle of the K10 knob. The ring 10 is deficient for the tip of 10S, the distal half of the K10 knob, and the short euchromatic tip. A crossover in a ring/rod heterozygote yielded a dicentric bridge at meiotic anaphase. Breaks at anaphase occur at different positions. One bridge broke in the euchromatic segment of 10L to produce a modified rod chromosome 10 (the K^m10) with a normal long arm and a short arm to which was attached a portion of 10L consisting of the K^m knob (= the proximal half of the K10 knob), the segment with the three small knobs and a short duplication of 10L. If we designate the three small knobs in the K10 chromosome as a, b, and c with a lying closest to the centromere, the linear order is centromere–a–b–c–K10 knob, and in the K^m10 chromosome the order is centromere–K^m knob–c–b–a; in other words, the order is reversed with respect to that in the K10 chromosome. The K^m10 chromosome does not cause preferential segregation in heterologous knobbed/knobless bivalents as does K10, but shares with it the ability to enhance recombination values in centric heterochromatin. Miles reconstituted a new ring 10 from intrachromosomal crossing over following synapsis of the duplicated regions of K^m10, and from this newly synthesized ring she obtained six rod derivatives that were studied for their ability to induce preferential segregation, neocentromeres, and enhanced recombination. Five of the new rod derivatives had all or part of the K^m knob, and the three small knobs in the same orientation, with respect to the centromere, as in the K10 chromosome. These five rod chromosomes, differing from the K^m10 chromosome in the orientation of the heterochromatic segment, induced the preferential segregation which K^m10 was unable to do. The variable activity of the knob may be considered a *position effect,* since the K^m10 and the five rod derivatives differ in the position of the heterochromatic piece derived from the K10 chromosome. Miles's rod derivatives varied among themselves in knob size. She found that the degree of preferential segregation was directly correlated with knob size. Her sixth rod derivative had none of the K10 knob. It was unable to induce either preferential segregation or crossover enhancement. Her studies indicate that preferential segregation and enhanced recombination are separable phenomena controlled by different genic products. The K10 knob is responsible for preferential segregation, since progressively smaller portions of it resulted in lower degrees of preferential segregation, and in the absence of any K10 knob heterochromatin there was no preferential segregation. The correlation between knob size and degree of preferential segregation suggests that the K10 knob is composed of repetitive sequences of DNA.

One missing link in the above work is the genetic effect of that portion of the K10 chromosome present in K^m10 and the five rod derivatives when it is incorporated into a ring chromosome. Since Miles had entered medical school and was unable to perform this experiment, we undertook it. Our results indicate that the same segment of K10 chromatin which enhanced recombination but did not cause preferential segregation when present in the K^m10 chromosome and which produced both preferential segregation and high crossover values in the five rod derivatives, did not induce either of these phenomena when incorporated into a ring chromosome. These rather remarkable observations lend themselves to a variety of interpretations with respect to gene regulation that are beyond the purview of this chapter. Suffice it to say that the position and orientation of the heterochromatic segment derived from the K10 chromosome play an unsuspected role in its activity.

HETEROCHROMATIN AND SEX DIFFERENCES IN RECOMBINATION

A problem of concern to maize geneticists for many years has been the relative amount of crossing over in male and female flowers. Sex differences in recombination were reported in the early days of maize genetics, but the values were erratic and inconsistent. No clear pattern was evident. Much of the data came from experiments involving the short arm of chromosome 9, which possesses a number of easily scored mutants affecting the color and composition of the kernel. Stadler (1926) reported significantly higher values in the male gametes for the $C-Wx$, $Sh-Wx$, and $C-Sh$ intervals in 9 S. Collins and Kempton (1927) found higher $C-Sh$ crossing over in the male gametes of four progenies, but in six others the rate was higher on the female side. According to Emerson and Hutchinson (1921), certain plants had higher female than male values, whereas in others no differences existed between the sexes for the $C-Wx$ region. In Eyster's (1922) data there was no sex difference for this interval. The varying and discordant results reported for the same segment and the lack of a sex difference in regions of other chromosomes led maize geneticists to accept the generalization that the rate of crossing over is essentially equal in male and female flowers. No distinction as to the heterozygous parent was usually made in determining map locations. However, in 1941 extensive data were published showing significantly higher crossover values in the male flowers for the $A2-Bt$,

A2–Bm, Bt–Pr, and *Bm–Pr* regions of chromosome 5 (Rhoades, 1941). All of these segments span or are adjacent to the centromere. Nel (1973) confirmed these results. Both Phillips (1969, 1971) and Nel found no sex difference in distal euchromatic regions of chromosome 5, indicating that the sex difference was restricted to regions in or flanking the deeply staining, diffuse chromomeres comprising the centric heterochromatin.

In addition to the well-documented variability observed in certain segments of chromosome 5, sex differences have been reported for the *Rg–lg2* and *Rg–d* regions of chromosome 3, the *la–su* interval of chromosome 4, the *gl–in* region of 7, and the *sh–gl*15 interval of chromosome 9 (Phillips, 1971). All have higher male than female crossing over and all of them include centric heterochromatin. When a sex difference exists, crossover values are usually higher in the male flowers but Phillips (1969, 1971) demonstrated that the *Y–su2* region in the long arm of chromosome 6 has higher *female* than male crossing over. The location of the *Y–su2* segment with respect to centric heterochromatin has not been established. According to the gross cytological positions given in Phillips (1969), the *Y–su2* region is situated in the proximal half of the long arm but is not close to the centromere since *Y* is placed cytologically at position 0.17. The role of centric heterochromatin in this instance of higher female recombination is uncertain. However, maize is not the only organism with higher male crossing over in one region and higher female in a second. DeWinton and Haldane (1935) reported such a situation in Primula. Dunn and Bennett (1967) also report sex differences in rate of crossing over in opposite directions in different chromosomes of the mouse where 54 regions have been studied. In 30 there was no sex difference, whereas of the remaining 24 regions, 19 had higher female than male percentages, and five had higher male than female crossing over.

Rhoades (1941) suggested that the sex difference arose as a consequence of the proximal heterochromatin being less tightly coiled in microsporocytes and that relaxation in coiling is conducive to more frequent exchange. No cytological observations exist in support of this hypothesis, which rests primarily on the fact that crossing over per unit length at pachynema is lower in heterochromatic regions than in euchromatic ones and that coiling is tighter in heterochromatin than in euchromatin (Du Praw, 1968).

An alternative hypothesis assumes that crossing over is equal in the sexes, but, following a crossover, the segregation of the two noncrossover and crossover chromatids of the tetrad is not random (Rhoades, 1955). Selective orientation of the tetrad on the MI spindle occurs, it is assumed, with the two noncrossover chromatids facing the spindle poles while the two crossover chromatids comprising the chiasma are directed toward the equatorial plate. If the orientation at MI persists to the second meiotic metaphase, the basal megaspore of the linear set of four would preferentially receive a noncrossover chromatid. The two crossover and one noncrossover chromatids of the tetrad are included in the three abortive megaspores. Inasmuch as the egg is derived from the basal megaspore, nonrandom segregation during meiosis gives eggs with fewer than expected crossovers. In contrast, all four microspores from each PMC form functional male gametophytes so no crossover chromatids are eliminated from the gametic pool. This hypothesis was laid to rest by Nel's (1975) determination of the frequency of crossover chromatids in the half-tetrads of diploid eggs produced by plants homozygous for the *el* gene. On the preferential exclusion hypothesis, the percent of crossing over calculated from diploid eggs should equal that found in haploid sperm cells, whereas unequal rates in the two kinds of gametes would indicate a real difference in exchange frequency. The latter was found to be true. Sex differences in recombination for the *A2–Bt–Pr* region of chromosome 5 cannot be ascribed to the mechanism of selective segregation. Although the mechanism responsible for sex differences in crossing over remains conjectural, the assumption of equivalence in all regions of the chromosome is clearly unjustified.

Since centric heterochromatin is responsible for sex differences in crossing over, other kinds of heterochromatin might also influence differentially the frequency of recombination in the male and female inflorescences. The specific effect would depend on the nature of the heterochromatin. Recent studies have disclosed a sex difference involving the discrete heteropycnotic knobs of chromosomes 9 and 3. These sharply delimited knobs are morphologically quite distinct from the diffuse chromomeres of centric heterochromatin.

Evidence that the heterochromatic knob terminating the short arm of chromosome 9 is responsible for a sex difference in crossing over was fortuitously discovered in an experiment designed to ascertain the amount of recombination in chromosomes bereft of their normal telomeres. Gillies (1975) observed that the ends of the leptotene chromosomes of maize are associated with the nuclear membrane, indicative of a possible presynaptic alignment of homologous chromosomes. Juxtaposition of the ends of homologues at this stage could facilitate pairing at zygonema and thus enhance crossing over. If the terminal chromomere, the telomere, is a specialized structure differing from the interstitial chromomeres, then a chromosome end consisting of an interstitial chromomere substituted for the true telomere might not be able to become associated with the nuclear membrane. Failure of presynaptic alignment could reduce pairing and crossing over. A chromosome without a normal telomere that is transmissible through both male and female gametophytes is available in maize. The *wd* chromosome 9 derived by McClintock (1941a) lacks the terminal knob, the thread-like stalk connecting the terminal chromomere to the knob and a portion of the ultimate chromomere. The lethal albino phenotype of homozygous *wd* plants would normally preclude the study of crossing over in these individuals were it not that the small *Wd* ring chromosome, which carries the genes deficient in the *wd* chromosome, can be introduced into *wd* plants. Plants with the *Wd* ring and homozygous *wd* have striped leaves of green and white tissue because of the instability of the ring chromosome during development. The ring chromosome seldom if ever pairs with homologous portions of the normal or deficient chromosomes 9 at pachynema and was found to have no effect on crossing over in the chromosome 9 bivalent. Consequently, plants with and without the ring chromosome (but varying in chromosome 9 constitution) were used in crossover studies.

The following four genotypes are produced by crossing

$$\frac{K^L \; Yg \; c \; sh \; wx}{wd \; c \; sh \; wx} \times \frac{K^L \; Yg \; C \; Sh \; Wx}{wd \; C \; Sh \; Wx} \text{ plus } Wd \text{ ring}$$

Class I $\dfrac{K^L \; Yg \; c \; sh \; wx}{K^L \; Yg \; C \; Sh \; Wx}$

Class II $\dfrac{wd \; c \; sh \; wx}{wd \; C \; Sh \; Wx}$

Class III $\dfrac{wd \; c \; sh \; wx}{K^L \; Yg \; C \; Sh \; Wx}$

Class IV $\dfrac{wd \; C \; Sh \; Wx}{K^L \; Yg \; c \; sh \; wx}$

All are heterozygous for the *C, Sh,* and *Wx* loci that lie in the distal half of the short arm of chromosome 9. Class I is homozygous for a large knob (K^L) on the end of the short arm; class II is homozygous knobless; classes III and IV are both heterozygous for a knobbed and knobless 9 but differ in the linkage phase of the knob with the marker genes. Class II plants with the *Wd* ring are green and white striped. Class I and II individuals segregated only for *C, Sh,* and *Wx.* Classes III and IV can be distinguished from class I because their heterozygosity for *wd* is revealed in testcross progenies. A number of plants of each of the four classes were testcrossed reciprocally. The fact that all four classes are sibs minimizes any variation in crossover values caused by genetic modifiers. Although all testcrossed individuals were segregating for the *Sh* and *sh* alleles, the shrunken phenotype was difficult to score on some ears. Crossover values for the *C–Sh* and *Sh–Wx* regions were obtained only when there was an accurate classification for the shrunken phenotype. The mean recombination values for the four classes are given in Table 4. In all cases the *C–Wx* value represents the percentage of recombination and not the sum of the *C–Sh* and *Sh–Wx* regions.

The data in Table 4 allow the following conclusions: First, in plants with two knobless chromosomes 9 (class II), the amount of recombination between *C* and *Wx* in the microsporocytes does not differ from that in the megasporocytes.

TABLE 4. Chromosome 9 recombination values in plants of varying knob constitution

Class I K^L *Yg c sh wx*
$\overline{K^L\ Yg\ C\ Sh\ Wx}$

Crossover data from eleven pairs of reciprocal testcrosses

	C–Sh		Sh–Wx		C–Wx		Population size	
	♂	♀	♂	♀	♂	♀	♂	♀
	4.25	3.02	26.25	19.43	29.75	22.34	3502	2950
	(41%)[a]		(35%)		(33%)			

Class II   *wd c sh wx*
$\overline{wd\ C\ Sh\ Wx}$

Crossover data from three pairs of reciprocal testcrosses

	C–Sh		Sh–Wx		C–Wx		Population size	
	♂	♀	♂	♀	♂	♀	♂	♀
	4.3	4.6	25.9	25.9	29.6	30.0	3464	841

(Equivalency of male and female recombination values)

Class III   *wd c sh wx*
$\overline{K^L\ Yg\ C\ Sh\ Wx}$

Crossover data from twelve pairs of reciprocal testcrosses

	C–Sh		Sh–Wx		C–Wx		Population size	
	♂	♀	♂	♀	♂	♀	♂	♀
	2.49	0.66	23.26	8.54	25.48	9.22	3704	3692
	(277%)[a]		(172%)		(176%)			

Class IV   *wd C Sh Wx*
$\overline{K^L\ Yg\ c\ sh\ wx}$

Crossover data from three pairs of reciprocal testcrosses

	C–Sh		Sh–Wx		C–Wx		Population size	
	♂	♀	♂	♀	♂	♀	♂	♀
	2.46	1.17	25.77	16.80	27.83	17.33	1555	991
	(105%)[a]		(53%)		(61%)			

[a] Percent increase in ♂ meiocytes.

However, plants containing K^L had higher male than female crossing over. The increase in male gametes was more pronounced in knobbed/knobless heterozygotes (classes III and IV) than in individuals homozygous for the knob (class I). The percent increase in class III plants was greater than that in class IV plants. This can probably be ascribed to sampling errors since class IV consisted of only three pairs of reciprocal crosses and all three had a sex difference comparable to pairs in the lower range of class III. Therefore, a sex difference in crossing over is associated with the terminal knob on 9S. Second, the amount of crossing over in microsporocytes where both homologues of a pair are without the usual telomere is equivalent to that in bivalents homozygous for the large terminal knob and is somewhat in excess of that in knob heterozygotes. Moreover, the amount of female crossing over in the $C-Wx$ region is significantly higher in homozygous knobless plants than in either of the knobbed classes. If pairing at zygonema and crossing over at pachynema are influenced by presynaptic alignment resulting from an association of telomeres with the nuclear membrane, it apparently matters little whether the telomere is one normally found at the end or consists of a formerly interstitial chromomere. A different situation exists in *Zea mays* than in *Drosophila,* where broken ends are never converted into stable ones. In a recent paper, Roberts (1975) argues for the reality of the telomere concept. He showed that some supposedly terminal losses in the X chromosome are capped by the distal X telomere. The difference in behavior of broken ends in the two organisms is not due to dissimilarities in chromosome structure but reflects a difference in specific cellular environments.

The unanticipated sex difference caused by the knob on chromosome 9 is confirmed by our analysis of the chromosome 9 data of Chang and Kikudome (1974) from K^s/K^s and K^L/K^s plants with different numbers of B chromosomes. Those authors were primarily concerned with the effect on crossing over of odd versus even numbers of B chromosomes and did not address the problem of a sex difference. Related plants of

$$\frac{K^s\ Yg\ Sh\ Bz\ Wx}{K^s\ yg\ sh\ bz\ wx} \quad \text{and} \quad \frac{K^s\ Yg\ Sh\ Bz\ Wx}{K^L\ yg\ sh\ bz\ wx}$$

genotypes, each with numbers of Bs ranging within 0–5, were reciprocally testcrossed. The four marker genes are in the short arm of 9 with the Yg locus near the distal end. The K^s knob is small, whereas K^L is a very large terminal knob. The amounts of recombination for the $Yg-Sh$ and $Bz-Wx$ regions in male and female inflorescences of plants with different numbers of Bs are given in Table 5. We did not include the $Sh-Bz$ interval because it is too short to provide meaningful crossover data.

In the K^L/K^s heterozygotes, recombination in the distal $Yg-Sh$ region is 82.8% higher in male flowers than in female, whereas in the proximal $Bz-Wx$ interval male crossing over is only 15.2% higher. Increased numbers of B chromosomes cause a progressively smaller sex difference in the $Bz-Wx$ region of K^L/K^s heterozygotes. Increasing the number of Bs may have a similar effect in the distal $Yg-Sh$ region, but the data are less convincing. The sex difference is the smallest in plants with 5 Bs, but there is no obvious trend in going from the 1B to the 5B class. The sex difference in both the $Yg-Sh$ and $Bz-Wx$ regions is greatest in plants with no Bs, suggesting that Bs reduce the sex differential.

The 8.4% increase in male over female recombination in the $Yg-Sh$ region of K^s/K^s homozygotes approximates the 10.1% found in the proximal $Bz-Wx$

TABLE 5.   Amounts of recombination for different numbers of Bs from data in Chang and Kikudome (1974)

$$\frac{K^s\ Yg\ Sh\ Bz\ Wx}{K^s\ yg\ sh\ bz\ wx} \times yg\ sh\ bz\ wx\ reciprocally$$

	Yg–Sh Region				Bz–Wx Region		
	Percent Recombination in $\male$	Percent Recombination in $\female$	$\male-\female$		Percent Recombination in $\male$	Percent Recombination in $\female$	$\male-\female$
0B	25.6	23.9	+1.7	0B	18.1	15.7	+2.4
1B	23.3	22.1	+1.2	1B	20.6	17.3	+3.3
2B	26.5	24.2	+2.3	2B	17.6	16.2	+1.4
3B	24.9	21.8	+3.1	3B	17.2	17.4	−0.2
4B	24.5	23.5	+1.0	4B	17.8	16.1	+1.7
5B	25.8	23.7	+2.1	5B	19.8	18.3	+1.5
(Average)	(25.1)	(23.2)	(+1.9)		(18.5)	(16.8)	(+1.7)

$$\frac{K^s\ Yg\ Sh\ Bz\ Wx}{K^L\ yg\ sh\ bz\ wx} \times yg\ sh\ bz\ wx\ reciprocally$$

	Yg–Sh region				Bz–Wx region		
	Percent Recombination in $\male$	Percent Recombination in $\female$	$\male-\female$		Percent Recombination in $\male$	Percent Recombination in $\female$	$\male-\female$
0B	20.5	8.4	+12.1	0B	25.2	19.1	+6.1
1B	18.2	10.9	+7.3	1B	25.3	21.4	+3.9
2B	15.9	8.1	+7.8	2B	25.5	21.1	+4.4
3B	19.2	10.4	+8.8	3B	25.5	23.4	+2.1
4B	18.5	9.0	+9.5	4B	24.1	21.9	+2.2
5B	16.7	12.6	+4.1	5B	23.9	22.8	+1.1
(Average)	(18.2)	(9.9)	(+8.3)		(24.9)	(21.6)	(+3.3)

interval indicating that the slight increase in male crossover values is evenly distributed along the short arm. This is in contrast to the K^L/K^s heterozygotes where there was a much greater excess of male crossing over in the distal Yg–Sh segment compared to the proximal Bz–Wx region. In the presence of the large terminal knob of K^L/K^s heterozygotes the sex differential is greatest for regions in proximity to the knob and is less pronounced in more distant regions. It should be noted that heterofertilization would give apparently higher male recombination if one trait (Yg) is expressed in the sporophyte and a second (Sh) in the endosperm, but Chang and Kikudome (1974) state that its infrequent occurrence did not affect their crossover values.

Plants homozygous and heterozygous for a large knob on 9S in our experiments had higher male crossover values, but there was no sex difference in plants homozygous for a knobless 9. Chang and Kikudome's data differ from ours in that their crossover values came from plants homozygous for a small knob and from those heterozygous for a large and small knob. Their K^L/K^s data with higher male values are in agreement with ours from K^L/k heterozygotes. Of interest is the comparison of male and female crossing over in their K^s/K^s homozygotes and in our knobless homozygotes. Slight but consistently higher male crossing over was found in the former and equivalent values in the latter. A consideration of the crossover data from our K^L/K^L, K^L/k, and k/k combinations and from Chang and Kikudome's K^L/K^s and K^s/K^s individuals permits the drawing of certain inferences regarding the role of knobs in a quantifiable sex difference. No difference between the sexes was found in knobless (kk) homozygotes, only slightly higher male recombination occurred in K^s/K^s homozygotes, and there was a much greater excess of male over female crossing over in K^L/K^L, K^L/K^s, and K^L/k combinations. The greater sex difference induced by a large knob is dominant to the lesser effect produced by a small one. However, the degree of crossover enhancement is not dose dependent since the sex difference in K^L/k heterozygotes exceeds that in K^L/K^L homozygotes. An apparent hierarchical arrangement exists for knob size with respect to the degree of sex differences. Large knobs produce a marked sex difference, small knobs a slight one, and knobless chromosomes no sex difference.

The discrepant results reported previously on sex differences in crossover values for the C–Wx region become intelligible on the assumption that knobbed chromosomes 9 were present in those strains giving a sex difference and were absent where no sex difference occurred. A plausible but unproved hypothesis has been advanced to account for sex differences in regions of the centric heterochromatin, but how knobs operate is wholly conjectural.

Studies on crossing over in plants heterozygous for $In3a$ also disclose a sex differential that is related to the presence of a large heteropycnotic knob at position 0.6 in the long arm of chromosome 3. Only a summary statement is given here because of space limitation. Inversion-3a ($In3a$) is a paracentric inversion with breakpoints at 0.4 and 0.95. The inverted segment, which is approximately 60 map units in length (Rhoades and Dempsey, 1953), includes the $Lg2$ and A loci as well as the knob at 0.6. The proximal $Gl6$ gene is not included in the inversion. There are 12 map units between the proximal breakpoint of the In and $Lg2$ and an equal distance between A and the distal break. In N3 N3, k3 k3 plants, 36% recombination is found in the Lg–A interval. The Lg locus and the knob are closely linked in K/k heterozygotes; no estimate of the frequency of crossing over between Lg and the knob is available from plants homozygous K3 K3 or k3 k3, but it is undoubtedly greater than that occurring in K3 k3 heterozygotes where asynapsis is frequently observed at pachynema in regions adjacent to the knob.

$$\text{The cross of } \frac{\text{K10}}{\text{N10}} \; \frac{\text{In } Gl \, A \, \text{K} \, Lg}{\text{N} \, gl \, lg \, \text{k} \, a} \times \frac{\text{N10}}{\text{N10}} \; \frac{\text{In } Gl \, A \, \text{k} \, Lg}{\text{N} \, gl \, lg \, \text{K} \, a}$$

individuals yielded four compounds heterozygous for the marker genes and varying in respect to the K10 chromosome and the K3 knob. These are:

$$\text{Class I} \qquad \frac{\text{K10}}{\text{N10}} \; \frac{\text{In } Gl \, A \, \text{K} \, Lg}{\text{N} \, gl \, lg \, \text{K} \, a}$$

$$\text{Class II} \qquad \frac{\text{K10}}{\text{N10}} \; \frac{\text{In } Gl \, A \, \text{k} \, Lg}{\text{N} \, gl \, lg \, \text{k} \, a}$$

$$\text{Class III} \qquad \frac{\text{N10}}{\text{N10}} \; \frac{\text{In } Gl \, A \, \text{K} \, Lg}{\text{N} \, gl \, lg \, \text{K} \, a}$$

$$\text{Class IV} \qquad \frac{\text{N10}}{\text{N10}} \; \frac{\text{In } Gl \, A \, \text{k} \, Lg}{\text{N} \, gl \, lg \, \text{k} \, a}$$

All are full sibs heterozygous for the inversion. Class I is heterozygous for K10, as is class II; they differ in that the K3 knob is homozygous in class I, whereas class II plants have knobless 3s. Class III is homozygous for the K3 knob and class IV is knobless. Both are homozygous N10. The four heterozygotes were testcrossed reciprocally and the ensuing progenies classified for the segregating traits. The summary of the recombination values presented in Table 6 is based on extensive data. A correction has been made in the data from heterozygous females for the occasional functioning of embryo sacs with deficient–duplicate chromosomes arising from breakage of the dicentric bridge at the meiotic anaphases (Rhoades and Dempsey, 1953). Dicentric bridges come from crossing over in the inversion loop. The crossover regions are indicated in Figure 7, which is a diagrammatic representation of the pachytene pairing in a plant heterozygous for *inversion*-3a.

The A allele could not always be classified in the sporophyte because the R^g allele was segregating. Consequently, all instances of heterofertilization could not be detected. Presumably its frequency was the same in the four genotypic classes and the observed amounts of crossing over were not markedly affected by its occurrence. The conclusions reached regarding recombination are valid but may be subject to minor revisions when the true amount of heterofertilization is accurately determined (see Table 6).

There is no sex difference in classes II and IV, which have only knobless chromosomes 3, whereas in classes I and III, where K3 is homozygous, male crossing over is higher than female. Since this is precisely the result found for homozygous knobbed and knobless chromosomes 9, it suggests that a similar sex difference will be found for other knobbed chromosomes. Obviously many questions remain to be answered. Is the sex difference produced by knobs restricted to the knob-bearing arm? Is it greatest in regions adjacent to the knob, as the data of Chang and Kikudome indicate? Is there any interaction between knobs on different chromosomes? Is there a sex difference in structurally normal chromosomes 3 which are knobbed but not in knobless 3s? This is an important matter to resolve, since it could be that the sex difference found for the structural heterozygote might not occur in homozygotes.

An unexpected feature of the *In*3a data is the increase in recombination for the inverted *Lg–A* region in K3 K3 compared to k3 k3 plants. That is, knob homozygosity gives significantly higher crossover values for this region in both sexes. This is confirmed by a comparison of class I with class II where K10 is held constant and the two classes differ by the presence and absence of the K3 knob and also in a comparison of class III with class IV where N10 is homozygous in both. The marked enhancement in crossing over within the inversion loop caused by knob homozygosity is surprising, since our data for the *Lg–A* region in knobbed and knobless normal chromosomes 3 show at most a very modest increase in K3 K3 plants.

A comparison of: (a) K10 N10, K3 K3 with N10 N10, K3 K3 and (b) K10 N10, k3 k3 with N10 N10, k3 k3 shows the enhancement in crossing over produced by K10. This is not an unanticipated finding inasmuch as the K10 effect on crossing over has been known for some time (Rhoades and Dempsey, 1966).

The most striking feature of the data is the synergistic action of K10 and K3. The crossover enhancement in K10 N10, K3 K3 plants is greater than the sum of the increases induced by each taken singly. The 1–2% of *Lg–A* crossing over found in N10 N10, k3 k3 inversion heterozygotes is a drastic reduction from the control value of 36% in normal chromosomes 3. All *Lg–A* crossovers come

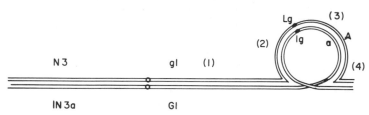

Figure 7. Diagrammatic representation of pachytene pairing in an *In*3a heterozygote showing the crossover regions. The K3 knob is homozygous.

TABLE 6. Percentages of recombination in *In*3a heterozygotes differing for K10 and K3

				Gl–Lg	Lg–A	Gl–A
Class I	K10 N10	K3 K3	♀	22.2	10.9	15.8
			♂	29.6	22.9	23.2
Class II	K10 N10	k3 k3	♀	25.2	4.0	23.9
			♂	25.5	2.7	25.3
Class III	N10 N10	K3 K3	♀	8.4	3.7	6.8
			♂	17.2	9.3	14.6
Class IV	N10 N10	k3 k3	♀	14.1	1.3	14.7
			♂	14.3	1.2	14.6

from double exchanges in regions 2–3 and 3–4 and from triple exchanges in regions 1–2–3, 1–3–4, and 2–3–4 (see Figure 7). From all of the double and triple exchanges, half of the chromatids are lost and half are recovered. The viable spores have equal numbers of *Lg–A* crossovers and noncrossovers. The relative numbers of *Lg–A* crossovers in the progenies of K10 N10, K3 K3; K10 N10, k3 k3; N10 N10, K3 K3; and N10 N10, k3 k3 individuals is a direct measure of the frequency of specific multiple exchanges in the microsporocytes of the four genotypes, since no *Lg–A* crossovers are recovered from single exchanges, which form dicentric bridges at AI. The frequency of *Lg–A* crossovers in K10 N10, K3 K3 plants, which is several times greater than in N10 N10, k3 k3 sibs, indicates that a striking increase in multiple exchanges has occurred within the inversion loop. Unless there is high negative chiasma interference, it appears that single, as well as multiple, exchange frequency in the chromatin of the inverted segment is very much greater than that taking place in structurally normal chromosomes 3. We have here a good example of a specific genetic effect (increased recombination) produced by the interaction of the heterochromatin of K10 and K3.

Incidentally, in these inversion heterozygotes the heterochromatic B chromosome had no effect on recombination. Sib plants with 0, 2, or 4 Bs in each of the four genotypes did not differ in rates of crossing over.

In summary, both the diffuse, deeply staining centric heterochromatin of several chromosomes and the discrete knobs of chromosomes 9 and 3 are associated with a sex difference in recombination. However, a distinction should be made between the activities of the two types of heterochromatin. The knobs of chromosomes 9 and 3 clearly act as inducers of the sex difference, since in their absence recombination occurs at the same rate in both mega- and microsporocytes. Whether centric heterochromatin also plays an inducing role is difficult to demonstrate, since strains of maize show little variation in appearance of centric heterochromatin in pollen mother cells at pachynema. No strains are known in which centric heterochromatin is missing. It seems more likely that changes in the physical state of centric heterochromatin, which modify exchange frequencies, may occur in response to dissimilar physiological conditions present in mega- and microsporocytes—that is, higher male crossing over in centric heterochromatin arises from changes in state induced by some unknown factor in the cellular environment. Another difference in the recombinational activities of the two kinds of heterochromatin involves the regions where the sex difference is expressed. The sex difference associated with centric heterochromatin is restricted to segments that consist, at least in part, of centric heterochromatin, whereas the sex difference due to the K9 knob is greatest in the euchromatic regions adjacent to the knob and is less in more distant regions.

In addition to the induction of sex differences in recombination, both the K10 and K3 knobs markedly increased crossing over in a heterozygous inversion. A synergistic interaction between K10 and K3 in inversion heterozygotes produced an increase of multiple exchanges in the PMCs of K10 N10, K3 K3 plants many times higher than that in N10 N10, k3 k3 sibs.

Knowledge of the location of centric heterochromatin with respect to the marked segments and of knob size and number would be highly desirable in any future study of a sex difference in crossing over. Dunn and Bennett (1967) are in sympathy with the view that sex differences come from localized modifications in the state of the chromatin that affect the probability of chiasma formation.

If the sex difference in recombination due to the K9 and K3 knobs holds for knobs in general, what adaptive role did knobs play in the evolutionary divergence of maize? The sex differential in recombination is the only characteristic that has been attributed to knobs alone. It is true that knobs interact with B chromosomes in chromatin elimination at the second microspore division, but this is surely a detrimental effect. Knobs also interact with the K10 chromosome in neocentromere formation and in preferential segregation. These phenomena are of interest to the cytogeneticist, but their adaptive significance in the economy of the species is questionable. Our inability to see what selective advantage a lower rate of female crossing over conferred to the maize plant as it evolved during the millennia may reflect a lack of perceptiveness. Possibly knobs are physiologically neutral relic organelles that have persisted because of tight linkage to favorable genetic factors. Be that as it may, knobs are an indispensable aid in arriving at an understanding of the architecture and evolution of the maize germplasm.

SUMMARY

The heterochromatin of knobs, B chromosomes, Abnormal 10, and the centric region is responsible for a variety of effects including preferential segregation, neocentromere formation, sex difference in recombination, chromatin elimination, and enhanced recombination. Some effects result from specific interaction between different kinds of heterochromatin; others require only the action of one source of heterochromatin.

REFERENCES

Carlson, W. 1969. Factors affecting preferential fertilization in maize. *Genetics* **62:** 543–554.

Carlson, W. 1970. Nondisjunction and isochromosome formation in the B chromosome of maize. *Chromosoma* **30:** 356–365.

Carlson, W. 1973a. Instability of the maize B chromosome. *Theor. Appl. Genet.* **43:** 147–150.

Carlson, W. 1973b. A procedure for localizing genetic factors controlling mitotic nondisjunction in the B chromosome of maize. *Chromosoma* **42:** 127–136.

Chang, C. C. and G. Y. Kikudome. 1974. The interaction of knobs and B chromosomes of maize in determining the level of recombination. *Genetics* **77:** 45–54.

Collins, G. N. and J. H. Kempton. 1927. Variability in the linkage of two seed characters in maize. *USDA Bull.* No. 1468.

DeWinton, D. and J. B. S. Haldane. 1935. The genetics of *Primula sinensis*. III. Linkage in the diploid. *J. Genet.* **31:** 67–100.

Dunn, L. C. and D. Bennett. 1967. Sex differences in recombination of linked genes in animals. *Genet. Res.* **9:** 211–220.

DuPraw, E. J. 1968. *Cell and molecular biology*. New York: Academic Press.

Emerson, R. A. and C. B. Hutchinson. 1921. The relative frequency of crossing over in microspore and in megaspore development in maize. *Genetics* **6:** 417–432.

Emmerling, M. H. 1959. Preferential segregation of structurally modified chromosomes in maize. *Genetics* **44:** 625–645.

Evans, G. M. and A. J. Macefield. 1973. The effect of B chromosomes on homoeologous pairing in species hybrids. I. *Lolium temulentum* × *Lolium pirenne*. *Chromosoma* **41:** 63–73.

Eyster, W. H. 1922. The intensity of linkage between the factors for sugary endosperm and for tunicate ears, and the relative frequency of their crossing over in microspore and megaspore development. *Genetics* **7:** 597–601.

Gillies, C. B. 1975. An ultrastructural analysis of chromosomal pairing in maize. C. R. Lab. Carlsberg **40:** 135–161.

Hanson, G. P. 1965. Some studies of crossing over in maize as influenced by accessory chromosomal elements. Ph.D. thesis. Indiana University, Bloomington.

Hanson, G. P. 1969. B-Chromosome-stimulated crossing over in maize. *Genetics* **63:** 601–609.

Heitz, E. 1928. Das Heterochromatin der Moose. I. *Jahrb. wiss. Bot.* **69:** 762–818.

Heitz, E. 1929. Heterochromatin, Chromocentren, Chromomeren. *Der Deuts. Bot. Gaz.* **47:** 274–284.

Jones, R. N. and H. Rees. 1969. Genotype control of chromosome behavior in rye. XI. An anomalous variation due to the presence of B-chromosomes in rye. *Heredity* **24:** 265–271.

Kikudome, G. 1959. Studies on the phenomenon of preferential segregation in maize. *Genetics* **44:** 815–831.

Kirk, D. and R. N. Jones. 1970. Nuclear genetic activity in B-chromosome rye, in terms of the quantitative interrelationships between nuclear protein, nuclear RNA and histone. *Chromosoma* **31:** 241–254.

Longley, A. E. 1937. Morphological characters of Teosinte chromosomes. *J. Agr. Res.* **54:** 835–862.

Longley, A. E. 1938. Chromosomes of maize from North American Indians. *J. Agr. Res.* **56:** 177–196.

Longley, A. E. 1945. Abnormal segregation during megasporogenesis in maize. *Genetics* **30:** 100–113.

McClintock, B. 1938. The fusion of broken ends of sister half-chromatids following chromatid breakage at meiotic anaphases. *Mo. Agr. Exp. Sta. Res. Bull.* No. 290, pp. 1–48.

McClintock, B. 1941a. The association of mutants with homozygous deficiencies in *Zea mays. Genetics* **26:** 542–571.

McClintock, B. 1941b. The stability of broken ends of chromosomes in *Zea mays. Genetics* **26:** 234–282.

Miles, J. H. 1970. Influence of modified K10 chromosomes on preferential segregation and crossing over in *Zea mays.* Ph.D. thesis. Indiana University, Bloomington.

Nel, P. M. 1973. The modification of crossing over in maize by extraneous chromosomal elements. *Theor. Appl. Genet.* **43:** 196–202.

Nel, P. M. 1975. Crossing over and diploid egg formation in the elongate mutant of maize. *Genetics* **79:** 435–450.

Phillips, R. L. 1969. Recombination in *Zea mays* L. II. Cytogenetic studies of recombination in reciprocal crosses. *Genetics* **61:** 117–127.

Phillips, R. L. 1971. A chromosomal region with more crossing over in megasporogenesis than in microsporogenesis. *Maize Genet. Coop. Newsl.* **45:** 123–125.

Rhoades, M. M. 1941. Different rates of crossing over in male and female gametes of maize. *J. Am. Soc. Agron.* **33:** 603–615.

Rhoades, M. M. 1942. Preferential segregation in maize. *Genetics* **27:** 395–407.

Rhoades, M. M. 1952. Preferential segregation in maize. In *Heterosis,* J. W. Gowen (ed.). Ames: Iowa State College Press.

Rhoades, M. M. 1955. The cytogenetics of maize. In *Corn and corn improvement,* G. F. Sprague (ed.). New York: Academic Press, Chapter 4.

Rhoades, M. M. 1968. Studies on the cytological basis of crossing over. In *Replication and recombination of genetic material,* W. J. Peacock and R. D. Brock (eds.). Canberra: Australian Academy of Science, pp. 229–241.

Rhoades, M. M. and E. Dempsey. 1953. Cytogenetic studies of deficient–duplicate chromosomes derived from inversion heterozygotes in maize. *Am. J. Bot.* **40:** 405–424.

Rhoades, M. M. and E. Dempsey. 1966. The effect of abnormal chromosome 10 on preferential segregation and crossing over in maize. *Genetics* **53:** 989–1020.

Rhoades, M. M. and E. Dempsey. 1972. On the mechanism of chromatin loss induced by the B chromosome of maize. *Genetics* **71:** 73–96.

Rhoades, M. M. and E. Dempsey. 1973a. Chromatin elimination induced by the B chromosome of maize. *J. Hered.* **64:** 12–18.

Rhoades, M. M. and E. Dempsey. 1973b. The functioning of the dissimilar sperm of high-loss plants in double fertilization. *Maize Genet. Coop. Newsl.* **47:** 58–62.

Rhoades, M. M. and E. Dempsey. 1975. Stabilization of freshly broken chromosome ends in the endosperm mitoses. *Maize Genet. Coop. Newsl.* **49:** 53–58.

Rhoades, M. M., E. Dempsey, and A. Ghidoni. 1967. Chromosome elimination in maize induced by supernumerary B chromosomes. *Proc. Nat. Acad. Sci. U. S. A.* **57:** 1626–1632.

Rhoades, M. M. and H. Vilkomerson. 1942. On the anaphase movement of chromosomes. *Proc. Nat. Acad. Sci. U. S. A.* **28:** 433–436.

Roberts, P. A. 1975. In support of the telomere concept. *Genetics* **80:** 135–142.

Roman, H. 1947. Mitotic nondisjunction in the case of interchanges involving the B-type chromosome in maize. *Genetics* **32:** 391–409.

Roman, H. 1948. Directed fertilization in maize. *Proc. Nat. Acad. Sci. U.S.A.* **34:** 36–42.

Stadler, L. J. 1926. The variability of crossing over in maize. *Genetics* **11:** 1–37.

Ward, E. J. 1973a. Nondisjunction: Localization of the controlling site in the maize B chromosome. *Genetics* **73:** 387–391.

Ward, E. J. 1973b. The heterochromatic B chromosome of maize: the segments affecting recombination. *Chromosoma* **43:** 177–186.

Ward, E. J. 1975. Further studies on the effects of accessory chromatin in maize. *Can. J. Genet. Cytol.* **17:** 124–126.

Chapter 41

CYTOGENETICS OF INTERCHANGES

Charles R. Burnham
Emeritus Professor,
Department of Agronomy and Plant Genetics,
University of Minnesota,
St. Paul

There are several kinds of changes in chromosome structure that involve the shift or transposition of chromosome segments. In most species the chromosome is a double-armed structure, since it has a single centromere that usually is not in a terminal position. Hence in a haploid cell, for instance, the cells in the pollen grain, the transposition of segments might be within the same chromosome arm, between the two arms of the same chromosome, or between nonhomologous chromosomes. In a diploid cell, such as the cells in a seed or growing plant, there is the additional possibility of transpositions between two homologs.

Chromosomal interchanges (translocations), the subject of this chapter, are exchanges of end segments of nonhomologous chromosomes. They have been useful for identifying the linkage group carried by a particular chromosome, locating new genes to a chromosome, determining the positions of genetic markers in relation to cytological markers, studying chromosome behavior at

Paper No. 1614, Miscellaneous Journal Series, Minnesota Agricultural Experiment Station.

meiosis, elucidating the genetics of complex characters, identifying kernels on the same ear with contrasting alleles for biochemical comparisons, and also as aids in plant breeding. Crosses between certain interchange stocks have also been used to produce duplications for particular chromosome segments.

In this chapter a general description of an interchange and its breeding and cytological behavior is followed by a discussion of the ways in which inter-crosses among a series of interchange lines that involve the same two chromosomes have been used to synthesize marker stocks that combine the two parental interchanges and to construct chromosomes with multiple duplications that might utilize genetic male sterility to produce "all male-sterile" progeny for production of maize hybrids without detasseling. Also explored are areas in which future research on interchanges in maize might be fruitful for basic information and applied research.

GENERAL DESCRIPTION AND BREEDING BEHAVIOR OF INTERCHANGES

The diagram in Figure 1 illustrates the chromosomal constitution in an interchange heterozygote. If homologous ends remain associated at diakinesis and metaphase-I of meiosis, the configuration is a ring of four (ring 4). Note that the interchange chromosomes, 8 with a piece of 9 (8^9) and 9 with a piece of 8 (9^8), alternate with the normal ones, 8 and 9, in the ring. From this ring there are six possible combinations with two chromosomes, only two of which have all the genetic material, one with both interchange chromosomes, and one with the two normal ones. The other four combinations have a duplication and a deficiency (Dp + Df) and in plants, spores containing these combinations usually abort. Hence the self-progeny of an interchange heterozygote include plants heterozygous for the interchange, ones with normal chromosomes, and ones homozygous for the interchange in a ratio of 2:1:1. Whereas the heterozygotes have an association of four chromosomes at meiosis, the others have only "pairs."

Occasionally three chromosomes in the ring pass to one pole. Referring to Figure 1, note that of the four possible 3/1 disjunctions, two of the 3-chromosome combinations have both interchange chromosomes, the other two have only one. The latter, when combined with n chromosomes from a normal stock, are known as *tertiary trisomics*.

If both interchange segments are long and if homologous parts are paired throughout in the interchange heterozygote, there will be a cross-shaped con-figuration at pachytene, the center of the "cross" theoretically indicating the original breakpoint positions. In heterozygotes for single interchanges in maize, the position of the "cross" varies, certain interchanges being more variable

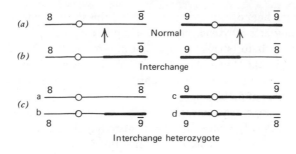

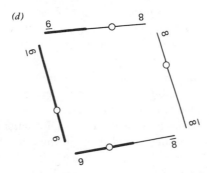

Figure 1. Diagrams illustrating the interchange hypothesis: (a) normal chromosomes, 8.8 and 9.9; (b) interchange chromosomes, 8.9 and 9.8 resulting from an exchange of terminal segments, 8 and 9; also designated as 8^9 and 9^8, read as 8 with a piece of 9, and 9 with a piece of 8; (c) chromosomal constitution of the interchange heterozygote; (d) the ring of four chromosomes formed at diakinesis and metaphase-I of meiosis.

than others. The segments adjacent to the "cross" may not be synapsed or they may be intimately associated. When the "cross" is not at the points of breakage, the segments along one axis on both sides of the new position of the "cross" are paired nonhomologously. Progeny with aberrant chromosomes found among the progeny of interchange heterozygotes may have been the result of illegitimate crossing over (McClintock, 1933). The average position of the "cross" may not be at the points of original exchange. If a breakpoint is near a centromere, this may lead to an error as to the arm in which the break occurred, and consequently there will be different combinations of ends in the interchange chromosomes. Pachytene analysis of homozygotes may furnish information on breakpoint positions relative to chromomere patterns and other cytological markers that would supplement and correct information obtained only from the heterozygotes. Another effect of the variable pairing is a reduction in recombination in regions on both sides of the breakpoints.

If both interchange segments are long, as meiosis proceeds into diakinesis and metaphase I, homologous ends usually remain paired to form a ring of four. If one interchange segment is very short, the short segment may not pair with its homolog at pachytene, and a T-shaped configuration results. At diakinesis there is a chain (string) of four chromosomes. If both interchange segments are short, the four chromosomes may always form two pairs at pachytene with one member of each pair longer than the other. Then there would be only pairs at diakinesis.

If both interchange segments are long, the sterility in maize interchange heterozygotes is usually about 50% (= semisterile). Usually pollen containing the deficiency is devoid of starch, partially filled, or filled but smaller than normal. If an interchange segment is very short or is not physiologically important in pollen development, the class of pollen deficient for it may be indistinguishable from normal pollen. With one such segment, pollen abortion is reduced to 25%. The Dp also present in such pollen grains may preclude their functioning, since pollen grains with extra genic material usually are less able to compete with normal pollen during germination and subsequent pollen tube growth. If this combination functions in the ovules, selfing such a heterozygote will include two additional classes of plants with normal-appearing pollen that are heterozygous for the Dp + Df. Plants in one of these classes have both normal chromosomes, and those in the other have both interchange chromosomes. The latter will produce, when selfed, only plants with normal pollen; and when testcrossed as ♂ on normal, only plants with low pollen sterility. As Dr. Earl Patterson has pointed out (personal communication), these results are similar to those for the tests regularly used to identify interchange homozygotes. Although all of the offspring from the above selfing will have normal pollen, some will be homozygous for the interchange and others will be like the parent, that is, with both interchange chromosomes plus the Dp + Df. If crossed as ♀ with normal, the former will produce all low sterile progeny, and the latter will produce progeny segregating for plants with normal pollen and ones with partial sterility. These results would distinguish such a line from one homozygous for the interchange.

If pollen carrying the Dp + short Df were visibly different, either partially filled with starch, or filled but smaller, then only the F_2 plants with normal pollen would be selfed and tested to establish the interchange homozygote.

The kinds of metaphase-I configurations and their frequencies have important effects on sterility and genetic recombination. Consider first the results when there is no crossing over in interstitial segments (between the centromeres and the breakpoints). The ring of four on the metaphase-I plate may have an open or a zigzag form. In the open ring two adjacent chromosomes pass to each pole. Two of the four spores produced have a Dp + Df, and the other two have a Df + Dp (complements of the Dp + Df in the other two). In plants all four usually abort. In animals, gametes carrying deficiencies may function, but the embryos receiving them usually do not complete development.

In the zigzag ring, two chromosomes in alternate positions pass to each pole. The four spores have complete chromosome complements, two of them have normal chromosomes, whereas the other two have both interchange chromosomes.

The theoretical possibilities are that the open ring may orientate on the metaphase plate so that homologous centromeres pass to opposite poles (adjacent-1) or pass to the same pole (adjacent-2). One might suppose that for each open configuration leading to adjacent-1 or adjacent-2 segregations, a zigzag one leading to alternate segregation should be equally likely (McClintock, 1930; Burnham, 1934). This would account for the 50% sterility. The zigzag configuration for alternate-1 segregation differs from that for alternate-2 segregation (John and Lewis, 1965). The products produced by the two alternate disjunctions are the same, but in one each centromere disjoins from its homolog, whereas in the other each disjoins from a nonhomolog. The two types of alternate disjunction have not as yet been recognized in maize, but there is random orientation, with a ratio of one open : one zigzag configuration.

In maize, chromosome segregation in ring configurations without crossing over in interstitial segments is two alternate : one adjacent-1 : one adjacent-2. In chain-forming interchange heterozygotes, using ones with one break in the satellite of chromosome 6, adjacent-2 segregations are rare regardless of the length of the interstitial segments. The segregation is one alternate : one adjacent-1 (Burnham, 1950, 1954).

In *Datura*, many of the interchange heterozygotes had normal fertility, but others ranged from less than 15% up to about 50%. Most of the configurations at metaphase-I were zigzag. If crossovers that occur in the interstitial segments are followed by alternate-1, alternate-2, or adjacent-1 segregations, two of the four spore products have a Dp + Df that causes them to abort. Therefore, the lines that possessed a ring of four and were fertile must have had either no interstitial segments or very short ones. The lines with sterility must have had at least one longer interstitial segment in which crossovers occurred. The fate of the two crossover and the two noncrossover chromatids that result when there is a crossover in the interstitial segment depends on the kind of segregation that follows. The two crossover chromatids are in the two Dp + Df spores when the segregation is alternate and in the functional spores when the segregation is adjacent-1. Hence a species with mostly alternate segregations will show greatly reduced recombination in the entire interstitial segment. This prediction has not been tested in *Datura*. If crossing over in interstitial segments were followed by equal frequencies of the four kinds of segregation, there should be some reduction in recombination in the entire interstitital segment, since only the adjacent-1 segregations would allow the recovery of those crossovers. The absence of such reduction in maize is indirect evidence that crossovers in interstitial segments in this species are probably followed by one alternate-1 : one adjacent-1 segregations. Adjacent segregations do occur following crossing over in an interstitial segment, but the tests thus far have furnished no information as to the types or their relative frequencies (Burnham, 1954).

The frequencies of the four types of segregation for three interchanges in cotton have been reported (Endrizzi, 1974). The difference in size of the two chromosomes involved (one from each genome), together with differences in length of the interchanged segments, made this possible. For at least one of the three lines, the ratios were 1:1 for alternate-1 : adjacent-1 and for alternate-2 : adjacent-2, but not 1:1 for the first two versus the last two. For one of the interchanges the ratio of alternate-1 : adjacent-1 was 1:1 but was 2:1 for alternate-2 : adjacent-2. Such behavior in maize would lower the sterility to a value below 50%.

In *Collinsia heterophylla* ($2n = 14$), a species with chiasmata localized toward the ends, random orientation is the rule for those induced translocations that had the breaks in the chiasma-forming regions (short interchange segments). Those that had the breaks in the nonchiasma-forming regions (longer interchange segments) show directed orientation (Garber and Dhillon, 1962; Soriano, 1957; Zaman and Rai, 1972). Therefore, the two types of behavior in *Collinsia* are determined mechanically, rather than directly by genetic factors.

INTERCROSSES BETWEEN INTERCHANGE STOCKS INVOLVING THE SAME TWO CHROMOSOMES

Intercrosses between stocks of interchanges involving the same two chromosomes have been used in our laboratory for studies of chromosome pairing, for generating duplications for between-break segments, and for producing special marker stocks for placing genes to chromosome.

The breakpoints in single interchanges may be in either the short (S) or the long (L) arm of each of the two chromosomes. If an interchange with the breakpoints in the short arms of both chromosomes is designated S–S, the other possible combinations of breakpoints are L–L, S–L, and L–S. The cytogenetic behavior of intercrosses among these depends on the relative positions of the breakpoints in the two parents. The different possible intercrosses have been grouped under four major types (Gopinath and Burnham, 1956).

In terms of the breakpoint positions in one parental interchange, designated c, relative to those in the other parental interchange, designated d, Type 1a crosses (Figure 2), referred to as "opposite arms," would be S–S × L–L or S–L × L–S; type 1b crosses, referred to as "one arm the same," would be S–S × S–S, S–S × L–S, L–L × S–L, and L–L × L–S; types 2a and 2b crosses, referred to as "same arms," would be S–S × S–S, L–L × L–L, S–L × S–L, and L–S × L–S. In type 2a both interchange segments from the c interchange

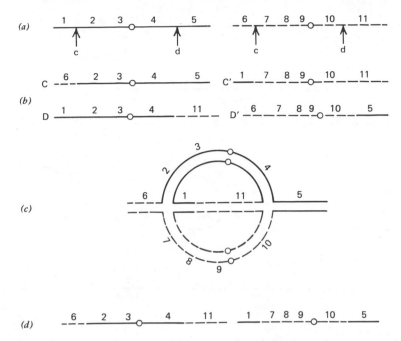

Figure 2. Diagrams illustrating the "opposite arms," Type 1a, intercross between two stocks of interchanges involving the same two chromosomes: (a) breakpoint positions in the two chromosomes; (b) the interchange chromosomes in the F_1, C, C^1 are the two from the c interchange, D, D^1, the two from the d interchange; (c) the pachytene configuration if pairing throughout is between homologous parts; (d) the two chromosomes that combine the two interchange chromosomes as a result of crossing over in both "between breaks," differential segments. The numbers used for the different chromosome segments in this and subsequent figures are the same as those used in the original table that listed the different intercross types, together with the chromosomal constitutions of the F_1 and of two of the nonparental combinations of interchange chromosomes (Gopinath and Burnham, 1956).

are shorter than those for the d interchange. In type 2b (shown in Figure 3) the interchange segment from c is shorter than the one for d in the first chromosome, but longer than the one for d in the second chromosome. Type 2a has been designated as a short–short × long–long intercross, and type 2b as a short–long × long–short intercross. Here "short" and "long" refer to relative lengths of the interchange segments.

In intercross types 1a, 2a, and 2b the interchange chromosomes in one parent have the same combinations of ends as those in the other parent, whereas in type 1b they have different combinations of ends. Using the assumption that homologous ends attract each other, one would predict rings of four for the type 1b intercrosses, and "pairs" for types 1a, 2a, and 2b crosses.

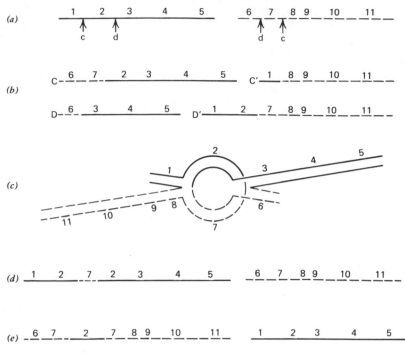

Figure 3. Diagrams illustrating the "same arms," Type 2b, intercross in which interchange c exchanged a segment of the first chromosome that is shorter and a segment of the second chromosome that is longer than the segments of corresponding chromosomes interchanged in d: (a) breakpoint positions in the two chromosomes; (b) the interchange chromosomes in the F_1 [the C + D' nonparental chromosome combination has a duplication for both "between breaks" differential segments, 2 and 7, a "double duplication"; the D + C' combination is deficient for both]; (c) the pachytene configuration if homologous parts are paired; (d) multiple duplication and normal chromosomes formed by a crossover at 7; (e) multiple duplication and normal chromosomes formed by a crossover at 2 (a ring fragment with no centromere results from crossing over at 2 and 7 between the same two strands).

In our laboratory we have obtained information on the types and frequencies of the configurations at diakinesis and also pollen-abortion values for all but a few of the 276 possible intercrosses between 24 T1-5 interchange stocks in maize, five of which are S–S interchanges, five L–L, six S–L, and eight L–S. The numbers of intercrosses of each type showing the different frequencies of "pairs" are given in Table 1. Few or no "pairs" were formed by "one arm the same," type 1b intercrosses. In the other types of intercrosses many had very high frequencies of "pairs." All but a few had more than 10%. Hence these observations are in agreement with the prediction.

TABLE 1. Number of T1-5 intercrosses with different frequencies of "pairs" at diakinesis

Type of intercross	Possible no.	Frequencies of "pairs" in percent							Total analyzed
		0	1-5	6-10	11-40	41-70	71-90	91-100	
1a, Opposite arms	73	1	6	7	24	14	9	10	71
1b, One arm the same	140	94	33	2	2	0	0	0	131
2a, Same arms	45	0	3	0	8	13	13	7	44
2b, Same arms	18	0	0	1	6	2	4	4	17
Totals	276								263

681

In the F_1s, nonparental combinations of the interchange chromosomes will be duplicate, with or without a deficiency, either for interchange segments, or for "between breaks" (differential) segments. If the deficiency for one of those segments is not physiologically important for gametophyte development, the spores with that deficiency may be normal in appearance and sterility will be low. Hence pollen-abortion data on the intercrosses will furnish a clue as to which ones generate transmissible duplications or deficiencies. The data on pollen abortion for the same T1-5 intercrosses are given in Table 2. Of the 266 intercrosses analyzed, 37 had sterility values lower than 35%. Each intercross type had some crosses with such low values. Eight intercrosses had very low values (< 10%), similar to those frequently observed in normal plants. The pachytene configurations for all the types and a more complete description is given elsewhere (Burnham, 1962).

For type 1a, opposite-arms intercrosses, Figure 2c is the pachytene configuration for an S–S × L–L intercross if homologous segments are associated throughout. Simultaneous crossing over in the two differential, "between breaks" segments, produces two new chromosomes which combine the two parental interchanges (Figure 2d). The stock homozygous for these two new chromosomes is designated a "2-chromosome double-interchange" stock. Since each arm of the two chromosomes is marked by an interchange breakpoint, its use as a stock for placing genes to chromosomes has been proposed (Burnham, 1968). A series of such stocks involving chromosome 9 and all but three of the other chromosomes has been synthesized by Kowles (1972). He also reported data for a T2-6 double interchange that did detect linkage with widely separated genes in chromosome 2.

In the type 1a intercross, each of the two nonparental chromosome combinations, C + D and C′ + D′, has a duplication for the centromere-bearing, between-breaks segment of one chromosome and a deficiency for the similar region in the other chromosome (Figure 2b). When the two breakpoints in each

TABLE 2. Number of T1-5 intercrosses with different percentages of pollen abortion

| Type of intercross | Possible no. | Percentages of pollen abortion | | | | Total analyzed |
		<10	14–34	35–65	65–75	>75	
1a, Opposite arms	73	5	8	28	25	7	73
lb, One arm the same	140	0	4	99	24	5	132
2a, Same arms	45	3	11	23	7	0	44
2b, Same arms	18	0	6	10	1	0	17
Totals	276						266

chromosome are close together, and still on opposite sides of the centromere, these Dp + Df spores may survive. The centromeres in the two chromosomes in each of these combinations will be from the same chromosome. Since there is no centromere from the other chromosome, there has been a centromere substitution (Kasha, 1968). The five intercrosses shown in Table 2 with less than 10% pollen abortion had the four breakpoints close to the centromeres. One of these, T1-5e × T1-5f, was crossed reciprocally with the brittle endosperm-1 (*bt*1) and brown midrib-1 (*bm*1) genetic markers (Kasha, 1968; Burnham, unpublished). These are closely linked, less than 1% recombination, with *bt* known to be in the long arm and *bm* in the short arm. The ears using the F_1 as female had 16% *bt* kernels, and none in the reciprocal cross. Hence the deficiency which included the chromosome-5 centromere was only female transmissible and included the *bt* locus. Tests of the same intercross with *bm* gave only normal plants, indicating that *bm* is probably distal to the breakpoint in 5S or the break supposedly in 5S was actually in the centromere. Intercrosses similar to this for other chromosome combinations might be used to screen for induced mutants potentially useful as markers for centromere regions.

In the types 1b and 2a intercrosses, crossing over in both differential segments also produces two new chromosomes that combine the two parental interchanges. The one from type 1b will have one breakpoint in one arm of each chromosome and two breakpoints in one arm of the other. The one from type 2a will have two breakpoints in one arm of each chromosome.

Type 2b Intercrosses

For the Type 2b, same arms, short–long × long–short intercross, note in Figure 3*b* that the C + D′ nonparental combination is duplicate for both "between-breaks" segments, 2 and 7; that is, a double duplication with no deficiency (Burnham, 1966). The D + C′ nonparental combination is deficient for both. The T1-5 Type 2b intercrosses given in Table 2 with 14–34% pollen sterility are likely candidates for producing plants carrying a double duplication for segments of chromosomes 1 and 5.

Type 2b intercrosses between interchange stocks involving chromosome 6 were used to produce duplications for portions of the nucleolar organizer in a study of the rRNA gene content of different portions of the organizer and the role of different segments in nucleolus formation (Givens, 1974).

Type 2b crosses have been used also to test the double-duplication method for using genetic male-sterile-1(*ms*1) on chromosome 6 in producing hybrid corn without detasseling (Phillips and Springer, 1972). The pachytene configuration for this type 2b intercross is shown in Figure 3*c*. Simultaneous crossing over in both differential segments (involving the same two strands, the ones shown in the diagram) will produce a ring fragment without a centromere (acentric). In the T1-5(070-12) × T1-5(8041) intercross, 3.5% of the meiocytes had such a fragment (Weinheimer, unpublished). Crossing over in only one differential segment in that configuration can generate a normal chromosome and a chromosome with "multiple duplications" that has no deficiency but has its own "between-breaks" segment in duplicate and the "between-breaks" segment from the other chromosome intercalated between them (Figure 3*d*). A single crossover in the other differential segment generates a similar-type product for the other chromosome (Figure 3*e*). These new chromosomes are the ones shown in Figure 4.14 in an earlier report (Burnham, 1966). I did not realize then that they could arise by crossing over in F_1 of the type 2b intercross. An intercross of this type that produces acentric fragments should produce also the chromosome with the multiple duplications just described. The problem would be to recover that chromosome as well as the normal one that is the complement of that crossover (Figure 3*d,e*).

Studies of Chromosome Pairing Using Types 1a and 2b Intercrosses

Studies of pachytene pairing in opposite arms type T1-5 intercrosses were used to ascertain where pairing is initiated. If pairing begins at the centromeres, the ends in those "pairs" would be nonhomologous. If pairing begins at the ends, the nontranslocated centromere-bearing regions would be nonhomologous. This is obvious from Figure 2 if "pairs" are formed first by holding homologous centromere-bearing regions together and then by holding homologous ends together. The studies demonstrated conclusively that in maize pairing is not initiated at the centromeres, but at or near the ends (Burnham, Stout, et al., 1972). The probability of pairing initiation decreases rapidly from a high value at the ends to almost zero at the middle of the arm in chromosomes 1 and 5.

Most of the type 2b intercrosses show high frequencies of "pairs" at diakinesis, as shown in Table 1. "Pairs" were frequent also at pachytene. As is obvious from Figure 3*c*, if homologous ends are associated in those "pairs," one member of the "pair" will be longer than the other, the difference being equal to the sum of the lengths of the two "between breaks" segments, numbered 2 and 7. Since pairing begins at or near the ends, these "pairs" will have a loop or buckle at pachytene. An extensive study of the position of the loop showed that it tended to be located more frequently at certain positions along the chromosome (Weinheimer, unpublished).

DUPLICATIONS, PRODUCTION, AND POSSIBLE USES

A duplication may have its origin in an insertional translocation or following an exchange of unequal end segments of the same arm of two homologs following treatments of sporophytic tissue with mutagenic agents. These duplications are unpredictable. Methods of screening for them using genetic markers

(Doyle, 1969–1972), or a gene marker that shows a dosage effect (Beard, 1960) have been proposed.

More interesting are the duplications that can be predicted to recur. Interchanges with one very short interchange segment may generate haplo-viable deficiency-duplications. These are useful for placing genes to chromosome (Phillips, Burnham, et al., 1971). They are being utilized also with male-sterile genes for producing "all male-sterile" progeny in the production of hybrid corn without detasseling the ♀ parent (Patterson, 1973). The method is subject to a patent held by the University of Illinois Foundation. Stocks needed to use the same method have been established for use with the polymitotic gene (po) for male sterility (Phillips, 1975).

In barley, tertiary trisomics are being used to furnish "all male-sterile" progeny to produce hybrid barley (Ramage, 1965). This method is not subject to the patent. In the desired tertiary trisomic stock the interchange chromosome carries the normal Ms allele and the two normal chromosomes carry ms. The only functional pollen has n chromosomes and the ms allele. When crossed on ms ms it should produce only male sterile progeny. The basic principle of the method was reported by Ramage and Tuleen (1964).

In my tests in maize, using this tertiary trisomic method the $n + 1$ from the T2-6a interchange source functioned through the pollen with a frequency of about 10% to give the unwanted nonms (fertile) plants, and the $n + 1$ from T4-6 (0 55-8) produced about 20% fertile plants (Burnham, 1975). Other interchanges as sources of tertiary trisomics may have much lower transmission rates, low enough to be usable for hybrid seed production.

Another method of producing "all male-sterile" progeny utilizes the double duplication produced by type 2b intercrosses between interchange stocks involving the same two chromosomes. Tests of the feasibility of this "double-duplication" method of producing "all male-sterile" progeny show that if the $Ms1$ locus is in the duplication, then crossovers within the duplication between either breakpoint and that locus will transfer the $Ms1$ allele to the normal chromosome and produce unwanted fertile (nonmale sterile) plants (Phillips and Springer, 1972; Springer, 1974). The same is true if the Ms locus is not in the duplication but closely linked to one breakpoint. One possible solution is to use a very short duplication that includes that locus, but a very long duplication from the other chromosome to preclude functioning of pollen carrying the duplication. An important feature was their use of the yellow versus white (Y–y) endosperm marker (closely linked to ms), which could be scored on self-pollinated ears and on yy ms ms ears crossed with putative double-duplication heterozygotes. This circumvented the necessity of growing the progeny in the field for classification for male steriles versus nonmale steriles. A good seedling marker linked with an ms locus might be used in feasibility tests if no closely linked endosperm marker were available.

Another possible method is the use of the chromosome with "multiple duplications" shown in Figure 3d or e that results from crossing over in one differential segment in a type 2b intercross. For example, this chromosome in Figure 3d is not deficient, but has a nontandem duplication for its own "between breaks" segment number 2 with the "between breaks" segment number 7 from the other chromosome intercalated between the two number 2 segments. If an MS allele were in the intercalated segment, crossovers between that segment and its homologs in the normal chromosome pair would be unlikely. Both members of the normal pair would carry the ms allele. The chromosome with the multiple duplications should not function through the male. The functional pollen should all carry male sterility. In Datura, pure-breeding plants with predicted characters from extra chromosomal material were reported, together with the methods to produce them (Blakeslee, Bergner, et al., 1933). The interchanges used as the source of tertiary trisomics were assumed to involve the transposition of entire chromosome arms. Certain of the ones used were described as simple translocations, one segment attached at the end of another chromosome, but were probably very unequal exchanges.

FUTURE RESEARCH NEEDED ON INTERCHANGES IN MAIZE

Certain information about interchanges in maize is needed if full use is to be made of their potentialities. For such studies there is a large number of interchange stocks that involve all possible combinations between the 10 chromosomes (Longley, 1961; Lambert, 1969) and also many that involve the supernumerary B chromosomes (Beckett, 1975). These have a wide range of breakpoints and furnish a wealth of material for physiological, biochemical, and developmental studies in addition to those mentioned in the paragraphs that follow.

Are there differences in fertility between heterozygotes for interchanges involving chromosomes with nearly median centromeres and those involving the other chromsomes, or between those involving short chromosomes versus those involving long ones? Does inbreeding have an effect on chiasma frequency, which in turn might affect disjunction types and frequencies? Although most interchange heterozygotes in maize show about 50% pollen and ovule abortion, there are some having breakpoints at positions where spores deficient for the interchange segment should abort, and yet they have significantly less than 50% sterility. Why?

There is very little information on the frequency of the different kinds of trisomics among the progeny of single-interchange heterozygotes. Do the frequencies vary for different interchanges, and if so, what affects the frequency?

Studies of stability and breeding behavior are needed for duplication and tertiary trisomic stocks. Are they sufficiently stable for use in breeding? If not, do those involving different chromosomes differ in stability?

Studies are needed to determine if a stock with the "multiple duplications" chromosome (Figure 3d,e) can be established from the type 2b intercrosses described earlier, and if so, are they stable? In *Neurospora,* restoration of the normal condition occurs frequently by spontaneous breakage, usually at the union in nontandem duplications. It has been stated that "stability of the duplications depends on both the specific rearrangement and the genetic background" (Perkins, Newmeyer, et al., 1972).

A study of the effect of deficiencies and duplications for different chromosome segments on gametophyte development, as well as on plants carrying them, should be rewarding. Interchanges between the normal and the B chromosomes may be used to avoid the gametophytic screen in the pollen and produce plants with a duplication or with a deficiency for the same segments (Roman, 1947).

Intercrosses between stocks of interchanges that involve the same two chromosomes would furnish plants with duplications for internal segments. An example of the information to be gained from this approach has been reported in *Drosophila* using translocations between the Y chromosome and the autosomes (Lindsley, Sandler, et al., 1972). They assessed the effect of aneuploidy for segments of the autosomes and determined the number and location of dosage-sensitive loci.

The duplications present in many diploid species may have been important in their evolution. Lines of maize homozygous for different duplications might be subjected to mutagens to produce changes not possible from treating material without the duplications. The rx-1 deficiency has been used to produce monosomic plants in maize (Weber, 1974). Could it be used with the B chromosome interchanges to survey the effects of nullisomy for segments of the maize chromosomes?

Seedlessness in certain fruits is or might be a desirable character. It is worth noting that in diploid species that have fruits with single seeds, single interchanges would be sufficient to produce some fruits with an aborted seed as a test of the effect on development and quality of the fruit.

Information is needed at the molecular level. Where do the interchange breakpoints occur? What are the limitations, if any, imposed by the structure and coding sequences on the kinds of chromosome changes that can be produced? Are certain changes more likely to be stable than others?

Any or all of the above studies could have long-term value for applications to studies of physiology, cytology, development, and plant breeding. The information would be valuable also for those working with other species.

SUMMARY OF THE CYTOGENETICS OF INTERCHANGES

Species Lacking Directed Segregation

Such species do not have an excess of alternate disjunctions. For example, in maize the interchange heterozygote shows partial sterility (PS), usually about

50%, in pollen and ovules. At meiosis it has an association of four chromosomes plus eight pairs.

The progeny from self pollination of the interchange heterozygote include plants heterozygous for the interchange that are partially sterile, plants with normal chromosomes that are fertile, and plants homozygous for the interchange that are also fertile in a ratio of $2:1:1$.

Breakpoint frequencies are in general proportional to arm length but higher than expected in a 10-μ region on either side of the centromere and also appear to be higher for certain chromosome combinations.

The configuration at diakinesis or metaphase I is a ring or a chain of four chromosomes or two "pairs," depending on lengths of interchanged segments.

The corresponding configurations at pachytene are cross-shaped, T-shaped, or two separate "pairs," respectively. The position of the cross or T is variable, resulting in nonhomologous associations. Products that could have come from illegitimate crossing over have been observed.

Recombination, as a result of the variable pairing, is reduced in regions adjacent to the breakpoints.

Spore abortion, depending on the viability of the two combinations that are duplicate (Dp) and deficient (Df) for the interchange segments, is 50% if both are not viable, 25% if one is viable, and similar to that in normals if both are viable. Those that form a Dp + short Df combination that may be transmissible only through the ovules may produce, following self-pollination, two additional classes of plants, one that has the two interchange chromosomes plus the Dp + short Df. Selfs of these have only fertile offspring and their test crosses on normals have only semisterile offspring. Although all of the offspring from those selfs have normal pollen, some will be homozygous for the interchange, and others will be like the parent. Crosses of the latter as ♀ with normals will segregate for plants with normal pollen and ones with partial pollen sterility, a behavior different from that for an interchange homozygote.

At metaphase-I of meiosis, the ring of four forms either an open or a zigzag configuration.

For the open rings, adjacent chromosomes pass to the same pole, but the orientation may be such that homologous centromeres disjoin and pass to opposite poles, that is, adjacent-1 segregation; or it may be such that homologous centromeres do not disjoin but pass to the same pole (i.e., adjacent-2). For the zigzag rings, alternate chromosomes in the ring pass to the same pole. Although not recognized in maize as yet, there is probably an alternate-1 and an alternate-2 segregation corresponding to adjacent-1 and adjacent-2 segregations, respectively.

Disjunction in ring configurations, without crossing over in the interstitial segments between the centromere and the breakpoint, is one alternate : one adjacent (probably one alternate-1 : one adjacent-1 : one alternate-2 : one adjacent-2). When crossing over occurs in interstitial segments it probably is one alternate-1 : one adjacent-1 in the meiocytes that have the crossover.

Disjunction in chain configurations, with or without crossing over in interstitial segments, is one alternate : one adjacent-1.

Alternate segregation in a meiocyte that had a ring of four with no crossing over in interstitial segments forms a spore quartet with two spores that have the normal chromosomes and two with the interchange chromosomes; all four are functional. Adjacent segregation in such a ring forms a spore quartet with four spores that are Dp + Df; all four abort.

The chromatids that arise from crossing over in an interstitial segment, when followed by alternate segregation, are in the two Dp + Df spores in the quartet of four, and when followed by adjacent-1 segregation, such chromatids are in the two genetically complete, functional spores. Hence those crossover chromatids can be recovered only from adjacent-1 segregations.

Disjunction from the association of four occasionally may be three chromosomes to one pole, and one to the other. The former may be of four types, two of them with both interchange chromosomes, and the other two with only one. When crossed with normal, the two with one interchange chromosome produce the tertiary trisomics with $2n + 1$ chromosomes.

Crosses between interchange stocks involving the same two chromosomes are of four major types, depending on the relative positions of the breakpoints. These have been used to establish stocks in which the two chromosomes are marked by four breakpoints, and also stocks with duplications for between breaks segments. Low percentages of pollen abortion characterize the intercrosses that produce viable duplications or deficiencies.

Species with Directed Segregation

For species with mostly zigzag orientations (e.g., *Oenothera* and *Datura*), sterility in the interchange heterozygote varies from an amount similar to that in normal plants up to 50%, depending on: (a) the kinds and frequencies of ring, chain, or "pairs" configurations, (b) the frequency of zigzag orientations; and (c) the frequency of crossing over in interstitial segments (Hanson and Kramer, 1949).

When there is an excess of zigzag orientations, and consequently an excess of alternate disjunctions, recombination in entire interstitial segments will be reduced, since the crossover chromatids pass to the spores with a Dp + Df that usually abort.

In a species with directed segregation, only the ones with a ring of chromosomes and short interstitial segments would be fertile. These would be overlooked if sterility were used as the criterion for selecting interchange heterozygotes.

ACKNOWLEDGMENTS

I wish to acknowledge the assistance of Dr. J. T. Stout, who constructed the original pachytene pairing diagrams for the intercrosses, and to thank him and Drs. Wm. H. Weinheimer, Jane Magill, R. V. Kowles, R. L. Phillips, and G. R. Stringam for assistance in gathering the data on the T1-5 intercrosses reported here.

REFERENCES

Beard, B. H. 1960. A homozygous heterozygote. *Barley Newsl.* **3:** 34.

Beckett, J. B. 1975. Genetic breakpoints of the B-A translocations in maize. *Maize Genet. Coop. Newsl.* **49:** 130–134.

Blakeslee, A. F., A. D. Bergner, and A. G. Avery. 1933. Methods of synthesizing pure-breeding types with predicted characters in the Jimson weed, *Datura stramonium*. *Proc. Nat. Acad. Sci. U. S. A.* **19:** 115–122.

Burnham, C. R. 1934. Cytogenetic studies of an interchange between chromosomes 8 and 9 in maize. *Genetics* **19:** 430–447.

Burnham, C. R. 1950. Chromosome segregation in translocations involving chromosome 6 in maize. *Genetics* **35:** 446–481.

Burnham, C. R. 1954. Studies of crossing-over and chromosome segregation in maize translocations. *Proc. IX Internat. Congr. Genetics, Bellagio Caryologia* (suppl.) **6:** 1171–1172 (abstr.).

Burnham, C. R. 1962. *Discussion in cytogenetics*, 1977 reprinting, pp. 383–385. Available only from the author.

Burnham, C. R. 1966. Cytogenetics and plant improvement. In *Plant breeding, a symposium held at Iowa State University*. Ames: Iowa Univ. Press, Chapter 4.

Burnham, C. R. 1968. A new method of using interchanges as chromosome markers. *Crop Sci.* **8:** 357–360.

Burnham, C. R. 1975. Progress report on three possible methods of producing an all male sterile progeny. *Maize Genet. Coop. Newsl.* **49:** 119–121.

Burnham, C. R., J. T. Stout, W. H. Weinheimer, R. V. Knowles, and R. L. Phillips. 1972. Chromosome pairing in maize. *Genetics* **71:** 111–126.

Doyle, G. G. 1969–1972. X-Ray induced duplications from translocations between homologous chromosomes. *Maize Genet. Coop. Newsl.* **43:** 129–134; **44:** 164–166; **46:** 138–142.

Endrizzi, J. E. 1974. Alternate-1 and alternate-2 disjunctions in heterozygous reciprocal translocations. *Genetics* **77:** 55–60.

Garber, E. D. and T. S. Dhillon. 1962. The genus *Collinsia* XVIII. A cytogenetic study of radiation-induced reciprocal translocations in *C. heterophylla*. *Genetics* **47:** 561–567.

Givens, Jean F. 1974. Molecular hybridization and cytological characterization of plants partially hyperploid for different segments of the nucleolar organizer region of *Zea mays* L. Ph.D. thesis. St. Paul: University of Minnesota.

Gopinath, D. M. and C. R. Burnham. 1956. A cytogenetic study in maize of deficiency duplication produced by crossing interchanges involving the same chromosomes. *Genetics* **41:** 382–395.

Hanson, W. D. and H. H. Kramer. 1949. The genetic analysis of two chromosome interchanges in barley from F_2 data. *Genetics* **34:** 687–707.

John, B. and K. R. Lewis. 1965. The meiotic system. *Protoplasmatologia* **6:** F1, 1–335.

Kasha, K. J. 1968. Centromere substitution in chromosomes of *Zea mays. Proc. XII Internat. Congr. Genet.* **1:** 172 (abstr.).

Kowles, R. V. 1972. The synthesis and behavior of two-chromosome double interchanges in maize and *Neurospora crassa*. Ph.D. thesis. St. Paul: University of Minnesota *Diss. Abstr. B.* **33:** 60–61.

Lambert, R. J. 1969. List of reciprocal translocation stocks maintained by the Maize Genetics Cooperative. *Maize Genet. Coop. Newsl.* **43:** 216–230.

Lindsey, D. L., L. Sandler, B. S. Baker, A. T. C. Carpenter, R. E. Denell, J. C. Hall, P. A. Jacobs, G. L. G. Milkos, B. K. David, R. C. Gethmann, R. W. Hardy, A. Hessler, S. M. Miller, H. Nozawa, D. M. Parry, and M. Gould-Somero. 1972. Segmental aneuploidy and the genetic gross structure of the Drosophila genome. *Genetics* **71:** 157–184.

Longley, A. E. 1961. Breakage points for four corn translocation series and other corn chromosome abberrations. *USDA* Agriculture Research Service. ARS-34-**16:** 1–40.

McClintock, B. 1930. A cytological demonstration of the location of an interchange between two non-homologous chromosomes of *Zea mays. Proc. Nat. Acad. Sci. U. S. A.* **16:** 791–796.

McClintock, B. 1933. The association of non-homologous parts of chromosomes in the mid-prophase of meiosis in *Zea mays. Zeits. Zellforsch. u. Mikr. Anat.* **19:** 191–237.

Patterson, E. B. 1973. Genic male sterility and hybrid maize production. *Proc. 7th Meeting Maize and Sorghum Sect. Eucarpia, Sept. 3-6, Zagreb, Yugoslavia.*

Perkins, D. D., D. Newmeyer, and B. C. Turner. 1972. Nontandem duplications in Neurospora and restoration of the euploid condition by chromosome breakage. *Genetics* **71:** 546–547 (abstr.).

Phillips, R. L. 1975. Nuclear male-sterility system of hybrid seed corn production. *Maize Gen. Coop. Newsl.* **49:** 118–119.

Phillips, R. L., C. R. Burnham, and E. B. Patterson. 1971. Advantages of chromosomal interchanges that generate haplo-viable deficiency–duplications. *Crop Sci.* **11:** 525–528.

Phillips, R. L. and W. D. Springer. 1972. A cytogenetic method for utilizing nuclear male sterility in hybrid corn production. *Maize Genet. Coop. Newsl.* **46:** 124–125.

Ramage, R. T. 1965. Balanced tertiary trisomics for use in hybrid seed production. *Crop Sci.* **5:**177–178.

Ramage, R. T. and N. A. Tuleen. 1964. Balanced tertiary trisomics in barley serve as a pollen source homogeneous for a recessive lethal gene. *Crop Sci.* **4:** 81–82.

Roman, H. 1947. Mitotic nondisjunction in the case of interchanges involving the 'B' type chromosome in maize. *Genetics* **32:** 391–409.

Soriano, J. D. 1957. The genus *Collinsia.* IV. The cytogenetics of colchicine-induced reciprocal translocation in *C. heterophylla. Bot. Gaz.* **118:** 139–145.

Springer, Warren D. 1974. A cytogenetic method for utilizing nuclear male sterility in hybrid corn production. M.S. thesis. St. Paul: University of Minnesota.

Weber, D. F. 1974. A monosomic mapping method. *Maize Genet. Coop. Newsl.* **48:** 49–52.

Zaman, M. A. and K. S. Rai. 1972. Cytogenetics of thirteen radiation-induced reciprocal translocations in *Collinsia heterophylla. Cytologia* **37:** 629–638.

Chapter 42

PROPERTIES AND USES OF DUPLICATE-DEFICIENT CHROMOSOME COMPLEMENTS IN MAIZE

E. B. Patterson

Department of Agronomy,
University of Illinois,
Urbana

The large collection of reciprocal translocations available in maize has enormous potential value both for increasing our cytogenetic knowledge of the maize genome and for developing the stocks and information needed for marking, manipulating, and controlling specific segments of it. The primary purpose of this chapter is to point out some of the diverse uses of reciprocal translocations in scientific investigations to develop a broader awareness of their potential and at the same time stimulate others to consider additional possibilities for innovative new applications to problems in their specific research areas.

Discussion here is limited to chromosome interchanges between chromosomes of the normal, or A, genome. Properties and transmission characteristics of reciprocal translocations are reviewed to the extent necessary to serve as a foundation for later portions of the chapter. The main emphasis is on the classification, transmission characteristics, and uses of duplicate–deficient chromosome complements.

BACKGROUND AND GENERAL CONSIDERATIONS

Two nonhomologous chromosomes in maize may interchange ends following chromosome breakage. The products of this event are termed a *reciprocal interchange* or *reciprocal translocation*. The two rearranged chromosomes that result are fully stable and, when transmitted together, contain all of the chromosomal material that was present in the two normally arranged chromosomes from which they arose. The rearranged chromosomes complement each other; spores that receive one of them must also receive the other in order to be chromosomally balanced.

In maize these physical points of interchange may· be determined by cytological observation of the chromosomes at pachynema of microsporogenesis in plants heterozygous for a reciprocal translocation. Pairing of the two rearranged chromosomes with homologous portions of their two normally arranged counterparts leads typically to formation of a cross configuration in which the center of the cross marks the points of interchange in the original parental chromosomes. Because regions adjacent to the interchange points may frequently fail to synapse or may show nonhomologous association, assignment of the physical positions of interchange points is frequently based on an average value obtained from measurements of several pachytene figures. Longley (1961) has published a listing of the cytological positions of 1003 reciprocal translocations. Of these, about 865 translocations have been perpetuated as part of the Maize Cooperation genetic stock collection.

The meiotic pairing relationship of rearranged chromosomes in a plant heterozygous for a reciprocal translocation is diagrammed in Figure 1. If, during microsporogenesis, homologous centromeres disjoin and nonhomologous centromeres segregate at random with respect to each other, then the pollen produced is expected to exhibit three distinguishable phenotypes that are ascribable to four different constitutions:

1. One-half of the pollen is expected to be normal in appearance and functioning and to consist in equal proportions of pollen grains carrying the chromosome combinations $6 + 9$ or $6^9 + 9^6$.

2. One-quarter of the pollen is expected to carry the chromosome combination $6^9 + 9$. Pollen of this constitution, since it is deficient for only a short tip segment of chromosome 6, is shown as exhibiting a subnormal, rather than clearly abortive, phenotype. Typically, pollen grains deficient for a short tip

segment of a chromosome arm are somewhat smaller than normal (chromosomally balanced) pollen grains produced by the same plant; usually they are also incompletely filled, or less densely filled, with starch.

3. The remaining one-quarter of the pollen is expected to carry the chromosome combination $6 + 9^6$. If the deficient segment is long, this pollen is ordinarily clearly abortive in appearance, being devoid, or nearly so, of starch.

The two products of adjacent-1 disjunction (disjunction in the plane of the centromeres) are duplicate–deficient chromosome complements. One complement is deficient for the chromosome 6 segment distal to the interchange point in chromosome 6 and is duplicated for the chromosome 9 segment distal to the interchange point in chromosome 9; the second complement is deficient for the chromosome 9 segment distal to the interchange point in chromosome 9 and is duplicated for the chromosome 6 segment distal to the interchange point in chromosome 6. In most translocations, neither interchange point is near the tip of a chromosome arm. Thus pollen carrying either of the two duplicate–deficient complements arising from adjacent-1 disjunction is usually clearly abortive, and such translocations typically exhibit about 50% abortion.

Burnham (1950) estimated frequencies of alternate, adjacent-1, and adjacent-2 disjunction in plants heterozygous for reciprocal translocations

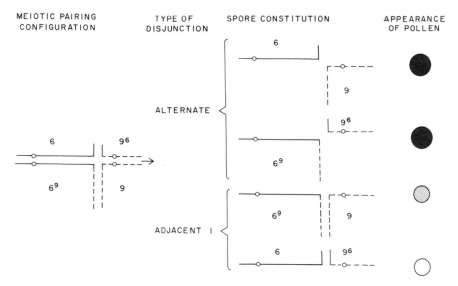

Figure 1.   Chromosome pairing configuration and products of meiosis in a maize plant heterozygous for a reciprocal translocation. Phenotypes of pollen grains with unbalanced chromosome complements reflect specific imbalances in different reciprocal translocations.

involving chromosome 6. Patterns of disjunction in individual instances were determined by observation of nucleolar constitution of microspore quartets and correlated frequencies of pollen abortion. Although individual exceptions are known, adjacent-2 and 3:1 disjunctional patterns from most translocation heterozygotes appear to be infrequent. Both arise as a result of centromere nondisjunction.

Duplicate–deficient chromosome complements that are egg-transmissible are produced by plants heterozygous for some reciprocal translocations. With respect to a particular rearrangement, these duplicate–deficient complements carry a translocated and a normal chromosome, rather than the usual combination of both translocated or both normal chromosomes. The abbreviation Dp–Df is used to refer to duplicate–deficient spores, pollen and gametes, or to duplicate–deficient kernels and plants.

In the type example diagrammed in Figure 1, if the Dp–Df complement carrying the chromosome constitution $6^9 + 9$ is egg-transmissible, then the fertilization of such eggs by pollen from normal-chromosome stocks will yield Dp–Df plants of the chromosome constitution shown in Figure 2. Plants of this constitution produce pollen largely of two genotypes (and phenotypes): (a) one-half of the pollen normal in appearance and functioning and having the constitution $6 + 9$ with respect to the chromosomes shown and (b) one-half of the pollen having the Dp–Df constitution $6^9 + 9$ and exhibiting the phenotype associated with pollen of this same chromosome constitution that is produced by plants heterozygous for the same reciprocal translocation (the subnormal

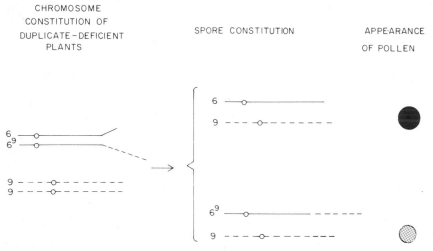

Figure 2. Chromosome constitution and spore products of a duplicate–deficient plant. The duplicate–deficient chromosome complement was derived from the reciprocal translocation shown in Figure 1.

pollen shown in Figure 1). Duplicate–deficient plants do not produce pollen of the clearly abortive type associated with the chromosomal constitution $6 + 9^6$, since these plants do not carry a 9^6 chromosome.

Phenotypes of pollen grains appear to reflect accurately their genetic contents, particularly with regard to deficient segments. Most reciprocal translocations display predominantly two unique subnormal or abortive pollen phenotypes, which in the case of translocations that yield egg-transmissible Dp–Df complements may be ascribed to the products of adjacent-1 disjunction. The rare occurrence of other abortive pollen types from most translocations argues against substantial frequencies of adjacent-2 or 3:1 disjunctional patterns.

In segregating progenies, plants heterozygous for a translocation are ordinarily identified by the pollen semisterility that they exhibit. These classifications may be confirmed by the ear semisterility shown by well-pollinated ears of the same plants. In crosses that generate or propagate egg-transmissible Dp–Df complements, resulting Dp–Df kernels are ordinarily somewhat smaller than sib kernels carrying balanced genomes. In planting, this differential kernel size may be used to advantage in selecting either for or against occurrence of Dp–Df plants. However, in stock maintenance or other types of crosses in which it is desirable to avoid transmission of unbalanced chromosome complements, it is advisable to use heterozygous translocation plants as male parents whenever there is an option.

If plants heterozygous for a translocation are self-pollinated, the progeny are of three types with regard to chromosome constitution, provided there is no transmission of unbalanced complements. These occur in the proportions one-half heterozygous translocation : one-fourth homozygous translocation : one-fourth homozygous normal. The latter two types are fully fertile and are ordinarily distinguished by testcrosses to plants carrying standard chromosomes. Testcrosses of homozygous translocation plants yield all semisterile plants; progeny from testcrossing plants with normal chromosomes are all fully fertile.

As mentioned previously, plants heterozygous for some translocations generate Dp–Df chromosome complements that are egg-transmitted. Most such complements are not pollen-transmitted under conditions of pollen competition. With regard to the translocation diagrammed in Figure 1, if the Dp–Df complement $6^9 + 9$ is egg-transmitted but not pollen-transmitted, self-pollination of plants heterozygous for the translocation yields five types of progeny plants. Three of these are balanced-chromosome types described above. In addition, two types of Dp–Df plants are produced: $6^9\ 9/6\ 9$ (type 1) and $6^9\ 9/6^6\ 9^6$ (type 2). Duplicate–deficient plants of these two types are identical in gene dosage but differ in chromosome structure and transmission characteristics. In testcrosses as pollen parents, type 1 plants yield all normal progeny, whereas type

2 plants yield all semisterile progeny. Since the behavior of type 2 Dp–Df plants in this respect is the same as plants that are homozygous for the translocation, it is necessary to exercise care in the isolation and confirmation of homozygous translocation stocks of those reciprocal translocations that transmit Dp–Df chromosome complements.

Since Dp–Df complements are ordinarily not pollen-transmitted in maize, the frequencies with which Dp–Df individuals occur in progenies depend on: (a) frequencies of Dp–Df megaspores and (b) viability of Dp–Df megaspores, megagametophytes, gametes, and zygotes. On ears borne by plants heterozygous for a translocation, Dp–Df kernels usually comprise less than one-third of the total kernels. On ears borne by Dp–Df plants, there is usually a low, but variable, level of ovule abortion, and Dp–Df kernels typically occur in frequencies less than one-half of total kernels.

The usual failure of Dp–Df complements to be pollen transmitted may result either from pollen inviability or the inability of Dp–Df pollen to compete successfully with chromosomally balanced pollen in effecting fertilization. To distinguish between these two possibilities, pollen competition may be reduced or eliminated through the use of sparse pollinations. A second approach may be used in instances in which there is a discrete size difference between normal and Dp–Df pollen grains. Through the use of a graded series of fine-mesh sieves, pollen grains of different sizes may be separated from mixtures, and pollen samples containing only small grains may be used in pollinations. Both approaches have been used in very limited tests of a half-dozen translocations, with negative results in transmission of Dp–Df complements to date. However, Dp–Df complements derived from three other translocations have apparently been pollen-transmitted in low frequencies from pollinations in which no precautions were taken to reduce pollen competition. Presumably, transmission frequencies in these latter instances may be greatly increased by reducing or eliminating pollen competition.

The only chromosomally balanced spores produced by plants heterozygous for a reciprocal translocation contain either both interchanged chromosomes or both normally arranged homologs. In crosses in which unbalanced chromosome complements are not transmitted, a complete artificial linkage is established between the two interchange points and between the corresponding points on their normal homologs. This complete linkage between points on nonhomologous chromosomes is a direct consequence of nontransmission of other combinations of these points. Because of their complete association in transmission, the two interchange points may be considered as one in inheritance. The symbol T is used to designate the combination of rearranged chromosomes, or alternatively, the interchange points themselves. The symbol + is commonly used to designate the alternative combination of normal chromosomes or their normal, or standard, structure. In crosses of standard maize by plants heterozygous for a reciprocal translocation, the translocation may be followed as if it were a dominant gene for semisterility. Since each reciprocal interchange involves two chromosomes, semisterility maps to two chromosomes and the linkage relations so derived may be diagrammed as a cross-shaped linkage map in which the two interchange points are simultaneously represented as the center of the cross configuration.

Because reciprocal translocations alter normal patterns of pairing and crossing over, values on crossing over frequencies obtained through the use of reciprocal translocations may not be safely incorporated in standard linkage maps. However, information on the sequential relationships of interchange points relative to gene loci is reliable.

Since the interchange points of reciprocal translocations in the physical chromosomes can be correlated with positions in the linkage maps, it is possible to make inferences regarding the positions of gene loci in the physical chromosomes. In general, positions of reciprocal translocations in the genetic maps as determined by direct genetic tests have shown good correlations with assigned cytological positions.

As a group, our current collection of reciprocal translocations permits complete artificial linkage of some 865 pairs of points in the maize genome. They add about 1730 genetic loci to the chromosome complement, all classified by the same character, semisterility.

Reciprocal translocations have been used extensively to locate genes for both qualitative and quantitative traits in maize. Many stocks have been developed and maintained for special purposes. Examples of some of the applications and proposed uses of translocations have been presented by Anderson (1956) and Burnham (1966).

USES OF DUPLICATE-DEFICIENT CHROMOSOME COMPLEMENTS

Patterson (1952a) identified Dp–Df plants in testcross progenies by their characteristic pollen phenotypes, and through associated marker gene phenotypes was able to specify the chromosome constitutions of individual female-transmissible Dp–Df chromosome complements. Duplicate–deficient plants derived from reciprocal translocations involving chromosome 2 were identified that were triplicated for *lg, gl2,* and *B,* and phenotypic effects of gene dosage in triplicated segments carrying these loci were noted. The deficient segment in the Dp–Df chromosome complement derived from T 2–3e was placed closely distal to *lg.*

Patterson (1952b) outlined the use of transmissible Dp–Df chromosome complements to locate genes within, or linked with, a deficient segment. Testcross data were reported showing that in Dp–Df plants derived from T 2–3e, recombination between *lg* and the point of deficiency was 0.1% (1 recombinant in 1219 progeny). Evidence on the cytological extent of deficient or triplicated segments in Dp–Df plants was used to estimate the cytological positions of several gene loci in the short arms of chromosomes 2 or 9.

Linkage information from crosses involving 72 reciprocal translocations in maize was summarized by Patterson (1958). From nine of these translocations, transmission of Dp–Df complements of designated chromosome constitutions was reported or predicted. Testcross data were presented on linkage of marker genes with the deficient segment in Dp–Df plants derived from two of the translocations.

Patterson (1959) reported that the locus of *po* was represented in the deficient segments of three Dp–Df complements; *yg2* was located in the deficient segments of two Dp–Df complements. Results of subsequent tests indicate that the deficient segments of four additional Dp–Df complements include the locus of *yg2* and deficient segments of 21 additional transmissible Dp–Df complements include the locus of *po*.

Phillips, Burnham, et al. (1971) described transmission characteristics and uses of interchanges in maize that transmit Dp–Df chromosome complements in substantial frequencies. In F_2 and testcross populations derived from heterozygous translocations, the specific patterns of distortion in segregation ratios of linked genes may be used to determine positions of interchange points and gene loci. About 50 reciprocal translocations known or predicted to transmit Dp–Df complements were listed. Distortion in segregation ratios may also be used to locate unplaced genes, and a special linkage-detection tester set of reciprocal translocations was proposed. The method has the advantage that no additional marker genes are needed and pollen classification is not required.

This chapter stresses particularly the kinds of information that may be further developed from the phenotypes and transmission characteristics of Dp–Df plants themselves. Since conclusions do not depend on distortion of segregation ratios caused by the presence of Dp–Df individuals in progenies, the procedures are also applicable for use with Dp–Df complements that are initially generated and transmitted from heterozygous translocations in low frequencies.

Figures 3 and 4 show the pairing configuration of the same heterozygous translocation shown as a type example in Figure 1. Chromosome constitutions and associated pollen phenotypes of spore products are also assumed to be the same. In the discussion that follows it is likewise assumed that the Dp–Df complement carrying the $6^9 + 9$ combination of chromosomes is egg-transmissible.

In Figure 3 a heterozygous translocation is used as female parent in cross (1) to allow transmission of a Dp–Df chromosome complement. In cross (2), Dp–Df plants so derived are used as male parents to prevent transmission of the same Dp–Df complement.

The diagram at the top of Figure 3 shows the pairing configuration and genotype of the female parent in cross (1). The male parent in this cross is a chromosome 6 tester stock carrying normally arranged chromosomes. Duplicate–deficient progeny derived from cross (1) are hemizygous for all loci in that portion of chromosome 6 distal to the point of interchange in chromosome 6. They thus show the recessive phenotype resulting from hemizygosity for any recessive allele of a gene in this segment that has been introduced from the male parent. This cross establishes that the loci of A and B are distal to the interchange point in chromosome 6. That is, they are carried on the 9^6 chromosome. This cross does not, however, provide any information on the positions or sequence of the A and B loci within this terminal segment.

If the female parent in cross (1) is homozygous for the dominant alleles A and B, all Dp–Df progeny show the hemizygous recessive phenotypes of a and b. Moreover, only Dp–Df progeny show these recessive phenotypes, since no viable chromosome complement derivable from the female parent by 3 : 1 disjunction lacks the A or B loci.

If the C, D, and E loci are known from independent evidence to be on chromosome 6, and if Dp–Df progeny show the dominant expression of each, then all three are located on the 6^9 chromosome.

Commonly, Dp–Df kernels that arise from cross (1) are somewhat smaller than sib kernels. If a recessive kernel phenotype occurs, that phenotype itself identifies Dp–Df kernels. In the absence of a recessive kernel phenotype, if smaller kernels are selected for planting, the frequency of Dp–Df plants is increased. Typically, Dp–Df plants are somewhat shorter than sib plants and shed pollen a few days later. Duplicate–deficient plants derived from different reciprocal translocations, however, may have additional or alternative distinctive characteristics. If no hemizygous recessive kernel or plant phenotypes occur in plantings, Dp–Df plants may be identified by their characteristic pollen phenotypes (as shown in Figure 2).

In cross (2), Dp–Df plants derived from cross (1) are used as pollen parents in testcrosses to determine the sequence of loci C, D, and E relative to the interchange point in the 6^9 chromosome. Pollen carrying the 6^9 chromosome does not function. Thus any locus carried on the chromosome 6 portion of the 6^9 chromosome can be transmitted only if through recombination it comes to be carried on a normal chromosome 6. The necessity for this unidirectional recombination is indicated by dotted lines. The frequencies with which the dominant alleles C, D, and E are transmitted are direct measures of recombination between each locus and the interchange point. This relationship allows the sequencing of loci that show different recombination values.

The procedures shown in Figure 3 can be used with translocations that show a low frequency of initial transmission of the Dp–Df complement from heterozygous translocation female parents. Conclusions drawn from cross (1)

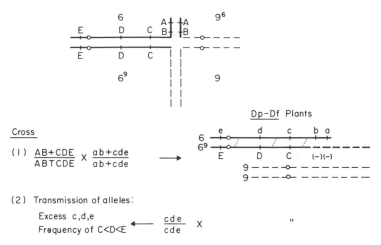

Figure 3. The use of duplicate–deficient plants derived from a reciprocal translocation between chromosomes 6 and 9 to determine the position of the interchange point and deficient segment in chromosome 6 relative to gene loci. The pairing configuration and genotype of the heterozygous translocation source stock is diagrammed at top (see text for further details).

are based not on the frequency with which Dp–Df progeny occur but on the phenotypes shown by those that do occur. Since Dp–Df plants are used as pollen parents in cross (2), a few such plants are sufficient.

The pairing diagram shown at the top of Figure 4 serves as a reference for crosses summarized below it. Together, these crosses develop information on the position of the interchange point in chromosome 9 relative to chromosome 9 gene loci. Cross (1) assigns genes to the 6^9 or 9^6 chromosomes. Cross (2) gives the sequence of genes distal to the interchange point (gene loci on the 6^9 chromosome). Cross (3) derives type 2 Dp–Df plants. Cross (4) gives the sequence of genes proximal to the interchange point and into the opposite arm of chromosome 9 (gene loci on the 9^6 chromosome).

The female parent in cross (1) has the pairing configuration and genotype shown in the diagram at top of Figure 4. It is heterozygous for the translocation and for chromosome 9 gene loci, with all dominant alleles carried on the interchanged chromosomes. It is necessary in the interpretation of the results of cross (1) that all recessive alleles be carried on the normal chromosome 9 in the female parent. The male parent is a tester stock for the same gene loci and carries normal chromosomes. If the male parent is also homozygous for the recessive allele of a chromosome 6 gene that is located in the deficient segment of Dp–Df plants (e.g., a) and the female parent is homozygous for the dominant allele (A), Dp–Df kernels or plants in the progeny may be recognized by their expression of the associated hemizygous recessive phenotype ($-/a$). In the absence of such a diagnostic marker, Dp–Df plants may be recognized by pollen phenotypes.

The question asked in cross (1) is whether a specific chromosome 9 gene locus is distal to the interchange point in chromosome 9 and thus carried on the 6^9 chromosome or is on the opposite side of the interchange point and is thus carried on the 9^6 chromosome. As extreme cases, if a gene locus were distal to the interchange point and completely linked with it, a Dp–Df megaspore would always carry a dominant allele of the locus; alternatively, if the gene locus were proximal to the interchange point and completely linked with it, a Dp–Df megaspore would never carry the dominant allele of the locus.

With respect to loci distal to the interchange point, the maximum frequency of homozygosis for a recessive allele in Dp–Df megaspores is one-sixth, regardless of the position of a locus in this segment; a minimum of five-sixths of all Dp–Df megaspores carry at least one dominant allele of any specific locus in this segment. Duplicate–deficient progeny carrying a dominant allele of a gene locus in this segment are correspondingly in excess.

With respect to chromosome 9 loci carried on the 9^6 chromosome, the proportion of Dp–Df megaspores that carry the dominant allele of a specific locus

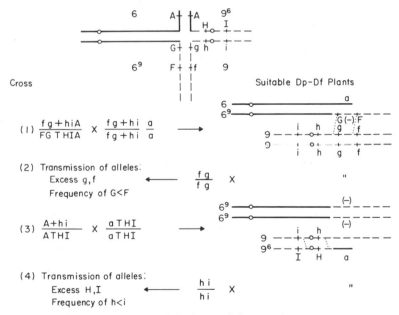

Figure 4. The derivation and use of duplicate–deficient plants to determine the position of the interchange point and triplicated segment in chromosome 9 relative to gene loci. The pairing configuration and genotype of one of the heterozygous translocation source stocks is diagrammed at top (the same reciprocal translocation as shown in Figure 3) (see text for further details).

reaches an upper limit of one-half for loci that assort independently of the interchange point.

In short, Dp–Df progeny derived from cross (1) usually exhibit the dominant phenotype of a chromosome 9 gene locus that is carried on the 6^9 chromosome and they usually show the recessive phenotype of a chromosome 9 gene locus that is linked to the interchange point and carried on the 9^6 chromosome. The Dp–Df genotype shown is a common one, but alternative ones occur as a consequence of crossing over. Most Dp–Df progeny carry the dominant G allele and are homozygous h; they thus appear phenotypically to be recombinants for the G–H interval. It is possible to determine the sequence of genes H and I by examining the frequency with which the dominant allele at each locus is transmitted to Dp–Df progeny. If both loci are closely linked to the interchange point, however, the sequence may be more easily determined in large progenies produced by cross (4).

Duplicate–deficient plants derived from cross (1) are suitable for use as male parents in cross (2) if they have the genetic constitution shown with regard to alleles at the G and F loci. In cross (2), since the 6^9 chromosome is not pollen-transmitted, alleles carried on the chromosome 9 portion of that chromosome must be transferred to a normal chromosome 9 by recombination before they can be transmitted. This requirement is indicated by dotted lines.

The G and F loci are carried on a chromosome segment that is triplicated in Dp–Df plants. In order that a dominant allele on this segment of the parental 6^9 chromosome be transmitted, it is necessary: (a) that the region between the locus and the interchange point pair with the homologous region of a normal chromosome 9, (b) that exchange occur in this region to produce a chromosome 9 carrying the dominant allele; and (c) that pollen carrying this recombinant chromosome 9 function in fertilization. Because of these requirements, transmission of a locus carried on the parental 6^9 chromosome at a position immediately distal to the interchange point may occur in very low frequency. In the linkage sequence diagrammed, $g\ F$ recombinants are transmitted in much higher frequency than $G\ f$ recombinants, since transmission of the latter combination requires that coincident recombination occur between G and the interchange point to yield a double-crossover chromatid. The sequencing of loci relative to the interchange point is straightforward, but rather large progenies may be required.

In cross (3), type 2 Dp–Df plants are derived by using a female parent heterozygous for the translocation and for genes that are carried on that segment of chromosome 9 that is present in a 9^6 chromosome. The pollen parent is homozygous for the translocation and for the dominant alleles of the same chromosome 9 gene loci. A marker gene, such as a, that is located in the deficient segment of chromosome 6 may be used to identify all Dp–Df progeny by their hemizygous recessive phenotype. The use of such a marker gene in this cross requires that the female parent be homozygous for the dominant allele and that the male parent carry the recessive allele on an interchanged chromosome. If the latter stock is not available, Dp–Df plants in the progeny may be identified by pollen phenotypes, which are the same as those shown in Figure 2. Most of the Dp–Df plants derived from cross (3) carry the recessive alleles h and i on the normal chromosome 9 if these gene loci are closely linked to the interchange.

In cross (4), type 2 plants are used as male parents in testcrosses. Pollen containing a normal chromosome 9 does not function due to chromosome imbalance. A locus carried on the parental chromosome 9 can be transmitted only if it becomes part of an interchanged chromosome as a result of recombination. The region of chromosome 9 that is of specific interest here is that region of chromosome 9 that is not represented in the triplicated segment (i.e., that portion of chromosome 9 represented in the 9^6 chromosome). Again, in transmission the recombination is unidirectional, as indicated by dotted lines. The frequencies of transmission of the h and i alleles measure recombination between each of these loci and the interchange point in the 9^6 chromosome.

In cross (4) all progeny receive the interchanged chromosomes from the pollen parent. Those progeny homozygous for a recessive marker gene received an interchanged chromosome carrying that recessive marker gene. In the translocation used here as a type example, this cross is an effective means of transferring chromosome 9 marker genes into the 9^6 chromosome. Plants of the same type 2 Dp–Df chromosome constitution may also be used to transfer genes from the remaining portion of chromosome 9 into a 6^9 chromosome. This transfer is simply selection of the reciprocal product of a recombinant event by which genes such as G and F are transferred from a 6^9 chromosome to a normal chromosome 9 in cross (2).

This technique is applicable to any translocation that generates a transmissible duplicate–deficient complement. This selective transfer procedure (transferring marker genes from a normal chromosome into interchanged chromosomes) is restricted to only one of the two normal chromosomes involved in a particular translocation—that normal chromosome that is present singly in type 2 Dp–Df plants.

Some additional points may be made regarding crosses outlined in Figures 3 and 4. In these crosses, dominant alleles are usually carried on interchanged chromosomes. With respect to most mutant alleles used by maize geneticists, wild-type alleles are dominant. Since reciprocal translocation stocks, as they are maintained, carry wild-type alleles at most loci, they ordinarily may be used directly without modification. In those crosses that are made to derive Dp–Df progeny, only Dp–Df individuals need be identified. Only a few Dp–Df plants are needed as male parents in testcrosses. Testcross progeny are classified only to determine frequencies with which gene alleles were transmitted; it is not necessary to classify for transmission of interchanged chromosomes to derive information on linkage of gene loci with interchange of points.

In maintenance, Dp–Df plants in each generation may be identified by pollen phenotypes. Alternatively, marker genes may be used. If two loci, A and B, are situated within the deficient segment of Dp–Df plants, stocks may be maintained by pollinating Dp–Df plants in alternate generations by an a tester stock, then by a b tester stock; Dp–Df plants in each generation may be identified by hemizygous recessive phenotype.

Duplicate–deficient stocks may also be maintained by use of a marker gene closely adjacent to an interchange point. There are two alternatives, which are illustrated in type 1 Dp–Df plants arising from cross (1), (Figure 3) and those arising from cross (1) (Figure 4). In the first case a dominant marker allele (e.g., C) is carried immediately proximal to the interchange point in the 6^9 chromosome. In the second case a dominant marker allele (e.g., G) is carried on the same chromosome immediately distal to the interchange point. In each case allelic positions in the genome carry the recessive allele. In the progeny from self-pollination, most individuals showing the dominant gene phenotype are Dp–Df; most of those showing the recessive phenotype have normal chromosomes.

Type 2 Dp–Df plants such as those that arise from cross (3) (Figure 4) may be marked in a similar fashion. Again, there are two alternatives. The two 6^9 chromosomes may be marked by a recessive gene located just distal to the interchange point. Alternatively, the 9^6 chromosome may be marked by a recessive gene located just proximal to the interchange point. In each case the normal chromosome 9 should carry the corresponding dominant allele. In the progeny from self-pollination, most individuals showing the dominant gene phenotype are type 2 duplicate–deficients, whereas most of those showing the recessive phenotype are homozygous for the interchanged chromosomes. Occasional exceptions occur in this linked-marker procedure as a result of crossing over between a marker gene and the adjacent interchange point, so it is necessary to monitor maintenance conducted in this way.

Duplicate–deficient complements may also provide information on the sequence of a centromere in a linkage map and in instances in which an interchange point is very near a centromere, may indicate whether the interchange point is in the short arm or the long arm of the chromosome. The argument may be followed more easily by reference to the pairing diagram and Dp–Df plants that arise from cross (1) (Figure 4). All genes represented on the triplicated segment in Dp–Df plants are located in the same chromosome arm. If any gene locus present in the triplicated segment is known to be located on the long arm of the chromosome, then the interchange point is in the long arm of the chromosome. Since gene loci may be sequenced with respect to an interchange point, and since an interchange point in Dp–Df plants may be sequenced with respect to a centromere, a centromere may be sequenced with respect to gene loci.

Patterson (1952b) used the waxy-staining reaction of Dp–Df pollen to assign interchange points of four reciprocal translocations to the short or long arm of chromosome 9. Linkage data indicated that the chromosome 9 interchange point in each of the four translocations was to the right of the wx locus in the linkage map. Duplicate–deficient pollen produced by two of the translocations was mostly Wx-staining, indicating that the wx locus was in the duplicated segment and that the interchange point was thus in the short arm. With regard to the other two translocations, Dp–Df pollen was mostly wx staining, indicating in each instance that the interchange point in chromosome 9 was in the long arm. Evidence from the staining reaction of Dp–Df pollen in the latter two instances served to correct the initial assignment of both interchanges to the short arm of chromosome 9 on the basis of cytological observations.

Several effective methods for locating genes to chromosome are available in maize. The crosses shown in Figure 3 might also be used in a general screening procedure to locate unplaced genes, but for this purpose other gene-locating techniques are more efficient.

The crosses that have been diagrammed are designed to determine positions of interchange points relative to mapped chromosome loci. The testcrosses measure recombination frequencies between individual gene loci and interchange points in Dp–Df plants. The latter information can be useful for its reference, or predictive, value in mapping new gene loci.

Duplicate–deficient complements can be used to "localize" the positions of genes that have already been assigned to chromosomes, but not mapped. The simplest procedure uses crosses of the type shown in Figure 3. Mutant plants are used as male parents in crosses to plants heterozygous for a translocation (or to Dp–Df plants). If the mutant gene locus is not found to be in the deficient segment, Dp–Df plants carrying the mutant allele are crossed as male parents to a tester stock for the mutant locus. Recombination frequency of the mutant locus with the interchange point is measured in the testcross progeny. This recombination frequency can be compared with previously measured recombination frequencies of various mapped loci in the same chromosome with the same interchange point in Dp–Df plants. This comparison would allow an approximate positioning of the new mutant locus in the chromosome map. This procedure, however, is subject to the objection that recombination frequencies derived from different tests may be somewhat variable and that estimates of gene position would be correspondingly uncertain.

This uncertainty may be avoided by using marker genes in the interchanged chromosome. Again, Dp–Df plants could be derived by a cross similar to that shown as cross (1), Figure 3. The female parent would be heterozygous for the translocation and homozygous for mutant alleles of appropriately spaced chromosome 6 marker genes. A plant carrying the new mutant allele would be used as male parent. In Dp–Df plants the 6^9 chromosome would carry marker genes derived from the female parent, and the chromosome 6 would carry the new mutant allele. If the new mutant gene locus were not found to be in the deficient segment, individual Dp–Df plants would be testcrossed both to a tester stock for the chromosome 6 marker genes and to a tester stock for the new mutant allele. Recombination of each locus with the interchange point would be determined in the two testcross progenies. Since both testcross progenies derive from the same male plant, recombination data may be combined to estimate the position of the mutant gene in the linkage map. The use of two separate tester stocks eliminates the necessity of deriving a chromosome tester stock that incorporates the recessive allele of the new mutant gene. If the new mutant allele is dominant to wild-type, the tester stock for the chromosome 6 marker genes suffices.

If available, two Dp–Df complements could be used for each chromosome—one deficient for a long arm tip segment and the second deficient for a short arm tip segment. In long chromosomes, a middle region might fail to show linkage with either deficiency.

In recent years dozens of genes in maize have been located to chromosomes but not mapped. Procedures for inducing new mutant genes and for assigning them to chromosomes are becoming increasingly effective. The use of an intermediate step to determine the approximate position of a gene, prior to mapping in standard stocks, may be a more efficient means of mapping large numbers of genes. If B–A translocations with interchange points at various positions in a chromosome arm are available, they can be used very effectively in this intermediate step.

A number of varied and specialized uses of Dp–Df complements may be mentioned. The Dp–Df plants carry a terminal chromosome segment in hemizygous condition. Since genes in this segment are present in single dose, this condition may be reflected in altered levels of specific chemical components in the plant. The presence of a terminal, unpaired region might also be used in studies of chromosome-pairing initiation. The unequal chromosomes present may be used in studies of preferential segregation. Transmissible terminal deficiencies might also be used in a screening procedure to detect mutagen-induced recessive mutations in tip segments of chromosomes.

Triplication of chromosome segments may be used in studies of gene dosage or to produce various combinations of alleles for studies of isozyme banding and hybrid enzyme formation. In these regards, their use is similar to that of trisomic plants. However, Dp–Df complements may be more easily derived and the triplicated segment is more limited in extent.

In male transmission the point of deficiency in Dp–Df plants has a selective effect similar to that of a male gametophyte factor. This permits the differential transmission of alleles at linked loci and the production from heterozygotes of progenies largely of a single, predictable genotype. The selective transmission of one recombinant type over its reciprocal also provides a useful method for determining the sequence of closely linked loci or of sites within a locus. It may also be used to extract recombinants of a specific desired type.

Chromosome deficiencies that are egg-transmissible, but not pollen-transmitted, may be used in conjunction with closely linked male sterile gene loci to effect differential male transmission of the male sterile gene alleles. In this way, progenies that are essentially all male-sterile may be produced for use as female rows in commercial production fields, and detasseling of such rows may be avoided. Screening studies to date have identified four specific Dp–Df chromosome complements, any one of which may be used to produce progenies consisting of at least 99.7% male-sterile plants. Details of the procedure for the derivation and use of the required stocks have been described by Patterson (1973).

Deficiencies may frequently be used as pollen size markers. A Dp–Df plant produces equal numbers of normal and deficient pollen grains, and these may be visually distinguished. Samples of such mixtures may be used in *in vitro* studies of pollen germination, and effects of varying regimes of media supplementation may be observed directly. The two pollen sizes may also be segregated from mixtures through the use of fine-mesh sieves. Since deficient pollen grains lack all loci in the deficient segment, absence of specific loci might be distinguished by associated alterations in electrophoretic banding patterns.

Techniques employing pollen markers in maize would be especially powerful, but options have been limited by the scarcity of good pollen markers that can be classified in combinations. One combination that can be exploited now is the waxy trait and small pollen traits.

An example of linkage employing these pollen traits is provided in the Dp–Df plant produced by cross (3) (Figure 4). If the locus designated I in this drawing were the wx locus, recombination between the locus and the interchange point could be determined directly in pollen. This linkage might then be used to test the influence of a wide array of variables on crossing over in this region. Recombination in this region could be used to study, or search for, genes affecting crossing over. Studies might also include the effects of chemicals or of genomic alterations (e.g., monosomics, Abnormal 10, B-chromosomes). Changes in crossing over in the same plant might be tested by checking pollen shed on successive days or by comparing crossing over in a parent plant with that in tillers that had been subjected to differential treatment. The same kind of plant can be used to sequence mutational sites within the wx locus itself through the association that is shown between the recombinant *Wx* phenotype and pollen size.

Most of the studies conducted with reciprocal translocations have been concerned with determining the positions of interchange points with respect to gene loci. Relatively little application has been made of this information to manipulate, control and monitor specific chromosome regions. Duplicate–deficient complements are especially useful for developing detailed information about positions of interchange points and gene loci and for deriving stocks that may be employed for additional purposes. In addition to the 50 or so known- or probably-transmissible Dp–Df complements mentioned earlier, there may be as many as 50 more that can be derived from reciprocal translocations now available in maize. It is evident that many opportunities exist for geneticists to collaborate with physiologists and biochemists to mutual benefit.

REFERENCES

Anderson, E. G. 1956. The application of chromosomal techniques to maize improvement. *Brookhaven Symp. Biol.* No. 9, *Genetics in plant breeding,* pp. 23–36.

Burnham, C. R. 1950. Chromosome segregation in translocations involving chromosome 6 in maize. *Genetics* **35**: 446–481.

Burnham, C. R. 1966. Cytogenetics in plant improvement. In *Plant breeding, a symposium.* Ames: Iowa State Univ. Press, Chapter 4.

Longley, A. E. 1961. Breakage points for four corn translocation series and other corn aberrations. *USDA Agr. Res. Serv. Crops Res. Bull.* No. 34-16, 40 pp.

Patterson, E. B. 1952a. Studies on crossing over in homozygous and heterozygous chromosome rearrangements in *Zea mays.* Ph.D. thesis. Pasadena: California Institute of Technology.

Patterson, E. B. 1952b. The use of functional duplicate–deficient gametes in locating genes in maize. *Genetics* **37**: 612–613 (abstr.).

Patterson, E. B. 1958. Genetic confirmation of chromosomes involved in reciprocal translocations. *Maize Genet. Coop. Newsl.* **32**: 54–66.

Patterson, E. B. 1959. Report on Maize Cooperative. *Maize Genet. Coop. Newsl.* **33**: 131.

Patterson, E. B. 1973. Genic male sterility and hybrid maize production. *Part 1, Proc. 7th Meeting Maize and Sorghum Sect., Eur. Assoc. Res. Plant Breeding (EUCARPIA).* Zagreb, Yugoslavia: Secretariat, Institute for Breeding and Production of Field Crops (ed.).

Phillips, R. L., C. R. Burnham, and E. B. Patterson. 1971. Advantages of chromosomal interchanges that generate haplo-viable deficiency–duplications. *Crop Sci.* **11**: 525–528.

Chapter 43

MOLECULAR CYTOGENETICS OF THE NUCLEOLUS ORGANIZER REGION

R. L. Phillips

Department of Agronomy and Plant Genetics,
University of Minnesota,
St. Paul

The nucleolus organizer region (NOR) offers many intriguing opportunities to the molecular, classical, and applied cytogeneticist. From the molecular viewpoint, the NOR contains the DNA coding for 18S and 28S rRNA, thereby permitting the study of transcription and genetic regulation of these important cellular macromolecules. The interaction of these ribosomal components (18S and 28S rRNA) with others, such as 5S RNA and ribosomal proteins, must be elucidated to understand the formation and role of the nucleolus in plant development. Classically, the region is interesting because the locus of the 18S and 28S rRNA genes, that is, the NOR, can be directly observed cytologically with the light or electron microscope and evaluated for various structural details. In addition, the functional capacity of the region can be revealed

Paper No. *1608*, Miscellaneous Journal Series, Minnesota Agricultural Experiment Station.

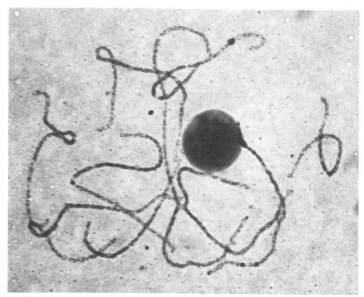

Figure 1. Pachynema of microsporogenesis displaying the 10 bivalent chromosomes including chromosome 6 associated with the nucleolus at the NOR. Magnification × 1600. Photograph courtesy of J. T. Stout.

cytologically as manifested by the presence, size, and structure of the nucleolus in various genetic and cytogenetic stocks. And since ribosomes are involved in the synthesis of all proteins in every cell of the plant, the applied cytogeneticist can manipulate the NOR to attempt to correlate the effects with economic traits. Finally, the ribosomal RNA system is controlled by genes reiterated thousands of times. Studies on such genetic systems should provide unique insights to maize genetics and breeding.

Information bearing on the molecular cytogenetics of the NOR in maize has not been extensively reviewed. This paper attempts to assess the current state of knowledge on the maize NOR, giving information on various molecular, classical, and applied cytogenetic aspects of the ribosomal RNA system along with our current interpretations and speculations on the NOR's genetic nature and function. A maize pachytene cell illustrating the chromosome-6 NOR–nucleolus assocation is shown in Figure 1.

THE MAIZE NUCLEOLUS

Nucleolar Cycle

The cell cycle represents a complex and highly controlled series of events that includes interactions between nuclear and cytoplasmic processes. The nucleolus, located within the nucleus, is central to many of these processes and interactions. The nucleolus functions in maize, as in other organisms, as the site of synthesis and maturation of a ribosomal precursor RNA into ribosomal components. The nucleolus undergoes cyclic changes that usually result in its disorganization in prophase and reorganization in telophase, but this varies with cell type. De la Torre and Clowes (1972) have shown that the dissolution and reorganization of the nucleolus is differentially timed relative to the chromosomal cycle in different root zones. In the root-cap initials, the cells that generate the cap cells, the nucleolus starts to disorganize before prophase and has completely reorganized by midtelophase. In contrast, nucleoli in cells of the quiescent center of the root meristem start to disorganize during prophase but do not reorganize until after telophase. Cells of the stelar region are more typical in that the nucleolus commences disorganization in prophase and undergoes reorganization in telophase.

In meiosis the nucleolus reaches its maximum size at midpachynema. Das (1965) and Das and Alfert (1966) found that nucleolar RNA synthesis had essentially ceased by this point of maize microsporogenesis. Autoradiography following a 2-h label with ^{3}H-cytidine or ^{3}H-uridine showed rapid RNA synthesis during premeiotic interphase, and as prophase progressed there was a decrease in RNA synthesis in the nucleolus. The nucleolus remained inactive in RNA synthesis from pachynema to the end of prophase (through diakinesis). Although inactive in RNA synthesis, the nucleolus remained prominent and intensely stainable with Azure B. The remainder of the genome continued to be involved in RNA synthesis at pachynema, but the rate decreased significantly at diplonema and diakinesis. Lin (1955) measured RNA amounts in the nucleolus by UV microspectrophotometry. The amount of nucleolar RNA doubled between leptonema and midpachynema. Nucleolar RNA synthetic rate, however, decreased during these stages and ultimately ceased. This elevation of the amount of RNA in the nucleolus while the RNA synthetic rate was decreasing led Das (1965) to suggest that there is a reduced rate of transport of nucleolar RNA to the cytoplasm from leptonema to midpachynema in microsporogenesis.

Nucleolar Composition

Pollister and Ris (1947), using microspectrophotometry, concluded that the pachytene nucleolus of maize contains 5.0×10^{-11} gram of protein. Comparing UV absorption before and after hot trichloroacetic acid (TCA) or ribonuclease treatment of maize pachytene nucleoli, Pollister and Leuchtenberger (1949) observed a reduction of 53% in the extinction value after treatment. Swift and Stevens (1966) reported that maize microspore nucleoli of the normal as well as diffuse (multiple, small, and irregularly sized nucleoli) types stain purple with Azure B, characteristic of the presence of RNA. These observations suggested that the nucleolus contains nucleic acid almost exclusively of the RNA type. Lin

(1955) estimated that maize early-pachytene nucleoli contain from 7.35–14.55 × 10^{-12} gm RNA, depending on the strain. This value for RNA coupled with the Pollister and Ris (1947) value for protein suggests that the maize pachytene nucleolus is approximately 80% protein. Lin's (1955) UV absorption spectrum of pachytene nucleoli also indicates a large amount of protein. McLeish (1964) reported the existence of DNA in isolated nucleoli of maize root tips even though they are Feulgen negative.

Maize ribosomal RNA is rich in guanine and cytosine. Three maize genotypes gave similar results (Pollard 1964) with the following average values (mols %):

$$28S \; rRNA - A = 21.2, \; U = 16.9, \; C = 28.0, \; G = 33.9 \; \frac{A + U}{C + G} = 0.62$$

$$18S \; rRNA - A = 22.0, \; U = 20.2, \; C = 25.0, \; G = 32.8 \; \frac{A + U}{C + G} = 0.73$$

Since the rRNA is G–C rich (61.9% for 28S and 57.8% for 18S) the complementary DNA would be G–C rich and more dense in a CsCl density gradient than the bulk DNA. Ingle, Timmis, et al. (1975) and Doerschug (1976) have shown this to be true. Ingle, Timmis, et al. list the buoyant density as 1.710–1.711 for maize rDNA and 1.701 for bulk DNA.

Ultrastructure of the Nucleolus and NOR

Swift and Stevens (1966) indicated that most of the nucleolar mass of the normal microspore nucleolus consists of a finely filamentous or nonparticulate region. The nucleolus is surrounded by a narrow ring of particles; the particulate component is especially prominent in a nucleolar cap that possesses thread-like nucleonemal structures. Nucleolar vacuoles also are observed. In early stages of microspore development, quartet or early interphase, the particulate component is not present. In diffuse nucleoli, generated in interchange T6-9a heterozygotes, particulate regions usually are not seen. The diffuse nucleolar structure resembles the inner component of normal nucleoli, the finely filamentous portion. Diffuse nucleoli are morphologically distinctive from normal nucleoli.

Stout (1973) noted that in maize anthers of normal inbred lines prepared for electron microscopy the nucleolus is associated with the nuclear membrane from leptonema through diakinesis of prophase I. During leptonema he observed small blebs of ribosome-sized particles extending from the inner membrane into the intramembrane space of the nuclear envelope. Cytoplasm with densely packed ribosome-sized particles appears to bud off and may be found near the plasma membrane. This phenomenon was not observed after

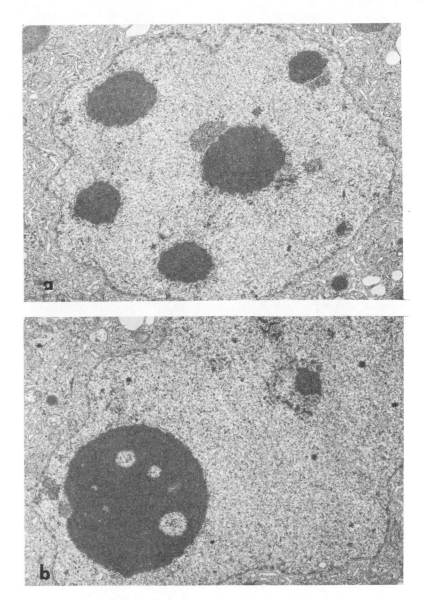

Figure 2. "Diffuse" (a) and normal (b) nucleoli in a 2-diffuse quartet of T5-6c heterozygote showing absence of lacunae (vacuoles) in diffuse nucleoli. Magnification × 11,400. Photograph courtesy of J. T. Stout.

715

zygonema. This would agree with Das's (1965) interpretation that reduced transport rate of nucleolar RNA to the cytoplasm could account for the increase in nucleolar RNA in prophase I found by Lin (1955) while RNA synthetic rate decreases. Stout also observed a lower cytoplasmic ribosome density in zygotene as compared with leptotene cells. The cytoplasmic ribosome density from diplonema to telophase II appeared to remain the same.

Stout (1973) also studied normal versus diffuse nucleoli in quartets of T5-6c heterozygotes. The diffuse nucleoli consistently lacked vacuoles or lacunae. Normal nucleoli in other cells of the same quartet contained vacuoles (Figure 2). He suggested that a functional NOR produces a vacuolated nucleolus. The diffuse nucleoli, which contained no vacuoles, may be aggregations of ribonucleoprotein normally produced by chromosome regions other than the NOR. Nonvacuolated diffuse nucleoli may indicate the absence of a functional NOR. Similar observations were made on diffuse nucleoli from T6-9a heterozygotes by Swift and Stevens (1966). The T6-9a diffuse nucleoli usually, but not always, lacked vacuoles. The T6-9a has a break in the NOR-heterochromatin, and microspores with diffuse nucleoli have a portion of the NOR present in duplicate. Perhaps a complete or partial NOR associated with the observed small vacuolated nucleoli obscured Stout's correlation of nonvacuolated nucleoli and absence of the NOR.

Stout also observed that the meiotic prophase nucleolus consists of a core of finely filamentous material and a cortex of more particulate nature. The core was not centrally located in the nucleolus but displaced to one side, such that a portion of the core became the outside surface of the nucleolus. Always at this point the NOR was associated with the nucleolus (Figure 3).

Nucleolus Organizer Region

The NOR, as observed ultrastructurally, protrudes into the core portion of the nucleolus (Figure 3). The association of the NOR and the nucleolus is complex and gives the impression of a large amount of chromatin entering the nucleolus from the NOR, with the satellite emerging nearby. The satellite in Figure 3 terminates at the nuclear envelope. Underbrink, Ting, et al. (1967) also noted a complex junction between the NOR and the nucleolus.

Gillies (1973) reconstructed the maize pachytene nucleus from electron micrographs of serial sections and also noted a complex NOR–nucleolus association. His techniques allowed the differentiation of the NOR into a dark-staining region corresponding to the NOR-heterochromatin and a lighter-staining region interpreted as corresponding to the NOR-secondary constriction. The lighter-staining region was associated with the nucleolus in a complex manner. The serial sectioning technique allowed the reconstruction of the synaptonemal complex path through the NOR and the satellite region between

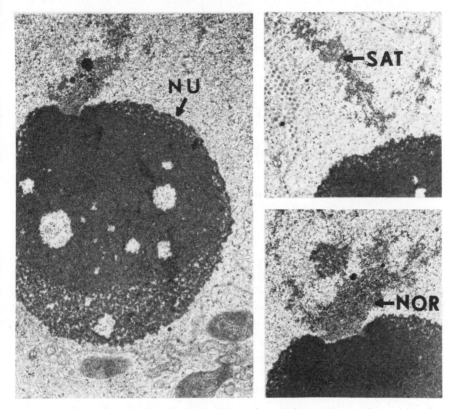

Figure 3. Three closely related sections through a pocket in the nucleus of a pollen mother cell at late zygonema containing the nucleolus (NU), NOR, and satellite (SAT). Magnification × 9200. Photographs courtesy of J. T. Stout.

the chromosome 6 homologs. The synaptonemal complex was present through the NOR-heterochromatin as well as the lighter-staining region and the satellite. An interesting point is that the synaptonemal complex traversed the lighter-staining region (secondary constriction), which has the complex junction with the nucleolus, without entering the nucleolus or undergoing significant morphological alterations. In light-microscopic observations of the NOR-secondary constriction, a fine thread of chromatin is often seen forming a direct and apparently uninterrupted connection between the NOR-heterochromatin and the satellite. This chromatin-thread connection seemed difficult to interpret if one assumes the argument that the NOR-secondary constriction represents a region where the chromatin is dispersed to some extent into the nucleolus. However, such a direct thread-like connection might be expected between the NOR-heterochromatin and the satellite by light microscopy if the synaptonemal complex does not enter the nucleolus but maintains a certain amount of attached chromatin while the remainder is dispersed into the nucleolus. Gillies also reported that the fibrillar chromatin of the lighter-staining region was finer in diameter suggesting a more active state. Although perhaps reasonable to assume, there is no direct evidence that the maize nucleolus contains DNA, with the exception of preliminary biochemical evidence on isolated nucleoli referred to by McLeish (1964).

RIBOSOMES AND POLYMERASES

The maize monoribosome has a sedimentation coefficient of 80S and can be dissociated into two subunits, 60S and 40S (Hsiao, 1964). The purified monoribosome is 43% RNA and the remainder consists of ribosomal proteins about which little is known. The RNA is characterized by high guanine and relatively high cytosine contents. The large subunit contains the 28S rRNA (Tang, 1971). The small subunit presumably contains the 18S rRNA; however, Tang (1971) obtained apparent RNA degradation products instead of the intact 18S rRNA. By using *E. coli* rRNAs as markers, Jacobson and Williams (1968) reported that maize cytoplasmic rRNAs have sedimentation coefficients of 16S and 26S. Since it is common to refer to eukaryotic ribosomal RNA subunits as 18S and 28S (Tang 1971), the more conventional nomenclature is used in this chapter. Loening (1968) gives the molecular weights for maize 18S and 28S rRNA as 0.7×10^6 and 1.3×10^6 daltons, respectively. Pring and Thornbury (1975) recently reported molecular weights of 0.67×10^6 and only 1.19×10^6 for maize 18S and 28S rRNA, respectively. The values of 0.7×10^6 and 1.3×10^6 have been used in this chapter for the rRNA gene-number determinations. Ruppel (1969) showed that maize ribosomes also contain 5S RNA. Whether the 5S RNA is located in the large ribosomal subunit is unknown.

Two DNA-dependent RNA polymerases have been shown to be present in maize nuclei (Stout and Mans, 1967; Strain, Mullinix, et al., 1971). Ribonucleic acid polymerase I is resistant to α-aminitin inhibition, suggesting that it is the nucleolar RNA polymerase and involved in 18S and 28S rRNA synthesis (Roeder and Rutter, 1969). Ribonucleic acid polymerase II is strongly inhibited by α-aminitin and can be resolved into two activities. Ribonucleic acid polymerase IIa is active with denatured nuclear DNA of maize, whereas IIb is more active with native DNA (Mullinix, Strain, et al., 1973). Ribonucleic acid polymerase IIa is composed of several polypeptide chains of different molecular weights.

Although the size and maturation processing of the ribosomal precursor RNA that gives rise to 18S and 28S rRNA has not been described for maize, the scheme in plants appears similar to other eukaryotes in that ribosomal RNAs of molecular weight 1.3×10^6 and 0.7×10^6 daltons both arise from the

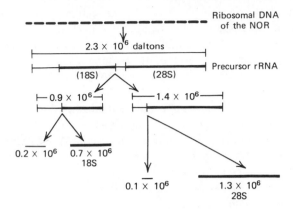

Figure 4. Processing of ribosomal precursor RNA to 18S and 28S rRNA. The 18S rRNA is complexed with proteins and rapidly transported to the cytoplasm where additional ribosomal proteins attach making the 40S subunit. The 28S rRNA is complexed with proteins and 5S RNA from genes in chromosome 2 (and perhaps 7S RNA also from the original 2.3×10^6 precursor rRNA) and transported to the cytoplasm where additional ribosomal proteins attach making the 60S subunit. Drawing is schematic.

selective cleavage of a larger precursor molecule. In plants, the rRNA genes transcribe a rRNA precursor molecule of about 2.3×10^6 daltons (Rogers, Loening, et al., 1970). This molecule is cleaved into two molecules of 1.4×10^6 and 0.9×10^6 molecular weights. The 0.9×10^6 molecule is rapidly cleaved to produce the 0.7×10^6 molecule (18S). The 1.4×10^6 molecule is cleaved to produce the 1.3×10^6 rRNA molecule (28S). During these maturation processes, the RNA becomes associated with characteristic proteins to produce the 60S (which also is expected to contain 5S and possibly 7S RNA) and 40S ribosomal subunits. The fate of the RNA in excess of the 18S and 28S produced by cleavage of the precursor molecule is unknown. The general scheme is illustrated in Figure 4.

In the maize cytoplasm the 60S and 40S subunits unite to form the 80S monoribosome. The 80S monosomes become associated with mRNA to form polyribosomes active in protein synthesis. Sucrose gradient profiles of maize polyribosomes reveal a peak of monosomes and several polymers of increasing size. Rapid changes in the polyribosome profile occur in maize as the result of altered physiological states, induced by such factors as water stress (Hsiao, 1970) and light patterns (Travis, Huffaker, et al., 1970).

LOCALIZATION OF 18S AND 28S rRNA GENES

In 1955 Lin showed that nucleolar RNA content could be elevated linearly by increasing the dosage of a portion of the NOR. He concluded that "the function of the nucleolar organizer involves the actual synthesis of nucleolar material; the organizer is doing something more than merely serving as a pump or a reservoir for the collection and organization of the matrix material, or some other material produced by the chromosomes, into a single body." This work pioneered the understanding of the NOR in terms of its synthetic capacity.

Now there are at least six lines of evidence indicating that the NOR of maize is the chromosomal site of the DNA complementary to 18S and 28S rRNA. The first demonstration, shown by our laboratory in 1971 (Phillips, Kleese, et al., 1971), involved the comparison of rRNA gene numbers by DNA/rRNA hybridization techniques of inbreds W23 and A188 versus a genetic marker strain possessing a NOR consisting of two heterochromatic portions separated by a secondary constriction. This NOR condition was termed "2NOR" (Figure 5). The nucleolar volume of the 2NOR strain at midpachynema was 64% larger than the average of the two inbreds. The large 2NOR–nucleolus segregated with the 2NOR–chromosome 6. The percent DNA hybridizable to 18S and 28S rRNA for the 2NOR strain was found to be twice the amount for A188 and more than twice W23. The NOR of maize was concluded to be the chromosomal site of the 18S and 28S rRNA genes. Further support for this conclusion was provided by showing that plants trisomic for chromosome 6

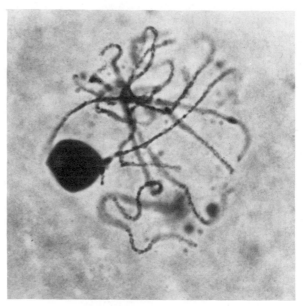

Figure 5. Pachytene cell of the 2NOR strain showing the two heterochromatic segments separated by the secondary constriction with attached nucleolus. Magnification × 1600.

possess approximately 50% more rRNA genes than their diploid sibs and that plants monosomic for chromosome 6 carry approximately half as many rRNA genes as the parent that contributed the chromosome to the monosomic progeny (Phillips, Weber, et al., 1974).

Perhaps the best evidence for any eukaryote that the rRNA genes are actually within the NOR and not in neighboring regions is that presented by Givens and Phillips (1976). The procedure involved the production of duplications by the method of Gopinath and Burnham (1956), where selfpollinations of intercrosses between appropriately chosen interchanges involving the same two chromosomes yield duplications of the between-breaks segments of each chromosome. Plants were produced carrying a duplication of either the NOR-heterochromatin or the site giving rise to the secondary constriction. These plants also carried a short duplication in either chromosome 1 or 2, but the previous evidence cited indicates that these chromosomes are not expected to carry rRNA genes. Hybridization of DNA from plants heterozygous for the duplication of the NOR-heterochromatin was approximately 50% higher than the controls. The heterochromatic portion of the NOR, therefore, contains most of the rRNA genes present in maize. As a check, similar experiments were performed using DNA from plants heterozygous for the duplication of the secondary constriction. Given that most of the rRNA genes are in the NOR-heterochromatin, the hybridization level using DNA from plants carrying the duplication of the secondary constriction would not be expected to be higher than the controls; this was observed. Such a result does not necessarily indicate that the secondary constriction carries no rRNA genes, but that it carries either none or a number below the limits of our error in the DNA/rRNA hybridization procedure. As this error is 10% or less, the conclusion is that the NOR-heterochromatin possesses 90% or more and the NOR-secondary constriction 10% or less of the total rRNA genes. Since maize has a high number of rRNA genes, 10% represents a rRNA gene multiplicity that is not dissimilar to that of many animals.

That the NOR of maize is the site of the 18S and 28S rRNA genes also is supported by DNA/rRNA hybridization experiments (Doerschug, 1976; Ramirez and Sinclair, 1972) with a translocation between the NOR of chromosome 6 and a B chromosome (TB-6a). Due to the property of nondisjunction of the B chromosome centromere at the second postmeiotic division in microsporogenesis (Roman, 1947), maize plants can be obtained with varying numbers of the interchanged B^6 chromosome. Since the break of TB-6a is near the center of the NOR-heterochromatin (Phillips, unpublished), plants may be obtained with multiple doses of the segment distal to the break. This segment includes the distal half of the NOR-heterochromatin and the secondary constriction, as well as the satellite. Based on DNA/rRNA hybridization results using DNA from hyperploid plants with from one to seven B^6 chromosomes, Doerschug (1976) estimates that the break occurred such that approximately half of the rRNA genes are in the 6^R chromosome and half in the B^6 chromosome.

The last line of evidence to be mentioned that the maize NOR is the site of rDNA comes from *in situ* hybridization experiments by Wimber, Duffey, et al. (1974). The placement of the ribosomal RNA genes to the distal end of the short arm of chromosome 6 was confirmed by annealing ^{125}I-labelled 18S and 28S rRNA to pachytene chromosomes. They state that "the only obvious labelled region was the nucleolus."

The placement of most if not all of the rRNA genes to the NOR appears to be more precise for maize than for any other higher organism. The well-known rDNA localization studies using DNA extracted from such stocks as the NOR-duplications and deficiencies in *Drosophila* (Ritossa and Spiegelman, 1965), the anucleolate mutant of *Xenopus* (Birnstiel, Wallace, et al., 1966) and the *bobbed* mutants of *Drosophila* (Ritossa, Atwood, et al., 1966) place the rRNA genes only in the vicinity of the NOR and not unequivocally within the structure. The *in situ* DNA/rRNA hybridization technique (Pardue, Gerbi, et al., 1970) also does not provide the resolution required for as precise a placement as has been accomplished for maize.

RIBOSOMAL RNA GENE MULTIPLICITIES

Variation in rRNA gene multiplicity is great among higher plant species. Cullis and Davies (1974) indicated that the range represents a 17-fold difference, from a DNA hybridization percentage of 0.022 or 1580 rRNA gene/2C nucleus for the artichoke, *Helianthus tuberosus* (Ingle and Sinclair, 1972), to a 3.1% hybridization value or 27,000 rRNA genes for cucurbit, *Cucurbita maxima* (Goldberg, Bemis, et al., 1972). We have found the range in rRNA gene multiplicities for maize to be nearly as great as the range reported for all higher plant species. Table 1 includes the rRNA gene multiplicities for various inbred lines utilized in the corn-breeding industry, and thus they may be considered as agronomically desirable lines. The lines were chosen from a list of the 25 most widely used publicly developed lines of corn (Horsfall, 1972). Seventeen of the 25 were tested along with four additional lines (Wf9, A188, A153, and W23). The range in rRNA gene multiplicities among the inbreds is 5,000–12,000 per 2C nucleus. A somewhat smaller range based on data from 10 of these inbreds was reported previously (Phillips, Wang, et al., 1973).

Feulgen microspectrophotometric DNA measurements of these 20 lines indicated that although there were differences in DNA values among the various lines, no correlation existed between DNA amount and rRNA gene multiplicity.

TABLE 1.   Variation in rRNA gene multiplicity among inbred lines of maize[a]

Inbred	Ribosomal RNA gene multiplicity ± S.E.[b]	Inbred	Ribosomal RNA gene multiplicity ± S.E.
W117	12,000 ± 450	A634	7,400 ± 370
Mo17	10,600 ± 240	W64A	7,300 ± 900
B37	10,100 ± 250	A619	7,200 ± 390
C103	9,900 ± 270	A188	7,000 ± 370
Oh43	9,400 ± 270	C123	6,500 ± 300
B14	8,500 ± 270	A153	6,400 ± 435
Wf9	8,400 ± 250	A635	6,200 ± 150
A632	8,300 ± 200	A554	6,100 ± 240
N28	8,000 ± 120	H84	5,500 ± 60
B57	7,600 ± 600	W23	5,000 ± 100
A239	7,400 ± 250		

[a] Deoxyribonucleic acid/rRNA hybridization followed previously published procedures (Phillips, Kleese, et al., 1971; Phillips, Weber, et al., 1974) using 6-day-old seedlings. Ribosomal RNA gene multiplicities based on minimum of six determinations per inbred. Hybridization done as one experiment using A188 ^{3}H-uridine-labeled rRNA under saturating conditions at 10 μg rRNA/ml. All gene multiplicities are per 2C nucleus.

[b] S.E. = standard error.

Certain inbreds in Table 1 with a high level of relatedness have retained comparable levels of rRNA gene multiplicities, while others have not. For example, A632 and B14 are closely related and have similar rRNA gene multiplicities, whereas A634 and B14 are related to the same degree but have some divergence in their rRNA gene multiplicities. An explanation cannot be offered without further study. This range in rRNA gene multiplicities among inbred lines of a single species should in the future provide the basis for relating the number of rRNA genes to various biochemical or agronomic characteristics.

The lowest rRNA gene multiplicity observed in maize is that of the *sticky chromosome* (*st*) mutant conditioned by a single locus in the short arm of chromosome 4 discovered by Beadle (1932). The mutant is generally characterized by the stickiness of the chromosomes at the meiotic anaphases giving chromatin bridges, but chromosomes of other stages also are abnormal in various ways. High ovule and pollen sterility, scarring of the endosperm in mature *st* kernels, and small plants with leaf striations were common characteristics. The *st* lines currently available give a variable expression of the *st* phenotype in terms of cytology, sterility, and endosperm and plant phenotypes (Stout, 1973).

Stout (1973) made several observations that bear on a possible relationship between the *st* locus in chromosome 4 and the NOR of chromosome 6. The

NOR of *st* plants was decidedly different from normal (Figure 6). The NOR-heterochromatin was abnormally small and often not particularly apparent. Instead, the locus of the normally heterochromatic portion of the NOR frequently possessed chromatin that appeared diffuse. The secondary constriction was quite large as was the nucleolus. Stout also observed a chromatin vesicle in the *st* homozygotes from the original seed and in certain F_3 families. The vesicle had a dark-staining reticulate core and a lighter-staining surrounding matrix. The vesicle was associated with the NOR or nucleolus more often than with any other part of the nucleus, either attached or unattached to other

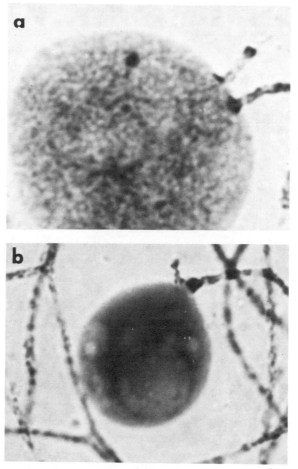

Figure 6. Comparison of the *st* NOR (a) and the NOR of the next lowest rRNA gene multiplicity line (W23) (b) at pachynema. Note small NOR-heterochromatin and large secondary constriction in *st* plant. Magnification × 4200. Photographs courtesy of J. T. Stout.

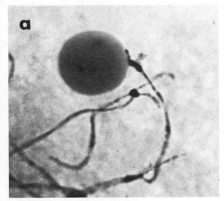
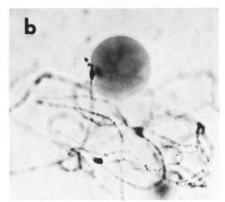

Figure 7.   Large NOR-heterochromatin of the University of Illinois reverse high-protein line (a) compared with the NOR-heterochromatin of the University of Illinois high protein line (b). Magnification × 1500.

chromosomes. The abnormal appearance of the *st* NOR-heterochromatin and the occurrence of a vesicle led us to perform DNA/rRNA hybridization experiments using DNA from the *st* homozygote (Stout, 1973). A value of about 3300 rRNA genes was obtained—the lowest rRNA gene multiplicity observed to date for any maize line. The cytologically anomalous NOR is probably a manifestation of this low rRNA gene multiplicity. Whether the observed vesicle contains rDNA and whether the NOR phenotype is somehow controlled by the *st* locus in chromosome 4 await further experimentation. Two other single gene variants of maize have been studied for rDNA content. Both mutants, *elongate* (*el*) and *ragged* (*rgd*) located in chromosome 6 near the NOR, have normal rRNA gene multiplicities (Stout 1973; Phillips, Weber, et al., 1974).

The highest rRNA gene multiplicity thus far observed for maize was discovered while investigating the protein lines selected by University of Illinois investigators since 1896 (see pp. 736–738). The reverse high protein (RHP) line was found to possess 23,100 rRNA genes. Since this value was unusually high, the strain was studied cytologically with special attention focused on the NOR. A strikingly large NOR was observed at pachynema; the NOR-heterochromatin was several times larger than comparable segments in other strains (Figure 7).

Another line with a high rRNA gene multiplicity is the 2NOR strain. In 1971 we compared the 2NOR strain with inbreds W23 and A188 and found that the 2NOR strain possesses twice the A188 number of rRNA genes (Phillips, Kleese, et al., 1971). Our current estimate of the rRNA gene multiplicity for A188 is lower than reported in 1971 (the W23 rRNA gene multiplicity was not reported) as the result of modifying our estimate of the amount of DNA/2C nucleus used for calculating rRNA gene numbers (Phillips, Weber, et al., 1974)

TABLE 2. Ribosomal RNA gene multiplicities of Wf9 lines with normal and T, C, S, and J male-sterile cytoplasms[a]

Wf9	Wf9 Tcms	Wf9 Ccms	Wf9 Scms	Wf9 Jcms
8100[b]	7600	6600	6600	6500

[a] Seed kindly provided by C. A. Laible, Funk Seed International, Bloomington, Illinois. The number of backcrosses to Wf9 for the T, C, S, and J cms lines is 12, 10, 10, and 7, respectively.

[b] Data average of two experiments each with eight determinations using saturating levels of A188 ^{3}H-rRNA. Deoxyribonucleic acid was extracted from 6-day-old seedlings. Duncan's multiple range test was used to determine significance. The cms lines are not significantly different; however, the Wf9 C, S, and J cms lines are significantly different from Wf9.

and a calibration problem with the spectrophotometer. Thus our estimate for the 2NOR rRNA gene multiplicity is 14,000. Genetic alterations in rRNA gene multiplicities have been demonstrated in animal species as the result of aneuploidy or gene mutation. Such changes in rRNA gene multiplicities, termed rDNA compensation (Tartof, 1973) for somatic alterations and rDNA magnification (Ritossa, 1968) for germ-line changes, were sought in maize by generating plants monosomic for chromosome 6 and determining their rRNA gene multiplicity as well as that of their progeny (Phillips, Weber, et al., 1974). The results were as expected assuming additivity and no disproportionate replication of the rDNA. Additiveness seems to be a general principle for rRNA gene multiplicities in maize. Hybrids between lines of known rRNA gene multiplicities (Phillips, unpublished) and various aneuploids (Doerschug, 1976, Phillips, Weber, et al., 1974) have been shown to behave in an additive fashion. To my knowledge, no well-documented exceptions to this general rule exist at the present time.

Another question of interest regarding rRNA gene multiplicity is whether it is influenced by the cytoplasm. Through extensive backcrossing, lines are available with the nuclear genetic constitution of inbred Wf9 and different cytoplasms. Beckett (1971) grouped the various sources of cytoplasms that confer male sterility into the S, C, and T groups. The J cytoplasmic male sterile (CMS) was included in the S group, according to the methods employed. Deoxyribonucleic acid from Wf9 T, C, S, and J CMS lines and normal cytoplasm Wf9 were hybridized to A188 rRNA. The results (Table 2) indicate that the sterile cytoplasms do not cause an increase in rRNA gene multiplicity. The various Wf9 CMS lines were not significantly different. However, the Wf9 C, S, and J CMS lines were significantly different from Wf9 with normal cytoplasm. Since these converted lines are the result of seven to twelve backcrosses to Wf9, it would be unlikely that the differences are simply a matter of not recovering the Wf9 rRNA gene constitution. Crossing over at the NOR during the conversion process could give rise to these differences.

The foregoing discussion focused on the multiplicity of 18S and 28S rRNA genes. Since 5S RNA is also part of the ribosome and is present in the nucleolus prior to its transport to the cytoplasm as part of the large subunit, knowledge on the 5S RNA gene multiplicity is important to an understanding of the NOR function. Based on *in situ* DNA/5S RNA hybridization, the genes for 5S RNA appear to be clustered near the end of the long arm of chromosome 2 (Wimber, Duffey, et al., 1974). Multiple copies of the 5S RNA genes are indicated by the autoradiographic observation of a heavily labeled chromosomal region. A high multiplicity would be expected, extrapolating from animal systems where the number of 5S RNA genes is usually greater than the number of 18S and 28S rRNA genes (Ford, 1973). Our preliminary DNA/5S RNA hybridization experiments suggest that 5S RNA genes of maize also are present in many thousands of copies; the data are not sufficient to give a gene multiplicity value at this time.

FUNCTIONAL MAP OF THE NOR

In 1931 Heitz reported on the physical relationship between the nucleolus and a particular chromosomal site. The nucleolus was shown to form in a similar manner in association with the same daughter chromosomes in daughter telophase nuclei. The nucleolus appeared to develop in association with the secondary constriction. The theory was advanced that the secondary constriction actually formed the nucleolus. McClintock (1934) challenged this idea in 1934 as a result of her study of nucleolus–chromosome associations in maize. She studied a chromosomal interchange which divided the NOR-heterochromatin in two parts (Figure 8). The break in this interchange (T6-9a) occurred such that the proximal two-thirds of the NOR-heterochromatin was separated from the distal one-third of the heterochromatin and the contiguous secondary constriction. In the homozygote, the proximal NOR-heterochromatic

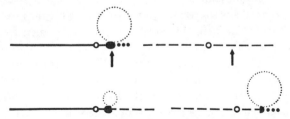

Figure 8. Diagram illustrating structurally normal chromosomes with typical NOR–nucleolus association (upper diagram) and chromosomes of an NOR-interchange with a break in the NOR-heterochromatin showing differential functional capacities of the NOR-segments (lower diagram).

segment formed a small nucleolus, whereas the distal part formed a larger nucleolus. Thus, the heterochromatic body located adjacent to the secondary constriction had the capacity to form a nucleolus. The fact that in T6-9a homozygotes a larger nucleolus is formed by the distal one-third of the NOR-heterochromatin while possessing 50% or less of the total rRNA genes (Doerschug, 1976), suggests that all of the rRNA genes may not be concomitantly functional in nucleolar formation. The question then is whether the secondary constriction has a nucleolus-forming capacity as believed by Heitz. Through elegant analyses of T6-9a and several other stocks, McClintock concluded that whereas the nucleolus formed as a result of activity of the NOR-heterochromatin, the secondary constriction formed as a passive result of nucleolar growth. The heterochromatin was referred to as the "nucleolar organizing body or element." In addition to the interchange strain, McClintock studied two other strains in which the nucleolus appeared to form at places other than the distal end of the NOR-heterochromatin. The nucleolus of one strain formed near the center of the heterochromatin. In another strain the nucleolus formed at the proximal end; in this case the NOR-heterochromatin was on the satellite side of the nucleolus creating a large satellite. These observations suggested that the functional site of the NOR in terms of nucleolar formation may be at any one of three locations— distal, central, or proximal.

That the NOR has three potential sites of activity also has been indicated from studies in our laboratory of additional maize interchanges with a break in the NOR. At least 20 interchanges (termed *NOR-interchanges* for the purposes of this chapter) are available today with a break in the NOR at various sites (Phillips and Wang, 1972). At pachynema in plants homozygous for NOR-interchanges, two bivalent chromosomes are associated with a nucleolus, making exact the placement of the break in the NOR. A break immediately proximal to the NOR in the short arm of chromosome 6 or immediately distal in the satellite results in only one chromosome pair associated with the nucleolus. The association of both interchanged bivalents with the nucleolus in NOR-interchange homozygotes also allows the precise placement of breakpositions in the other chromosome involved in the interchange. Thus the cytological placement of the breaks is more accurate than is possible for most maize interchanges. When the break is in the NOR, indicated by two pachytene bivalents associated with a single nucleolus or each with a separate nucleolus, there may or may not be an obvious segment of the NOR-heterochromatin translocated. When a portion of the heterochromatin is translocated, the result is apparent and a breakposition can be assigned in the NOR-heterochromatin by measuring the two heterochromatic portions. When there is no obvious heterochromatin translocated, but there are two bivalents with nucleolar associations, the break is assumed to have occurred in the distal site that forms the secondary constriction in these strains. The two resultant interchanged chromosomes may each form an obvious secondary constriction. The average lengths of the two secondary constrictions were taken to indicate the position of the break in the site giving rise to the secondary constriction. Although the measurement is subject to error, we have assigned the breaks to the proximal part of the secondary constriction near the heterochromatin, about 0.25 of the heterochromatin-satellite distance, or midway (Table 3). There were no indications of any interchange break beyond midway. Seven of the interchanges have a break in the NOR-heterochromatin and 12 in the secondary constriction.

Nucleolar volumes were determined using pachytene cells with two nucleoli. The data are presented (Table 3 and Figure 9) as the volume of the nucleolus associated with the proximal part of the NOR expressed as a percentage of the total nucleolar volume. Apparent is the fact that nucleolar size and, therefore, the functional capacity of the NOR portion, depends on the position of the break in the NOR-heterochromatin or the secondary constriction. Perhaps the most striking observation is for interchange T2-6(8786). The break is close to the distal end of the heterochromatin, resulting in chromosome bivalent 6^2 possessing nearly all of the NOR-heterochromatin. An extremely small amount

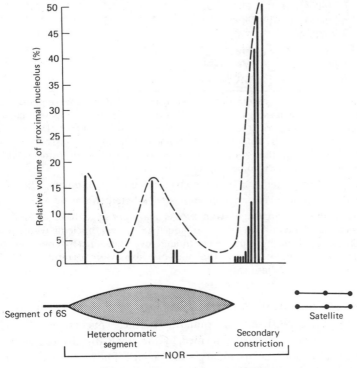

Figure 9. Functional map of NOR.

TABLE 3. Functional capacities of portions of the NOR when separated in homozygous NOR-interchange stocks

Interchange	Breakpoints		Relative proximal nucleolar volume (%)[a]
	Chromosome 6	Other	
1-6Li[b]	S.C.[c]—prox.	1L.81	0.2
1-6(4986)	S.C.—prox.	1S.11	0.8
1-6(6189)	Het. 0.10	1S.50	17.8
1-6(8415)	S.C.—prox.	1L.31	0.01
2-6(5419)	S.C.—0.25	2L.82	13.7
2-6(8441)	S.C.—prox.	2L.95	0.1
2-6(8786)	Het. 0.88	2S.97	0.1
2-6(027-4)	S.C.—prox.	2L.04	0.6
3-6(030-8)	S.C.—0.25	3S.05	1.6
3-6(032-3)	S.C.—midway	3S.34	50.0[d]
4-6(4341)	Het. 0.50	4S.36	15.5
5-6f	S.C.—midway	5S.23	47.3[d]
5-6(8696)	S.C.—midway	5L.79	42.0[d]
6-7(4964)	Het. 0.32	7L.67	1.3
6-7(5181)	Het. 0.71	7L.85	2.6
6-7(035-3)	S.C.—0.25	7L.59	5.8
6-9a	Het. 0.67	9L.32	2.5
6-9d	Het. 0.46	9L.84	3.3
6-10(5519)	S.C.—prox.	10L.10	8.0

[a] Nucleolar volumes were determined using diameter measurements from camera lucida drawings of three cells from each homozygous interchange and converting to volumes assuming a spherical shape. The relative proximal nucleolar volume is the volume of the nucleolus associated with the proximal portion of the NOR when two nucleoli were formed divided by total nucleolar volume (expressed in percentage).

[b] Originally reported as a T4-6 interchange (Stout and Burnham, 1968).

[c] S.C. = Secondary constriction, Het. = NOR-heterochromatin.

[d] In all three interchanges with breaks that appear to have divided the secondary constriction into two equal parts, two nucleoli were never observed at pachynema, even though over 2000 pachytene cells per interchange were studied. The nucleolar sizes were determined at the quartet stage in these cases.

of nucleolar material, 0.1% of the total nucleolar volume, is associated with nearly the entire NOR-heterochromatin. In these pachytene cells with two nucleoli, the large nucleolus is formed by the small distal piece of the NOR-heterochromatin and the secondary constriction. The nearly complete NOR-heterochromatin has little functional capacity in pachytene cells in which the small distal piece of the NOR-heterochromatin and secondary constriction also are present. Confirmation of this conclusion comes from an independent study of the previously mentioned stock carrying a duplication of the NOR-heterochromatin generated in the F_2 of intercrosses of interchanges involving the same two chromosomes (Givens and Phillips, 1976). The bivalent chromosome carrying only the NOR-heterochromatin (and none of the original secondary constriction) in homozygous condition had no associated nucleolar material in 71% of the pachytene cells ($N = 324$). When that chromosome was associated with a separate nucleolus (13% of the cells), the nucleolus was extremely small representing only 0.6% of the total nucleolar volume. The remaining cells (16%) had both chromosome bivalents associated with one nucleolus.

In contrast to the T2-6(8786) interchange, others with a break in the NOR-heterochromatin revealed a greater functional capacity of the proximal portion of the NOR-heterochromatin. The functional capacity, however, was not a linear function of the NOR breakposition. Maximum activities occurred when the break was either near the midpoint of the NOR-heterochromatin or near the proximal end (Figure 9). Interchanges with breaks between these sites revealed lower activities associated with the proximal NOR segment.

Interchanges with a break in the secondary constriction gave activities that may be a linear function of the breakposition. Little activity was associated with the proximal NOR segment when the break was near the proximal end of the secondary constriction. A larger nucleolus was associated with the proximal portion when the break was at 0.25 in the secondary constriction. Pachytene cells of T3-6 (032-3), T5-6f, and T5-6 (8696) with breaks about midway in the secondary constriction contained only one nucleolus; two nucleoli were not observed in more than 2000 pachytene cells of each interchange. However, two nucleoli of approximately equal volumes were observed at the quartet stage in homozygotes of all three interchanges.

Our interpretation of these results is that the NOR has three potential sites of activity. We hypothesize that these interchanges arose in maize strains in which the distal site was active, at least as expressed in prophase I of microsporogenesis. When the break occurred in this site, activity was divided proportionately between the two segments and formed appropriately sized secondary constrictions. The other two potential sites of activity in these stocks are inactive in meiosis of microsporogenesis unless a break occurs in or near one of them. Such a break, then, allows a small amount of activity and reveals their existence. This conclusion is in basic agreement with McClintock's theory regarding the existence of three potentially functional sites. The difference is that these results suggest that the secondary constriction does have a functional capacity. This idea also is supported by observations on nucleolus–chromosome associations in plants heterozygous for a duplication of the secondary constriction (Givens and Phillips, 1976). In these plants, all three of the secondary constrictions were associated with a nucleolus in 97–100% of the pachytene cells.

This result is unlike that for the duplication of the NOR-heterochromatin where all heterochromatic segments were associated with a nucleolus in only 29% of the pachytene cells. Therefore, the secondary constriction appears to possess a functional capacity, even when duplicated in a cell and physically separated from the NOR-heterochromatin in a chromosome other than chromosome 6.

Coupling the cytological observations on the interchange and duplication stocks with the previously mentioned DNA/rRNA hybridization data on the duplication stocks, the interpretation is advanced that although most of the rRNA genes are in the NOR-heterochromatin the secondary constriction appears to be the site primarily responsible for nucleolus formation (at least in pachynema of microsporogenesis). Also, two secondary sites appear to be located in the NOR-heterochromatin in these strains—one near the midpoint and one at the proximal end.

Electron-microscopic observations of transcribing rRNA genes in primary spermatocyte nucleoli of *Drosophila hydei* show that a maximum of one-half of the rRNA genes are transcribing at any one time and the number varies with developmental stage (Meyer and Hennig, 1974). Regulation of rRNA synthesis appears to occur primarily by the activation or inactivation of groups of adjacent rRNA genes. The NOR primary and secondary sites suggested for maize may represent groups of rRNA genes coordinately controlled and active at various developmental stages.

The question arises as to the purpose of the thousands of rRNA genes present in maize nuclei. The argument can be made that most of them are not active in any one strain since (a) animal species usually possess rRNA genes in the hundreds rather than the thousands, (b) maize strains differ so vastly in rRNA gene multiplicities, and (c) the region that appears primarily responsible for nucleolus formation, the site giving rise to the secondary constriction, contains relatively few rRNA genes. However, one can argue that evolution does not generate an inefficient genetic system and that all of the rDNA must be useful at certain developmental stages or under certain stress conditions. A supporting fact is that all plant genera studied thus far have more than 1500 rRNA genes per 2C nucleus, and, therefore, high rRNA gene multiplicities must be important to plant development and survival. One view is that plants meet peak rRNA demands by transcribing more of the rRNA genes already present in the cell rather than amplifying their rDNA (Phillips, Weber, et al., 1974). The final answer to this question will provide important information on genetic regulation in higher plants.

Considerable information can also be gained on the functional capacity of NOR segments from plants *heterozygous* for NOR-interchanges. Meiotic products of these plants principally are of four chromosome constitutions. For example, T6-9a (breakpoints: NOR-heterochromatin 0.67, 9L.32) produces microspores with chromosome combinations $6 + 9$, $6^9 + 9^6$, $6^9 + 9$, and $6 + 9^6$.

McClintock (1934) showed that chromosome 6^9 in $6^9 + 9$ microspores formed a normal sized nucleolus at prophase of the first postmeiotic division, whereas it formed a small nucleolus in $6^9 + 9^6$ microspores. Therefore, the proximal 67% of the NOR-heterochromatin has the potential of forming a normal-sized nucleolus and exhibits this potential when the distal portion is not present in the cell. This observation is consistent with the idea of three potential sites of NOR-activity (proximal, central, and distal). The secondary site(s) may become active on deletion of the primary site of that strain. Now that several NOR-interchanges are available, the functional capacity of various proximal portions of the NOR can be assayed in the absence of the distal portion. The results (Table 4) show that the proximal portion of the NOR-heterochromatin can form a normal nucleolus in every case when the distal portion is absent, even in the extreme case of T1-6(6189) in which the break is at 0.10 in the NOR-heterochromatin. The segment proximal to 0.10 in the heterochromatin can function to apparently normal capacity in the absence of the remainder of the NOR. This supports the idea of a site of potential activity at the proximal end of the NOR-heterochromatin.

Microspores from T6-9a heterozygotes of the $6 + 9^6$ chromosome constitution possess a diffuse nucleolus at prophase of the first postmeiotic division as reported by McClintock (1934) and reconfirmed in our laboratory (Table 4). McClintock suggested that the diffuse nucleolus may have been the result of deleting a part of 9L; that is, 9L may carry a gene(s) necessary for normal nucleolus formation. If this interpretation is correct, the gene must be between 9L.32 and 9L.84, since T6-9d does not produce microspores with diffuse nucleoli. Another such gene may reside between 7L.59 and 7L.85, since T6-7(035-3) heterozygotes produce diffuse nucleolated microspores whereas none occur in T6-7(5181) heterozygotes. The T6-10(5519) heterozygotes also produce microspores with diffuse nucleoli at a relatively high frequency (11%), but mostly in the microspore type $6^{10} + 10$, deficient for a portion of the NOR and duplicate for a portion of chromosome 10. Although not understood, it is important to note that either of the two unbalanced chromosome combinations can lead to a diffuse nucleolus.

Another important observation is that the appearance of a diffuse nucleolus at prophase of the first postmeiotic division does not mean that a diffuse nucleolus existed at the quartet stage (Table 4—see T6-9a for example). A nucleolus apparently can form immediately following meiosis in certain cases and then become diffuse. The reverse also is observed as in T1-6 Li in which 16% of the quartet cells possessed a diffuse nucleolus while diffuse nucleoli were not observed at the first postmeiotic prophase. In these cases, a diffuse nucleolus appears able to coalesce into a single nucleolus of normal appearance. The homozygous anucleolate mutant of *Xenopus* also may form nucleoli that appear normal in older embryos (Barr, 1966). The genetic cause of diffuse

TABLE 4. Nucleolar constitution of microspores in the first postmeiotic prophase or quartet stages of NOR-interchange heterozygotes

Interchange	Breakpoints		Spore types				Total cells	Prophase microspores with diffuse nucleolus (%)	Quartet microspores with diffuse nucleolus (%)	Total cells (no. quartets × 4)
	Chromosome 6	Other	$6 + X$	$6^x + X$[b]	$6^x + X$	$6 + X$[b]				
1-6 Li	S.C.[a]—prox.	1L.81	1N(60)[c]	1N(50)2U(2)	1N(52)	1N(60)	224	0	16.0	1680
1-6(4986)	S.C.—prox.	1S.11	1N(54)	1N(41)2U(3)	1N(42)D(8)	1N(40)	192	4.2	12.5	2128
1-6(5495)	S.C.—prox.		1N(51)	1N(35)2U(1)	1N(60)	1N(74)	221	0	1.1	2488
1-6(6189)	Het.[b] 0.10	1S.50	1N(30)	1N(30)	1N(34)	1N(25)	119	0	0.8	2816
2-6(027-4)	S.C.—prox.	2L.04	1N(46)	1N(35)2U(13)	1N(48)	1N(38)2U(1)D(5)	186	2.7	8.5	2152
2-6(8786)	Het. 0.88	2S.97	1N(80)	1N(26)2U(14)	1N(45)	1N(55)	220	0		2392
2-6(5419)	S.C.—0.25	2L.82							17.4	2160
3-6(030-8)	S.C.—0.25	3S.05	1N(52)	1N(35)2U(6)	1N(46)	1N(48)D(2)	189	1.1	1.2	2276
3-6(032-3)	S.C.—midway	3S.34							2.2	2328
4-6(4341)	Het. 0.50	4S.36	1N(65)	1N(60)	1N(54)	1N(29)	208	0	9.2	2116
5-6f	S.C.—midway	5S.23	1N(70)	1N(69)	1N(70)	1N(10)	219	0	9.7	2292
5-6(8696)	S.C.—midway	5L.79	1N(45)	1N(45)2U(9)	1N(31)	1N(45)	175	0	0.1	2768
6-7(4964)	Het. 0.32	7L.67	1N(28)	1N(22)2U(11)	1N(24)	1N(23)D(1)	109	0.9	18.6	2416
6-7(035-3)	S.C.—0.25	7L.59	1N(49)	1N(36)2U(2)	1N(34)	1N(16)D(24)	161	14.9	9.7	2276
6-7(5181)	Het. 0.71	7L.85	1N(38)	1N(34)2U(2)	1N(24)	1N(28)	126	0	0.9	2116
6-9a	Het. 0.67	9L.32	1N(54)	1N(60)2U(4)	1N(50)D(2)	1N(8)D(42)	220	20.0	1.0	2192
6-9d	Het. 0.46	9L.84	1N(47)	1N(32)2U(10)	1N(45)	1N(40)	174	0	0.6	1912
6-10(5519)	S.C.—prox.	10L.10	1N(60)	1N(50)2U(12)D(4)	1N(42)D(24)	1N(55)	247	11.3	2.8	

[a] S.C. = Secondary constriction.

[c] Number of microspores given in parentheses; 1N = normal nucleolus, 2U = two unequal-sized

nucleoli at quartets but not at the first postmeiotic prophase is unknown, but it is not the result of a deficiency for a segment of the nonorganizer chromosome involved in the interchange. Weber (1975 and personal communication) observed that plants monosomic for chromosome 1, 2, 4, 7, 8, 9, or 10 did not produce quartet microspores with diffuse nucleoli. Half of the microspores in these plants would be deficient for the respective chromosome. Therefore, the cause of diffuse nucleoli at the quartet stage in NOR-interchange heterozygotes appears to be due to either the duplication of a portion of the nonorganizer chromosome or a duplication or deletion of NOR segments. Diffuse nucleoli at the quartet versus first postmeiotic prophase stages may have different genetic bases.

THE NOR–SYNTHESIZER AND ORGANIZER

Our finding that the NOR is the site of DNA complementary to 18S and 28S rRNA (Phillips, Kleese, et al., 1971; Phillips, Weber, et al., 1974) shows that the maize NOR is functioning in a synthetic capacity—synthesizing 18S and 28S rRNA. McClintock (1934) showed that the NOR serves to organize nucleolar material into a nucleolus. When the NOR is absent in maize microspores, nucleolar material is present as small nucleoli or what was termed a "diffuse nucleolar condition." This observation, along with several other considerations, demonstrated the organizing capacity of the NOR.

The hypothesis being tested in our laboratory is that the NOR is a *synthesizer* of 18S and 28S rRNA and an *organizer* of other nucleolar components, such as 5S RNA and ribosomal proteins. The 5S RNA genes of maize are localized in a chromosome other than chromosome 6, most likely chromosome 2 (Wimber, Duffey, et al., 1974). Therefore, 5S RNA is synthesized by a chromosome different from that to which the nucleolus is attached and must move to the nucleolus. No information is available for the genes coding for the various ribosomal proteins, but they also may be in chromosomes other than chromosome 6. The ribosomal proteins are expected to be synthesized on cytoplasmic polyribosomes and transported to the nucleolus within the nucleus. The diffuse nucleolar condition, therefore, may not be the result of all nucleolar material simply remaining dispersed and forming small nucleoli. But based on present knowledge, the nucleolar material observed in diffuse-nucleolated microspores may represent collections of 5S RNA and ribosomal and other nucleolar proteins. These bodies would appear as nucleolar material but be lacking 18S and 28S rRNA. According to the hypothesis, when a cell is deficient for the NOR, 18S and 28S rRNA is not synthesized and the 5S RNA and ribosomal proteins do not become organized into a single nucleolar structure associated with chromosome 6.

Perry (1973) reviewed the evidence that 5S RNA and ribosomal proteins are synthesized independent of 18S and 28S rRNA synthesis in eukaryotes. Persistent synthesis of 5S RNA, for example, occurred in mammalian cells in which low doses of actinomycin D suppressed 18S and 28S rRNA synthesis. Miller (1974) showed that the anucleolate mutant of *Xenopus* synthesizes 5S RNA while unable to synthesize 18S and 28S rRNA; also, the 5S RNA has a shorter half-life than 5S RNA synthesized by normal embryos. Hay and Gurdon (1967) described the existence of multiple nucleolar bodies and pseudo-nucleoli in embryonic cells of the anucleolate mutant. Thus maize microspores deficient for 18S and 28S rRNA genes (NOR-deficient microspores) may continue to synthesize 5S RNA, ribosomal and other nucleolar proteins that fail to coalesce into a single nucleolus due to the absence of 18S and 28S rRNA synthesis and instead form many small nucleolar bodies.

RIBOSOMAL RNA GENE MULTIPLICITY
AND PROTEIN PRODUCTION

A potential application of the knowledge gained on the NOR is the possible effect of varying numbers of rRNA genes on protein levels. A preliminary

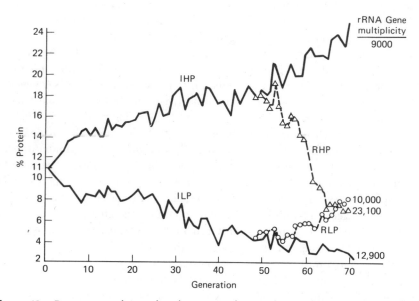

Figure 10. Response to forward and reverse selection for modified protein levels and rRNA gene multiplicities of the resultant lines (IHP—Illinois high protein, ILP—Illinois low protein, RHP—reverse high protein, RLP—reverse low protein). Modified from Dudley (1974).

TABLE 5. Ribosomal RNA gene multiplicities of University of Illinois protein and oil lines[a]

Protein line	Ribosomal RNA gene multiplicity	Oil line	Ribosomal RNA gene multiplicity
Illinois high protein	9,000(a)[b]	Illinois high oil	10,000(a)[b]
Reverse low protein	10,000(a)[b]	Reverse low oil	9,000(a)[b]
Reverse high protein	23,100(c)[b]	Reverse high oil	9,500(a)[b]
Illinois low protein	12,900(b)[b]	Illinois low oil	5,400(b)[b]

[a] Hybridization performed using DNA from 6-day-old seedlings. Each value for the protein lines based on average of two experiments each with eight determinations per line using A188 ^{3}H-rRNA at saturating levels. Oil lines evaluated in one experiment with eight determinations per line.

[b] a,b,c Significance relationships using Duncan's multiple range test.

report is presented here on research directed toward this goal. Since many factors influence the quantity of protein in the vegetative portion or in the grain of a plant, a system was needed where protein levels had been extensively selected. In the selection process, perhaps many of the genetic factors influencing protein synthesis would have been modified including the rRNA gene multiplicity.

As an initial step, we decided to survey the University of Illinois protein lines selected for high as well as low protein from one initial open-pollinated variety since 1896 (Dudley, 1974). Four protein lines were tested: (a) Illinois high protein (IHP); (b) Illinois low protein (ILP); (c) Illinois reverse high protein (RHP); and (d) Illinois reverse low protein (RLP). The RHP line was developed by selecting for low protein from the IHP line, and the RLP line was the result of selecting for high protein from the ILP line (Figure 10). Seed selected in 1969 and 1971 was used for rRNA gene-number determinations. Since rRNA gene-number estimates are from 6-day-old seedlings, protein values of the lines were determined by the *microkjeldahl* technique using vegetative tissue from 6-day-old seedlings. The protein values for the vegetative tissue reflect the same rankings as for grain protein, although not of the same magnitude. The lines were found to differ considerably in rRNA gene multiplicities from an average of 9000 for IHP to 23,100 for the RHP line (Table 5). As mentioned previously, cytological examination of the NOR of the RHP line revealed a greatly enlarged NOR-heterochromatic segment, supporting the rRNA gene-number estimate. The relationship of these rRNA gene multiplicities with protein level is not obvious. The variation, however, may be related to protein selection, since rRNA gene multiplicities for the University of Illinois oil lines (selected from the same open-pollinated variety) did not display the same degree of variation. Based on one experiment, three of the oil lines were not significantly different and possessed 9,000–10,000 rRNA genes, whereas the low oil line possessed fewer (Table 5).

An additional interesting observation is that the RHP line had a dramatically rapid response to selection that was greater than for any of the other protein lines (Figure 10). Perhaps sometime after the reverse selection program was initiated, the RHP line developed a greatly elevated number of rRNA genes as the result of unequal crossing over. Considering the current rRNA gene multiplicity of IHP, an unequal crossover event must have occurred at least twice to achieve the RHP level of 23,100 rRNA genes. If one assumes that the rapid selection response is related to the elevated rRNA gene multiplicity, the unequal crossover event must have occurred soon after the initiation of reverse selection (Figure 10). On the other hand, the selection response has been slow in recent years (Figure 10), and perhaps the sudden slow down in response is related to the high rRNA gene multiplicity. The data presented here show only that variation exists among the protein lines; further tests are underway in an attempt to determine the genetic relationship between protein level and rRNA gene multiplicity.

ACKNOWLEDGMENTS

The dedicated and careful technical assistance of Dr. S. S. Wang in the studies reported herein is gratefully acknowledged. Graduate research assistants P. J. Buescher, T. C. Murphy, and W. D. Springer also contributed importantly to the maize NOR investigations.

REFERENCES

Barr, H. J. 1966. Problems in the developmental cytogenetics of nucleoli in *Xenopus*. *Nat. Cancer Inst. Monogr.* **23:** 411–424.

Beadle, G. W. 1932. A gene for sticky chromosome in *Zea mays. Zeitschr. Ind. Abstam. u. Vererbungsl.* **63:** 195–217.

Beckett, J. B. 1971. Classification of male-sterile cytoplasms in maize (*Zea mays* L.). *Crop Sci.* **11:** 724–727.

Birnstiel, M. L., H. Wallace, J. L. Sirlin, and M. Fischberg. 1966. Localization of the ribosomal DNA complements in the nucleolar organizer region of *Xenopus laevis. Nat. Cancer Inst. Monogr.* **23:** 431–447.

Cullis, C. and D. R. Davies. 1974. Ribosomal RNA cistron number in a polyploid series of plants. *Chromosoma* **46:** 23–28.

Das, N. K. 1965. Inactivation of the nucleolar apparatus during meiotic prophase in corn anthers. *Exp. Cell Res.* **40:** 360–364.

Das, N. K. and M. Alfert. 1966. Nucleolar RNA synthesis during mitotic and meiotic prophase. *Nat. Cancer Inst. Monogr.* **23:** 337–352.

De la Torre, C. and F. A. L. Clowes. 1972. Timing of nucleolar activity in meristems. *J. Cell Sci.* **11:** 713–721.

Doerschug, E. B. 1976. Placement of the genes for ribosomal-RNA within the nucleolar organizing body of *Zea mays. Chromosoma.* **55:** 43–56.

Dudley, J. W. (ed.). 1974. *Seventy generations of selection for oil and protein in maize.* Madison, Wisc.: Crop Science Society of America.

Ford, P. J. 1973. The genes for ribosomal ribonucleic acid. *Biochem. Soc. Symp.* **37:** 69–81.

Gillies, C. B. 1973. Ultrastructural analysis of maize pachytene karyotypes by three dimensional reconstruction of the synaptonemal complexes. *Chromosoma* **43:** 145–176.

Givens, J. F. and R. L. Phillips. 1976. The nucleolus organizer region of maize (*Zea mays* L.): Ribosomal RNA gene distribution and nucleolar interactions. *Chromosoma* **57:** 103–117.

Goldberg, R. B., W. P. Bemis, and A. Siegel. 1972. Nucleic acid hybridization studies within the genus *Cucurbita. Genetics* **72:** 253–256.

Gopinath, D. M. and C. R. Burnham. 1956. A cytogenetic study in maize by deficiency-duplication produced by crossing interchanges involving the same two chromosomes. *Genetics* **41:** 382–395.

Hay, E. D. and J. B. Gurdon. 1967. Fine structure of the nucleolus in normal and mutant *Xenopus* embryos. *J. Cell Sci.* **2:** 151–162.

Heitz, E. 1931. Die Ursache der gesetzmässigen Zahl, Lage, Form, and Grösse pflanzlicher Nukleolen. *Planta* **12:** 775–844.

Horsfall, J. G., Ch. 1972. Genetic vulnerability of major crops. *Monogr. Nat. Acad. Sci. U. S. A.,* p. 106.

Hsiao, T. C. 1964. Characteristics of ribosomes isolated from roots of *Zea mays. Biochim. Biophys. Acta.* **91:** 598–605.

Hsiao, T. C. 1970. Rapid changes in levels of polyribosomes in *Zea mays* in response to water stress. *Plant Physiol.* **46:** 281–285.

Ingle, J. and J. Sinclair. 1972. Ribosomal RNA genes and plant development. *Nature* **235:** 30–32.

Ingle, J., J. N. Timmis, and J. Sinclair. 1975. The relationship between satellite deoxyribonucleic acid, ribosomal ribonucleic acid gene redundancy, and genome size in plants. *Plant Physiol.* **55:** 496–501.

Jacobson, A. B. and R. W. Williams. 1968. Sedimentation studies on RNA from proplastids of *Zea mays. Biochim. Biophys. Acta* **169:** 7–13.

Lin, M. 1955. Chromosomal control of nuclear composition in maize. *Chromosoma* **7:** 340–370.

Loening, U. E. 1968. Molecular weights of rRNA in relation to evolution. *J. Molec. Biol.* **38:** 355–365.

McClintock, B. 1934. The relation of a particular chromosomal element to the development of the nucleoli in *Zea mays*. *Z. Zellforsch. u. Mikr. Anat.* **21:** 294–328.

McLeish, J. 1964. Deoxyribonucleic acid in plant nucleoli. *Nature* **204:** 36–39.

Meyer, G. F. and W. Hennig. 1974. The nucleolus in primary spermatocytes of *Drosophila hydei*. *Chromosoma* **46:** 121–144.

Miller, L. 1974. Metabolism of 5S RNA in the absence of ribosome production. *Cell* **3:** 275–281.

Mullinix, K. P., G. C. Strain, and L. Bogorad. 1973. RNA polymerases of maize: Purification and molecular structure of DNA-dependent RNA polymerase II. *Proc. Nat. Acad. Sci. U. S. A.* **70:** 2386–2390.

Pardue, M. L., S. A. Gerbi, R. A. Eckhardt, and J. G. Gall. 1970. Cytological localization of DNA complementary to ribosomal RNA in polytene chromosomes of *Diptera*. *Chromosoma* **29:** 268–290.

Perry, R. P. 1973. Regulation of ribosome content in eukaryotes. *Biochem. Soc. Symp.* **37:** 105–116.

Phillips, R. L., R. A. Kleese, and S. S. Wang. 1971. The nucleolus organizer region of maize (*Zea mays* L.): Chromosomal site of DNA complementary to ribosomal RNA. *Chromosoma* **36:** 79–88.

Phillips, R. L. and S. S. Wang. 1972. Cytological localization of interchange breakpoints to the nucleolus organizer region or satellite. *Maize Genet. Coop. Newsl.* **46:** 123.

Phillips, R. L., S. S. Wang, D. F. Weber, and R. A. Kleese. 1973. The nucleolus organizer region (NOR) of maize: a summary. *Genetics* **74**(supp.)**2:** s212.

Phillips, R. L., D. F. Weber, R. A. Kleese, and S. S. Wang. 1974. The nucleolus organizer region of maize (*Zea mays* L.): Tests for ribosomal gene compensation or magnification. *Genetics* **77:** 285–297.

Pollard, C. J. 1964. The specificity of ribosomal ribonucleic acids of plants. *Biochem. Biophys. Res. Commun.* **17:** 171–176.

Pollister, A. W. and C. Leuchtenberger. 1949. Nucleotide content of the nucleolus. *Nature* **163:** 360–361.

Pollister, A. W. and H. Ris. 1947. Nucleoprotein determination in cytological preparations. *Cold Spring Harbor Symp. Quant. Biol.* **12:** 147–157.

Pring, D. R. and D. W. Thornbury. 1975. Molecular weights of maize mitochondrial and cytoplasmic ribosomal RNA's under denaturing conditions. *Biochim. Biophys. Acta* **383:** 140–146.

Ramirez, S. A. and J. H. Sinclair. 1972. Variation of gene redundancy in *Zea mays*. *Am. Soc. Cell Biol.* 211a: (abstr.).

Ritossa, F. M. 1968. Unstable redundancy of genes for ribosomal RNA. *Proc. Nat. Acad. Sci. U. S. A.* **60:** 509–516.

Ritossa, F. M., K. C. Atwood, and S. Spiegelman. 1966. A molecular explanation of the bobbed mutants of *Drosophila* as partial deficiencies of "ribosomal" DNA. *Genetics* **54:** 819–834.

Ritossa, F. M. and S. Spiegelman. 1965. Localization of DNA complementary to ribosomal RNA in the nucleolus organizer region of *Drosophila melanogaster. Proc. Nat. Acad. Sci. U. S. A.* **53:** 737–745.

Roeder, R. G. and W. J. Rutter. 1969. Multiple forms of DNA-dependent RNA polymerase in eukaryotic organisms. *Nature* **224:** 234–237.

Rogers, M. E., U. E. Loening, and R. S. S. Fraser. 1970. Ribosomal RNA precursors in plants. *J. Molec. Biol.* **49:** 681–692.

Roman, H. 1947. Mitotic non-disjunction in the case of interchanges involving the "B" type chromosome in maize. *Genetics* **32:** 391–409.

Ruppel, H. G. 1969. Nucleic acids in chloroplasts. III. Detection of a low molecular weight RNA fraction in the chloroplast ribosomes of *Allium porrum* and *Zea mays. Z. Naturforsch. B.* **24:** 1467–1475.

Stout, J. T. 1973. The biochemical cytogenetics of a meiotic mutant in maize. Ph.D. thesis. St. Paul: University of Minnesota.

Stout, J. and C. R. Burnham. 1968. T4-6 (C.H.Li). *Maize Genet. Coop. Newsl.* **42:** 121.

Stout, E. R. and R. J. Mans. 1967. Partial purification and properties of RNA polymerase from maize. *Biochim. Biophys. Acta* **134:** 327–336.

Strain, G. C., K. P. Mullinix, and L. Bogorad. 1971. RNA polymerases of maize: Nuclear RNA polymerases. *Proc. Nat. Acad. Sci. U. S. A.* **68:** 2647–2651.

Swift, H. and B. J. Stevens. 1966. Nucleolar–chromosomal interaction in microspores of maize. *Nat. Cancer Inst. Monogr.* **23:** 145–166.

Tang, C. L. 1971. Dissociation of ribosomes and some characteristics of ribosomal RNA of *Zea mays.* Ph.D. thesis. Davis: University of California.

Tartof, K. D. 1973. Unequal mitotic sister chromatid exchange and disproportionate replication as mechanisms regulating ribosomal RNA gene redundancy. *Cold Spring Harbor Symp. Quant. Biol.* **38:** 491–500.

Travis, R. L., R. C. Huffaker, and J. L. Key. 1970. Light-induced development of polyribosomes and the induction of nitrate reductase in corn leaves. *Plant Physiol.* **46:** 800–805.

Underbrink, A. G., Y. C. Ting, and A. H. Sparrow. 1967. Note on the occurrence of a synaptinemal complex at meiotic prophase in *Zea mays* L. *Can. J. Genet. Cytol.* **9:** 606–609.

Weber, D. F. 1975. The template for 5S ribosomal RNA is not necessary for formation of a nucleolus. *Maize Genet. Coop. Newsl.* **49:** 38–39.

Wimber, D. E., P. A. Duffey, D. M. Steffensen, and W. Prensky. 1974. Localization of the 5S RNA genes in *Zea mays* by RNA-DNA hybridization *in situ. Chromosoma* **47:** 353–359.

Two additional papers confirming aspects of rRNA gene localization and variation in maize were published after this chapter was written: Ramirez, S. A., and J. H. Sinclair. 1975. Intraspecific variation of ribosomal gene redundancy in *Zea mays. Genetics* **80:** 495–504. Ramirez, S. A., and J. H. Sinclair. 1975. Ribosomal gene localization and distribution (arrangement) within the nucleolar organizer region of *Zea mays. Genetics* **80:** 505–518.

Chapter 44

IDENTIFICATION OF GENETIC FACTORS CONTROLLING CENTROMERIC FUNCTION IN MAIZE

Wayne R. Carlson

Department of Botany,
University of Iowa,
Iowa City

The B chromosome of maize is a supernumerary, subtelocentric chromosome that frequently undergoes nondisjunction at the second pollen mitosis. Nondisjunction changes the number of B chromosomes in the sperm from the expected and aids in maintaining and increasing the number of B chromosomes present in a population (Roman, 1947, 1948). Roman utilized translocations between A chromosomes and the B to analyze nondisjunction. One translocation, TB-4a, divided the B chromosome approximately in half. Roman found that the B^4 chromosome with the B centromere retained the ability to undergo nondisjunction, whereas the 4^B did not. Nevertheless, the presence of 4^B in the pollen was required for nondisjunction of the B^4 (Roman, 1949). Thus a proximal and a distal segment on the B are necessary for nondisjunction to occur. Ward (1973) showed with further analysis of A–B translocations that a factor on the distal

743

tip of the B is required for nondisjunction. Carlson (1973) found that modification of the B centromere by misdivision could prevent the occurrence of nondisjunction. The findings of Ward and Carlson may be interpreted as a more accurate localization of factors previously identified by Roman. However, the proximal and distal halves of the B may each contain more than a single region that controls nondisjunction.

In a previous paper (Carlson, 1973), a method for selecting mutant B chromosomes was presented. (The mutant phenotype is an inability to undergo nondisjunction at the second pollen mitosis.) Selection was carried out on the A–B translocation TB-9b, since useful genetic markers are associated with it. The translocation has a slightly more distal exchange point in the B than TB-4a, and its properties were described by Robertson (1967) and Carlson (1969b). A partial analysis of 11 mutant B-9b translocations is given here. In addition, the properties of certain B^9 chromosomes with modified centromeres are reported. The findings suggest that at least three separate regions of the B chromosome control nondisjunction.

MATERIALS AND METHODS

Genetic Markers

Unless otherwise indicated, the mutant B translocations and the standard TB-9b carry dominant alleles for genes in 9S. Chromosome 9 in the translocation heterozygotes may carry various recessive alleles. Genes that mark the B^9 chromosome include, from distal to proximal regions of 9S, Yg, C, Sh, and Bz. The Wx locus of 9S is present on 9^B. The C, Sh, Bz and Wx genes have endosperm phenotypes: C—purple, c—white or yellow; Sh—normal, sh—collapsed (shrunken); Bz—purple, bz—brown or bronze; Wx—endosperm with normal starch, and wx—endosperm starch that lacks amylose. The Wx versus wx phenotypes are determined by staining the starch with an iodine–potassium iodide solution. The kernel is chipped and the exposed region stained: Wx—black or purple, wx—brown. The Wx versus wx phenotypes may also be identified by staining in the haploid pollen, so that heterozygotes and homozygotes for the dominant allele may be distinguished in the sporophyte. The seedling marker Yg is classified on the sand bench as Yg—green, yg—yellow-green.

Ethylmethanesulfonate Treatment

The selection of spontaneous TB-9b mutants was described previously (Carlson, 1973). A second selection of mutant B translocations was carried out after treatment with EMS. Seeds of a homozygous W22 TB-9b line were treated according to Briggs (1969). The seeds were immersed in a 0.008 M aqueous solution of EMS maintained at pH 7.5 with phosphate buffer. After incubation for 7 hs at room temperature, the seeds were rinsed with distilled water and planted wet in the field. Plants from the treated kernels were crossed as male parents to a $bzbz$ tester and mutant B translocations selected. Several unselected mutations of endosperm and plant characters were recovered in subsequent generations, indicating successful infiltration of EMS.

Determination of Rates of Nondisjunction

In homozygous 9^B 9^B B^9 B^9 plants, one type of microspore (9^B B^9) is produced. Nondisjunction of the B^9 at the second pollen mitosis produces 9^B and 9^B B^9 B^9 sperm. (In the absence of nondisjunction, both sperm have one B^9). If following nondisjunction the 9^B sperm fertilizes the polar nuclei, a recessive endosperm phenotype (bz, c, or sh) appears in testcrosses. If the 9^B sperm fertilizes the egg, a recessive plant phenotype (yg) results. For example, the cross $ygygbzbz$ $\times$ 9^B 9^B B^9 B^9 gives three classes of endosperm and plant phenotypes: (a) bronze endosperm and green plant, (b) purple endosperm and yellow–green plant, and (c) purple endosperm and green plant. The frequency of (a) and (b) equals the rate of nondisjunction. Fertilization of the polar nuclei by the 9^B sperm occurs approximately twice as frequently as fertilization of the egg, so that the bronze endosperm, green plant phenotype is more common than purple endosperm, yellow–green plant. If only the bz marker is available in a testcross, the rate of nondisjunction can be approximated by multiplying the frequency of bz by 1.5.

The rate of nondisjunction may also be determined for 9 9^B B^9 heterozygotes, if chromosome 9 is marked by wx and the 9^B by Wx (Carlson, 1969b). The Wx marker is rarely transferred to chromosome 9 by crossing over, since the locus is very close to the translocation exchange point (Robertson, 1967). For practical purposes all Wx pollen from 9^{wx} 9^{BWx} B^9 plants arises from 9^B B^9 microspores. Nondisjunction of the B^9 may, therefore, be calculated in crosses of heterozygotes if Wx progeny are selected and classified for the appropriate B^9 markers.

In the crosses of Table 3 the rate of nondisjunction was estimated for heterozygous translocations in which the 9 and 9^B both carried the dominant Wx allele. Since translocation heterozygotes produce about 50% of functional pollen grains from 9^B B^9 microspores (Table 4), the rate of nondisjunction was estimated by multiplying the observed nondisjunction by 2. In addition, nondisjunction was determined utilizing an endosperm marker (bz) so that the frequency of bronze kernels was multiplied by 3. A different estimation of nondisjunction was made for plants carrying extra B chromosomes. In this case, fertilization of the egg by 9^B sperm occurs equally as often as fertilization of the polar nuclei (Carlson, 1969a), and the frequency of bronze kernels was multiplied times 4.

RESULTS

B^9 chromosomes with modified centromeric regions were recovered from a B^9 isochromosome through misdivision (Carlson, 1973). Four telocentric derivatives of the isochromosome were found to be virtually incapable of nondisjunction, and they are included here as mutants of nondisjunction. (The term *mutant* is used in its broader sense to include chromosomal aberrations). The isolations of the four telocentrics are numbered 1852, 1853, 1854, 1855. Two screenings for mutant B translocations that lack the ability to carry out nondisjunction yielded 11 cases. In the first screening, spontaneous mutants were selected and two such translocations, numbered 1866 and 1867, were identified (Carlson, 1973). The second selection involved pretreatment of seeds with the EMS mutagen according to Briggs (see Materials and methods section). A combination of spontaneous and induced mutations may be expected from this screening. Nine modified translocations were recovered, numbered 8, 9, 10, 11, 12, 13, 14, 2010, and 2150.

Several tests are routinely carried out in the analysis of mutant B-9b translocations. First, the inability of the B^9 to undergo nondisjunction is tested for several progeny of the original isolate. Second, the modified region of the translocation is located by genetic tests either to the B^9 or the 9^B. Third, translocations with mutant B^9 chromosomes are tested for nondisjunction of the B^9 in the presence of extra B chromosomes. Fourth, the presence or absence of chromosomal aberrations in the mutant B translocations is determined cytologically. Fifth, transmission of the modified translocations through the male parent is determined for the T/N heterozygotes.

The first step, confirmation that the translocations are severely inhibited in nondisjunction, was previously reported for the telocentrics 1852, 1853, 1854, 1855, and the mutants 1866 and 1867 (Carlson, 1973). (Recent crosses with the telocentrics have occasionally produced nondisjunction at a higher rate than originally reported. The variability in results with the telocentrics has not been found with any of the other modified translocations and is being studied; data are too preliminary to be reported here.) Tests of nondisjunction for the remaining mutant B translocations are given in Table 1. In each case, a *bzbzwxwx* tester was crossed as female parent to a translocation heterozygote, 9^{Bzwx} $9^{B\,Wx}$ B^{9Bz}. Kernels with the dominant *Wx* phenotype were selected and classified for the purple (*Bz*) versus bronze (*bz*) endosperm phenotype. From the ratio *bzWx*/total *Wx* an approximate rate of nondisjunction was calculated (see Materials and Methods). Rates of nondisjunction in Table 1 are all below 1%

TABLE 1. Estimated rates of nondisjunction for nine mutant translocations (results for a standard TB-9b are included for comparison)

Mutant translocation	Family numbers of crosses	Endosperm classification for Wx progeny of bzbzwxwx $\times$ $9^{Bz\,wx}$ $9^{B\,Wx}$ $B^{9\,Bz}$		Percent bzWx per total Wx	Estimated percent of nondisjunction (frequency bzWx $\times$ 1.5)
		BzWx	bzWx		
8	2343 × 2331	855	3	0.3	0.5
9	2358 × 2333	748	0	0.0	0.0
10	2358 × 2336	789	1	0.1	0.2
11	2358 × 2337	808	0	0.0	0.0
12	2358 × 2338	836	1	0.1	0.2
13	2344 × 2339	743	1	0.1	0.2
14	2343 × 2341	851	0	0.0	0.0
2010	2009, 2086 × 2010	7471	12	0.2	0.2
2150	2358 × 2400	1160	2	0.2	0.3
Standard TB-9b	2028 × 2043	3451	4149	54.6	81.9

compared to normal rates of 50–100% (Roman, 1948; Bianchi, Bellini, et. al., 1961; Carlson, 1969a).

Next, the mutant site on the translocations was localized to the 9^B or B^9 chromosome. All of the modified translocations, except for the telocentrics, whose origin was known, were crossed as female parents to a TB-9b stock. One of the following crosses was made:

1. 9^{BWx} 9^{BWx} B^{9C} B^{9C} $\times$ 9^{wx} 9^{BWx} B^{9c} B^{9c}
   (homozygous modified translocation)　(hyperploid normal translocation with c marker)

2. 9^{wx} 9^{BWx} B^{9C} $\times$ 9^{wx} 9^{BWx} B^{9c} B^{9c}
   (heterozygous modified translocation)　(hyperploid normal translocation with c marker)

Plants with only purple (Wx)-staining pollen grains and 50% pollen abortion were selected from the progeny. Due to the very small amount of crossing over between Wx and the translocation exchange point (Robertson, 1967), these plants were almost certainly homozygous for 9^B. The 50% pollen sterility indicated the presence of one B^9. The 9^B 9^B B^9 plants selected arose by transmission of 9^{BWx} B^{9C} through the female and 9^{BWx} from the male. (Nondisjunction in the male produced the 9^B sperm.) The chromosomes may be designated

9^{B*} 9^B B^{9*}, with the asterisk signifying origin from a mutant B translocation. If the translocation being analyzed is mutant in 9^B constitution but has a normal B^9, the chromosomal combination 9^R B^{9*} should be capable of nondisjunction at the second pollen mitosis. Alternately, if the B^9 is mutant and the 9^B is unchanged, nondisjunction should not occur in either the 9^B B^{9*} or 9^{B*} B^{9*} combinations. (Theoretically, both the 9^B and the B^9 could be mutant, with the result that nondisjunction does not occur in pollen of 9^{B*} 9^B B^{9*} plants. However, this is unlikely and, as shown below, only possible for translocation 1866). Two or more plants of the 9^{B*} 9^B B^{9*} type were selected for the translocations 8, 9, 10, 11, 12, 13, 14, 1866, 1867, 2010, and 2150. The plants were crossed as female parents to a cc tester to confirm absence of the normal B^{9c}. Each plant was also crossed as male parent to a $bzbzwxwx$ tester. The $WxWx$ genotype of the plants was confirmed in these crosses. In addition, an approximate rate of nondisjunction was calculated from the frequency of $bronze$ kernels (Table 2). Translocation 1866 remained incapable of significant levels of nondisjunction in this test, indicating mutation of the B^9 chromosome. The remaining ten translocations were restored to high rates of nondisjunction and must contain normal B^9s and mutant 9^Bs.

Translocations with altered B^9 chromosomes include 1866, mentioned above, and the telocentrics 1852, 1853, 1854, 1855. These translocations may be further characterized by their response to the presence of additional B chromosomes. If a centromeric alteration, for example, is cis-dominant, it would be unaffected by the presence of extra B chromosomes. However, if the mutation behaves as a simple recessive, extra B chromosomes would induce nondisjunction. One of the telocentrics (1852) and translocation 1866 were crossed as female parents to two Black Mexican strains with and without B chromosomes. The two Black Mexican lines were isogenic, differing only in the presence or absence of B chromosomes. From the F_1, plants with semisterile pollen (9 9^B B^9 ± Bs) were selected and crossed as male parents to a $bzbzwxwx$ tester (results are given in Table 3). The addition of B chromosomes to 1866 restores nondisjunction to high levels, indicating a recessive mutation or deletion. For the telocentric 1852, however, nondisjunction is increased to only 8% of the pollen, and the chromosome appears to be destabilized by the presence of extra B chromosomes. Kernels with recessive sectors of bronze on a purple background appear to increase in numbers in the presence of B chromosomes, as shown in Table 3. The rate of sectoring in the absence of Bs is 2.3%, and in the presence of Bs, 10.6%. (The rate equals the frequency of sectoring times 2, since the male parent is heterozygous for the translocation and frequently transmits the stable chromosome 9 rather than the translocation chromosomes.) The increase in sectoring may result if extra B chromosomes initiate an early step in nondisjunction of the B^9 that often cannot be brought to completion. Resolution of the incomplete nondisjunction may damage the centromere,

TABLE 2. Estimated rate of nondisjunction for mutant translocations in the presence of a standard 9^B. Male parents were of 9^{B*} 9^B B^{9*} constitution, in which the asterisk indicates origin from a mutant translocation; each cross listed represents testing of a separate male parent

Mutant translocation	Family numbers of crosses	Endosperm classification of bzbz × 9^B 9^B B^{9Bz}		Percent bz	Estimated rate of nondisjunction[a] (frequency bz × 1.5)
		Bz	bz		
8	2358 × 2318 A	228	73		
	2358 × 2318 B	204	97		
	2361 × 2318 C	268	129		
		700	299	30	45%[a]
9	2343 × 2319 A	295	135		
	2358 × 2319 B	319	163		
	2358 × 2319 C	265	115		
		879	413	32	48%[a]
10	2358 × 2320 A	175	135		
	2358 × 2320 B	220	119		
		395	254	39	58%[a]
11	2361 × 2366 A	192	97		
	2358 × 2366 B	210	88		
	2358 × 2366 C	301	165		
		703	350	33	50%[a]
12	2502 × 2515 A	474	249		
	2502 × 2515 B	810	366		
		1284	615	32	49%[a]
13	2343 × 2321 A	305	183		
	2343 × 2321 C	221	126		
		526	309	37	56%[a]

TABLE 2. (*Continued*)

Mutant translocation	Family numbers of crosses	Endosperm classification of $bzbz \times 9^B\ 9^B\ B^{9Bz}$		Percent bz	Estimated rate of nondisjunction[a] (frequency bz × 1.5)
		Bz	bz		
14	2343 × 2322 A	258	133		
	2343 × 2322 B	203	108		
	2344 × 2322 C	346	174		
		807	415	34	51%[a]
1866	2345 × 2324 A	416	2		
	2344 × 2324 B	429	2		
	2344 × 2324 C	443	2		
	2345 × 2325 A	396	4		
	2344 × 2325 B	408	3		
	2358 × 2325 C	359	5		
		2451	18	0.7	1.1%
1867	2583 × 2633 A	151	121		
	2583 × 2633 B	129	85		
	2583 × 2634 A	135	138		
	2583 × 2634 B	143	117		
		558	461	45	68%
2010	2344 × 2323 A	317	150		
	2343 × 2323 B	310	146		
	2344 × 2323 C	309	138		
		936	434	32	48%[a]
2150	GH-75-2 × 1 A	152	66		
	GH-75-2 × 1 B	215	120		
	GH-75-2 × 1 C	191	69		
	GH-75-2 × 1 D	313	141		
		871	396	31	47%[a]

[a] Rates shown are the combined values for $9^B\ B^{9*}$ and $9^{B*}\ B^9$ pollen. When nondisjunction of the B^{9*} is induced by 9^B, the rate of nondisjunction in $9\ B^{9*}$ is approximately twice the value shown for all cases except mutant 1867. The 9^B of translocation 1867 carries a duplication of a 7L segment that decreases transmission of $9^{B*}\ B^{9*}$ pollen in competition with $9^B\ B^{9*}$.

perhaps by misdivision and may produce sectoring. The data in Table 3 indicate that telocentric 1852 is deficient in a *cis*-dominant manner for one step in nondisjunction. (The findings do not, in fact, discriminate between true dominance and *cis*-dominance, but the appearance of a true dominant seems unlikely).

Cytological observations have thus far been made on two of the modified translocations, 1866 and 1867. The translocation 1867 has an apparently normal B^9, but the 9^B is modified. The mutant 9^B lacks almost all of the distal B segment carried by the normal 9^B and contains in its place a terminal segment of 7L (Figure 1). The reciprocal half of the 9^B-7 translocation was not found so that mutant 1867 contains the normal 7, a highly modified 9^B, and an apparently normal B^9. For 1866, a structural change in the B^9 was found (Carlson, 1973), but none in the 9^B. Preliminary observations were made in pachynema of a heterozygote between the B^9 of 1866 and a standard B^9. The

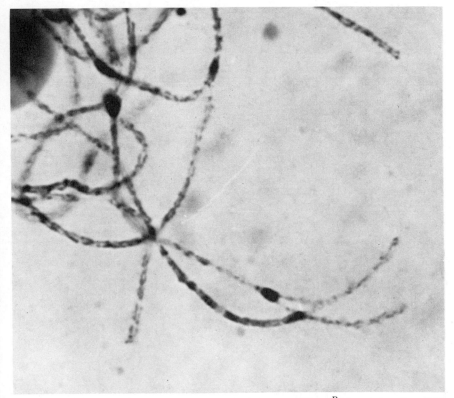

Figure 1. Pairing between chromosome 7 and the modified 9^B of translocation 1867. The lower chromosome is the modified 9^B.

TABLE 3. Influence of extra B chromosomes on nondisjunction of the B^9 in translocations 1852 and 1866: male parents with mutant translocations were $9^{Bz}\ 9^B\ B^{9Bz} \pm Bs$; female parents were $bzbz$

Mutant translocation	Family numbers of crosses	Number of extra Bs	Endosperm classification			Frequency bz	Estimated rate of nondisjunction (frequency bz × 3 for no B plants; frequency bz × 4 for B-containing plants)
			Bz	bz	Sectored[a]		
1852 (Telocentric)	2009 × 2071 A	0	1460	3	21	0.2%	
	2009 × 2071 B	0	1391	7	12	0.5%	
	2086 × 2071 C	0	1554	4	19	0.2%	
	2086 × 2071 D	0	1283	2	15	0.2%	
			5688	16	67	0.3%	0.8%
	2009 × 2073-4	3	1414	3	22	0.2%	
	2009 × 2073-7	4–5	1072	22	41	1.9%	
	2009 × 2074-7	7	1803	50	162	2.4%	
	2009 × 2073-6	7–8	1259	47	91	3.5%	
			5548	122	316	2.0%	8.2%

752

	Cross	Sector	Count	Not recorded: small in number	%	%
1866	2040 × 2033 A	0	912	1	0.1%	
	2040 × 2033 B	0	950	4	0.4%	
	2028 × 2033 C	0	878	0	0.0%	
	2042 × 2033 D	0	933	1	0.1%	
	2042 × 2033 E	0	876	0	0.0%	
			4549	6	0.1%	0.4%
	2030 × 2034-11	2	914	240	21%	
	2030 × 2034-15	4	634	181	22%	
	2030 × 2034-4	6–7	772	235	23%	
	2030 × 2034-9	8	624	239	28%	
			2944	895	23%	93%

[a] Kernels with multiple recessive sectors or a large single sector on the endosperm are recorded here. A sector was considered large when it covered one-quarter or more of the endosperm.

TABLE 4.   Transmission of the 9^B B^9 (Wx) pollen type for the mutant B translocations and standard TB-9b

Mutant translocation	Family numbers of crosses	Endosperm classification of $wxwx \times 9^{wx} 9^{B^{Wx}}$ B^9		Percent Wx
		Wx	wx	
8	2343 × 2331	858	653	57
9	2358 × 2333	748	810	48
10	2358 × 2336	790	684	54
11	2358 × 2337	808	675	54
12	2358 × 2338	837	657	56
13	2344 × 2339	744	648	53
14	2343 × 2341	851	681	56
1852 (telocentric)	2009 × 2012	1722	1970	47
1854 (telocentric)	2009 × 2013	1922	2149	47
1866	2009, 2086 × 2076	4074	4838	46
1867	2086 × 2078	3137	6646	32
2010	2009, 2086 × 2010	7161	7471	49
2150	2358 × 2400	1162	907	56
Standard TB-9b	2028 × 2043	7600	7743	50

mutant B^9 was found to be deficient for a segment of proximal euchromatin and, perhaps, some of the centric heterochromatin.

The possibility of genetic imbalance in the mutant translocation chromosomes was checked through pollen competition. The cross $wxwx$ ♀ × 9^{wx} $9^{B^{Wx}}$ B^9 ♂ was performed for each translocation, including two of the telocentrics. Transmission of Wx through the pollen is a measure of genetic balance for the $9^{B^{Wx}}$ B^9 chromosomal combination. Duplication or deletions of A chromosome material, associated with the mutation of TB-9b, should reduce the frequency of Wx progeny. Table 4 gives results for male transmission of the modified translocations and the standard TB-9b. Only translocation 1867 shows evidence of a reduced Wx frequency in comparison to the control value of standard TB-9b. The finding of selection against the 1867 translocation agrees with the cytological observation of duplication of a 7L segment.

DISCUSSION

The B^9 chromosome of translocation B-9b undergoes nondisjunction at the second pollen mitosis at a very high frequency. Derivatives of TB-9b have been found in which the B^9 is incapable of nondisjunction. The first such translocations identified contained telocentric derivatives of the B^9. Four telocentrics that lack nondisjunction, numbered 1852, 1853, 1854 and 1855, arose by misdivision from a B^9 isochromosome (Carlson, 1973). A procedure for selecting additional translocations which lack B^9-nondisjunction was developed and two spontaneous translocation mutants (1866 and 1867) isolated (Carlson, 1973). The screening procedure was subsequently modified to include pretreatment of TB-9b kernels with EMS and selection in the next generation for mutant translocations. Both spontaneous and induced mutations should be recovered by this procedure. Nine mutant translocations (8, 9, 10, 11, 12, 13, 14, 2010, 2150) were identified in the second screening. The mutant nature of each of the nine translocations was confirmed by progeny tests recorded in Table 1.

A thorough analysis of the selected B-9b translocations should lead to an understanding of the number of loci on the B chromosome that control nondisjunction and the functions of these loci. As a first step, the chromosome affected in the mutant translocations (B^9 vs. 9^B) was identified. The B^9 chromosome from a mutant translocation was combined with the 9^B from a standard TB-9b as described. If nondisjunction occurred, the mutation was localized to the 9^B. If nondisjunction did not occur, a mutant B^9 was present. Data of Table 2 show that translocation 1866 is a B^9 mutant, whereas 8, 9, 10, 11, 12, 13, 14, 1867, 2010 and 2150 have mutant 9^Bs. The test was not applied to the telocentrics, since their origin through B^9 alteration was known.

The two translocation types with modified B^9s, 1866 and the telocentrics, were further tested for the influence of extra B chromosomes on B^9 nondisjunction. It was considered possible that B^9 mutations could be *cis*-dominant, particularly centromeric alterations. Extra B chromosomes should not induce nondisjunction of a *cis*-dominant mutation. Results of Table 3 show that the telocentric 1852 behaves as a *cis*-dominant with little effect of extra B chromosomes, whereas translocation 1866 acts as a recessive, giving a high frequency of nondisjunction in the presence of extra Bs. Apparently, two functionally distinct regions on the B^9 control nondisjunction. Cytological localization of the region missing from mutant 1866 is not precise, but preliminary observations reveal a deletion of proximal euchromatin and possibly some centric heterochromatin from the B^9. Dr. Bor-Yaw Lin (personal communication) has also obtained evidence that a region in the proximal euchromatin or centric heterochromatin is required for nondisjunction. Dr. Lin constructed a series of B-10 translocations and made various combinations of the B^{10} from one translocation with the 10^B from another. The translocations were selected to yield deletions of B chromatin but not chromosome 10 material. One of the 10^B–B^{10} hybrid translocations lacked approximately the same region which was absent from mutant 1866, and the B^{10} chromosome failed to undergo nondisjunction at the second pollen mitosis.

Since chromosomal aberrations are involved in at least some of the mutant B translocations, an analysis of genetic imbalance in 9B B^9 spores was made. All of the modified translocations were crossed as 9^{wx} $9B^{Wx}$ B^9 males to a $wxwx$ tester. Transmission of Wx through the pollen is sensitive to imbalance of the $9B^{Wx}$ B^9 chromosomal combination. The frequency of Wx progeny was comparable to that of the standard TB-9b for all mutant translocations except 1867 (Table 4). The reduction in Wx transmission by translocation 1867 was expected from the cytological observation of a 7L duplication.

SUMMARY

Nondisjunction of the maize B chromosome was analyzed with derivatives of the A–B translocation B-9b. Several modified B-9b translocations were available in which the B^9 lacks the ability to undergo nondisjunction. Four translocations with telocentric B^{9}s (1852, 1853, 1854, and 1855) were reported to be incapable of nondisjunction (Carlson, 1973). In addition, a screening procedure for selecting mutants of nondisjunction yielded eleven TB-9b derivatives that lack nondisjunction, numbered 8, 9, 10, 11, 12, 13, 14, 2010, 2150, 1866, and 1867. In the present chapter the chromosomal site of mutation of the modified TB-9bs was determined and certain properties of the translocations analyzed.

The telocentric chromosomes, due to their origin, were known to be modified in the centromeric region of the B^9. In addition, it was shown (Table 2) that translocation 1866 has a modified B^9, whereas 8, 9, 10, 11, 12, 13, 14, 1867, 2010 and 2150 have mutant 9Bs.

A telocentric B^9 and the B^9 of 1866 were tested for their response to the presence of extra B chromosomes. Nondisjunction was restored to the 1866 B^9 by extra Bs but the telocentric B^9 retained a low rate of nondisjunction. It was concluded that two functionally distinct regions on the B^9 are required for nondisjunction. One of the regions is *cis*-dominant and located in or near the centromere. The other region is recessive and occupies a site within the 1866 deletion. Analysis of the 9^B mutant translocations should similarly distinguish the presence of one or more distal factors required for nondisjunction.

ACKNOWLEDGMENT

Supported in part by National Science Foundation Grant BMS74-19034.

REFERENCES

Bianchi, A., G. Bellini, M. Contin, and E. Ottaviano. 1961. Nondisjunction in presence of interchanges involving B-type chromosomes in maize and some phenotypical consequences of meaning in maize breeding. *Z. Vererbungsl.* **92:** 213–232.

Briggs, R. W. 1969. Induction of endosperm mutations in maize with ethyl methanesulfonate. *Maize Genet. Coop. Newsl.* **43:** 23–31.

Carlson, W. R. 1969a. Factors affecting preferential fertilization in maize. *Genetics* **62:** 543–554.

Carlson, W. R. 1969b. A test of homology between the B chromosome of maize and abnormal chromosome 10, involving the control of nondisjunction in B's *Molec. Gen. Genet.* **104:** 59–65.

Carlson, W. R. 1973. A procedure for localizing genetic factors controlling mitotic nondisjunction in the B chromosome of maize. *Chromosoma* **42:** 127–136.

Carlson, W. R. 1977. The Cytogenetics of Corn. *Corn and corn improvement.* Chapter 5. Am. Soc. Agron. Madison.

Robertson, D. S. 1967. Crossing over and chromosomal segregation involving the B[9] element of the A–B translocation B-9b in maize. *Genetics* **55:** 433–449.

Roman, H. 1947. Mitotic nondisjunction in the case of interchanges involving the B-type chromosome in maize. *Genetics* **32:** 391–409.

Roman, H. 1948. Directed fertilization in maize. *Proc. Nat. Acad. Sci. U. S. A.* **34:** 36–42.

Roman, H. 1949. *Rec. Gen. Soc. Am.* **18:** 112. (abstr.).

Ward, E. J. 1973. Nondisjunction: Localization of the controlling site in the maize B chromosome. *Genetics* **73:** 387–391.

ABSTRACTS OF
DEMONSTRATION PAPERS

BREEDING OF MAIZE FOR IMPROVED PROTEIN QUALITY AND HIGH YIELD OF GRAIN

M. Misović,
Danica Jelenić, N. Pesev,
V. Trifunović

Maize Research Institute,
Zemun, Yugoslavia

There is a growing necessity for improved protein quality and high-yield hybrids in maize. To realize this aim an extensive breeding program including backcrossing (BC_1–BC_5) of 64 of the best normal inbred lines, development of new lines by recombination of genes, a broad-based germplasm pool of five synthetics, and a local population of *opaque*-2 (*o*2) genes has been initiated. Original, Italian, and Yugoslav sources of *o*2 genes were used.

The results of this research and its implications to the development of high-yielding maize hybrids with modified protein quality are discussed.

Among about 1500 *o*2 inbred lines analyzed from a series of widely different germplasms, broad ranges of protein (5.27–16.64%) and lysine content (2.49–5.73 g/100 of protein) were found. Thus variability in the protein and lysine content of many genotypes exceeds original *o*2 mutant, having other good agronomic properties.

A broad testcross for combining ability of different sources and between groups of inbreds was included.

The best *o*2 hybrids were tested at two or three densities and different locations for 3–5 years, and its yield was nearly that (86.8–94.8%) of normal maize of same maturity group. The best yielded hybrids had satisfactory protein level (9.32–11.56%) and lysine (3.44–4.70 g/100 protein).

On the basis of official 3 years of testing by the Federal Commission for Variety Recognition in Yugoslavia, a large quantity of seed of the best three single crosses, maturity group 700, was produced for commercial use in 1975.

STALK QUALITY IMPROVEMENT IN MAIZE

M. S. Zuber and Terry Colbert

USDA Agricultural Research Service and the
University of Missouri,
Columbia

Although corn breeders have increased stalk lodging resistance during the past two decades, stalk breakage continues to be a major production problem. The progress that has been made in developing strains with better stalk quality has been offset in part by higher plant densities and greater application of fertilizers.

Future emphasis will be on increased food production. To meet this challenge breeders will be giving a high priority to developing hybrids with greater genetic yield potential, and other researchers will be changing practices to maximize yield. Higher corn yields will require hybrids with superior stalk quality.

The purpose of this demonstration is fourfold:

1. Show the effectiveness of cyclic selection for stalk-quality improvement.
2. Compare the contribution of several morphological components to stalk quality.
3. Demonstrate and evaluate different methods for measuring stalk strength.
4. Report the interrelationship of several chemical components involved with stalk strength.

THE SUBAPICAL DWARFISM LOCATED IN THE LINE W401 AND ITS HEREDITY

M. Pollacsek

INRA Plant Breeding Station
Clermond-Ferrand, France

Dwarfism would bring the advantages linked to shortness, namely, ease of transport of pipes for irrigation, detasseling, harvest, and increase in resistance to lodging. An increase in productivity could be hoped for if the photosynthetic products not wasted in the stalk would go to the ear. Because the *brachytic*-2 gene has some depressive effect on the yield and a low ear insertion, the subapical dwarfism located above the ear is tentatively used.

A new source of subapical dwarfism was found within the line W401 following a natural mutation. The stability of its expression is better than those of the first source studied (K257), particularly as the character can be seen during winter culture.

Within crosses, between W401 subapical dwarfism and normal lines, some of them issued of Wisconsin (W617, W374) or not (Oh51A), show in the F_1 generation a good level of dwarfism similar to the dwarf parent. With other lines, the F_1s have an intermediate level (F 195 . . .) or are normal, according the presence of a few genes needed for the expression of the dominant mutation.

Heredity and comparative studies for the effect on yield are in progress.

ESTIMATION OF GENETIC ADVANCE AT VARIOUS PLANT-POPULATION DENSITIES IN AN opaque-2 MAIZE COMPOSITE

B. N. Singh and Joginder Singh

Cummings Laboratory,
Indian Agricultural Research Institute,
New Delhi, India

Investigations were undertaken to determine the estimates of genetic variances at various plant-population densities at selection and evaluation. Information thus gathered was used to determine the most appropriate plant population density for realizing optimum yield and genetic advance.

Half-sib family groups (North Carolina Design-1) were developed in *Shakti* and *opaque*-2 composite at three-plant population densities, specifically, 25,000, 50,000, and 100,000 plants per hectare (A, B, and C), and each group of progenies was evaluated at similar levels of plant densities (a, b, and c) for grain yield, agronomic traits, and quality characters.

The group of progenies developed and evaluated at high plant density (Cc) gave significantly highest grain yield. The group of progenies developed at low and tested at high density (Ac) was next in order. Relative magnitude of additive and dominance variance showed marked variation at different levels of selection and evaluation. Marked increase in the estimates of dominance variance at increasing levels of plant densities was recorded. This was particularly true at increasing levels of densities at evaluation. Different levels of selection did not exhibit any consistent pattern.

Estimates of genetic advance at 5% selection intensity through full-sib family selection were estimated to be highest for grain yield for the progenies developed at high density, irrespective of the plant-population level of evaluation. High mean performance and genetic advance suggested the selection and evaluation of progenies at high plant densities for grain yield.

In contrast to grain yield, higher estimates of genetic advance for protein and tryptophan percent in endosperm were realized at medium to low levels of selection.

Graphs relating to mean performance and genetic advance through full-sib family selection, at various levels of selection and evaluation, comprise the main exhibit.

INDUCED DOMINANT MUTANTS

M. G. Neuffer

Division of Biological Sciences,
University of Missouri,
Columbia

Mutation from normal to dominant, whether spontaneous or induced, is a rare occurrence. From a population of 7738 M_1 plants produced by the treatment of pollen with ethyl methanesulfonate, 10 viable, transmissible dominant mutants were obtained. They included one striped virescent, one tiny dwarf, one yellow–green, two different disease lesion mimics, one yellow-streaked type, one shrunken endosperm, one collapsed endosperm, and two etched endosperm types. A 3461-plant subset of this population produced eight of the above mutants, 12 probable dominant lethal cases, 855 recessive endosperm mutants, and 751 recessive seedling mutants. From other treatments five additional viable dominant mutants—one yellow–green, two different intermediate dwarfs, one speckled leaf, and one lazy mutant—were obtained, bringing the total induced viable dominant mutants to 15. Data comparing frequencies and mutant types are presented in tabular form. Photographs and plant materials presented demonstrate the phenotypes observed.

EMBRYO CULTURE OF DEFECTIVE MAIZE MUTANTS

W. F. Sheridan, W. J. Caputo, and M. G. Neuffer

*Division of Biological Sciences,
University of Missouri,
Columbia*

Included in a large number of ethylmethylsulfonate-induced maize mutants are many with defective endosperm development that do not germinate or survive beyond the coleoptile state. We believe that the defective mutants may in some cases be auxotrophs, and that these mutants might be rescued if their embryos were cultured on approximately supplemented media prior to embryo abortion. Normal 8-, 10-, and 12-day-old (days postpollination) embryos were successfully grown on a modified Linsmaier and Skoog medium (Sheridan, 1975) containing either no amino acids, 1 g/liter casamino acids, or a mixture at 1 mM concentration of all 20 amino acids at either pH 4.9 or 5.6. The four mutants examined in this study (E738, E783, E792, and E873) are all recessive, being expressed in the homozygotes as small, nonviable kernels. Mutant E873 was more extreme, having very little endosperm tissue in the mature seed. The mutant kernels could be readily distinguished from normal kernels at an immature stage because of their reduced size and lighter color. Ears showing a 3:1 segregation of normal:defective kernels were used. Normal and defective 13–18-day-old embryos from each ear were cultured on media with no amino acids, or containing 1 g/liter of casamino acids. Embryos from normal kernels from all ears grew into plants on both kinds of media. Embryos from defective kernels of mutants E738, E783, and E792 did not grow on either medium. Embryos from defective kernels of mutant E873 developed into normal-appearing embryos on both kinds of media. Fourteen days after their initial culturing 22 of the mutant embryos were transferred to new media containing no supplements. These embryos grew into plants, some of which grew to 4 inches in height, but after 21 days following transfer all were dead or dying. Since the E873 embryos had initially developed equally well on both media, it was believed that the casamino acids were not essential for embryo development and plant growth. However, of the 10 most vigorous mutant plants among the 22 mutant plants cultured, eight of these developed from embryos initially cultured on the casamino-acid-supplemented medium. This observation together with the death of all the E873 mutant plants following transfer to basal medium suggests that the casamino acids were beneficial and that, in fact, this mutant may be auxotrophic for one or more amino acids. Of the four mutants studied, the one with the least promising phenotype, E873, responded to embryo culture.

PHYLLOTAXY—A FACTOR IN CORN PRODUCTIVITY?

R. I. Greyson and D. B. Walden

Department of Plant Sciences,
University of Western Ontario,
London, Ontario, Canada

Theory suggests that different leaf arrangements might alter plant productivity in certain contexts. The validity of this theory has yet to be tested in corn due to the apparent rarity of phyllotaxy differing from the normal distichous arrangement. A few heritable variables do occasionally arise, however, and these materials represent a unique resource of genetic variability with which to test the relationship between phyllotaxy and productivity. A bibliography of known reports is included in our demonstration.

In an attempt to encourage discussion and exchange of information on this topic, our demonstration focuses on the ABPHYL material using living specimens, models, and photographic records. The following points are illustrated and discussed.

1. ABPHYL exhibits phyllotactic expression from spiral to decussate.

2. The expression of the syndrome can commence at any time from embryo to tassel formation.

3. In decussate specimens, having twice the normal number of leaves, leaves are approximately one-half the width of those of comparable siblings due to fewer cells.

4. ABPHYL shoot apices are broader than those of normal siblings.

5. The character is transmitted through either the male or female parent and has been maintained for 10 generations.

6. Several dozen cultivars have been established, some of which yield progeny that fall into discrete segregation classes. Among the progeny of other stocks, variable expressivity or penetrance is encountered.

Clearly, this material along with that of other reports, should provide opportunities to test the original hypothesis. The phenotypic variability exhibited in these strains in combination with other well-known genetic material should assist in fashioning corn plants better suited to either forage or grain and also could influence the improvement of photosynthetic efficiency.

MEMBRANE POLYPEPTIDES AND CHLOROPHYLL–PROTEIN COMPLEXES OF MAIZE MESOPHYLL CHLOROPLASTS

**D. B. Hayden, W. G. Hopkins,
and B. C. Saurino**

*Department of Plant Sciences,
University of Western Ontario,
London, Ontario, Canada*

Polypeptide profiles of maize mesophyll chlorplast membranes have been obtained using SDS-acrylamide gel electrophoresis. Molecular weights have been assigned to the major polypeptides using the same technique. Chlorophyll-protein complexes from these membranes have been resolved using both SDS-acrylamide gel electrophoresis and chromatography on hydroxylapatite. Chlorophyll a is associated with a polypeptide having a molecular weight of 69 kilodaltons. This chlorophyll–protein complex probably serves in both photosystems I and II. A polypeptide with a molecular weight of 23 kilodaltons is associated with both chlorophylls a and b. This complex is presumed to serve in the light harvesting network serving mainly PS II, but also PS I.

ANALYSIS OF GREENING IN VIRESCENT MUTANTS OF MAIZE BY *IN VIVO* SPECTROPHOTOMETRY

**W. G. Hopkins, D. B. Hayden
and D. B. Walden**

*Department of Plant Sciences,
University of Western Ontario,
London, Ontario, Canada*

We have begun a study of the genetics and physiology of chloroplast development and pigment accumulation in virescent mutants of maize. Virescence is by definition a condition wherein the level of pigmentation changes with time. Moreover, we have observed in preliminary experiments that both the rate and pattern of pigment accumulation may be significantly modified by environmental factors. To adequately describe the changing status of pigmentation in localized regions of the seedling or leaf, we have developed a rapid, *in vivo* method for quantifying chlorophyll content of maize tissue. The method to be described is based on absorbance characteristics of leaf tissue. Samples consist of a 5-mm disk, and chlorophyll content is expressed as the difference (ΔA) between peak absorbance at 675 nm and the long wavelength minimum at 735 nm. There is significant correlation between ΔA and extractable chlorophyll. Using this method one is able to construct a sensitive, quantitative profile of pigment distribution within a single seedling leaf. The application of this technique for defining environmental influence on expression of virescence is described. In addition, the technique permits definition of the status of greening in plants prior to harvesting chloroplasts for subsequent analysis of photosynthetic electron transport and fluorescence emission characteristics.

SELECTION AND CHARACTERIZATION OF PHOTOSYNTHETIC MUTANTS IN *ZEA MAYS*

Kenneth Leto and C. D. Miles

Biology Division,
University of Missouri,
Columbia

A simple screening technique used for the detection of photosynthetic mutants in algae can also be used for the detection of photosynthetic mutants in higher plants. Corn seedlings at the two- or three-leaf stage can be rapidly screened in the sand bench at night by illuminating the seedlings with a hand-held longwave UV mineralight and viewing the chlorophyll fluorescence through UV protective goggles containing red cutoff filters. Plants showing high levels of chlorophyll fluorescence are defective in photosynthesis. After selection, defects in photosystem II can be distinguished from those in photosystem I by observing the kinetics of fluorescence induction in whole leaves and isolated chloroplasts. Using these simple screening techniques over the past year we have been able to select 43 F_2 families from selfed ethylmethanesulfonate-treated material that are photosynthetically mutant. These mutants have been designed *hcf* for high chlorophyll fluorescence.

Extensive biochemical analysis has been carried out on four mutant lines. All segregate 3 : 1 and have normal or near normal levels of chlorophyll. The *hcf*1, a protosystem I mutant, grows to a height of 40 cm under optimal conditions and exhibits reduced NADP diaphorase activity. Both *hcf*2 and *hcf*3 are seedling lethal photosynthesis mutants that are missing cytochromes; plastoquinone and cytochrome f are missing in *hcf*2, whereas the high-potential form of cytochrome b-559 is missing in *hcf*3. Work with *hcf*4 indicates that although these chloroplasts pump protons in nonycyclic electron transport, photophosphorylation is altered.

A limitation in the use of the fluorescence screening technique is that mutants involving the oxidizing side of photosystem II cannot be detected. Work is under way to investigate the suitability of the photodynamic herbicide diquat as a selection agent for mutants on the oxidizing side of photosystem II. Diquat kills wild-type plants in the light, whereas mutant plants bearing defects on the reducing side of photosystem II or on the oxidizing side of photosystem I are not killed in the light. Thus plants surviving diquat treatment and showing low levels of fluorescence will have mutations localized on the oxidizing side of photosystem II.

LOCALIZED SYNTHESIS OF ZEIN IN MAIZE ENDOSPERM

Frances A. Burr and Benjamin Burr

Biology Division,
Oak Ridge National Laboratory,
Oak Ridge, Tennessee

Electron micrographs show that protein bodies, distributed throughout the internal layers of the developing endosperm, are entirely limited by a single membrane and have polyribosomes at their surface. Protein bodies isolated from other cytoplasmic organelles and matrix protein contain zein as their major protein component and continue to be associated with RNA. The polyribosomes can be separated from the protein bodies by detergent treatment. They are active as the particulate component in *in vitro* protein synthesis, but will not incorporate radioactive amino acids in the absence of added soluble factors or in the presence of the inhibitor puromycin. Zein is the only protein formed because lysine, which is absent from zein, is not incorporated and because completed chains migrate with zein in SDS gel electrophoresis.

This work demonstrates a specialized apparatus for the synthesis of the largely insoluble zein protein close to its site of deposition in the protein bodies. There are probably a number of processes peculiar to this system. It is conceivable that mutations blocking any one of these steps could prevent the accumulation of zein and thus explain why a number of loci condition the opaque phenotype.

STUDY OF GENE ACTION IN SYNTHESIS OF STORAGE PROTEINS IN MAIZE

M. Denić, J. Dumanović, R. Simić,
K. Konstantinov, Danica Jelenić,
and Vesna Sukalović

Department of Genetics, INEP, and
Maize Research Institute,
Zemun, Yugoslavia

Due to increased demands for plant proteins, an extensive breeding program to improve the quality of maize protein has been initiated. To achieve this goal, the influence of the *opaque*-2 gene on phenotype and some biochemical properties of the endosperm was investigated.

The variation of endosperm texture, resulting from modifier genes within a number of homozygous *opaque*-2 converted materials of different genetic backgrounds, was studied. It was found that in modified *opaque*-2 endosperm the amount of protein and glutamic acid, leucine, phenylalanine, proline, and alanine was somewhat increased, whereas the amount of lysine, histidine, arginine, aspartic acid, and glycine was decreased. A tendency was indicated for RNase activity and lysine : histidine ratio to be decreased with increasing proportions of modified endosperm.

Using an Osborn-Mendel procedure modified by introducing 2-mercapto ethanol, a higher accumulation of zein-2 fraction was found in normal size than in the mutant during endosperm development. Both absolute and relative amounts of glutelins were higher in the mutant as compared to the normal genotype. Detailed analysis of material showed a higher absolute amount of all 19 free amino acids in mutant endosperm than in normal maize 14 days after pollination. However, such differences between the genotypes were not found in the relative amounts of most of the free amino acids. This would suggest that the lower amounts of some protein-bound amino acids in the mutant or normal maize are not due to reduced concentrations of free amino acids.

Possible involvement of some cell components in *opaque*-2 gene action is also discussed.

VARIABILITY OF CONTENT AND FATTY-ACID COMPOSITION OF OIL IN MAIZE AND BREEDING POTENTIAL

M. Misović, S. Ratković, Mirjana Mihajlović, S. Kapor, V. Trifunović, and J. Dumanović

Maize Research Institute and INEP,
Zemun, Yugoslavia

Due to an increased demand for vegetable oil in recent years and the already high quality of maize oil, an extensive breeding program has been undertaken to increase the content of oil in maize kernels and further improve its quality.

A number of inbred lines (ca. 500) with oil contents ranging from 3.2% to 12.5%, which had previously been partly screened for other agronomically important traits, were studied in respect to: (a) variation of oil content, (b) variation of fatty-acid composition of oil, (c) relationship of oil content to fatty-acid composition, (d) interrelationships among the major fatty acids, and (e) distribution pattern of fatty acids within triglyceride molecules.

The range of single-kernel variability in oil content within samples was different. The observed differences among the lines might be ascribed at least partly to different intensities of selection during their development and maintenance. It is suggested that simple kernel selection could still be effective (if necessary) in lines in which the range of oil content is broad enough. On the other hand, further individual selection could be expected to be more or less limited and practically ineffective in the lines that exhibit a narrow range in oil content.

Great variability in the four major fatty acids was found (palmitic, 7.64–18.7%; stearic, 0.72–3.82%; oleic, 16.36–43.35%; linoleic, 39.20–68.26%). From a breeding point of view it is of interest that a satisfactory variation of linoleic acid content is present. The very broad variability in fatty-acid composition, which is probably heritable, to a considerable degree provides possibilities of breeding corn with a desired fatty-acid composition.

However, it should be pointed out that with a change in oil level a considerable change in fatty-acid pattern is observed. In general, with increased oil content the relative content of oleic acid (probably stearic, too) is increased, and the content of linoleic acid decreased, and the rate of increase of oleic acid is much the same as the rate of decrease of linoleic acid, resulting in more-or-less constant values for their sum at all oil levels.

The overall results are further discussed with regard to breeding potential and distribution pattern of fatty acids within triglyceride molecules.

SELECTION SYSTEM FOR LYSINE, THREONINE, AND METHIONINE MUTANTS IN MAIZE

C. E. Green, R. L. Phillips, and B. G. Gengenbach

University of Minnesota, St. Paul

Isolation of regulatory mutants in maize based on *in vivo* and *in vitro* manifestations of feedback inhibition in the lysine–threonine–methionine biosynthetic pathway is displayed via tissue culture and seedling responses. Also demonstrated are the: (a) use of mutagenized callus and seedling populations and (b) molecular basis of lysine–threonine inhibition at the enzyme level.

Simultaneous addition of lysine and threonine severely inhibits growth of maize callus and germinating seedlings from embryos or whole kernels. Reversal of this inhibition by low levels of methionine or homoserine implicated feedback inhibition as the molecular mechanism. Isolated enzyme studies confirmed that aspartokinase (E.C. 2.7.24) activity was inhibited by lysine and that homoserine dehydrogenase (E.C. 1.1.1.3) was inhibited by threonine. This endproduct inhibition of two enzymes required in methionine biosynthesis presumably results in growth inhibition due to methionine deficiency.

Lysine–threonine inhibition should provide an effective selection system for mutants insensitive to feedback inhibition that are potential "overproducers" of lysine, threonine, or methionine. Such enzyme regulatory-site mutations are expected to behave as dominants and, therefore, can be isolated as heterozygotes in diploids, reducing the required time and population size.

Eight lysine–threonine-resistant callus clones have been selected from mutagenized maize callus cultures. Two of these presumptive mutants grow normally in the presence of lysine–threonine. Various maize genotypes and mutagenized populations are being screened for seedling growth in the presence of lysine–threonine. Screening of 130 Corn Belt inbred lines indicated little variability for this character, except for B37. B37 Seedlings from whole kernels grew but were inhibited if from isolated embryos. B37 Endosperm apparently prevents lysine–threonine inhibition, perhaps because it contains higher free methionine. Screening EMS-treated W23 M-1 seed is in progress.

MAIZE HISTONES

J. T. Stout and J. L. Kermicle

Laboratory of Genetics,
University of Wisconsin,
Madison

A genetic characterization of variation in lysine-rich histones (Hl) is in progress. Apparatus, sample gels, charts, and diagrams are exhibited to illustrate the application of various segregation and mapping methods available in maize for use in conjunction with high-resolution polyacrylamide gel electrophoresis. Particular areas of inquiry include:

1. *Gene number and arrangement.* Analysis of haploids from plants heterozygous for two dissimilar histone I electrophoretic subfraction systems has provided data concerning the question of histone-gene clustering in plants. Thus so far three variant forms of histone Ia (the slowest migrating, most abundant lysine-rich subfraction of each system) have been identified. They segregate in crosses as alleles of one locus. Variants of the remaining Hl subfractions (Ib, Ic) are likewise simply inherited, but as a single unit. The Ib, c cluster is inherited independently of Ia, confirming the initial report of Ia and Ic independent (*PNAS* **70**: 3043, 1973). The fourth and fastest subfraction (Id) is visible only in some inbreds. Its behavior in crosses has not yet been determined.

2. *Chromosome location.* By the combined use of monosomics and A–B translocations, histone gene Ia has been placed to the short arm of chromosome 1. The position of the Ib, c cluster is not yet definite. Chromosomes 1 and 6–10 have been excluded.

Our evidence concerning histone gene organization in maize contrasts with the single cluster model inferred from investigations with *Drosophila* and sea urchin. The genes for maize histone I are clustered only in part, involving two chromosomes. Variation described previously involving histone II subfractions has yet to be characterized with respect to histone I variation.

THE *F-cu* TWO-UNIT CONTROLLING ELEMENT SYSTEM

Jaime Gonella and Peter A. Peterson

Agronomy Department
Iowa State University,
Ames

A new two-unit controlling element system in maize, *F–cu* plus *r–cu*, has been identified. This is an addition to the previously described systems, *Ac–Ds*, *a1–Dt*, and I-Spm-En. The *F–cu* system was found in a maize accession (race Chococeno) that originated from the Cuna Indian Tribe in northwestern Colombia, South America. That this is unrelated to previously described systems has been confirmed by crosses to the various aleurone color testers (*a1/a1*, *a2/a2*, *r/r*, and *c/c*), as well as standard testers for each of the controlling element systems (*C1–Ds*, *a1dt*, *a1m(r)/a1m-1*). The new system consists of a receptive allele *r–cu* at the *r* locus that responds to a second factor identified as *F–cu*. The *F–cu* system is specific for *r–cu*, not affecting the other controlling element receptive loci such as *a1m(r)*, *a1dt*, or *C1–Ds*. The *r–cu* system does express gene activity in the absence of *F–cu*, giving a variable range of lightly pigmented coloration. The crown area of *r–cu* kernels is generally nonpigmented, the extent of the nonpigmented area varying between kernels.

This Cuna Indian accession from Colombia also contains *En* and *Dt*. The particular *En*, identified as *En–cu*, affects and *A1* allele (colored aleurone) in two ways. It causes a partial suppression of color expression and induces the *A1* allele to mutate to a higher level of pigmentation, producing very dark sectors on a dark background.

KNOB NUMBER AND POSITION IN THE STUDY OF EVOLUTIONARY RELATIONSHIP OF CLASSIFIED YUGOSLAV MAIZE FLINT TYPES

Lj. Zecević and Jelena Pavlicić

Institute for Biological Research (Belgrade),
Maize Research Institute (Zemun),
Yugoslavia

Flint types studied are considered as the eldest ones of maize introduced in Yugoslavia. Preliminary investigations showed some differences in their genetical characters.

The aim of the present investigation was to use pachytene analysis of the number and position of knobs to find parameters for establishing the relationship, origin, and evolution of autochthonous maize flint types.

The classified maize flint types were examined as follows: Montenegrin, Kosovo, Macedonian, and Mediterranean flints. Those are early flints, grown in hilly areas, where they were preserved mostly unchanged after their introduction.

Pachytene analysis showed up that the number of knobs in examined types ranged within 0–4. The smallest average number of knobs was by Macedonian flints (1.61), followed by Montenegrin flints (2.07) and Kosovo flints (2.50), and the largest one was by Mediterranean flints (2.66).

In all these types the most frequent knob position was 9ST, and also a high frequency was observed in positions 7L3, 2L, and 1L3.

Our results indicate the existence of noticeable differences in the number and position of knobs among these flint types and confirm that their classification corresponds also to their genetical characteristics.

The differences noticed among them may be caused by the different origin of these types, although we can suppose they all came from South America.

A DIFFERENTIAL EFFECT ON KERNEL SIZE OF MATERNAL AND PATERNAL FORMS OF A CHROMOSOME REGION

Bor-yaw Lin

Department of Genetics,
University of Wisconsin,
Madison, Wisconsin

B Translocations with a breakpoint proximal to *du* in 10L condition a 50% reduction in size of kernels of the hypoploid-endosperm class. Translocations with a more distal breakpoint, such as TB-10(14) and Roman's TB-10a, are associated with only a 5% reduction in this class. Thus a factor or factors markedly affecting endosperm growth is situated in the chromosome region delineated by the breakpoint of TB-10(14) and the locus of *du*. Small seed size could be based in either of two effects on gene action, either a reduction in total number of the critical chromosome region from three to two (i.e., a dosage effect) or, more specifically, due to the absence of the paternal form of this segment (i.e., an imprinting effect). To distinguish between the two, kernel classes possessing endosperm of the same dosage level but differing in parentage of the critical region were synthesized using TB-10(19), one of three translocations with a breakpoint proximal to *du*. The control series utilized TB-10(14) in parallel tests. The 10 10 B^{10} partial trisomic was crossed as female with the balance translocation homozygote as male. Chromosome 10 and the two B^{10}s were marked differentially by alleles of R. Kernels of endosperm constitution 10 10 B^{10} B^{10} from this cross have two origins. Either the B^{10}s are maternal (a $4\male:0\female$ combination for the critical region), or the B^{10}s are paternal (a $2\male:2\female$ combination). Both classes were near normal in the TB-10(14) control series. With TB-10(19), kernels of the $2\male:2\female$ endosperm class also were comparable to normal ($2\male:1\female$). Kernels of $4\male:0\female$ composition, however, exhibited a 50% size reduction, not unlike the hypoploid endosperm ($2\male:0\female$) class. Thus a paternal form of the region in question evidently is necessary for normal endosperm growth.

B–A TRANSLOCATIONS IN MAIZE

J. B. Beckett

USDA Agricultural Research Service and
University of Missouri,
Columbia

B–A Translocations result from reciprocal interchanges between a supernumerary (B) chromosome and members of the basic (A) set of maize chromosomes. Since B–A exchanges are now available on 18 of the 20 maize chromosome arms, it seems an appropriate time to demonstrate some of the advantages these translocations offer for a wide variety of studies. Each translocation generates functional male gametes carrying zero and two doses, respectively, of a particular chromosome arm (that part beyond the breakpoint of the translocation), making it possible to manipulate the ploidy level of that arm. The ability of B–A translocations to locate recessive endosperm and seedling mutants in the F_1 are demonstrated. Other uses of the translocations are shown; living material is used as much as possible.

IN VITRO POLLEN GERMINATION AS A TECHNIQUE IN GENETIC ANALYSIS

D. B. Walden, S. J. Gabay,
E. B. Patterson, and J. R. Laughnan

University of Western Ontario,
London, Ontario, Canada, and
University of Illinois, Urbana

The *in vitro* pollen germination system, as refined by Cook and Walden, has a potential for genetic analysis that has not been fully exploited. Germination and tube growth of pollen of most inbred lines and genetic strains of maize is sufficient to afford its use in genetic studies. The demonstration illustrates: (a) variation in growth responses among various inbred lines and genetic strains, (b) the inhibitory effect of toxin from *Helminthosporium maydis,* race T, on germination of pollen from restored *cms*T plants, (c) *in vitro* growth responses of pollen carrying balanced and duplicate–deficient genomes derived from certain translocation heterozygotes, and (d) germination of small pollen grains from inbred line M825L.

Applications of the pollen sieving technique are also considered.

POLLEN METHOD FOR DETERMINATION OF CHROMOSOMAL INTEGRATION SITES OF THE CMS S RESTORER EPISOME IN MAIZE

S. J. Gabay and J. R. Laughnan

Department of Genetics and Development
University of Illinois,
Urbana

Tests for allelism, available for six of the ten new restorers of S male-sterile cytoplasm, indicate that these six restorers are neither allelic with the standard restorer RF3 carried by inbred line CE1 nor with each other. We have developed a new technique that takes advantage of the fact that S restoration is gametophytic, rather than sporophytic, and affords the assignment of the fertility episome (F) to linkage group on the basis of pollen analysis. Each restorer strain is crossed with the series of *waxy* translocation stocks, and pollen of the F_1 interchange heterozygotes is examined for the proportion of blue- and red-staining phenotypes among the 25% nonaborted pollen grains. Since equal numbers of blue- and red-staining grains are expected, if F assorts independently of *waxy*, a distorted ratio favoring blue-staining over red-staining grains is taken as preliminary evidence of linkage. By this means, restorer IV has been assigned to chromosome 3, and restorer I has been assigned to chromosome 8, and these assignments have been confirmed recently by data from testcross progenies. In addition, restorers VII and VIII have tentatively been assigned to chromosomes 3 and 8, respectively, and in the restorer IX and X strains, the fertility episome appears to have integrated into chromosome 1. The procedure is illustrated with slide preparations and diagrams, as well as with testcross data.

THE GENETIC CONTROL OF ANTHOCYANIN PIGMENTATION

E. H. Coe, Jr.

*USDA Agricultural Research Service and
University of Missouri,
Columbia*

Major loci that affect anthocyanins and related pigments, as displayed in closely similar backgrounds, are demonstrated, summarized, and interpreted. Each locus is either universal (i.e., pigmentation is determined by the locus in all tissues) or limited (pigmentation is determined by the locus in some tissues but not in others). Certain loci (*C* and *Pl*) have alleles whose recessive effects are contingent on darkness; at the *C* locus other recessive alleles are null (i.e., noncontingent). Dominant inhibition of pigmentation is found at *C* (*C–I*), *C2* (*C2–Idf*), *in* (*In–D*), and *a3* (*A3*), but of these only *C–I* is pigmentless when homozygous in those tissues (aleurone and embryo) over which it is determinative. Duplicate dominant interactions are found in aleurone tissue and in plant tissues between specific representatives in the *B* and *R* regions. Variants for tissue specifically are not common at a given locus—in fact, the examples that have been studied at *R* are accompanied by segmental duplication rather than cistronic variation, implying that changes in tissue specificity may occur only with structural changes. Although any or all of the universal loci might be regulatory for the battery of functions in the pathway, the limited loci are better candidates for a regulatory role as they include members that are responsive only to particular conditions or to particular tissues.

ACTION OF INTENSIFIER GENE (*in*) IN FLAVONOID SYNTHESIS OF MAIZE

A. R. Reddy and G. M. Reddy

*Department of Agronomy,
Iowa State University,
Ames*

The intensifier gene (*in*), when homozygous recessive, in the appropriate genetic background, accentuates the formation of anthocyanin pigments in the aleurone tissue of maize. A comparative study was undertaken with *In* versus *in* combinations with five loci controlling anthocyanin pigmentation, *A, A2, Bz, Bz2,* and *Pr* for qualitative and quantitative differences in the accumulated pigments to determine the role of *in* in the pathway for anthocyanin formation. Data were obtained from paper chromatography and spectrophotometric studies. With *Pr, pr, bz,* and *bz2* tissue, it enhances the quantity of anthocyanins and 3-deoxyanthocyanidin by approximately 5 times. In the *a2* tissue a threefold increase of the leucoanthocyanidin was observed. Thus *in* acts differentially with various anthocyanin loci in enhancing the formation of different pigments. No qualitative changes in the pigment composition of aleurone can be ascribed to recessive *in*. Presumably, the dominant allele *In* is an amorph, having no action. Although the basis of *in* action is not yet fully known, it is proposed that it may be involved in the regulation of availability of precursors of anthocyanins and leucoanthocyanidins.

MAIZE IN ART AND CULTURE

J. R. Harlan and J. M. J. de Wet

Agronomy Department,
University of Illinois,
Urbana

This is a 30-minute tape-and-slide show prepared by the Crop Evolution Laboratory, Agronomy Department, University of Illinois. The slides and story feature maize in the arts, religions, and mythologies of American Indians from pre-Columbian times to the present. A few modern artifacts are also available.

POSSIBLE MUTAGENIC ACTIVITY OF ATRAZINE DEGRADATION PRODUCTS

M. J. Plewa and J. M. Gentile

Department of Agronomy,
University of Illinois, Urbana and
Department of Human Genetics,
Yale University School of Medicine,
New Haven, Connecticut

We have devised a plant-microbe bioassay by which we can test pesticides for possible mutagenic activity under conditions normally encountered in the environment. This system is superior to other commonly used mutagen test systems because the mutagenicity of a pesticide as well as any plant-mediated metabolite can be determined rapidly and accurately. We have studied the herbicide atrazine as there is a controversy in the literature concerning the mutagenicity of s-triazine compounds. Our results indicate that there is an increase of an order of magnitude in the reversion at the *waxy* locus of *Zea mays* microsporocytes from plants treated with atrazine as compared to sibling controls not exposed to the herbicide. Also, a water-soluble extract from maize seedlings exposed to atrazine increases the frequency of reversion and gene conversion in *Saccharomyces cerevisiae*. However, atrazine does not induce mutation in *S. cerevisiae* if it is directly applied to the microbe. The data from our studies indicate that:

1. atrazine is converted by the maize plant into a product that can induce reversion at the *waxy* locus in maize microsporocytes, the *ade*2 locus in haploid yeast, and induces mitotic gene conversion at the *ade*2 and *trp*5 loci in diploid yeast.

2. The molecular nature of the lesion is probably a base-pair substitution in the DNA.

3. The mutagenic agent is water-soluble.

4. The agent probably remains active for a relatively long period of time in the maize plant, at least until anthesis of the plant.

MORPHOLOGY OF MAIZE X TRIPSACUM DERIVATIVES

J. R. Harlan and J. M. J. de Wet

Agronomy Department,
University of Illinois,
Urbana

Plant specimens from maize × *Tripsacum* hybrid derivatives are displayed to show a part of the range of morphological variation that can be generated in these materials. Some of the characters that are known to be "tripsacoid" are pointed out in plants that carry a few *Tripsacum* chromosomes. Similar characteristics also occur in $2n = 20$ plants. The specimens were prepared by the Crop Evolution Laboratory.

THE PRODUCTION AND UTILIZATION OF MONOSOMIC *ZEA MAYS*

David F. Weber

Department of Biological Sciences,
Illinois State University,
Normal

We have produced plants monosomic for chromosomes 1–4 and 6–10 using the system described in this demonstration. In addition, two plants genetically monosomic for chromosome 5, but not cytologically confirmed, have been isolated. This is the first time that most (probably all) of the monosomic types have been produced in a diploid organism. The monosomics are produced using the $r–x1$ deficiency, a deficiency of the R locus on chromosome 10. Megaspores carrying this deficiency are normal, but we have found that chromosomal nondisjunction occurs in deficiency-carrying gametes during the mitotic megagametophyte divisions. Over 11% of the progeny carrying the deficiency are monosomic, and an equivalent frequency are trisomic. If the male parent in the cross carries a recessive mutation that is expressed in the sporophyte, progeny expressing this mutation are monosomic for the chromosome carrying this gene. Plants monosomic for different chromosomes are shown.

A few of the many uses for monsomics are presented. First, the behavior of univalent chromosomes is poorly understood. Monosomics provide an ideal opportunity to study the behavior of univalent chromosomes because each meiotic cell in a monosomic contains a univalent chromosome. In addition, doubly monosomic plants (plants monosomic for two different chromosomes) have been recovered, and with these plants one can test for interactions between nonhomologous univalent chromosomes. Second, quartets of microspores from monosomic plants have two haploid cells and two cells that are nullisomic for the monosomic chromosome. A comparison of these nullisomic and haploid cells is a comparison of zero and one copy of all genes on a given chromosome. Third, by comparing a plant monosomic for a specific chromosome with its diploid siblings, one compares one and two copies of all genes on a given chromosome. If a gene is present on the monosomic chromosome that exhibits dosage effects, a difference will be found between these two plant types. This is a new method of analyzing the genome, and with it one analyzes all gene loci on a given chromosome without inducing or utilizing gene mutations. Using this rationale, we are exploring the genome for several types of genetic factors. Examples where each of these experimental approaches has been used are presented.

SELECTED STAGES OF POLLEN DEVELOPMENT EXAMINED WITH LIGHT AND ELECTRON MICROSCOPES

C. E. Hall, J. R. Laughnan, and S. J. Gabay

*Department of Genetics and Development
University of Illinois,
Urbana*

Selected stages of normal pollen development are demonstrated employing coordinated light and electron microscope techniques.

Anthers were fixed in gluteraldehyde, postfixed in osmium tetroxide, and embedded in either Epon 812 or Spurr's low-viscosity medium. Thick sections (1.5μ) were cut from the epoxy embedded anthers, transferred to glass slides, stained with Paragon-1721, and examined with the light of the microscope. Following this examination, blocks containing anthers of special interest were retrimmed and thin-sectioned for examination with the electron microscope.

The use of a single fixation procedure for tissue to be examined with both light and electron microscopes has two significant advantages: (a) fixation artifacts are more easily detected when a single specimen is examined with both the light and electron microscopes and (b) individual stages in the development of the anther can be determined via a light-microscope examination before the block is sectioned with the electron microscope. Since sectioning and staining of a tissue sample for electron microscopy is technically more difficult and time-consuming than preparation for light microscopy, examination with the light microscope of large numbers of anthers fixed and embedded for examination with the electron microscope affords the investigator a relatively rapid screening technique.

RACES OF MAIZE DEMONSTRATION

Major M. Goodman
North Carolina State University,
Raleigh

Specimen ears of 130 races of Latin American maize were arranged by country or region of origin. The ears were chosen as being reasonably representative of their respective races. The races were chosen so as to span the range of variability of the maize of Latin America. With only a few exceptions, all ears on display were open-pollinated and were grown in winter nurseries near Homestead, Florida. (The few exceptions were grown in Raleigh, North Carolina.) All ears were from collections designated in the series of races of maize monographs as being typical of the races represented.

INDEX